Anti-Angiogenic Functional and Medicinal Foods

NUTRACEUTICAL SCIENCE AND TECHNOLOGY

Series Editor

FEREIDOON SHAHIDI, PH.D., FACS, FCIC, FCIFST, FIFT, FRSC
University Research Professor
Department of Biochemistry
Memorial University of Newfoundland
St. John's, Newfoundland, Canada

Anti-Angiogenic Functional and Medicinal Foods

Edited by

Jack N. Losso
Fereidoon Shahidi
Debasis Bagchi

CRC Press
Taylor & Francis Group
Boca Raton London New York

CRC Press is an imprint of the
Taylor & Francis Group, an **informa** business

CRC Press
Taylor & Francis Group
6000 Broken Sound Parkway NW, Suite 300
Boca Raton, FL 33487-2742

First issued in paperback 2019

© 2007 by Taylor & Francis Group, LLC
CRC Press is an imprint of Taylor & Francis Group, an Informa business

No claim to original U.S. Government works

ISBN-13: 978-1-57444-445-2 (hbk)
ISBN-13: 978-0-367-38927-7 (pbk)

Visit the Taylor & Francis Web site at
http://www.taylorandfrancis.com

and the CRC Press Web site at
http://www.crcpress.com

DEDICATION

To My mentors, the late John T. Edward and Patrick G. Farrell
-Fereidoon Shahidi

To My children, Mariel, Boriss, MerryJean, and my wife Jane
-Jack N. Losso

To My beloved professor, Dr. Amalendu Banerjee, Ph.D.
-Debasis Bagchi

Table of Contents

Part II The *"Ohmics" Technologies and Functional Foods*

Part III *Angiogenesis and Chronic Degenerative Diseases*

Part IV Angiogenesis, Functional, and Medicinal Foods

Preface

The concept of angiogenesis or the growth of new blood vessels was proposed by Dr. Judah Folkman in 1971. The hallmark of angiogenesis is the proliferation and migration of endothelial cells that form the primitive tools that become blood vessels. Physiological angiogenesis is a precisely regulated process that occurs as brief burst of capillaries blood vessel growth and lasts days or weeks. Physiological angiogenesis is important for body development, reproduction, healing wounds and restoring blood flow to tissues after ischemic injury or insult. The angiogenic process "switches off "after achieving the appropriate biological end point (e.g. healed wound or menstrual cycle). Pathological angiogenesis, however, can be excessive or insufficient. Excessive angiogenesis continues unabated and paves the way to aberrant tissue growth and contributes to more than seventy chronic degenerative diseases such as chronic inflammation, rheumatoid arthritis, psoriasis, obesity, diabetes retinopathy, macular degeneration, plaque formation, and cancer. Insufficient angiogenesis may lead to infertility, heart disease, stroke, ulcers, scleroderma, and delayed wound healing.

Research on pathological angiogenesis is bringing together molecular biologists, bio/chemists, biomedical researchers, pharmacologists, medical doctors, pathologists, medicinal chemists, nutritionists, and food scientists to build a working framework for promoting the utility and scientific merit of this new field.

Although, the clinical benefits of anti-angiogenic therapy have been slow to realize, it has generated a body of information that is being harnessed to develop and identify new bioactive compounds for safely inhibiting pathological angiogenesis.

There are several advantages to consider anti-angiogenic therapy over conventional therapies. Since angiogenesis in healthy individuals is normally restricted, only about 0.01% of adult endothelial cells undergo the process of angiogenesis at any given time, the side effects of inhibitors of pathological angiogenesis on normal tissues should be negligible. These inhibitors affect tumor growth indirectly by targeting the steps involved in the process leading to the formation of aberrant new blood vessels and are mainly cytostatic. As a result, the FDA has approved the use of anti-angiogenic drugs for cancer and macular degeneration treatments.

Experimental and clinical evidence show that several bioactive compounds from foods, also known as functional foods are among the most efficacious and investigated anti-angiogenic molecules. These bioactive compounds can serve as the basis for developing anti-angiogenic foods or drugs.

This book discusses functional foods research in some major parts of the world and some recent advances on dietary bioactive compounds and the mechanisms by which these compounds inhibit pathological angiogenesis in areas as diverse as inflammation, cardiovascular disease, brain tumorigenesis, obesity, and diabetes. The book concludes with a chapter on developing and delivering anti-angiogenic functional food products, an aspect that could be of value to anyone working in the field of functional foods.

We have assembled authors from the six continents who have provided inputs that have the potential to significantly advance research on anti-angiogenic functional foods. We are grateful to all the contributors for taking the time to provide up to date information in subject matters of their expertise.

Editors

Jack N. Losso, PhD is an associate professor in the department of Food Science at Louisiana State University Agricultural Center, Baton Rouge, Louisiana. His current research interests include angiogenesis, macular degeneration, chronic inflammation, cancer, and functional foods. Dr. Losso has published more than 80 peer-reviewed papers, book chapters, proceedings, and journal magazine articles. He is an active professional member of the Institute of Food Technologists and the American Chemical Society.

Fereidoon Shahidi, PhD, FACS, FCIC, FCIFST, FIAFoST, FIFT, FRSC is a university research professor at Memorial University of Newfoundland in Canada. Dr. Shahidi is the author of over 600 research papers and book chapters, has also authored or edited 40 books. His research contributions have led to several industrial developments worldwide. Dr. Shahidi's current research interests include different areas of nutraceuticals and functional foods, natural antioxidants, marine foods and aquaculture. Dr. Shahidi serves as the editor-in-chief of the Journal of Food Lipids, an editor of Food Chemistry as well as an editorial board member of a number of journals, including the Journal of Food Science, Journal of Agricultural and Food Chemistry, Nutraceuticals and Food, International Journal of Food Properties, Current Nutrition and Food Science and Inform. Dr. Shahidi has received numerous awards, including the 2005 Stephen Chang Award from the Institute of Food Technologists, for his outstanding contributions to science and technology. Between 1996 and 2006, Dr. Shahidi was the most published and most frequently cited scientist in the area of food, nutrition, and agricultural science as listed by the ISI.

Debasis Bagchi received his PhD in Medicinal Chemistry in 1982. He is a professor in the Department of Pharmacy Sciences at Creighton University Medical Center, Omaha, Nebraska. He is also the Senior Vice President of R&D of Inter Health Nutraceuticals, Benicia, California. Dr. Bagchi is a Member of the Study Section and Peer Review Committee of NIH, Bethesda, Maryland, and Editorial Board Members of numerous journals. His research interests include oxidative stress, angiogenesis, metabolic syndrome, gene expression, and medicinal and functional foods. Dr. Bagchi has 225 research publications, edited numerous books and special issues, and organized and chaired several workshops and symposiums.

Contributors

O. I. Aruoma
Faculty of Health and Social Care
London South Bank University
London, United Kingdom

Debasis Bagchi
Dept. of Pharmacy Sciences
Creighton University Medical Center
Omaha, Nebraska

Manashi Bagchi
InterHealth Nutraceuticals Incorporated
Benicia, California

T. Bahorun
Department of Biosciences
Faculty of Science
University of Mauritius
Réduit, Mauritius

Snigdha Banerjee, PhD
Cancer Research Unit
The University of Kansas Medical Center
Kansas City, Missouri

Sushanta K. Banerjee, PhD
Cancer Research Unit
The University of Kansas Medical Center
Kansas City, Missouri

Narayan Bhaskar
Department of Meat, Fish and Poultry
    Technology
Central Food Technological Research
    Institute
Mysore, India

Antoni Caimari
Laboratori de Biologia Molecular
    Nutrició i Biotecnologia
Universitat de les Illes Balears
Palma de Mallorca, Spain

Maurizio Canavari
Department of Agricultural Economics
    and Engineering
Alma Mater Studiorum-University
    of Bologna
Bologna, Italy

Alessandra Castellini
Department of Agricultural Economics
    and Engineering
Alma Mater Studiorum-University
    of Bologna
Bologna, Italy

Shampa Chatterjee
Institute for Environmental Medicine
University of Pennsylvania
    Medical Center
Philadelphia, Pennsylvania

Christopher H. K. Cheng
Department of Biochemistry
The Chinese University of Hong Kong
Hong Kong, China

Thomas Paul Devasagayam, PhD
Radiation Biology and Health
    Sciences Division
Bhabha Atomic Research Centre
Mumbai, India

Kelley Fitzpatrick
NutriTech Consulting
Winnipeg, Manitoba, Canada

Jennifer M. Fostel
National Center for Toxicogenomics
National Institute of Environmental
    Health Sciences
Durham, North Carolina

Mike Gidley
Centre for Nutrition and Food Sciences
University of Queensland
Brisbane, Australia

Roland H. Goldbrunner
Department of Neurosurgery
University of Munich
Munich, Germany

Pramod Gopal
Fonterra Research Centre
Palmerston North
New Zealand

Gayle M. Gordillo
Laboratory of Molecular Medicine
Dorothy M. Davis Heart and Lung
    Research Institute
The Ohio State University
Columbus, Ohio

Soumyadipta Hazra
Department of Psychiatry
Brookdale University Hospital
    Medical Center
Brooklyn, New York

Nanba Hiroaki
Department of Microbial Chemistry
Graduate School of Kobe
    Pharmaceutical University
Kobe, Japan

C. A. Hornick
Department of Physiology
Louisiana State University
    School of Medicine
New Orleans, Louisiana

V. Ivanov
Matthias Rath Research Institute
Cancer Division
Santa Clara, California

Lianji Jin
Rieveschl Laboratories for Mass
    Spectrometry
Department of Chemistry
University of Cincinnati
Cincinnati, Ohio

T. Kalinovsky
Matthias Rath Research Institute
Cancer Division
Santa Clara, California

Min Jung Kang
Bioanalysis and Biotransformation
    Research Center
Korea Institute of Science
    and Technology
Seoul, Korea

Asami Kawakami
Pharmaceutical Research
    Laboratories
Ajinomoto Co., Inc.
Kawasaki, Japan

Dong-Hyun Kim
Bioanalysis and Biotransformation
    Research Center
Korea Institute of Science
    and Technology
Seoul, Korea

Paterna Kotzekidou, PhD
Department of Food Science
    and Technology
Aristotle University of Thessaloniki
Thessaloniki, Greece

Satoru Koyanagi
Department of Biochemistry
Faculty of Pharmaceutical
    Sciences
Fukuoka University
Fukuoka, Japan

Franco M. Lajolo, PhD
Departamento de Alimentos e Nutrição
  Experimental
Faculdade de Ciências
  Farmacêuticas Universidade
  de São Paulo
São Paulo, Brazil

Hyong Joo Lee
Department of Agricultural
  Biotechnology
College of Agriculture
  and Life Sciences
Seoul National University
Seoul, South Korea

Patrick A. Limbach
Department of Chemistry
University of Cincinnati
Cincinnati, Ohio

Jack N. Losso
Department of Food Science
Louisiana State University
  Agricultural Center
Baton Rouge, Louisiana

V. A. Luximon-Ramma
Department of Biosciences
Faculty of Science
University of Mauritius
Réduit, Mauritius

Saptarshi Mandal
Research Fellow, Stanford University
  Medical Center and Department
  of Biological Sciences
Stanford University
Palo Alto, California

Alexandra-Maria Michaelidou
Department of Food Science
  and Technology
Aristotle University of Thassaloniki
Thessaloniki, Greece

K. Miyashita
Dept. Bioresources Chemistry
Graduate School of Fisheries
  Sciences
Hokkaido University
Hakodate, Japan

Eliane Miyazaki
FoodStaff Assessoria
  em Alimentos Ltda.
São Paulo, Brazil

Paul J. Moughan
Riddet Centre
Massey University
New Zealand

V. S. Neergheen
Department of Biosciences
Faculty of Science
University of Mauritius
Réduit, Mauritius

A. Niedzwiecki
Matthias Rath Research Institute
Cancer Division
Santa Clara, California

Giuseppe Nocella
Alma Mater Studiorum - University
  of Bologna
Faculty of Agriculture - Food Science
  and Technology
Cesena, Italy

Kodama Noriko
Department of Microbial Chemistry
Graduate School of Kobe
  Pharmaceutical University
Kobe, Japan

Paula Oliver
Department of Agricultural Economics
  and Engineering
Alma Mater Studiorum-University
  of Bologna
Bologna, Italy

Andreu Palou
Catedràtic de Bioquímica i Biologia
  Molecular
Universitat de les Illes Balears (UIB)
Palma de Mallorca, Spain

Carlo Pirazzoli
Department of Agricultural Economics
  and Engineering
Alma Mater Studiorum-University
  of Bologna
Bologna, Italy

M. Rath
Matthias Rath Research Institute
Cancer Division
Santa Clara, California

Gibanananda Ray
Cancer Research Unit
The University of Kansas
  Medical Center
Kansas City, Missouri

Smriti Kana kundu-Raychaudhuri
Department of Genetics
Stanford University School
  of Medicine
Palo Alto, California

Siba P. Raychaudhuri
Department of Medicine,
School of Medicine, University
  of California
Davis, California

Lenore Polo Rodicio
Department of Natural Sciences
Miami-Dade College
Miami, Florida

Ana Mª Rodríguez
Laboratori de Biologia Molecular,
  Nutrició i Biotecnologia
Universitat de les Illes Balears
Palma de Mallorca, Spain

M. Waheed Roomi
Matthias Rath Research Institute
Cancer Division
Santa Clara, California

Sashwati Roy
Department of Surgery
Heart and Lung Research Institute
The Ohio State University
  Medical Center
Columbus, Ohio

Maria Saarela
VTT Technical Research Centre
  of Finland
Tietotie, VTT, Finland

Kenji Sato, PhD
Department of Food Sciences
  and Nutritional Health
Kyoto Prefectural University
  Shimogamo Hangicho
Kyoto, Japan

Chandan K. Sen
Department of Surgery
Heart and Lung Research Institute
The Ohio State University
  Medical Center
Columbus, Ohio

Fereidoon Shahidi, PhD
Department of Biochemistry
Memorial University
  of Newfoundland
St. John's, NF, Canada

Kazuhiro Shichinohe
Department of Laboratory
  Animal Science
Nippon Medical School
Bunkyo-ku Tokyo, Japan

Hiroshi Shimeno, PhD
Department of Biochemistry
Faculty of Pharmaceutical
  Sciences
Fukuoka University
Fukuoka, Japan

Makoto Shimizu, PhD
Department of Applied
  Biological Chemistry
The University of Tokyo
Tokyo, Japan

R. W. Siggins
Department of Physiology
Louisiana State University School
  of Medicine
New Orleans, Louisiana

Rekha Singhal, PhD
Food and Fermentation Department
University Institute of
  Chemical Technology
Mumbai, India

Harjinder Singh
Riddet Centre
Massey University
New Zealand

Shinji Soeda
Department of Biochemistry
Faculty of Pharmaceutical
  Sciences
Fukuoka University
Fukuoka, Japan

M. A. Soobrattee
Department of Biosciences
Faculty of Science
University of Mauritius
Réduit, Mauritius

Gottumukkala V. Subbaraju
Laurus Labs Private Limited
Hyderabad, India

Masumi Suganuma, PhD
Department of Laboratory
  Animal Science
Nippon Medical School
Bunkyo-ku Tokyo, Japan

Young-Joon Surh
College of Pharmacy
Seoul National University
Seoul, South Korea

Linda Tapsell
National Centre of Excellence in
  Functional Foods
University of Wollongong
Wollongong, Australia

Jai C. Tilak
Radiation Biology
  and Health Sciences
  Division
Bhabha Atomic Research Centre
Mumbai, India

Robert J. Tomanek
Professor
University of Iowa Medical
  School
Iowa City, Iowa

Golakoti Trimurtulu
Laila Research Center
Vijayawada, India

Peter Van Veldhuizen
Cancer Research Unit
The University of Kansas
    Medical Center
Kansas City, Missouri

Somepalli Venkateswarlu
Laila Research Center
Vijayawada, India

Michael D. Waters
National Center for Toxicogenomics
National Institute of Environmental
    Health Sciences
Durham, North Carolina

Anthony Y. H. Woo
Department of Biochemistry
The Chinese University
    of Hong Kong
Hong Kong, China

Fatih Yildiz
Food Engineering Department
Middle East Technical University
Ankara, Turkey

Shirley Zafra-Stone
InterHealth Nutraceuticals
Incorporated
Benicia, California

R. Zhang
Department of Pharmacology
Kunming Medical College
Kunming, China

Y. Zhao
Dept. of TCM
    and Natural Drug Research
College of Pharmaceutical
    Sciences
Zhejiang University
Hangzhou, China

C. Zhou
Dept. of TCM
    and Natural Drug Research
College of Pharmaceutical
    Sciences
Zhejiang University
Hangzhou, China

Part I

History and Scope of Functional Foods Around the World

1 Scope of Conventional and Functional Foods in the U.S.A.

Jack N. Losso

CONTENTS

1.1   INTRODUCTION

The U.S.A. has one of the safest food systems and best health care in the world. People are living longer than ever before. Life expectancy averages 75 and 78.5 years for males and females, respectively. However, these long-living individuals are susceptible to debilitating chronic and long-term illnesses such as vision impairment, diabetes blindness, multiple sclerosis, arthritis, Parkinson's, Alzheimer's, and cancer. There are few effective treatments and no real cures, and health care costs for these diseases are rising. At the same time, most Americans understand the link between diet and health. Individuals seeking cost-effective and improved health status also understand that choosing a good diet and using dietary supplements are two of the most important achievable options. The majority of these individuals are turning to some special types of food.

1.2   DEFINITIONS

In the United States, different definitions are used by scientific, non-profit and federal organizations for functional foods.

The Institute of Food Technologists (IFT) defines *functional foods* as "foods and food components that provide a health benefit beyond basic nutrition (for the intended population) [1]. Examples may include conventional foods; fortified, enriched or enhanced foods; and dietary supplements. These substances provide essential nutrients."

The Institute of Food Technologists defines *nutrients* as "traditional vitamins, minerals, essential fatty acids for which recommended intakes have been established and other components that include phytonutrients, zoochemicals, and fungochemicals present in foods for which a substantial body of evidence (physical and physiological) exists for a plausible mechanism, but for which a recommended intake and function have not been definitively established." [1].

The *International Life Sciences Institute* of North America (ILSI) defines functional foods as "foods that, by virtue of physiologically active food components, provide health benefits beyond basic nutrition."

The *Institute of Medicine of the National Academy of Sciences* uses the following definition: "foods in which the concentration of one or more ingredients has been manipulated or modified to enhance their contribution to a healthful diet" [2].

The *American Dietetic Association* (ADA) classifies all foods as "functional" at some physiologic level [2]. The American Dietetic Association considers that by using moderation and a variety of foods, all foods can be incorporated into a healthful eating plan. The term "functional food" should not imply that there are good foods and bad foods.

The *National Institutes of Health* (NIH) and the *National Cancer Institute* (NCI) use the words "bioactive compounds" in foods which in fact translate into compounds in foods that deliver a health benefit beyond basic nutrition, as suggested by the ILSI.

The *Food and Drug Administration* (FDA) has no statutory definition for functional foods.

The U.S. Congress defines *dietary supplement* as a product taken by mouth that contains a "dietary ingredient" intended to supplement the diet. The "dietary ingredients" in these products may include vitamins, minerals, herbs or other botanicals, amino acids, enzymes, organ tissues, glandulars, metabolites, extracts, concentrates, tablets, capsules, softgels, gelcaps, liquids, or powders. Dietary Supplement Health and Education Act (DSHEA) classifies dietary supplements and ingredients as food.

A *dietary supplement* is defined as "a product (other than tobacco) intended to supplement the diet by increasing the total daily intake. " The product is intended for ingestion in pill, capsule, tablet, or liquid form.

A *"new dietary ingredient"* is one that meets the definition for a "dietary ingredient" but was not sold in the U.S. as a dietary supplement before October 15, 1994.

A *"fortified food"* contains a significant quantity of a nutrient that was not originally present in that food or was present only in nutritionally insignificant amount.

Medical food: a prescription food for a patient with special nutrient needs in order to manage a disease or health condition. The patient is under the physician's ongoing care. Examples include enteral foods for hospitalized patients and phenylalanine-free foods for use by persons with phenylketonuria.

Nutraceutical: any substance that may be considered a food or part of a food and provides medical and health benefits as well as prevention and treatment of disease.

A *drug* is a substance used to prevent, treat, cure or mitigate a disease. Most drugs are designed to bind with high affinity, often irreversibly, specifically to a protein, an enzyme, or receptor in a specific tissue, and with half-maximal activity at nanomolar to micromolar concentration. Drugs are designed to be very specific, but functional foods often are pleiotropic in their effects.

1.3　HISTORICAL EVOLUTION OF CONVENTIONAL FOODS IN THE U.S.A.

1.3.1　CONVENTIONAL FOODS IN THE U.S.A. BETWEEN 1400 AND 1699

This paragraph covers the evolution of foods in America before and after Christopher Columbus' visit. We acknowledge Key Ingredients, Wikipedia, and Food History Timeline on the internet as some of the major sources of our information. Long before Christopher Columbus came to America, Native Americans in and around North Dakota preserved *pemmican*, a sun-dried meat jerky (buffalo, venison, moose, or caribou) containing lean meat, rendered animal fats and berries, for months without refrigeration. *Pemmican* is still manufactured today by companies like Conagra Foods and used as high energy food by mountain climbers and other high altitude sports aficionados, as well as being promoted as low-carbohydrate food. When Christopher Columbus sailed in 1492 in the new world, he found that corn, potato, berries, tomato, bell pepper, chili pepper, vanilla, tobacco, beans, pumpkin, cassava root, avocado, peanut, pecan, cashew, pineapple, sunflower, petunia, black-eyed susan, dahlia, marigold, quinine, wild rice, cocoa (chocolate), gourds and squash—all were foods then unknown to Europeans. When Columbus returned to America in 1493, he and his crew enjoyed native wild fruits and vegetables, including pineapple. Green and red chili peppers were introduced in New Mexico in 1589 by a Mexican businessman, Don Juan de Onate and his crew. These peppers are still around and used as ingredients in many famous Southern dishes such as "chile con care" and the "El Diablo" hot spicy sauce. Corn porridge was a favorite dish of both Native Americans and colonialists. In the 16th century, Native American communities had a highly diverse diet that included plants, vegetables, meats, chili, locusts, spiders, lice, and insects. At the Thanksgiving celebrated in 1621, the Mayflower Pilgrims of Plymouth and the native Wampanoag community shared native foods and recipes from Europe. The dishes included boiled lobster, roasted goose, boiled turkey, rabbit stew, Indian corn pudding with dried berries, boiled cod, roast duck, stewed pumpkin, roasted venison with mustard sauce, wild fowl, a dish of fruit and Dutch cheese. Cranberry was a wild fruit grown in New Jersey. In 1690, rice was grown from Louisiana to South Carolina for use as a staple food and cash crop.

1.3.2   Conventional Foods in the U.S.A. Between 1700 and 1899

Between 1700 and1800, a mixture of cultures and cuisines that included Natives, Europeans, and African slaves was established. Slaves brought yams, watermelons, okra, several varieties of beans, and cultivated them in secret in the plantations' backyards. The slaves who cooked for their white masters in the plantations introduced new dishes such as candied yams and sweet potato pie by combining ingredients from their native foods with those of their masters' recipes. The Southern cuisine was born. African slaves cooked unused parts of pigs, including the snout, ears, feet, thighs, stomach, and intestines to flavor stews and green vegetables such as kale, collards, turnip, spinach, and mustard greens. Cornbread, grits, and hush puppies were slaves' daily recipes that have become part of today's Southern cooking. Barbecue originated from the West Indies. In the South, African slaves basted pork meat with sauce and cooked it slowly on open fire. "Barbecue" was born. At the same time, in the Southwest, beef was pit-cooked and was also named "barbecue." In the South, Cajuns from Nova Scotia and Creoles from the Caribbean combined their cooking skills to serve spicy hearty meats, seafood or soup together over rice, creating *gumbo* and *jambalaya*. In 1763, oyster, which was then eaten only by rich people and fishermen, made its entry in New York City and became the food of ordinary men and women. During the American Revolutionary War (1775–1783), soldiers ate their meals (cornmeal, bread, beef, pork, or fish) sweetened with maple sugar and molasses and flavored with dried apples or cranberries. In California, monks and priests from Mexico introduced the first "hot chocolate" by combining a popular corn-based hot drink known as "*atole*" with cinnamon, sugar, or bitter chocolate. Soybeans arrived in America in the early 1800s and farmers began to plant it around 1879. In 1804, Lewis and Clark's expedition encountered thousands of buffalo herds in what is now part of Montana. Meat became the focus of diet and these men feasted on 9 lbs of buffalo meat daily in this region. As early as 1828, "chili," a combination of pork, beef, native chiles, oregano, and garlic, became popular among cowboys in the Southwest. The pioneers on the Oregon Trail preserved food that could withstand 2000 mile journeys. They carried bacon, salted pork and beef jerky and dry beans but could not carry fresh fruits, vegetables or dairy products. When Chinese immigrants came to California around 1850 to work on railroad, in mines and on crop fields, they brought and drank lukewarm tea and ate abalone, cuttlefish, dried bamboo sprouts, dried oysters, dried mushrooms, dried seaweed and rice. The snack food industry was born in 1853 when George Crum, a chef at the Moon Lake Resort in Saratoga Springs, NY, created the first potato chip. During the Civil War (1861–1865), the mainstay of the Union soldier's diet was a flour biscuit called "*hardtack* " whereas the Confederate soldiers, had fewer choices, and ate bacon, cornmeal, tea, sugar or molasses, and occasionally vegetables. The prospectors of the California gold rush in the Sierra Nevada ate mostly meat and bread, drank coffee and tea with lots of sugar, and suffered of scurvy. The gold rush also saw the rise of the restaurant business around the California mining towns. Men did not want to cook; women took the opportunity and ran restaurants. Gold prospectors celebrated with expensive French food, champagne, and oysters when they hit it rich. The first modern chewing gum was developed in 1871 when Thomas Adams, then secretary to exiled president of Mexico living in Staten Island, NY, attempted to combine "chicle," a milky plant juice chewed by Mexicans, with rubber to make carriage tires. The result was not a tire, but the world's first chewing gum. Adams sold his idea to pharmacists who flavored the chewing gum with mint and licorice. In 1886, Dr. John Pemberton, a pharmacist, formulated a drink that contained cocaine, caffeine, sugar, citric acid and fruit oils for use as a soda fountain drink and tonic for digestion, along with other medicinal uses. He named the drink after two of its main ingredients: cocaine and caffeine. Coca–Cola was born. The drink started as a functional food [3] and was an immediate hit. Cocaine is no longer an ingredient, but some parts of the cocoa plant are still part of Coca–Cola formulation which has remained a closely-guarded secret even until this day. The world's first dry whole grain cereal breakfast was invented in 1863 by James Caleb Jackson who named it Granula. At the same time, Charles William Post introduced an easier to chew cold cereal known today as "Grape-Nuts." In 1869, Henry J. Heinz was selling grated horseradish, a rich source of glucosinolate, in clear jars; he also marketed the first sweet pickles. Tomato ketchup is a Nineteenth Century creation made from unripe tomatoes, with a light vinegar touch. Heinz argued with then United States Department of Agriculture (USDA) chief chemist that by adding more vinegar to ketchup, the product would be superior, safer, and tastier than the original watery product. The thick, viscous ketchup as we know it today was born. The rise of health food began when Dr. John Harvey Kellogg and his brother William Keith invented a cold breakfast cereal known as "corn flakes" in 1894, to replace meat for breakfast. Dr. Kellogg was a Seventh Day Adventist who imposed a strict vegetarian diet on his patients and family. The original corn flakes were an unpalatable gruel.

1.3.3 CONVENTIONAL FOODS IN THE U.S.A. BETWEEN 1900 AND 2006

George Washington Carver, considered by many to be the father of the peanut industry, began his peanut research in 1903. Peanut butter, hamburgers, hot dogs and ice cream cones were introduced to the public for the first time at the St. Louis World's Fair, also known as Louisiana Purchase Exposition, in 1904. Dr. John Harvey Kellogg was awarded the first patent that described peanut butter as a meat alternative. The Popsicle was invented by San Franciscan Frank Epperson in 1905; he was 11 years old at the time. Frank left a fruit drink outside, with a stirrer in it, on a very cold night. The next day, he found it frozen and named it the "Ep-cicle." At age 29, he patented "Ep-cicle" and named it "Popsicle." Frank also created the "Twin Popsicle" (for two children to share), as well as the Fudgsicle, Creamsicle, and Dreamsicle. Later, William Kellogg sweetened his flakes with malt, bought out his brother and started the Battle Creek Toasted Corn Flake Company in 1906, the parent company of Kellogg's. Crisco shortening was invented during World War I as a replacement for butter and eggs, which were hard to find. At the same time, the first grocery stores, such as Piggly Wiggly, appeared in Memphis, TN. Moon Pie, made of two graham cookies dipped in marshmallow and coated in chocolate, and RC Cola, both of which became Southern traditions, began in Chattanooga, TN. Pizza was introduced in 1905 when Gennaro Lombardi applied to the New York

City government for the first license to make and sell the food in his Italian–American restaurant. By 1925, pizza had spread to Connecticut and by the 1930s it was sold in Boston, MA and San Francisco, CA. By 1920 in California, *"chop suey,"* a Cantonese combination of bean sprouts, bamboo shoots, water chestnuts, mushrooms, meat or chicken, soy sauce and rice became one of the trendiest dishes of the Jazz Age. In 1921, hamburgers were served commercially for the first time at White Castle restaurant in Wichita, KS, at a price of $0.05 apiece. Years later, employees eliminated the need to flip the burger over during cooking by adding five holes to the patty which also helped it cook faster and more evenly. Fast food restaurants were born and quickly began improving. By the late 1920s convenience foods (canned tuna and pineapple, Kool-Aid, Jell-O, Velvetta, and peanut butter) were on sale. Ruth Wakefield, owner of the Toll House Inn in Whitman, MA invented chocolate chips cookies by substituting baker's chocolate with a semi-sweet chocolate. Thanksgiving was signed into law as a national holiday in 1941 by President Franklin D. Roosevelt. Cranberry sauce has always been included in the menu. The peanut butter and jelly (PB&J) sandwich was popular at home and for GIs during World War II. It has remained a favorite for both adults and children ever since. Casseroles and barbecues entered the food eating habits in the 1950 and have remained popular. In the 1950s in New York, Birdseye introduced a "quick–freeze method" to preserve and maintain the flavor and taste of meat, poultry, fruits, and vegetables during frozen storage. Carl Swanson introduced fast food at home with his brands of "TV dinners." Turkey, corn-bread stuffing and gravy, sweet potatoes, and peas packed in an aluminum tray could be heated in an oven and eaten while watching TV. The era of convenience was springing up. The sunflower, a species native to the U.S., began as an important agronomic crop in the U.S. in the 1950s, starting in North Dakota and Minnesota. In 1955, the first McDonald's restaurant opened in suburban Chicago, IL, Ray Kroc, the founder of McDonald's epitomized fast food. Cheap and tasty meals ranging from breakfasts to dinners (15 cents for a hamburger, 4 cents for extra cheese, 10 cents for fries, and 20 cents for a shake) could be served in less than 30 seconds. McDonald's was the "The All-American Meal" that has transcended cultures around the world. During the 1960s, the "Green Revolution" which combined pesticides, irrigation, and genetic engineering to develop high yield "seeds," promised to end world hunger. Scientists promised to double or triple harvest around the world using the same size of plots. The Green Revolution came and went away but world hunger did not. Astronauts' space missions revolutionized food science research and convenience significantly. "Space foods" were healthy and developed in freeze–dried forms that included drinks, dried fruit, PB&J, candy bars, nuts, sandwiches, liquid salt, and Tabasco sauce. In 1984, FDA made it mandatory to add sodium content on nutrition labels. Dietary fiber became a buzz word in nutrition circles. Fat substitutes began to be synthesized in research laboratories [4]. Fat substitutes are chemically modified carbohydrates (dextrins, maltodextrins, modified starch, cellulose, and various gums), microparticulated proteins (Simplesse by NutraSweet), or fat-based replacers (Olestra by Proctor and Gamble) that were developed to mimic the taste and mouthfeel of natural fat and provide 1–4 kcal/g. Fat substitutes are FDA approved for human consumption and are available in foods like potato chips. In the 1990s, saturated fatty acids, cholesterol, dietary fiber, and calories from fat content were all required by FDA on nutrition labels of certain foods sold in the U.S. As the 1990s went by, despite advances in medical, biomedical, and pharmaceutical research, age-related and chronic degenerative diseases were on the rise, health care costs were skyrocketing, and standard therapies for certain diseases were still limited. At the same time, consumers began to understand the relationship between diet and disease and the positive health benefits of non-nutrient bioactive compounds in foods. The changing regulatory environment and the technical advances that allowed the food industry to market health-promoting products to health-conscious consumers stimulated the functional foods industry to grow. Today the industry is at more than $39 billion annual business.

1.4 U.S. LEGAL STANDARDS AND REGULATIONS OF FOODS AND DIETARY SUPPLEMENTS

1.4.1 INTRODUCTION

The United States public policy process that affects the regulation of foods includes the legislative, executive, and judiciary branches. The lawmakers in Congress identify societal problems, identify options, and implement a policy to solve the problems. With regard to food, the Department of Health and Human Services (DHHS) and U.S. Department of Agriculture (USDA) are two cabinet-level departments in the executive branch that regulate foods. The judicial branch adjudicates disputes arising from regulations and enforces implementation of regulations governing health messages made by food and dietary supplement manufacturers. Federal agencies and commissions involved in regulating foods include the FDA of the DHHS, the food safety inspection service (FSIS) of the USDA, the federal trade commission (FTC), and the environmental protection agency (EPA). The FDA regulates food labeling, animal feeds, pet foods, shell eggs, seafood, dietary supplements, cosmetics, drugs, and medical devices. The USDA regulates meat, poultry products, and egg products. The FTC regulates advertising, including infomercials for dietary supplements and most other products sold to consumers. All these agencies work very closely with one another but may operate under different laws.

1.4.2 FEDERAL STANDARDS TO SUPPORT CLAIMS FOR CONVENTIONAL FOODS AND DIETARY SUPPLEMENTS

Congress has delegated to FDA the authority to establish the procedures and standards for health claims for conventional foods and dietary supplements. All foods (conventional foods, food additives, dietary supplements) are regulated by the Federal Food, Drug, and Cosmetic Act (FDCA) of 1938 as amended. U.S.C. 343 (r) (6) makes clear that no claim should be made that the food is intended to cure, mitigate, treat, or prevent a specific disease or class of disease. Manufacturers are required to provide pre-market notice and evidence of safety for any supplement the manufacturer plans to sell that contains a "new dietary ingredient." Since the FDCA does not provide a statutory definition for functional foods, the FDA has no authority to establish a formal regulatory category for these foods [5].

Three categories of claims can be used on the labels of foods and dietary supplements sold in the U.S. These are *health claims*, *nutrient content claims*, and *structure/function claims*. The responsibility for ensuring the validity of these claims rests with the manufacturer, the FDA, and the FTC.

1.4.2.1 Health Claims

Health claims characterize the relationship between a food substance (ingredient or food) and a disease (macular degeneration, cancer) or health-related condition (high cholesterol). Evidence supporting a health claim must be presented to FDA for review before the claim may appear on food labels. There are 3 levels of standards that the FDA has allowed to support health claims for foods and dietary supplements, namely (1) significant scientific agreement (evidence), (2) weight of the scientific evidence (emerging evidence), and (3) competent and reliable scientific evidence (overwhelming evidence of efficacy) for the bioactivity of the component.

Food and Drug Administration can exercise its oversight in determining which health claims may be appropriate on a food or dietary supplement label by using (1) the 1990 Nutrition Labeling and Education Act (NLEA), (2) the 1997 Food and Drug Administration Modernization Act (FDAMA) or (3) the 2003 FDA *Consumer Health Information for Better Nutrition Initiative*. Food and Drug Administration requirements for approved health claims can be found in 21 CFR Section 101.72 through 21 CFR Section 101.83.

1.4.2.1.1 NLEA Authorized Health Claims

Nutrition Labeling and Education Act (1990), the Dietary Supplement Act of 1992, and the DSHEA (1994) allow health claims to be made as a result of an extensive review of the scientific literature on a health claim petition, using the significant scientific agreement standard to determine that the nutrient/disease inverse relationship is well established. Food and Drug Administration may authorize health claim by regulation (21 U.S.C. 343 (r) (3) (B) (i)) when the agency determines that there is significant scientific agreement (SSA), among experts qualified by scientific training and experience to evaluate such claims, that the claim is supported by the totality of publicly available scientific evidence including epidemiological (relation between consumption and disease), mechanistic (in vitro and in vivo animal models interactions with the bioactive compound), and clinical trials (randomized controlled clinical trials). All available data support and qualified experts agree on the inverse relationship between the bioactive compound and disease. Health claims that meet the SSA (21 CFR 101.14), also known as approved health claims, are presented in Table 1.1.

1.4.2.1.2 Health Claims Based on Authoritative Statements

Food and Drug Administration Modernization Act of 1997 allows the use of an expedited process that authorizes manufacturers to use "authoritative statement-based health claims" from a scientific body of the U.S. Government with official

TABLE 1.1
Bioactive Compounds with Food and Drug Administration (FDA)-Approved Health Claims Based on Strong Evidence from Clinical Studies

Bioactive Compound	Food Sources	Health Claims	Recommended Amount or Frequency of Intake
β-Glucan	Whole grains	⇓ total and LDL cholesterol	3 g/day
Soluble fiber	Psyllium	⇓ total and LDL cholesterol	1 g/day
Phytosterols and phytostanols	Fortified margarines	⇓ total and LDL cholesterol	1.3 g/day for sterols and 1.7 g/day for stanols
Protein	Soy	⇓ total and LDL cholesterol	25 g/day

Source: From Hasler, C. M, Bloch, A. S., Thomson, C. A., Enrione, E., and Manning, C., *J. Am. Diet. Assoc.*, 104 (5), 814–826, 2004.

responsibility related to human nutrition (such as National Institute of Health [NIH] or Center for Disease Control [CDC]) or the National Academy of Science, when there is competent and reliable scientific evidence. Thus, foods high in whole grains are allowed to claim efficacy in preventing cardiovascular disease (CVD) and certain types of cancer. These claims cannot be used for dietary supplements. Manufacturers notify FDA 120 days before use of claims and if FDA fails to act, the claim is authorized by statute. Food and Drug Administration regulations in 21 CFR Section 101.14(c)(6) require that foods that are not dietary supplements and bear health claim should contain at least 10% or more of the reference daily intake (RDI) of daily reference value (DRV) for vitamin A, vitamin C, iron, calcium, protein or fiber before any nutrient addition unless it is exempted by FDA. For instance, diets containing foods that are good sources of potassium and whole grains qualify for health claims authorized under FDAMA (1997).

1.4.2.1.3   Qualified Health Claims
When the scientific evidence is not sufficiently established to meet the SSA standards, the FDA' 2003 *Consumer Health Information for Better Nutrition Initiative* allows manufacturers to use *qualified health claims* when there is emerging evidence for a relationship between a food, food component, or dietary supplement and reduced risk of a disease or health-related condition. Examples include FDA-approved qualified health claims for nuts and heart disease, and Se and cancer. Qualified health claims can be used for any food or dietary supplement that demonstrates that the "weight of scientific evidence" supports the proposed claims.

1.4.2.2   Nutrient Content Claims

Nutrient content claims characterize the level of a nutrient or dietary substance in a product, using terms such as *free*, *high* (hi in vitamin C), and *low* (low in sodium), or compare the level of a nutrient in a food to that of another food, using terms such as *more*, *reduced*, and *lite*. Percentage claims may be used to describe a percentage or comparative percentage level of a dietary ingredient for which there is no established daily value (e.g., 40% omega-3 fatty acids, 10 mg/capsule or twice the omega-3 fatty acids per capsule (80 mg) as in 100 mg of menhaden oil (40 mg)). Health claims and nutrient content claims must be reviewed by the agency prior to product marketing unless authorized by regulation. When FDA issues a new nutrient content claim regulation, such regulation can be used by any person who is not the petitioner and whose product contains the referenced nutrient at the required level. A manufacturer is not allowed to make a claim on labels (even if it was truthful and not misleading) about a food as "good source of a nutrient" for which no FDA nutrient claim regulation already exists. This also applies to nutrients that do not have a RDI.

Medical foods are regulated by 21 CFR Section 101.9(j) (8). These foods cannot claim to cure, mitigate, treat or prevent a disease. However, since the number of individuals who need medical foods is increasing, FDA is exploring ways to better regulate these products including establishing safety evaluations, standards for claims, and specific information required on the labels.

1.4.2.3   Structure/Function Claims for Conventional Foods

Structure/function claims consider the effect of a food on the structure or function of the body, such as "calcium builds strong bones," "fiber maintains bowel regularity" or "antioxidants maintain cell integrity." These claims are not health claims and not subject to FDA review and authorization but must be accurate, truthful, and not misleading (21 U.S.C. 343(r) (6); 21 CFR 101.93). Manufacturers have 30 days to notify FDA of their marketing plan but do not have to indicate the scientific evidence on the label. Several foods are being sold using structure/function claims. However, when FDA recognizes the use of non GRAS (Generally recognized as safe) ingredients, the agency can easily issue a warning letter with potentials of closing production or facilities.

1.4.2.4   Position of IFT vis-à-vis the FDA Regulations

The IFT 2004 expert panel does support the concept of qualified health claims [1]. However, the panelists believe that current FDA policies on functional foods specifically (1) limit the scope and accuracy of consumer information about functional foods; (2) hinder the development and marketing of innovative functional foods; (3) deny ascribed health benefits of functional foods to consumers; and (4) tie health benefits to very limited concept of "nutritive value." Rather, the panel recommends (1) the use of science-based policies to enhance the development and marketing of functional foods; (2) product labeling that accurately reflects the scientific evidence to avoid drug classification for functional foods; (3) use of broad-based scientific criteria that accurately addresses the links between health/nutrition and physiology, endocrinology, biochemistry, neurology, and genetics; and (4) that FDA prohibits claims that rely on very limited and preliminary information and protects consumers from limited scientific information.

1.4.2.5   Position of American Dietetic Association vis-à-vis FDA Regulations

The FDA regulations remain confusing. The ADA does not support the FDA acceptance of qualified health claims approach to labeling and believes that "*health and nutrient content claims authorized for foods and dietary supplements should be based on the totality of the publicly available scientific evidence ... and should not be preliminary and speculative*" [2]. The ADA recommends (1) an evaluation of the body of available scientific evidence prior to the development of consumer diet-health messages, (2) regulation of functional foods and dietary supplements to ensure that the products are safe, have been manufactured using recognized good manufacturing practices and all labels and claims are truthful, based on significant scientific agreement and not misleading, (3) the food industry should provide specific guidelines for future research and development in functional foods, and (4) manufacturers should disclose information on any ingredients (nutrients, phytochemicals, zoochemicals, fungochemicals, botanicals, amount available in a serving size) used to market the product.

1.4.2.6   Position of the Grocery Manufacturers Association vis-à-vis FDA Regulations

The Grocery Manufacturers Association (GMA), in a letter to FDA dated January 17, 2006 showed that member companies are interested in using health claims on food products but would like to see both unqualified and qualified health claims defined by FDA in a way that consumer can understand without confusion. The Grocery Manufacturers Association suggests alternative ways of conveying health claims on food labels. For instance, instead of the use of terminologies such as "significant scientific evidence," GMA suggests, for instance, the use of different approaches that help consumers understand health claims in context by using three levels of scientific evidence such as "conclusive scientific evidence shows that calcium reduces the risk of osteoporosis," "strong scientific evidence shows that nuts may reduce the risk of heart disease," or "limited scientific evidence shows that green tea may help reduce the risk of prostate and breast cancer." GMA also agrees with IFT which suggested the use of independent expert evaluations of proposed health claims. Food and Drug Administration-authorized wording for the difference between qualified and unqualified health claims cannot be easily understood.

1.5   STEPS TO BRING FUNCTIONAL FOODS ON THE MARKET

The IFT Expert Panel [1] suggested seven steps to successfully bring potential functional foods on the market. We suggest one additional step that we consider very essential for food manufacturers to sell and maintain their products on the market: that is taste, convenience, and trust.

1.5.1 Step 1. Identify Relationship Between Food Component and Health Benefits

The target for a functional food can be the reduction of CVD, hypertension or delay of macular degeneration. Once a link has been established between the functional food and body tissue, controlled studies using the functional food need to be carried out to confirm the observations.

1.5.2 Step 2. Demonstrate Efficacy and Determine intake Level Necessary to Achieve Desired Effect

Demonstrating efficacy is the ultimate "raison d'être" of functional foods and one that every researcher, product developer, regulatory agent, and consumer is searching for. The stability of functional foods during processing and storage, their bioavailability, and organ tropism need to be established. The efficacy of functional foods can be measured using biological end points or biomarkers such as physical performance, cognitive, behavioral, physical, or organ function (bone, gastrointestinal). C-reactive protein may be a biomarker associated with inflammatory conditions such as heart, obesity, perivascular, cancer, and many other chronic diseases [1]. The burden for demonstrating efficacy lies on the manufacturer and the burden for confirming efficacy lies on independent peer reviewers. The efficacy of functional foods can be evaluated using a combination of the Hill criteria [1] and some others. These include (1) the strength of association, (2) consistency of the observed association, (3) the specificity of association, (4) temporal relationship of the observed association, (5) dose-response relationship, (6) biological plausibility, (7) coherence of evidence, (8) the amount and type of evidence, (9) the quality of evidence, (10) the totality of the evidence, and (11) the relevance of the evidence to the specific claim. These criteria have been used with success by the Food and Nutrition Board of the Institute of Medicine to propose the Dietary Reference Intake and by the FDA to design criteria for structure/function claims for specific foods. For instance, daily intake of 1 g of β-glucan from whole oat products or 3 g of ω-3 fatty acids (from fish) may reduce the risk of CVD. Lutein is bioavailable and consumption of 6 mg daily may delay the onset and/or progression of macular degeneration. At 200 μg/d, Se may reduce cancer risk and mortality in humans [6]. The Selenium and Vitamin E Cancer Prevention Trial (SELECT), sponsored by the NCI, is a twelve year (2001–2013) intergroup Phase III, randomized, double-blind, placebo-controlled, population-based clinical trial, using African Americans of at least 50 years old and Caucasians of at least 55 years, designed to test the efficacy of selenium and vitamin E alone and in combination in the prevention of prostate cancer [7].

1.5.3 Step 3. Demonstrate Safety of the Functional Component at Efficacious Levels

GRAS components and pre-approved food additives must be safe at projected levels of use. However, they can be assessed by estimating the ranges of intakes among consumers before and after product introduction [1]. When the bioactive components of a food matrix are not known, safety data obtained from epidemiological observations of populations that regularly consume the functional food may be used to substantiate the safety of the bioactive compound. The safety levels of Se, ω-3 fatty acids, and β-glucan have been established. The assessment of safety should include bioactive compound bioavailability, estimated half-life in vivo, estimated dose-response for a range of potential effects, known pharmacologic/toxic effects, evidence of non allergenicity using in vitro systems or experimental animal models. The potential interactions between the compound and P-glycoproteins and cytochrome P540s need to be determined.

1.5.4   Step 4. Develop a Suitable Food Vehicle
for Bioactive Ingredients

In order to protect these bioactive compounds from degradation or unwanted interactions, suitable matrix or carriers are needed during processing, storage, and consumption. For compounds targeting the gastrointestinal tract, encapsulation techniques using cyclodextrin are ideal. For compounds targeting tissues such as the breast or prostate, encapsulation, and absorption enhancers may be needed. Details can be found in Chapter 23.

1.5.5   Step 5. Demonstrate Scientific Sufficiency of Evidence
for Efficacy

The responsibility for ensuring evidence for efficacy rests on the manufacturer and this needs to be confirmed by a generally recognized as efficacious (GRAE) panel made of scientists with excellent track records of scientific publications in the area of food science, nutrition, pharmacology, toxicology, or internal medicine. The panel could be an independent body or proposed by the manufacturers. The resulting GRAE report would be submitted to FDA for public review. Drake [8] has hypothesized that sodium selenite and methylselenic acid may be more potent than selenomethionine and Se-methylselenocysteine because in sodium selenite Se is in the $+4$ oxidation state and both compounds (sodium selenite and methylseleninic acid) exert their cancer chemopreventive effects by directly oxidizing critical thiol-containing cellular substrates compared to selenomethionine and Se-methylselenocysteine that lack oxidation capability. This hypothesis is supported by the work of Zhao et al. [9] who reported that methylselenic acid at 3–30 μM modulated expression of many androgen-regulated genes, suppressed androgen receptor (AR) expression at both mRNA and protein level, and decreased levels of secreted prostate specific antigen. However, Unni et al. [10] suggested that Se-methylselenocysteine inhibits DNA synthesis, total protein kinase C and cyclin-dependent kinase 2 activities, leading to prolonged S-phase arrest and elevation of growth-arrested DNA damage genes, followed by caspase activation and apoptosis in a synchronized TM6 mouse mammary tumor model. See also targets osteopontin gene. Mounting preclinical and clinical evidence indicate that indole-3-carbinol, a key bioactive food component in cruciferous vegetables, inhibits cancer by targeting p21, p27, cyclin-dependent kinases, retinoblastoma, Bax/Bcl-2, cytochrome P-450 1A1 and GADD153 [11].

1.5.6   STEP 6. TASTE, CONVENIENCE, AND TRUST

Before launching a functional food product, the manufacturer has to make sure that the product tastes good. Otherwise, it must be reformulated it until the taste is right. Although there is a shift toward sacrificing taste for nutrition, most consumers consider taste very essential. Products must fit consumers' busy life-styles and manufacturers should be aware that ready-to-eat, ready-to-heat, and on-the-go products are becoming more popular [3]. Consumers want to trust claims and get guidance from brand names they trust [3]. *"For consumers today, taste is king and pleasure is queen… If you don't get the taste right it doesn't matter what else you do… If taste is missing, the consumer is missing too"* [12]. *"It is the consumer who will decide whether a product is successful on the marketplace, and the most successful companies consider the consumer very early on in the product development process."*

1.5.7   STEP 7. COMMUNICATE PRODUCT BENEFITS TO CONSUMERS

Following validation of the health claim, manufacturers can communicate the benefits of the functional food for which evidence exists by using simple and easy to understand language on the labels. Labels represent a very good vehicle for educating the public on the health benefits of dietary components. The Institute of Food Technologists and the International Food Information Council (IFIC) have developed "Guidelines for Communicating the Emerging Science on Dietary Components for Health" [13].

1.5.8   STEP 8. CONDUCT IN-MARKET SURVEILLANCE TO CONFIRM EFFICACY AND SAFETY

Once the functional food has been introduced into the market place, it is important to determine whether the intended intake has been achieved and the bioactive compound is being absorbed and utilized. Bioavailability can be measured by analyzing blood and body fluid samples from the consumers and measuring the levels of the compound, or its bioactive metabolites, in circulation or target tissue, as well as the health effects. Once information has been obtained on compound bioavailability, efficacy can be determined on a short and long term basis. The responsibility rests on the manufacturer but FDA should be part of the process.

1.6   FUNCTIONAL FOODS ON THE MARKET AND CURRENT TRENDS

Most functional foods on the market are products derived from discoveries in nutritional science or epidemiological observations and few are from deliberate

TABLE 1.2
Food and Drug Administration (FDA)-Approved Conventional Foods Based on Weak or Moderate Evidence from Clinical Studies

Bioactive Compound(s)	Food Sources	Health Claims	Recommended Amount or Frequency of Intake
Resveratrol	Grape, red wine, peanut	⇓ Platelet aggregation	8–16 oz/day
Proanthocyanidins	Cranberry	⇓ Urinary tract infection	300 ml/day
Prebiotics	Jerusalem artichoke, garlic	⇓ Blood pressure	3–10 g/day
Probiotics	Fermented dairy products	Support GI health	1–2 billion CFU/day
n-3 Fatty acids	Egg with ω-3 fatty acids	⇓ Cholesterol	Unknown
Organosulfur	Garlic	⇓ Cholesterol	600–900 mg/day

Source: From Hasler, C. M, Bloch, A. S., Thomson, C. A., Enrione, E., and Manning, C., *J. Am. Diet. Assoc.*, 104 (5), 814–826, 2004.

research strategy to develop functional foods [3]. These include (1) Foods with very strong evidence of efficacy and health claims status through clinical trials (Table 1.1). (2) Food and Drug Administration-approved conventional foods with weak evidence or moderate evidence of efficacy through clinical studies (Table 1.2).

TABLE 1.3
Food and Drug Administration (FDA)-Approved Conventional Foods Based on Weak or Moderate Evidence from Epidemiological, Cell and/or Animal Studies

Bioactive Compound(s)	Food Sources	Health Claims	Recommended Amount or Frequency of Intake
Catechins	Green tea, cranberry	⇓ risk of certain types of cancer	4–6 cups/day
Lutein/Zeaxanthin	Spinach, kale, collar green, sweet potato leaves	⇓ risk of macular degeneration	6 mg/day
Lycopene	Tomato/processed tomato products	⇓ risk of prostate cancer	½ cup/day (30 mg or 10 servings/week)
Conjugated linoleic acid	Milkfat, beef, lamb, turkey	⇓ risk of breast cancer	Unknown
Glucosinolates	Cruciferous vegetables	⇓ risk of certain types of cancer	½ cup/day

Source: From Hasler, C. M, Bloch, A. S., Thomson, C. A., Enrione, E., and Manning, C., *J. Am. Diet. Assoc.*, 104 (5), 814–826, 2004.

(3) Foods with weak to moderate evidence of efficacy through epidemiological observations, cell cultures, and animal studies (Table 1.3); and (4) Foods or bioactive compounds with emerging evidence of efficacy through epidemiological observations, cell cultures, and animal studies, but no extensive clinical evidence of efficacy. These include (1) functional foods from animal and avian sources such as lactoferrin, conjugated linoleic acid, vitamin D, butyrate, prebiotics (lactulose, inulin, fructooligosaccharide, and oligosaccharides), carnosine, taurine, coenzyme Q10, collagen, vitamin K2, phosvitin, and protamine; (2) functional foods from plant sources such as isoflavones (genistein, daidzein), phytoalexins (resveratrol), protein protease inhibitors (Bowman–Birk inhibitor), inositol, saponins (diosgenin, soyasaponins), phytosterols (β-sistosterol), anthocyanins, procyanidins, vitamins (C, E, K1, K2), lignans (enterodiol, enterolactone), glucans (β-glucan), carotenoids (lutein, zeaxanthin, lycopene), glucosinolates (sulforaphane), indole carbinols, flavonoids (kaempferol, quercetin), terpenes (limonene), quinones (thymoquinone), sulfoxides (alliin, allicin), sulfides (diallyl sulfides), catechins (epigallocatechin gallate), coumarin (scopoletin), pectin, fiber, gingerol; tocopherols, tocotrienols, waxes (policosanols, octosanol); (3) functional foods from fungi such as β-glucan, selenomethionine, mucoprotein; (4) functional foods from marine food sources (collagen, protamine, omega 3 fatty acids). Clinical trials are expensive and time-consuming and may not be necessary for every functional food. Epidemiological observations of individuals in communities where consumption of the bioactive compounds is regular may help further the acceptability of these foods as being functional.

1.7 FUTURE TRENDS AND RESEARCH NEEDED

Top ten trends in functional foods [14] are presented in Table 1.4. The outlook for the future of functional food is good. The ever-rising cost of health care, the rising proportion of elderly people, and the burden of lifestyle diseases such as diabetes, cancer, osteoporosis, CVD, and the well known link between diet and health are cause for need for more research on functional foods. Research on functional foods is needed to identify healthy raw materials, preserve and improve their positive attributes throughout the manufacturing process, ensure stability during storage and gastrointestinal transit (for some products), and organ tropism after consumption. A toxicological and pharmacological approach that considers the safety issues of functional foods is needed even though FDA does not consider conventional foods to be drugs. An approach that integrates insights into consumer needs, demands, and acceptance, and a structured scientific research process using proteomics, nutrigenomic, metabolomic, and bioinformatics, genetic susceptibility, individualized health traits, food processing and preservation, bioavailability and organ tropism is needed. Research needs involve two major areas: products providing physiology-based or disease reduction benefits (heart, osteoporosis, cancer) and psychology/behavior-based or daily benefits (stress, lack of energy, mental, and physical appearance).

TABLE 1.4
Top 10 Trends in Functional Foods

Trend #	Trend	Target Consumer Group	Comments
1	Addressing the kids	Kids pre-school to adolescent	Adapt kids early to bitter and less sugared functional foods [15]. Formulate balance diets for children (more nutrients, less fat, calories, sugar, and sodium) [14]
2	Light and lower calories products	All age groups	Start early
3	Phytochemicals	Baby boomers	Phytochemicals rich in antioxidants should be part of healthy eating for all ages
4	Multi-benefits foods	Baby boomers, people age 50 and above	Products that target more than one risk factor are in high demand. Heart disease, obesity, cholesterol, cancer are targets
5	Healthy fats	Baby boomers, all ages	Omega-3s, low in saturated fat, cholesterol-reducing spreads are the most appealing fat-directed label claims
6	Mature matters	Baby boomers	Preferences given to food products that aid against joint pain/arthritis, osteoporosis, digestive problems, vision, blood glucose, and pressure
7	Glycemic, gluten and whole grains	All age groups	All age groups should have interest in these products
8	Natural solutions	Organic users	Only 10% of consumers consume these products
9	Boosting performance	Adolescents, younger and older boomers	All need extra energy and want to stay healthy
10	Fun favorites	Human and pets	Taste, convenience, quality, premium ingredients, and value remain important despite healthy choices.

1.7.1 CONSUMER AND CHILDREN EDUCATION

Consumer and child education about the link between nutrition and health is important and the design of such education should introduce the benefits of whole grains and other functional foods to school-age children and parents as early as possible. Healthy choices will reduce the need for FDA approval of functional food benefits. When consumer becomes well informed, manufacturers can launch products without specific FDA approved health claims and expect consumers to shift their purchasing desires to functional foods.

Children can easily adapt to new diets, tastes, flavors, and eating habits if exposed early and continuously. The purpose of the functional foods revolution should not be to sugar-coat existing nutritionally poor foods with functional foods for the children. Efforts should be made by manufacturers to develop future foods using healthy raw ingredients and preserving their contents and attributes throughout the manufacturing process until consumption. These newly-developed tastes should be part of the healthy foods for children.

1.7.2 THE BENEFITS OF WHOLE GRAIN FOODS AND FRUITS AND VEGETABLES

Most ancient civilizations recognized the benefits of whole grain foods (which include the bran, germ, and endosperm in their naturally occurring proportions) and used them to produce foods with health enhancing properties. Epidemiological observations show that diets rich in vegetables, fruits, grains, and fish, lean meat, and certain oils are associated with sustained health benefits. Explaining the mechanism(s) of activities of each molecule in such foods is difficult and tedious even with today's available technologies. However, it has been recognized that whole grain food constituents together may create a "hurdle effect" or "synergistic effect" that disease stimulators may not overcome and this may explain the health-enhancing attributes of these foods. When it comes to chronic diseases and functional foods, the pharmaceutical-inspired approach of looking for "one magic molecule" is not working, because most chronic diseases use several pathways to proliferate and the magic bullet that can counteract all the known and unknown pathways of chronic diseases has not yet been identified. More research is needed to understand the synergistic and matrix effects of whole foods on healthy living. Whole grains increase the bioavailability of folate, reduce homocysteine plasma level as well as systolic blood pressure by increasing potassium and magnesium levels [16,17]. The fact that whole wheat flour promotes insulin sensitivity relative to white flour and yet has a near identical glycemic index suggests that certain nutrients such as magnesium or phytochemicals in whole wheat, depleted by the refining process, promote preservation of insulin sensitivity [18]. Whole grains deliver as many if not more phytochemicals and antioxidants than do fruits and vegetables [19]. Bioactive compounds in garlic, i.e., alliin, steroidal saponins, lectins, S-allylmercaptocysteine, diallyl disulfide, and oligosaccharides act synergistically to produce the health benefits of garlic [20,21]. While Healthy People 2010 (DHHS) recommends 3 servings of whole grains per day, others have suggested 5–10 servings a day for our busy society. The ADA recognizes that whole foods as well as other foods (functional, fortified, enriched, or enhanced foods) have a potentially beneficial effect on health when consumed as part of a varied diet on a regular basis, at effective levels [2].

1.7.3 FUNCTIONAL FOODS FOR CARDIOVASCULAR DISEASE

By 2013, the following chronic diseases will top major health issues in the U.S.: osteoporosis, heart disease, cancer, high blood pressure, high cholesterol, diabetes, arthritis, joint chronic pain, memory loss, vision loss, and bladder infection [22]. Cardiovascular disease has been the leading cause of death in the U.S. since 1918. Risk factors include obesity, type 2 diabetes, rheumatoid arthritis, cholesterol, overconsumption of energy-rich foods with little or no regular physical activity, hypertension, circulating oxidized low-density lipoprotein, high sodium intake, low intake of potassium and calcium, low level of nitric oxide, homocysteine, occlusive thrombosis, inflammation, and oxidative stress. Decreased intakes of sodium alone, and increased intakes of potassium, calcium, and magnesium each decreases elevated blood pressure. High intake of fruits and vegetables (500 g/day) has been reported to cause about 11% decrease in homocysteine while increasing folate by as much as 15%. Functional foods that help decrease arterial blood pressure (quercetin, allicin) are needed. The benefits of whole foods or a combination of foods has often been better that individual elements studied in isolation which show no efficacy in vivo in humans. Phytosterols (mainly campesterol and sitosterol) and their respective stanols (5 alpha-saturated derivatives) lower serum cholesterol but increase serum plant sterol levels, providing up to 20%–25% reduction of serum cholesterol over a short period of time [23]. Whole grains containing magnesium, potassium, vitamin E, folate, fiber, and melatonin may protect better against CVD than refined foods to which folate and other micronutrients are added back [17]. Effective functional foods may have a pleitropic effect against several of the pathways associated with CVD including activation of antioxidant defenses and anti-inflammatory activities through inhibition of oxidative stress-induced transcription factors (e.g., NF-κB, AP-1), cytotoxic cytokines and cyclooxygenase-2.

Fermented milk products containing angiotensin converting enzyme inhibitors which may lower blood pressure may be a good strategy to produce a functional food with antihypertensive activity. Increasing consumption of marine n-3 fatty acids and decreasing certain saturated fatty acids may reduce hemostatic CVD risk factors [24].

1.7.4 FUNCTIONAL FOODS FOR CANCER

The six essential alterations in cell physiology that collectively dictate malignant growth include (1) self-sufficiency in growth signals, (2) insensitivity to growth-inhibitory signals, (3) evasion of programmed cell death (apoptosis), (4) limitless replicative potential, (5) sustained angiogenesis (6) and tissue invasion and metastasis [25]. Cancer like many other angiogenic diseases is associated with aberrantly regulated apoptotic cell death that leads to inhibition of apoptosis and disease propagation. The disease is multifactorial and has multi origins. When detected early some cancers (colon, breast, prostate, bladder, skin) have a good prognosis. However, despite advances in detection technologies, there are still cases where false negative diagnostics can reach 20% and when the disease is detected late, disease-free survival rate is low and prognosis often not good. Since most cancers involve multiple pathways with multiple and redundant stimulators, it is often late to return the body to the initial homeostatic stage. Cancer-related anorexia/ cachexia syndrome is associated with tissue wasting, weight loss, reduced food intake, systemic inflammation, and poor performance [26,27]. At the time of diagnosis, 80% of patients with upper gastrointestinal cancers and 60% of patients with lung cancer have already experienced substantial weight loss and the prevalence of cachexia increases from 50 to >80% before death, and in more than 20% of patients, cachexia is the main cause of death [26]. The benefits of functional foods, when initiated early, are that regular consumption at level that can keep the body under homeostatic stage may delay the onset and progression of cancer. Several functional foods have been identified mostly by epidemiological observations. The list includes soybean isoflavones, fenugreek, curcumin, green tea catechins, lactoferrin, milk vitamin D and butyrate, and many others. This paragraph is described in details in several chapters in Section 4.

1.7.5 FUNCTION FOODS FOR DIABETES

Diabetes is a disorder of chronic hyperglycemia, and glucose participates in diabetic complications such as the formation of advanced glycation end (AGE) products, hypertension, atherosclerosis, cataract, blindness, proteinuria, kidney disease and failure, heart disease and stroke, high blood pressure, nervous system disease, amputation, dental disease, complications of pregnancy and high susceptibility to deadly pneumonia or influenza. Chronic hyperglycemia accelerates the reaction between glucose and proteins and leads to the formation of AGE products, which form irreversible cross-links with many macromolecules such as collagen. In diabetes, these AGE accumulate in tissues at an accelerated rate. In diabetes, AGE products act as agonists to advanced glycation end product receptors (RAGE) of various cell types, which stimulate the release of profibrotic growth factors, promote collagen deposition, increase inflammation, and ultimately lead to tissue fibrosis. In the heart, large vessels, and kidney, these reactions produce diastolic dysfunction, atherosclerosis, and renal fibrosis. Type 1 diabetes accounts for about 5%–10% of diagnosed diabetes in the United States. Treatment for Type 1 diabetes includes taking insulin shots, making wise food choices, exercising regularly, taking aspirin daily (for some), and controlling blood pressure and cholesterol. Type 2 diabetes accounts for 90%–95% of people with diabetes and is associated with older age, obesity, family history of diabetes, previous history of gestational diabetes, physical inactivity, and ethnicity. About 80% of people with type 2 diabetes are overweight. Treatment includes using diabetes medicines, making wise food choices, exercising regularly, taking aspirin daily, and controlling blood pressure and cholesterol. It appears that, except for insulin, for types 1 and 2 diabetes, treatments are similar. However, when proper treatment is not given or followed, diabetes complications can be devastating. Diabetes is widely recognized as one of the leading causes of premature death and disability in the United States. In 2000, diabetes was the sixth leading cause of death. About 65% of deaths among those with diabetes are attributed to heart disease and stroke. The U.S. government statistics show that, in 2002, diabetes cost the country $132 billion. Indirect costs, including disability payments, time lost from work, and premature death, totaled $40 billion; direct medical costs for diabetes care including hospitalizations, medical care, and treatment supplies, totaled $92 billion.

Since diabetes is a multifactorial disease with multiorgan consequences, a multidrug approach that would maximize the blockade of several pathways at different levels, without toxic side effects, may slow down the progression of the disease. Modern medicine is trying to turn diabetes into a manageable chronic disease. At the same time there is growing interest and evidence concerning the role played by functional foods for the prevention of the progression of diabetes complications. Functional foods that contain bioavailable bioactive compounds effective against AGE product accumulation, oxidative stress, RAS, and pathological angiogenesis may find applications in reducing or delaying the onset and/or progression of diabetes complications. Fenugreek has been reported for its anti-diabetic effects which derive from a combinatorial effect of several bioactive compounds such as galactomannans, saponins, flavonoids, amino acid hydroxy isoleucine, selenium, and others [28]. Curcumin inhibits AGE products and oxidative stress, matrix metalloproteinases, as well as promoting wound healing [29–32]. It also modulates several important molecular targets, including transcription factors (e.g., NF-κB, AP-1, Egr-1, beta-catenin, and PPAR-γ), enzymes (e.g., COX-2, 5-LOX, iNOS, and hemeoxygenase-1), cell cycle proteins (e.g., cyclin D1 and p21), cytokines (e.g., TNF, IL-1, IL-6, and chemokines), receptors (e.g., EGFR and Her-2), and cell surface adhesion molecules which are biomarkers of diabetes progression and complications [33]. Six-gingerol from the root of ginger (*Zingiber officinalis*) exhibits a biologic activity profile similar to that of curcumin. *Chapter 22* of the book provides more insights into functional foods and diabetes.

1.7.6 FUNCTIONAL FOODS FOR THE ELDERLY

As the elderly population increases, so will the prevalence of age-related diseases [33]. Normal aging is accompanied by loss of natural teeth (often replaced by uncomfortable dentures), declines in motor, cognitive, and physiological performances which in turn require specific foods to address this group [34]. Lactose intolerance due to lactase decline, increased triglyceride and blood pressure levels, arrhythmia, risk of thrombosis, declined vision and vision loss and cardiovascular problems are common health problems that affect elderly people. Older people have metabolic needs different from growing younger populations and malnutrition is a common problem in the elderly in need of long-tem care. Lactose intolerance is widening among non-Caucasians as well as Caucasian Americans and may be a risk factor for developing osteoporosis. Aging impairs mitochondrial function causing depletion of endogenous antioxidant defense enzymes, overproduction of reactive oxygen species and biomarkers of peroxidation, and induction of heat shock proteins. As a result, reactive oxygen species (ROS) overload also causes DNA mutations that in turn can cause mitochondrial failure, the latter being associated with myocardium ischemic injury, Alzheimer's and Parkinson's diseases, epilepsy, and Wilson's disease. Mitochondrial dysfunction also causes opening of mitochondrial membrane and release of apoptotic biomarkers such as caspases, and cytochrome *c*. With aging, the gastrointestinal tract undergo physiological and biochemical changes such as development of atrophic gastritis and the inability to secrete gastric acid which affect one third of older people in the U.S. [35]. *Helicobacter pyloris* has been identified as risk factor in the majority of cases and is associated with malabsorption of micronutrients such as iron, folate, calcium, vitamin K and B12, and lactose intolerance.

Several functional foods have been identified as potential inhibitors of mitochondrial oxidation. Foods rich in antioxidants that can suppress radical buildup and inhibit mitochondrial oxidation, protect endogenous antioxidant enzymes, forestall and reverse the deleterious physiological effects of aging include flavonoids, melatonin, coenzyme Q10, L-carnitine, carnosine, iron, and zinc chelators. The use of staple foods to develop enriched and novel foods as potential vectors for these functional ingredients can increase dietary intakes of needed functional foods

Age-related macular degeneration (ARMD) is a painless disease that affects the macula and blurs the sharp, central vision. The etiology of ARMD is complex; some risk factors have been identified and include light exposure, oxidative stress, the pathologic growth of new blood vessels in the eye of premature infants and diabetic and sickle cell anemia patients, smoking, age, age-related accumulation of lipofuscins in the retinal pigment epithelial cells (RPE), Stargardt's disease, genetic and environmental factors [36–38]. Age-related macular degeneration rarely affects people less than 55 years old, but causes reduced visual acuity and legal blindness in 10%–25% of people at age 65 and over. Over three million Americans aged 40+ are visually impaired and these numbers are expected to double by the year 2030 [39].

There are two types of ARMD, the dry form and the wet form. The progressive loss of vision resulting from alterations of pigment distribution, loss of RPE cells and photoreceptors, and diminished retinal function due to an overall atrophy of the cells has been termed "atrophic (or dry) ARMD." Among the risk factors, the accumulation of lipofuscin *N*-retinyl-*N*-retinylidene (A2E) in the RPE cells of the eyes over time has been identified as one of the major factors. There is no generally accepted treatment for dry ARMD, except for some palliative treatment such as visual aids and rehabilitation of low vision [40]. Isotretinoin (13-*cis*-retinoic acid or Allitane), a drug reported to cure A2E-related ARMD causes several side effects which include depression, suicide and birth defects, which preclude long-term therapy with this product [41]. The other form of ARMD is the wet, exudative, or neovascular ARMD. Wet ARMD involves proliferation of abnormal choroidal vessels, which penetrate the Bruch's membrane and RPE layer into the subretinal space, thereby forming extensive clots and/or scars. Treatment options for wet ARMD include (1) anti-angiogenic compounds that inhibit the choroidal neovas-cularization, (2) photodynamic therapy, and (3) the NIH Age-Related Eye Disease Study (AREDS) antioxidant supplementation strategy [42]. Several angiogenic factors such as the VEGF, IL-8, and MCP-1 have been associated with the formation of blood vessels in the eye; however, angiogenesis inhibition alone is not sufficient to restore the function of a macula affected by ARMD [43]. Since ARMD is a multistep disease with multiple signaling pathways that develops over several years, there is a growing interest in and evidence for the role of dietary compounds that can be used metronomically (low dose over a long period of time) to (1) inhibit A2E accumulation in RPE cells, (2) inhibit choroidal neovascularization, and (3) suppress oxidative stress-mediated RPE apoptosis for early disease prevention. Lutein at 6 mg/day has been reported to delay the onset and/or progression of macular degeneration by as much as 43% [44]. Lutein and zeaxanthin were more potent than tocopherol in protecting retinal pigment epithelial cells against lipofuscin photooxidation [45]. Anthocyanins from bilberry showed ability to protect against A2E permeabilization through RPE cells [46]. Other functional foods with ability to permeate the eye-brain barrier and interfere with A2E or other lipofuscins accumulation during normal aging or diabetes-enhanced retinal angiogenesis need to be identified.

1.7.7 FUNCTIONAL FOODS FOR ATTENTION DEFICIT HYPERACTIVITY DISORDER

Attention deficit hyperactivity disorder (ADHD) is a common disorder that afflicts 3%–7% of school age children [47,48]. The prevalence of current adult ADHD has been estimated at 4.4% [49]. Catecholamines dopamine dysfunction or excess dopaminergic activity in particular dopamine D2 receptor and norepinephrine dysfunction are both important in the pathophysiology of ADHD. Purported risk factors include genetic predisposition, iron deficiency [50], deficiencies or imbalances in certain highly unsaturated fatty acids [47], exposure to non-nutritional food additives (L-glutamic acid, quinoline yellow, aspartame, brilliant blue) during the critical development window [51], exposure to salicylates, sugar, food colors and certain flavors [48]. Dietary interventions such as the Feingold Diet (removal of salicylates, artificial colors, flavors, preservatives), supplementation with high doses of vitamins and/or minerals, avoidance of sugars, use of essential fatty acids EPA and DHA (show promising results), flax oil in combination with vitamin C (showed promising results) have been suggested [47,48]. Results have been conflicting because ADHD is not a one size fits all disease. Additional research is needed to find out foods that may prevent ADHD and enhance learning capability and academic performance while identifying dietary factors that stimulate the onset of the disorder.

1.7.8 FUNCTIONAL FOODS FOR SATIETY

The world-wide epidemic of obesity has brought appetite control to the forefront of functional foods research. Dietary restriction and increased physical activity can help individual lose weight, but these lifestyle changes are difficult to implement and maintain. The feeling of satiety help individuals eat less and store fewer calories. Patterns of neural activity related to food motivation begin in childhood implying that obese children and adults may carry abnormal hunger and satiation mechanisms early in life [52]. Functional foods that can affect energy metabolism and fat partitioning, or control appetite by triggering gut hormones such as cholecystokinin (CCK) and GLP-1 or inhibit ghrelin are needed as dietary approach to weight control [53,54]. Dietary fats such as medium chain triacyglycerols and diacylglycerols may increase the feeling of satiety. Medium chain triacylglycerols are catabolized to 2 free fatty acids and a 2-monoacylglycerol molecule. Diacylglycerols are catabolized to 2 fatty acids and a glycerol moiety and are more thermogenic than triacylglycerols. Diacylglycerols have a greater effect on loss of fat mass due to greater fat oxidation, energy expenditure, and reduction of appetite. Cotton oil is a natural source of diacylglycerol and contains 9.5% of diacylglycerol [55]. Medium chain triacylglycerols in combination with plant sterols may increase satiety and at the same time prevent adverse cardiovascular side effects of medium chain triacylglycerols, the latter being associated with increase in cholesterol and LDL-cholesterol. Most spices are flavoring ingredients that do not add calories to foods. Consumption of thermogenic ingredients such as spiced foods can lead to greater thermogenesis, satiety and prevention of obesity. Capsaicin (hot peppers), piperine (black pepper), gingerols (ginger), and mixed spices (black pepper, coriander, turmeric, red chili, cumin, and ginger) may bind to receptor potential vanilloid receptor 1 (TRPV1) which may contribute the their bioactivities [56,57]. Clinical trials with capsaicin showed that oral ingestion of 0.9 g of red pepper (0.25% capsaicin) in capsules or tomato juice over two days reduced daily energy intake by 10% and 16%, respectively [58]. Habitual high intake of green tea containing caffeine was associated with greater weight maintenance, thermogenesis and fat oxidation [59]. The anti-obesity activity of green tea, which is controversial, may be associated with catechin inhibition of O-methyl-transferase, the enzyme that degrades norepinephrine [60]. Increase in thermogenesis induced by caffeine is probably based on inhibition of adipocyte phosphodiesterase-induced degradation of intracellular cyclic AMP. However, caffeine stimulated less thermogenic response in obese subjects, but caused a high and sustained response in lean women [61]. Clinical trials with coffee showed that long-term high coffee intake can lead to higher thermogenesis and fat oxidation and greater weight loss [58]. A combination caffeine-catechin induced increase in daily energy expenditure but was associated with increased systolic and diastolic blood pressure [62]. However it was suggested that a combination of functional ingredients such as caffeine and catechin with exercise would promote a decrease in blood pressure and stimulate metabolism while sustaining weight loss.

1.7.9 FUNCTIONAL FOODS AND IMMUNE ENHANCEMENT

Deficiency of one or more essential nutrients impairs immune function and increases susceptibility to infectious pathogens. Basic research and clinical research on probiotics have suggested and proven the beneficial effects of probiotics on different intestinal and extraintestinal diseases. Most probiotics fall into the group of organisms known as lactic acid-producing bacteria and are normally consumed in the form of yogurt, fermented milks or other fermented foods. Diarrhea, gastroenteritis, irritable bowel syndrome, inflammatory bowel disease (Crohn's disease and ulcerative colitis), depressed immune system, lactose indigestion, infant allergies, *H. pylori* infection, hypertension, cancer, and genitourinary tract infections may be prevented with probiotics [63–65]. Most probiotics belong to the *Lactobacillus* and *Bifidobacterium* genera. Among the Lactobacillus, Actimel®, Yakult®, ProViva®, Actifit$^{Plus®}$, Gefilus®, LC1®, and various brand names of *Bifidobacterium lactis* have been used in clinical trials and dominate the more than $1 billion probiotic industry [64]. Functional foods that can cause delayed-type hypersensitivity responses, increase the proportions of helper T cells, activate B cells, enhance neutrophil phagocytosis and attenuate declines in cytotoxic/suppressor lymphocytes need to be identified. Intake of 100–400 µg Se/day is safe and may optimize immune function [66].

1.7.10 FUNCTIONAL FOODS FOR ENDURANCE PERFORMANCE

The use of performance-enhancing supplements by children and adolescents to fight tiredness, stress, and increase alertness is on the rise. There is interest in the ergogenics such as arginine aspartate, taurine, caffeine, hydroxycitrate, and creatine that are used to enhance endurance and performance [67–69]. Most ergogenics may work by promoting fatty acid utilization as an energy source and glycogen sparing. More of these products are highly in demand as many people try to overcome tiredness and fatigue.

1.7.11 FUNCTIONAL FOODS FOR JOINT PAIN AND RELATED CHRONIC DISEASES

Beside cancer, diabetes, and CVD, there are several other chronic degenerative diseases characterized by excessive inflammation and regular joint pain. Multiple sclerosis, osteoporosis, arthritis and rheumatoid arthritis are typical cases. Functional foods that can be pluripotent in alleviating the pain caused by these diseases are needed. Fenugreek, omega-3 fatty acids, curcumin, capsaicin, gingerol, soybean isoflavones, green tea catechins, garlic sulfides, glucosamine broccoli glucosinolates and butyrate are typical of products that may fill the need [33,70–74].

1.7.12 FUNCTIONAL FOODS FOR SKIN HEALTH

Functional foods for healthy skin has been a popular topic in recent years. Name like cosmeceuticals have been used to indicate functional food for cosmetic purposes. Plants bioactive compounds are commonly used in cosmetics. The idea of eating well for a healthy skin is an extension of the potential of functional on overall human health. Carotenoids are suitable candidates for photoprotection because of their perceived protective role in the macula lutea [75]. Diets rich in vitamin D and carotenoids and low in alcohol may reduce the risk for melanoma [76].

1.7.13　Delivery for Bioactive Ingredients

Discussed in Chapter 33.

1.7.14　New and Existing Biomarkers for Functional Foods

The identification of biomarkers of health and the development of food ingredients and technologies that address the issues of biomarkers will be keys to the success of functional foods. Whereas biomarkers have often been biological endpoints associated with health, it would be preferable to identify markers associated with disease development rather than disease endpoint. Exposure markers are needed to assess intake, bioavailability and utilization of bioactive compounds in foods. Proteinase levels in the body can be used as biomarkers of health and performance. Ghrelin, CCK and GLP-1 may be used as biomarkers of satiety. The effects of diets on genes need to be studied. Clinical trials that evaluate the effects of diets on biomarkers are needed.

1.7.15　Research on Behavior

This will encompass the physiology of taste and food choice, the psychology of food selection and consumption to motivate consumer to shift behavior to healthy foods.

1.7.16　Food Composition and Dietary Intake Database

There is a need to update and expand the USDA food component database. These databases have in the past assisted in identifying the relationship between diet and health. New database with information on composition of newly identify functional foods would serve consumers well.

1.8　CONCLUSION

The U.S. has one of the safest food and best health care systems and lifespan in the world. The long-living individuals are facing health care costs that are on the rise. Disease prevention through better utilization of functional foods can reduce the costs of health care. Interest and opportunities in functional foods research have identified functional foods that may help prevent or delay the progression of degenerative diseases such as cancer, diabetes, macular degeneration, osteoporosis, and Alzheimer's. Research and funding are keys to the future of functional foods. Children should also be the focus for present and future functional foods research. Research is needed to provide the molecular basis of functional foods' efficacy and safety in young as well as old people. Research is also needed to determine the molecular mechanisms of functional foods or their metabolites. A toxicological and pharmacological approach that considers the safety issues of functional foods is needed, even though FDA does not consider conventional foods as drugs. An approach that integrates insights into consumer needs, demands and acceptance, and a structured scientific research process using proteomics, nutrigenomics, bioinformatics, genetic susceptibility, individualized health traits, food processing and preservation, clinical evidence, bioavailability, metabolomics, and organ tropism is needed. Similarly, more research and clinical evidence are needed to support the benefits of products (single or mixed compounds or whole foods) providing physiology-based or disease reduction benefits (heart, osteoporosis, cancer) and psychology/behavior-based or daily benefits (stress, lack of energy, mental and physical appearance). Partnerships among the food industry, food science and nutrition academics, medical and biomedical researchers and practitioners and the federal agencies (FDA, USDA, EPA, NOAA) will be vital to realize the promise of functional foods endeavor. These agencies must embrace changes by using the findings from food science and nutrition research to encourage efforts to investigate and use functional foods for health and wellness.

REFERENCES

1. Clydesdale, F., Bidlack, W. R., Birt, D. F., Bistrian, B. R., Borzelleca, J. F., Clemens, R. A., Dreher, M. L. et al., *Functional Foods: Opportunities and Challenges*, IFT Expert Report, Institute of Food Technologists, 2005.
2. Hasler, C. M., Bloch, A. S., Thomson, C. A., Enrione, E., and Manning, C., Position of the American dietetic association: functional foods, *J. Am. Diet. Assoc.*, 104 (5), 814–826, 2004.
3. Weststrate, J. A., van Poppel, G., and Verschuren, P. M., Functional foods, trends and future, *Br. J. Nutr.*, 88 (suppl. 2), S233–S235, 2002.
4. Akoh, C. C., Lipid-based fat substitutes, *Crit. Rev. Food Sci. Nutr.*, 35 (5), 405–430, 1995.
5. Milner, J. A., Functional foods and health: a US perspective, *Br. J. Nutr.*, 88 (suppl. 2), S151–S158, 2002.
6. Finley, J. W., Proposed criteria for assessing the efficacy of cancer reduction by plant foods enriched in carotenoids, glucosinolates, polyphenols and selenocompounds, *Ann. Bot. (Lond.)*, 95 (7), 1075–1096, 2005.
7. Klein, E. A., Selenium and vitamin E cancer prevention trial, *Ann. NY Acad. Sci.*, 1031, 234–241, 2004.
8. Drake, E. N., Cancer chemoprevention: selenium as a prooxidant, not an antioxidant, *Med. Hypotheses*, 67 (2), 318–322, 2006.
9. Zhao, H., Whitfield, M. L., Xu, T., Botstein, D., and Brooks, J. D., Diverse effects of methylseleninic acid on the transcriptional program of human prostate cancer cells, *Mol. Biol. Cell.*, 15 (2), 506–519, 2004.
10. Unni, E., Kittrell, F. S., Singh, U., and Sinha, R., Osteopontin is a potential target gene in mouse mammary cancer chemoprevention by Se-methylselenocysteine, *Breast Cancer Res.*, 6 (5), R586–R592, 2004.
11. Kim, Y. S. and Milner, J. A., Targets for indole-3-carbinol in cancer prevention, *J. Nutr. Biochem.*, 16 (2), 65–73, 2005.
12. Gilbert, L., Functional foods: what consumers want, http://www.nutraingredients-usa.com/news/news-ng.asp?id=59890-functional-foods-what, 2005.
13. Sloan, E. A., Top trends functional foods, *Food Technol.*, 60 (4), 22–40, 2006.
14. Losso, J. N., Preventing degenerative diseases by anti-angiogenic functional foods, *Food Technol.*, 56 (6), 78–88, 2002.

15. McKeown, N. M. and Jacques, P., Whole grain intake and risk of ischemic stroke in women, *Nutr. Rev.*, 59 (5), 149–152, 2001.
16. McKeown, N. M., Meigs, J. B., Liu, S., Wilson, P. W., and Jacques, P. F., Whole-grain intake is favorably associated with metabolic risk factors for type 2 diabetes and cardiovascular disease in the Framingham Offspring Study, *Am. J. Clin. Nutr.*, 76 (2), 390–398, 2002.
17. McCarty, M. F., Magnesium may mediate the favorable impact of whole grains on insulin sensitivity by acting as a mild calcium antagonist, *Med. Hypotheses*, 64 (3), 619–627, 2005.
18. Jones, J. M., Reicks, M., Adams, J., Fulcher, G., and Marquart, L., Becoming proactive with the whole-grains message, *Nutr. Today*, 39 (1), 10–17, 2004.
19. Amagase, H., Clarifying the real bioactive constituents of garlic, *J. Nutr.*, 136 (3 suppl.), 716S–725S, 2006.
20. Losso, J. N. and Nakai, S., Molecular size of garlic fructooligosaccharides and fructopolysaccharides by matrix-assisted laser desoprtion ionization mass spectrometry, *J. Agric. Food Chem.*, 45 (11), 4342–4346, 1997.
21. Sloan, E. A., Top trends functional foods, *Food Technol.*, 58 (4), 1–14, 2004.
22. Gylling, H. and Miettinen, T. A., Cholesterol absorption: influence of body weight and the role of plant sterols, *Curr. Atheroscler. Rep.*, 7 (6), 466–471, 2005.
23. Lefevre, M., Kris-Etherton, P. M., Zhao, G., and Tracy, R. P., Dietary fatty acids, hemostasis, and cardiovascular disease risk, *J. Am. Diet. Assoc.*, 104 (3), 410–419, 2004.
24. Hanahan, D. and Weinberg, R. A., The hallmarks of cancer, *Cell*, 100 (1), 57–70, 2000.
25. Mantovani, G., Madeddu, C., Maccio, A., Gramignano, G., Lusso, M. R., Massa, E., Astara, G., and Serpe, R., Cancer-related anorexia/cachexia syndrome and oxidative stress: an innovative approach beyond current treatment, *Cancer Epidemiol. Biomarkers Prev.*, 13 (10), 1651–1659, 2004.
26. Fearon, K. C., Voss, A. C., and Hustead, D. S., Definition of cancer cachexia: effect of weight loss, reduced food intake, and systemic inflammation on functional status and prognosis, *Am. J. Clin. Nutr.*, 83 (6), 1345–1350, 2006.
27. Kim, J. S., Kwon, C. S., and Son, K. H., Inhibition of alpha-glucosidase and amylase by luteolin, a flavonoid, *Biosci. Biotechnol. Biochem.*, 64(11), 2458–2461, 2000.
28. Sinha, R., Anderson, D. E., McDonald, S. S., and Greenwald, P., Cancer risk and diet in India, *J. Postgrad. Med.*, 49 (3), 222–228, 2003.
29. Sidhu, G. S., Mani, H., Gaddipati, J. P., Singh, A. K., Seth, P., Banaudha, K. K., Patnaik, G. K., and Maheshwari, R. K., Curcumin enhances wound healing in streptozotocin induced diabetic rats and genetically diabetic mice, *Wound Repair Regen.*, 7 (5), 362–374, 1999.
30. Altavilla, D., Saitta, A., Cucinotta, D., Galeano, M., Deodato, B., Colonna, M., Torre, V. et al., Inhibition of lipid peroxidation restores impaired vascular endothelial growth factor expression and stimulates wound healing and angiogenesis in the genetically diabetic mouse, *Diabetes*, 50 (3), 667–674, 2001.
31. Osawa, T. and Kato, Y., Protective role of antioxidative food factors in oxidative stress caused by hyperglycemia, *Ann. NY Acad. Sci.*, 1043, 440–451, 2005.
32. Shishodia, S., Sethi, G., and Aggarwal, B. B., Curcumin: getting back to the roots, *Ann. NY Acad. Sci.*, 1056, 206–217, 2005.
33. Lau, F. C., Shukitt-Hale, B., and Joseph, J. A., The beneficial effects of fruit polyphenols on brain aging, *Neurobiol. Aging*, 26 (suppl. 1), S128–S132, 2005.

34. Saltzman, J. R. and Russell, R. M., The aging gut. Nutritional issues, *Gastroenterol. Clin. North Am.*, 27 (2), 309–324, 1998.

35. Bailey, T. A., Kanuga, N., Romero, I. A., Greenwood, J., Luthert, P. J., and Cheetham, M. E., Oxidative stress affects the junctional integrity of retinal pigment epithelial cells, *Invest. Ophthalmol. Vis. Sci.*, 45 (2), 675–684, 2004.

36. King, A., Gottlieb, E., Brooks, D. G., Murphy, M. P., and Dunaief, J. L., Mitochondria-derived reactive oxygen species mediate blue light-induced death of retinal pigment epithelial cells, *Photochem. Photobio.*, 79 (5), 470–475, 2004.

37. Sparrow, J. R., Therapy for macular degeneration: insights from acne, *Proc. Natl Acad. Sci. U.S.A.*, 100 (8), 4353–4354, 2003.

38. National Institutes of Health, http://www.nei.nih.gov/health/maculardegen/armd_-facts.asp, accessed on July 21, 2006.

39. Figueroa, M., Schocket, L. S., DuPont, J., Metelitsina, T. I., and Grunwald, J. E., Effect of laser treatment for dry age related macular degeneration on foveolar choroidal haemodynamics, *Br. J. Ophthalmol.*, 88 (6), 792–795, 2004.

40. Wolf, G., Lipofuscin and macular degeneration, *Nutr. Rev.*, 61 (10), 342–346, 2003.

41. Liu, M., A review of treatments for macular degeneration: a synopsis of currently approved treatments and ongoing clinical trials, *Curr. Opin. Ophthalmol.*, 15 (3), 221–226, 2004.

42. Schlingemann, R. O., Role of growth factors and the wound healing response in age-related macular degeneration, *GRAEFES Arch. Clin. Exp. Ophthalmol.*, 242 (1), 91–101, 2004.

43. Seddon, J. M., Ajani, U. A., Sperduto, R. D., Hiller, R., Blair, N., Burton, T. C., Farber, M. D. et al., Dietary carotenoids, vitamins A, C, and E, and advanced age-related macular degeneration. Eye Disease Case-Control Study Group, *JAMA*, 272 (18), 1413–1420, 1994.

44. Kim, S. R., Nakanishi, K., Itagaki, Y., and Sparrow, J. R., Photooxidation of A2-PE, a photoreceptor outer segment fluorophore, and protection by lutein and zeaxanthin, *Exp. Eye Res.*, 82 (5), 828–839, 2006.

45. Jiang, Y. L., Escano, M. F., Sasaki, R., Fujii, S., Kusuhara, S., Matsumoto, A., Sugimura, K., and Negi, A., Ionizing radiation induces a p53-dependent apoptotic mechanism in ARPE-19 cells, *Jrn. J. Ophthalmol.*, 48 (2), 106–114, 2004.

46. Joshi, K., Lad, S., Kale, M., Patwardhan, B., Mahadik, S. P., Patni, B., Chaudhary, A., Bhave, S., and Pandit, A., Supplementation with flax oil and vitamin C improves the outcome of attention deficit hyperactivity disorder (ADHD), *Prostaglandins Leukot. Essent. Fatty Acids*, 74 (1), 17–21, 2006.

47. Marcason, W., Can dietary intervention play a part in the treatment of attention deficit and hyperactivity disorder? *J. Am. Diet. Assoc.*, 105 (7), 1161–1162, 2005.

48. Kessler, R. C., Adler, L., Barkley, R., Biederman, J., Conners, C. K., Demler, O., Faraone, S. V. et al., The prevalence and correlates of adult ADHD in the United States: results from the National Comorbidity Survey Replication, *Am. J. Psychiatry*, 163 (4), 716–723, 2006.

49. Konofal, E., Cortese, S., Lecendreux, M., Arnulf, I., and Mouren, M. C., Effectiveness of iron supplementation in a young child with attention-deficit/hyperactivity disorder, *Pediatrics*, 116 (5), e732–e734, 2005.

50. Lau, K., McLean, W. G., Williams, D. P., and Howard, C. V., Synergistic interactions between commonly used food additives in a developmental neurotoxicity test, *Toxicol. Sci.*, 90 (1), 178–187, 2006.

51. Holsen, L. M., Zarcone, J. R., Thompson, T. I., Brooks, W. M., Anderson, M. F., Ahluwalia, J. S., Nollen, N. L., and Savage, C. R., Neural mechanisms underlying food motivation in children and adolescents, *Neuroimage*, 27 (3), 669–676, 2005.

52. St-Onge, M. P., Dietary fats, teas, dairy, and nuts: potential functional foods for weight control? *Am. J. Clin. Nutr.*, 81 (1), 7–15, 2005.

53. de Graaf, C., Blom, W. A., Smeets, P. A., Stafleu, A., and Hendriks, H. F., Biomarkers of satiation and satiety, *Am. J. Clin. Nutr.*, 79 (6), 946–961, 2004.

54. Flickinger, B. D. and Matsuo, N., Nutritional characteristics of DAG oil, *Lipids*, 38 (2), 129–132, 2003.

55. Szallasi, A., Piperine: researchers discover new flavor in an ancient spice, *Trends Pharmacol. Sci.*, 26 (9), 437–439, 2005.

56. Dedov, V. N., Tran, V. H., Duke, C. C., Connor, M., Christie, M. J., Mandadi, S., and Roufogalis, B. D., Gingerols: a novel class of vanilloid receptor (VR1) agonists, *Br. J. Pharmacol.*, 137 (6), 793–798, 2002.

57. Westerterp-Plantenga, M., Diepvens, K., Joosen, A. M., Berube-Parent, S., and Tremblay, A., Metabolic effects of spices, teas, and caffeine, *Physiol. Behav.*, Mar 28, [Epub ahead of print], 2006.

58. Westerterp-Plantenga, M. S., Lejeune, M. P., and Kovacs, E. M., Body weight loss and weight maintenance in relation to habitual caffeine intake and green tea supplementation, *Obes. Res.*, 13 (7), 1195–1204, 2005.

59. Borchardt, R. and Huber, J. A., Catechol *O*-methyltransferase. 5. Structure-activity relationships for inhibition by flavonoids, *J. Med. Chem.*, 18 (1), 120–122, 1975.

60. Bracco, D., Ferrarra, J. M., Arnaud, M. J., Jequier, E., and Schutz, Y., Effects of caffeine on energy metabolism, heart rate, and methylxanthine metabolism in lean and obese women, *Am. J. Physiol.*, 269 (4 Pt 1), E671–E678, 1995.

61. Berube-Parent, S., Pelletier, C., Dore, J., and Tremblay, A., Effects of encapsulated green tea and Guarana extracts containing a mixture of epigallocatechin-3-gallate and caffeine on 24 h energy expenditure and fat oxidation in men, *Br. J. Nutr.*, 94 (3), 432–436, 2005.

62. Parvez, S., Malik, K. A., Ah Kang, S., and Kim, H. Y., Probiotics and their fermented food products are beneficial for health, *J. Appl. Microbiol.*, 100 (6), 1171–1185, 2006.

63. Saxelin, M., Tynkkynen, S., Mattila-Sandholm, T., and de Vos, W. M., Probiotic and other functional microbes: from markets to mechanisms, *Curr. Opin. Biotechnol.*, 16 (2), 204–211, 2005.

64. Stanton, C., Ross, R. P., Fitzgerald, G. F., and Van Sinderen, D., Fermented functional foods based on probiotics and their biogenic metabolites, *Curr. Opin. Biotechnol.*, 16 (2), 198–203, 2005.

65. Broome, C. S., McArdle, F., Kyle, J. A., Andrews, F., Lowe, N. M., Hart, C. A., Arthur, J. R., and Jackson, M. J., An increase in selenium intake improves immune function and poliovirus handling in adults with marginal selenium status, *Am. J. Clin. Nutr.*, 80 (1), 154–162, 2004.

66. Paddon-Jones, D., Borsheim, E., and Wolfe, R. R., Potential ergogenic effects of arginine and creatine supplementation, *J. Nutr.*, 134 (10 suppl.), 2888S–2894S, discussion 2895S, 2004.

67. Tokish, J. M., Kocher, M. S., and Hawkins, R. J., Ergogenic aids: a review of basic science, performance, side effects, and status in sports, *Am. J. Sports. Med.*, 32 (6), 1543–1553, 2004.

68. Lim, K., Ryu, S., Suh, H., Ishihara, K., and Fushiki, T., (−)-Hydroxycitrate ingestion and endurance exercise performance, *J. Nutr. Sci. Vitaminol. (Tokyo)*, 51 (1), 1–7, 2005.

69. Payne, A., Nutrition and diet in the clinical management of multiple sclerosis, *J. Hum. Nutr. Diet.*, 14 (5), 349–357, 2001.

70. Losso, J. N. and Bawadi, H. A., Hypoxia inducible factor pathways as targets for functional foods, *J. Agric. Food Chem.*, 53 (10), 3751–3768, 2005.

71. Bruyere, O., Pavelka, K., Rovati, L. C., Deroisy, R., Olejarova, M., Gatterova, J., Giacovelli, G., and Reginster, J. Y., Glucosamine sulfate reduces osteoarthritis progression in postmenopausal women with knee osteoarthritis: evidence from two 3-year studies, *Menopause*, 11 (2), 138–143, 2004.

72. Aggarwal, B. B. and Shishodia, S., Suppression of the nuclear factor-kappaB activation pathway by spice-derived phytochemicals: reasoning for seasoning, *Ann. NY Acad Sci.*, 1030, 434–441, 2004.

73. Stahl, W. and Sies, H., Bioactivity and protective effects of natural carotenoids, *Biochem. Biophys. Acta*, 1740 (2), 101–117, 2005.

74. Ross, S. A., Finley, J. W., and Milner, J. A., Allyl sulfur compounds from garlic modulate aberrant crypt formation, *J. Nutr.*, 136 (suppl. 3), 852S–854S, 2006.

75. Whitehead, A. J., Mares, J. A., and Danis, R. P., Macular pigment: a review of current knowledge, *Arch. Ophthalmol.*, 124 (7), 1038–1045, 2006.

76. Millen, A. E., Tucker, M. A., Hartge, P., Halpern, A., Elder, D. E., Guerry, D. IV, Holly, E. A., Sagebiel, R. W., and Potischman, N., Diet and melanoma in a case-control study, *Cancer Epidemiol. Biomarkers Prev.*, 3 (6), 1042–1051, 2004.

2 Functional Foods, Nutraceuticals and Natural Health Products in Canada

Kelley Fitzpatrick and Fereidoon Shahidi

CONTENTS

2.1 INTRODUCTION

There is widespread recognition that diet can significantly impact the development of chronic disease. Estimates in the United States suggest that diet plays a role in 5 of the 10 leading causes of death [1]. Many people intend to eat a healthy diet, but fail to do so because they associate healthy food with poor taste and extensive preparation time [2]. This has created a market demand for foods that do not sacrifice taste and convenience in order to be healthy.

The ability to link a food or food component to health is based on sound scientific evidence, with the desired standard being replicated, randomized, placebo-controlled, intervention trials in human subjects [1]. Once a sound scientific basis has been established for a food, the process of obtaining a health claim for the product can be initiated.

There are three basic types of health claims: structure/function (e.g., calcium helps to build strong bones), risk reduction (e.g., calcium helps reduce the risk and progression of osteoporosis), and therapeutic (e.g., product X is indicated for the treatment of osteoporosis).

Currently, the ability to utilize health claims to promote the value-added attributes associated with functional foods varies internationally. Comparing this sector across regions is also challenging, due to the various definitions and thus scope of functional foods as well as nutraceuticals. "Functional foods" is essentially a marketers' or analysts' term that for most countries is not recognized in law or defined in the dictionary. In fact, only Japan has a legal definition for functional food, while other jurisdictions focus on the application of health claims rather than establishing a definition that may have difficulty encompassing a range of innovative food products.

In 1998, Health Canada proposed that a functional food be defined as being similar in appearance to a conventional food, be consumed as part of the usual diet, have demonstrated physiological benefits, and/or to reduce the risk of chronic disease beyond basic nutritional functions. In the same paper, a nutraceutical was defined as a product that has been isolated or purified from foods and generally sold in a medicinal form not usually associated with food and that exhibits a physiological benefit or provides protection against chronic disease [3]. Since this proposal, several legislative developments have occurred in Canada, resulting in neither definition being formally passed into law. However, both terms are used extensively in reference to research and company activity in the area.

Significantly, Canada is the only global jurisdiction that has legislation related to natural health products (NHP) that includes homeopathic preparations, substances used in traditional medicine, a mineral or trace element, a vitamin, an amino acid, an essential fatty acid (EFA) or other botanical, animal or microorganism-derived substance [4]. These products are generally sold in a medicinal or "dosage" form. Since early 2004, the product category of nutraceuticals has been legally included within NHP regulations.

Of importance to building successful domestic companies is a well-established regulatory environment that will support research, innovation and product development. While it can be said that such a climate exists in Canada for NHP, the same situation does not exist for functional foods and nutraceutical ingredients of interest to food manufacturers. The reasons for this are twofold: outdated federal legislation and hesitancy on behalf of federal bureaucrats to aggressively pursue regulatory reform.

2.2   CANADIAN FOOD AND DRUGS ACT AND REGULATIONS

The Canadian Food and Drugs Act and Regulations were passed into law in 1953. The definition of food under the Food and Drugs Act includes "... any article manufactured, sold or represented for use as food or drink for human beings, chewing gum, and any ingredient that may be mixed with food for any purpose whatever." Drugs are defined as "...any substance or mixture of substances manufactured, sold or represented for use in: the diagnosis, treatment, mitigation or prevention of a disease, disorder or abnormal physical state, or its symptoms, restoring, correcting or modifying organic function."

Through its definitions of "food" and "drug," this legislation currently restricts health-related statements ("claims") on product labels for foods and food ingredients. Since 1953, these products have been considered as either foods or drugs depending on the type and concentration of the "active ingredient" and whether claims are made. When a product is regulated as a food, there are no provisions in the legislation to make claims of a "health" or "therapeutic" nature regarding its use or possible side effects.

Further limiting the development of health claims for foods in Canada are statements in Section 3 of the Act and Regulations which prohibit the sale or advertisement to the general public of any food, drug, cosmetic or device which indicates a treatment, cure or preventive role for diseases or disorders referred to in Schedule A, which include heart disease, diabetes, cancer, hypertension, obesity, and arthritis. These are the most common causes of morbidity and mortality in Canada and are diseases for which functional foods have the potential to be most beneficial. Physiological effects that relate to these conditions, such as lowering of serum cholesterol or glucose, are also considered under the umbrella of Schedule A and are precluded from appearing on labeling and in advertising. In order for health claims to appear on food labels in Canada, changes would have to be made to allow claims about Schedule A diseases or physiological effects related to these diseases. Progress has been made in this regard for NHP, as described later in this chapter.

The regulatory environment in Canada is believed to have stifled innovation, competition, and investment in the food industry. A report released as early as 1996 concluded that the burden of Canadian regulations has lead to significant lost opportunities and sales for Canadian companies [5]. Canada's regulatory system has recently been described as the most cumbersome and time-consuming in the Western world [6].

The average time for a regulatory amendment under the current system, after a good deal of departmental analysis, is at least 18–24 months and can take much longer. There are nine essential steps for any regulatory change under the Act and Regulations: conception and development of the regulation; departmental drafting of a draft regulation and the preparation of a detailed Regulatory Impact Analysis Statement and related documents; examination by Department of Justice and review by Privy Council Office Regulatory Affairs Directorate; Ministerial Approval for prepublication; prepublication review by the Regulatory Affairs and Orders in Council Secretariat (RAOIC) and by Treasury Board (TB) Committee of Cabinet; prepublication in the Canada Gazette, Part I with comment period; departmental preparation of Regulatory Proposal for Final Submission including analysis of comments; final review by RAOIC and return a second time to Cabinet (TB); making, registering, and publishing in Canada Gazette, Part II including approval by the Governor General [6].

This process is appropriate for large new regulatory initiatives, especially if new regulations are imposing major new requirements on industry. This is currently not the situation with the Canadian food industry, in which most regulatory amendments are required to refine existing regulations in order to allow the law to respond to new science, new nutritional trends, new international standards, new manufacturing processes, new products, new claims to better inform consumers, new functional foods, new NHP, etc. If it cannot respond quickly to these forces, Canada's regulatory system may undermine innovation and competitiveness.

Recognizing these constraints, Health Canada began a series of regulatory reviews in 1996. Regulatory developments for NHPs progressed swiftly, and new legislation came into effect in Canada on January 4, 2004. However, functional food regulatory initiatives have been less than speedy.

2.3   REGULATORY INITIATIVES IN FUNCTIONAL FOODS

Beginning in early 1996, the Therapeutic Products Programme (that regulates drug products) and Food Directorate (that regulates food products) of the Health Protection Branch of Health Canada initiated a joint project, with the formation of a government Working Group and the support of an External Advisory Panel, to develop a policy framework for health claims for foods. These deliberations led to a Policy Options paper entitled "Recommendations for defining and dealing with Functional Foods" [7].

As noted earlier, the proposed definition in the Policy Options paper for a nutraceutical was a product that has been isolated or purified from foods and generally sold in medicinal forms not usually associated with food and that exhibits a physiological benefit or provides protection against chronic disease. With the establishment of the Natural Health Products Directorate (NHPD) and a definition for NHP, the proposed category of "nutraceuticals" has been included in the regulatory framework for NHP.

Health Canada also proposed a definition for a functional food as being similar in appearance to a conventional food, consumed as part of the usual diet, with demonstrated physiological benefits, and/or the ability to reduce the risk of chronic disease beyond basic nutritional functions [7]. In late 2001, Health Canada announced that a new regulatory definition for functional foods would not be required under the current Canadian Food and Drugs Act in order to permit health claims for foods. However, this term is used extensively in Canada to describe foods with health benefits beyond basic nutrition.

In 1998, a Final Policy paper dealing specifically with health claim provisions was released [8]. In this document, Health Canada proposed "… to permit structure/function and risk reduction claims for food products, and continue to regulate remaining health claims as drugs."

The proposal defined "structure/function" claims as those that "assert the role of a nutrient or other dietary component intended to affect the structure or physiological function in humans" and risk reduction claims as those that "assert a relationship between a nutrient in a diet and reduced risk of a disease." Therapeutic claims, those intended to cure, prevent, treat or mitigate a disease, would continue to be regulated as drugs. The proposal also indicated that product-specific disease risk-reduction claims that assert a relationship between a specific food product and reduced risk of a disease or a specific health outcome would also be considered. Release of this proposed framework showed that progress was being made. However, any statement regarding disease reduction could only be allowed with specific amendments to the Act and Regulations and Schedule A.

The implementation of health claim allowance in Canada has involved three distinct steps.

2.3.1 ADOPTION OF CERTAIN OF THE INITIAL TEN DIET-BASED DISEASE RISK REDUCTION CLAIMS APPROVED IN THE UNITED STATES UNDER THE NUTRITION LABELING AND EDUCATION ACT

At the time of the Canadian Final Policy paper, ten generic health claims had been approved in the United States under the Nutrition Labeling and Education Act (NLEA). Such claims apply to a food or a group of foods that have compositional characteristic(s) that contribute to a dietary pattern associated with reducing the risk of a disease or health condition. Once the claim is authorized, any food that meets the specified conditions for composition and labeling may carry the claim without further assessment [9].

In 2000, it was announced that five of the NLEA claims were considered valid in a Canadian context; these were:

- Sodium and hypertension
- Calcium and osteoporosis
- Saturated and trans fat and cholesterol and Coronary Heart disease
- Fruits and vegetables and cancer
- Sugar alcohols and dental caries

Each of these health claims required an individual amendment to the Foods and Drug Act. Details of the specific wording of the claims and criteria for the nutrient content of the food and other factors were published in January 2003 in Canada Gazette Part II, "Regulations Amending the Food and Drug Regulations (Nutrition Labeling, Nutrition Claims and Health Claims)" as part of mandatory nutrition labeling regulations [10].

For the other five generic U.S. claims considered, Health Canada concluded that issues remained that did not allow their immediate approval. In the case of fat and cancer, there had been considerable new evidence to question the role of dietary fat in cancer risk, such that Health Canada decided not to proceed with this claim. For the following four claims, decisions are still pending:

- Folate and neural tube defects
- Fiber-containing grain products, fruits, and vegetables, and cancer
- Fruits, vegetables and grain products that contain fiber, particularly soluble fiber and risk of Coronary Heart disease
- Soluble fiber and risk of Coronary Heart disease

A possible sixth generic health claim, which would link a diet rich in vegetables, fruits, and whole grains with a reduction in the risk of coronary heart disease, is currently being considered by Health Canada; however, it has not yet been sent out for public consultation [11]. Health claims based on authoritative statements from organizations (e.g., "the Canadian Medical Association recommends...") and qualified health claims used in the United States were not considered for use in Canada at this time.

2.3.2   DEVELOPMENT OF STANDARDS OF EVIDENCE AND A GUIDANCE DOCUMENT ON DATA REQUIREMENTS FOR SUPPORTING THE VALIDATION OF NEW HEALTH CLAIMS FOR FOODS

For other health claims, the Health Products and Foods Branch of Health Canada, using an internal working group, and in consultation with an Expert Advisory Panel, developed a framework for the standards of scientific evidence required for health claims and a guidance document for industry. In June 2000, Health Canada published a Consultation Document entitled "Standards of Evidence for Evaluating Foods with Health Claims: A proposed framework" [12].

The principles governing the proposed standards have three important elements. The first is product safety, which Health Canada interprets to be reasonable assurance of no adverse health effects. Second is claim validity, as determined by demonstration of product efficacy and effectiveness, based upon establishing an etiologic link between the desired effect and consumption of the food or bioactive substance at the recommended level of intake in the target population that will most likely benefit. Third, is quality assurance, meaning that foods bearing health claims should be able to identify, measure and maintain a consistent level of the bioactive substance to ensure efficacy without jeopardizing safety. Health Canada proposes that initially only claims with the potential for major public health benefit and for which there is sufficient acceptable scientific evidence should be given priority for evaluation.

2.3.3   DEVELOPMENT OF AN APPROPRIATE REGULATORY FRAMEWORK TO ALLOW PRODUCT-SPECIFIC HEALTH CLAIMS FOR FOODS

In October 2001, Health Canada released a draft approach to regulating product-specific health claims for foods [13]. The authorization was proposed for specific foods having a direct, measurable metabolic effect beyond normal growth, development or health maintenance, reducing disease risk or aiding in the dietary management of a disease or condition. In the November 1998 policy options paper, Health Canada recognized that a claim concerning the effect of a food or its ingredient(s) or component(s) should not be generalized to other similar products unless acceptable supporting evidence was provided. This is the rationale for product-specific authorization, whereby each product with the intended claim is evaluated on its own merits. This concept recognizes that food matrices and processing conditions could have an effect on the physiological property of foods. Therefore, an application containing product-specific evidence would be required for a similar claim on another product, unless generally accepted or specific nutritional or food science theory or knowledge would indicate otherwise.

Given the current definitions of "food" and "drug" within the Food and Drugs Act and Regulations, as well as the restrictions imposed under Schedule A, new health claims on foods still require individual amendments be made to the Act and Regulations. This process is laborious, and as evidenced with the five generic claims recently approved, can take upwards of three years to complete. It is of note that no new health claims have been approved in Canada since the initial five generic claims announced in 2000.

A potential solution to the impediments of the Act would be to allow foods with health claims to be regulated as a subset of drugs, the route that new NHP regulations are taking in Canada [14]. However, this is very unlikely to occur. Instead, Health Canada has announced that it will review its regulatory process under the "Smart Regulation" initiative, a government-wide project meant to improve the Government of Canada's regulatory performance [15]. Health Canada announced in the fall of 2005 that it intended to develop a new regulatory framework for the use of food labels, including a new regulatory framework for product-specific health claims for foods and disease risk reduction claims currently considered drug claims. This consultation will be based upon an earlier guidance document (*Interim Guidance Document on Preparing a Submission for Foods with Health Claims*) released in 2001 which described proposed requirements [16]. Health Canada has targeted 2007 for the completion of this regulatory framework.

2.4   NATURAL HEALTH PRODUCTS

The Office of NHP was formed in March, 1999, to establish and implement a new regulatory framework for NHPs. In 2001, the Office became the NHPD, a parallel and equal department to Health Canada's Food Directorate and Therapeutic Products Programme. The Directorate moved quickly to develop NHP regulations to allow health claims and increased consumer access.

Proposed regulations for NHPs were initially published in *Canada Gazette Part I*, the first step toward amendments to the Foods and Drug Act and Regulations, on December 21, 2001 [4]. Final regulations were published in *Canada Gazette II* on Wednesday, June 18, 2003 [17]. The main components of the regulations include NHP definitions, product licensing, adverse reactions reporting, site licensing, good manufacturing practices (requirements for product specifications [identity, purity, potency], premises, equipment, personnel, sanitations program, operations, quality assurance, stability, records, sterility, lot or batch samples and recall reporting), standards of evidence for safety and health claims, and labeling and packaging.

2.5 DEFINITION AND HEALTH CLAIMS

In Canada, the definition of a NHP, as noted earlier, includes a wide range of materials [4]. Regarding health claims, NHP products can be manufactured, sold or represented for use in: (i) the diagnosis, treatment, mitigation or prevention of a disease, disorder, or abnormal physical state or its symptoms in humans (i.e., claims previously allowed in Canada only for drug products); (ii) resorting or correcting organic functions in humans, or (iii) maintaining or promoting health or otherwise modifying organic functions in humans [4].

In order to allow "drug type" claims within the Foods and Drug Act and Regulations, these products are regulated under a subsection of the Drug Regulations, the vehicle used to manage the process without requiring the opening of the Food and Drugs Act and Regulations, or the time-consuming steps involved in establishing a new regulation for each new health claim. The standards of evidence for safety and a health claim correspond to the strength of the claim. For example, traditional references are required for a product to carry "self care" claims such as "Traditionally used to…" This is in contrast to a "disease related" claim such as "Clinical trials show…" which requires the submission of data from a meta-analysis of randomized controlled trials, or at least one well-designed (multi-center) randomized, controlled trial.

One area of debate continues to be the definition of NHPs. Even though the proposal does not include conventional foods and is not intended to capture a product in a food medium, the rule is unclear as to whether foods bearing health claims or structure/function claims relating to a nutrient (e.g., a beverage with the statement "contains calcium to help build strong bones") are outside the scope of the NHP regulation. For example, the proposed regulations state, "Although 'dosage form' is not an express part of the definition, the NHPD recognizes that NHPs are usually sold in capsule, pill, tablet or liquid form. As well, certain other forms, such as gum or bars, have come to considered acceptable dosage forms."

2.6 NATURAL HEALTH PRODUCT REGULATIONS—CANADA GAZETTE II

Effective January 1st, 2004, product licenses for NHPs were available through the NHPD. For NHPs already licensed as drugs (with a Drug Identification Number—DIN), a Product License application and market authorization from the NHPD will be required by December 31, 2009. An abbreviated application process will be used to issue a market authorization in the form of a Product License. During the transition period, DIN products will continue to be regulated under the Food and Drug Regulations until they obtain NHP product licenses.

The transition period is four years (i.e., by December 31, 2007) for products that do not hold DINs. Applications for a product license must provide specific information about the NHP, including the quantity of the medicinal ingredients it contains, the specification it complies with, the recommended use or purpose for which the NHP is intended to be sold, and the supporting safety and efficacy data.

Once a license is granted, a natural product number (NPN) is issued. Requirements for Adverse Event Reporting (AER) come into effect as products obtain their NHP licenses.

There is a "default licensing" provision whereby products containing a single active ingredient for which Health Canada has created a monograph will be automatically granted a license within 60 days. The Compendium of Monographs can be found on the NHPD website [18]. The NHPD is working on a combination product policy that will allow products with multiple active ingredients to use this default licensing provision.

In August 2005, the NHPD published its first list of NHP that have been authorized for sale in Canada [19]. This list, which is updated on a monthly basis, provides the name of each approved product; the name of the product license holder; the dosage form; the route of administration, and the NPN or Homeopathic Medicine Number (HMN) (DIN–HM) issued to the product.

Significant proposals to modify Schedule A of the Act and Regulations have recently been published. A Committee established by Health Canada in February 2003 [20] made several recommendations to address regulatory impediments associated with this statute. The results of the Committee's deliberations appeared in *Canada Gazette, Part 1* [21]. The publication described proposed amendments that would allow NHPs to make Schedule A health claims. Specifically NHPs are proposed to be exempt with respect to advertisement to the general public as a treatment or preventative for any of the diseases, disorders or abnormal physical states referred to in Schedule A of the Act.

According to Health Canada, the proposed amendment is consistent with evolving scientific knowledge and the overall health and safety protection provided to Canadians through the Food and Drugs Act and Regulations, while supporting informed choice for non-prescription drugs and NHPs. Unfortunately, the proposed regulatory amendment does not include food.

The NHPD has been challenged with the implementation of these new regulations. Insufficient resources for product evaluation and licensing are leading to backlogs, particularly for new ingredients and for products with more than one active ingredient. In fact, the NHPD has received a total of 7983 product, site, and clinical trial applications since January 1, 2004. By early 2006, a little over 900 product/site licenses and notices of authorization have been issued. As a result, an analysis of current operating resources and processes has been initiated to address weaknesses in the Directorate's application review operations.

Extensive guidance documents are available on the NHPD website that provide a complete description of the regulatory requirements [22].

2.7 IMPACT ON NEW NATURAL HEALTH PRODUCT REGULATIONS ON THE MARKET

Canadians want safe and effective products but they also want access to as many options as possible. The NHPD of Health Canada enforces the new NHP regulations, which address five key areas: product licensing, site licensing, good manufacturing practices, labeling, and packaging provisions, and adverse reaction reporting. Natural health products will have to be proven to be safe and effective. Although this is similar to the current drug system, the standards of proof for NHP will be lower. Many NHPs will be able to simply rely on standardized NHPD monographs rather than have to conduct clinical trials. Further, licenses will be granted for vitamins, minerals, homeopathic preparations, and botanicals if they have a long history of safe use. NHPs now have the regulatory benefits of drugs (such as the ability to have health claims) but with a less costly and lengthy approval process.

Consumers benefit from new NHP regulations as they now have access to products that they know are safe, effective, and of high quality. They can now make more informed decisions when it comes to shopping for NHP. Health Canada predicts that with consumers having more confidence in the safety and efficacy of NHP, there will be an increase in the use of such products.

The new regulatory regime for NHP is also expected to soften the medical profession's longstanding resistance to NHP. The industry is now bestowed with greater legitimacy. A historical hands-off approach to NHPs by the medical profession may be the reason that two-thirds of NHP users do not tell their physicians that they are doing so [23].

The ability for consumers to know that what is on the label is also in the bottle and to have all health claims supported by evidence comes at a cost. NHP manufacturers who also manufacture drugs will face minimal new costs, as they already hold valid establishment licenses and are in compliance with good manufacturing practices. Those companies that do not manufacture drugs will necessarily face some new costs, which eventually will be passed on to the consumer. The additional costs that manufacturers face appear to be outweighed by the benefits mentioned above. The increase in long-term stable demand will benefit the industry. Consumers may have to pay a little more for their products, but consumers have continually demonstrated that they are willing to absorb cost increases in return for increased quality and safety of the goods that they purchase.

Companies that can provide Canadian consumers with safe NHP of an exceptional level of quality will benefit the most.

2.8   THE CANADIAN MARKET

Canadian sales figures for functional foods and NHPs are difficult to interpret, as much of the data is extrapolated from U.S. sales and adjusted downwards [24]. Estimates are that Canadians purchased close to U.S. $5 B worth of natural personal care products, NHP, organic, and functional food products in 2003. This translates into nearly $150 per capita spending, a 130% increase in only four years.

NBJ data show that Canadian sales of natural/organic and functional foods in 2003 were valued at U.S. $3.5B, a steady increase over the last four years. While functional food sales in the U.S. represent approximately 4.0% of total food sales, Canada's portion of total food sales is only 2.2%, representing a significant growth potential for the domestic industry.

Over the past two decades, Canadian consumers have steadily increased their usage of NHP for general nutrition, for enhancement of health or for preventive measures [25]. Surveys have reported that more than 80% of Canadians regularly take vitamin supplements and believe vitamins can help prevent a number of diseases. Additionally, Canadians aged 65 + appear to be the heaviest supplement users [26].

Canadian sales figures for sport, meal, homeopathic and specialty supplements, vitamins and minerals, herbs and botanicals were estimated to be $1.3 B U.S. in 2003, a 10% increase over 2002 figures [24]. Parallel to the interest in NHP, the expectations by healthcare professionals and consumers that these products are safe, and deliver the health benefits promised, is increasing.

2.9   THE CANADIAN INDUSTRY

A recent study conducted in 2003 by Statistics Canada for Agriculture and Agri-Food Canada (AAFC) provides insight into the Canadian industry (Table 2.1). The survey of 576 companies in the food industry provided the key findings. The reader is cautioned, though, that some of this data may capture other segments of the NHP industry that are not included under the traditional nutraceutical "umbrella," such as vitamins, minerals, amino acids and hormones, as well as food categories not considered by Health Canada as included within functional foods such as fortified sports bars and fortified sports drinks.

In 2002, KPMG Consultants reported 215 Canadian companies involved in functional foods and nutraceuticals. Of importance, these were companies that "self-identified" as working almost exclusively in the functional food and nutraceutical market [28].

Canadian companies produce a wide range of NHP, nutraceutical and functional food products. Canada's prairie and forested regions offer an abundant source of wild plants and large areas of fertile land that make the country an ideal location for the cultivation of a wide variety of commodity, specialty, and medicinal crops. Along with enhancing the nutritive value and functional properties of common crops, there has been a trend in Canada towards value-added processing and the extraction of nutritionally valuable constituents.

Grains such as wheat, oats, and barley are mainstays of the North American diet. These products are excellent sources of dietary fiber, carbohydrate, and vitamins. Canadian companies, such as Saskatoon based InfraReady Products and Edmonton based Cevena BioTech, have developed specialized fractionation technologies for the processing of raw materials such as legumes, oats, and other cereals into starch, protein and fiber, which are used as functional food additives. In addition, specialty crops such as fenugreek, produced by a Regina company, Emerald Seed Products, are increasingly being cultivated to meet the demands of manufacturers seeking specific raw materials for functional food and NHP products.

The range of herbs produced by Canadian companies is diverse. Saskatchewan growers, for example, reported production of over 70 different herbs and spices, principally echinacea, ginseng, garlic, milk thistle, feverfew, goldenseal, St. John's

TABLE 2.1
Survey of the Canadian Functional Food and Nutraceutical Industry

Factor	Description
Size	161 companies are involved in the functional food sector only
	264 companies are involved in the nutraceutical sector only
	144 companies stated that they are involved in both the functional food and nutraceutical industry
	40 companies have more than 100 employees devoted only to functional food and nutraceuticals
	7 of the companies employed more than 100 workers
	More than half the companies had fewer than 10 employees
Revenues	11% of the industry has revenues in excess of $5M
	18% reported revenues of between $1M and $5M
	30% had revenues of less than $1M
Export markets	In priority
	U.S.
	Japan
	North and South Korea
	Taiwan
	Australia/New Zealand
Export materials	77% exported finished functional food or nutraceutical products for sale at the wholesale or retail level
	44% exported raw materials or ingredients for use in functional foods or nutraceuticals
	33% exported semi-finished products for further processing before sale
	8% exported technology relating to production

Source: Statistics Canada, *Functional Foods and Nutraceuticals Survey*, The Daily, Monday, October 6, 2003.

wort, valerian, ginseng, astragalus, and cayenne. Other herbs include seabuckthorn, anise, fireweed, senega root, sarsasparilla, milk thistle, chamomile, yarrow, calendula, and stinging nettle. These herbs are also common across Canada. Canadian companies specialize in the standardization of herb and plant extracts and have developed the extraction, isolation, and purification expertise necessary to manufacture herbal products to pharmaceutical standards. Also, companies have developed and refined analytical methods to verify the potency and bio-activity of herbal extracts and other compounds.

Spice and fruit crops under production across the country include caraway, coriander, mustard, dill, peppermint, cumin, seabuckthorn, blueberry, Saskatoon berry, chokecherry, and buffalo berry. Canadian companies, including New Brunswick's Vaccinium Technologies Inc., have developed technologies and expertise in the extraction, characterization, stabilization, modification and enhancement of the flavonoid constituents of fruits.

Canadian companies such as Montreal based Institute Rosell–Lallemand produce micro organisms for the dairy, meat, and brewing industries. Micro organisms are also being manufactured as sources of pre- and probiotic supplements and food ingredients. NHP and cosmetics derived from elk antler, such as elk velvet capsules, powders and tinctures as well as emu oil, are produced and processed in various parts of Canada.

Expertise in the formulation and manufacturing of single and complex vitamins, minerals and antioxidants is available from a number of Canadian manufacturers. In addition to consumer brands, Canadian companies also offer full-service contract manufacturing of private label vitamin and mineral supplements as well as herbals, specialty and combination products.

The Canadian industry is a leader in the development and manufacturing of EFA products from plant and marine sources, including evening primrose oil, flaxseed, borage, hemp and marine animal oils, as well as herbal/EFA condition-specific combination products. Such companies such as Saskatoon's Bioriginal Food and Science Corporation and Halifax's Ocean Nutrition are producing EFA oils for the global market. Further, Canadian companies have developed specialized encapsulation and other packaging technologies that preserve the integrity and bio-activity of EFA products. Canola, tall and soy sterols and stanols produced by Vancouver-based Forbes Medi-Tech, and flaxseed lignans from Winnipeg based Pizzey's Milling are also sold into the health food market in the form of capsules, blended with oil or as part of foods.

The food and food ingredient sector is also a very important part of the Canadian nutrition industry. The types of food and food ingredient products produced by Canadian companies are quite diverse, and include milk and eggs with increased levels of omega-3 fatty acids, cereals, and grains including wheat, oat, barley, and fenugreek products with enhanced amounts of dietary fiber (soluble and insoluble), modified fatty acid vegetable oils, vegetable proteins from soy, canola and hemp, legumes and fruit products.

The growth of the Canadian functional food and NHP sector reflects the demand for nutritional products based on increasing scientific evidence linking diet to the quality of health. Consumer interest in self-care and alternative medicine is on the rise. According to a recent study conducted for AAFC, it is estimated that up to Can$1B of farm production value is devoted to supplying the functional foods and NHP sector [29]. This estimate does not include the marine industry, which contributes to the sector through the production of omega-3 fatty acids and other marine-based products.

A significant opportunity exists in Canada for functional food and NHP products to positively impact health care costs. In another study conducted for AAFC [30], it was noted that lifestyle-related chronic disorders are a major component of increasing health care expenditures in this country. The proportion of disease onset attributable to diet is estimated to be approximately 40%–50% for cardiovascular disorders and diabetes, while 35%–50% of all cancers are directly related to dietary factors. Approximately 20% of osteoporosis is diet-related.

Strong evidence supports the role of functional foods and NHPs in reducing the prevalence of chronic disease in Canada and providing impressive savings in health

TABLE 2.2
Examples of Natural Health Products Available in the Canadian Marketplace

Item	Examples
Herbals	Echinacea, milk thistle, black cohosh, saw palmeto, ginkgo biloba, panax ginseng
Fruits and by-products	Cranberry extract, noni juice, blueberry powder, apple cider vinegar capsules
Specialty lipids/lipid by-product	Omega-3, omega 3.6.9, evening primrose oil, flax oil, lecithin, phosphatidylserine, co-enzyme Q10
Carotenoids	Lycopene, supervision formula with Lutein
Other plant sources	Soy isoflavone, garlic, grapeseed
Marine products	Omega-3 oils, glucosamine, glucosamine-chondroitin sulfate, cod liver oil, Alaskan salmon oil
Others	Acidophilus with bifiders, Diabetex, Lakota

care costs without significant overall dietary changes. Based on the degree to which various chronic disorders are diet-related, and using the current direct medical costs of these disorders, the author estimated potential annual savings to the health care system could be on the order of $20 billion per year. Canadian companies are focusing their product research and development, production, and wholesaling in the areas of (by priority): general well being; immune system; vascular and heart health; energy; diabetes and weight control.

A survey of the range of NHP in the Canadian marketplace is shown in Table 2.2. This table is by no means comprehensive, but provides a cursory account of what is currently available. The health benefits of products are generally recognized by consumers' knowledge, developed through a myriad of advertisement strategies.

2.10 CONCLUSIONS

The relationship between regulations and industry competitiveness is an extremely critical issue. The Canadian regulatory system can provide a competitive advantage and encourage companies to do business there, or it can be seen as a regulatory system that is unclear and unresponsive, creating an unduly heavy regulatory burden that undermines competitiveness and discourages innovation and investment. As this chapter has described, on the whole, the regulation of health claims for functional food products has fallen behind that of NHP in Canada. The regulatory divide between the two is of prominent concern for industry stakeholders on both sides, as is the need for greater clarity regarding the nature and extent of scientific evidence required to support claims. For the benefit of all involved in these sectors, Canadian regulators need to clarify the boundaries and improve the interface between functional foods and NHP.

REFERENCES

1. Hasler, C. M., Functional foods: benefits, concerns and challenges—a position paper from the American council on science and health, *J. Nutr.*, 132 (12), 3772–3781, 2002.
2. Guthrie, J. F., Derby, B. M., and Levy, A. S., *America's Eating Habits: Changes and Consequences*, Department of Agriculture, Washington, 1999.
3. Health Canada, Standards of Evidence for Evaluating Foods with Health Claims—Fact Sheet, 2000, http://www.hc-sc.gc.ca/food-aliment/ns-sc/neen/health_claims-allegations_sante/e_soe_fact_sheet.htm (accessed on November 2000).
4. Health Canada, Natural Health Products Regulations, Canada Gazette Part 1, http://canadagazette.gc.ca/partI/2001/20011222/pdf/g1-13551.pdf (accessed on December 2001).
5. Smith, B. L., Marcotte, M., and Harrison, G., A Comparative Analysis of the Regulatory Framework Affecting Functional Food Development and Commercialization in Canada, Japan, the European Union and the United States of America, Commissioned by Agriculture and Agri-Food Canada to Intersect Alliance, Ottawa, Ontario, 1996.
6. Doering, R. L., A Duty to Do it Well: Regulations and Food Industry Competitiveness and Innovation under the Food and Drugs Act, Commissioned by Agriculture and Agri-Food Canada to Gowling LaFleur Henderson LLP, Ottawa, Ontario, 2005.
7. Health Canada, Discussion Document, Functional Foods and Nutraceuticals, 1997, http://www.hc-sc.gc.ca/main/drugs/zfiles/english/ffn/ffdscdoc_e.html.
8. Health Canada, Final Policy Paper on Nutraceuticals/Functional Foods and Health Claims on Foods, 1998, Internet: http://www.hc-sc.gc.ca/food-aliment/ns-sc/ne-en/health_claims-allegations_sante/e_nutra-funct_foods.html (accessed on November 1998).
9. Food and Drug Administration, Department of Health and Human Services, Guidance for Industry: Significant Scientific Agreement in the Review of Health Claims for Conventional Foods and Dietary Supplements, Docket No. 99D-5424, Vol. 64, No. 245, 1999, http://www.cfsan.fda.gov/~dms/ssaguide.html (accessed on December 1999).
10. Health Canada, Regulations Amending the Food and Drug Regulations (Nutrition Labeling, Nutrition Claims and Health Claims), Canada Gazette Part II, 2003, http://canadagazette.gc.ca/partII/2003/20030101/pdf/g2-13701.pdf (accessed on January 2003).
11. Personal Communication with Nora Lee, Health Canada, November 2005.
12. Health Canada, Consultation Document: Standards of Evidence for Evaluating Foods with Health Claims: A Proposed Framework, 2000, Internet: http://www.hc-sc.gc.ca/food-aliment/english/subjects/health_claims/standards_of_evidence (accessed on June 2000).
13. Health Canada, Product-Specific Authorization of Health Claims for Foods: A Proposed Regulatory Framework, 2001, Internet: http://www.hc-sc.gc.ca/food-aliment/ns-sc/ne-en/health_claims-allegations_sante/e_finalproposal01.html (accessed on October 2001).
14. Stephen, A. M., Liston, A. J., Anthony, S. P., Munro, I. A., and Anderson, G. H., Regulation of foods with health claims: a proposal, *Can. J. Public Health*, 93, 328–331, 2002.
15. Government of Canada, Smart Regulation: Report on Actions and Plans, Fall 2005, http://www.regulation.gc.ca/ (Fall Update 2005).

16. Health Canada, Interim Guidance Document: Preparing a Submission for Foods with Health Claims Incorporating Standards of Evidence for Evaluating Foods with Health Claims, 2001. http://www.hc-sc.gc.ca/food-aliment/ns-sc/neen/health_claims-allegations_sante/pdf/e_guidance_doc_interim.pdf

17. Health Canada, Natural Health Products Regulations, Canada Gazette Part 2, 2003, Internet: http://canadagazette.gc.ca/partII/2003/20030618/html/sor196-e.html (accessed on June 2003).

18. Health Canada, Compendium of Monographs, Natural Health Products, 2005, http://www.hc-sc.gc.ca/dhp-mps/prodnatur/applications/licen-prod/monograph/mono_list_e.html (accessed on March 2005).

19. Health Canada, List of Approved Natural Health Products, 2005, http://www.hc-sc.gc.ca/dhp-mps/prodnatur/applications/licen-prod/lists/listapprnhp-listeapprpsn_e.html (accessed on August 2005).

20. Health Canada, Health Canada Reviews Schedule A to the Food and Drugs Act, 2003, http://www.hc-sc.gc.ca/hpfb-dgpsa/sched_a_review_e.html (accessed on February 2003).

21. Health Canada, Proposed Amendments to Schedule A of the Foods and Drug Act and Regulations, Canada Gazette Part 1, Vol. 139, No. 47, 2005, http://canadagazette.gc.ca/partI/2005/20051119/html/regle1-e.html (accessed on November 19, 2005).

22. Health Canada, Natural Health Products: Guidance Documents, 2003, http://www.hc-sc.gc.ca/dhp-mps/prodnatur/legislation/docs/index_e.html (accessed on November 2003).

23. International Business Strategies, The Canadian Natural Health Products Market, 2004, www.internationalbusinessstrategies.com (accessed on June 2004).

24. Ferrier, G., Nutrition Business Journal, NBJ Industry Overview, Webcast, 2005, www.nutritionbusiness.com (accessed on April 2005).

25. Canadian Health Foods Association, Results of a survey on Canadians' perceptions and attitudes towards natural health products and whole foods, May 2002. Prepared for the CHFA by Rotenberg Research, Markham, Ontario, 2002.

26. Roche vitamin Information Program, Gallup Survey of Canadians, 2003.

27. Statistics Canada, *Functional Foods and Nutraceuticals Survey*, The Daily, Monday, October 6, 2003.

28. KPMG, Canadian Technological Road Mapping on Functional Foods and Nutraceuticals: MARKET report commissioned by the National Research Council of Canada, Ottawa, ON and University of Laval, QC, 2002.

29. Scott Wolfe Management, Potential Benefits of Functional foods and Nutraceuticals to the Agri-Food Industry in Canada, report submitted to Agriculture and Agri-Food Canada, Food Bureau, March 2002.

30. Holub, B., Potential benefits of functional foods and nutraceuticals to reduce the risk and costs of diseases in Canada, Report submitted to Agriculture and Agri-Food Canada, Food Bureau, June 2002.

3 History and Scope of Functional Foods in Japan

Makoto Shimizu and Asami Kawakami

CONTENTS

3.1 HISTORY OF FUNCTIONAL FOOD SCIENCE AND FOODS FOR SPECIFIED HEALTH USES IN JAPAN

Since the end of World War II, the food industry of Japan has been changing its objectives in response to shifts in social demand; e.g., economical food with sufficient nutrients in the 1940s–1950s, hygienic food with better taste in the 1950s to 1960s, convenient food, such as precooked products in the 1970s, and, most recently, gourmet food. A new objective appeared for the Japanese food industry in the 1980s: health-oriented and disease-preventive food.

The concept of this health-oriented food was born from the national research project "Systemic analysis and development of food functions" supported by the Ministry of Education, Science, and Culture (MESC) in 1984 [1]. In this project, the health benefits of foods were studied by many experts specializing in food science, nutrition, pharmacology, and medical science. The project also proposed for the first time the new concept of "functional food" and defined food functions as primary (nutritional), secondary (sensory) and tertiary (physiological). Food with physiological functions was of particular interest, because such food would be useful for improving the health of the general public, thus reducing the national medical costs that had been predicted to increase rapidly in the near future.

Since the first MESC project was started in 1984, a large number of food scientists in universities and institutes have been involved in research on the physiological functions of food substances [2–4]. Many food companies have also been showing great concern about the applied aspects of functional food science, exploring the possibility of developing "functional food" that is distinctly different from common food. After the first MESC project ended in 1987, the concept of "preventing life-style-related diseases by specific foods" became widely recognized by society, and the project was followed by two more large-scale grant-aided studies sponsored by MESC, namely "Analysis of body-modulating functions of foods (1988–1991)" and "Analysis and molecular design of functional food (1992–1995)."

In 1991, the Ministry of Health and Welfare established a policy of officially approving some selected functional foods in terms of "foods for specified health uses (FOSHU)," that led to the approval of the first two FOSHU products in 1993 [1–4]: hypoallergenic rice for patients with rice allergy [5] and low-phosphorus milk for people with kidney problems. Although originally classified as FOSHU products, these first two were later reclassified as "foods for the ill." The number of FOSHU items has progressively increased since then, and 590 FOSHU products were available on the Japanese market as of August, 2006. The market size for FOSHU was almost 630 billion yen in the year 2005.

Chinese tradition states that medicine and food are isogenic. FOSHU in Japan has revived this principle, based on scientific evidence. Since FOSHU products are scientifically evaluated by the government in terms of their effectiveness and safety, a large volume of scientific data is required to gain approval: the food and its functional constituents should be safe to eat; the physicochemical properties of the constituents should be well-defined; and the claimed health benefits of the food or its constituents should be based on available data concerning the relationship between the foods/food constituents and health benefits that have been obtained on a clear medical/nutritional basis [6].

During the basic studies on food functions, scientists have found various biomarkers or functional markers for evaluating the effectiveness of the food and food constituents. These markers are useful for designing and developing FOSHU products. To evaluate food for improving a gastrointestinal condition, e.g., the population of intestinal microflora, the number of bowel movements and the shape of feces are used as markers. A variety of biomarkers, such as the oral pH value and demineralization of enamel slabs (for dental health), bone mineral density and Gla-modified osteocalcin in blood (for bone health and strength), systolic and diastolic blood pressure (for blood pressure regulation), blood triglyceride level and body mass index (for obesity), and blood glucose, glycohaemoglobin, and insulin levels (for blood glucose regulation) have been established, these being used to evaluate the psysiological functions of foods and food ingredients [6].

3.2 FOSHU PRODUCTS CURRENTLY AVAILABLE IN JAPAN

FOSHU products available as of September, 2006 can be roughly classified into eight categories, according to their specific functions that are permitted to be shown on the label as health claims:

1. Foods to regulate gastrointestinal conditions.
2. Foods for those with slight hypertension.
3. Foods for those with slight hypercholesterolemia.
4. Foods for those who are concerned about their blood glucose levels.
5. Foods for those who are concerned about fat accumulation.
6. Foods to help mineral absorption.
7. Foods for those who are concerned about bone strength.
8. Foods less likely to cause tooth decay.

These foods contain effective ingredients such as probiotics, oligosaccharides, dietary fibers, proteins, peptides, minerals, and polyphenols. FOSHU products are supplied as a wide variety of foods, and consumers can freely select an appropriate item carrying the health claim label to improve the dietary practices of daily life. Categories of FOSHU are summarized in Table 3.1.

3.2.1 Foods to Regulate Gastrointestinal Conditions

This category contains the largest number of FOSHU products and comprises the vast majority of sales. Most are dairy foods containing either probiotics or prebiotics to promote gut health. These components favorably influence the indigenous intestinal microflora.

Probiotics are defined as "viable, non-pathogenic microorganisms having beneficial properties for hosts by improving the balance of intestinal microflora." Bifidobacterium species and lactic bacteria are examples, and are frequently provided as constituents of fermented drinks and yogurt. Daily consumption of these products plays a role in treating or preventing intestinal disorders. The most common and well-studied probiotic in Japan is the *Lactobacillus casei* shirota strain. Several studies have demonstrated that the *L. casei* shirota strain exerts a multifunction effect on gut health; e.g., it prevents the proliferation of pathogenic bacteria [7,8] and enhances the function of the immune system [9].

Prebiotics are "indigestible food substances that pass through the small intestine and are fermented by endogenous microflora." Dietary fibers (indigestible dextrin,

TABLE 3.1
Categories and Examples of FOSHU Products Currently Available in Japan (as of August, 2006)

Categories[a]	Functional Ingredients	Mechanisms of Action	Types of Food	No. of Approval
Foods to regulate gastrointestinal conditions	Bifidobacterium, lactic bacteria, dietary fibers, oligosaccharides	Improve the balance of intestinal microflora	Fermented drinks, yoghurt, cereals, soft drinks, snacks, noodles	244
Foods for those with slight hypertension	Lactotripeptides (VPP, IPP), fish peptides (VY, LKP), Tochu tea extracts	Inhibit ACE activity Activate parasympathetic system	Fermented milk, freshener drinks, powdered soup, tea	75
Foods for those who with slight hypercholesterolemia	Soybean proteins/peptides, chitosan, depolymerized sodium alginate, plant sterols	Increase excretion of cholesterol and bile acids Prevent intestinal cholesterol absorption	Freshener drinks, soy milk, meatball powdered soup, edible oil, margarine, mayonnaise	87
Foods for those who are concerned about blood glucose levels	Indigestible dextrin, guava leaf polyphenols, wheat albumin, L-arabinose, Touchi-extract	Delay the rate of glucose uptake Inhibit α-amylase activity Inhibit α-glucosidase activity	Freshener drinks, tea, powdered soup, table sugar	79
Foods for those who are concerned about fat accumulation	Globin digests, diacylglycerol, medium-chain triglycerides, tea catechins	Reduces blood triglyceride level Inhibit pancreatic lipase Suppress intestinal fat absorption, activate fat burn	Soft drinks, edible oil, tea	34
Foods to help mineral absorption	Calcium citrate malate, casein phosphopeptide (CPP), fructooligosaccharide, polyglutamic acid (PGA) Heme iron	Increase intestinal Ca solubility Provide soluble form of iron Increase intestinal Ca absorption	Freshener drinks, table sugar, snacks	10

Foods for those who are concerned about bone strength	Calcium, vitamin D, soybean isoflavone, vitamin K, MBP (milk basic protein)	Increase the rate of bone formation Inhibit bone resorption	Soft drinks, soy milk, natto (fermented soybeans)	22
Foods less likely to cause tooth decay	Maltitol and erythritol, xylitol, casein phosphopeptide-amorphous calcium phosphate (CPP-ACP), phosphoryl oligo-saccharides of calcium (POs-Ca)	Inhibit the proliferation of the bacteria Suppress demineralization and/or promote remineralization	Chewing gum, candies, chocolate	39
			Total	590

[a] Categories by health claim of FOSHU products.

polydextrose, guar gum, psyllium seed coat, etc.) and certain oligosaccharides (fructo-, xylo-, and galacto-oligosaccharides, soybean oligosaccharides, raffinose, lactulose, arabinose, etc.) are examples. Adding or increasing prebiotics in the diet stimulates the growth of specific bacterial populations. It is important to keep consuming these prebiotics to maintain an appropriate number of probiotic colonies in the intestinal mucosa.

3.2.2 FOODS FOR THOSE WITH MILD HYPERTENSION

Hypertension is associated with such common diseases as myocardial infarction and cerebral apoplexy. Several food factors that have blood-pressure-moderating properties are the subjects of intense interest, as the mortality rates of these diseases are becoming higher in Japan.

Blood pressure is controlled by the renin-angiotensin system. The angiotensin I-converting enzyme (ACE) in this system plays an important role in increasing the blood pressure as it cleaves angiotensin I (a decapeptide), transforming it into angiotensin II (an octapeptide) which raises the blood pressure by causing the smooth muscle cells of the blood vessels to constrict or narrow.

Several food-derived peptides with ACE inhibitory activity have been isolated and are used as functional factors in FOSHU products for individuals with mild hypertension. A fermented sour milk drink containing casein-derived ACE-inhibitory peptides is one of the most popular FOSHU products in this category. Two kinds of lactotripeptides (LTP) that have an antihypertensive effect have been isolated and identified as Val-Pro-Pro (VPP) and Ile-Pro-Pro (IPP). These LTPs were made from skim milk during its fermentation by *Lactobacillus helveticus* and *Saccharomyces cerevisiae*. They have been found to inhibit ACE activity by an in vitro assay that showed IC_{50} values for VPP and IPP of 9 and 5 μM, respectively, [10]. Their anti-hypertensive effects have also been demonstrated in vivo in an experiment with spontaneously hypertensive rats [11]. A single oral administration of VPP (0.6 mg/kg of BW) or IPP (0.3 mg/kg of BW) significantly decreased the systolic blood pressure (SBP) for several hours. In contrast, their administration did not change SBP of normotensive WKY rats, suggesting that their anti-hypertensive effect was specific to the hypertensive state [11,12]. Another placebo-controlled study has demonstrated that a daily 95-ml consumption of the LTPs for 8 weeks significantly decreased the blood pressure in hypertensive patients [13].

Several other food-derived peptides with ACE inhibitory activity are also used as functional components of FOSHU products. For example, a dipeptide with the Val-Tyr sequence, originally found in the alkaline protease hydrolysate of sardine muscle [14], is added as a supplement to sports drinks. Katsuobushi (dried bonito), a traditional Japanese seasoning made from bonito muscle, contains a peptide with the five amino acid sequence of Leu-Lys-Pro-Asn-Met. It is converted into Leu-Lys-Pro in the digestive organs or blood and exhibits ACE inhibitory activity [15]. Katsuobushi oligopeptides are used as functional components in such FOSHU products as powdered soup, tea, and tablets.

There is another functional ingredient that exerts an anti-hypertensive effect, but by a different mechanism. "Tochu tea," an aqueous extract of *Eucommia ulmoides*

Oliv. leaves, lowers blood pressure without inhibiting ACE activity. One of the functional components in Tochu tea is supposed to be a geniposidic acid, which activates the parasympathetic system by regulating the muscarinic receptor, resulting in enhanced endothelium-dependent vasodilation. A soft drink containing Tochu tea has been approved as a FOSHU product for people with mild hypertension.

3.2.3 FOODS FOR THOSE WITH MILD HYPERCHOLESTEROLEMIA

A high plasma level of low-density lipoprotein-cholesterol (LDL) is a risk factor for the development of cardiovascular diseases. The plasma level of LDL is dependent on both genetic factors and non-genetic factors such as diet and exercise. A growing number of people with hypercholesterolemia is now becoming a serious problem in Japan.

A number of foods are known to have cholesterol-lowering effects, as modern anti-hyperlipidemic medicines do. These foods could be promising substitutes for medicines, as they have no adverse effects. Soybean protein, chitosan, depolymerized sodium alginate and plant sterols are examples.

The cholesterol-lowering effect of soybean protein has long been known, although the mechanism and components responsible for its effect have not been fully established. One hypothesis is that soy proteins associate with dietary cholesterol and/or bile acids in the intestinal tract and suppress their absorption and reabsorption. Several experiments have demonstrated that a diet containing soy protein increased the fecal excretion of cholesterol and bile acids [16]. There are other mechanisms that have been proposed to explain the hypocholesterolemic effect of soy proteins. These are associated with: increased activity of the hepatic LDL receptor [17], increased rates of hepatic cholesterol and bile acid synthesis, and changes in the endocrine status [18]. It is believed that soy protein itself, together with such other components as isoflavones, is responsible for these beneficial results. There are many types of FOSHU products containing soy proteins as functional ingredients, for example: instant powdered soup containing soy proteins, nutrient drinks supplemented with soybean globulin, processed meat products like hamburger and sausage, soy milk, and yogurt made from soy milk.

Chitosan, a kind of dietary fiber derived from crab and prawn chitin, has been found to have a cholesterol-lowering effect. A possible mechanism for this is that the positive charge of chitosan interacts with the negative charge of cholesterol, thus interrupting cholesterol absorption in the intestines [19]. An instant cup-of-noodles product supplemented with 1000 mg of chitosan is now available, and is claimed to be beneficial for those concerned about hypercholesterolemia.

Depolymerized sodium alginate is used as an effective component of plasma cholesterol-reducing FOSHU products. Sodium alginate is a polysaccharide originally derived from seaweed. It binds and absorbs cholesterol or bile acids in the small intestine, thereby preventing cholesterol absorption [20].

Dietary cholesterol absorption can also be reduced by plant sterols and stanols. The mechanism of action is believed to be the competitive inhibition of cholesterol absorption. Cholesterol is solubilized within micelles by bile acids and phospholipids in the small intestine. Plant sterols and stanols are very similar in structure to cholesterol and have greater affinity to the micelles, enabling them to displace cholesterol from the micelles. Cholesterol absorption is therefore, prevented and it is instead excreted in the feces. Plant sterols, as well as cholesterol from the micelles, are absorbed via NPC1L1, which is believed to be a selective transporter of cholesterol expressed at the surface of enterocytes. However, almost all of the absorbed plant sterols are taken back to the gut by the efflux transporters, ABCG5/G8. This mechanism results in plant sterols showing a lower absorption rate (0.02%–3.5%) than that of cholesterol (35%–70%) [21,22].

Cooking oil made from soybean kernels has been approved as a FOSHU product. This oil contains about 1700 mg of plant sterols in 100 g of oil. An experiment using healthy adult males with a slightly high cholesterol level (220 mg/dl and higher) has demonstrated that, after 11 g/day consumption of this oil, plasma levels of LDL cholesterol decreased by 10–15 mg/dl on average. A new mayonnaise, another cholesterol-lowering product containing plant sterols, has also been approved as a FOSHU product.

3.2.4   FOODS FOR THOSE CONCERNED ABOUT THEIR BLOOD GLUCOSE LEVEL

Hyperglycemia, when continued for a long period, is prone to trigger such microvascular complications as diabetes, insulin resistance and ketoacidosis. It is important for those who have a relatively high plasma concentration of glucose to attenuate any increase in the postprandial blood glucose level. Postprandial hyperglycemia is often due to impaired insulin-mediated glucose uptake by peripheral tissues.

Indigestible dextrin is the most common principal ingredient of products in this category. Indigestible dextrin is a soluble dietary fiber that can remain in the stomach with food for a while and delay its movement to the intestines. It therefore, physically ameliorates the glucose uptake and prevents a rapid increase in the postprandial blood glucose level [23].

Several food-derived compounds with inhibitory effects on ,intestinal sugar-degrading enzymes are also used as functional ingredients. Guava tea manufactured from a hot-water extract of the leaves of *Psidium guajava* L. has been confirmed to inhibit the activity of starch-degrading enzymes, especially that of α-amylase. It has also been demonstrated by experimental studies on animals and human subjects that the oral administration of guava tea can successfully prevent an increase in the postprandial glucose level [24]. A tea beverage containing guava leaf polyphenols has been approved as a FOSHU product. Another food-derived functional ingredient with inhibitory activity against α-amylase is wheat albumin (WA). Wheat albumin can be extracted from the water-soluble fraction of wheat kernel and comprises several proteins, the principal one being 0.19-WA [25]. Wheat albumin showed strong inhibitory activity against gastrointestinal α-amylases in both in vitro and in vivo experiments, and the administration of 1.5 g of WA significantly suppressed the postprandial hyperglycemic response and serum insulin level in the borderline and mildly diabetic groups [26,27]. Wheat albumin is now available in the form of

FOSHU-labeled powdered soup. Regular administration of this product has demonstrated its usefulness not only for preventing postprandial hyperglycemia, but also for lowering the fasting blood glucose level of patients with mild diabetes.

L-Arabinose, a natural pentose that is manufactured as functional table sugar, can also delay carbohydrate digestion through its inhibitory activity against intestinal sucrase [28]. In a study using both healthy volunteers and type 2 diabetes patients, 3% L-arabinose added to sucrose suppressed the maximal increase in plasma glucose after sucrose ingestion.

The fermented soybean-derived "touchi" extract is known to exert a strong inhibitory effect on intestinal α-glucosidase. Touchi is a traditional Chinese foodstuff mainly used for seasoning. The touchi extract is obtained from soybeans that have been steamed and fermented with *Aspergillus* sp. Powdered tea supplemented with this touchi extract has been approved as a FOSHU product. In a study using subjects with borderline and mild type 2 non-insulin-dependent diabetis mellitus patients, three months' consumption of powdered tea supplemented with 0.3 g of this touchi extract three times daily with meals significantly decreased the fasting blood glucose level and glycated hemoglobin level [29].

3.2.5 FOODS FOR THOSE CONCERNED ABOUT THEIR FAT ACCUMULATION

Obesity is now becoming one of the most critical health problems in modern life. It is linked to several serious medical conditions, including type 2 diabetes, heart disease, high blood pressure, and stroke. Obesity is also linked to higher rates of certain types of cancer. To decrease the risk of contracting these disorders, it is most important to manage hypertriglyceridemia, as this condition is closely associated with coronary diseases that can provoke many serious conditions. Dietary control to ingest less fat and carbohydrate and more dietary fiber is the first choice to improve the serum triglyceride level, but it does not alone have sufficient effect.

Cooking oil formulated to reduce the triglyceride concentration has been approved as a FOSHU product. This product contains diacylglycerol (DG), particularly 1,3-diacylglycerol, as the main component. Triacylglycerol (TG), the main component of conventional edible oil, is hydrolyzed in the duodenum by lipase to form 2-monoacylglycerol and free fatty acids which are easily absorbed in the small intestine. 2-Monoacylglycerol and fatty acids are reformed into TG in the intestinal wall and then flow into the bloodstream as chylomicrons. On the other hand, 1,3-diacylglycerol is neither absorbed nor resynthesized into TG in the small intestine. As a result, this oil can prevent postprandial lipidemia. It has been demonstrated that the postprandial level of TG, four hours after ingesting DG, was significantly less than that after ingesting a TG emulsion in healthy human subjects [30]. Long-term use of this oil can also prevent the increase of body fat, especially fat that is deposited in the internal organs [31,32].

Globin digest (GD) is another functional ingredient that suppresses the postprandial rise in the serum neutral fat concentration. Globin digest is an acidic protease hydrolysate of globin, and its effective constituent is the tetrapeptide, Val-Val-Thr-Pro. It is known that VVTP suppresses fat absorption by inhibiting the activity of pancreatic lipase. It has also been suggested that GD can activate internal neutral fat-digesting lipases like lipoprotein lipase and hepatic lipase [33]. Several soft drinks containing GD are now commercially available as FOSHU products in Japan.

Body fat accumulation can also be inhibited by activating fat burn. Tea catechins have the beneficial effect of activating lipid metabolism in the liver. It has been demonstrated that long-term consumption of a high-fat diet supplemented with a high dose of tea catechins significantly increased the β-oxidation activity in the liver. Tea catechins enhance the gene expression of β-oxidation enzymes like acyl-CoA oxidase and medium-chain acyl-CoA dehydrogenase [34]. A green tea FOSHU product fortified with catechins is now available to consumers concerned about obesity.

3.2.6 FOODS TO HELP MINERAL ABSORPTION

Calcium and iron are among the minerals which are often in short supply in modern diets. Calcium is the main constituent of hydroxyapatite, the principal component of bones and teeth, while iron is required for the formation of hemoglobin in red blood cells and various enzymes. A deficiency of these minerals leads to serious physical disorders, such as osteoporosis and anemia.

FOSHU products that help mineral absorption supply these minerals either in a highly absorbable form or with an enhancer that facilitates their absorption.

Calcium citrate malate (CCM) supplies calcium in its most bioavailable form, and is used in some FOSHU products to aid calcium absorption [35]. Calcium is generally absorbed by two distinct mechanisms. Active, transcellular absorption by calcium channels in the duodenum occurs when the calcium intake is relatively low. This process is greatly enhanced by vitamin D. Ionized calcium, on the other hand, diffuses through tight junctions into the basolateral spaces. This paracellular passive absorption occurs in the jejunum and ileum, and, to a much lesser extent, in the colon, when the dietary calcium level has been moderate or high. Calcium citrate malate is taken up by the second mechanism as it produces a highly soluble type of calcium in the small intestine.

Some FOSHU products contain casein phosphopeptides (CPPs), which promote calcium absorption. Casein phosphopeptides is a peptide fragment produced during casein digestion and has specific phosphorylated parts capable of binding to calcium. Casein phosphopeptides can therefore, maintain more soluble calcium in the small intestine by forming a soluble complex with calcium, thus facilitating its passive absorption [36].

Fructooligosaccharide (FOS) is also known to promote calcium absorption and is used as an ingredient of FOSHU products. The mechanism by which FOS facilitates calcium absorption is not fully understood, but an increase in the relative amount of calcium-binding protein (calbindin) in the large intestine of FOS-fed rats has been reported [37]. Calbindin acts as a carrier protein of calcium in the intestinal epithelial cells and is involved in the rate-limiting step of active transcellular absorption.

Polyglutamic acid (PGA), a large peptide of glutamic acid, is another functional ingredient that supports mineral absorption. A granulated FOSHU product is available containing calcium, together with PGA, as effective components. Polyglutamic acid increases Ca solubility because of the negative charges in its molecule. It also inhibits the formation of insoluble complexes of Ca with other substances like phosphate in the lower part of small intestine, where the pH is relatively high. An experiment to verify the effect of PGA on Ca absorption indicated that about 70% of Ca was soluble in the lower half of the small intestine of rats fed with a PGA-supplemented diet, as compared to only about 40% in the control group [38].

Iron absorption is affected by the physical state of iron entering the duodenum. There are two major forms of dietary iron: heme iron, a complex of iron with porphyrin, is abundant in such animal foods as meat and liver, while inorganic iron ions (non-heme iron) are contained in vegetable foods such as soybeans and spinach. Heme is the most easily absorbed form of iron, the absorptivity being about ten times higher than that of inorganic iron [39]. Heme iron is the form used in FOSHU products.

3.2.7 FOODS FOR THOSE CONCERNED ABOUT BONE STRENGTH

The skeleton has multiple functions, such as supporting body weight, protecting the internal organs, providing a primary site for hematopoiesis, and storing minerals. Bone contains 99% of the whole calcium in the body, in the form of hydroxyapatite. Bone is in fact a dynamic organ that undergoes an active renewal process called remodeling [40]. There are three types of cell involved in bone remodeling: osteoblasts, osteocytes, and osteoclasts. Osteoblasts and osteocytes are responsible for synthesizing the bone matrix. These cells produce collagen and other proteins indispensable for crystallization. Osteoclasts are involved in the resorption of bone tissue. These cells secrete bone-reabsorbing enzymes which digest the bone matrix. The two processes in bone remodeling, resorption, and formation, are usually coupled so that is no net bone loss. However, as we get older, the rate of breakdown increases and exceeds the rate of formation. Overall, bone is lost, and the skeleton becomes more fragile and vulnerable to fracture. More than 50% of Japanese women age 60 or over are considered to be suffering from osteoporosis.

There are several FOSHU products useful to strengthen our bones. Products supplemented with calcium and vitamin D are well known to help bone health. Calcium is indispensable as a major component of bone, and vitamin D promotes calcium absorption in the intestines [41].

There are some other functional materials that act directly on the mechanism for bone remodeling. FOSHU products containing a large amount of soybean isoflavone are now available. Isoflavone has a weak estrogen-like action, and is referred to as phytoestrogen. Estrogen plays an important role in controlling bone mass by maintaining the proper balance between bone formation and resorption. It increases osteoblast activity and simultaneously inhibits osteoclast activity. Soybean isoflavone is believed to act like estrogen and increases the rate of bone formation [42]. It has been demonstrated that consumption of 60 mg a day of isoflavone for three months resulted in a significant decrease in the concentration of a urinary bone absorption marker (pyridinoline) [43].

Vitamin K is required to produce an active form of osteocalcin, a non-collagenous protein accounting for one of the components of the extracellular matrix of bone. Osteocalcin has high affinity for bone mineral constituents and is believed to be involved in bone formation. The mineral-binding capability of osteocalcin requires vitamin K-dependent carboxylation of its molecule [44]. "Natto" is a traditional Japanese food made of fermented soybeans. A special type of "natto" containing a large amount of vitamin K2 has been approved as a FOSHU product.

Milk is widely known as an excellent source of calcium and is good for bone health. It has recently been revealed that milk contains not only nutrients such as calcium but also functional components that affect bone remodeling. A functional component was found in whey [45] and designated as MBP (milk basic protein) because it was recovered in the alkaline fraction. MBP is a mixture of basic proteins and peptides that increases the number of osteoblasts or prevents the activity of osteoclasts, thereby promoting bone formation and inhibiting bone resorption. It has been demonstrated that a daily MBP supplementation of 40 mg for six months to healthy adult women significantly increased their bone mineral density compared to the placebo group [46]. In the same experiment, bone resorption markers in urine were significantly lower in the MBP-supplemented group. Cystatin C has recently been identified as one of the factors in MBP that inhibit bone resorption. Cystatin C is a protease inhibitor that inhibits cathepsin K, a major protease secreted by osteoclasts to digest bone matrix proteins [47]. Several types of dairy products containing MBP have been approved and are now commercially available.

3.2.8 FOODS LESS LIKELY TO CAUSE TOOTH DECAY

Tooth decay is induced by acids created when the bacteria that normally reside in the mouth react with dietary sugars. These acids cause dissolution of the tooth structure by depleting it of calcium and phosphate, the process known as demineralization, which finally leads to the cariogenic cavity. Based on this mechanism, foods less likely to cause tooth decay can be categorized into two types: (1) those that inhibit the proliferation of bacteria and (2) those that suppress demineralization and/or promote remineralization, the reverse process.

FOSHU products that are categorized into the former type include a chewing gum containing sugar alcohols like maltitol and erythritol, the sweeteners used as sugar substitutes. The sugar-free nature of the gum deprives the bacteria of nutrients, thus suppressing their proliferation [48]. Tea polyphenols, which are known to act directly on the bacteria to prevent their proliferation, are also used as ingredients. The latter type of FOSHU product includes another type of chewing gum containing xylitol and casein phosphopeptide-amorphous calcium phosphate (CPP-ACP).

Phosphopeptide-amorphous calcium phosphate delivers calcium and phosphate to the tooth enamel in a stable manner, then promotes remineralization [49]. The peptide part CPP maintains calcium and phosphate in an amorphous, soluble form, and binds to the surface of the tooth, thus presenting this soluble calcium phosphate at a high concentration to promote remineralization. A seaweed polysaccharide, funoran, and calcium phosphate are also used as ingredients of FOSHU products to promote remineralization of the enamel subsurface lesions.

Another FOSHU product features phosphoryl oligosaccharides of calcium (POs-Ca), which suppress demineralization by keeping the pH value high in the plaque. Pos-Ca also stabilizes calcium and phosphate in a soluble form, even in the neutral condition, and is used as an ingredient in FOSHU products that promote remineralization [50].

3.3 FUTURE VIEWS ON FUNCTIONAL FOOD SCIENCE AND FOSHU PRODUCTS

One of the major target organs for current FOSHU products is the intestines. More than half of the present FOSHU products are in the category of "foods to regulate gastrointestinal conditions by promoting an increase in beneficial bacteria in the intestinal microflora." The intestines are responsible for the absorption of nutrients. Modulation of the nutrient absorption in the intestinal tract can also be expected to be effective for regulating the physiological conditions [4,51]. As already described, food substances that inhibit intestinal carbohydrate digestion and/or glucose absorption are effective in reducing the postprandial glucose level. Several FOSHU products helpful for people with high blood cholesterol utilize functional substances that suppress the intestinal absorption of dietary cholesterol and bile acids. Substances inhibitory to gastrointestinal lipase are useful for anti-obesity FOSHU products. Stimulating the intestinal absorption of minerals is also an effective means for better mineral supplementation. Intestine-modulating foods have therefore, constituted the primary FOSHU products up to the present. However, research on the functional foods modulating other body systems such as the blood pressure, lipid metabolism and bone formation has been progressing very rapidly in recent years, creating some new FOSHU products.

In addition to the above-mentioned FOSHU products, foods with other functions are eagerly awaited by the public. These may include foods for people suffering from allergy, foods to prevent cancer development, foods to modulate brain-nerve functions and foods for people who feel fatigued.

3.3.1 IMMUNITY-MODULATING FOSHU

Food factors that modulate the immune system are being paid increased attention in Japan because of the recently increasing number of patients with such allergies as asthma and dermatitis. According to certain statistics, one third of the Japanese population has some kind of allergy. Hay fever is becoming a significant problem in Japan, and autoimmune diseases such as inflammatory bowel disease are also increasing. There have also been recent outbreaks of pathogenic bacterial infection, such as that by *Escherichia coli* O157 in 1996, which may suggest that the body defense ability of Japanese people has become attenuated. Many scientists are therefore, interested in studying immunity-modulating food factors, and such factors as polyphenols, nucleotides, oligosaccharides, and peptides have already been reported to modulate the immune-system from both in vivo and in vitro experiments [3,4]. The functions of probiotics and prebiotics in modulating the immune-system have already been recognized by many researchers [52] and also been demonstrated by epidemiological studies. However, no immunity-modulating food has yet been approved as a FOSHU product, because the immune system is highly complex and its regulation by food is difficult to verify. Convenient and reliable biomarkers for evaluating immunity-modulating food substances have not yet been established.

Japanese scientists began working to solve these problems in 2001, by founding the Japanese Association for Food Immunology (JAFI). Food scientists and medical scientists from universities and research institutes, as well those from major food manufacturing companies, joined together for this purpose. The first objective for the members was to construct a database of food substances with immune functions. The working group read 16,000 papers published in scientific journals during the past decade, to clarify the relationships between food factors and the immune system. JAFI formally started work in October, 2004 and is expected to give scientific advice regarding immunological issues on food to the administrative organs and public. Getting public recognition of the immunity-modulating functions of foods based on scientific evidence will be important in creating a new FOSHU category of immunity-modulating products.

3.3.2 CANCER-PREVENTING FOSHU PRODUCTS

The search for cancer-preventing food has long been carried out in many countries. The Designer Food Program (U.S.A., 1990) is a well-known cancer-preventing food project [53].

Cancer develops in several stages. Damage to DNA caused by such problems as oxidative stress and carcinogenic substances is the initiating step. It is becoming evident that various antioxidative food factors, such as carotenoids and polyphenolic compounds, eliminate reactive oxygen species (ROS) and are beneficial for inhibiting oxidative stress to cell membranes, proteins, and DNA [54]. These compounds would be promising constituents of prospective cancer-preventing food. The induction of apoptosis in cancerous cells, inhibition of the promotional stage of cancer, inhibition of cancer metastasis, and activation of such immune cells as natural killer (NK) cells would also be effective means by which food substances might prevent cancer.

Chemical compounds with anti-cancer promotion activity isolated from tropical plants represent examples of promising anti-carcinogenic food substances [55]. Anti-angiogenic food factors such as terpenes, carotenoids, and polyphenols are also being paid increased attention [56]. In spite of these many studies on the anti-cancer functions of food and food constituents, it remains difficult for these foods to be approved as FOSHU products because of the highly complex aspects of regulating cancer formation, as is also the case with immunity-modulating food products.

3.3.3 ANTI-FATIGUE FOSHU PRODUCTS

According to recent statistical data from the Ministry of Health, Labor, and Welfare of Japan, 59% of the Japanese population feel fatigued, with 37% of those reporting chronic fatigue. In 2003, a national project was established with the title of "Fatigue evaluation and anti-fatigue food & drug development." Researchers from seven universities, nine pharmaceutical/chemical companies, seven food companies and two general trading firms are involved in this project. Its goals include (1) evaluating the type (physical or mental) and degree of fatigue, as well as susceptibility and resilience to fatigue, (2) developing anti-fatigue FOSHU products, and (3) providing self-monitoring tools for evaluating fatigue. Since the definition of "fatigue" itself is still obscure, it is difficult to establish good biomarkers for its evaluation. However, this project has already found several interesting physiological, biochemical, and immunological parameters that could be applicable as markers for fatigue. A computer-based testing instrument to examine fatigue during research trials has been developed by this project and should be useful for objectively and quantitatively evaluating fatigue. It may be possible to approve anti-fatigue FOSHU products in the future if this project continues to make the progress already begun.

3.3.4 SCIENCE AND TECHNOLOGY ESSENTIAL FOR THE DEVELOPMENT OF **FOSHU** PRODUCTS

A large number of studies on food and nutrition during the past two decades have demonstrated that the relationship between food and body systems is highly complex. Predicting the effects of dietary substances on body systems is therefore, not easy. To elucidate food–body interactions in more detail, studies based on the latest advances in science and technology are essential.

3.3.4.1 Approaches Using Genomics and Proteomics

It is known that food substances sometimes affect gene expression in various organs, thereby modulating body systems. However, one food substance may elicit the expression of multiple genes. Furthermore, food is a mixture of diverse chemical substances. It is therefore, essential to develop methods to comprehensively analyze the multiple responses of body systems to various food factors. Genomics and proteomics provide valuable means of gathering a large amount of information on body–food interactions. The recent development of microarray technology has made it possible to use this method for searching and evaluating functional food substances [57,58].

3.3.4.2 Approaches Using Metabolomics

The fate or movement of food substances after oral administration is, in many cases, not well understood. Most food substances are digested and absorbed in the gastrointestinal tract, followed by further cleavage in the intestinal cells. Some components may be conjugated with other molecules in the intestinal and/or liver cells. Modification of the components may occur in the blood circulation and also in the peripheral tissues. Without knowing details of these complicated changes that occur to food substances, it is difficult to accurately evaluate these substances in terms of their functions in the body. Metabolomics should therefore, be applied to functional food science [58,59].

3.3.4.3   Approaches Using Bioinformatics

The application of computer-aided analysis to functional food science is essential to answer the question of "how do heterogeneous factors (food) affect the heterogeneous system (body)?" Informatics drawing on a comprehensive database of functional food substances [60], their properties and their content in food materials would be a powerful tool for evaluating and designing functional food products.

3.3.4.4   Establishment of Biomarkers

Many biomarkers have already been used to evaluate food/food constituents that would be beneficial for preventing life-style-related diseases [6]. The foregoing approaches using new technologies would be helpful to find other biomarkers, some of which may be useful for developing the new FOSHU categories just described.

3.4   CONCLUSION

The value of any FOSHU product is always justified by scientific evidence. This is the distinct difference between FOSHU products and traditional foods, including so-called health foods. The development of FOSHU products with new health claims requires more scientific information on the interaction between food substances and body systems. For example, it is important to reveal how food factors are recognized by body systems and to express the functions thereof. The molecular mechanism for the recognition of any food factor should be investigated in detail. The mechanisms by which dietary substances interact with metabolism in different individuals must be defined through biochemical and physiological studies. Information provided by such basic studies will help to reinforce clinical proof and also to confirm the safety of the food factors. The remarkable increase in the number of FOSHU products in Japan seems to promise rapid expansion of the FOSHU market in the near future. However, concerted research into the basic aspects of functional foods must be continued to keep pace with the development of the FOSHU market, because the success of the FOSHU concept is essentially dependent on its scientific basis.

REFERENCES

1. Arai, S., Studies on functional foods in Japan—state of the art, *Biosci. Biotechnol. Biochem.*, 60 (1), 9–15, 1996.
2. Arai, S., Global view on functional foods: Asian perspectives, *Br. J. Nutr.*, 88 (suppl. 2), S139–S143, 2002.
3. Arai, S., Osawa, T., Ohigashi, H., Yoshikawa, M., Kaminogawa, S., Watanabe, M., Ogawa, T. et al., A mainstay of functional food sciences in Japan—history, present status and future outlook, *Biosci. Biotechnol. Biochem.*, 65 (1), 1–13, 2001.
4. Arai, S., Morinaga, Y., Yoshikawa, T., Ichiishi, E., Kiso, Y., Yamazaki, M., Morotomi, M., Shimizu, M., Kuwata, T., and Kaminogawa, S., *Biosci. Biotechnol. Biochem.*, 66 (10), 2017–2029, 2002.

5. Watanabe, M., Hypoallergenic rice as a physiologically functional food, *Trend. Food Sci. Technol.*, 4 (5), 125–128, 1993.
6. Shimizu, T., Health claims on functional foods: Japanese regulations and an international comparison, *Nutr. Res. Rev.*, 16, 242–252, 2003.
7. Ogawa, M., Shimizu, K., Nomoto, K., Tanaka, R., Hamabata, T., Yamasaki, S., Takeda, T., and Takeda, Y., Inhibition of in vitro growth of Shiga toxin-producing *Escherichia coli* O157:H7 by probiotic Lactobacillus strains due to production of lactic acid, *Int. J. Food Microbiol.*, 68 (1–2), 135–140, 2001.
8. Ogawa, M., Shimizu, K., Nomoto, K., Takahashi, M., Watanuki, M., Tanaka, R., Tanaka, T., Hamabata, T., Yamasaki, S., and Takeda, Y., Protective effect of *Lactobacillus casei* strain Shirota on Shiga toxin-producing *Escherichia coli* O157:H7 infection in infant rabbits, *Infect. Immun.*, 69 (2), 1101–1108, 2001.
9. Yasui, H., Nagaoka, N., Mike, A., Hayakawa, K., and Ohwaki, M., Detection of Bifidobacterium strains that induce large quantities of IgA, *Micro. Ecol. Health*, 5, 155–162, 1992.
10. Nakamura, Y., Yamamoto, N., Sakai, K., Okubo, A., Yamazaki, S., and Takano, T., Purification and characterization of angiotensin I-converting enzyme inhibitors from sour milk, *J. Dairy Sci.*, 78, 777–783, 1995.
11. Nakamura, Y., Yamamoto, N., Sakai, K., and Takano, T., Antihypertensive effect of sour milk and peptides isolated from it that are inhibitors to angiotensin I-converting enzyme, *J. Dairy Sci.*, 78, 1253–1257, 1995.
12. Itakura, H., Ikemoto, S., Terada, S., and Kondo, K., The effect of sour milk on blood pressure in untreated hypertensive and normotensive subjects, *J. Jpn Soc. Clin. Nutr.*, 23 (3), 26–31, 2001.
13. Hata, Y., Yamamoto, M., Ohni, M., Nakajima, K., Nakamura, Y., and Takano, T., A placebo-controlled study of the effect of a sour milk on blood pressure in hypertensive subjects, *Am. J. Clin. Nutr.*, 64, 767–771, 1996.
14. Matsui, T., Matsufuji, H., Seki, E., Osajima, K., Nakashima, M., and Osajima, Y., Inhibition of angiotensin I-converting enzyme by Bacillus licheniformis alkaline protease hydrolyzates derived from sardine muscle, *Biosci. Biotechnol. Biochem.*, 57 (6), 922–925, 1993.
15. Fujita, H. and Yoshikawa, M., LKPNM: a prodrug-type ACE-inhibitory peptide derived from fish protein, *Immunopharmacology*, 44 (1–2), 123–127, 1999.
16. Greaves, K. A., Wilson, M. D., Rudel, L. L., Williams, J. K., and Wagner, J. D., Consumption of soy protein reduces cholesterol absorption compared to casein protein alone or supplemented with an isoflavone extract or conjugated equine estrogen in ovariectomized cynomolgus monkeys, *J. Nutr.*, 130 (4), 820–826, 2000.
17. Lovati, M. R., Manzoni, C., Gianazza, E., Arnoldi, A., Kurowska, E., Carroll, K. K., and Sirtori, C. R., Soy protein peptides regulate cholesterol homeostasis in HepG2 cells, *J. Nutr.*, 130 (10), 2543–2549, 2000.
18. Potter, S. M., Overview of proposed mechanisms for the hypocholesterolemic effect of soy, *J. Nutr.*, 125 (suppl. 3), 606S–611S, 1995.
19. Maezaki, Y., Tsuji, K., Nakagawa, Y., Kawai, Y., Akimoto, M., Tsugita, T., Takekawa, W., Terada, A., Hara, H., and Mitsuoka, T., Hypocholesterolemic effect of chitosan in adult males, *Biosci. Biotechnol. Biochem.*, 57, 1439–1444, 1993.
20. Kimura, Y., Watanabe, K., and Okuda, H., Effects of soluble sodium alginate on cholesterol excretion and glucose tolerance in rats, *J. Ethnopharmacol.*, 54 (1), 47–54, 1996.

21. Igel, M., Giesa, U., Lutjohann, D., and von Bergmann, K., Comparison of the intestinal uptake of cholesterol, plant sterols, and stanols in mice, *J. Lipid Res.*, 44 (3), 533–538, 2003.

22. Turley, S. D. and Dietschy, J. M., Sterol absorption by the small intestine, *Curr. Opin. Lipidol.*, 14 (3), 233–240, 2003.

23. Wakabayashi, S., Kishimoto, Y., and Matsuoka, A., Effects of indigestible dextrin on glucose tolerance in rats, *J. Endocrinol.*, 144 (3), 533–538, 1995.

24. Yusof, R. M. and Said, M., Effect of high fibre fruit (Guava—*Psidium guajava* L.) on the serum glucose level in induced diabetic mice, *Asia Pac. J. Clin. Nutr.*, 13 (suppl.), S135, 2004.

25. Oneda, H., Lee, S., and Inouye, K., Inhibitory effect of 0.19 alpha-amylase inhibitor from wheat kernel on the activity of porcine pancreas alpha-amylase and its thermal stability, *J. Biochem.*, 135 (3), 421–427, 2004.

26. Choudhury, A., Maeda, K., Murayama, R., and DiMagno, E. P., Character of a wheat amylase inhibitor preparation and effects on fasting human pancreaticobiliary secretions and hormones, *Gastroenterology*, 111 (5), 1313–1320, 1996.

27. Koike, D., Yamadera, K., and DiMagno, E. P., Effect of a wheat amylase inhibitor on canine carbohydrate digestion, gastrointestinal function, and pancreatic growth, *Gastroenterology*, 108 (4), 1221–1229, 1995.

28. Seri, K., Sanai, K., Matsuo, N., Kawakubo, K., Xue, C., and Inoue, S., L-Arabinose selectively inhibits intestinal sucrase in an uncompetitive manner and suppresses glycemic response after sucrose ingestion in animals, *Metabolism*, 45 (11), 1368–1374, 1996.

29. Fujita, H., Yamagami, T., and Ohshima, K., Long-term ingestion of a fermented soybean-derived Touchi-extract with alpha-glucosidase inhibitory activity is safe and effective in humans with borderline and mild type-2 diabetes, *J. Nutr.*, 131 (8), 2105–2108, 2001.

30. Tada, N., Watanabe, H., Matsuo, N., Tokimitsu, I., and Okazaki, M., Dynamics of postprandial remnant-like lipoprotein particles in serum after loading of diacyl-glycerols, *Clin. Chim. Acta*, 311, 109–117, 2001.

31. Maki, K. C., Davidson, M. H., Tsushima, R., Matsuo, N., Tokimitsu, I., Umporowicz, D. M., and Dicklin, M. R., Consumption of diacylglycerol oil as part of a reduced-energy diet enhances loss of body weight and fat in comparison with consumption of a triacylglycerol control oil, *Am. J. Clin. Nutr.*, 76, 1230–1236, 2002.

32. Nagao, T., Watanabe, H., Goto, N., Onizawa, K., Taguchi, H., Matsuo, N., Yasukawa, T., Shimasaki, H., and Itakura, H., Dietary diacylglycerol suppresses accumulation of body fat compared to triacylglycerol in men in a double-blind controlled trial, *J. Nutr.*, 130, 792–797, 2000.

33. Kagawa, K., Matsutaka, H., Fukuhama, C., Watanabe, Y., and Fujino, H., Globin digest, acidic protease hydrolysate, inhibits dietary hypertriglyceridemia and Val-Val-Tyr-Pro, one of its constituents, possesses most superior effect, *Life Sci.*, 58 (20), 1745–1755, 1996.

34. Murase, T., Nagasawa, A., Suzuki, J., Hase, T., and Tokimitsu, I., Beneficial effects of tea catechins on diet-induced obesity: stimulation of lipid catabolism in the liver, *Int. J. Obes. Relat. Metab. Disord.*, 26 (11), 1459–1464, 2002.

35. Patrick, L., Comparative absorption of calcium sources and calcium citrate malate for the prevention of osteoporosis, *Altern. Med. Rev.*, 4 (2), 74–85, 1999.

36. Tsuchita, H., Suzuki, T., and Kuwata, T., The effect of casein phosphopeptides on calcium absorption from calcium-fortified milk in growing rats, *Br. J. Nutr.*, 85 (1), 5–10, 2001.

37. Takahara, S., Morohashi, T., Sano, T., Ohta, A., Yamada, S., and Sasa, R., Fructooligosaccharide consumption enhances femoral bone volume and mineral concentrations in rats, *J. Nutr.*, 130 (7), 1792–1795, 2000.
38. Tanimoto, H., Mori, M., Motoki, M., Torii, K., Kadowaki, M., and Noguchi, T., Natto mucilage containing poly-gamma-glutamic acid increases soluble calcium in the rat small intestine, *Biosci. Biotechnol. Biochem.*, 65 (3), 516–521, 2001.
39. Layrisse, M. and Garcia-Casal, M. N., Strategies for the prevention of iron deficiency through foods in the household, *Nutr. Rev.*, 55 (6), 233–239, 1997.
40. Erlebacher, A., Filvaroff, E. H., Gitelman, S. E., and Derynck, R., Toward a molecular understanding of skeletal development, *Cell*, 80 (3), 371–378, 1995.
41. Suda, T., Ueno, Y., Fujii, K., and Shinki, T., Vitamin D and bone, *J. Cell. Biochem.*, 88 (2), 259–266, 2003.
42. Setchell, K. D. and Lydeking-Olsen, E., Dietary phytoestrogens and their effect on bone: evidence from in vitro and in vivo, human observational, and dietary intervention studies, *Am. J. Clin. Nutr.*, 78 (suppl. 3), 593S–609S, 2003.
43. Uesugi, T., Toda, T., Okuhira, T., and Chen, J. T., Evidence of estrogenic effect by the three-month-intervention of isoflavone on vaginal maturation and bone metabolism in early postmenopausal women, *Endocr. J.*, 50 (5), 613–619, 2003.
44. Miki, T., Nakatsuka, K., Naka, H., Kitatani, K., Saito, S., Masaki, H., Tomiyoshi, Y., Morii, H., and Nishizawa, Y., Vitamin K_2 (menaquinone 4) reduces serum undercarboxylated osteocalcin level as early as 2 weeks in elderly women with established osteoporosis, *J. Bone Miner. Metab.*, 21, 161–165, 2003.
45. Francis, G. L., Regester, G. O., Webb, H. A., and Ballard, F. J., Extraction from cheese whey by cation-exchange chromatography of factors that stimulate the growth of mammalian cells, *J. Dairy Sci.*, 78 (6), 1209–1218, 1995.
46. Yamamura, J., Aoe, S., Toba, Y., Motouri, M., Kawakami, H., Kumegawa, M., Itabashi, A., and Takada, Y., Milk basic protein (MBP) increases radial bone mineral density in healthy adult women, *Biosci. Biotechnol. Biochem.*, 66 (3), 702–704, 2002.
47. Matsuoka, Y., Serizawa, A., Yoshioka, T., Yamamura, J., Morita, Y., Kawakami, H., Toba, Y., Takada, Y., and Kumegawa, M., Cystatin C in milk basic protein (MBP) and its inhibitory effect on bone resorption in vitro, *Biosci. Biotechnol. Biochem.*, 66 (12), 2531–2536, 2002.
48. Van Loveren, C., Sugar alcohols: what is the evidence for caries-preventive and caries-therapeutic effects? *Caries Res.*, 38 (3), 286–293, 2004.
49. Cai, F., Shen, P., Morgan, M. V., and Reynolds, E. C., Remineralization of enamel subsurface lesions in situ by sugar-free lozenges containing casein phosphopeptide–amorphous calcium phosphate, *Aust. Dent. J.*, 48 (4), 240–243, 2003.
50. Kamasaka, H., Uchida, M., Kusaka, K., Yoshikawa, K., Yamamoto, K., Okada, S., and Ichikawa, T., Inhibitory effect of phosphorylated oligosaccharides prepared from potato starch on the formation of calcium phosphate, *Biosci. Biotechnol. Biochem.*, 59 (8), 1412–1416, 1995.
51. Shimizu, M., Modulation of intestinal functions by food substances, *Nahrung*, 43 (3), 154–158, 1999.
52. Kaminogawa, S., Intestinal immune system and prebiotics, *Biosci. Microflora*, 21, 63–68, 2002.
53. Caragay, A. B., Cancer-preventive foods and ingredients, *Food Technol.*, 46 (6), 65–68, 1992.
54. Uchida, K. and Osawa, T., Cellular response to the lipid peroxidation products. Oxygen radicals, In *Redox Regulation of Signalling and Disease Control in*

Antioxidants, Yodoi, J. and Packer, L. Eds., Marcel Dekker Inc., New York, pp. 105–113, 1999.

55. Murakami, A., Ohigashi, H., and Koshimizu, K., Chemoprevention: insights into biological mechanisms and promising food factors, *Food Rev. Int.*, 15 (3), 335–395, 1999.

56. Losso, N. J., Preventing degenerative diseases by antiangiogenic functional foods, *Food Technol.*, 56 (6), 78–88, 2002.

57. Muller, M. and Kersten, S., Nutrigenomics: goals and strategies, *Nat. Rev. Genet.*, 4 (4), 315–322, 2003.

58. Davis, C. D. and Milner, J., Frontiers in nutrigenomics, proteomics, metabolomics and cancer prevention, *Mutat. Res.*, 551 (1–2), 51–64, 2004.

59. German, J. B., Roberts, M. A., and Watkins, S. M., Genomics and metabolomics as markers for the interaction of diets and health: lessons from lipids, *J. Nutr.*, 133 (suppl. 1), 2078S–2083S, 2003.

60. Kuipers, O. P., Genomics and bioinformatics enhance research on food and gut bacteria, *Curr. Opin. Biotechnol.*, 15 (2), 83–85, 2004.

4 Functional Foods in India: History and Scope

Thomas Paul Devasagayam, Jai C. Tilak, and Rekha Singhal

CONTENTS

4.1   INTRODUCTION

The issues relating to the use of natural products for prevention and therapy of human diseases are gaining momentum in recent years. Plant products have been used for food and fiber throughout history. In India, they have also been used to cure and prevent diseases, as stated in many traditional Indian systems of medicine. The modulation of diseased states such as cardiovascular ailments, neurological disorders, cancer, diabetes, etc., by using dietary components including fruits and vegetables, natural products, medicinal plants, etc., as a possible therapeutic measure has become a subject of recent scientific investigations [1–4]. This basic concept has existed in the ancient Vedic scripture, the *Ayurveda*, and has been practiced in Indian traditional medicine for many centuries. Apart from *Ayurveda*, there are other established systems of medicine, such as *Siddha* and *Unani* prevalent in India that also have similar approaches. Among these, the *Ayurveda* system of medicine is the most popular [5].

In the West, natural products were widely viewed as templates for structure optimization programs designed to make perfect new drugs, referred to by industry as new chemical entities (NCEs). Unlike the Western NCE paradigm, traditional medicinal systems of the East always believed that complex diseases are best treated with complex combinations of botanical and non-botanical remedies that should be further adjusted to the individual patient and to the specific stage of the disease. This approach, best articulated and developed in traditional Ayurvedic medicinal system, emphasizes the synergistic effect of different components of complex medicinal mixtures [6].

4.2   HISTORY OF FUNCTIONAL FOODS IN INDIA—CONCEPT OF AYURVEDA, HEALTH, AND NUTRITION

Ayurveda, an ancient system of medicine that evolved in India thousands of years ago, probably represented the first record of scientific medicine in the history of the world. The story goes that the sages converged on the snow-clad slopes of the Himalayas to discuss ways to alleviate human misery. They then approached the gods, who taught them this elaborate science in a highly simplified way. The two main approaches to illness in *Ayurveda* are preventive and curative. The main reason for the sages' request to the gods was to avoid diseases and remain fit for a long time. Hence, the major approach to therapy in *Ayurveda* emphasizes prevention. The second approach is the curative one [7–9].

According to the Ayurvedic theory, a harmonious balance between three humors of the body, "Vayu," "Pitta" and "Kaf" is needed for positive health; imbalance of these may cause disease. It aims to promote positive health. "Rasayanas" are devoted to enhance the body's resistance. The prescribed procedures include not only drugs but also a daily routine including exercise, diet, and nutrition, besides mental attitude and discipline [8]. This is achieved by using extracts of various plant materials, the *rasayanas*. By using *Rasayana* therapy, one obtains longevity, regains youth, and achieves sharp memory and intellect, as well as freedom from diseases, lustrous complexion and the "strength of a horse."

In order to understand the concept of *Rasayana* and draw parallels in contemporary science, we have to become familiar with some of the basic principles of *Ayurveda*. It rests on the *dosha–dhatu–mala* tripod. There are three *doshas*, namely *vata*, *pitta*, and *kapha*, each with specific characteristics and functions. There are seven *dhatus*, or tissues. Formed by assimilation of dietary items, the tissues influence the characteristics and behavior of the *doshas*. The dhatus are arranged in hierarchical fashion—*rasadhatu* being the primordial tissue. The *rasadhatu* has been likened to the plasma, although its exact interpretation remains elusive. The tissues receive nutrients from the *rasadhatu*, picking up the components they need. It is obvious then, that the quality of the *rasadhatu* is very important and also follows that it would influence the working of subsequent tissues. Drugs that improve the quality of this *rasadhatu* and thereby of the entire body are the *Rasayanas*. On ingesting the "appropriate formulation" of the *Rasayana* in the "appropriate season" by the "appropriate person," the beneficial effects of *Rasayana* are seen [7]. Employing a unique holistic approach, Ayurvedic medicines are usually customized to an individual's constitution. Ayurvedic Indian and traditional Chinese systems are living "great traditions" and have important roles in bioprospecting of new medicines [10], as well as in rediscovering the functionality of different food components.

4.2.1 HEALTH BENEFITS OF INDIAN MEDICINAL PLANTS: THE MAIN CONSTITUENTS OF *AYURVEDA*

Indian medicinal plants are rich sources of substances that have several therapeutic properties such as being cardioprotective and chemopreventive, among other effects [11,12]. Please see Table 4.1 for details. These can form the basis for a healthy and "curative nutrition." These plants can provide useful ingredients for functional foods.

4.2.2 BENEFITS OF TRADITIONAL MEDICINE—*AYURVEDA*

Considerable research on biopharmacognosy, chemistry, pharmacology, and clinical therapeutics has been carried out on Ayurvedic medicinal plants. Many medicines have been derived from the Ayurvedic experimental base. Examples include *Rauwolfia* alkaloids for hypertension, guggulsterons as hypolipidaemic agents, *Mucuna pruriens* for Parkinson's disease, baccosides in aiding mental retention,

TABLE 4.1
Indian Medicinal Plants Based on their Therapeutic Effects

Therapeutic Properties	Plant (Common Name)
Cardioprotective properties	*Curcuma longa* (Turmeric), *Emblica officinalis* (Amla), *Ginkgo biloba*, *Panax ginseng*, *Panax pseudoginseng*, *Vitis vinifera* (Grapes)
Hepatoprotective activities	*Allium cepa* (Onion), *Andrographis paniculata* (Kalmegh, Bhunimba), *Artemisia campestris*, *C. longa* (Turmeric), *E. officinalis* (Amla), *Hibiscus sabdariffa*, *Picrorhiza kurroa* (Kutki), *Premna tomentosa*, *Trianthema portulacastrum*
Chemopreventive/ anticarcinogenic and antimutagenic properties	*Aegle marmelos*, *Allium sativum* (Garlic), *A. cepa* (Onion), *Aloe vera* (Ghritakumari), *A. paniculata* (Kalmegh, Bhunimba), *Azadirachta indica* (Neem), *Brassica juncea* (Mustard), *Camellia sinensis* (Tea), *C. longa* (Turmeric), *Cymbopogan citrates* (Lemon grass), *E. officinalis* (Amla), *Garcinia atroviridis*, *Glycyrrhiza glabra*, *Ocimum sanctum* (Tulsi), *P. kurroa* (Kutki), *T. portulacastrum*, *V. vinifera* (Grapes), *Withania somnifera* (Ashwagandha)
Anti-inflammatory activity	*Aglaia roxburghiana*, *A. indica* (Neem), *C. longa* (Turmeric), *E. officinalis* (Amla), *Emilia sonchifolia*, *Hemidesmus indicus* (Anantamul), *Nigella sativa*, *W. somnifera* (Ashwagandha)
Antimicrobial/antifungal and antiviral properties	*A. indica* (Neem), *E. officinalis* (Amla), *G. atroviridis*, *G. glabra*, *O. sanctum* (Tulsi)
Antidiabetic properties	*Capparis desidua*, *C. longa* (Turmeric), *P. kurroa* (Kutki), *Salacia oblonga*, *Syzigium cumini* (Jamun), *Trigonella foenum—graecum* (Fenugreek)
Immunomodulatory properties	*Asparagus racemosus* (Shatavari), *W. somnifera* (Ashwagandha)
Anti-ulcerogenic activities	*A. roxburghiana*, *C. sinensis* (Black tea), *P. kurroa* (Kutki)
Neuroprotective activity	*G. biloba*
Thyroid stimulatory properties	*Achyranthes aspara*, *Bauhania purpurea*, *O. sanctum* (Tulsi), *W. somnifera* (Ashwagandha)
Wound healing property	*Cantella asiatica* (Chandan)
Antidermatophytic activity	*C. tamala* (Tejpat, Dalchini)
Hypolipidaemic activity	*Commiphora mukul*, *P. kurroa* (Kutki), *Tinospora cordifolia*
Radioprotective activity	*C. longa* (Turmeric), *O. sanctum* (Tulsi)
Antipyretic activity	*A. paniculata* (Kalmegh, Bhunimba), *H. indicus* (Anantamul)
Anti-venom activity	*H. indicus* (Anantamul)
Antidepressant activity	*Hypericum perforatum*
Anti-allergic activity	*P. kurroa* (Kutki). *Tinospora cordifolia*
Anti stress activity	*A. racemosus* (Shatavari), *W. somnifera* (Ashwagandha)
Cholinergic activity	*G. biloba*, *Mellissa officinalis*
Memory improving capacity	*Bacopa monniera* (Brahmi), *M. officinalis*
Antiageing effects	*W. somnifera* (Ashwagandha)

curcumin in reducing inflammation, withanolides and many other steroidal lactones and glycosides as immunomodulators [10]. Current estimates indicate that about 80% of people in developing countries still rely on traditional medicine, based largely on various species of plants and animals for their primary healthcare [6,10]. *Ayurveda* remains one of the most ancient and yet living traditions practiced widely in India. Indian healthcare reflects medical pluralism, and *Ayurveda* still remains dominant over modern medicine, particularly for treatment of a variety of chronic disease conditions. *Ayurveda* also provides details of various healthy diets for patients afflicted with diseased conditions [5,10].

4.3   FUNCTIONAL FOODS AND THEIR RELEVANCE TO INDIA

4.3.1   CONCEPT OF FUNCTIONAL FOODS

The very concept of food is changing, from a past emphasis on health maintenance to the promising use of foods to promote better health and prevent chronic illnesses. The advent of functional foods may allow us to improve public health. "Functional foods" are those that provide more than simple nutrition; they supply additional physiological benefit to the consumer. This broad definition is the one used in this chapter. Currently there is no precise, universally accepted definition of functional foods. The term is defined and used differently in various countries. Whole foods represent the simplest example of a functional food. Broccoli, carrots, and tomatoes are considered functional foods because of their high contents of physiologically active components (sulforaphen, β-carotene, and lycopene, respectively). Green, leafy vegetables, and spices like mustard and turmeric, used extensively in India, also can fall under this category [13].

4.3.2   INDIAN DIET AND FUNCTIONAL FOODS

Because dietary habits are specific to populations and vary widely, it is necessary to study the disease-preventive potential of functional micronutrients within the context of regional diets. There are various regions with their characteristic diets in India. For instance, Southern and Eastern parts consume more rice, while Western and Northern parts consume more wheat as the staple food. There are also significant differences between urban and rural populations in the different regions. Indian food constituents such as spices, as well as medicinal plants, provide a rich source of compounds that can be used in functional foods. There is a need to create a legal category for dietary supplements and functional foods in India. At present, they are regulated as either foods or medicine. India's rapid increase in diet-related non-communicable diseases calls for new approaches to diet [14,15].

In many developing countries, such as India, the lack of availability of a balanced diet leads to micronutrient deficiencies. Antioxidants are important micronutrients in this context. Some of the other ingredients that make a food functional are: dietary fibers, vitamins, minerals, oligosaccharides, essential fatty acids like omega-3, lactic acid bacteria cultures and lignins [13]. Modern biotechnologies such as genomics, genetic expression and biomarkers of health and performance can help to a large extent in making functional food science a success that also has immense commercial potential in India.

4.3.3 Special Requirements for India

Under-nutrition is being rapidly reduced in India. The diet is shifting toward higher fat and lower carbohydrate content. Distinct features are intakes of dairy products and added sugar. The proportion of overweight individuals is increasing rapidly, and the shift is more pronounced among urban and high-income rural residents. India's rapid increase in diet-related non-communicable diseases projects similar economic costs for both under-nutrition and over-nutrition by 2025. Hence, it is very important to address these issues in the coming years [14]. People in India still eat too much fat and take in too little fiber within the framework of their daily diet. The supply of dietary fiber did not even meet 50% of the recommended daily amount of 40 g. The result was a dramatic increase in the diseases of the heart, circulatory complaints, constipation, and diabetes. Vegetables, roots, fruits, and cereals are the main sources of dietary fibers [16].

4.4 MOLECULES BEHIND THE BENEFICIAL EFFECTS AND MODE OF ACTION OF SOME FUNCTIONAL INGREDIENTS FROM INDIAN DIETARY COMPONENTS

The commonly consumed foods have various chemical moieties, many of which have been recognized for some biological activity or other. For instance, terpenoids have an essential function in cellular membranes in shear stress, indicating evolutionary significance. Similar findings for other biomolecules are also likely to emerge. Genetic manipulation of such important biomolecules through biotechnological approaches may be the guiding force in years to come [17]. We list below the functionalities of some Indian food constituents.

4.4.1 Grains, Legumes, and Oilseeds

Rice, wheat, jowar, bajra, ragi, and maize are the major grains consumed in India. There are also many types of minor grains consumed, especially by the rural poor. *Paspalum scorbiculatam*, a minor grain consumed in India, has been characterized by its various constituents, which have demonstrated antifungal, tranquilizing, and anti-food-poisoning properties [18]. A similar report is available for Job's tears (*Coix lacrym Jobi* L.), growing in the northeastern hilly regions of India [19]. Anti-tumor potential of *Asteracantha longifola* seeds against chemically induced hepatocarcinogens (2-acetylaminofluorene) and antioxidant potential have been shown in Wistar rats [20].

Although soy is not of Indian origin, its consumption in this part of the world has been on the increase due to its health benefits. Being a storehouse of several phytochemicals, its nutraceutical value has assumed unprecedented importance. The isoflavones, genistein, and daidzein do function as estradiol agonists, but can have antiesterogenic qualities as well. The antiesterogenic potential is due to its ability to bind estrogen receptors and by stimulating sex hormone- binding globulin production, to decrease the free and active hormone in the blood. Genistein is also effective in the inhibition of both androgen-dependent and androgen-independent prostate cancer cells in vitro and in vivo [21]. In another study, the interaction between anti-mitotic vitamin D and isoflavones in soy has been shown to be of great value in reducing the incidence of not only prostate cancer, but also other types of sporadic epithelial cancers, such as colonic and mammary tumors [22]. Soy proteins per se, in particular the 7S globulins, are effective in regulating the intake and degradation of low density lipoprotein (LDL) in human hepatoma cells, $\alpha + \alpha'$ subunits being more effective than the β subunit. The effect is attributed to the peptides derived from soy protein during digestion [23]. Lecithin, a component of soy lipids, has been shown to act as an antistress and adaptogenic functional food. It improved resistance during exposure to cold, hypoxia restraint and enhanced recovery from hypothermia [24].

Legumes play a major role in different food preparations of the vegetarian diet. They also constitute important sources of protein for a major part of the population. A diet rich in heated chick pea (*Cicer arietinum*) has been recommended for humans with altered lipid profiles, such as type IIa hyperlipoproteinemia. It has also been demonstrated to have a positive effect in diabetic therapy [25]. Germinated legumes are an integral part of the Indian diet. Interestingly, germination increases the levels of melatonin, the hormonal product of the pineal gland, reported to be an antioxidant. Melatonin production in the pineal gland declines progressively with age such that its levels in the elderly are a fraction of those found in young individuals. Supplemental administration of melatonin is hence thought to be beneficial in delaying age-related degenerative conditions. Very few studies related to foodstuffs derived from plants as a possible source of melatonin are available. The presence of immunoreactive melatonins in germinated legume seeds is another lesser-known beneficial effect, and needs study, along with the commonly used legumes [26].

Oilseeds and whole grains, as well as fruits and vegetables, are known to contain lignans, secoisolariciresinol and matairesinol, which after ingestion are converted by intestinal bacteria to mammalian lignans, enterodiol, and enterolactone. Lignans have protective properties against breast, prostate, and colon cancers, attributed partly to their antioxidant activity as shown in model systems, e.g., inhibition of lipid peroxidation, scavenging of hydroxyl radicals, and inhibition of DNA strand scissions [27]. Enrichment of bakery products by supplementation with non-wheat flours will add to their nutritional value. The flours and protein products of legumes, oil seeds, other cereals, tubers, corn gluten, and germ, rice bran, vegetable protein, etc., can be used for nutritional enrichment of the bakery products being used in India [28].

4.4.2 SPICES, HERBS, FLAVOURANTS, AND CONDIMENTS

Indian food preparations contain several spices to give them specific flavour and taste. They contain several nutritionally important constituents, including antioxidants [29]. The spices having antioxidant properties are *Allium cepa* (onion), *Allium sativum* (garlic), *Brassica juncea* (mustard), *Capsicum annum* (red chilli), *Curcuma longa* (turmeric), *Cinnamomum tamala* (dalchini), *Cinnamomum verum* (cinnamon), *Crocus sativus* (saffron), *Murayya koenigii* (curry leaf), *Trigonella foenum graecum* (fenugreek), and *Zingiber officinalis* (ginger). Curcumin, eugenol, capsacicin are some of the active ingredients in spices such as turmeric, black pepper, asafoetida, pippali, coriander, and garlic.

India is the home of many herbs and spices, being both the largest producer and consumer in the world. In a recent study, many herbs have been shown to contain dimethyl furan fatty acids, a group of minor components of lipid fractions that is reported to have antioxidant properties (by hydroxyl radical scavenging and peroxidase inhibitor activity) [30]. Similar research on spices produced and consumed in India is well warranted. Many ingredients from Indian spices have been shown to have health benefits. We detail below some such examples. Spice principles like eugenol lowered inflammation [31]. Capsaicin from red peppers (chillies) and curcumin, in combination with dietary fatty acids lowered generation of reactive oxygen species (ROS) in macrophages [32]. The salt–spice herbal mixture "amrita bindu" prevented carinogen-induced depletion of antioxidant enzymes, glutathione, and vitamins A, C, and E [33]. While the seeds of *Bixa orellana* L. are well known as a source of food color, the other parts of its plant are reported to possess antipoeriodic, antipyretic, diuretic, antidysenteric, hypotensive, hypoglycaemic, antitumor, and antihepatitis properties, one or all the effects being attributed to the presence of carotenoids, flavonoids, tannins, amino acids, and trace amounts of alkaloids. This has been recently reviewed [34].

Adding spices such as rosemary, marjoram, and sage to fried meat balls (pork and beef) has been observed to have an inhibitory effect on the formation of cholesterol oxidation products during subsequent storage [35]. Work on these aspects of traditional Indian spices needs to be done. Many essential oils from aromatic herbs and their constituents are known for their antiseptic, carminative, and antispasmodic activities [36]. A 1:1 water: alcoholic extract of the commonly used spices showed a dose-dependent inhibition of linoleic acid in the presence of soy lipoxygenase. The maximum effect was seen with cloves, and thereafter the order of effectiveness was cinnamon > pepper > ginger > garlic > mint > onion. The same study showed a synergistic effect of spices on the antioxidant activity. The activity was also resistant to thermal treatments, and was unaffected by boiling for 30 min at 100°C [37]. Extracts of black pepper, asafoetida, pippali, and garlic could increase life span. Garlic and asafoetida extracts can inhibit 2-stage chemical carcinogenesis [38]. Mace (*Myristica fragrans*) causes significant increase in acid soluble sulfhydryl (SH)-content in the livers of mice [39]. Coriander (*Coriandrum sativum*) appears to play a protective role in colon cancer [40]. *Cinnamomum cassia* exerted significant antimutagenic effects and these can be attributed to its modulatory effect on metabolism of xenobiotics [41]. The commonly used spices garlic, ginger, onion, mint, cloves, cinnamon, and pepper inhibit lipid peroxidation. They show "synergistic" antioxidant activity. Antioxidant activities were retained even after boiling extracts for 30 min [37]. Curry leaves show anticarcinogenic effects [42]. Asafoetida affords protection against free radical mediated diseases such as cancer [43].

4.4.2.1 Fenugreek

Among the various herbs and spices, none has attracted greater attention than fenugreek seed. It is one of the most important and time-tested folk medicinal plants in India. Dietary administration at 1%–2% has been shown to increase the levels of glutathione as well as the glutathione S-transferase activity in the liver [44]. Oral administration of fenugreek powder before eating at 25–50 g twice a day has been shown to have a hypolipidemic effect in hypercholesterolemic patients [45]. A hypoglycemic effect of fenugreek when incorporated with millets and legumes in the form of traditionally processed products, has been confirmed by measurements of glycemic index [46]. Isolated fenugreek fractions have been shown to act as hypoglycemic and hypocholesterolemic agents in both animal and human studies. The unique dietary fiber composition and high saponin content in fenugreek appears to be responsible for these therapeutic properties [47]. Inclusion of fenugreek in the diet significantly decreased lipid peroxidation with simultaneous enhancement of circulating antioxidants and thus exerts chemopreventive effects [48].

4.4.2.2 Turmeric

Turmeric has been consumed as a spice in Indian cuisine since ancient times. Curcumin, one of its major components, has been used for centuries as a naturally occurring medicine and as an anti-inflammatory agent. This effect is also attributed to inhibition of prostaglandin synthesis [21]. Antibacterial activity of the turmeric oleoresin after extraction of the yellow color, curcumin, has also been shown effective against organisms such as *Bacillus cereus*, *Bacillus coagulans*, *Bacillus subtilis*, *Staphylococcus aureus*, *Escherichia coli*, and *Pseudomonas aeruginosa* [49]. Xanthorrhizol, another compound present in a different species of turmeric, *Curcuma xanthorrhiza* Roxb., has also been shown to have antibacterial activity against the cariogenic bacterium, *Streptococcus mutans* [50]. Turmeric is used in Chinese and Ayurvedic systems of medicine as an anti-inflammatory agent, as well as for treatment of flatulence, jaundice, menstrual difficulties, haemauria, haemorrage and colic. Current research focuses on antioxidant, hepatoprotective, anti-inflammatory, anticarcinogenic, and antimicrobial properties in addition to its use in cardiovascular disease and gastrointestinal disorders [51]. Turmeric also enhances the detoxifying capacity of xenobiotics [52]. It contains a water-soluble antioxidant peptide "turmerin" which protects membranes and DNA against oxidative injury [53]. Curcumin and related compounds from turmeric inhibit cancer at initiation, promotion, and progression stages [54]. It also inhibits free radical-induced damage to DNA [55]. Curcumin (combined with garlic) reduced chromosomal aberrations induced by mutagens [56,57]. It causes tumor cell death by up-regulating proto-oncoprotein, Bax [58]. It also reduced oxidative damage and amyloid pathology in an Alzheimer's transgenic mouse [59].

4.4.2.3 Saffron

Saffron (*C. sativus*), a highly valued spice in India, has been shown to exert modulatory effects on the in vivo genotoxicity of test compounds such as cisplatin, cyclophosphamide, mitomycin C and urethane. The efficacy was also confirmed by the increased levels of hepatic glutathione S- transferase as compared to its levels in the absence of saffron extract [60].

4.4.2.4 Garlic

Garlic (*A. sativum*) is used in various Indian food preparations. In low doses, it has been shown to enhance the endogenous antioxidant status. Care must be taken, however, in using this spice, since higher doses can reverse the effect and also cause morphological changes in the liver and kidney [61]. The protective effect of fresh garlic against the clastogenic effects of known genotoxicants has also been reported [62]. Aged garlic extract is a powerful antioxidant derived from prolonged extraction of fresh garlic at room temperature—the effect being attributed to S-allylcysteine and S-allylmercaptocysteine, and lipid soluble compounds such as diallyl sulfide, diallyl disulfide, triallyl sulfide and diallyl polysulfides. This extract has been shown to increase cellular glutathione in a variety of cells, thereby exhibiting a protective effect against cell damage and the risk of cancer development. This extract has shown a synergistic action with pharmacologically-used cancer drugs [21]. It is probable that garlic and its associated allyl sulfur compounds influence several key molecular targets in disease prevention [63].

4.4.2.5 Ginger

Ginger is another important ingredient in various Indian curry preparations. The active components in ginger stimulate digestion and absorption and also relieve constipation and flatulence by increasing the muscular activity in the digestive tract. One of the features of rheumatic disease is polyarthritis inflammation. The biochemical manifestation of inflammation is increased oxygenation of arachidonic acid, which results in production of prostaglandin and leukotrienes. In *Ayurveda*, ginger is reported to be useful in treating inflammation and rheumatism, possibly through inhibition of prostaglandin and leukotriene biosynthesis [64]. Organophosphate pesticides as contaminants in foods possibly involve the generation of free radicals, as indicated by enhanced levels of malonaldehyde and the key enzymes. This effect has been shown to be counteracted by ginger [65]. [6]-Gingerol, the pharmacologically active component of ginger, inhibits xanthine oxidase, an enzyme responsible for the generation of ROS such as superoxide. It is reported to possess substantial antioxidant and antitumor activity. [6]-Gingerol, shogaol and other structurally related substances in ginger inhibit prostaglandin and leukotriene biosynthesis through suppression of the enzymes 5-lipoxygenase and prostaglandin synthetase, respectively. This explains the anti-tumor effect of ginger at the molecular level [21]. Its antimicrobial activity against *E. coli*, *Proteus sp.*, *Staphylococci*, *Streptococci*, *Salmonella* and the aflatoxin producing *Aspergillus* has been reported. Its potential to provide relief from migraine and to protect against the toxic effects of xenobitics has also been suggested. The latter effect has been confirmed by the increased levels of glutathione-S-transferases in the liver [64].

A list of essential oil-yielding plants (118 species belonging to 92 genera and 45 families) of the Indian Himalayan Region has been prepared, in order to explore, identify and prepare an inventory of potential oil-yielding plants. Various parts, such as roots, seeds, leaves, flowers, fruits, rhizomes, and bulbs have been used by local people to cure a variety of ailments, such as ulcers, skin diseases, cold, cough, fever, stomach ache, chicken pox, diarrhea, asthma, bronchitis, headache, snake bite, jaundice, rheumatism, cardiovascular, and other problems. Their oils are also traded. However, many of these plant species are recognized as critically rare, endangered, vulnerable or threatened. Efforts to develop a strategy for the conservation, utilization, and management of such species are urgently under way [66]. In a field study carried out on the ethnomedicinal practice of tribal and non-tribal people of coastal Karnataka, 43 herbs have been recognized as being administered as herbal emetics, pastes, and medicated foods for activities ranging from antibiotic, antiseptic, antiinflammatory, and ulcer healing. Some of these herbs are used in the Ayurvedic system of medicine, while others in the *Siddha* system for a similar purpose [67].

4.4.3 FRUITS AND VEGETABLES

Among the Indian fruits, *Aegle marmelos* (bilva), *Mangifera indica* (mango), *Punica granatum* (pomegranate), *Syzigium cumini* (jamun), and *Vitis vinifera* (grapes) have functional properties such as antioxidant activity. Other dietary components, used as vegetables or for seasoning, with possible antioxidant effects are *Cicer arientium* (Bengal gram), *Emblica officinalis* (amla), *Momordica charantia* (bitter gourd), *Sesamum indicum* (til), *Psophocarpus tetragonolobus* (winged bean), *Cyamopois tetragonolobus* (guar), *Spirulina platensis* (blue green algae), *Helianthus annus* (sunflower), *Vigna sinensis* (cowpea), *Linus usitatissimum* (linseed), *Arachis hypogea* (ground nut), *Hibiscus cannabinus* (kenaf), *Carthamus tinctorus* (safflower), *Brassica campestris* (Rape seed), *Doichos biflorus* (horse gram), rice bran, pearl millet, wheat grass, sorghum, soybean etc., *Amaranthus hypochondriacus*, *Coccinia indica* (ivy guard), *Brassica dorada* (cabbage), cassava, sweet potato, and yams [11,12].

Epidemiological evidence associating higher intake of fruits with lower risk of coronary heart disease has been attributed to antioxidants that can inhibit radical mediated oxidative damage to lipids and proteins. Consumption of five portions of fruits and vegetables per day may provide significant physiological effects for individuals with relatively high levels of oxidative stress. This has been demonstrated in smokers (>10 cigarettes for >1 year) consuming less than three servings of fruits and vegetables [68]. Green leafy vegetables and yellow fruits like mango and papaya are good sources of provitamin A. Carrot is a source of α-carotene and papaya is a good source of β-cryptoxanthin. Incorporation of the above in the diet can combat vitamin A deficiency.

Anthocyanins and other dietary polyphenols have been the focus of studies on their effects in prevention of heart diseases and cancers [69]. Anthocyanins have been reported to have several therapeutic properties, ranging from vasoprotection and anti-inflammation, maintenance of normal vascular permeability, anticancer properties and even treatment of age-related decline of neurological functions. In particular, studies on delphinidin and cyanidin and their glycosilated derivatives have demonstrated cyanidin to be very effective in human fibroblast cell lines against DNA damage, with the presence of sugar moiety shown to reduce their efficacy (Bianchi et al. 2001). Several other factors, such as metal chelation, interaction with emulsifiers, and other antioxidants and distribution in oil and water play an important role in antioxidant action [70].

Enzymatic enhancement of anthocyanins and other natural co-existing polyphenolic antioxidants in black currant juices by pectinase treatments have been standardized and could be marketed based on its nutraceutical importance [71]. A similar approach could be used for anthocyanin-rich exotic and minor fruits such as kokum, jamun, pomegranate and karvanda, used extensively and popularly in India. Anthocyanin-tailored foods by co-pigmentation, i.e., intermolecular hydrogen bonds, has been shown to stabilize anthocyanins in the plants in vivo [72]. Red berries of the family *Rosaceae* such as raspberries, cloudberries and strawberries are rich in ellagitannins, complex polyphenolics that are hydrolyzed to ellagic acid in the gut. Both elagitannins and ellagic acid have been shown to possess in vitro antitumor, antimicrobial and antioxidative effects [73].

There is great interest in plant foods for the management of diabetes mellitus, especially some vegetables viewed as potential hypoglycaemic agents. Both bitter gourd (*M. charantia*) and ivy gourd (*C. indica*) are useful. Hypoglycaemic effects are mediated through an insulin secretagogue effect or through influence on enzymes involved in glucose metabolism. There are also similar studies on cabbage (*Brassica oleracia*), green leafy vegetables, beans, and tubers [74].

Garcinia cambogia has flavonoids, administration of which has been shown to significantly lower the lipid levels in rats fed normal and cholesterol-containing diets. The effect has been explained to be due to a lower rate of lipogenesis and higher rate of degradation of cholesterol caused by the flavonoids [75]. Flavonoids that have an inhibitory activity on lipid peroxidation, and are not affected by heating up to 100°C/10 min., have been identified in another *Garcinia* species, i.e., *Garcinia kola* [76].

Ethnomedicinal, pharmaceutical and food values of drumstick (*Moringa* sp.) are known and reviewed [77]. Its leaves are very popular as leafy vegetables in various parts of India. Its ethanolic extract has been shown to enhance the recovery from hepatic damage induced by antitubercular drugs. The effect was confirmed from levels of glutamic oxaloacetic transaminase, glutamic pyruvic transminase, alkaline phosphatase and bilirubin in the serum, as well as from levels of lipids and lipid peroxides in the liver [78]. The seeds of this plant are shown to possess antimicrobial properties against a number of organisms [79].

Isothiocyanates present in cabbage, broccoli, and cauliflower inhibited cancer of the esophagus, colon, and lung in experimental animals inhibiting cytochrome P450 IIE4 [21]. Indole-3-carbinol, a compound present in cruciferous vegetables as a glucosinolate, has shown a decrease in both viral and chemically-induced mammary carcinogenesis by a similar mechanism in human subjects in both acute and long term studies.

The pods of *Phaseolus vulgaris* are commonly consumed as a vegetable in the Indian subcontinent. In an experimental study, consumption of aqueous extracts for 45 days showed a significant reduction in blood glucose, serum triglycerides, free fatty acids, phospholipids, total cholesterol, very low density lipoprotein cholesterol, and low density lipoprotein cholesterol. A lowering effect on thiobarbituric acid reactive substances (TBARS) and hydroperoxides was also recorded [80].

4.4.4 DAIRY PRODUCTS

India is one of the World's leading producers and consumers of dairy products. Milk protein has many bio-active peptides which are latent until released and activated by enzymatic proteolysis, such as in gastrointestinal digestion or food processing. The proteolytic activity of lactic acid bacteria as seen in widely consumed products in India such as curds or yogurt can contribute to the liberation of bioactive peptides. It should be noted that all the known biologically active peptides can also be cleaved by the lactic acid bacteria. Activated peptides are potential modulators of various regulatory processes in the body: opoid peptides are the receptor ligands which can modulate the absorption processes in the intestinal tract; angiotensin-1 converting enzyme (ACE)-inhibitory peptides are hemodynamic regulators and exert an hypertensive effect; immunomodulating casein peptides stimulate the activities of cells of the immune system; antimicrobial peptides kill sensitive microorganisms; antithrombotic peptides inhibit aggregation of platelets; caseinophosphopeptides may function as carriers for different minerals, especially calcium, and bioactive peptides can interact with target sites at the luminal side of the intestinal tract. Food-derived peptides, hence, can be used in functional foods and pharmaceutical preparations [81].

4.4.5 INDIAN MEDICINAL PLANTS

Indian medicinal plants (as listed below) are some of the best sources of beneficial phytonutrients, such as polyphenols, flavonoids, sterols, quinones, saponins, sulphides/thiols, and tannins having antioxidant activities besides having potential as therapeutic agents. Some commonly used Indian medicinal plants, with common or local names in brackets, are *Achyranthes aspera* (latjira), *Aegle marmelos* (bilva), *Aglaia roxburghiana* (priyangu), *Allium cepa* (onion), *A. sativum* (garlic, lasun), *Aloe vera* (korphad), *Amomum subulatum* (bari elaichi), *Andrographis paniculata* (kiryat), *Asparagus racemosus* (shatavari), *Azadirachta indica* (neem), *Bacopa monniera* (brahmi), *Bauhinia purpurea* (rakta chandan), *Brassica campastris, B. juncea* (mustard), *Butea monosperma* (palas), *Camellia sinensis* (tea), *C. annum* (chilli), *Centella asiatica, C. arietinum* (Bengal gram), *C. tamala* (tejpat, dalchini), *C. verum* (cinnamom), *Commiphora mukul* (guggulu), *Corylifoliae fructus, C. sativus* (saffron), *C. longa* (turmeric, haldi), *Cymbopogan citrates* (lemon grass), *E. officinalis* (amla), *Erycibe obtusifolia, Fagopyrum esculentum* (buckwheat), *Garcinia atroviridis, Garcinia indica* (kokum), *Ginkgo biloba, Glycyrrhiza glabra, Hemidesmus indicus* (anantamul), *Hibiscus sabdariffa* (lalambari), *Hypericum perforatum* (basant), *Indigofera tinctoria*, *Lycopus europaes*, *Mangifera indica* (mango, amra), *Mimordica charantia* (bitter gourd), *Murraya koenigii* (curry leaf), *Nigella sativa* (kalajira), *Ocimum sanctum* (tulsi), *Origanum vulgare* (common marjoram), *Panax ginseng*, *Panax pseudoginseng*, *Petroselinum crispum* (parsley), *Picrorhiza kurroa* (katuki), *Pinus maritime* (pine), *Piper beetle*, *Plumbago zeylanica* (chitrak), *P. granatum* (pomegranate), *Rubia cordifolia* (manjit), *Salacia oblonga*, *S. indicum*, *Sida cordifolia*, *Spirulina fusiformis*, *Terminalia arjuna* (arjun), *Tinospora cordifolia* (guduchi), *Trianthema portulacastrum* (bishkapra), *Triogenella foenum-graecum* (fenugreek, methi), *V. vinifera* (grapes), *Withania somnifera* (ashwagandha), and *Zingiber officinale* (ginger, sunth) [82,83].

There are several detailed studies on the mechanisms behind the beneficial properties of these plants. One such study shows that the acetone and methanol extracts of *T. arjuna* has demonstrated antimutagenic effects against 2-amino-fluorene and 4-nitro-*o*-phenylenediamine in a TA 98 Frameshift mutagen tester strain of *Salmonella typhimurium* [84]. In another study, an equal combination of 75% extracts of *Terminalia chebula*, *Terminalia belerica*, and *E. officinalis* (named "triphala") reduced blood sugar in normal and diabetic rats. This extract was found to inhibit lipid peroxide formation and scavenge hydroxyl and superoxide radicals in vitro [85].

Essential oils of various medicinal plants such as *Ocimum gratissimum*, *Zingiber cassumunar*, *Cymbopogon citratus*, and *Caesulia axilliaris* have shown strong antifungal activity against a dominant storage fungi, *Aspergillus flavus*. The essential oil of *O. gratissimum* has shown its curative potency and applicability as an antidermatophytic agent, when used in the form of an ointment prepared in polyethylene glycol [86].

Withania somnifera, commonly known as ashwagandha, has been shown to have adaptogenic, cardiotropic, cardioprotective, and anticoagulant properties [9,11,12,82,83]. Many Indian medicinal plants also help in cancer therapy by inducing apoptosis [87].

4.4.6 PLANTATION CROPS (TEA, COFFEE, AND COCOA)

Consumption of black tea and coffee is fairly common in India. Both have many essential components, including antioxidants. Tea is a pleasant, popular, socially accepted, economical, and safe drink that is enjoyed every day by millions of people across all continents. Black tea, along with milk, is a popular beverage in India. Tea also provides a dietary source of biologically active compounds that help prevent a wide variety of diseases. It is the richest source of a class of antioxidants called flavonoids and contains many other biological compounds, such as vitamins, and fluoride. Moderate consumption of tea may protect against several forms of cancer, cardiovascular diseases, the formation of kidney stones, bacterial infections, and dental cavities [88]. Tea is particularly rich in catechins, of which epigallocatechin gallate (EGCG) is the most abundant. These may contribute to the beneficial effects ascribed to tea. They are effective scavengers of ROS in vitro and may also function through their effects on transcription factors and enzyme activities [89]. 1,2-Dimethylhydrazine induced oxidative DNA damage in the colon mucosa of rats, eventually leading to cancer, can be prevented by the consumption of the polyphenols from black tea [90].

Coffee, along with milk, is another popular beverage, especially in South India. One of the main constituents, caffeine, is known to have significant antioxidant effect [91]. The coffee used in India mostly contains caffeine. Our studies show that, at low concentrations, similar to those used in India, caffeine has significant antioxidant properties in vitro using rat liver microsomes and mitochondria besides DNA [91–94].

Cocoa, being rich in polyphenolic antioxidants, is recognized as an important element in neutralizing the pathology of high blood pressure and cardiovascular disorders. It also possesses the ability to modulate immune functions. India has significant consumption of chocolate as a part of healthy and balanced diet [95].

4.4.7   ROOTS AND TUBERS

Root tuber crops, including cassava, sweet potato, yams, and aroids enjoy considerable importance as vegetables, staple foods or sources of raw materials for small-scale industries. They are also sources of "lectins" [11,12]. Some wild tubers from Jammu and Kashmir like *Pueraria* tubers can be consumed without cooking, while *Droscoreus bellophyla* tubers need some preparation prior to consumption. Wild tubers also are rich sources of vitamin C, Ca, iron, and carbohydrates [96]. Potato is another major tuber used in India in considerable amounts by almost all sections of the society. A transgenic potato with genes from *A. hypochondriacus* has increased total protein content and many essential amino acids [97]. Such an "enriched" potato holds promise as a possible functional food in India also.

4.4.8   MEAT, FISH, AND POULTRY

Meats from goats, buffalo, and poultry are major sources of protein in India, especially for those consuming a non-vegetarian diet. A study showed that minced buffalo meat can be blended with 2% sodium chloride and pH increased to improve water holding capacity, emulsifying capacity, emulsion stability and yield of patties [98]. This will increase the utility of meat from old buffalo, which form the major part of animals used for meat. Buffalo meat is generally consumed by poorer sections of Indian society while poultry is consumed by all sections of the society.

Fish is one of the cheapest sources of animal protein. It is also a major source of healthy omega 3 and 6 fatty acids in the Indian diet. In India, fish production from coastal areas contributes to about 50% of the total production. Value-added production, such as fish oils, provides a good source of nutrients and vitamins that are not found in adequate quantities in a cereal-based diet. Their use will significantly contribute to the improvement of the health status of the Indian population [99]. Omega 3 fatty acids can be obtained from marine sources, which are produced by phytoplanktons and algae and then transferred eventually via a nutrition chain in membrane phospholipids and fat depots in fish and other marine animals. In India, also, these n-3 fatty acids in fatty fish including herring, mackerel, trout, other marine animals like shrimp, as well as fish oil and cod liver oil, are included in the diet. These are known to protect against cardiac arrythmias, rheumatoid arthritis, and insulin resistance [100,101]. These fatty acids also can form important constituents of various functional foods.

Among the different products delivering essential nutrients to the body, the egg has arguably a special place, being a rich, and balanced source of essential amino, and fatty acids, as well as some minerals and vitamins. It also is a major component of the Indian diet, and due to its fairly low cost, consumed in high amounts. Enrichment of eggs with vitamin E, carotenoids, selenium, and omega-3 fatty acids include better stability of polyunsaturated fatty acids during egg storage and cooking, high availability of such nutrients as vitamin E and carotenoids, absence of off-taste and an improved antioxidant and n-3 status of people consuming these eggs. Such "designer eggs" can be considered a new type of functional foods [102]. Such designer eggs are making their appearance in the Indian market.

4.4.9 HONEY

Honey is used as a home remedy in various households in India. It is also used as a vehicle for many Ayurvedic drug formulations. It possesses immunosuppressive activity on induction of murin humoral antibody responses against different allergens [103]. Honey, on induction of antigen-specific IgE antibody response, induces immunosuppression as determined by enzyme-linked immunosorbent assay (ELISA) and T cell proliferation. Hence it confirms its antiproliferative activity [104]. Honey has a beneficial effect on the physiological constitution of animals fed with it. Lactic acid bacterial counts increased 10–100-fold in the presence of honey as compared to sucrose [105]. In India, mothers of children under five years of age who had a lower respiratory tract infection and pneumonia were interviewed, and the most common home remedy used for relief of cough was honey (25%) and ginger (27%) [106]. In India, petals or whole flowers are employed in medicine along with honey, which is known as "gulkand." It is particularly useful during the summer season since it cools the body temperature [107].

4.4.10 FERMENTED FOODS

Many types of fermented foods are used in India. Idli, made from rice and legumes, is one such favorite South Indian dish. Studies indicate that dietary changes in lunches in the workplace and at home could bring about a behavioral change and improvement in iron-deficiency anemia, which is prevalent in poorer sections. Increasing consumption of idli enhanced haemogobin status in women significantly [108]. A probiotic fermented food mixture was developed by fermentation of autoclaved and cooled slurry of pearl millet flour, chickpea flour, skim milk powder and fresh tomato pulp with *Lactobacillus acidophilus*. It inhibits the growth of pathogens, namely shigella, salmonella, *E. coli*. It is also a cost-effective drink, available for not more than a rupee per 200 ml glassful [109]. Barley rabadi is an indigenous fermented food in India. Rabadi fermentation of barley flour and buttermilk mixture enhances titratable acidity at temperatures below 40°C and was acceptable in terms of taste [110].

4.4.11 PREBIOTICS AND PROBIOTICS

Probiotics, defined as microbial food supplements that beneficially affect the host by improving its intestinal microbial balance, are used therapeutically to prevent diarrhea, improve lactose tolerance, modulate immunity and even prevent cancer and lower serum cholesterol. *Lactobacilus*, *Bifidobacterium* and several other microbial species are perceived to exert such effects by changing the composition of gut microflora. The use of non-oligosaccharide prebiotics fortifies the intestinal bacteria and stimulate their growth [111]. They are bifidogenic at a dose of 2.75 g/day and the effect lasts for about 7 days [112]. The inulin type fructans, exemplified by chicory, inulin, and oligofructose, have effects on the colonic microflora, the gastrointestinal physiology, and the immune functions. While improvement in bowel habits is widely known, the evidence for calcium bioavailability is promising [113,114]. Non-digestible oligosaccharides have been shown to have a strong prebiotic effect in human subjects. After successful identification of a prebiotic, issues such as regulation, labeling, safety, and effective scale-up, stability, and shelf life need to be addressed [115].

4.4.12 MISCELLANEOUS

Capsules prepared from spray dried *Spirulina* are a rich source of antioxidants, gamma-linolenic acid, amino acids, and fatty acids. In human subjects, these have been shown to reduce the increased levels of lipids in patients with hyperlipidemic nephrotic syndrome. Its ability to prevent accumulation of cholesterol in the body has been attributed to gamma-linolenic acid [116]. Some crude extracts from agro-industrial waste such as wheat bran, potato peel extract, shrimp shell waste, red grape pomace peels, apple pomace, lemon seeds, sweet orange seeds, lentil seed coat, grape mare, grape pomace, lemon peel, grape seeds, orange essential oil, and grape seed extract also contain antioxidants that can be commercially exploited as constituents for functional food [117].

4.5 ANGIOGENESIS AND FUNCTIONAL FOODS

Angiogenesis is involved in many diseases such as cancer, diabetic retinopathy, and chronic inflammation. Several polyphenols, especially those derived from tea, grapes or red wine, inhibit angiogenesis when administered orally. Several Indian dietary components are rich in related polyphenols. These include grapes, peanuts, fruits, vegetables, herbs, soybeans, citrus fruits, beans, tea, and spices like fenugreek. Many more studies are needed to characterize such ingredients from Indian plants. The discovery of these compounds (polyphenols, etc.) as angiogenesis inhibitors has shed light on the mechanisms behind the health benefits of natural products, which are rich in these components. The possible therapeutic advantage of these small natural molecules over large protein compounds is that they can be administered orally without causing severe side effects [118,119]. They also can form important ingredients of functional foods with potential health effects in prevention of diseases.

4.6   EFFECT OF FOOD PROCESSING ON FUNCTIONALITY OF FOODS

Food processing technologies are increasingly being used in India, especially in urban areas. The consequences of food processing and preservation for phytochemicals may greatly differ in relation to their concentration, chemical structure, and oxidation state, localization in the system, possible interactions with other food components and the type of intervention applied. In foods containing vastly different phytochemicals, the physiological activity due to food processing may be a result of more than one mechanism. Consequently, there may be a decrease, increase, or a slight change in the content and functionality of phytochemicals [120].

No/slight effect: carotenoids comprising of β-carotene and lycopene are generally stable to heat treatments encountered in blanching, cooking, and pasteurization/sterilization. Interactions between polyphenols and ascorbic acid may slow the degradation of the latter during storage, while α-tocopherol can be regenerated from its radical form by reduction with ascorbic acid.

Decrease in content and activity of phytochemicals: a classic example is that of the "technological indicator," ascorbic acid, which is by far the most sensitive nutrient, and can be damaged during most treatments. Chemical and/or enzymatic oxidations are reported to decrease the antioxidant efficacy of polyphenolics, while leaching into the cooking water is mainly responsible for loss of folates. Heat processing of *Brassica* vegetables of the *Cruciferae* family greatly reduces their functionality [121]. Manufacture of black tea causes a higher degree of enzymatic aerobic oxidation of flavonoids, resulting in lower antioxidant activity. Some processing operations such as peeling and juice clarification can remove the polyphenolics [122].

Increase in content and activity of phytochemicals: partially oxidized polyphenolics that result during food processing, have been recently shown to exhibit higher antioxidant activity than the corresponding non-oxidized forms, due to increased ability to donate a hydrogen atom from the aromatic hydroxyl group to a free unpaired electron. A moderate increase in carotenoid bioavailability and enhanced phytochemical nutrient function in cereal processing are the other instances where processing can increase the activity of phytochemicals [123]. A similar situation arises for the carotenoid-rich carrot pomace during processing of carrots for juices [124]. Enzymatic treatments can effect the solubilization of antioxidants from wastes such as blueberry skin (Stanley and Miller 2001). A deeper understanding of the effect of processing on the bio-active compounds in foods is an essential factor for designing functional foods that can deliver functional phytochemicals better or equal to the raw produce [125].

4.7   INGREDIENTS AVAILABLE FOR USE IN FUNCTIONAL FOODS IN INDIA

Functional Food ingredients available in the Indian market include: vitamin antioxidant & mineral premixes; tomato powder, garlic powder, onion powder, spice mixes; amino acids, chitosan; Omega-3-fatty acids (fish and flax seed); whey protein powder; Guarana extract, *G. biloba* extract, ginseng extract, rosemary probiotics; natural antioxidants (from tea); "shield" liquid antioxidants; vegetable peptones; essential fatty acids concentrates; performance proteins; natural fruit based flowering compounds; natural colours; total extracts of medicinal plants "antioxidants"; soy ingredient, soy proteins, soy protein hydrolysate; soya protein isolate & concentrate; super critical extracts of spices; and herbs; glutamine peptides; lactoferrin, milk calcium; lycopene, *garcinia*, raw herbs; whey protein concentrate; wheat fiber, β-carotene; *A. vera* gel powder, etc. [16].

4.8   INDIAN CONSUMERS AND FUNCTIONAL FOODS

The link between diet and health in the consumer's mind is now a worldwide phenomenon. Among all the shoppers surveyed world-wide by Health Focus International in 2003, Indian shoppers were the most health-active, with 99% choosing foods, and beverages for health reasons at least sometimes. Seven out of ten Indian shoppers surveyed (mostly from the affluent groups) maintain predominantly vegetarian diets and heart-healthy diets. The connection between food choices and health is very important to Indian shoppers and they consider their natural food diets to be very healthy. They place high priority on freshness, purity, natural ingredients, and minimal processing but are still open to the idea of fortification, with two-thirds showing an interest in functional foods. More than half are willing to pay extra for them. Indian awareness of the benefits of many nutrients is low, but shoppers are interested in learning more about foods that enhance health, foods that reduce the risk of disease, vegetarian eating, and high-energy foods [126]. Many fruit and vegetable juices rich in antioxidants are being marketed at present.

4.9   NUTRACEUTICALS, FUNCTIONAL FOODS, AND FUTURE PROSPECTS IN INDIA

Nutraceuticals are derived from nutritional compounds having therapeutic value. They need not be derived solely from plants. Even vitamins, minerals, and some synthetic compounds are classified as nutraceuticals. One of the segments includes vitamins, minerals, and supplements (VMS) as well as herbals/botanicals. Growth in nutraceuticals between 2003 and 2007 in India is likely to be driven by the functional food market. It includes brain builders such as *Gingko biloba*, brahmi for smooth mental functions, and ashwagandha, used to treat insomnia and senile debility. Lutein, zeaxanthin, carotenoids, and vitamin C are antioxidants with potential beneficial effects on eye health and are considered nutraceuticals for better vision. Vitamin A, vitamin E, lycopene, selenium, resveratrol, and soy isoflavones are nutraceuticals against cancer. Polyphenols from green tea, phytic acid in cereals, and legumes, arginine and glycine inhibit angiogenesis, a key step in the development of degenerative diseases. α-Lipoic acid, conjugated linolenic acid, creatin monohydrate, γ-linolenic acid, inulin, fenugreek, L-glutathione, omega-3, and omega-6 fatty acids, gelatin hydrolysate, and vitamin E are some of the multinational nutraceuticals.

India is relatively a new market for nutraceuticals. The size of the Indian nutraceutical market is estimated to be about Rs. 1600 crores in 2001. The companies supplying ingredients for functional foods are given in Table 4.2. *Ayurveda* gives India tremendous leads in finding newer applications in this area. The country also has a historical predisposition towards medicinal herbs. India can become a significant part of this industry if it develops clinical documentation and scientific bases to support claims of safety and efficacy. Companies have to evolve more adaptable varieties suitable for cultivation in order to survive in this business. To achieve a competitive edge, Indian companies should develop superior products, trimmer preparations, bioenhancers, entirely new herbal preparations, food products, and special formulations of traditional systems. India should develop knowledge-based products [16].

TABLE 4.2
Indian Companies/Manufacturers Supplying Ingredients for Functional Foods

Alok International	Mumbai
Chaitanya Chemicals	Malkapur
Chandrakant Enterprises	Mumbai
Croda Chemicals	Mumbai
Danisco	Gurgaon, Haryana
Darmom Food Additives Pvt. Ltd.	Thane
Eltech Fine Chemicals Pvt. Ltd.	Mumbai
Fine Organics	Mumbai
Hexagon Group	Mumbai
Hi Media Laboratories Pvt. Ltd.	Mumbai
Lucid Colloids Ltd.; Lucid Flavitalia Pvt. Ltd.	Mumbai
M/s Kushalchand & Sons	Mumbai
Mahaan Healthcare	New Delhi
Naram's Food Products	Mumbai
Novartis Medical Nutrition	Thane
Sieber Hegner India Pvt. Ltd.	Mumbai
The Solae Company	Gurgaon, Haryana
Vedic Supercriticals & Biotechnologies (i) Pvt. Ltd.	Mumbai
Zytex Arun & Co.	Mumbai

4.10   REGULATORY AGENCIES IN INDIA

A meeting in October, 2003 involving more than 250 Indian food supplement and botanical industry heads, along with legislators, has highlighted the need for improved manufacturing processes and regulatory changes to meet changing domestic and international demands. There is a need to create a legal category for dietary supplements. At present they are regulated as either foods or medicine. The modernization of India's supplements sector has been gaining momentum. The government and industry realize they have to improve manufacturing processes and gather the kind of science the developed world demands if they are to compete with the likes of Chinese traditional medicines. At present, there is a Good Manufacturing Practices program in India [15].

4.11   CONCLUSIONS

Indian traditional systems of medicine and home remedies have identified many forms of health foods. Mainly whole foods are consumed as functional foods, rather than supplements or processed foods. Many dietary constituents provide ingredients for possible use in functional foods. The mechanisms behind their beneficial effects have been reviewed. There are specific requirements for the Indian population in terms of functional foods. These pertain to the adverse effects of over-nutrition in the affluent population and micronutrient deficiencies in the economically weaker sections. With proper support from appropriate government agencies in terms of suitable legislation and help from food scientists, there is a tremendous potential for processed functional foods in India in the future.

REFERENCES

1. Block, G., Patterson, B., and Subar, A., Fruit, vegetables, and cancer prevention: a review of the epidemiological evidence, *Nutr. Cancer*, 18, 1–29, 1992.
2. Hertog, M. G. L., Feskens, E. J. M., Hollman, P. C. H., Katan, M. B., and Kromhout, D., Dietary antioxidant flavonoids and risk of coronary heart disease: the Zutphen elderly study, *Lancet*, 342, 1007–1011, 1993.
3. Broekmans, W. M. R., Klopping-Ketelaars, I. A. A., Schuurman, C. R. W. C., Verhagen, H., van den Berg, H., Kok, F. J., and van Poppel, G., Fruits and vegetables increase plasma carotenoids and vitamins and decrease homocysteine in humans, *J. Nutr.*, 130, 1578–1583, 2000.
4. Surh, Y.-J., Cancer chemoprevention with dietary phytochemicals, *Nat. Rev. Cancer*, 3, 768–780, 2003.
5. Lele, R. D., *Ayurveda and Modern Medicine*, Bharatiya Vidya Bhavan, Mumbai, 2001.
6. Raskin, I., Ribnicky, D. M., Komarnytsky, S., Ilic, N., Poulev, A., Borisjuk, N., Brinker, A. et al., Plants and human health in the twenty-first century, *Trends Biotechnol.*, 20, 522–530, 2002.
7. Upadhyay, S. N., *Immunomodulation*, Narosa Publishing House, New Delhi, 1997.

8. Agarwal, S. S. and Singh, V. K., Immunomodulators: a review of studies on Indian medicinal plants and synthetic peptides. Part I: medicinal plants, *Proc. Indian Natl Sci. Acad. B*, 65, 179, 1999.

9. Devasagayam, T. P. A. and Sainis, K. B., Immune system and antioxidants, especially those derived from Indian medicinal plants, *Indian J. Exp. Biol.*, 40, 639–655, 2002.

10. Patwardhan, B., Vaidya, A. D. B., and Chorghade, M., Ayurveda and natural products drug discovery, *Curr. Sci.*, 86, 789–799, 2004.

11. Kirtikar, K. R., Basu, B. D., Blatter, E., Caius, J. F., and Mhaskar, K. S., *Indian Medicinal Plants*, Vols. 1 & 2, Lalit Mohan Basu, Allahabad, 1984.

12. Thakur, R. S., Puri, H. S., and Husain, A., *Major Medicinal Plants of India*, Central Institute of Medicinal and Aromatic Plants, Lucknow, 1989.

13. Krishnaswamy, K., Indian functional foods: role in prevention of cancer, *Nutr. Rev.*, 54, S127–S131, 1996.

14. Popkin, B. M., Horton, S., Kim, S., Mahal, A., and Shuigao, J., Trends in diet, nutritional status, and diet-related non-communicable diseases in China and India: the economic costs of nutrition transition, *Nutr. Rev.*, 59, 379–390, 2001.

15. Starling, S., India rethinks supplements sector, *Funct. Foods Neutraceutic*, 10, 138–150, 2004.

16. Souvenir and abstract book: *Prof. P. J. Dubash Memorial Seminar on Functional Additives and Ingredients in Foods—Present Trends and Newer Developments*, Institute of Chemical Technology, University of Mumbai, Mumbai, 2004.

17. Khanuja, S. P. S., Molecular basis of phylogenic variation in plant terpenoid biosynthesis, *J. Med. Aromat. Plant Sci.*, 22, 340–345, 2000.

18. Mishra, M., Shukla, Y. N., and Kumar, S., Chemistry and biological activity of *Paspalum scorbiculatum*: a review, *J. Med. Aromat. Plant Sci.*, 22, 288–292, 2000.

19. Jimo, P. V., Singh, R. A., Singh, N. P., and Gupta, R. C., Job's tears: an under-exploited medicinal plant of north-east India, *J. Med. Aromat. Plant Sci.*, 22–23, 696–698, 2000–2001.

20. Ahmed, S., Rahman, A., Mathur, M., Athar, M., and Sultana, S., Anti-tumour promoting activity of *Asteracantha longifolia* against experimental hepatocarcino-genesis in rats, *Food Chem. Toxicol.*, 39, 19–28, 2001.

21. Hollis, D. M. and Wargovich, M. J., Phytochemicals as modulators of cancer risk, In *Biologically-Active Phytochemicals in Food*, Pfannhauser, W., Fenwick, G. R., and Khokhar, S. Eds., Royal Society of Chemistry, Cambridge, pp. 13–20, 2001.

22. Farhan, H. and Cross, H. S., Genistein action in prostate cancer cells, In *Biologically-Active Phytochemicals in Food*, Pfannhauser, W., Fenwick, G. R., and Khokhar, S. Eds., Royal Society of Chemistry, Cambridge, pp. 28–31, 2001.

23. Arnoldi, A., D'Agostina, A., Boschin, G., Lovati, M. R., Manzoni, C., and Sirtori, C. R., Soy protein components active in the regulation of cholesterol homeostasis, In *Biologically-Active Phytochemicals in Food*, Pfannhauser, W., Fenwick, G. R., and Khokhar, S. Eds., Royal Society of Chemistry, Cambridge, pp. 103–106, 2001.

24. Kumar, R., Divekar, H. M., Gupta, V., and Srivastava, K. K., Antistress and adaptigenic activity of lecithin supplementation, *J. Altern. Complement. Med.*, 8, 487–492, 2002.

25. Zulet, M. A., Macarulla, M. T., Portillo, M. P., Noel-Suberville, C., Higuerert, P., and Martinez, J. A., Lipid and glucose utilization in hypercholesterolemic rats fed a diet containing heated chickpea (*Cicer arietinum* L.): a potential functional food, *Int. J. Vitam. Nutr. Res.*, 69, 403–411, 1999.

26. Zielinski, H., Lewczuk, B., Przybylska-Gornowicz, B., and Kozlowska, H., Melatonin in germinated legume seeds as a potentially significant agent for health, In *Biologically-Active Phytochemicals in Food*, Pfannhauser, W., Fenwick, G. R., and Khokhar, S. Eds., Royal Society of Chemistry, Cambridge, pp. 110–117, 2001.

27. Niemeyer, H. B. and Metzler, M., Antioxidant activities of lignans in the FRAP assay (ferric reducing/antioxidant power assay), In *Biologically-Active Phytochemicals in Food*, Pfannhauser, W., Fenwick, G. R., and Khokhar, S. Eds., Royal Society of Chemistry, Cambridge, pp. 394–395, 2001.

28. Chavan, J. K. and Kadam, S. S., Nutritional enrichment of bakery products by supplementation with nonwheat flours, *Crit. Rev. Food Sci. Nutr.*, 33, 189–226, 1991.

29. Peter, K. V., *Handbook of Herbs and Spices*, CRC Press, Boca Raton, FL, 2001.

30. Sigrist, I. A., Manzardo, G. G. G., and Amadò, R., Analysis of furan fatty acids in dried green herbs using ion trap GC–MS/MS, In *Biologically-Active Phytochemicals in Food*, Pfannhauser, W., Fenwick, G. R., and Khokhar, S. Eds., Royal Society of Chemistry, Cambridge, pp. 237–240, 2001.

31. Reddy, A. C. and Lokesh, B. R., Studies on anti-inflammatory activity of spice principles and dietary n-3 polyunsaturated fatty acids on carrageenan-induced inflammation in rats, *Ann. Nutr. Metab.*, 38, 349–358, 1994.

32. Joe, B. and Lokesh, B. R., Role of capsaicin, curcumin and dietary n-3 fatty acids in lowering the generation of reactive oxygen species in rat peritoneal macrophages, *Biochem. Biophys. Acta*, 1224, 255–263, 1994.

33. Shanmugasundaram, K. R., Ramanujam, S., and Shanmugasundaram, E. R., Amrita Bindu—a salt–spice–herbal health food supplement for the prevention of nitrosamine induced depletion of antioxidants, *J. Ethnopharmacol.*, 42, 83–93, 1994.

34. Srivastava, A., Shukla, Y. N., Jain, S. P., and Kumar, S., Chemistry, pharmacology and uses of *Bixa orellana*: a review, *J. Med. Aromat. Plant Sci.*, 21, 1145–1154, 1999.

35. Schlemmer, R., Razzazi, H. W., Hulan, H. W., Bauer, F., and Luf, W., The effect of the addition of different spices on the development of oxysterols in fried meat, In *Biologically-Active Phytochemicals in Food*, Pfannhauser, W., Fenwick, G. R., and Khokhar, S. Eds., Royal Society of Chemistry, Cambridge, pp. 471–473, 2001.

36. Singh, J., Bagchi, G. D., Srivastava, R. K., and Singh, A. K., Pharmaceutical aspects of aromatic herbs and their aroma chemicals, *J. Med. Aromat. Plant Sci.*, 22, 732–738, 2000.

37. Shobana, S. and Naidu, K. A., Antioxidant activity of selected Indian spices, *Prostaglandins-Leukot. Essent. Fatty Acids*, 62, 107–110, 2000.

38. Unnikrishnan, M. C. and Kuttan, R., Tumour reducing and anticarcinogenic activity of selected spices, *Cancer Lett.*, 51, 85–89, 1990.

39. Kumari, M. V. and Rao, A. R., Effects of mace (*Myristica fragrans*, Houtt.) on cytosolic glutathione S-transferase activity and acid soluble sulfhydryl level in mouse liver, *Cancer Lett.*, 46, 87–91, 1989.

40. Chitra, V. and Leelamma, S., *Coriandrum sativum*—effect on lipid metabolism in 1,2-dimethyl hydrazine induced colon cancer, *J. Ethnopharmacol.*, 71, 457–463, 2000.

41. Sharma, N., Trikha, P., Athar, M., and Raisuddin, S., Inhibition of benzo[a]pyrene- and cyclophoshamide-induced mutagenicity by *Cinnamomum cassia*, *Mutat. Res.*, 480–481, 179–188, 2001.

42. Khanum, F., Anilkumar, K. R., Sudarshana, K. K. R., Viswanathan, K. R., and Santhanam, K., Anticarcinogenic effects of curry leaves in dimethylhydrazine-treated rats, *Plant Foods Hum. Nutr.*, 55, 347–355, 2000.

43. Saleem, M., Alam, A., and Sultana, S., Asafoetida inhibits early events of carcinogenesis: a chemopreventive study, *Life Sci.*, 68, 1913–1921, 2001.

44. Choudhary, D., Chandra, D., Choudhary, S., and Kale, R. K., Modulation of glyoxalase glutathione S-transferase and antioxidant enzymes in the liver, spleen and erythrocytes of mice by dietary administration of fenugreek seeds, *Food Chem. Toxicol.*, 39, 989–997, 2001.

45. Prasanna, M., Hypolipidemic effect of fenugreek: a clinical study, *Indian J. Pharmacol.*, 31, 34–36, 2000.

46. Pathak, P., Srivastava, S., and Grover, S. S., Development of food products based on millets, legumes and fenugreek seeds and their suitability in the diabetic diet, *Int. J. Food Sci. Nutr.*, 51, 409–415, 2000.

47. Madar, Z. and Stark, A. H., New legume sources as therapeutic agents, *Br. J. Nutr.*, 88, S287–S292, 2002.

48. Devasena, T. and Menon, V. P., Enhancement of circulatory antioxidant by fenugreek during 1,2-dimethylhydrazine-induced rat colon carcinogenesis, *J. Biochem. Mol. Biol. Biophys.*, 6, 289–292, 2002.

49. Negi, P. S., Jayaprakasha, G. K., Rao, L. J. M., and Sakariah, K. K., Antibacterial activity of turmeric oil: a by-product from curcumin manufacture, *J. Agric. Food Chem.*, 47, 4297–4300, 1999.

50. Shim, J., Na, G., Kim, J., Pyun, Y., and Hwang, J., Xanthorrhizol: a novel antibacterial compound from *Curcuma xanthorrhiza*, In *Biologically-Active Phytochemicals in Food*, Pfannhauser, W., Fenwick, G. R., and Khokhar, S. Eds., Royal Society of Chemistry, Cambridge, pp. 77–80, 2001.

51. Devasagayam, T. P. A., Kamat, J. P., and Sreejayan, N., Antioxidant action of curcumin, In *Micronutrients and Health: Molecular Biological Mechanisms*, Nesaretnam, K. and Packer, L. Eds., AOCS Press, U.S.A., pp. 42–59, 2001.

52. Goud, V. K., Polasa, K., and Krishnaswamy, K., Effect of turmeric on xenobiotic metabolising enzymes, *Plant Foods Hum. Nutr.*, 44, 87–92, 1993.

53. Srinivas, L., Shalini, V. K., and Shylaja, M., Turmerin: a water soluble antioxidant peptide from turmeric [*Curcuma longa*], *Arch. Biochem. Biophys.*, 292, 617–623, 1992.

54. Nagabhushan, M. and Bhide, S. V., Curcumin as an inhibitor of cancer, *J. Am. Coll. Nutr.*, 11, 192–198, 1992.

55. Subramanian, M., Sreejayan, N., Rao, M. N. A., Devasagayam, T. P. A., and Singh, B. B., Diminution of singlet oxygen-induced DNA damage by curcumin and related antioxidants, *Mutat. Res.*, 311, 249–255, 1994.

56. Shukla, Y., Arora, A., and Taneja, P., Antimutagenic potential of curcumin on chromosomal aberrations in Wistar rats, *Mutat. Res.*, 515, 197–202, 2002.

57. Shukla, Y. and Taneja, P., Antimutagenic effects of garlic extract on chromosomal aberrations, *Cancer Lett.*, 176, 31–36, 2002.

58. Pal, S., Choudhuri, T., Chattopadhyay, S., Bhattacharya, A., Datta, G. K., Das, T., and Sa, G., Mechanisms of curcumin-induced apoptosis of Ehrlich's ascites carcinoma cells, *Biochem. Biophys. Res. Commun.*, 288, 658–665, 2001.

59. Lin, G. P., Chu, T., Yang, F., Beech, W., Frautschy, S. A., and Cole, G. M., The curry spice curcumin reduces oxidative damage and amyloid pathology in an Alzheimer transgenic mouse, *J. Neurosci.*, 21, 8370–8377, 2001.

60. Premkumar, K., Abraham, S., Santhiya, S. T., Gopinath, P. M., and Ramesh, A., Inhibition of genotxicity by saffron (*Crocus sativus* L.) in mice, *Drug Chem. Toxicol.*, 24, 421–428, 2001.

61. Banerjee, S. K., Maulik, M., Manchanda, S. C., Dinda, A. K., Das, T. K., and Maulik, S. K., Garlic-induced alteration in rat liver and kidney morphology and associated changes in endogenous antioxidant status, *Food Chem. Toxicol.*, 39, 793–797, 2001.

62. Das, T., Choudhury, A. R., Sharma, A., and Talukder, G., Effects of crude garlic extract on mouse chromosome in vivo, *Food Chem. Toxicol.*, 34, 43–47, 1996.

63. Spigelski, D. and Jones, P. J. H., Efficacy of garlic supplementation in lowering serum cholesterol levels, *Nutr. Rev.*, 59, 141–236, 2001.

64. Polasa, K. and Nirmala, K., Ginger—its role in xenobiotic metabolism, *Nutr. News*, 23, 1–6, 2002.

65. Ahmed, R. S., Seth, V., Pasha, S. T., and Banerjee, B. D., Influence of dietary ginger (*Zingiber officinale* Rosc.) on oxidative stress induced by malathion in rats, *Food Chem. Toxicol.*, 38, 443–450, 2000.

66. Samant, S. S. and Palni, L. M. S., Diversity, distribution and indigenous uses of essential oil yielding medicinal plants of the Indian Himalayan region, *J. Med. Aromat. Plant Sci.*, 22, 671–684, 2000.

67. Bhandary, M. J. and Chandrashekar, K. R., Treatment of poisonous snake-bites in the ethnomedicine of coastal Karnataka, *J. Med. Aromat. Plant Sci.*, 22–23, 505–510, 2000–2001.

68. Gordon, M. H., Walker, A. F., and Roberts, W. G., A human study investigating the effects of increased fruit and vegetable consumption in smokers, In *Biologically-Active Phytochemicals in Food*, Pfannhauser, W., Fenwick, G. R., and Khokhar, S. Eds., Royal Society of Chemistry, Cambridge, pp. 21–23, 2001.

69. Knuthsen, P., Ma, H., and Leth, T., Occurrence and analysis of anthocyanins in foods and feeds, In *Biologically-Active Phytochemicals in Food*, Pfannhauser, W., Fenwick, G. R., and Khokhar, S. Eds., Royal Society of Chemistry, Cambridge, pp. 206–208, 2001.

70. Kähkönen, M., Heinämäki, J., Hopia, A., and Heinonen, M., In *Biologically-Active Phytochemicals in Food*, Pfannhauser, W., Fenwick, G. R., and Khokhar, S. Eds., Royal Society of Chemistry, Cambridge, pp. 357–359, 2001.

71. Landbo, A. K. and Meyer, A. S., Enzymatic enhancement and antioxidant activities of anthocyanins and other phenolic compounds in black currant juice, In *Biologically-Active Phytochemicals in Food*, Pfannhauser, W., Fenwick, G. R., and Khokhar, S. Eds., Royal Society of Chemistry, Cambridge, pp. 354–356, 2001.

72. Heins, A., Stöckmann, H., and Schwarz, K., Designing 'anthocyanin-tailored' food composition, In *Biologically-Active Phytochemicals in Food*, Pfannhauser, W., Fenwick, G. R., and Khokhar, S. Eds., Royal Society of Chemistry, Cambridge, pp. 378–381, 2001.

73. Kähkönen, M., Vainionpää, M., Hopia, A., and Heinonen, M., In *Biologically-Active Phytochemicals in Food*, Pfannhauser, W., Fenwick, G. R., and Khokhar, S. Eds., Royal Society of Chemistry, Cambridge, pp. 360–362, 2001.

74. Platel, K. and Srinivasan, K., Plant foods in the management of diabetes mellitus: vegetables as potential hypoglycaemic agents, *Nahrung*, 41, 68–74, 1997.

75. Koshy, A. S., Anila, L., and Vijaylakshmi, N. R., Flavonoids from *Garcinia cambogia* lower lipid levels in hypercholesterolemic rats, *Food Chem.*, 72, 289–294, 2001.

76. Adegoke, G. O., Kumar, M. V., Sambaiah, K., and Lokesh, B. R., Inhibitory effect of Garcinia kola on lipid peroxidation in rat liver homogenate, *Indian J. Exp. Biol.*, 36, 907–910, 1998.

77. Mughal, M. H. S., Srivasta, P. S., and Iqbal, M., Drumstick (*Moringa pterygosperma* gaertn.): a unique source of food and medicine, *J. Econ. Taxonomic Bot.*, 23, 47–61, 1999.

78. Pari, L. and Kumar, A. A., Hepatoprotective activity of *Moringa olifera* on antitubercular drug-induced liver damage in rats, *J. Med. Food*, 5, 171–177, 2002.

79. Spiliotis, V., Lalas, S., Gergis, V., and Dourtoglou, V., Comparison of antimicrobial activity of seeds of different *Moringa olifera* varieties, *Pharm. Pharmacol. Lett.*, 8, 39–40, 1998.

80. Venkateswaran, S., Pari, L., and Sarvanan, G., Effect of *Phaseolus vulgaris* on circulatory antioxidants and lipids in rats with streptozotocin-induced diabetes, *J. Med. Food*, 5, 97–103, 2002.

81. Meisel, H. and Bocklemann, W., Bioactive peptides encrypted in milk proteins: proteolytic activation and thropho-functional properties, *Antonie-Van Leeuwenhoek*, 76, 207–215, 1999.

82. Tilak, J. C. and Devasagayam, T. P. A., Indian medicinal plants as sources of antioxidants: biological significance, *Assoc. Food Scientists Technologists (India) Newslett.*, July, 3–7, 2003.

83. Tilak, J. C., Devasagayam, T. P. A., and Lele, R. D., Antioxidant activities from Indian medicinal plants: a review—current status and future prospects, *Res. Commun. Pharmacol. Toxicol.*, in press.

84. Kaur, K., Arora, S., Kumar, S., and Nagpal, A., Antimutagenic activities of acetone and methanol fractions of *Terminalia arjuna*, *Food Chem. Toxicol.*, 40, 1475–1482, 2000.

85. Sabu, M. C. and Kuttan, R., Anti-diabetic activity of medicinal plants and its relationship with their antioxidant property, *J. Ethnophramacol.*, 81, 155–160, 2002.

86. Dubey, N. K., Tripathi, P., and Singh, H. B., Prospects of some essential oils as antifungal agents, *J. Med. Aromat. Plant Sci.*, 22, 350–354, 2000.

87. Taraphdar, A. K., Roy, M., and Bhattacharya, R. K., Natural products as inducers of apoptosis: implication for cancer therapy and prevention, *Curr. Sci.*, 80, 1387–1396, 2000.

88. Trevisanato, S. I. and Kim, Y.-I., Tea and health, *Nutr. Rev.*, 58, 1–10, 2000.

89. Higdon, J. V. and Frei, B., Tea catechins and polyphenols: health effects, metabolism and antioxidant functions, *Crit. Rev. Food Sci. Nutr.*, 43, 89–143, 2003.

90. Lodovici, M., Casalini, C., Filippo, C. De., Copeland, E., Xu, X., Clifford, M., and Dolara, P., Inhibition of 1,2-dimethylhydrazine-induced oxidative DNA damage in rat colon mucosa by black tea complex polyphenols, *Food Chem. Toxicol.*, 38, 1085–1088, 2000.

91. Devasagayam, T. P. A., Kamat, J. P., Mohan, H., and Kesavan, P. C., Caffeine as an antioxidant: inhibition of lipid peroxidation induced by reactive oxygen species in rat liver microsomes, *Biochim. Biophys. Acta*, 1282, 63–70, 1996.

92. Kamat, J. P., Boloor, K. K., Devasagayam, T. P. A., and Kesavan, P. C., Protection of superoxide dismutase by caffeine in rat liver mitochondria against γ-irradiation, *Curr. Sci.*, 77, 286–289, 1999.

93. Kamat, J. P., Boloor, K. K., Devasagayam, T. P. A., Jayashree, B., and Kesavan, P. C., Differential modification of oxygen-dependent and -independent effects of γ-irradiation in rat liver mitochondria by caffeine, *Int. J. Rad. Biol.*, 76, 1281–1288, 2000.

94. Kumar, S. S., Devasagayam, T. P. A., Jayashree, B., and Kesavan, P. C., Mechanism of protection against radiation-induced DNA damage in plasmid pBR322 by caffeine, *Int. J. Rad. Biol.*, 77, 617–623, 2001.

95. Patairiya, M., Food of the Gods, chocolate, *Sci. Rep.*, September, 10–15, 2000.

96. Atal, C. K., Sharma, B. M., and Bhatia, A. K., Search of emergency foods through wild flora of Jammu and Kashmir state (India): Sunderbani area, *Indian Forester*, 106, 211–219, 2002.

97. Chakraborty, S., Chakraborty, N., and Datta, A., Increased nutritive value of transgenic potato by expressing a nonallergenic seed albumin gene from *Amaranthus hypochondriacus*, *Proc. Natl Acad. Sci. U.S.A.*, 97, 3724–3729, 2000.

98. Anjaneyelu, A. S. R., Sharma, N., and Kondaiah, N., Evaluation of salt polyphosphate and their blends at different levels on physicochemical properties of buffalo meat and patties, *Meat Sci.*, 25, 293–306, 1989.

99. Pandian, T. J., Sustainable Indian fisheries, *Curr. Sci.*, 82, 753–754, 2002.

100. Wildman, R. E. C., *Handbook of Neutraceuticals and Functional Foods*, CRC Press, Boca Raton, FL, 2001.

101. Din, J. N., Newby, D. E., and Flapen, A. D., Omega 3 fatty acids and cardiovascular disease-fishing for a natural treatment, *Br. Med. J.*, 328, 30–35, 2004.

102. Surai, P. F. and Sparks, N. H. C., Designer eggs: from improvement of egg composition to functional food, *Food Sci. Technol.*, 12, 7–16, 2001.

103. Duddukuri, G. R., Kumar, P. S., Kumar, V. B., and Athota, R. R., Immunosuppressive effect of honey on the induction of allergen-specific humoral antibody response in mice, *Int. Arch. Allergy Immunol.*, 114, 385–388, 1997.

104. Duddukuri, G. R., Rao, D. N., and Athota, R. R., Suppressive effect of honey on antigen/mitogen stimulated murine T cell proliferation, *Pharm. Biol.*, 40, 39–44, 2002.

105. Shamala, T. R., Shri-Jyoti, Y., and Saibaba, P., Stimulatory effect of honey on multiplication of lactic acid bacteria under *in vitro* and *in vivo* conditions, *Lett. Appl. Microbiol.*, 30, 453–455, 2000.

106. Mishra, S., Kumar, H., and Sharma, D., How do mothers recognize and treat pneumonia at home? *Indian Pediatr.*, 31, 15–18, 1994.

107. Bhatt, D. C., Mitaliya, K. D., Mehta, S. K., and Patel, N. K., Flowers employed as gulkand in medicine, *Adv. Plant Sci.*, 13, 539–542, 2000.

108. Gopaldas, T., Iron-deficiency anemia in young working women can be reduced by increasing the consumption of cereal-based fermented food or gooseberry juice at the workplace, *Food Nutr. Bull.*, 23, 94–105, 2002.

109. Rani, B. and Khetarpaul, N., Probiotic fermented food mixtures: possible applications in clinical antidiarrhoea usage, *Nutr. Health*, 12, 97–105, 1998.

110. Gupta, M., Khetarpaul, N., and Chauhan, B. M., Preparation nutritional value and acceptability of barley–rabadi an indigenous fermented food in India, *Plant Food Hum. Nutr.*, 42, 351–358, 1992.

111. Kaur, I. P., Chopra, K., and Saini, A., Probiotics: potential pharmaceutical applications, *Eur. J. Pharm. Sci.*, 15, 1–9, 2002.

112. Roberfroid, M. B., Prebiotics and synbiotics: concepts and nutritional properties, *Br. J. Nutr.*, 80, S197–S202, 1998.

113. Roberfroid, M. B., Functional effects of food components and the gastrointestinal system: chicory fructo-oligosaccharides, *Nutr. Rev.*, 54, S38–S42, 1996.

114. Roberfroid, M. B., Concepts in functional foods: the case of inulin and oligofructose, *J. Nutr.*, 129 (7 suppl.), 1398S–1401S, 1999.

115. Sanders, M. E. and Huis-in't-Veld, J., Bringing a probiotic containing functional food to the market: microbiological, product, regulatory and labeling issues, *Antoine Van Leeuwenhoek*, 76, 293–315, 1999.

116. Samuels, R., Mani, U. V., Iyer, U. M., and Nayak, U. S., Hypocholesterolemic effect of spirulina in patients with hyperlipidemic nephrotic syndrome, *J. Med. Food*, 5, 91–96, 2002.

117. Moure, A., Cruz, J. M., Franco, D., Dominguez, J. M., Seneiro, J., Dominguez, H., Nunez, M. J., and Parajo, J. C., Review—natural antioxidants from residual sources, *Food Chem.*, 72, 145–171, 2001.

118. Jiang, C., Agarwal, R., and Lu, J., Antiangiogenic potential of a cancer chemopreventive flavonoid antioxidant, silymarin: inhibition of key attributes of vascular endothelial cells and angiogenic cytokine secretion by cancer epithelial cells, *Biochem. Biophys. Res. Commun.*, 276, 371–378, 2000.

119. Cao, Y., Cao, R., and Brakenhielm, E., Antiangiogenic mechanisms of diet-derived polyphenols, *J. Nutr. Biochem.*, 13, 380–390, 2002.

120. Gibson, G. R. and Williams, C. M., *Functional Foods: Concept to Product*, CRC Press, Boca Raton. FL, 2000.

121. Cejpek, K., Valušek, J., and Velíšek, J., Decomposition of allyl isothiocyanate in model systems containing nucleophilic reagents, In *Biologically-Active Phytochemicals in Food*, Pfannhauser, W., Fenwick, G. R., and Khokhar, S. Eds., Royal Society of Chemistry, Cambridge, pp. 480–484, 2001.

122. Galaverna, G., Sforza, S., Di Silvestro, G., and Marchelli, R., Variation of the antioxidant activity in fruit juices during technological treatments, In *Biologically-Active Phytochemicals in Food*, Pfannhauser, W., Fenwick, G. R., and Khokhar, S. Eds., Royal Society of Chemistry, Cambridge, pp. 474–476, 2001.

123. Schieber, A., Keller, P., Hendreb, H., Rentschler, C., and Carle, R., Recovery and characterization of phenolic compounds from by-products of food processing, In *Biologically-Active Phytochemicals in Food*, Pfannhauser, W., Fenwick, G. R., and Khokhar, S. Eds., Royal Society of Chemistry, Cambridge, pp. 502–504, 2001.

124. Stoll, T., Schieber, A., and Carle, R., Carrot pomace—an underestimated by-product? In *Biologically-Active Phytochemicals in Food*, Pfannhauser, W., Fenwick, G. R., and Khokhar, S. Eds., Royal Society of Chemistry, Cambridge, pp. 525–527, 2001.

125. Anese, M. and Nicoli, M. C., Optimising phytochemical release by process technology, In *Biologically-Active Phytochemicals in Food*, Pfannhauser, W., Fenwick, G. R., and Khokhar, S. Eds., Royal Society of Chemistry, Cambridge, pp. 455–470, 2001.

126. Starling, S., Key European and Asian consumer data unveiled, *Funct. Foods Neutraceutic*, October, 6–20, 2003.

5 History and Scope of Functional Foods in China

Anthony Y. H. Woo, Y. Zhao, R. Zhang, C. Zhou, and Christopher H. K. Cheng

CONTENTS

5.1 THE ORIGIN AND DEVELOPMENT OF FUNCTIONAL FOODS IN CHINA

China is one of the ancient cultures with the earliest development of written languages. Some of the Chinese literature on functional foods available today can be dated back 3000 years. These written records are unquestionably the most valuable sources of information for understanding the origin and historical development of functional foods in China. In the first part of this article, a concise history of the origin and development of functional foods in China, we will highlight the representative written works which appeared in different dynasties.

5.1.1 THE ORIGIN OF FUNCTIONAL FOODS IN CHINA

In the earliest Chinese encyclopedia—"Shan Hai Jing" (The Classic of Mountains and Seas, eleventh to eighth century BC)—as many as 110 food items from animal and plant sources were described, and most of them have functional properties. Historical records suggest that wine and vinegar had been widely used in medicine as early as the dynasty of Xia Yu[*] (twenty-third to eighteenth century BC). Another

---

[*] It is a Chinese custom to name a person with the family name first followed by the given names. In this case, "Xia" is the family name and "Yu" is the given name.

fermented product—sauce—was produced in the Zhou dynasty. As described in "Zhou Li" (The Rites of Zhou), the different kinds of sauces were referred to as Bai-zhang (a hundred sauces). The production of bean curd was first documented in "Huai Nan Zi" (Gentleman from the South of River Huai) written by Liu An around the second century BC. The production of sauce, wine, sugar, vinegar, and bean curd manifests the evolution from a simple way of direct consumption of animals or plants as food to a greater variety enriched by food processed by different physicochemical or biological procedures. These advances enhanced nutrient availability and enriched the diversity of food species. It indicates that the Chinese acquired much knowledge in food and nutrition in the early days of their history.

The earliest emergence of functional foods in China can be dated back several thousand years. In China, the earliest medicine is food and the earliest therapeutic method is diet therapy. In the earliest anthology of civilian poetries, "Shi Jing" (The Classic of Odes), which appeared around the eleventh to the sixth century BC (from the Western Zhou dynasty to the middle of the Spring and Autumn period in Chinese history), more than a hundred herbs such as Radix Puerariae (pueraria root), Radix Paeoniae (peony root), Fructus Trichosanthis (snackgourd fruit), Fructus Lycii (Chinese wolfberry), Radix Glycyrrhizae (licorice root), and Fructus Mume (dark plum) were recorded as both medicine and food. According to "Zhou Li—Tian Guan Chapter," the practical needs of clinical medicine prompted the appearance of Shi-yi, dietary physicians, around the fifth century BC. Shi-yi at that time was "in charge of the preparation of the King's six foodstuffs, six beverages, six meals, a hundred delicacies, a hundred sauces and the adequacy of the eight exotic food." In this sense, Shi-yi in Zhou can be regarded as the first dietitians in the world.

5.1.2 THE DEVELOPMENT OF FUNCTIONAL FOODS IN CHINA

Since ancient times, the Chinese people have discovered the medicinal properties of foods, and that the effects of different food species vary. Over a long period of practical use of these agents to treat illnesses and to maintain health, the special properties of different food species were recognized and theories of medical practice developed. According to the description of medicinal foods in different dynasties, the development of functional foods in China can be divided into the following eight phases—and the milestone classics that appeared during the development of functional foods in Chinese history are shown in Table 5.1.

5.1.2.1 The Pre-Qin Era (475–221 BC)

More than two thousand years ago in the Warring States period, the first Chinese work on medical theory, "Huang Di Nei Jing" (The Yellow Emperor's Classic of Internal Medicine) was compiled. This "Nei Jing," in short, describes many fundamental theories of nutrition and diet therapy in traditional Chinese medicine (TCM) including the declaration of an important concept that "medicine and food are of the same origin." This concept formed the basis of the use of food in TCM as agents for the prevention and treatment of diseases.

The therapeutic principles of *"Nei Jing"* utilize the special properties of each food substance or medicinal material, viz. the Four Natures, the Five Tastes, the

TABLE 5.1
Milestone Classics on Functional Foods in China

Name of the Classic	Author (Publication Year)	Importance
Huang Di Nei Jing	Author unknown (475–221 BC)	First classic on functional foods and medicine
Sheng Nong Ben Cao Jing	Author unknown (First century AD)	First classic on materia medica (at least half of the 365 items described can be used both as food and as medicine)
Shang Han Lun and Jin Kui Yao Lue	Zhang Zhongjing (around AD 200)	First classic on disease diagnosis and treatment (containing about 110 prescriptions with functional foods)
Qian Jin Fang	Sun Simiao (around AD 600)	The therapeutic effects of 154 food items were described
Yin Shan Zheng Yao	Hu Sihui (1330)	First systematic classic on nutritional principles
Ben Cao Gang Mu	Li Shizhen (1596)	More than 500 kinds of foods for diet therapy were collected

Channel Tropism, and the Four Bearings to regulate the human body in an integral and holistic manner, i.e., by "tonifying viscera (major organs) and bowels (organs subsidiary to viscera)" (strengthening the organs), "purging repletions and dispelling evils" (eliminating toxins and pathogens), and "rectifying yin and yang"* (regulating physiological functions) so that the objectives of preventing diseases and maintaining health could be achieved. This work greatly promoted the advancement of TCM both in theory and in practice and can be regarded as the first classic on functional foods in China.

In "Wu Shi Er Bing Fang" (Prescriptions for Fifty-two Kinds of Diseases), which was engraved from the Han Tomb at Mawangdui in Hunan province, one-fourth of the medicinal materials documented were foodstuffs and about half of the 52 kinds of diseases mentioned could be treated or managed by the judicious choice of foods in diet therapy.

5.1.2.2 From the Qin Dynasty to the Northern and Southern Dynasties (221 BC to AD 589)

At around the first century AD in the Han dynasty, the first classic of materia medica, "Sheng Nong Ben Cao Jing" (The Divine Farmer's Classic of Materia Medica) was compiled. This book recorded 365 different kinds of medicinal materials categorized into three grades: superior, intermediate, and low. Superior grade materials have

---

* Some of the terms used in TCM do not have the same meaning as in modern medicine. For example, "spleen" in TCM represents the entire digestive system in modern medicine, and "kidney" has extensive functions not limited to excretion. In addition, some TCM terms such as "yin" (essential materials of the body), "yang" (essential activities of the body), "essence" (basis of the body) and "qi" (vital energy of the body) do not have synonyms in modern medicine.

good health promoting (or tonic) effects and low toxicity. Frequent consumption is recommended for the promotion of health and longevity. A considerable number of foodstuffs were classified as superior grade, such as Fructus Ziziphi (jujube), Rhizoma Dioscoreae (yam), Rhizoma Nelumbinis (lotus root), Semen Euryales (gordon euryale seed), Semen Coicis (coix seed), and Mel (honey). The intermediate grade materials are mostly medicinal agents with therapeutic effects but they may produce toxic side effects if the materials were used inappropriately. The low grade materials are mostly toxic substances and should be used with caution to treat some diseases. This grading system reflects the traditional Chinese concept that promotion of health is superior to disease therapy and food occupies an important position in the maintenance of health.

In the Eastern Han dynasty, a distinguished physician, Zhang Zhongjing (AD 168–196), completed two medical works, "Shang Han Lun" (Treatise on Cold-induced Febrile Diseases) and "Jing Kui Yao Lue" (Synopsis of the Golden Chamber). Considered classics on disease pattern diagnosis and treatment in TCM, these works contain prescriptions for the treatment of various disorders. Among the 112 prescription formulae in "Shang Han Lun," about 60 were composed of foodstuffs; and among the 262 formulae in "Jing Kui Yao Lue," about 50 contained food components. In addition, some effects of diet on the treatment of diseases were also described. For example, to treat "externally contracted diseases" (flu or the common cold), Zhang proposed that patients should drink some hot gruel after taking medicines to enhance the effects of the medicines. He also claimed that during the course of flu, patients should avoid raw, cold, greasy, and spicy food. This shows that the importance of diet in clinical medicine had already been recognized at that time.

In addition, physicians such as An Qisheng of the Qin dynasty, Li Shaojun of the Han dynasty, and Ge Hong of the Jin dynasty had made various advances on the nutritional value of food. For example in "Zhou Hou Bei Ji Fang" (A Handbook of Prescriptions for Emergency) by Ge Hong (281–341), the treatment of goiter with seaweed liquor and diabetes with pig pancreas was first documented. The Eastern Jin physician Zhi Facun, a specialist in "leg qi" (beriberi), developed a number of formulae to treat this disease. Incidentally, most of the ingredients in his formulae are rich in vitamin B_1. Later from Zhi's theory, Sun Simiao found that beriberi usually appeared in the rice-eating region and further demonstrated that it could be prevented by consuming rice gruel with the husks.

5.1.2.3   The Sui-Tang Era (581–907)

In the Sui dynasty and the Tang dynasty, the development of medicinal foods in China entered a new era with the emergence of a number of works on functional foods. In the Sui dynasty, in his work on etiology and semiology, "Zhu Bing Yuan Hou Lun" (General Treatise on the Causes and Symptoms of Diseases), Chao Yuanfang introduced the use of sheep dimple to treat goiter, toad venom to treat wounds, and antelope horn to treat apoplexy. During the turn of the seventh century, Sun Simiao wrote "Qian Jing Fang" (Prescriptions Worth a Thousand Pieces of Gold). Quoting Bian Que, Sun, a famous ancient physician in the second century BC, wrote, "A doctor should look into the origin of an ailment, know how it is contracted, and cure it with food; only when food therapy fails, then resort to medicine." A chapter in that book known as "Qian Jing Shi Zhi" (Dietary Treatment Worth a Thousand Pieces of Gold) was entirely devoted to diet therapy. The use of pig liver to treat night blindness was first recorded in this work. From that time onward, the TCM principle of "supplementing viscera with viscera" (treating the weakness of an organ with the corresponding organ of an animal) was established.

The first monograph on diet therapy in China, "Shi Liao Ben Cao" (Dietetic Materia Medica) by Meng Shen was also completed at that time. Moreover, in "Shan Fu Jing Shou Lu" (The Manuscript of the Diet Minister's Classic) written by Yang Yezhuan, the author introduced a number of formulae for diet therapy. In addition, the characteristics and dining customs of different kinds of foods were also described. In particular, a detailed account of the characteristics of different species of tea and their places of origin was given.

5.1.2.4 The Song Dynasty (960–1279)

In the Song dynasty, food was commonly used for the prevention and treatment of diseases. Diet therapy made significant progress during that period. In the imperial pharmacopoeia, "Tai Ping Shen Hui Fang" (Holy Benevolence Prescriptions of the Taiping Period), compiled by 12 imperial physicians of the Song dynasty, diet therapies for 28 diseases were documented. A collection of articles about the detailed dietary treatment methods for 30 diseases were also included in a special chapter on diet therapy in "Shen Ji Zong Lu (The General Catalog of Holy Relief)." Lin Hong's "Shan Jia Qing Gong" (Vegetable Meals of the Hill Inhabitants), containing a collection of 102 kinds of vegetable dishes for disease treatment and health maintenance, was another monograph on diet therapy written at that time. Chen Zhi wrote another monograph, "Yang Lao Feng Qing Shu" (Health Care of Parents and Old People), that contained 162 formulae of diet therapy for disease treatment and health care of aged people.

5.1.2.5 The Jin (1115–1234) and Yuan Dynasties (1271–1368)

Li Gao, a key medical scholar during this period, greatly emphasized the importance of diet therapy. He proposed the use of sweet and warm medicines such as Radix Ginseng (ginseng root) and Radix Astragali (astragalus root) to "tonify the spleen and the stomach," so as to "nurture the original qi" (build up the vital energy of the body). His thoughts were embodied in the book "Pi Wei Lun" (Treatise on the Spleen and Stomach). Zhang Congzheng of the Jin dynasty wrote in "Ru Men Shi Qin" (The Literati's Care of Their Parents), "to prolong life, one should count on dietary supplementation" and "in case of deficiencies in essence and blood, food should be supplemented." He suggested the principle of "supplementing vacuity by food nourishment" (strengthening the weakness of the body by nutrient supplementation). Hu Sihui, an imperial dietary physician of the Yuan dynasty, began his arguments in his famous work "Yin Shan Zheng Yao" (Orthodox Essentials of Dietetics) on the essential diet for a healthy person. This work is considered the first systematic monograph on nutritional principles in China. In addition, "Yin Shi Xu Zhi" (Must Know for Diet) by Jia Ming and "Ri Yong Ben

Cao" (Materia Medica for Daily Use) by Wu Rui were also famous works on dietetics in China written during that time.

5.1.2.6 The Ming Dynasty (1368–1644)

"Ben Cao Gang Mu" (The Compendium of Materia Medica) written by Li Shizhen was a representative work during this period. In this great work, a total of 1892 medicinal materials were collected and among them 347 were described for the first time. Food constituted a significant portion of the materials and biotransformation procedures were required during the preparation of some of the food items such as butter, fermented bean curd, and salted fish. This work greatly enriched the species of food items used in diet therapy.

In "Jiu Huang Ben Cao" (Materia Medica for the Relief of Famine) written by Zhu Su, most of the edible plants documented were newly mentioned. During this period, two works of similar names, but by different authors: "Shi Wu Ben Cao" (Food Materia Medica) by Wang Ying and "Shi Jian Ben Cao" (Food Guide for Materia Medica) by Ning Yuan also made important contributions to the development of medicinal foods. In addition, some attention had been devoted to the development of diet therapy for febrile diseases. For example, a special chapter on dietetics was included in "Wen Yi Lun" (Treatise on Pestilence) by Wu Youxing.

5.1.2.7 The Qing Dynasty (1636–1911)

A number of important publications on diet therapy appeared during this period. The following are the representative ones: "Shi Wu Ben Cao Hui Zhuan" (A Compilation of Food Materia Medica) a monograph with 12 chapters and illustrations compiled by Shen Lilong; the distinctive "Sui Yuan Shi Dan" (Cooking Menu of Suiyuan) by Yuan Zicai with particular emphasis on the skill of cooking; "Yin Shi Shi Er He Lun" (Twelve Comments on Dietetics) by Zhang Ying; "Tiao Zhi Yin Shi Bian" (Identification of Diets for the Rectification of Diseases) by Zhang Xingyu; "Sui Xi Ju Yin Shi Pu" (Recipes for an Easy Living) by Chen Xiuyuan; "Shi Jian Ben Cao" (Food Guide for Materia Medica), "Ben Cao Yin Shi Pu" (Recipes for Materia Medica), and "Shi Yang Liao Fa" (Diet Therapy), collectively known as "Fei Shi Shi Yang" (Fei's Dietetics), by Fei Boxiong who first used the term "diet therapy"; and "Zhou Pu" (Gruel Recipes)—with a supplement, "Guang Zhou Pu" (Gruel Recipes of the Guang Provinces), the first monograph on medicinal gruel compiled by Huang Haoji with a collection of 200 recipes for medicinal gruel. In addition, "Chuan Ya Nei Bian" (Internal Treatise on Folk Medicine) and "Ben Cao Gang Mu Shi Yi" (Omissions Due in the Compendium of Materia Medica) compiled by Zhao Xuemin also contained many records on diet therapy.

Particular note should be given to the establishment of the Qing theory of febrile diseases. A great deal of precious experience on the diet therapy of febrile diseases was accumulated during this period. For example, Wu-zhi-ying (Five Juices Drink) in "Wen Re Lun (Treatise on Warm Heat) compiled by Ye Tianshi was a typical prescription of diet therapy given during the recovery period of febrile diseases for the "nourishment of the stomach yin" (sustaining the function of the stomach).

5.1.2.8   After the Fall of Imperial China (Since 1911)

After the fall of the Qing dynasty, development of medicinal food continued to progress and diet therapy was further promoted. Representative works include "Shi Wu Zhi Bing Xin Shu" (A New Book on the Treatment of Diseases by Food) edited by Zhang Zi, "Shi Wu Liao Bing Chang Shi" (Common Knowledge in the Treament of Diseases by Food), "Bu Pin Yan Jiu" (Research on Tonics) edited by Yang Zhiyi, "Jia Ting Shi Wu Liao Fa" (Family Diet Therapy) by Zhu Renkang, and "Ji Bing Yin Shi Zhi Nan" (A Quick Guide on Diets for Diseases) edited by Cheng Guoshu. All these works had inherited the previous experience in diet therapy and made some advancement in their specialized fields.

5.1.3   CURRENT STATUS OF CHINESE FUNCTIONAL FOODS

The domestic functional food market in China enjoyed a huge growth since the early 1980s. "The Measures of Functional Food Administration" published by the Ministry of Health have been enforced since July 1, 1996 to regulate the use, examination, production, marketing, and sales, labeling, specification, advertising, and monitoring of functional foods in China. The definition of functional food, as stipulated in Article 2 of Chapter 1 in this document, "is the food with specific function for health care. Functional food is suitable for use by specific group of people to improve their body functions but it is not aimed at treating disease."

As yet, over 2000 Chinese functional food items have been approved. The qualification of a material to be either a food or a medicine, or both, has been much debated. To resolve this problem, the Ministry of Health published on March 5, 2002 a list of materials approved or prohibited in functional food products. The materials are categorized into three groups: Materials that are both medicine and food, materials approved in functional foods, and materials prohibited in functional foods. Among them, eighty-seven materials were classified as both medicine and food (Table 5.2).

5.2   EXAMPLES OF RECENT STUDIES ON CHINESE FUNCTIONAL FOODS

Systematic functionality studies of some selected Chinese functional foods have been carried out in China since the 1960s. These studies not only allow us to understand the underlying mechanisms of actions of these materials and to relate them to their traditional use, they also contribute towards the discovery of many new functions in these foods. Recent studies on some of the food items widely consumed in China are described below (see also the summary in Table 5.3).

5.2.1   CHINESE WOLFBERRY

Fructus Lycii (or Chinese wolfberry) is the ripen fruit of *Lycium barbarum* L. or *Lycium chinenses* Mill. (Scolanaceae). Its synonyms in Chinese include Qizi

TABLE 5.2
List of Materials Recognized as Both Food and Medicine in China

Latin Name	Common Name	Species Name
Agkistrodon	Pallas pit viper	*Agkistrondon halys*
Arillus Longan	Longan aril	*Dimocarpus longan*
Bulbus Allii Macrostemi	Longstamen onion bulb	*Allium macrostemon*
Bulbus Lilii	Lily bulb	*Lilium lancifolium*; *L. brownii* var. *virdulum*; *L. pumilum*
Colla Corii Asini	Ass-hide glue	*Equus asinus*
Concha Ostreae	Oyster shell	*Osterea gigas*; *O. talienwhanensis*; *O. rivularis*
Cortex Cinnamomi	Cassia bark	*Cinnamommum cassia*
Endothelium Corneum Gigeriae Galli	Chicken's gizzard-skin	*Gallus domesticus*
Exocarpium Citri Rubrum	Red tangerine peel	*Citrus reticulate*
Flos Caryophylli	Clove	*Eugenia caryophylata*
Flos Chrysanthemi	Chrysanthemum flower	*Chrysanthemum morifolium*
Flos Lablab Album	White hyacinth bean flower	*Dolichos lablab*
Flos Lonicerae	Honeysuckler flower	*Lonicera japonica*; *L. hypoglasuca*; *L. confusa*; *L. dasystyla*
Flos Sophorae	Japanese pagoda tree flower	*Sophora japonica*
Folium Mori	Mulberry leaf	*Morus alba*
Folium Nelumbinis	Lotus leaf	*Nelumbo nucifera*
Folium Perillae	Perilla leaf	*Perilla frutescens*
Fructus Alpiniae Oxyphyllae	Sharpleaf glangal fruit	*Alpinia oxyphylla*
Fructus Amomi	Villous amomum fruit	*Amomum villosum*; *A. villosum* var. *xanthioides*; *A. longilaigulare*
Fructus Anisi Stellati	Chinese star anise	*Illcium verum*
Fructus Aurantii	Orange fruit	*Citrus aurantium*
Fructus Canarii	Chinese white olive	*Canarium album*
Fructus Cannabis	Hemp seed	*Cannabis sativa*
Fructus Chaenomelis	Common flowering quince	*Chaenomeles speciosa*
Fructus Citri	Citron fruit	*Citrus medica*; *C. wilsonii*
Fructus Citri Sarcodactylis	Finger citron	*Citrus medica* var. *sarcodactylis*
Fructus Crataegi	Hawthorn	*Crataegus pinnatifida* var. *major*; *C. pinnatifida*
Fructus Foeniculi	Fennel	*Foeniculum vulgare*
Fructus Gardeniae	Cape jasmine fruit	*Gardenia jasminoides*
Fructus Hippophae	Seabuckthorn fruit	*Hippophae rhamnoides*
Fructus Hordei Germinatus	Germinated barley	*Hordeum vulgare*
Fructus Hoveniae	Japanese raisin tree fruit	*Hovenia dulcis*
Fructus Jujubae	Chinese date/jujube	*Ziziphus jujuba*
Fructus Lycii	Chinese wolfberry	*Lycium barbarum*

(continued)

TABLE 5.2 *(Continued)*

Latin Name	Common Name	Species Name
Fructus Momordicae	Grosvenor momordica fruit	*Siraitia grosvenori*
Fructus Mori	Mulberry fruit	*Morus alba*
Fructus Mume	Dark plum	*Prunus mume*
Fructus Perillae	Perilla fruit	*P. frutescens*
Fructus Phyllanthi	Emblic leafflower	*Phyllanthus emlica*
Fructus Rubi	Palmate raspberry	*Rubus chingii*
Fructus Sophorae	Japanese pagoda tree fruit	*Sophora japonica*
Herba Cichorii	Cichory herb	*Cichorium glandulosum*; *C. intybus*
Herba Cirsii	Common cephalanoplos	*Cephalanoplos segetum*
Herba Houttuyniae	Heartleaf houttuynia herb	*Houttuynia cordata*
Herba Lophatheri	Lophatherum herb	*Lophatherum gracile*
Herba Menthae	Peppermint	*Mentha haplocalyx*
Herba Moslae	Haichow elsholtzia herb	*Mosla chinensis*
Herba Pogostemonis	Cablin patchouli herb	*Pogostemon cablin*
Herba Portulacae	Purslane herb	*Portulaca oleracea*
Herba Taraxaci	Dandelion	*Taraxacum mongolicum*; *T. sinicum*
Mel	Honey	*Apis cerana*; *A. mellifera*
Pericarpium Citri Reticulatae	Dried tangerine peel	*Citrus reticulata*
Pericarpium Zanthoxyli	Pricklyash peel	*Zanthoxylum schinifolium*; *Z. bungeanum*
Poria	Indian bread	*Poria cocos*
Radix Angelicae Dahuricae	Dahurian angelica	*Angelica dahurica*; *A. dahurica* var. *formosana*
Radix Glycyrrhizae	Liquorice	*Glycyrrhiza uralensis*; *G. inflata*; *G. glabra*
Radix Platygodon	Balloonflower root	*Platygodon grandiflorum*
Radix Puerariae	Kudzuvine root	*Pueraria lobata*; *P. thomsonii*
Rhizoma Alpiniae Officinarum	Lesser galangal rhizome	*Alpinia officinale*
Rhizoma Dioscoreae	Common yam	*Dioscorea opposita*
Rhizoma Imperatae	Lalang grass rhizome	*Imperata cylindrica* var. *major*
Rhizoma Phragmitis	Reed rhizome	*Phragmites communis*
Rhizoma Polygonati	Wirledear solomonseal	*Polygonatum sibiricum*
Rhizoma Polygonati Odorati	Fragrant solomonseal	*Polygonatum odoratum*
Rhizoma Zingiberis	Ginger	*Zingiber officinale*
Seman Ziziphi Spinosae	Spine date seed	*Ziziphus jujuba* var. *spinosa*
Semen Armeniacae Amarum	Apricot seed	*Prunus armeniaca* var. *ansu*; *P. sibirica*; *P. mandshurica*; *P. armeniaca*
Semen Brassicae Junceae	Yellow mustard seed	*Brassica juncea*
Semen Canavaliae	Jack bean	*Canavalia gladiata*
Semen Cassiae	Cassia seed	*Cassia obtusifolia*; *C. tora*
Semen Coicis	Coix seed	*Coix lacrymajobi* var. *ma-yuen*
Semen Euryales	Gordon euryale seed	*Euryale ferox*

(continued)

TABLE 5.2 *(Continued)*

Latin Name	Common Name	Species Name
Semen Ginkgo	Ginkgo seed	*Ginkgo biloba*
Semen Lablab Album	White hyacinth bean	*Dolichos lablab*
Semen Myristicae	Nutmeg	*Myristica fragrans*
Semen Nelumbinis	Lotus seed	*Nelumbo nucifera*
Semen Persicae	Peach seed	*Prunus persica*; *P. davidiana*
Semen Phaseoli	Red bean	*Phaseolus calcaratus*; *P. angularis*
Semen Piper	Black pepper	*Piper nigrum*
Semen Pruni	Chinese dwraf cherry seed	*Prunus humilis*; *P. japonica*; *P. pedunculata*
Semen Raphani	Radish seed	*Rhaphamus sativus*
Semen Sesami Nigrum	Black sesame	*Sesamum indicum*
Semen Sojae Preparatum	Fermented soyabean	*Glycine max*
Semen Sterculiae	Boat-fruited sterculia seed	*Sterculia lychnophora*
Semen Torreyae	Grand torreya seed	*Torreya grandis*
Thallus Laminariae	Kelp/sea-tangle	*Laminaria japonica*
Zaocys	Black snake	*Zaocys dhumnades*

(lycium berry), Xueguo (blood fruit), Tianjing (heaven essence), and Dixian (earth spirit). Due to its anti-aging effect, medical practitioners traditionally call it Quelao (anti-aging). This fruit is often used to tonify the body and to promote longevity.

Chinese wolfberry is sweet in taste and mild in nature according to TCM principles. It is used to treat "the vacuity of liver and kidney yin" (general weakness), dizziness, deterioration of eyesight, soreness, and weakness of the back and knees, and abnormal seminal emission. It was first documented in "Shen Nong Ben Cao Jing" and was classified as a superior grade material. "Ben Cao Gang Mu" describes its functions as "tonifying kidney, moistening lung, engendering essence and benefiting qi." Its chemical constituents include Lycium barbarum polysaccharides (LBP), alkaloids, carotenoids, vitamin C, cyclopeptides, amino acids, and trace elements [1]. Pharmacological studies indicate that Chinese wolfberry had anti-aging effects [2]. Recently, the immunomodulatory effects of Chinese wolfberry have attracted much attention. LBP are thought to be the active principle. Its immunomodulatory effects can be categorized into the following three aspects.

5.2.1.1 Effects on the Immune System

In one study [3], intraperitoneal injection of 5 mg/kg LBP into normal mice enhanced the NK cell-mediated killing, with the percentage of cytotoxicity increased from 12.4 to 18%. In addition, LBP antagonized the immunosuppressive effect of cyclophosphamide on mice NK cells. It was shown that LBP

TABLE 5.3
Functionalities of the Food Items Described in this Chapter

Function of Food[a]	Chinese Wolfberry	Tremella	Hawthorn	Walnut	Jujube	Common Yam	Pueraria	Perilla
Immune regulation	X	X			X	X		X
Postponement of senility	X	X			X	X		
Memory improvement				X				X
Promotion of growth and development				X				X
Anti-fatigue				X	X			
Body weight reduction								
Oxygen deficit tolerance		X		X				
Radiation protection								
Mutation inhibition								
Antitumor		X			X			X
Blood-lipid regulation			X	X		X		X
Sexual potency improvement							X	
Blood glucose regulation						X		
Gastrointestinal function improvement			X					
Sleep improvement				X				X
Improvement of nutritional anemia	X							
Protection of liver from chemical damage								
Lactation improvement								
Enhancement of beauty								
Vision improvement							X	X
Promotion of metal removal								
Throat soothing								
Blood pressure regulation			X				X	
Enhancement of bone calcification							X	X

[a] The functions listed here are the 24 tests on functional food products required by the Chinese authority [75]. The symbol X indicates that the food item described in this chapter possesses that particular function.

(5–10 mg/kg) significantly increased the percentage of cytotoxicity by NK cells in cyclophosphamide-treated mice from 9.5 to 15%–16%. These results indicate that LBP can increase the activity of NK cells in normal and cyclophosphamide-treated mice.

An extract of Chinese wolfberry exhibited biphasic immunomodulatory effects on mouse T lymphocytes, with stimulatory effect at low concentration (10^{-5} mg/ml) and suppressive effect at high concentration (1 mg/ml). Further studies indicated that 50–400 µg/ml LBP could increase the cGMP and cAMP levels in mouse lymphocytes in a dose-dependent manner [3]. Moreover, 100 µg/ml LBP has been shown to increase the plasma membrane protein kinase C (PKC) activity in Con A-activated mouse splenocytes [4]. It was postulated that LBP might produce immunostimulatory effects through the regulation of cAMP/cGMP levels and PKC activity. The immunostimulatory effects of LBP suggest its possible application in supportive therapy during cancer treatment.

5.2.1.2 Effects on Hematopoiesis

Intraperitoneal injection of 10 mg/kg LBP for three consecutive days increased the proliferation of bone marrow hematopoietic stem cells in normal mice. The content of granulocyte/macrophage (GM) progenitor cells was significantly increased and differentiation from the GM progenitor cells to granulocytes was also promoted. The erythrocyte burst-forming cells and erythrocyte colony-forming cells were also increased to 342 and 192% of the control levels, respectively. After 6 days of treatment, the peripheral reticulocyte count was increased to 218% of the control. Injection of LBP also promoted the secretion of colony-stimulating factor by mouse splenocytes and increased the colony-stimulating activity of the mouse serum [5,6]. The results suggest that LBP possesses hematopoietic stimulatory activities which may relieve the condition of bone marrow suppression during cancer chemotherapy.

5.2.2 TREMELLA

Tremella fuciformis Berk (Tremellaceae) is an epiphyte commonly known as tremella, or white jelly fungus, or "king of fungi." According to TCM, it is slightly sweet in taste and mild in nature and is traditionally used as a superior grade tonic and precious food alongside ginseng root, antelope horn, and bird's nest. Records of this epiphyte could be found in TCM classics from the earliest "Shen Nong Ben Cao Jing" to the recent "Shi Yong Jun (Edible Fungi)."

The chemical composition of tremella is complex. Preliminary analysis indicated that tremella contains 6.6% of protein, 78.3% of carbohydrate, 0.6% of fat, 2.6% of crude fibers, and many different kinds of vitamins [7]. Since the 1970s, studies on tremella have mainly focused on its polysaccharides, the putative major active ingredient [8]. The pharmacological effects of Tremella fuciformis polysaccharides (TFP) include the following.

5.2.2.1 Immunomodulatory Effects

Intraperitoneal injection of 200 mg/kg of TFP significantly increased the weight of the spleen in mice. Administration of 100 mg/kg of TFP increased the production of hemolysin in normal and cyclophosphamide-treated mice immunized with sheep erythrocytes by 92.9 and 112.9%, respectively. TFP (50–200 µg/ml) also significantly increased Con A-stimulated proliferation of mouse splenocytes in vitro [9]. The immunostimulatory effects of TFP might be the basis of its anti-cancer effects as described below.

5.2.2.2 Anti-Cancer Effects

TFP (100 mg/kg) significantly inhibited the growth of Ehrlich ascites tumor in mice and inhibited DNA synthesis of the tumor cells in vivo, but there was no observed cytotoxic effect on the cancer cells in vitro. Further studies indicated that the anti-cancer action of TFP is mediated through the stimulation of the immune function, i.e., by enhancing the phagocytic activity of the reticuloendothelial system and the production of interferon and tumor necrosis factor (TNF) [10]. Clinical studies showed that a formula of TFP and other medicines produced beneficial effects on late-stage lung cancer and during post-operation management in liver cancer. In addition, TFP has been shown to reduce the side effects of chemotherapy and radiotherapy, increase the therapeutic effects of the treatment, improve the quality of life, and prolong the survival of patients. These results are consistent with the traditional view regarding tremella as a superior grade medicine for the nourishment of the body and "upholding the right and securing the root" (enhancement of the internal forces to guard against illness).

5.2.2.3 Anti-Aging Effects

Experimental evidence suggested that TFP could significantly prolong the life expectancy of fruit flies and decrease the lipofuscin level of the flies by 24%. It could also decrease the lipofuscin level in mouse cardiomyocytes, increase the superoxide dismutase (SOD) activity in mouse brain and liver, inhibit mouse brain monoamine oxidase B (MAO-B) activity, and extend the survival time of mice undergoing anoxia [11]. These results suggest that TFP possesses anti-aging effects.

5.2.3 HAWTHORN

Fructus Crataegi (or hawthorn) is the dried ripened fruit of *Crataegus pinnatifida* Bge. var. *major* N. E. Br or *Crataegus pinnatifida* Bge. (Rosaceae). It is regarded in TCM as sweet and sour in taste and slightly warm in nature. Hawthorn has the function of "dispersing food accumulation and stagnation" (easing indigestion); "supplementing spleen and fortifying stomach" (promoting digestion and assimilation), "quickening blood and transforming stasis" (promoting blood circulation). Recent studies have demonstrated that hawthorn possesses the following pharmacological actions.

5.2.3.1 Promoting Digestion

Hawthorn contains lipases which can promote the digestion of fat. It also increases the secretion of pepsin and therefore facilitates the digestion of proteins. Hawthorn possesses certain regulatory effects on gastrointestinal function. It can inactivate the hyperactive rabbit duodenal smooth muscles and slightly stimulate the contraction of the relaxed smooth muscle of rat stomach. Both the ethanol extract and the water extract of hawthorn have been shown to noticeably inhibit the contraction of gastrointestinal smooth muscles of rabbits and rats induced by acetylcholine and negative ions [12].

5.2.3.2 Effects on the Cardiovascular System

It was found that intravenous administration of the ethanol extract and the total flavonoids of hawthorn increased the coronary flow in dogs by 37.5%. The ethanol extract, the total flavonoids, and especially the flavonoid hydrolysate of hawthorns enhanced ^{86}Rb-uptake in the mouse myocardium. These results indicate that hawthorn can increase blood supply to the mouse myocardium, increase coronary blood flow, and concomitantly decrease oxygen consumption, thus increasing the utilization of oxygen by the myocardium [13]. Moreover, the ethanol extract of hawthorn has been shown to produce protective effects against acute myocardial ischemia induced by pituitrin and isoproterenol in rats. The total flavonoids of hawthorn also antagonized aconitine-induced arrhythmias in rabbits and reestablished the normal rhythm of the heart in a shorter period [13]. Hypotensive effects of the extract of hawthorn have been demonstrated in mice, rabbits, and cats. Intravenous administration with the ethanol extraction of hawthorn lowered the blood pressure of anaesthetized rabbits slowly. In cats, intravenous injection of 10 mg/kg total flavonoids of hawthorn lowered the blood pressure by 40% [12].

5.2.3.3 Regulatory Effects on Lipid Metabolism

In an in vivo study in rats, hawthorn, and hawthorn flavones have been shown to greatly lower the malondialdehyde (MDA) content in serum and liver, enhance the SOD activity in red cells and liver, and also enhance the serum glutathione peroxidase activity. Moreover, they significantly suppressed the rise in the concentrations of total cholesterol (TC), low-density lipoprotein-cholesterol (LDL-C), and ApoB in the serum of rats fed with a high-fat and high-cholesterol diet—while increasing the concentrations of high-density liporpotein-cholesterol (HDL-C) and ApoAl. However, they had no obvious effect on the serum triglyceride level. RT-PCR studies demonstrated that a high-fat and high-cholesterol diet could decrease the expression of LDL receptors (LDLR) mRNA. Treatment with hawthorn and hawthorn flavones greatly increased the protein level of LDLR and augment the number of LDLR in rat liver with no obvious effect on the affinity of the receptor. This indicates that hawthorn and hawthorn flavones could prevent lipid metabolic disorders in rats by upregulating the transcription level of LDLR and enhancing the antioxidant potential [14].

5.2.3.4   Anti-Bacterial Effects

The water extract and ethanol extract of hawthorn possess anti-bacterial effects on many bacteria such as *Shigella dysenteria, Shigella flexneri, Shigella sonnet, Proteus* species, *Escherichia coli, Pseudomonas aeruginosa, Staphylococcus aureaus, Bacillus anthracis,* and *Corynebacterium diphtheria* [15]. This is consistent with the use of hawthorn to treat sore throat in Chinese folk medicine.

5.2.4   WALNUT

Semen Juglandis (or walnut) is the kernel of a tall deciduous tree *Juglans regia* L. (Juglandaceae). It is one of the oldest cultured nuts. Walnut, together with almond, cashew, and hazelnut, are collectively known as "the four main dried fruits of the world." The global yield of walnut is only slightly less than that of almond, and its yield in China is the largest in the world. In 2002, the yield of walnut in China was over 320,000 tons [16]. Walnut is not only nutritious but it also possesses good therapeutic values. Li Shizhen of the Ming dynasty claimed that walnut "benefits qi and enriches blood, moistens dryness and transforms phlegm, benefits life gate and triple burners,* and moistens lung and intestine" and could be used to treat "cough, low-back pain and leg pain, abdominal colic, and dysenteric diarrhea." According to TCM understanding, walnut is sweet in taste and warm in nature. It can "supplement kidney essence (vital substance)" and "fortify brain and increase intelligence." Therefore, it is also entitled Wanshuizi (long life seed) and Changsouguo (longevity fruit).

Walnut is rich in nutrients. It contains 16.7% of protein, 15.9% of carbohydrate, and 66.9% of fat [17]. The main types of fat in walnut are linoleic acid, linolenic acid, and oleic acid. Apart from the high contents of the eight essential animo acids, walnut also contains plenty of arginine [17]. Arginine can stimulate the secretion of growth hormone from the pituitary gland and inhibit the accumulation of excess fat [18]. Apart from fat and protein, walnut also contains appreciable amounts of vitamins and minerals. A hundred grams of walnut contain 2.5 mg vitamin E, 316 mg of phosphorus, and 2.5 mg of iron [19]. Recent pharmacological studies reveal the following beneficial effects of walnut.

5.2.4.1   Effects on Learning and Memory

Behavioral studies have shown that walnut can improve the learning ability and memory in mice [19,20]. Oral administration with 200 mg/kg of a walnut extract for 30 days significantly improved the response of mice in one-trail passive avoidance and Morris water maze experiments as compared with the control, while the response at higher or lower dosages was insignificant. The levels of acetylcholine

---

* Life gate is the right kidney. According to TCM theory, kidney deficiency may lead to low-back pain and leg weakness. Collectively "triple burners" refer to a virtual organ in the thoracic and abdominal cavities which constitute the passage for water. Its obstruction may result in cough, abdominal colic and diarrhea.

and NO in the brain increased with increasing dosage of walnut extract and a significant difference could be observed in NO content at the high dosage group (400 mg/kg). The increase in NO level can be attributed to the presence of its precursor, arginine, in the extract. It is concluded that extract of walnut can improve learning and memory of mice within a certain dosage range [20]. Another study has also shown that oral administration with walnut oil and vitamin E-fortified walnut oil for 8 weeks significantly improved the learning ability and memory of mice in the Morris water maze experiment [21].

5.2.4.2 Effects on Tolerance to Anoxia and Tiredness

Oral administration with walnut oil and especially vitamin E-fortified walnut oil has been shown to significantly extend the survival time of mice subjected to anoxia and brain ischemia, and could also prolong the swimming time of weight-loaded mice. Walnut oil and vitamin E-fortified walnut oil could also significantly enhance the pentobarbital-induced sleeping of mice. It was concluded that walnut increases tolerance to anoxia and tiredness in mice [21].

5.2.4.3 Effects on Lipid Profiles in Type 2 Diabetes

In a parallel randomized controlled study in patients with type 2 diabetes, the effects of the addition of 30 g of walnut to a modified low fat diet was compared with two other low fat diet groups [22]. The walnut group achieved a significantly greater increase in HDL-C as compared to other groups and a 10% reduction in LDL-C, with no significant difference in body weight, percent body fat, total antioxidant capacity, or HbA_{1c} level. The authors concluded that the delivery of a substantial amount of polyunsaturated or omega-3 fatty acids could be important for the improvement of the lipid profiles of these patients in the walnut group.

5.2.5 JUJUBE

Fructus Jujubae (or jujube), also called common jujube, Chinese jujube, or Chinese date, is the fruit of *Zizyphus jujuba* Mill. (family: Rhamnaceae). Its yield in China is more than 90% of the total global yield. Jujube is widely distributed in China with a high abundance in Hebei, Henan, Shandong, and Shanxi provinces. For ages, jujube has been considered a superior grade tonic. Because of its remarkable therapeutic effects, jujube is often combined with other medicines to improve their efficacy. Therefore, jujube is also entitled Hungliang (a matchmaker—a woman who tries to arrange marriages). In "Ben Cao Gang Mu," Li Shizhen regarded jujube as sweet in taste and warm in nature. It has "middle burner-calming and spleen-fortifying" effects. He also commented that jujube is "edible when ripe and tonifying when dried; can be stored as medicine in good harvest and relieve the situation during times of famine; and assists the food supplies for the well-being of the people."

Jujube is rich in nutrients and is considered a good tonic. Fresh jujube is rich in vitamins. The content of vitamin C in the fresh fruit is 400–600 mg/100 g, 70–100 times that of apple [23]. However, the content of vitamins especially vitamin C in fresh jujube will greatly decrease after certain processes occur, such as drying inside an oven or under the sun. All types of jujube contain the eight essential amino acids and most of them also contain two semi-essential amino acids for infants: arginine and histidine [23]. There are many studies reporting the nutritional values and the health enhancement functions of jujube. Most of these reports focus on the following pharmacological actions.

5.2.5.1 Anti-Cancer Effects

Many reports have demonstrated the anti-cancer effects of jujube. The neutral polysaccharides of jujube have been shown to induce the secretion of TNF from mice macrophages, thereby increasing the immunity of the animals [23]. In addition, pentacyclic triterpenoids in jujube such as betulinic acid, betulonic acid, and especially maslinic acid (Figure 5.1) have been shown to inhibit the growth of S120 sarcoma in mice [24].

5.2.5.2 Anti-Aging Effects

Many classical works have recorded the anti-aging effects of jujube. This effect is now confirmed by modern pharmacological studies and functional evaluation. In a mouse aging model, mice orally administered with jujube polysaccharides for a long period were found to have bigger spleens as compared with the control [23].

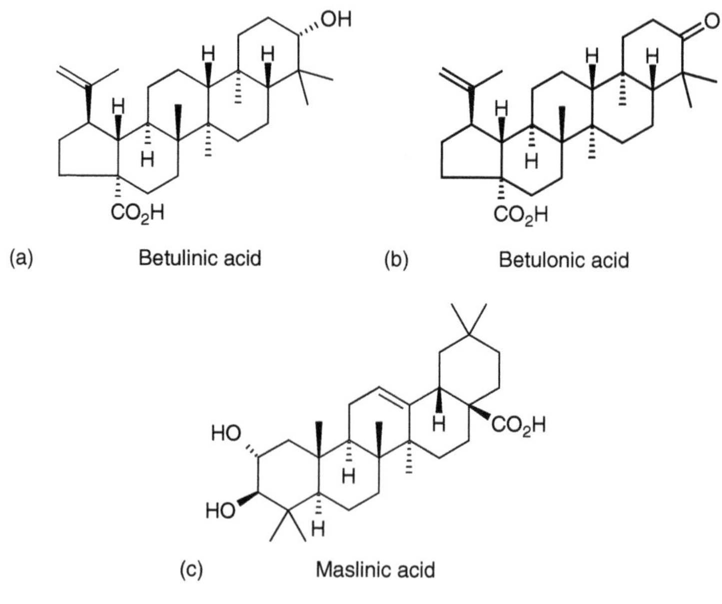

(a)   Betulinic acid   (b)   Betulonic acid

(c)   Maslinic acid

FIGURE 5.1 Chemical structures of pentacyclic triterpenoids from jujube.

Histological examination of tissues from the treated animals showed that the splenic lobules, medulla, and cortex became more distinctive, with enlarged white medulla, and dilated and filled sinus in the red medulla. The lymphoid tissues around the splenic arterioles became bigger, and the number of lymphocytes increased significantly as well.

All these indicate that jujube polysaccharides can significantly reduce the atrophy of the immune organs and delay the process of aging. Treatment of mice with an extract of jujube and wolfberry for 30 days also increased SOD activity in red cells and concomitantly decreased the MDA level in plasma [25]. These results show that jujube has free radical scavenging activity and can enhance the antioxidant potential against lipid peroxidation. The abundant amounts of antioxidants such as vitamin C and flavone glycosides in jujube could be responsible for the scavenging of free radicals and slowing down the process of aging.

5.2.5.3 Immunostimulatory Effects

Jujube polysaccharides have been reported to increase the proliferation of lymphocytes and splenocytes. The crude polysaccharides, the neural polysaccharides, and the acidic polysaccharides in jujube all possess obvious effects, but the activity of the neural polysaccharides seems to be higher than that of the acidic polysaccharides [26].

5.2.6 Common Yam

Rhizoma Dioscoreae (Chinese yam or common yam) is the tuber of *Dioscorea opposita* Thunb. (family: Dioscoreaceae). It was first recorded in "Shen Nong Ben Cao Jing" and was classified as a superior grade tonic. It is regarded in TCM as sweet in taste and mild in nature. Common yam was originally consumed as food. Its therapeutic effects were later recognized. According to "Ben Cao Gang Mu," common yam can be used to treat "all kinds of consumptive diseases" and "five types of exhaustion and seven types of wounds." The most outstanding function of common yam is its therapeutic effect on diabetes. Common yam contains saponins, glycoproteins, mannan, phytic acid, amino acids, coarse fibres, pectin, and minerals [27]. The following pharmacological actions of common yam have been described.

5.2.6.1 Immunostimulatory Effects

Common yam is rich in polysaccharides which can stimulate the immune system. Common yam has been shown to increase the number of blood leucocytes and their phagocytic activity. Moreover, common yam polysaccharides were reported to antagonize the cyclophosphamide-induced suppression of cellular immunity and cause a partial or complete recovery of the suppressed immune functions [28].

5.2.6.2   Hypoglycemic Effects

An extract of *Dioscorea* species has been shown to produce a hypoglycemic effects on fasting rats and rabbits and control hyperglycemia induced by alloxan. The water-soluble components in the ethanol extract may be related to these activities. In addition, a crude preparation of common yam polysaccharides significantly lowered blood glucose in mice [29,30].

5.2.6.3   Hypolipidemic Effects

Oral administration of purified common yam starch could lower the concentration of blood lipid in atherosclerotic mice. Common yam could also lower the serum cholesterol concentration in mice fed with free cholesterol or cholesterol-containing food [31].

5.2.6.4   Anti-Aging Effect

Common yam has been shown to significantly inhibit the activities of enzymes such as MAO which are related to aging. This effect is consistent with its traditional application in "invigorating liver and spleen" (promoting general physiological functions) and prolonging life and may thus explain the traditional use of common yam as the major ingredient in some tonic formulae such as Liu-wei-di-huang-wan (Six-ingredient Rehmannia Pill), Ba-wei-di-huang-wan (Eight-ingredient Rehmannia Pill) and Si-da-huai-yao-he-ji (Combination of the Four Medicines of Huai) [32].

5.2.7   PUERARIA

Radix Puerariae (or pueraria) is the tubers of *Pueraria lobata* (Willd.) Ohwi or *P. thomsonii* Benth. It was first recorded in "Shen Nong Ben Cao Jing" and was classified as an intermediate grade material. In TCM, it is regarded as sweet and pungent in taste and mild in nature and was traditionally used to relieve fever, headache, and stiffness of the nape in externally contracted disease, measles with inadequate eruption, acute dysentery or diarrhea, and alcohol intoxication.

The fresh tuber contains 68.6% of water, 27.8% of carbohydrate, 2.1% of protein, and trace elements. Pueraria is also rich in essential amino acids. The root tuber starch can be made into various kinds of foods, beverages, and wines. The major bioactive constituents of pueraria include flavonoids, triterpenes, and alkaloids. The isoflavones diadzein, diadzin, and puerarin (Figure 5.2) have been isolated from the herb; yet, puerarin is considered the most important because it is found only in this herb. Studies have revealed some new pharmacological actions of the herb, especially the cardiovascular effects of puerarin, as described below.

5.2.7.1   Effects on Hypertension

Pueraria flavonoids and puerarin can reduce the blood pressure in both normotensive and hypertensive animals [33]. Puerarin (100 mg/kg, i.p.) significantly reduced the blood pressure and heart rate in conscious spontaneous hypertensive rats [34]. The

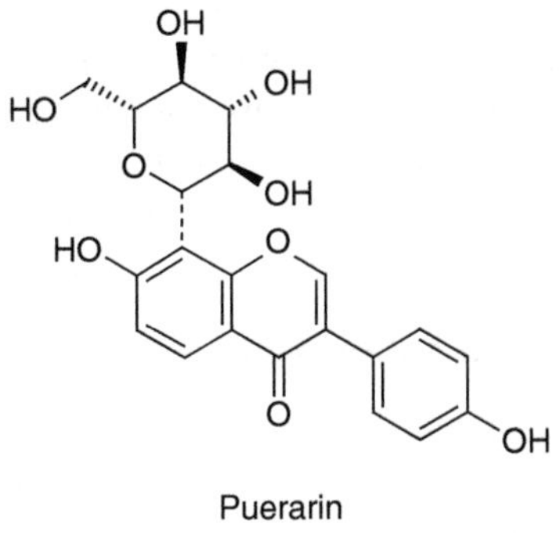

Puerarin

FIGURE 5.2 Chemical structure of puerarin.

plasma renin activity was found to decrease significantly, suggesting that its hypotensive effect may be related to the lowering of the catecholamine level and the suppression of the renin-angiotensin system. Puerarin has vasodilating properties. Huang et al. [35] have studied the effect of puerarin on plasma NO and endothelin (ET) levels in patients with hypertension. The results showed that treatment with puerarin could improve the low NO/ET ratio and regulate the function of the endothelium—and this may be one of the mechanisms involved in the stabilization of blood pressure.

5.2.7.2 Effects on Coronary Heart Diseases

Intravenous injection of puerarin could increase coronary flow and decrease vascular resistance in both conscious and anaesthetized dogs in a dose-dependent manner. Though puerarin could reduce the heart rate and contractility of the hearts, it did not result in a reduced blood flow to the lateral coronary arteries in the ischemic zone. This suggests that the effect of puerarin on coronary flow was produced by decreasing vascular resistance in the lateral coronary arteries [36]. In addition, puerarin could also antagonize experimental myocardial ischemia induced by pituitrin [37]. Wang et al. [38] have studied the effect of puerarin on coronary heart disease and found significant improvements in the hemodynamic parameters and electrocardiogram as compared with the control. Puerarin has also shown efficacy in treating angina pectoris [39] and acute myocardial infarction [40].

5.2.7.3 Effects on Microcirculation

In a clinical study on patients with cerebral ischemia, significant improvement in the hemorrheological parameters was observed in the group treated with puerarin [41]. Puerarin could dilate cerebral blood vessels, reduce resistance in the peripheral vessels and platelet aggregation, and improve microcirculation. The efficacy of intravenous puerarin in treating diabetic retinopathy was also studied [42]. Comparing before and after treatment, hemorrheological parameters such as the erythrocyte aggregation index, whole blood viscosity, fibrinogen, and erythrocyte sedimentation rate had all improved significantly. Comparing before and after treatment, the peak systolic velocity of arteria centralis retinae, their end diastolic volume, the acceleration, and the central retinal vein reflux velocity were all improved. Compared with the control group, naked eye visions were also improved. It is concluded that puerarin can improve microcirculation and may play a positive role in a number of microcirculation disorders including brain ischemia and diabetic retinopathy.

5.2.7.4   Hypolipidemic and Antioxidant Effects

Puerarin effectively reduced blood cholesterol, triglyceride, and LDL levels in rabbits with hyperlipidemia. By increasing SOD activity and reducing lipid peroxidation, it could decrease MDA level and LDL oxidation in blood [43]. Puerarin has been shown to increase SOD activity and reduce lipid peroxidation in patients with coronary heart disease. It also increased tissue plasminogen activator activity and promoted thrombolysis [44]. The results suggest that puerarin can relieve the condition of hyperlipidemia and reduce the incidence of atherosclerosis.

5.2.7.5   Antipyretic Effect

Puerarin was found to produce hypothermia in both the basal body temperature and pyrogenic fever in unanesthetized rats [45]. The peurarin-induced hypothermia was abolished by co-treatment with a 5-HT(2A)-receptor agonist or a 5-HT(1A)-receptor antagonist, while 5-HT(2A)-receptor antagonist or 5-HT(1A)-receptor agonist co-treatment had a potentiating effect. The results indicate that puerarin exerts its hypothermic and antipyretic effect by activating the 5-HT(1A)-receptor or antagonizing the 5-HT(2A)-receptor in the hypothalamus.

5.2.8   Perilla

Perilla frutescens (L.) Britt. (or perilla) is a herb regarded in TCM as pungent in taste and warm in nature. Perilla leaf has the property of dissipating wind cold and is used to treat cold, snuffling nose, headache, cough, nausea, and diarrhea caused by fish and crab consumption. Perilla fruit is indicated in cough, asthma, and constipation. Perilla stem has the function of rectifying qi and relieving the dullness of the stomach. It is indicated in distention and fullness of the abdomen, morning sickness, and abnormal fetal movement.

Perilla leaves may be eaten cold and dressed with sauce or soup. The whole herb and especially its fruits contain oil, an important raw material in the food industry. Perilla fruit contains 35%–50% of oil in which over 90% is unsaturated fatty acids such as α-linolenic acid (63%–84.5%) and linoleic acid (7%–10.2%) [46,47]. It is the richest plant source of α-linolenic acid. Recent studies showed that omega-3 polyunsaturated fatty acids such as α-linolenic acid is metabolized inside the body to the essential fatty acids docosahexaenoic acid (DHA) and eicosapentaenoic acid (EPA) which are essential for brain development and the well-being of the cardiovascular system. The chemical composition of perilla herb includes essential oil, terpenoids, sterols, phenolics, organic acids, flavonoids, glycosides, fatty acids, and pigments [48]. The essential oil content of perilla can be as high as 2.7%, in

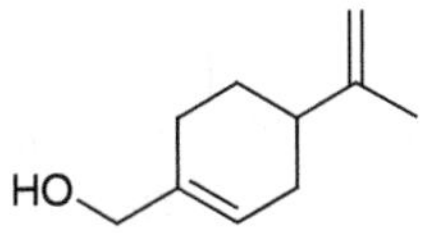

Perillyl alcohol

FIGURE 5.3 Chemical structure of perillyl alcohol.

which perillaldehyde, limonene, β-caryophyllene, perillaketone, perillyl alcohol, and myristicin are some of the major compounds isolated [48]. Major studies on the pharmacological actions of the herb are listed below.

5.2.8.1 Anti-Cancer Effects

Dietary perilla oil could decrease the incidence of carcinogen-induced tumorigenesis in Sprague–Dawley rats [49]. It was found to significantly inhibit tumor growth in rats with mammary and colon cancer [49]. Pulmonary metastasis of ascites tumor cells in rats was also inhibited by oral treatment with perilla oil [50].

The monoterpene perillyl alcohol (POH) (Figure 5.3) found in perilla has been shown to possess anti-cancer and cancer prevention properties. Its antitumor effects have been shown to be of broad spectrum with high efficacy and low toxicity. The compound is currently undergoing phase I and phase II clinical trials in the United States [51,52]. Perillic acid (PA) is the major metabolite found in the circulation in human and other mammals after oral ingestion of POH, suggesting that the antitumor effect of POH is mediated through PA and other metabolites [53,54]. In vitro and in vivo studies have shown that POH has antitumor effects on leukemia and carcinoma of the mammary, pancreatic, prostatic, ovarian, gastrointestinal, hepatic, cerebral, and pulmonary origins [55]. It can induce apoptosis and differentiation, and suppress proliferation and metastasis of the tumor cells. It also acts as a chemopreventive agent in some animal cancer models such as pancreatic cancer and esophageal cancer [56,57]. Induction of pro-apoptotic proteins [57], phase I and phase II detoxification enzymes [54], inhibition of ubiquinol synthesis [54], inhibition of angiogenesis [58], and regulation of Ras [59] and the transforming growth factor-β [60] signal transduction process have been suggested to be involved.

5.2.8.2 Sedative Effect

Perillaldehyde has been shown to produce a sedative effect in mice, as suggested by an increase in barbital-induced sleep time [61].

5.2.8.3 Antipyretic, Anti-Inflammatory, Anti-Allergic, Antitussive, and Anti-Asthmatic Effects

Perilla extract, perilla oil, and essential oil extracted from perilla have also been shown to produce antipyretic effects in rabbits [62]. A perilla leaf extract has been shown to relieve arachidonic acid-induced inflammation. These actions were attributed to the inhibition of the production of TNF-α [63]. Triterpene acids isolated

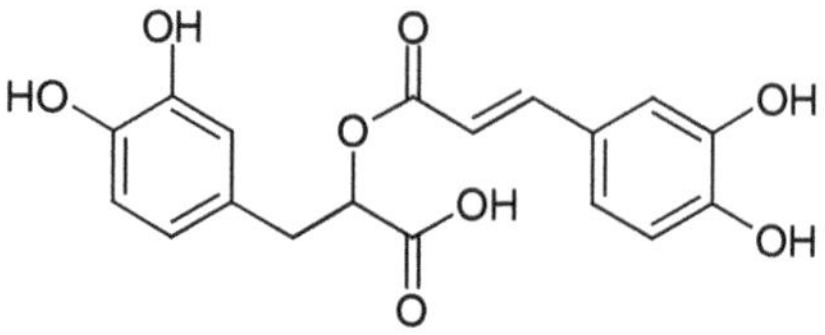

Rosmarinic acid

FIGURE 5.4 Chemical structure of rosmarinic acid.

from perilla leaf also inhibited 12-O-tetradecanoylphorbol-13-acetate-induced inflammation in mice [64]. In addition, rosmarinic acid (Figure 5.4) in perilla extract has been identified to inhibit allergic inflammation [65,66], and dietary α-linolenate from perilla seed oil has been shown to significantly reduce leukotriene B$_4$ (LTB$_4$) and platelet-activation factor production in polymorphonuclear leukocytes in rats [67,68].

Okamoto et al. have studied the effect of dietary perilla oil and corn oil on bronchial asthma in asthmatic subjects [69]. It was found that the generation of LTB$_4$ and LTC$_4$ by leukocytes was increased in the group supplemented with corn oil and decreased in the group supplemented with perilla oil. Pulmonary functional parameters were also significantly improved in the perilla oil group. The authors concluded that perilla oil is useful for the treatment of asthma. Perilla extracts, perilla oil, and essential oil extracted from perilla have also been shown to produce antitussive, expectorants, and anti-asthmatic effects in mice and guinea pig [62,70]. Oral administration of these preparations significantly reduced the total number of coughs and extended the latency period in animals treated with nebulized ammonia or asthmogenic drugs.

5.2.8.4   Effects on Vision, Learning, and Memory Function

Zhuo et al. [71] have reported that oral treatment with perilla oil could improve the learning and memory function of mice in the platform jumping test and the Morris water maze experiment. The perilla oil treatment could increase the DNA, protein, and monoamine neurotransmitter levels in the mouse brain. Watanabe et al. [72] have studied the effect of supplementation with α-linolenate-rich perilla oil on electroretinographic responses in rats. Results showed that perilla oil supplementation significantly increased the response as compared with the control group without supplement and the group supplemented with linoleate-rich safflower oil. The DHA level in the retina was also increased by perilla oil supplementation. These data suggest that dietary perilla oil can improve vision and memory function.

5.2.8.5   Hypotensive Effect

Shimokawa et al. [73] have studied the effects of α-linolenate-rich perilla oil and linoleate-rich safflower oil on the survival time and blood pressure of stroke-prone spontaneous hypertensive rats and found that both male and female rats fed with perilla oil lived longer than those fed with safflower oil. Post-mortem examination showed that apoplexy was the major cause of death for most of the rats. Diastolic blood pressure in the perilla oil group was also found to be lower than the safflower oil group by 10%. It was concluded that the increase in survival time in the rats treated with perilla oil was due to the lowering of blood pressure.

5.2.8.6 Hypolipidemic Effect

Guo et al. [74] have studied the effects of different high fat diets on the lipid profile in rats. High fat diets were prepared by supplementing the standard diet with 6% of lard, perilla oil, or pinon oil so that fat constituted 32.6% of the total calorie supply. After treatment for 3 weeks, total triglyceride, TC, LDL-c, LDL-c/HDL-c, and atherosclerotic index in the perilla group was found to be significantly lower than the lard group, while HDL-c was increased. These data show that perilla oil could improve the lipid profile in rats fed with a high fat diet, implicating its prophylactic effect on cardiovascular diseases.

5.3 CONCLUSION

The use of functional foods in China has a long history. The development of functional foods in China is summarized in the literature written by contemporary TCM scholars in different dynasties who succeeded their predecessors in the collection of new materials and prescriptions, and in the development of new theories based on clinical practice, finally synthesizing their knowledge into new works. This paradigm, originated from the concept that "medicine and food are of the same origin," has resulted in the growing number of materials used both as medicine and as food in the Chinese culture. In addition, recent studies have also elucidated some of the active principles and action mechanisms. Further studies on Chinese functional foods should focus on the integration of these scientific findings with traditional theories to achieve the objective of rendering them in the traditional language as "edible when ripe and tonifying when dried; can be stored as medicine in good harvest and relieve the situation during times of famine; and assists the food supplies for the well-being of the people," as aptly described by Li Shizhen for jujube.

REFERENCES

1. Qian, Y. and Yu, W., Recent progress in the research of the chemical constituent and the pharmacology of Chinese wolfberry [Chinese], *Acta Chin. Med. Pharmacol.*, 28 (4), 33–35, 2000.
2. Shi, R., Liu, Y., Zhao, S., Liu, J., Yung, Y., and Cao, Y., The regulatory action of ginseng (*Panax ginseng*), Huangqi (*Astragalus membranaceus*) and Chinese wolfberry (*Lycium chinenses*) on heart β-adrenoceptor of aged rats [Chinese], *Trad. Chin. Herb. Drugs*, 29, 389–391, 1998.
3. Xu, Y., He, L., Xu, L., and Liu, Y., Advances in the immunopharmacological research of *Lycium chinensis* [Chinese], *J. Chin. Med. Mater.*, 23, 295–298, 2000.

4. Zhang, X., Xiang, S., Cui, X., and Qian, Y., Effects of *Lycium barbarum* polysaccharide (LBP) on the lymphocyte signal transduction system in mice [Chinese], *Chin. J. Immunol.*, 13, 289–292, 1997.

5. Zhou, Z., Zhou, J., and Xing, S., Effects of *Lycium barbarum* polysaccharides on proliferation and differentiation of hematopoietic stem cells and progenitors of granulocytes and macrophages in mouse bone marrow [Chinese], *Chin. J. Pharmacol. Toxicol.*, 5, 44–46, 1991.

6. Zhou, Z., Zhou, J., and Xing, S., Effects of *Lycium barbarum* polysaccharides on hematopoiesis and colony-stimulating factors in normal mice [Chinese], *Chin. J. Hematol.*, 12, 409–411, 1991.

7. Hu, Z., Qi, Z., and Wu, Y., Research and development of instant food products of *Tremella fuciformis* [Chinese], *Food Sci. (Beijing)*, 17 (8), 35–40, 1996.

8. Yang, S., Yin, C., and Mou, S., Advances research on Yiner (*Tremella fuciformis*) polysaccharides and its pharmacological actions [Chinese], *Trad. Chin. Herb. Drugs*, 24, 153–157, 1993.

9. Nei, W., Zhang, Y., and Zhou, J., Phramacological research of *tremella fuciformis* polysaccharides [Chinese], *Pharmacol. Clin. Trad. Chin. Med.*, 16 (4), 44–46, 2000.

10. Zhou, A., Wu, Y., and Hou, Y., Study on the anti-tumor effect of *Tremella fuciformis* polysaccharides [Chinese], *J. Beijing Med. Univ.*, 19, 150, 1987.

11. Chen, Y., Xia, E., and Wang, S., The anti-aging effects of the polysaccharides of *Auricularia auricula* and *Tremella fuciformis* spores, *Modern Appl. Pharm.*, 6 (2), 9–11, 1989.

12. Shen, Y. J., *Pharmacology of Chinese Medicinal Materials* [Chinese], People's Medical Publishing House, Beijing, pp. 574–578, 2000.

13. Shen, Y. J., *Pharmacology of Chinese Medicinal Materials* [Chinese], Shanghai Scientific and Technical Publishers, Shanghai, p. 115, 1997.

14. Lin, Q. and Chen, J., Molecular mechanism of hawthorn and its flavonoids in prevention of lipid metabolism disorder in rats [Chinese], *Acta Nutr. Sin.*, 22, 131–136, 2000.

15. Liu, W., Nutritional chemical ingredients and prophylactic actions of hawthorn [Chinese], *Food Res. Dev.*, 23 (5), 65–66, 2002.

16. United States Department of Agriculture, *World Horticultural Trade and U.S. Export Opportunities*, Circular Series FHORT, 11-02, November 2002, http://ffas.usda.gov/htp/hort_circular/2002/02-11/toc.htm

17. Sze-Tao, K. W. and Sathe, S. K., Walnuts (*Juglans regia* L.): Proximate composition, protein solubility, protein amino acid composition and protein *in vitro* digestibility, *J. Sci. Food Agric.*, 80, 1393–1401, 2000.

18. Barbul, A., Sisto, D. A., Wasserkrug, H. L., Levenson, S. M., Efron, G., and Seifter, E., Arginine stimulates thymic immune function and ameliorates the obesity and the hyperglycemia of genetically obese mice, *J. Parenter. Enteral Nutr.*, 5, 492–495, 1981.

19. Wardlaw, G. M., *Perspective in Nutrition*, 4th ed., McGraw-Hill, New York, 1999.

20. Zhao, H., Li, X., and Xiao, R., Effect of Semen Juglands extract on improving learning and memory in mice [Chinese], *J. Shanxi Med. Univ.*, 35, 20–22, 2004.

21. Wang, Z., Yang, S., Li, W., and Yang, S., Effects of walnut oil and vitamin E-fortified walnut oil on animal behavior [Chinese], *Shanxi Med. J.*, 29, 325–326, 2000.

22. Tapsell, L. C., Gillen, L. J., Patch, C. S., Batterham, M., Owen, A., Bare, M., and Kennedy, M., Including walnuts in a low-fat/modified-fat diet improves HDL cholesterol-to-total cholesterol ratios in patients with type 2 diabetes, *Diabet. Care*, 27, 2777–2783, 2004.

23. Wang, J., Zhang, B., and Chen, J., A research review on the nutritive elements and function of *Zizyphus jujuba* [Chinese], *Food Res. Dev.*, 24 (2), 68–72, 2003.

24. Zhang, Y. and Yang, C., Study on the chemical constituents of *Zizyphus jujuba* [Chinese], *Trad. Chin. Herb. Drugs*, 609, 580, 1992.

25. Yu, S. Y., Cui, H. B., *Progress of Chinese Health Products* [Chinese], People's Medical Publishing House, Beijing, pp. 49–50, 2001.

26. Zhang, X., Lin, S., Lin, Q., Sun, L., and Yang, S., In vitro anticomplementary activity and mouse spenocytes proliferating effect of jujube polysaccharide [Chinese], *Pharmacol. Clin. Trad. Chin. Med.*, 14 (5), 19–21, 1998.

27. Ni, S. and Song, X., Analysis of the nutritional constituent of *Dioscorea opposita* [Chinese], *Jiangsu Pharm. Clin. Res.*, 10 (2), 26–27, 2002.

28. Nei, G., Dong, X., and Zhang, C., An overview of the research on common yam [Chinese], *Trad. Chin. Herb. Drugs*, 24, 158–160, 1993.

29. Wu, M. M., Okunji, C. O., Akah, P., Tempesta, M. S., and Corley, D., Dioscoretine: the hypoglycemic principle of *Dioscorea dumetorum*, *Planta. Med.*, 56, 119–120, 1990.

30. Hikino, H., Konno, C., Takahashi, M., Murakami, M., Kato, Y., Karikura, M., and Hayashi, T., Isolation and hypoglycemic activity of dioscorans A, B, C, D, E, and F; glycans of *Dioscorea japonica* rhizophors, *Planta Med.*, 49, 168–171, 1986.

31. Zhang, Y. and Wang, A., Advances in the pharmacological study of yam [Chinese], *Trad. Chin. Med. Res.*, 13 (5), 49–51, 2000.

32. Cao, K. and Liu, Y., Effects of the Four Medicines of Huai (Radix Rehmanniae Preparata, Flos Chrysanthemi, Rhizoma Dioscoreae and Radix Achyranthis Bidentatae) on mouse brain microsomal monoamine oxidase activity [Chinese], *Chin. J. Gerontol.*, 18, 102–104, 1998.

33. Fan, L., The effects of pueraria flavonoids on coronary and cardiac hemodynamics and cardiac metabolism in dogs [Chinese], *Chin. Med. J.*, 108, 724–725, 1995.

34. Song, S., Chen, P., and Cai, X., Effects of puerarin on blood pressure and plasma renin activity in spontaneous hypertensive rats [Chinese], *Acta Pharmacol. Sin.*, 9, 55–58, 1988.

35. Huang, S., Ye, W., and Cheng, Z., Effects of puerarin on plasma endothelin and NO levels in hypertensive patients [Chinese], *Chin. J. Contem. Appl. Pharm.*, 16 (3), 13–16, 1999.

36. Guo, J., Sun, Q., and Zhou, Q., Advances in the study of the pharmacological effects of puerarin [Chinese], *Trad. Chin. Herb. Drugs*, 26, 163–166, 1995.

37. Zhou, Y., Su, X., Cheng, B., Jiang, J., and Chen, H., Comparative study on pharmacological effects of various species of *Peuraria* [Chinese], *Acta Pharmacol. Sin.*, 20, 619–621, 1995.

38. Wang, Y. and Hu, W., The observation of the clinical efficacy of puerarin in coronary heart disease and its effect on hemodynamics [Chinese], *J. Anhui Trad. Chin. Med. Coll.*, 18 (3), 9–10, 1999.

39. Shu, L., Li, Z., and Chou, Y., A 114 case clinical observation of the treatment of angina pectoris by puerarin injection [Chinese], *Guangxi Med. J.*, 22, 1067–1069, 2000.

40. Zhao, H., Wei, J., and Li, Y., Effects of puerarin injection on myocardial infarct area and total ischemic burden in patients with acute myocardial infarction [Chinese], *Chin. J. Integr. Trad. West Med.*, 8, 367–369, 2001.

41. Li, S., Lei, Z., and Chui, Z., Analysis of a 36 case study of the treatment of acute cerebral ischemia by puerarin [Chinese], *J. Guangdong Sch. Pharm.*, 15, 313–315, 1999.

42. Ren, P., Hu, H., and Zhang, R., Observation on efficacy of puerarin in treating diabetic retinopathy [Chinese], *Chin. J. Integr. Trad. West. Med.*, 20, 574–576, 2000.

43. Wang, J., Gao, M., and Sun, S., Effects of emulsified puerarin on the lipid profile and the antioxidant potential in a rabbit hyperlipidemia model [Chinese], *J. Weifang Med. Coll.*, 23 (1), 6–9, 2001.

44. Chen, J., Xu, J., and Li, J., Effect of puerarin on thrombolytic function and lipid peroxidation in patients with coronary heart disease [Chinese], *Chin. J. Integr. Trad. West. Med.*, 19, 649–650, 1999.

45. Chueh, F. S., Chang, C. P., Chio, C. C., and Lin, M. T., Puerarin acts through brain serotonergic mechanisms induce thermal effects, *J. Pharmacol. Sci.*, 96, 420–427, 2004.

46. Tan, Y., Lai, B., Yan, Y., Wang, Y., Zheng, C., and Lu, P., Analysis of fatty acids in *Perilla frutescens* seed oil [Chinese], *Chin. Pharm. J.*, 33, 400–402, 1998.

47. Zhang, W., Liu, X., and Wang, H., Study of the chemical constituents of the fruits of *Perilla frutescens*, *Chin. Wild Plant Resou.*, 17, 42, 1998.

48. Wang, Y., Yang, J., Zhao, Y., and Zhu, X., An overview of the research in the chemistry and pharmacology of the Chinese medicine perilla, *Chin. Pharm. J.*, 38, 250–253, 2003.

49. Hirose, M., Masuda, A., Ito, N., Kamano, K., and Okuyama, H., Effect of dietary perilla oil, soybean oil and safflower oil on 7,12-dimethylbenz(α)anthracene (DMBA)- and 1,2-dimethylhydrazine (MDH)-induced mammary gland and colon carcinogenesis in female SD rats, *Carcinogenesis*, 11, 731–735, 1990.

50. Hori, T., Moriuchi, A., Okuyama, H., Sobujima, T., Tamiya-Koizumi, K., and Kojima, K., Effect of dietary essential fatty acid on pulmonary metastasis of ascites tumor cells in rats, *Chem. Pharm. Bull.*, 35, 3925–3927, 1987.

51. Morgan-Meadows, S., Dubey, S., Gould, M., Tutsch, K., Marnocha, R., Arzoomanin, R., Alberti, D. et al., Phase I trial of perillyl alcohol administered four times daily continuously, *Cancer Chemother Pharmacol*, 52, 361–363, 2003.

52. Meadows, S. M., Mulkerin, D., Berlin, J., Bailey, H., Kolesar, J., Warren, D., and Thomas, J. P., Phase II trial of perillyl alcohol in patients with metastatic colorectal cancer, *Int. J. Gastrointest Cancer*, 32, 125–128, 2002.

53. Crowell, P. L., Prevention and therapy of cancer by dietary monoterpenes, *J. Nutri.*, 129, 775S–778S, 1999.

54. Belanger, J. T., Perillyl alcohol: applications in oncology, *Altern. Med. Rev.*, 3, 448–457, 1998.

55. Hu, D. and Chen, Y., Anti-cancer effects of the dietary monoterpene perillyl alcohol [Chinese], *J. Clin. Hematol.*, 14, 141–143, 2001.

56. Liston, B. W., Nines, R., Carlton, P. S., Gupta, A., Aziz, R., Frankel, W., and Stoner, G. D., Perillyl alcohol as a chemopreventive agent in *N*-nitrosomethylbenzylamine-induced rat esophageal tumorigenesis, *Cancer Res.*, 63, 2399–2403, 2003.

57. Burke, Y. D., Ayoubi, A. S., Werner, S. R., McFarland, B. C., Heilman, D. K., Ruggeri, B. A., and Crowell, P. L., Effects of the isoprenoids perillyl alcohol and farnesol on apoptosis biomarkers in pancreatic cancer chemoprevention, *Anticancer Res.*, 22, 3127–3134, 2002.

58. Loutrari, H., Hatziapostolou, M., Skouridou, V., Papadimitriou, E., Roussos, C., Kolisis, F. N., and Papapetropoulos, A., Perillyl alcohol is an angiogenesis inhibitor, *J. Pharmacol. Exp. Ther.*, 311, 568–575, 2004.

59. Stayrook, K. R., Mackinzie, J. H., Barbhaiya, L. H., and Crowell, P. L., Effects of the antitumor agent perillyl alcohol on H-Ras vs. K-Ras farnesylation and signal transduction in pancreatic cells, *Anticancer Res.*, 18, 823–828, 1998.

60. Ariazi, E. A., Satomi, Y., Ellis, M. J., Haag, J. D., Shi, W., Sattler, C. A., and Gould, M. N., Activation of the transforming growth factor beta signaling pathway and induction of cytostasis and apoptosis in mammary carcinomas treated with the anticancer agent perillyl alcohol, *Cancer Res.*, 59, 1917–1928, 1999.

61. Honda, G., Koezuka, Y., Kamisako, W., and Tabata, M., Isolation of sedative principles from *Perilla frutescens*, *Chem. Pharm. Bull.*, 34, 1672–1677, 1986.

62. Wang, J., Tao, S., Xing, Y., and Zhu, Z., Pharmacological effects of zisu and baizu [Chinese], *Chin. Pharm. J.*, 22, 48–51, 1997.

63. Uede, H. and Yamazaki, M., Inhibition of tumor necrosis factor-α production by orally administering a perilla leaf extract, *Biosci. Biotechnol. Biochem.*, 61, 1292–1295, 1997.

64. Banno, N., Akihisa, T., Tokuda, H., Yasukawa, K., Higashihara, H., Ukiya, M., Watanabe, K., Kimura, Y., Hasegawa, J., and Nishino, H., Triterpene acids from the leaves of *Perilla frutescens*: their anti-inflammatory and anti-tumor-promoting effects, *Biosci. Biotechnol. Biochem.*, 68, 85–90, 2004.

65. Sanbongi, C., Takano, H., Osakabe, N., Sasa, N., Natsume, M., Yanagisawa, R., Inoue, K. I., Sadakane, K., Ichinose, T., and Yoshikawa, T., Rosmarinic acid in perilla extract inhibits allergic inflammation induced by mite allergen, in a mouse model, *Clin. Exp. Allergy*, 34, 971–977, 2004.

66. Takano, H., Osakabe, N., Sanbongi, C., Yanagisawa, R., Inoue, K., Yasuda, A., Natsume, M., Baba, S., Ichiishi, E., and Yoshikawa, T., Extract of *Perilla frutescens* enriched for rosmarinic acid, a polyphenolic phytochemical, inhibits seasonal allergic rhinoconjunctivitis in human, *Exp. Biol. Med.*, 229, 247–254, 2004.

67. Hashinoto, A., Katogiri, M., Torii, S., Dainaka, J., Ichikawa, A., and Okuyama, H., Effect of dietary α-linolenate/linoleate balance on leukotriene production and histamine release in rats, *Prostaglandins*, 36, 3–16, 1998.

68. Horii, T., Satouchi, K., Kabayashi, Y., Saito, K., Watanabe, S., Yoshida, Y., and Okuyama, H., Effect of dietary α-linolenate on platelet-activating factor production in rat peritoneal polymorphonuclear leukocytes, *J. Immunol.*, 147, 1607–1613, 1991.

69. Okamoto, M., Mitsunobu, F., Ashida, K., Mifune, T., Hosaki, Y., Tsugeno, H., Harada, S., and Tanizaki, Y., Effects of dietary supplementation with n-3 fatty acids compared with n-6 fatty acids on bronchial asthma, *Intern. Med.*, 39, 107–111, 2000.

70. Wang, Y., Xing, F., and Liu, F., Antituissive, expectorants and antiasthmatic effects of *Perilla frutescens*, *Central South Pharm.*, 1, 135–138, 2003.

71. Zhou, D., Hang, D., and Wang, Y., Effects of perilla seed oil on learning and memory function in mice, *Chin. Trad. Herb. Med.*, 25, 251, 1994.

72. Watanabe, S., Kato, M., Aonuma, H., Hashimoto, A., Naito, Y., Moriuchi, A., and Okuyama, H., Effect of dietary alpha-linolenic acid balance on the lipid composition and ERG responses in rats, *Adv. Biosci.*, 62, 563–570, 1987.

73. Shimokawa, T., Moriuchi, A., Hori, T., Saito, M., Naito, Y., Kabasawa, H., Nagae, Y., Matsubara, M., and Okuyama, H., Effects of dietary α-linolenate/linoleate balance on mean survival time, incidence of stroke and blood pressure of spontaneous hypertensive rats, *Life Sci.*, 43, 2067–2075, 1988.

74. Guo, Y., Cai, X., Li, H., and Su, Z., Effect of Perilla oil and pinon oil on lipid and lipid peroxidation in rats, *Acta Nutri. Sin.*, 18, 268–273, 1996.

75. Dang, Y., Peng, Y., and Li, W., *Chinese Functional Food*, New World Press, Beijing, 1999.

6 History and Current Status of Functional Foods in Korea

Hyong Joo Lee and Young-Joon Surh

CONTENTS

6.1 TRADITIONAL CONCEPTS OF FUNCTIONAL FOOD IN KOREA

6.1.1 DRUGS AND FOOD SHARE THE SAME ORIGIN

A functional food is usually depicted as a product that has properties of both drugs and food. In Asian countries, there is an old saying that "drugs and food share the same origin." This philosophy is compatible with what Hippocrates addressed about 25 centuries ago: "Let food be thy medicine and medicine be thy food." Oriental medicine has been often considered not only as a source of nutrients but also as a valuable component of physiological activities. It has been common to consider food as maintaining health and preventing diseases, and this is a very similar concept to that of today's functional food.

One of the most renowned classical texts of oriental medicine in Korea is *Dongui bogam*, meaning "Mirror of Korean Medicine." Twenty-five volumes of this book were written by Heo Jun, and first published in 1613. The work comprises five parts: the internal organs, the external organs, miscellany, drugs, and acupuncture and moxibustion. He discussed the invisible internal organs at the beginning of the book, before dealing with other physical organs or treatments. He also emphasized that the natural promotion of health is much more important than that of curing after a disease onset. This concept is very similar to the aim of today's functional food: the best way of maintaining health is prevention of disease. The *Dongui bogam* lists more than 6000 prescriptions using more than 900 herbal materials. Many of these materials (including phytochemicals) are applied and utilized in modern functional food materials as well as in oriental medicine.

6.1.2　EFFICACY OF DRUGS AND FOOD VARIES BETWEEN INDIVIDUALS

Another important concept of Korean traditional medicine is *Sasang euihak*, meaning "Four-constitution medicine," as developed by Lee Je-Ma in 1894. In this theory, the physical and mental constitutions of people are divided into four categories, and different physiological, pathological, and recuperative treatments are proposed for each. In the practice of this medicine, the physical constitution and personality of a person is first determined, before diagnosis and prescription can occur. This approach may offer a different diagnosis and prescription for the same disease to people of different physical constitutions. In other words, both the physical and mental characteristics of the person are taken into account in the treatment [1]. The emphasis is on curing an individual who has a disease, rather than curing a disease that a patient has, as in modern medical science.

The completion of the human genome project will make it possible to characterize each individual by genomic differences that also govern responses to various diseases. An understanding of this genetic diversity will make it possible to develop "custom-made treatments" according to these differences between individuals. This means that future medical treatment could be improved by providing different treatments for the same disease depending upon the genetic makeup of each person, thereby adopting the concept of *Sasang euihak*.

6.2　TRADITIONAL FUNCTIONAL FOODS IN KOREA

Numerous traditional foods have been developed throughout the long history of Korea, many of which have been identified as health-promoting functional foods. Among these, kimchi, fermented soy foods (doenjang, chonggukjang, and ganjang), jeotgal (fish sauce), and ginseng deserve special attention, because they are unique products to Korea and are produced on a large commercial scale for both domestic and foreign consumers, and also because considerable amounts of scientific data have accumulated that demonstrate their functional properties.

6.2.1 KIMCHI

6.2.1.1 General Characteristics of Kimchi

Kimchi is a vegetable food fermented by lactic acid bacteria (LAB). Although any type of vegetable can be used as a raw material for kimchi, Chinese cabbage is the most common. Other common materials include radish, cucumber, leaf mustard, scallion, and green onion. A typical kimchi-making procedure includes the salting and washing of Chinese cabbage, followed by mixing with ingredients such as sliced radish, green onion, red pepper powder, garlic, ginger, and fish sauces made from shrimp, anchovy, launce, or others. Raw kimchi is usually aged for several days to several weeks to achieve the proper stage of ripening, depending upon temperature and season.

Kimchi was reportedly first produced in Korea around the 7th century, and is mentioned in the book of *Samkuksagi*, a Korean history book published in 1145 A.D. [2]. The name kimchi is believed to have originated from chimchae (salted vegetable), and salting and fermenting was used to provide vegetable products with vitamins and minerals during the cold Korean winter.

Heterofermentative LAB are the major microorganisms responsible for kimchi fermentation. The dominant microflora is *Leuconostoc mesenteroides,* although other microorganisms such as *Lactobacillus plantarum, Leuconostoc citreum,* and *Pediococcus pentosaceus* are also involved. Lactic fermentation results in the formation of organic acids including lactic acid, as well as acetic acid, ethanol, and carbon dioxide. These metabolites from heterofermentation together with many esters give kimchi its distinct refreshing flavor. Kimchi is also a very good source of vitamin B_1, B_2, B_{12}, C, and niacin, the content of which increases during the optimum stage of ripening [3].

6.2.1.2 Health Promoting Effects of Kimchi

Various health-promoting effects of kimchi have been reported, including antimutagenic, anticarcinogenic, immunostimulating, hypocholesterolemic, anti-hypertensive, and antithrombotic activities. These health-promoting effects are attributable not only to the metabolites produced during the fermentation but also to the components from the raw vegetable materials, spices, the other ingredients, and LAB.

Solvent extracts of kimchi exhibit antimutagenic and anticarcinogenic activities [4], with these extracts being shown to inhibit the growth or to induce apoptosis of several cancer cell lines [5]. Kimchi extract was also shown to enhance the immune system in mice by increased phagocytic activity of the peritoneal macrophages [6]. Vegetable raw materials of kimchi, such as Chinese cabbage and radish, are good sources of the dietary fiber that protect against constipation and colon cancer, and also lower the cholesterol level to prevent circulatory diseases such as hypertension and arteriosclerosis. Also, isothiocyanates and indole-3-carbinol, which are components of the cruciferous family (which includes cabbage, radish, and broccoli), are well known to have cancer chemopreventive activities [7].

Another important contribution to the efficacy of kimchi comes from spices. In particular, red pepper, garlic, ginger, and green onion are indispensable in the preparation of kimchi. Many different chemicals, including phenolics, alkaloids, and sulfur compounds in the spices possess diverse physiological activities such as antioxidant, antifungal, anti-inflammatory, and cancer chemopreventive activities. The chemopreventive properties of spices have been well documented [7,8].

Red pepper is one of the most important spices used in kimchi, and has a high nutritional value with a relatively high content of carotenoids, vitamin C, and vitamin E. It has been suggested that red pepper improves circulation, promotes sweating, stimulates digestive secretions, increases appetite, and relieves pain and inflammatory symptoms [9]. Capsaicin—a pungent component of red pepper—has been shown to act as a carcinogen or a cocarcinogen in experimental animals, but other studies indicate that the compound has chemopreventive effects. Capsaicin inhibited constitutive and induced activation of nuclear factor-κB in human malignant-melanoma cells, leading to inhibition of melanoma-cell proliferation. Similarly, capsaicin was shown to cause apoptotic death in other types of transformed or cancerous cells [7].

Allium spices used in kimchi such as garlic, green onions, leeks, and chives have been reported to protect against stomach, esophageal and colorectal cancers. Their chemopreventive effects have been attributed to sulfur compounds, mainly as allyl derivatives. Organosulfur compounds have been found to modulate the activity of several enzymes that activate or detoxify carcinogens. In addition, an antiproliferative activity has been observed in several tumor cell lines, which is possibly mediated by the induction of apoptosis and alterations of the cell cycle. Garlic has been used both as a spice and for medicinal purposes. Garlic has been reported to be effective in preventing many chronic diseases including cardiovascular disease, arteriosclerosis, and cancer [9]. Onions are a common ingredient in many foods worldwide. There have been numerous reports that quercetin (the major flavonol in onion) exhibits strong antioxidant, anticancer, and chemopreventive properties. The mechanisms underlying the antitumorigenic effects of quercetin include cell-cycle arrest, inhibition of kinases, and inhibition of carcinogen-activating enzymes [9].

Ginger is used both as a spice and in oriental medicine, which recommends it for colds, fever, chills, rheumatism, motion sickness, and leprosy. Ginger exerts antiemetic, diuretic, anti-inflammatory, analgesic, carminative, stimulant, antioxidative, and antipyretic effects. Gingerol, a phenolic substance that is responsible for the spicy taste of ginger, has been reported to inhibit tumor promotion.

It is well known that LAB exert probiotic effects such as the alleviation of lactose intolerance, immune modulation, reduction of diarrhea, lowering of serum cholesterol, lowering of blood pressure, and reduction of some cancer recurrences [10]. Lactic acid bacteria has been utilized for thousands of years in the making of fermented dairy and vegetable products. Kimchi, as one of the most complicated fermented vegetable products, also benefits from the beneficial effects of LAB. When the antiproliferative effects of whole cells, peptidoglycans, and cytoplasm fractions of ten different LAB were measured against 11 human cancer cell lines, the cytoplasm fractions derived from *L. plantarum* (one of the kimchi LAB) were most effective for the retardation of proliferation of a human colon cell line [11].

This cytoplasmic fraction of *L. plantarum* exhibited strong antitumor activity against teratocarcinoma bearing mice in vivo [12].

6.2.2 FERMENTED SOY FOODS (DOENJANG, CHONGGUKJANG, AND GANJANG)

6.2.2.1 General Characteristics of Doenjang, Chonggukjang, and Ganjang

Doenjang is a Korean soy paste made from cooked soybean mass (meju), usually without using other cereals. Typical microorganisms in traditional meju include *Aspergillus oryzae*, *Aspergillus sojae*, and *Bacillus subtilis* [13], although other microorganisms such as *Penicillium*, *Mucor*, and *Rhizopus* are active therein. In the past, doenjang was obtained by separating the soy sauce (ganjang) after fermentation of meju in brine for several months. However, commercial doenjang is now produced as a sole product without separation of ganjang, and other cereals are often added to soybeans as substrate. Modern meju is prepared on an industrial scale by inoculating *A. oryzae* and *Bacilllus* rather than depending upon natural contaminant flora.

Chonggukjang is another type of soy paste consumed as a basis of soup in Korea. It was first mentioned in *Sanlim gyongje*, an agricultural encyclopedia published by Hong Man-Sun in 1715. Unlike doenjang, which takes several months to ferment, chonggukjang can be fermented in only a few days. Traditionally, fully cooked soy bean was placed on rice straw in a vessel, and was kept at 40°C–43°C for 3–4 days. White slime and fibrous materials are developed after the fermentation with microorganisms such as *Bacillus subtilis*, *Bacillus natto*, and *Bacillus licheniformis*. The fermented soybean are usually mixed with salt, garlic, ginger, and red pepper before being used as a soup base.

6.2.2.2 Health-Promoting Effects of Doenjang, Chonggukjang, and Ganjang

Doenjang reportedly shows diverse bioactivities such as antimutagenic, antic-arcinogenic, hypocholesterolemic, antihypertensive, and antithrombotic activities [3]. Some of these health-promoting effects originate from constituents of soybean such as isoflavones, protease inhibitors, phenolics, and globulins. Others are contributed by products formed during the fermentation process, such as peptides. Solvent extracts of traditional and modern commercial doenjang (and also chonggukjang) exhibit antimutagenic activities [14,15]. Solvent and water extracts of doenjang also exhibited anticarcinogenic activity in vivo: inhibiting tumor formation in sarcoma-180 transplanted mice [16], and increasing survival times [15]. Modifications of enzyme activities, including glutathione *S*-transferase or the cell cycle of the cancer cells, were involved in the anticarcinogenic activities [16].

It has been reported that peptides fractionated from fermented soy foods such as ganjang and doenjang are cytotoxic to several tumor cell lines [17,18]. Various peptide fractions were obtained under more controlled conditions by hydrolyzing soy proteins with 16 different proteases and extracted them using five different solvents [17]. Among the many extracts, thermoase/ethanol extract showed the highest anticancer activity and an isolated anticancer peptide was sequenced as X-Met-Leu-Pro-Ser-Tye-Ser-Pro-Tyr, which was determined to be a fragment of soy glycinin [19]. The anticancer activity was attributed to the arresting the G_2/M phase of the DNA cell cycle in a mouse lymphoma cell line. A peptide fraction isolated from ganjang exerted a dose-dependent antiproliferative effect on P388D1 mouse lymphoma, F9 mouse teratocarcinoma, and DLD-1 human colon cancer cells [20]. Tumor growth in F9 teratocarcinoma-bearing BALB/c mice, orally administered with 80 and 200 mg/kg/day of the peptides, was inhibited by 10.3% and 52.4%, respectively, compared to a control group.

It has been well documented that many peptides exhibit anti-hypertensive activity by inhibiting angiotensin converting enzyme (ACE). Angiotensin converting enzyme converts antiotensin I to angiotensin II (a vasoconstrictor), and inhibits the bradykinin (a vasodilator). Many food peptides originating from casein, gelatin, tuna, zein, and fermented milk have been reported as ACE inhibitors [21]. Angiotensin converting enzyme-inhibiting peptides have also been isolated from doenjang [22]. An ACE-inhibiting dipeptide of Arg-Pro was isolated from a water extract of doenjang; the IC_{50} value of this peptide was 92 µg [23]. Another peptide of His-His-Leu exhibited strong ACE inhibition, with an IC_{50} value of 2.2 µg/ml [24]. Moreover, the synthetic His-His-Leu significantly decreased of ACE activity in the aorta and lowered the systolic blood pressure in spontaneously hypertensive rats compared to control. Triple injections of the peptide (at 5 mg/kg of body weight per injection) also significantly decreased the systolic blood pressure: by 61 mmHg ($p < 0.01$) after the third injection [24]. Some peptides such as those containing Arg-Gly-Asp-Ser sequence act as antithrombotic peptides that inhibit fibrinogen binding to specific receptor sites of platelets. An antithrombotic peptide fraction was isolated from water extract of doenjang and His, Arg, and Ala were major amino acids in the peptide [25].

6.2.3   JEOTGAL (FISH SAUCE)

Various types of jeotgal (fish sauce) are manufactured on a large scale in Korea. The most common sources are shrimp, anchovy, and launce, although other materials including clam, oyster, small octopus, and the roe, intestine, and gills of fish are also used to make jeotgal. These raw materials are usually washed, salted, and aged for about 1 year to complete the fermentation process. Some products are filtered to obtain a clear liquid sauce. Jeotgals are very good sources of amino acids and peptides, and these constituents provide good background flavoring in the making of kimchi, salad, and soup.

Many different physiological activities of jeotgal have been reported, some of which must originate from the high content of peptides, given that many of food peptides exert diverse physiological functions such as anticancer, antihypertensive, hypocholesterolemic, and immunostimulating activities [26]. Butanol and aqueous fraction of anchovy sauce exhibited strong antioxidative activity, the active component of which is a methionine derivative [27].

The addition of anchovy, shrimp, and launce jeotgals has been shown to increase the ACE-inhibiting activities in kimchi samples [28], and some types of jeotgal reduced the nitrite content in kimchi during fermentation [29]. A microbial strain of potential fibrinolytic enzyme was isolated from shrimp jeotgal and identified as *Bacillus sp.* [30]. Fish sauces are also reported to exhibit anticancer activities. When a human lymphoma cell line (U937) was exposed to a hydrophobic peptide fraction of an anchovy sauce, a sub-G_1 peak representing the apoptotic cell population was found by the cell cycle analysis. The apoptosis in the peptide-treated U937 cell was also confirmed by internucleosomal DNA fragmentation and increased caspase-3 and -8 activities [31].

6.2.4 GINSENG

Korean ginseng (*Panax ginseng* C.A. Meyer) has been used extensively as an important ingredient of traditional medicines for thousands of years in Asia. The scientific name *Panax ginseng* is derived from "pan (all)" combined with "axos (medicine)," and hence means "cure-all." Many in vitro and in vivo studies and clinical trials have suggested that ginseng has various physiological properties, including antidiabetic, antistress, opioid, antioxidant, immunostimulating, antithrombotic, and anticancer activities [32]. The anticancer activity of ginseng is attributed to gensenosides and polyacetylene compounds. Ginsenoside Rh_2 inhibits the growth of human ovarian cancer cells [33] and B16 melanoma cells, and induces the apoptosis of human hepatoma SK-Hep-1 cells [34–36]. Ginsenoside Rh_1 and Rh_2 induce reverse transformation of cancer cells [37].

The *Panax ginseng* extracts of certain organic solvents, such as ethyl acetate and petroleum ether have also been reported to exert antiproliferative effects on various cancer cell lines, including human gastric, colon, and uterus carcinoma, human erythroleukemia, murine melanoma, and mouse fibroblast-derived tumor cells [38–40]. Oligopeptides from ginseng with physiological activity also have been reported. A peptide with a sequence of Gly-Arg-Glu-Val was purified from an alkaline fraction of ginseng extracts, which was found to stimulate the proliferation of baby hamster kidney-21 cells [41]. A protein fraction from ginseng also inhibited protein kinase C, which is activated by tumor-promoting phorbol 12-myristate 13-acetate in vitro [42]. Anti-cancer peptides isolated from ginseng were analyzed as either Trp-Trp or Trp-Trp-Trp sequences. These peptides appeared as to arrest the G_0/G_1 phase of DNA cell cycle in a mouse lymphoma cell line [43].

Korean red ginseng (the steamed and dried root of ginseng) has been reported to have physiological effects such as antioxidant, immunostimulating, antithrombotic, and anticancer activities [36,40,44]. Polyacetylene compounds such as panaxynol, panaxydol, and panaxytriol reportedly inhibit cancer cell growth in vitro [38,39,45]. Panaxytriol was very effective for suppressing the growth of B16 melanoma transplanted into mice [46], inhibiting the cellular respiration and energy balance of a human breast carcinoma cell line [47], and enhancing the cytotoxicity of mitocycin C in a human gastric adenocarcinoma cell line [48]. Other studies have focused on the inhibitory effects of panaxytriol on tumorigenesis and tumor growth in vivo and in vitro [46,49–51]. Panaxytriol showed both significant cytotoxicity and inhibition of DNA synthesis of a mouse lymphoma cell line by inducing cell cycle arrest at the G_2/M phase [45].

6.3 REGULATION OF MODERN FUNCTIONAL FOODS IN KOREA

So called "health foods" were first regulated in Korea in 1973: they were categorized as "nutrient supplement food" and regulated under the Food Hygiene Act. In 1989, 21 categories of "health supplement food" were added, and the categories were expanded to 24 in 2002 when a new regulation system was implemented. A new regulatory framework on functional food in Korea was promulgated by the Health Functional Food Act in 2002, and subsequent ordinances and regulations were enforced in 2003.

According to the new law, health functional food (HFF) refers to processed and manufactured goods in the form of tablets, capsules, powder, granules, liquid, and pills that help enhance and preserve the health of the human body using nutritional or functional ingredients. These are defined by the commissioner of the Korea Food and Drug Administration (KFDA), who provides the standards and specifications for the manufacturing, usage, and maintenance of HFF products. For products that are not specified by the commissioner of the KFDA, a business operator can provide the standards, specifications, safety and other relevant data to the designated inspection agency to obtain approval for the production of HFF.

The new code of HFF contains 32 categories of food sources that include health food supplements, nutrient supplements, and ginseng products: (1) nutrient supplements including protein, vitamin, mineral, amino acid, fatty acid, and dietary fiber (2) eel oil products (3) EPA and/or DHA products (4) royal jelly products (5) yeast products (6) pollen products (7) squalene products (8) enzyme products (9) LAB products (10) chlorella products (11) spirulina products (12) γ-linolenic acid products (13) germ oil products (14) germ products (15) lecithin products (16) octacosanol products (17) alcoxyglycerol products (18) grape seed oil products (19) fermented plant extract products (20) mucopolysaccharide /protein products (21) chlorophyll products (22) mushroom products (23) aloe products (24) plum extract products (25) snapping turtle products (26) β-carotene products (27) chitosan products (28) chito-oligosaccharide products (29) glucosamine products (30) propolis extract products (31) ginseng products and (32) red ginseng products. Health functional food is different from regular food in that its function can be expressed in physiological terms. If a company wishes to advertise or place labels on its HFF products, they must be reviewed based on the standards, methods, and procedures established by the commissioner of the KFDA. The commissioner can also assign the reviewing process to a suitable organization.

ACKNOWLEDGMENTS

This work was supported by the research grant from the Korea Institute of Science and Technology Evaluation and Planning (KISTEP), Ministry of Science and Technology, for functional food research and development.

REFERENCES

1. http://www.seoulnow.net/SITE/data/html (accessed on April 2004).
2. Cheigh, H. S. and Park, K. Y., Biochemical, microbiological, and nutritional aspects of kimchi, *Crit. Rev. Food Sci. Nutr.*, 34, 175–203, 1994.
3. Kwon, H. and Kim, Y. K. L., Korean fermented foods: Kimchi and Doenjang, In *Handbook of Fermented Functional Foods*, Farnworth, D. R. Ed., CRC Press, Boca Raton, FL, pp. 287–304, 2003.
4. Cho, E. J., Rhee, S. H., Lee, S. M., and Park, K. Y., In vitro antimutagenic and anticancer effects of kimchi fractions, *J. Korean Assoc. Cancer Prev.*, 2, 113–121, 1997.
5. Cho, E. J., Rhee, S. H., and Park, K. Y., In vitro anticarcinogenic effect of kimchi fractions, *J. Korean Assoc. Cancer Prev.*, 4, 79–85, 1999.
6. Choi, M. W., Kim, K. H., and Park, K. Y., Effects of kimchi extracts on the growth of sarcoma-180 cells and phagocytic activity of mice, *J. Korean Soc. Food Sci. Nutr.*, 26, 254–260, 1997.
7. Surh, Y. J., Cancer chemoprevention with dietary phytochemicals, *Nature Rev. Cancer*, 3, 768–780, 2003.
8. Surh, Y. J., Lee, C. J., Park, K. K., Mayne, T., Liem, A., and Miller, J. A., Chemopreventive effects of capsaicin and diallyl sulfide against mutagenesis or tumorigenesis by vinyl carbamate and *N*-nitrosodiethylamine, *Carcinogenesis*, 16, 2647–2741, 1995.
9. Surh, Y. J., Na, H. K., and Lee, H. J., Chemopreventive effects of selected spice ingredients, In *Phytopharmaceuticals in Cancer Chemoprevention*, Bagchi, D. Ed., CRC Press, Boca Raton, FL, pp. 575–598, 2004.
10. Salminen, S., Ouwehand, A., Benno, Y., and Lee, Y. K., Probiotics: how should they be defined?, *Trends Food Sci. Technol.*, 10, 107–110, 1999.
11. Kim, J. Y., Woo, H. J., Kim, Y. S., and Lee, H. J., Screening for antiproliferative effects of cellular components from lactic acid bacteria against human cancer cell lines, *Biotechnol. Lett.*, 24, 1431–1436, 2002.
12. Kim, J. Y., Woo, H. J., Kim, K. H., Kim, E. R., Jung, H. K., Juhn, H. N., and Lee, H. J., Antitumor activity of *Lactobacillus plantarum* cytoplasm on teratocarcinoma-bearing mice, *J. Microbiol. Biotechnol.*, 12, 998–1001, 2002.
13. Snyder, H. E. and Kwon, T. W., *Soybean Utilization*, Van Nostrand Reinhold, New York, p. 346, 1987.
14. Hong, S. S., Chung, K. S., Yoon, K. D., and Cho, T. J., Antimutagenic effects of solvent extracts of Korean fermented soybean products, *Foods Biotechnol.*, 5, 263–267, 1996.
15. Park, K. Y., Son, M. H., Moon, S. H., and Kim, K. H., Cancer preventive effects of doenjang in vitro and in vivo: antimutagenic and in vivo antitumor effects of doenjang, *J. Korean Assoc. Cancer Prev.*, 4, 68–78, 1999.
16. Son, M. H., Moon, S. H., Choi, J. W., and Park, K. Y., Cancer preventive effects of doenjang in vitro and in vivo: effect of doenjang extracts on the changes of serum and liver enzyme activities in sarcoma-180 transplancted mice, *J. Korean Assoc. Cancer Prev.*, 4, 143–154, 1999.
17. Kim, S. E., Pai, T., and Lee, H. J., Cytotoxic effects of the peptides from traditional Korean soy sauce on tumor cell lines, *Food Sci. Biotechnol.*, 7, 75–79, 1998.
18. Kim, J. Y., Woo, H. J., Ahn, C. W., Nam, H. S., Shin, Z. I., and Lee, H. J., Cytotoxic effects of peptides fractionated from bromelain hydrolyzates of soybean protein, *Food Sci. Biotechnol.*, 8, 333–337, 1999.

19. Kim, S. E., Kim, H. H., Kim, J. Y., Kang, Y. I., Woo, H. J., and Lee, H. J., Anticancer activity of hydrophobic peptides from soy proteins, *BioFactors*, 12, 151–155, 2000.

20. Lee, H. J., Lee, K. W., Kim, K. H., Kim, H. K., and Lee, H. J., Antitumor activity of peptide fraction from traditional Korean soy sauce, *J. Microbiol. Biotechnol.*, 14, 628–630, 2004.

21. Okamoto, A., Hanagata, H., Matsumoto, E., Kawamura, Y., Koizumi, Y., and Yanagida, Y., Angiotensin-I converting enzyme inhibitory activities of various fermented foods, *Biosci. Biotechmol. Biochem.*, 59, 1147–1149, 1995.

22. Shin, Z. I., Ahn, C. W., Nam, H. S., Lee, H. J., Lee, H. J., and Moon, T. H., Fractionation of angiotensin converting enzyme (ACE) inhibitory peptides from soybean paste, *Korean J. Food Sci. Technol.*, 27, 230–234, 1995.

23. Kim, S. H., Lee, Y. J., and Kwon, D. Y., Isolation of angiotensin converting enzyme inhibitor from Doenjang, *Korean J. Food Sci. Technol.*, 31, 848–854, 1999.

24. Shin, Z. I., Yu, R., Park, S. A., Chung, D. K., Ahn, C. W., Nam, H. S., Kim, K. S., and Lee, H. J., His-His-Leu, an angiotensin I converting enzyme inhibitory peptide derived from Korean soybean paste, exerts antihypertensive activity in vivo, *J. Agric. Food Chem.*, 49, 3004–3009, 2001.

25. Shon, D. H., Lee, K. A., Kim, S. H., Ahn, C. W., Nam, H. S., Lee, H. J., and Shin, Z. I., Screening of antithrombotic peptides from soybean paste by the microplate method, *Korean J. Food Sci. Technol.*, 28, 684–688, 1996.

26. Mills, E. N. C., Alcocer, M. J. C., and Morgan, M. R. A., Biochemical interactions of food-derived peptides, *Trends Food Sci. Technol.*, 3, 64–68, 1990.

27. Park, J. O., Yoon, M. S., Cho, E. J., Kim, H. S., and Ryu, B. H., Antioxidant effects of fermented anchovy, *Korean J. Food Sci. Technol.*, 31, 1378–1385, 1999.

28. Park, D. C., Park, J. H., Gu, Y. S., Han, J. H., Byun, D. S., Kim, E. M., Kim, Y. M., and Kim, S. B., Effects of salted-fermented fish products and their alternatives on angiotensin converting enzyme inhibitory activity of kimchi during fermentation, *Korean J. Food Sci. Technol.*, 32, 920–927, 2000.

29. Park, D. C., Park, J. H., Gu, Y. S., Han, J. H., Byun, D. S., Kim, E. M., Kim, Y. M., and Kim, S. B., Effects of salted-fermented fish products and their alternatives on nitrite scavenging activity of kimchi during fermentation, *Korean J. Food Sci. Technol.*, 32, 942–948, 2000.

30. Jang, S. A., Kim, M. H., Lee, M. S., Lee, M. J., Jhee, O. H., Oh, T. K., and Sohn, C. B., Isolation and identification of fibrinolytic enzyme producing strain from shrimp jeotgal, a tiny salted shrimp, and medium optimization for enzyme production, *Korean J. Food Sci. Technol.*, 31, 1648–1653, 1999.

31. Lee, Y. G., Kim, J. Y., Lee, K., Kim, K. H., and Lee, H. J., Peptides from anchovy sauce induce apoptosis in human lymphoma cell (U937) through the increase of caspase-3 and -8 activities, *Ann. NY. Acad. Sci.*, 1010, 399–404, 2003.

32. Kim, H., Lee, Y. H., and Kim, S. I., A possible mechanism of polyacetetylene: membrane toxicity, *Korean J. Toxicol.*, 4, 95–105, 1998.

33. Kikuchi, Y., Sasa, H., Kita, T., Hirata, J., Tode, T., and Nagata, I., Inhibition of human ovarian cancer cell proliferation in vitro by ginsenoside Rh2 and adjuvant effects to cisplatin in vivo, *Anticancer Drugs*, 2, 63–67, 1990.

34. Ota, T., Maeda, M., Odashima, S., Ninomiya-Tsuji, J., and Tatsuka, M., G_1 phase-specific suppression of the Cdk2 activity by ginsenoside Rh2 in cultured murine cells, *Life Sci.*, 60, PL39–PL44, 1997.

35. Odashima, S., Ohta, T., Kohno, H., Matsuda, T., Kitagawa, I., Abe, H., and Arichi, S., Control of phenotypic expression of cultured B16 melanoma cells by plant glycosides, *Cancer Res.*, 45, 2781–2784, 1985.

36. Park, J. A., Lee, K. Y., Oh, Y. J., Kim, K. W., and Lee, S. K., Activation of caspase-3 protease via a Bcl-2-insensitive pathway during the process of ginsenoside Rh2-induced apoptosis, *Cancer Lett.*, 121, 73–81, 1997.

37. Odashima, S., Nakayabu, U., Honjo, N., Abe, H., and Arichi, S., Induction of phenotypic reverse transformation by ginsenosides in cultured *Morris* hepatoma cells, *Eur. J. Cancer*, 15, 885–892, 1979.

38. Matsunaga, H., Katano, M., Yamamoto, H., Fujito, H., Mori, M., and Takata, K., Cytotoxic activity of polyacetylene compounds in *Panax ginseng* C.A. Meyer, *Chem. Pharm. Bull. (Tokyo)*, 38, 3480–3482, 1990.

39. Lee, S. H. and Hwang, W. I., Inhibitory effect of petroleum ther extract of *Panax ginseng* root against growth of human cancer cells, *Korean J. Ginseng Sci.*, 10, 141–150, 1986.

40. Sohn, J., Lee, C. H., Chung, D. J., Park, S. H., Kim, I., and Hwang, W. I., Effect of petroleum ether extract of *Panax ginseng* roots on proliferation and cell cycle progression of human renal cell carcinoma cells, *Exp. Mol. Med.*, 30, 47–51, 1998.

41. Yagi, A., Akita, K., Ueda, T., Okamura, N., and Itoh, H., Effect of a peptide from *Panax ginseng* on the proliferation of baby hamster kidney-21 cells, *Planta Med.*, 60, 171–174, 1994.

42. Park, H. J., No, Y. H., Rhee, M. H., Park, K. M., and Park, K. H., Effects of protein fractions and ginsenosides from *Panax ginseng* C.A. Meyer on substrate phosphorylation by a catalytic fragment of protein kinase, *Korean Biochem. J.*, 27, 280–283, 1994.

43. Kim, S. H., Kim, J. Y., Park, S. W., Lee, K. W., Kim, K. H., and Lee, H. J., Isolation and purification of anticancer peptides from Korean ginseng, *Food Sci. Biotechnol.*, 12, 79–82, 2003.

44. Kim, Y. S., Jin, S. H., Kim, S. I., and Hahn, D. R., Studies on the mechanism of cytotoxicities of polyacetylenes against L1210 cells, *Arch. Pharmacol. Res.*, 12, 207–213, 1989.

45. Kim, J. Y., Lee, K. W., Kim, S. H., Wee, J. J., Kim, Y. S., and Lee, H. J., Inhibitory effect of tumor cell proliferation and induction of G2/M cell cycle arrest by panaxytriol, *Planta Med.*, 68, 119–122, 2002.

46. Katano, M., Yamamoto, H., Matsunaga, H., Mori, M., Takara, K., and Nakamura, M., Cell growth inhibitory substance isolated from *Panax ginseng* root: panaxytriol, *Jpn J. Cancer Chemother.*, 17, 1045–1049, 1990.

47. Matsunaga, H., Saita, T., Nagumo, F., Mori, M., and Katano, M., A possible mechanism for the cytotoxicity of a polyacetyleneic alcohol, panaxytriol: inhibition of mitochondrial respiration, *Cancer Chemother. Pharmacol.*, 35, 291–296, 1995.

48. Matsunaga, H., Katano, M., Saita, T., Yamamoto, H., and Mori, M., Potentiation of cytotoxicity of mitocycin C by a polyacetylenic alcohol, panaxytriol, *Cancer Chemother. Pharmacol.*, 33, 291–297, 1994.

49. Katano, M., Matsunaga, J., and Yamamoto, H., A tumor inhibitory substance isolated from *Panax ginseng*, *J. Jpn Surg. Soc.*, 89, 971–977, 1988.

50. Nakano, Y., Matsunaga, H., Saita, T., Mori, M., Katano, M., and Okabe, H., Antiproliferative constituents in Umberliferae plants II. Screening for polyacetylenes in some Umbelliferae plants, and isolation of panaxynol and falcarindiol from the root of *Heracleum moellendorffii*, *Biol. Pharm. Bull.*, 21, 257–261, 1998.

51. Kim, J. S., Lim, Y. J., Im, K. S., Jung, J. H., Shim, C. J., Lee, C. O., Hong, J., and Lee, H., Cytotoxic polyacetylenes from the marine sponge *Petrosia* sp., *J. Nat. Prod.*, 62, 554–559, 1999.

7 Australia and New Zealand

Harjinder Singh, Linda Tapsell, Mike Gidley, Pramod Gopal, and Paul J. Moughan

CONTENTS

7.1 INTRODUCTION

Toward the end of last century a major lifestyle trend emerged in the Western world, emphasizing wellness and disease prevention in the human population rather than treatment therapies—a development not lost on approaches to food and nutrition [1]. This focus is expected to strengthen as the relatively affluent and well educated "baby boomers" of the 1950s and 1960s progressively reach middle age and begin to contemplate their retirement years. It has been calculated that in the U.S. alone, in the period from 1996 through to 2006, a "baby boomer" (defined as someone born between 1946 and 1964) will turn 50 years of age every seven and a half seconds [2]. Such individuals are acutely aware of health and lifestyle issues. A further significant driver of this trend is the escalating cost of traditional disease treatment. Allied to this, there has been a growing realisation of the pivotal role of nutrition in disease prevention and the maintenance of human health [3]. Epidemiological studies have exposed statistical correlations between the intake of certain dietary constituents and the development or prevention of various non-communicable diseases. In many cases, controlled human intervention studies have supported the epidemiological observations and a considerable body of knowledge has developed around the nexus of nutrition, health, and longevity. It has been postulated (the Barker hypothesis), and considerable supportive evidence has been amassed, that a propensity to develop certain diseases in later life may be related to maternal nutrition and subsequently modified by the nutritional habits of the individual [4,5].

The food industry has been quick to capitalise upon the changes in societal attitudes and today pays far more attention to nutrient contents and bioavailability than was the case in the past. This growing emphasis on lifestyle has also influenced the emergence and vigorous marketing of a completely new category of food, the so-called "functional foods" and "nutraceuticals." Functional foods are foods that, when consumed as part of a normal diet elicit beneficial effects on target functions in the body beyond nutritional effects in a way that is relevant to health and well-being and, or the reduction of disease [6]. Certain foods, when consumed on a sustained basis, may have quite subtle longer-term effects on aspects of physiological function, whereas other foods and natural food extracts may have acute pharmacological effects. Moreover, as the human genome project gains momentum, the very genetic basis of the so-called lifestyle diseases is being unravelled and described. This will soon allow the early identification (pre-symptomatic) of individuals prone to developing certain diseases, thus heralding specific dietary and lifestyle prescriptions. Functional foods and their place in a balanced diet are going to become increasingly important. The balancing of different food types and the role of functional foods and nutraceuticals has long been central to Eastern culture, but in the Western world has only been highlighted recently. In the West the functional foods revolution began in earnest in the early 1980s. The health-related benefits of materials such as plant fibre, fish oil, calcium, and probiotics attained respectability following publication of clinical studies in the medical literature and with physicians beginning to publicly promote their use. Examples are diverse, including the use of garlic (allicin) to reduce atherosclerosis, cranberry juice to prevent urinary tract infections, the role of calcium in treating osteoporosis, phytoestrogens, lycopene, antioxidants, and many others [7]. An example of one of the earlier and now well established functional foods is oat bran, whose soluble oat fibre acts in the alimentary canal to bring about a reduction in blood cholesterol. For hypercholesterolaemic individuals, the inclusion of oat bran in the diet (e.g., oat bran enriched cereals) can be used as part of an overall strategy to lower blood cholesterol, without resorting to prescription drugs. A further example of a cholesterol-lowering functional food, and one that has been successful commercially in Australasia, is that of phytosterol-enriched margarines and spreads. A recent study in Australia has demonstrated their utility in the clinical practice context [7]. There are many other examples of functional foods and nutraceuticals and many more products can be expected to enter the market as product development and clinical testing increase apace.

The increasing appreciation of the diverse physiological roles of food constituents is also having a profound influence on the contemporary view of nutritional science itself and of the very definition of "a nutrient." In the past, for example, volatile fatty acids were viewed as products of the fermentative breakdown of fibre, acting as an energy source for the host, whereas now their role in the development and regulation of gut function is being emphasised [8]. Proteins were once viewed as simply supplying amino acids for body protein synthesis, whereas now the distinct and diverse physiological effects of dietary peptides are being discovered and documented with an increasing frequency [9]. Fatty acids, far from acting solely as a source of energy or as body energy stores, are now known to profoundly affect red blood cell membrane composition, with consequent physiological effects [10].

7.2   THE DEVELOPMENT OF FUNCTIONAL FOODS
IN AUSTRALIA AND NEW ZEALAND

The production of high quality foods is central to the Australian and New Zealand economies, and both countries have well developed science infrastructures particularly relating to primary production and food processing. Not surprisingly, there is a strong interest in the development of functional foods. Key issues for the development of functional foods in Australia and New Zealand today concern their definition, how they are linked to food innovation, public health and safety issues, the substantiation of claims and communication of their value to consumers [11]. From a government perspective, this has implications for management of public health and safety aspects, consumer information and facilities for food innovation.

The rapidly ageing population in both New Zealand and Australia is also driving the demand for functional foods in these countries. In New Zealand, current statistics predict that the population aged 65 and over will grow by 100,000 within this decade, reaching 552,000 by the year 2011. After 2011, this population is expected to increase more rapidly, growing by approximately 200,000 between the years 2011 and 2021. By 2051, an estimated 25% of New Zealand's population will be over 65. A similar population-ageing trend is observed in Australia, due to increased life expectancies and low birth rates. Growth rate percentage predictions for the years 2000–2020 indicate Australia will have significant growth in its population 65 or older—higher than many Western European countries, the U.S., and Japan, but slightly lower than Canada. By 2050, population experts estimate, the percentage of Australia's population aged 65 and over will reach 25% [12].

Issues for the regulatory management of functional foods entering the market place revolve around protection of public health and safety, and ensuring that information given to consumers is meaningful and not false or misleading. Food Standards Australia New Zealand (FSANZ) is the statutory authority responsible for development of food standards to address these issues [13].

Food Standards Australia New Zealand's approach to the management of functional foods focuses on the key elements that characterise functional foods and utilises a "horizontal" approach across all food standards. The key characteristics that have been identified include: novel foods, high levels of fortification, inclusion of non-culinary herbs and other bioactive substances, and health and related claims. Novel foods have been regulated since 2001, and work is in progress on a number of other areas. However, non-culinary herbs and certain bioactive substances are yet to be formally addressed.

Underlying this approach is a regulatory paradigm that considers the history of safe use of a food or its components in the domestic food supply. Foods with a history of safe consumption may enter the food supply on the basis of minimal legislative requirements around safety and suitability. Foods with a more limited history of safe use require pre-market approval by FSANZ, which is determined through a science-based risk analysis processes.

The regulatory systems for label and advertising claims are currently under development. Nutrition, health, and related claims are seen to include a spectrum from nutrient content through to links between nutrients, or the food, and health outcomes. Claims linking foods to prevention of or curing diseases (i.e., therapeutic claims) are definitively not permitted on foods. For regulatory purposes, health, and related claims are grouped into "general level" and "high level" claims. This model overlays an increasing degree of regulation, which in turn, is related to the increasing degree of importance of the claim in relation to health.

Under consideration in the development of a policy framework, the term "health claim" covers assertions along the following lines:

- The food is a slimming food or has intrinsic weight-reducing properties;
- A claim for therapeutic or prophylactic action or a claim described by words of similar import;
- The word "health" or any word or words of similar import as a part of or in conjunction with the name of the food;
- Any word, statement, claim, express or implied, or design that directly or by implication could be interpreted as advice of a medical nature from any person; or
- Name of or a reference to any disease or physiological condition.

A new food standard, considered under proposal P293 currently under review, will allow health claims at low and high levels, both requiring substantiation, but with only high level claims needing pre-approval. Part of the consultation process has involved consideration of the early adoption of claims approved in other countries. Workshops in this area have suggested that there may be some preference for claims relating calcium to the prevention of osteoporosis, low saturated fat and cholesterol levels with reduced risk of heat disease, the benefits of fibre, fruit, and vegetables as protective from cancer, and lower energy values for obesity prevention [14].

In summary, interest in functional foods is multi-sectoral, at both the government and non-government level, and includes industry, research, and development sectors, public health interests and regulators. Key considerations for functional foods focus on definition, innovation by industry, information for consumers and management of public health and safety issues, with emphasis on the development and utilisation of a rigorous scientific framework.

7.3   SPECIALISED RESEARCH CENTRES

In both Australia and New Zealand there is substantial strength in a range of areas to support the science of functional food development. This is conducted on the basis of studies of food components, whole foods and whole diets. The scientific framework for developing and establishing the evidence for the effects is comprehensive, based on key questions and acknowledging the range of methodologies required to provide a suitable depth of evidence of effects for consumers.

As may be expected in a burgeoning area of science, specialist research centres focusing on functional foods have been established in both Australia and New Zealand. The Australian National Centre of Excellence in Functional Foods (NCEFF) was established in 2003 to integrate the capabilities of four of the largest food and nutrition research organisations in the country (CSIRO—Health Sciences and Nutrition, Food Science Australia, Department of Primary Industries (DPI)-Victoria, and the Australian Research Council (ARC) Key Centre for Smart foods [15]. The mix of capabilities is organised under four clusters of activity: Administration and Communication, Market Intelligence and Innovation Management, Substantiation Research and Development and Regulatory Affairs. The programmes developing in the Centre are working toward support along all stages of the innovation cycle. The Centre uses a working definition of functional foods with reference to the three main influencers of science and technology, regulation, and market and aims to consolidate the R&D effort across Australia.

The commercial successes of functional foods have been predominantly in the processed food sector, where opportunities to add value to current processed food categories has been a strong driver. This trend has also led to a reawakened interest in the potential health-promoting properties of "natural" foods, such as fruits, vegetables, nuts, grains, fish, meat, and milk. In many cases there is reasonable epidemiological evidence for likely health benefits, but molecular mechanisms of action for such complex foods are difficult to disentangle. In order to better understand such "naturally-functional" foods, and to bring the required breadth of scientific skills to bear, a new joint initiative between the University of Queensland and the Queensland Department of Primary Industries and Fisheries has resulted in the creation of a Centre for Nutrition and Food Sciences that brings together expertise in biological, chemical, physical, engineering, and health sciences with primary production and food technology skills. Such a wide scope maximises opportunities for both public good and commercial benefit, but presents challenges in working across many organisational groupings. The Queensland initiatives in food research, however, have provided for a useful link with the National Centre of Excellence in Functional Foods.

A parallel development in New Zealand to that of Australia is the recent establishment of the Riddet Centre, devoted to innovation in food and biologicals. The Riddet Centre [16] is a collaborative initiative between the University of Auckland, Massey University and the University of Otago and brings together leading scientists in food technology and human nutrition. The Centre acts as a focal point for the New Zealand food and health industries, and has a substantial research programme in science underpinning the development of functional foods.

A new National Centre of Excellence in Nutrigenomics has recently been established in New Zealand. The centre brings together a large multi-disciplinary team from three Crown Research Institutes and the University of Auckland. The major aim of this Centre is to determine how foods and food components affect health at a molecular genetic level by using nutritional genomic methods. These include gene expressions (microassays), genotyping in humans, metabolomics, and proteomics. An initial target is gut health, focussing on Crohn's disease.

7.4   SELECTED EXAMPLES OF FUNCTIONAL FOOD DEVELOPMENT IN AUSTRALIA AND NEW ZEALAND

As is the case in other countries, dairy products constitute a major market category of functional foods in Australia and New Zealand. These products include fat-reduced modified milk, the replacement of saturated fat with other types of fats (such as omega-3 fatty acids) in products including infant formulas, milk, and yoghurt. The use of functional dairy ingredients, such as whey protein isolates, casein, and whey protein hydrolysates in numerous foods, especially in the sports nutrition market, has increased markedly in recent years.

One of the most active areas of development has been that relating to the fortification of dairy products, particularly liquid milk, milk powders and yoghurt. This has been mainly in the area of fortification with calcium, iron, probiotic cultures, vitamins, omega-3 fatty acids and soluble fibres. For example, Fonterra Cooperative Ltd New Zealand's largest food company, has been particularly successful with its calcium-fortified milk powders sold under the brand name "Anlene." Anlene brand was launched in 1991 in Malaysia, Taiwan, Hong Kong, and Singapore and is now available in 23 countries. "Anlene" has three variants, each formulated to meet the needs of different consumers. Anlene is available in various forms such as powder, liquid (ultra high temperature treated and Pasteurised) and yogurt, and offers different variants to suit different individual needs. This includes Anlene for consumers between the ages of 19 and 50 years, and Anlene Gold for consumers 51 years old and above. There is also a low lactose format for lactose intolerance.

A unique approach of Fonterra in development of these products has been to use sound scientific evidence to support bone health claims. Well designed clinical trials using target population and final product, were carried out by reputed scientists and clinicians to obtain the evidence [17–19]. Bioavailability of calcium from fortified powders was also confirmed by studies in appropriate model systems [20]. Strength of clinical evidence has convinced regulatory authorities in Malaysia and several other South East Asian countries to allow a "clinically proven" claim for appropriate formulations sold under the Anlene brand.

From Australia, the Devondale brand of reduced fat dairy spread enriched with calcium was developed with Murray Goulbourn (with MG Nutritionals NatraCal milk minerals), and Dairy Farmers' created Farmers Best milk, replacing saturated fat with monounsaturated and omega-3 fatty acids from vegetable oils.

Lactic acid bacteria have long been a focus of research in New Zealand due to their importance to the dairy industry. The research into the health-promoting potential of lactic acid bacteria did not receive much attention in New Zealand until the last decade. A formal programme in probiotic research was established at what was then the New Zealand Dairy Research Institute (NZDRI) in late 1995. The main objective of the probiotic programme was to develop credible probiotic strains based on the highest quality science. The approach was multi-disciplinary and involved collaboration between New Zealand Dairy Research Institute and Universities and Research Institutions outside New Zealand. This resulted not only in the development of commercial probiotic strains supported by good science, but also gave impetus to further research in this ever-developing area. Strain *Bif. lactis* HN019 was trade marked as DR10™ and strain *L. rhamnosus* HN001 as DR20™.

Most of the research in the past five to six years on these strains has been focused on demonstrating their safety and efficacy. The main focus of health benefit from probiotic strains in New Zealand has been immune system enhancement. A number of clinical trials have been carried out to demonstrate the impact of the consumption of *Bif. lactis* HN019 and/or *L. rhamnosus* HN001 on various indices of the immune system [21,22].

These strains have been successfully commercialised in through two channels. Firstly New Zealand Milk has used these strains in its own consumer products such as a growing up milk powder (GUMP) sold under Fernleaf brand (Fernleaf "Defens"). This product was introduced in Taiwan in 1999. A cheddar cheese containing DR20 was launched in Australia under Mainland brand (Mainland "Inner Balance") in 2001. A second channel of commercialisation has been through third party licensing deal. Danisco, a Danish Ingredient Company, has developed a brand HOWARU, and these strains have been successfully commercialised in a number of countries across the world as HOWARU "Rhamnosus" and HOWARU "Bifidus."

There are several genome projects ongoing in New Zealand involving probiotic bacteria. Investigation of the gene content of probiotic strains holds great promise for both the discovery of important genes and understanding the mechanisms through which probiotic bacteria elicit beneficial activities. At Fonterra Research Centre (formerly NZDRI), the complete gene sequence of *L. rhamnosus* HN001 has been determined. The four-step genomic strategy includes (i) acquisition of gene sequences, (ii) identification and selection of relevant genes (iii) validation of selected genes and (iv) testing of biological relevance. This approach will lead to an understanding of the correlation of gene and gene networks to phenotypic behaviours, and analysis of products of gene expression will be critically important in unravelling the functional properties and behaviour of probiotic strains. Australia has also made functional foods marketing inroads with pro- and pre-biotics. Yakult introduced its yoghurt drink for gut health seven years ago in Australia, and as of 2002, 2% of Australia's population reports drinking the product every day.

Several Australian and New Zealand dairy companies are manufacturing functional dairy ingredients, such as lactoferrin, colostrum products, lactoperoxidase, glycomacropeptide, immunoglobulins, protein hydrolysates and phospholipid fractions. Recaldent™, derived from casein and patented by the School of Dental Science at the University of Melbourne, Australia, is used in chewing gum to increase resistance to tooth decay.

There have been a number of recent developments of functional food products and marketing campaigns in the cereals field in Australia. The concept of the Glycemic Index enjoys a high level of awareness and acceptance by the general public, reinforced by labelling information and advertising messages that emphasise "long-lasting energy" or satiety as a benefit, rather than insulin and blood glucose management. The fact that there are probable health benefits underlying a diet that emphasises low Glycemic Index foods appears to be a secondary, but reinforcing, benefit at the consumer level. Australia has also begun to develop and market "resistant" starches as ingredients for breads, noodles, cereals, and snack foods. Resistant starch escapes digestion in the upper gastrointestinal tract, and so acts more like dietary fibre than regular starches, In addition, fermentation of resistant starch in the colon delivers high levels of butyrate which is considered important for colonic health. It is encouraging that the concept of "healthy" starch-containing has achieved a level of acceptance, despite the recent publicity surrounding very low carbohydrate diets. Opportunities with resistant starch have been recognised [23] and commercialised through ingredients such as GEMSTAR R70 produced by Manildra and Hi Maize from Penford Australia, used in products in such as breads, breakfast cereals, bars, and liquid breakfast products. Successful product development based on Glycemic Index and Resistant Starch concepts has been made possible by internationally-recognised research efforts in these two areas within Australia [24,25].

Research from the Crop & Food Research and Industrial Research Ltd. In New Zealand has given rise to a product called Glucagel™, which is an ingredient derived from barley. This is a gelling form of barley β-glucan (cereal soluble fibre) with interesting nutritional and fat-imitating properties. Research has also indicated that it reduces blood sugars, (important for control of diabetes) as well as blood cholesterol [26]. A further trend in the cereals sector in Australia is the fortification of bread. Over the last few years, the amount of shelf space afforded to fortified breads (compared to non-fortified) has increased significantly, despite the fact that they ask for a 20%–30% price premium. A wide range of fortifications are available including folate, calcium, omega-3 fatty acids, iron, and fibre. The major advantages of bread as a carrier for functional ingredients are that it is usually eaten on a daily basis, and that it has a solid form and short shelf life, thereby easing formulation challenges. A disadvantage of current products is that they are largely based on processed white bread which has a high Glycemic Index and negligible Resistant Starch. Whilst wholegrain breads would be a better nutritional choice as a product vehicle, the fortification of white bread exemplifies a current emphasis on making ingredient-based food "healthier," rather than utilising the natural benefits of molecular and structural complexity in less-refined ingredients.

Other Australian functional foods successes are products designed to lower cholesterol, including spreads enriched with phytosterols. However, FSANZ requires that phytosterol margarines have information labels stating that they are not intended to help cholesterol management in infants, pregnant or lactating women, or children, except under medical supervision. Their utility in assisting adults with hypercholesterolemia, however has been demonstrated [27].

7.5 CONCLUSIONS

Australia and New Zealand has served as a test market for many functional food categories. Both with the same recognisable health issues as other Westernised countries, yet set apart geographically, Australia and New Zealand constitute an ideal learning market. While the manufactured functional foods market is quite innovative in Australia and New Zealand, whole food promoted for its health benefits beyond nutrition is an emerging category.

The development of functional foods provides real opportunities for the Australasian food industry. However, precisely defining these opportunities requires a working definition of functional foods that triangulates market trends with the science base and regulatory issues, where relevant. The heterogeneity of the food industry in Australia and New Zealand suggests that this will mean different things to different sectors, but the process will still apply. Commercialising science is the weakest step in the development process and experience shows that a long term vision and a focus on proven starting points works best [28]. The respective research centres established in Australia and New Zealand and that are dedicated to a study of functional foods are working towards this goal in setting up infrastructure and communication frameworks to produce benefits for the Australasian food industry. It is a complex task, but a first pass suggests that all the ingredients are there.

REFERENCES

1. Schneeman, B., Evolution of dietary guidelines, *J. Am. Diet. Assoc.*, 103 (suppl.), S5–S9, 2003.
2. Sloan, E. S., Food industry forecast: consumer trends to 2020 and beyond, *Food Tech.*, 52, 37, 1998.
3. WHO, *Diet, Nutrition and Prevention of Chronic Disease*, World Health Organisation, Geneva, 2003.
4. Barker, D. J. P., *Mothers, Babies, and Disease in Later Life*, BMJ Publishing Group, London, 1994.
5. Barker, D. J. P., Commentary: intrauterine nutrition may be important, *Br. Med. J.*, 318, 1471, 1999.
6. Diplock, A. T., Aggett, P. J., Ashwell, M., Bornet, F., Fern, E. B., and Roberfroid, M. B., Scientific concepts for functional foods in Europe. Consensus document, *Br. J. Nutr.*, 81, S1–S27, 1999.
7. Patch, C. S., Tapsell, L. C., and Williams, P. G., Plant sterol/stanol prescription is an effective treatment strategy for managing hypercholesterolemia in outpatient clinical practice, *J. Am. Diet Assoc.*, 105, 46–52, 2005.
8. Sakata, T., Stimulatory effect of short-chain fatty acids on epithelial cell proliferation in the rat intestine: a possible explanation for trophic effects of fermentable fibre, gut microbes and luminal trophic factors, *Br. J. Nutr.*, 58, 95, 1987.
9. Rutherford-Markwick, K. J. and Moughan, P. J., Bioactive peptide derived from food, *J. AOAC Int.*, 88, 955–966, 2005.
10. Storlien, L. H., Baur, L. A., Kriketos, A. D., Pan, D. A., Cooney, G. J., Jenkins, A. B., Calvert, G. D., and Campbell, L. V., Dietary fats and insulin action, *Daibetologia*, 39, 621–631, 1996.
11. Tapsell, L. C., Functional foods: definition and commercialisation, *Food Aus.*, 57, 384–386, 2005.
12. www.just-food.com/articlearchive.aspx
13. www.foodstandards.gov.au/newsroom/publications/annualreport/fsanzanimalreport20052006/index.cfm
14. www.nceff.com.au/pdf/Australia.pdf
15. www.nceff.com.au/about/index.html
16. http://riddetcentre.massey.ac.nz/research.html
17. Lau, E. M., Woo, J., Lam, V., and Hong, A., Milk supplementation of the diet of postmenopausal Chinese women on low calcium intake retards bone loss, *J. Bone Miner. Res.*, 16, 1704–1709, 2001.
18. Lau, E. M., Chan, L. H., and Woo, J., Milk supplementation prevents bone loss in postmenopausal Chinese women over 3 years, *Bone*, 4, 536–540, 2002.
19. Chee, W. S. S., Suriah, A. R., Chan, S. P., Zaitun, Y., and Chan, Y. M., The effect of milk supplementation on bone mineral density in postmenopausal Chinese women in Malasia, *Osteoporos Int.*, 14, 828–834, 2003.

20. Kruger, M., Gallaher, B., and Schollum, L., Bioavailability of calcium is equivalent from milk fortified with either calcium carbonate or milk calcium in growing male rats, *Nutr. Res.*, 23, 1229–1237, 2003.
21. Arunachalam, K., Gill, H. S., and Chandra, R. K., Enhancement of natural immune function by dietary consumption of *Bifidobacterium lactis* (HN019), *Eur. J. Clin. Nutr.*, 54, 263–267, 2000.
22. Cahiang, B. L., Sheih, Y. H., Wang, L. H., Liao, C. K., and Gill, H. S., Enhancing immunity by dietary consumption of probiotic lactic acid bacterium (*Bifidobacterium lactis* HN019): optimisation and definition of cellular responses, *Eur. J. Clin. Nutr.*, 54, 849–855, 2000.
23. Tapsell, L. C., Diet and metabolic syndrome: where does resistant starch fit in?, *J. AOAC Int.*, 87, 756–760, 2004.
24. Foster-Powell, K., Holt, S. H., and Brand-Miller, J. C., International table of glycemic index and glycemic load values, *Am. J. Clin. Nutr.*, 76, 5, 2002.
25. Topping, D. L. and Clifton, P. M., Short-chain fatty acids and human colonic function: roles of resistant starch and non-starch polysaccharides, *Physiol. Rev.*, 81, 1031, 2001.
26. www.glucagel.com/product.html
27. Patch, C. S., Tapsell, L. C., and Williams, P. G., Dietetics and functional foods, *Nutr. Diet*, 61, 22–29, 2004.
28. Mellentin, J., Functional ingredient strategy—commercialisation skills are the missing element, *New Nutr. Bus.*, 9, 24, 2004.

8 Prophylactic Phenolic Antioxidants in Functional Foods of Tropical Island States of the Mascarene Archipelago (Indian Ocean)

T. Bahorun, V. S. Neergheen, M. A. Soobrattee, V. A. Luximon-Ramma, and O. I. Aruoma

CONTENTS

8.1 INTRODUCTION

Plant resources have been emphasized worldwide as precious sources of raw materials. They not only provide the basic needs of life such as food, feed, fiber, fuel, and shelter, but they have always been a valuable source of prophylactic phytochemicals, flavors, and fragrances and other industrial products. A large number of studies provide convincing evidence of the beneficial role of plant foods and their nutraceuticals for the maintenance of health [1,2]. Epidemiological studies show a protective effect of fruits, vegetables, traditional plant preparations and beverages like teas against the risk of chronic diseases such as cancer, atherosclerosis, cardiac dysfunctions, diabetes, hypertension, and neurodegenerative disorders [3–6]. The benefits that plant-rich diets confer are believed to be ascribed to various antioxidants, especially carotenoids and antioxidant vitamins, including ascorbic acid and tocopherols. However, the antioxidant capacity of a particular fruit, vegetable or tea may originate from compounds other than β-carotene, vitamin C or vitamin E. The significance of phenolics such as catechins, phenylcarboxylic acids, phenylpropanoids, anthocyanins, and proanthocyanidins as dietary antioxidants in fruits, vegetables, medicinal plant extracts and teas is attracting considerable attention [7–10]. The antimutagenic, antibacterial, anti-inflammatory, antithrombotic, and vasodilatory actions of polyphenolics are well characterized [11] and accumulating chemical, biochemical, clinical, and epidemiological evidence supports the chemopreventive effects of phenolic antioxidants against oxidative stress-mediated disorders [12,13]. The pharmacological actions of phenolic antioxidants are strongly suggested to stem mainly from their free radical scavenging and metal chelating properties, as well as from their effects on cell signaling pathways and on gene expression [14–18].

The Mascarene Archipelago, consists of three islands: the Republic of Mauritius, Réunion Island (which is part of France) and Rodrigues Island (a dependency of Mauritius). All are located in the southwestern part of the Indian Ocean, between latitudes 10°S and 20°S and longitude 55°E and 65°E.

The area is enormously rich in plant wealth, with more than 955 native flowering plants, of which 72% are endemics. The island of Mauritius is endowed with 681 native species, of which 47% are endemic to Mauritius and a further 21% endemic to the Mascarenes [19] while in Réunion island 550 native flowering species exist, of which 30% endemic to the island and a further 25% endemic to the Mascarenes have been reported [20]. In Rodrigues Island, 133 native species comprised of 39% Rodriguan endemics and a further 15% endemic to the Mascarenes have so far been recorded [19]. It is also estimated that more than 2100 and 1675 species have been introduced in Réunion [21] and Mauritius [22], respectively. No estimates are currently available for Rodrigues, but the number of introduced species may reach several hundred, at least. In Mauritius more than 119 tropical fruit and 62 vegetable species are currently cultivated on a large scale or in backyard gardens—for local consumption, exports or for the emerging processing industries. The most common fruit and vegetable species commonly cultivated in the Mascarene islands or exploited specifically in the individual island states are shown in Box 8.1 and Box 8.2. Tea is a major crop of the Island of Mauritius with a production of 1482 tn. of manufactured tea mostly in the fermented form in 2004 [23] aimed for both local use and export.

These tropical island states, with a population nearing 2 million people, are also faced with a difficult health situation: several non-communicable diseases, such as cardiovascular diseases, diabetes, and cancers, usually associated with industrialized countries with a relatively fast life style, are steadily increasing. Data for the island of Mauritius alone indicate that cardiovascular diseases and cancer account for 50% and 10% of annual deaths, respectively, and that 20% of the Mauritian population suffer from diabetes [24]. These figures are of great concern and have triggered interest in the study of the Mauritian diet with particular emphasis on the

BOX 8.1   MAJOR TROPICAL FRUITS CULTIVATED AS COMMERCIAL SPECIES OR BACKYARD GARDEN SPECIES IN THE MASCARENE ISLANDS

Commercial Spps

Ananas comosus (Pineapple), *Annona hybrid* (Atemoya), *Annona squamosa* (Custard apple), *Artocarpus heterophyllus* (Jackfruit), *C. papaya* (Pawpaw), *Citrus aurantifolia* (Lime), *Citrus aurantium* var. *bergamia* (Bergamot), *Citrus aurantium* var. *bigaradia* (Seville orange), *Citrus hybrid* (Tangelo), *Citrus hystrix* (Combava), *Citrus limonia* (Lemon), *Citrus medica* (Citron), *Citrus grandis* (Pamplemousses), *Citrus mitis* (Calamondin), *Citrus deliciosa* (Tangerine), *Citrus paradisi* (Grapefruit), *Citrus reticulata* (Mandarin), *Citrus sinensis* (Orange), *Cocos nucifera* (Coconut), *Eriobotrya japonica* (Bibasse), *E. longana* (Longan), *Feijoa sellowiana* (Pineapple guava), *Ficus carica* (Fig), *Fragaria x ananassa* (Strawberry), *Hylocereus undatus* (Dragon fruit), *L. chinensis* (Litchi), *Malpighia glabra* (Acerola), *Malus sylvestris* (Apple), *M. indica* (Mango), *Manilkara zapota* (Sapodilla), *Morus alba* var *indica* (Indian mulberry), *Musa spp* (Banana), *P. edulis* f. *edulis* (Passion fruit purple), *P. edulis* f. *flavicarpa* (Passion fruit yellow), *Passiflora molissima* (Banana passion fruit), *Passiflora quadrangularis* (Grenadine carri), *Persea Americana* (Avocado pear), *Phyllanthus acidus* (Star gooseberry), *Prunus persica* (Peach), *Pyrus communis* (Pear), *P. guajava* (Guava), *P. cattleianum* (Strawberry guava red). *P. cattleianum* var.*lucidum* (Strawberry guava yellow), *Punica granatum* (Pomegranate), *Richardella campechiana* (Egg fruit), *Rubus rosifolius* (Framboise), *Spondias cytherea* (Hog's plum), *S. cumini* (Java Plum), *Syzygium malaccense* (Malay apple), *S. samarangense* (Java apple), *Tamarindus indica* (Tamarind), *Vitis vinifera* (Grape), *Zizyphus mauritiana* (Indian jujube), *Zizyphus jujuba* (Jujube)

Backyard Spps

Aberia gardneri (Ceylon gooseberry), *Achras zapota* (Sapodilla plum), *Aegle marmelos* (Bael), *Anacardium occidentale* (Cashew nut), *Annona cherimola* (Cherimoya), *Annona muricata* (Soursop), *Annona reticulata* (Bullock's heart), *Artocarpus altilis* (Breadfruit). *Averrhoa bilimbi* (Pickle fruit), *A. carambola* (Starfruit), *Bombax edulis* (Pistache malgache), *Carica cauliflora* (Mountain Pawpaw), *Carisa carandas* (Caranda), *Carya oliviformis* (Pecan nut), *Citrus aurantium* (Sour orange), *Citrus vangassay* (Vangasaille), *Clausena lansium* (Sapote de Chine), *Diospyros nigra* (Sapote Negro), *Diospyros discolor* (Mabolo), *Doryalis hebecarpa* (Groseille de ceylan), *Durio zibethinus* (Durian), *Elaeocarpus serratus* (Olive), *Eugenia brasilensis* (Brazilian cherry), *Eugenia cauliflora* (Jaboticaba), *Eugenia uniflora* (Surinam cherry), *Flacourtia ramontchi* (Governor's plum), *Fragaria vesca* (Fraisier des bois), *Garcinia mangostana* (Mangoustan), *Lucuma nervosa* (Egg fruit), *Mamea americana* (Mammey apple), *Mimusops coriacea* (Pomme jacot), *Mimusops elengi* (Spanish cherry), *Phoenix dactylifera* (Datepalm), *Phyllanthus emblica* (Aonla), *Prunus amygdalus* (Almond), *Prunus armeniaca* (Apricot), *Prunus domestica* (Plum), *Psidium araca* (Guava), *Psidium guineense* (Goyave Georges), *Psidium friedrichsthalianum* (Wild guava), *Rubus alceufolius* (Mauritian raspberry), *Sandoricum koetjape* (Faux mangoustan), *Syzygium jambos* (Rose apple), *Terminalia catappa* (Indian Almond), *Vangueria madagascariensis* (Chinese Tamarind).

Source: From Agricultural and Extension Unit, Food and Agricultural Research Council, Mauritius.

BOX 8.2   VEGETABLE SPECIES AND CULTIVARS COMMONLY GROWN IN THE MASCARENE ISLANDS AND MORE SPECIFICALLY PLANTED IN THE ISLANDS OF MAURITIUS*, RODRIGUES** AND RÉUNION***.

MASCARENE ISLANDS

Cynara scolymus (Artichoke), *Colocasia antiquorum* (Arouille Violet), *Beta vulgaris* (Beetroot), *Brassica oleracea var botrytis* (Brocoli), *Brassica oleracea var. capitata* (Cabbage), *Daucus carota* (Carrot), *Brassica oleracea var botrytis* (Cauliflower), *Capsicum frutescens* (Chilli), *Brassica chinensis* (Chinese cabbage), *Lycopersicon esculentum* (Cooking and salad tomato), *Cucumis sativus* (Cucumber), *Solanum melongena* (Eggplant), *Phaseolus vulgaris* (French Bean), *Allium sativum* (Garlic), *Lactuca sativa* (Lettuce), *Cucumis melo* (Melon), *Solanum tuberosum* (Potato), *Cucurbita maxima* (Pumpkin), *Raphanus sativus* (Radish), *Pisum sativum* L. Macrocarpon (Snap pea), *Capsicum annuum* (Sweet pepper), *Cucurbita pepo* (Zucchini)

IN MAURITIUS*

Amaranthus tricolor (Amaranthus), *Cajanus cajan* (Ambrevade), *Maranta arundinacea* (Arrow root), *Asparagus officinalis* (Asparagus), *Momordica charantia* (Bittergourd), *Lageeneria siceraria* (Bottlegourd), *Brassica chinensis* (Brède baton blanc, Brède de chine frisée, Brède Hamchoy, Brède Pak choy, Brède Tom pouce), *Moringa oleifera* (Brède mouroum), *Colocasia esculenta* (Brède songe), *Cichorium endivia* (Bringelle anguive), *Vicia faba* (Broad bean), *Juglans cinerea* (Butternut), *Benincasa hispida* (Calebasse chinois), *Manihot esculenta* (Cassava), *Sechium edule* (Chayotte), *Chicorium intybus* (Chicorée frisée), *Brassica rapa* (Chou navet), *Brassica oleracea var. gongylodes* (Chou rave), *Lycopersicon esculentum* (Cooking and salad tomato), *Vigna unguiculata* (Cowpea), *Cucumis anguria* (Gherkin), *Zingiber officinale* (Ginger), *Passiflora quadrangularis* (Grenadine carri), *Apios tuberosa* (Roundnut), *Dolichos lablab* (Lab lab), *Abelmoschus esculentus* (Ladies Finger), *Zea mays* (Maize), *Allium cepa* (Onion), *Ipomea purpurea* (Patate chinois), *Pisum sativum* (Pea), *Pleurotus sajor caju* (Pleurotus), *Tetragonolobus purpupeus* (Pois carré), *Brassica oleracea var. capitata f. rubra* (Red Cabbage), *Luffa acutangula* (Ridgegourd), *Trichosanthes anguina* (Snakegourd), *Glycine max* (Soya bean), *Spinacia oleracea* (Spinach), *Cucurbita pepo* (Squash), *Zea mays var. saccarata* (Sweet corn), *Ipomoea batatas* (Sweet potato), *Brassica napus* (Turnip), *Citrillus lanatus* (Water melon), *Rorippa nasturtium-aquaticum* (Watercress)

IN RODRIGUES**

Amaranthus tricolor (Amaranthus), *Momordica charantia* (Bittergourd), *Lageeneria siceraria* (Bottlegourd), *olanum nigrum* (Brède martin), *Manihot esculenta* (Cassava), *Sechium edule* (Chayotte), *Brassica pekinensis* (Chou de Chine), *Brassica rapa* (Chou navet), *Vigna unguiculata* (Cowpea), *Liriomyza sativae* (Dry beans), *Vigna radiata* (Emberique), *Zingiber officinale* (Ginger), *Apios americana* (Groundnut), *Abelmoschus esculentus* (Ladies Finger), *Zea mays* (Maize), *Allium cepa* (Onion), *Pisum sativum* (Pea), *Luffa acutangula* (Ridgegourd), *Trichosanthes anguina* (Snakegourd), *Spinacia oleracea* (Spinach), *Cucurbita pepo* (Squash), *Ipomoea batatas* (Sweet potato), *Colocasia esculenta* (Brède songe), *Brassica napus* (Turnip), *Citrillus lanatus* (Watermelon).

IN REUNION***

Liriomyza sativae (Dry beans), *Lactuca sativa* (Lettuce&—Batavia, Blonde, Feuille de chêne), *Brassica oleracea var. capitata f. rubra* (Red Cabbage)

Source: From Agricultural and Extension Unit, Food and Agricultural Research Council, Mauritius.

phytochemistry and antioxidant capacity of a number of fruits, vegetables, and teas commonly consumed by the population of these islands.

8.1.1   FREE RADICALS IN BIOLOGY AND MEDICINE

Free radicals are generated in vivo as part of normal cellular functioning and can be significantly increased beyond the threshold for normal physiological function by exposure to certain risk factors [25,26]. A free radical is any species capable of independent existence with one or more unpaired electrons in an atomic or molecular orbital. These include the hydroxyl radical ($OH^{\cdot}$), superoxide radical ($O_2^{\cdot-}$), peroxyl radical ($LOO^{\cdot}$), alkoxyl radical ($LO^{\cdot}$) and nitric oxide ($NO^{\cdot}$). Hydrogen peroxide (H_2O_2), hypochlorous acid ($HOCl$), singlet oxygen ($^1O_2$), ozone (O_3) and peroxynitrite ($ONOO^-$) are not free radicals but can easily mediate damage to biological molecules. The terms reactive oxygen species (ROS) and reactive nitrogen species (RNS) are often used to include both the radical and the non-radical species. In vivo sources of free radicals arise from the reduction of molecular oxygen during respiration and from the synthesis of complex biochemical compounds. Moreover, living organisms have evolved a system that uses ROS to defend them against invading foreign organisms. Within the body, the production of ROS and antioxidant defenses are approximately balanced. When this balance is tipped in favor of ROS production, oxidative stress results [26–29]. Oxidative stress contributes to cellular dysfunctioning by damaging DNA, proteins, lipids, and other biomolecules. Fortunately, the body has evolved intricate defense systems to reduce the cumulative load of ROS and RNS within cells [25]. These include protection afforded by the antioxidant enzymes superoxide dismutase, catalase, and glutathione peroxidase; the antioxidant response elements (ARE) and protection through low molecular weight antioxidants, some endogenously produced (glutathione, NADH, carnosine, uric acid, melatonin, α-lipoic aicd, bilirubin) [27,28] and some provided through dietary intake (ascorbic acid, tocopherols, ergothioneine, carotenoids, ubiquinol, quinones, phenolics) (Table 8.1) [26–31]. ARE, also referred to as electrophile response element (EpRE) is found in the promoter regions of several genes and is involved in gene expression. Gene induction through the ARE is emerging as an important mechanism for activation of cytoprotective antioxidant genes, such as heme oxygenase (HO) or phase II detoxifying enzymes (GSH transferase and quinone oxido reductase) [32].

There appears to be a consensus that oxidative stress plays a key role in the etiology and development of several pathologies. Several pathogens, including rheumatoid arthritis [43], cancer [44], neurodegenerative diseases [45], cardiovascular disorders [46], stroke [47], diabetes [48], and the process of aging [49], as well as impacting on the signal transduction systems leading to a predisposition to disease of both acute and a chronic nature (Figure 8.1). For instance enhanced free radicals produced in a diabetic state function as signaling molecules to activate a number of stress-sensitive pathways namely nuclear factor-κB (NF-κB), NH_2-terminal Jun kinases/stress activated protein kinases (JNK/SAPK), p38 mitogen-activated protein (MAP) and hexosamine that can chronically lead to the complications of diabetes including embryopathy, retinopathy, neuropathy,

TABLE 8.1
Low Molecular Weight Antioxidants

Low Molecular Weight Antioxidants	Function	Reference
Vitamins		
α-Tocopherol (vitamin E)	Scavenger of peroxyl and alkoxyl radicals	[33]
Ascorbate (vitamin C)	Scavenge peroxyl, thyil, sulfenyl radicals, HOCl acid, $^1O_2$, O_3	[34,35]
	Regenerate α-tocopherol from its radical	
Glutathione	Substrate for glutathione peroxidase	[27,36]
	Scavenger of O_2^-, $OH^·$, $^1O_2$	
Ubiquinol	Recycling of α-tocopherol from α-tocopheroxyl radical	[28]
Metal sequestrators (e.g., lactoferrin, caeruloplasmin)	Bind transition metal ions	[37,38]
Ergothioneine	Powerful scavenger of $ONOO^-$, $OH^·$, HOCl acid	[39–41]
	Quencher of $^1O_2$	
Polyphenol (e.g., flavonoids, phenolic acid, proanthocyanidins)	Exhibit free radical scavenging activity Chelate metal ions	[14,17,31,42]

nephropathy, and cardiovascular diseases [50–54]. The implication of free radicals in low-density lipoprotein (LDL) oxidation and in foam cell formation during atherogenesis has been greatly advocated [55]. ROS may also induce acute alterations in cellular functions via specific covalent modifications of target molecules, for example, key proteins involved in myocardial excitation-contraction coupling, such as sarcolemmal ion channels, sarcoplasmic reticulum calcium release channels, and contractile proteins, can all undergo redox sensitive alterations in activity [56]. Redox signaling has been suggested in vascular smooth muscle proliferation, atherosclerosis, angiogenesis, cardiac hypertrophy, fibrosis [57]. Modulation of intracellular signaling pathways (MAPKs), and the subsequent activation of downstream redox sensitive transcription factors like NF-κB, HIF-1, AP-1 results in alterations in gene and protein expression [58] that significantly enhance cardiac dysfunction.

The brain is particularly sensitive to oxidative stress due to its high content in unsaturated fatty acid, elevated oxygen consumption rate and chemical reactions involving dopamine and glutamate oxidation [59,60]. Moreover, alterations in the cell signaling pathways, mainly the activation of MAPK may lead to the upregulation of apoptosis-associated gene expression and trigger a caspase cascade leading to neuronal cell death [61]. Similarly the disrupted intracellular signaling cascades are central to cancer biology where alteration in cell signaling pathways (MAPK, NF-κB, AP-1, PLA, ASK 1) leads to modulation of proto-oncogene and tumor suppressor gene expression that affects the cell growth and proliferation [16,62]. A common link

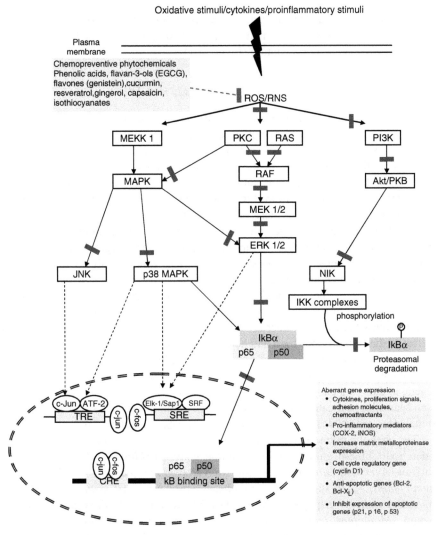

FIGURE 8.1 Schematic representation of the intracellular signal transduction cascades activated by reactive oxygen species (ROS) and converging on downstream transcription factors NF-κB and AP-1. Phosphorylation by the protein kinases causes the translocation of NF-κB into the nucleus. The latter binds to the corresponding promoter enhancer region of various target genes, for example, cytokines (TNF-α, IL-1), chemoattractants (MCP-1); adhesion molecules (ICAM-1, VCAM-1) involved in cardiac dysfunction; pro-inflammatory mediators (COX-2, iNOS) that enhance inflammation in neurodegenerative condition and cancer; anti-apoptotic proteins (cIAP 1, cIAP2, Bcl-2, Bcl-X$_L$) and matrix metalloproteinase (MMP-9) that contribute to cancer. The activation of AP-1 is mediated predominantly via the MAPK and association of c-fos and c-Jun forms a heterodimer that binds to AP-1 response element located in the promoter region of target genes up-regulating pro-inflammatory gene and cell cycle regulatory gene expressions (cyclin D1) contributing to the pathogenesis of neurodegeneration, cancer, diabetes, and cardiovascular diseases. (━━) indicates the direct scavenging effect of the phenolic acids and flavonoids on ROS/RNS while (━━) indicates the sites where phenolic compounds have been reported to modulate and suppress the cell signal transduction cascades.

between free radicals and the aforementioned pathological condition is the disrupted intracellular signal transduction networks, which suggests a rationale for targeting these pathways in chemoprevention (Figure 8.1).

8.1.2 PHYTOPHENOLICS IN FRUITS, VEGETABLES, AND TEAS

Phenolic compounds or polyphenols constitute one of the most numerous and ubiquitously distributed groups of plant secondary metabolites. Natural polyphenols can range from simple molecules (quinones, phenolic acids,) to highly polymerized compounds (lignins, melanins, tannins), with flavonoids such as flavonols, flavones, isoflavones, flavonones, flavanols, and anthocyanins representing the most common and widely distributed sub-group (Figure 8.2) [63]. Phenolics are widely distributed in the plant kingdom and are therefore an integral part of the diet, with significant amounts being reported in vegetables, fruits, and teas [7,9,64]. Although the dietary intake of phenolics varies considerably among geographic regions, it is estimated that daily intake ranges from about 20 mg to 1 g, which is higher than that for Vitamin E [65]. Phenolics are particularly attractive as prophylactic agents due to their prevalence in the diet and also due to their pluripharmacological effects. Bioactive food components are known to influence multiple biological processes. Determining which is most instrumental in bringing about a phenotypic change is critical to the future of nutrition and health, as it will assist in determining who may benefit and who may be placed at risk by intervention strategies.

Data obtained from a comprehensive phytophenolic study of 17 exotic fruits, 10 vegetable cultivars and 9 tea brands from the Mascarene islands, more particularly from Mauritius, indicated a wide distribution of phenolics (Table 8.2) that correlated to their free radical scavenging and antioxidant capacities [64,123].

The phenolic richness of the fruits investigated were in the following decreasing order: red Chinese guava > yellow Chinese guava (*Psidium cattleianum*) > sweet starfruit (*Averrhoa carambola*) > white guava (*Psidium guajava*) > jamblon (*Syzygium cumini*) > acid starfruit (*A. carambola*) > pink guava (*P. guajava*) > hogplum (*Spondias dulcis*) > papaya (*Carica papaya*) > passion fruit (*Passiflora edulis*) > mango (*Mangifera indica*) > pineapple (*Ananas comusus*) > Jamalac (*Syzygium samarangense*) > Litchi (*Litchi chinensis*) > Avocado (*Persea americana*) > banana (*Musa acuminata*) > Longanberry (*Euphoria longan*). Their antioxidant activities as assessed by two independent methods, the TEAC and FRAP assays are presented in Figure 8.3. The data summarized show that Mauritian fruits exhibit a wide range of antioxidant potentials associated with high levels of phenolics of which flavanols, flavones, and flavonols were major components. Fruit proanthocyanidin contents were much higher than those of any phenolic subclass assayed, and therefore appear to account significantly for the antioxidant capacity of fruits. It is known that proanthocyanidins comprising B dimers and their derivatives and oligomers are effective antioxidants due to the presence of hydroxyl functions attached to their ring structures (Figure 8.4) [66–68].

For comparison with data obtained previously for similar or different types of fruits using the same assays, the antioxidant capacities of Mauritian red and yellow Chinese guava seem to be very close to the antioxidant activity of blueberries, which

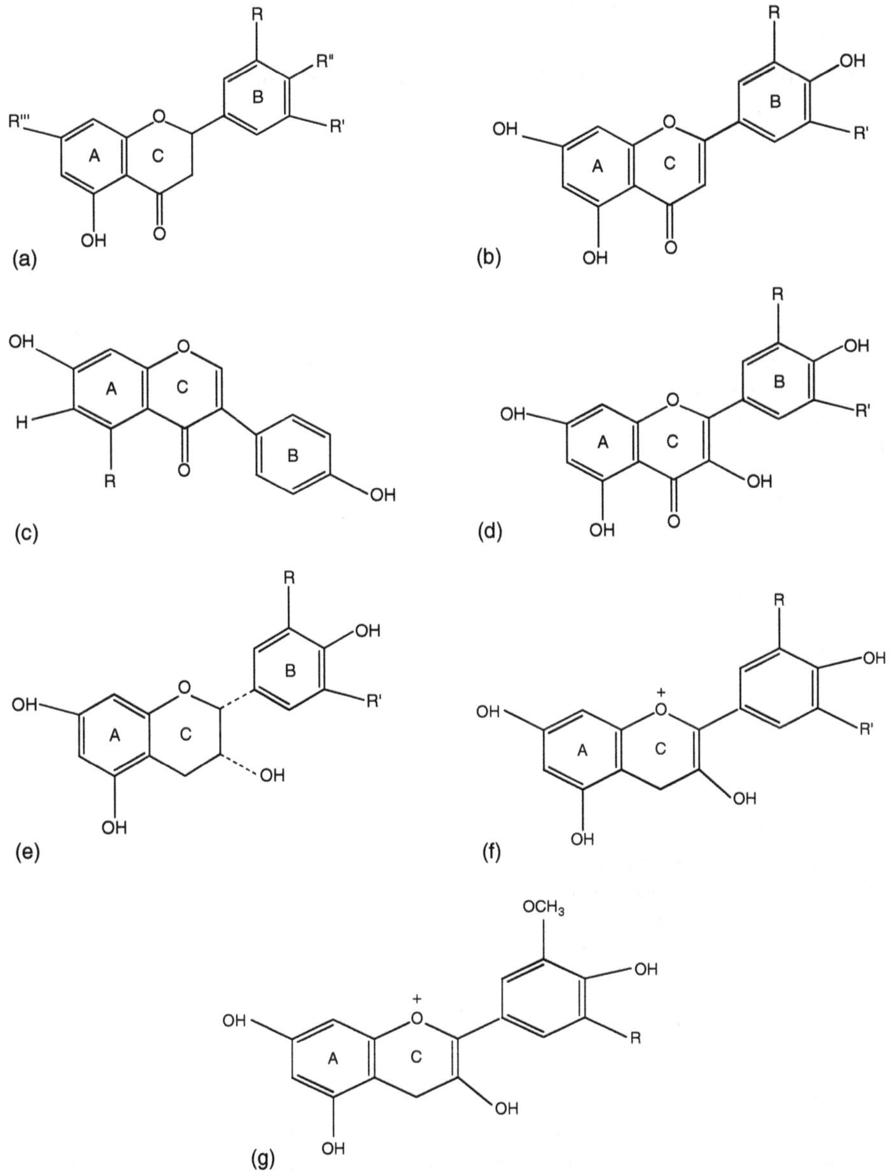

FIGURE 8.2 (a) Flavanones: $R=R^1=H$, $R^{11}=R^{111}=OH$; Naringenin; $R=OH$, $R^1=H$, $R^{11}=R^{111}=OH$; Eriodyctiol; $R=R^1=OH$, $R^{11}=R^{111}=OH$; 5^1-OH-Eriodyctiol. (b) Flavones: $R=R^1=H$; Apigenin; $R=OH$, $R^1=H$; Luteolin; $R=R^1=OCH_3$; Tricetin. (c) Isoflavones: $R=H$; Daidzein; $R=OH$; Genistein. (d) Flavonols: $R=R^1=H$; Kaempferol; $R=OH$, $R^1=H$; Quercetin; $R=R^1=OH$; Myricetin. (e) Flavanols: $R=OH$, $R^1=H$; (+)-Epicatechin; $R=R^1=OH$; (+)-Epigallocatechin; $R=OH$, $R^1=H$; (−)-Catechin; $R=R^1=OH$; (−)-Gallocatechin. (f) Anthocyanidins: $R=OH$, $R^1=H$; Cyanidin; $R=R^1=OH$; Delphinidin; $R=R^1=OCH_3$; Malvidin; $R=R^1=H$; Pelargonidin. (g) Anthocyanidins: $R=H$; Peonidin; $R=OH$; Petunidin.

TABLE 8.2
The Phenolics Content and Antioxidant Activities of Mauritian Exotic Fruits, Vegetables, and Teas

	Fruits[a]	Vegetables[a]	Teas[b]
Total phenols	118–5,638	132–1,189	62–107[c]
Total flavonoids	21–712	45–944	15–26[c]
Quercetin derivatives	—	15–390	1,074–3,288
Myricetin derivatives	—	1–32	339–1,467
Kaempferol derivatives	—	6–125	269–693
Other flavonoid derivatives (Apigenin, Luteolin)	—	2–45	—
Total proanthocyanidin	7–2,561	4–116	25–74[c]
Procyanidin B1	—	—	2,464–4,993
Procyanidin B2	—	—	1,208–3,139
Catechin index	—	—	14,612–30,749
Gallic acid	—	—	5,503–10,942
Trolox equivalent antioxidant capacity (TEAC)	1–47[d]	0.4–3.7[d]	423–1,147[d]
Ferric reducing antioxidant power (FRAP)	0.3–34[e]	0.8–8.5[e]	357–927[e]

Ranges are indicated.

Source: Bahorun, T., Luximon-Ramma, A., Crozier, A., and Aruoma, O. I., *J. Sci. Food Agric.*, 84, 1553–1561, 2004; Luximon-Ramma, A., Bahorun, T., and Crozier, A., *J. Sci. Food Agric.*, 83, 496–502, 2003; Luximon-Ramma, A., Bahorun, T., Crozier, A., Zbarsky, V., Datla, K. K., Dexter, D. T., and Aruoma, O. I., *Food Res. Int.*, 38, 357–367, 2005.

[a] Results are expressed as μg/g fresh weight.
[b] Results are expressed as μg/g dry weight.
[c] Results are expressed as indicates results in mg/g dry weight.
[d] Results are expressed as μmol Trolox equivalent/g fresh weight.
[e] Results are expressed as μmol Fe(II) equivalent/g fresh weight.

constitute one of the richest sources of antioxidants studied so far [69]. Rubus species (with antioxidant capacities measured as TEAC ranging from 0 to 25.3 μmol g^{-1} fresh weight), recommended for the improvement of nutritional value through germplasm enhancement programmes [70] have a lower antioxidant potency than the Mauritian fruits with TEAC values ranging from 1 to 47 μmol g^{-1} fresh weight.

Vegetables exhibiting the highest antioxidant activities were Chinese cabbage, onion, mugwort, broccoli, chilli pepper, and cauliflower (Figure 8.5). FRAP and TEAC values were similar in the vegetable extracts, with the following antioxidant profiles in this order: onion > Chinese cabbage > broccoli > mugwort > cauliflower > white cabbage > lettuce > tomato > carrot. The levels of total phenols in the vegetables varied between 132 μg g^{-1} fresh weight in carrot and 1189 μg g^{-1} fresh weight in Chinese cabbage. Due to this large phenolic content variation,

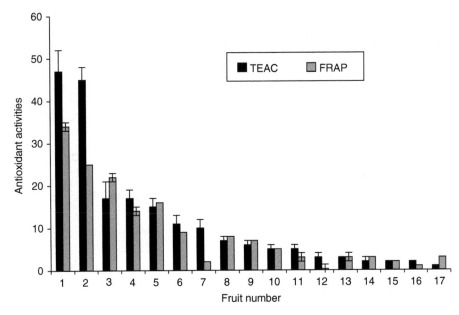

FIGURE 8.3 Antioxidant activities in 17 Exotic fruits as assessed by TEAC (µmol Trolox g^{-1} fresh weight) and FRAP (µmol FeII g^{-1} fresh weight) assays. (1) Red *Psidium cattleianum*; (2) Yellow *P. cattleianum*; (3) Sweet *Averrhoa carambola*; (4) White *Psidium guajava*; (5) *Syzygium cumini*; (6) Acid *A. carambola*; (7) *Carica papaya*; (8) Pink *P. guajava*; (9) *Spondias dulcis*; (10) *Mangifera indica*; (11) *Litchi chinensis*; (12) *Euphoria longan*; (13) *Passiflora edulis*; (14) *Ananas comosus*; (15) *Syzygium samarangense*; (16) *Persea americana* and; (17) *Musa acuminata*. (From Bahorun, T., Luximon-Ramma, A., Crozier, A., and Aruoma, O. I., *J. Sci. Food Agric.*, 84, 1553–1561, 2004; Luximon-Ramma, A., Bahorun, T., and Crozier, A., *J. Sci. Food Agric.*, 83, 496–502, 2003; Luximon-Ramma, A., Bahorun, T., Crozier, A., Zbarsky, V., Datla, K. K., Dexter, D. T., and Aruoma, O. I., *Food Res. Int.*, 38, 357–367, 2005.)

Mauritian vegetables were categorized in three groupings: (i) high phenolic level: >800 µg g^{-1} fresh weight including Chinese cabbage, mugwort, broccoli; (ii) medium level: 275–425 µg g^{-1} fresh weight including chilli pepper, tomatoes, cauliflower and (iii) low level: <275 µg g^{-1} fresh weight comprising white cabbage, lettuce, and carrot. Flavonoids were the predominant phenolic sub-class in the vegetables studied, with their distribution pattern being analogous to that of total phenols in most cases. The amounts of flavonoids were in the following order: Chinese cabbage > onion > mugwort > broccoli > chilli pepper > cauliflower > white cabbage. Mauritian lettuce, tomato, and carrot cultivars are relatively poor in flavonoids. Free flavonoids were not detected in the vegetable extracts; however, after hydrolysis, HPLC analysis showed that quercetin was the dominant flavonol in the hydrolyzed vegetable extracts, with maximum levels in Chinese cabbage, onion, and mugwort. Kaempferol, myricetin, apigenin, and luteolin levels were less pronounced in Mauritian vegetables. Contents of flavonoid derivatives in Mauritian vegetables were generally high. Factors including differences in variety and high sunlight conditions (a characteristic feature of

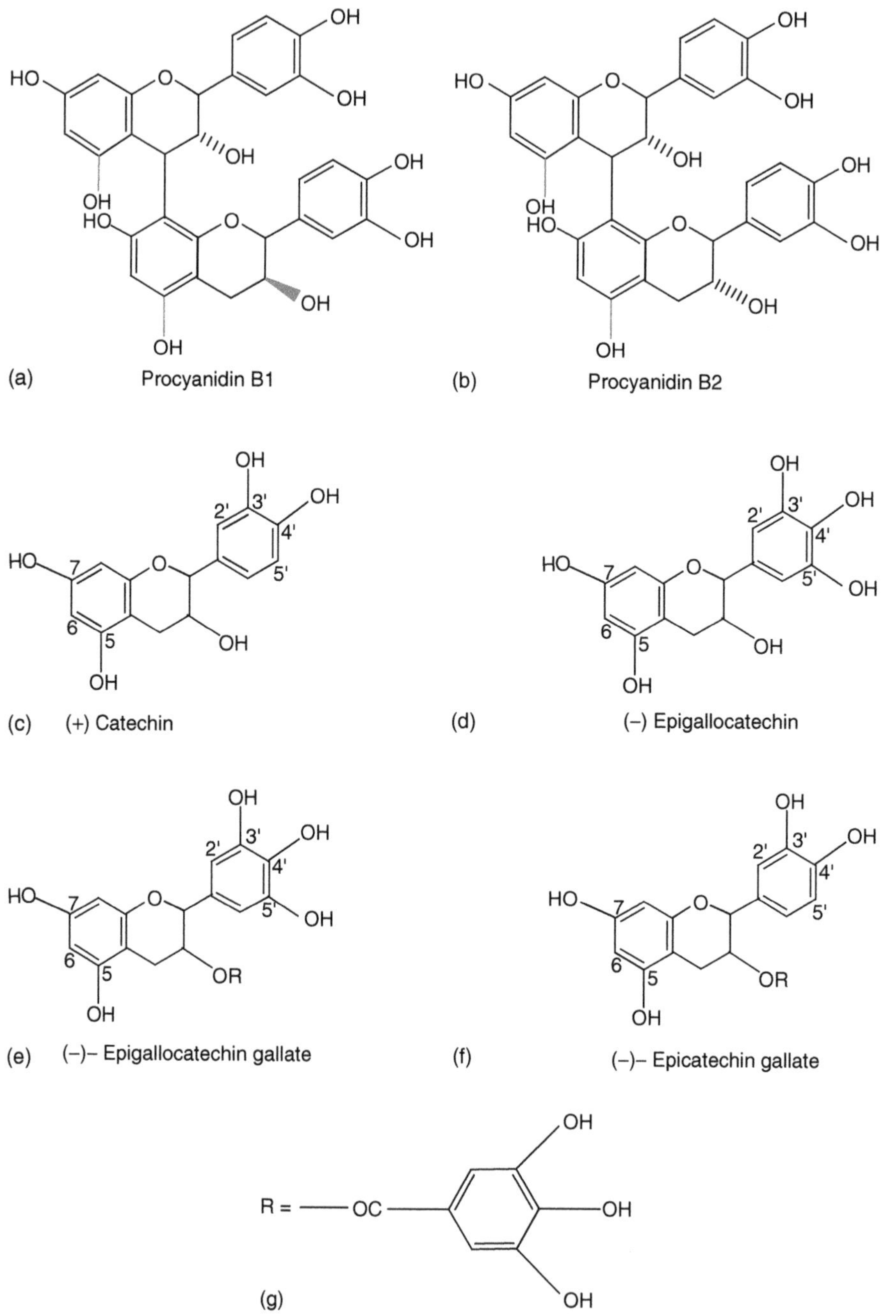

(a)   Procyanidin B1

(b)   Procyanidin B2

(c)   (+) Catechin

(d)   (−) Epigallocatechin

(e)   (−)− Epigallocatechin gallate

(f)   (−)− Epicatechin gallate

$R = \;\;\; OC$

(g)

FIGURE 8.4 Structure of the multiply-hydroxylated procyanidins and catechins.

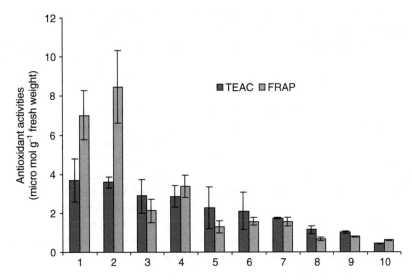

FIGURE 8.5 Antioxidant activities of vegetables as assessed by TEAC and FRAP assay: (1) chinese cabbage; (2) onion; (3) mugwort; (4) broccoli; (5) chilli pepper; (6) cauliflower; (7) white cabbage; (8) lettuce; (9) tomatoes; (10) carrot. (From Bahorun, T., Luximon-Ramma, A., Crozier, A., and Aruoma, O. I., *J. Sci. Food Agric.*, 84, 1553–1561, 2004; Luximon-Ramma, A., Bahorun, T., and Crozier, A., *J. Sci. Food Agric.*, 83, 496–502, 2003; Luximon-Ramma, A., Bahorun, T., Crozier, A., Zbarsky, V., Datla, K. K., Dexter, D. T., and Aruoma, O. I., *Food Res. Int.*, 38, 357–367, 2005.)

tropical countries like Mascarene islands), which can induce the accumulation of flavonoids [71] are most probably responsible for the high yields. Although geographical differences in the levels of phenolics are widely discussed for teas [64], it is suggested that this may also be widespread for other food items, including vegetables.

The antioxidant data are consistent with literature data for Chinese cabbage, broccoli, and onion, which are generally reported to have high antioxidant activities [72–74] and vegetables such as carrot and lettuce, which are shown to be weak antioxidants in in vitro assays [72,74]. However, high antioxidant activities have been reported in tomato, cabbage, and carrot and lower activities in onion, chilli pepper, and cauliflower, using a model system consisting of β-carotene and linoleic acid [75]. The emerging consensus is that use of one method to assess antioxidant capacity does not give universal answers. It is clear that the antioxidant efficacy of an extract should be evaluated by a multitude of methods, rather than depending on the results of one method [76,77]. In addition to phenolics, vegetables contain other bioactive components of prophylactic interest. Cruciferous vegetables are excellent sources of antioxidants and glucosinolates, precursors to a group of isothiocyanates known to possess anticarcinogenic property [78]. Interestingly, it is the hydrolysis products that are thought to protect against cancer. For example, sulforaphane, the hydrolysis product of the glucosinolate glucoraphanin, is highly potent at upregulating detoxification enzymes in cell cultures and is reported to prevent mammary cancer in rodents [79]. Epidemiological studies have revealed that the ingestion of three or more half-cup servings of cruciferous vegetables such as broccoli and cabbage, per week significantly lowered the risk of prostate cancer by 40%, compared to ingestion of one or fewer servings per week [80]. The anticarcinogenic effect of broccoli may therefore be ascribed to its high phenolic and glucosinolate content.

The constituents of green and black teas have been the subject of intensive investigations for a long time [64,81,83,122,123]. Recent trends, partially in response to claims of health benefits associated with the beverage, show an increased preference for fruit and herbal teas in continental Europe and to a lesser degree in the United Kingdom; however, black tea with added milk (white tea) remains by far the most common form consumed in the United Kingdom [82] and in Mauritius, where tea exports were the second source of national income, after sugar, in the early seventies. Interest in determining the phytophenolic profile and antioxidant propensities of Mauritain teas was triggered by its intended use in clinical trials to study the modulation of risk factors of cardiovascular diseases, diabetes, cancer, and neurodegenerative diseases.

The data generated on Mauritian teas indicated high total phenol content, ranging from 62 ± 9 to 107 ± 22 mg/g dry weight, total flavonoid content varying between 15 ± 2 and 26 ± 3 mg/g dry weight and total proanthocyanidin content between 25 ± 2 and 74 ± 10 mg/g dry weight in the black tea infusates (Table 8.2). The total phenol data for Mauritian tea infusates are consistent with those obtained by HPLC for Chinese Pu'er tea infusates, which ranged from 64 to 126 mg/g [84]. It was also observed that the phenolic content of one of the tea brands analyzed was 25% higher than U.K. PG-Tips black tea, 9% higher than Yorkshire Gold tea and 39% higher than Japanese Green tea (Bandia) studied by Khokhar and Magnusdottir [82]. HPLC quantification of the individual compounds revealed very high levels of (-)-EC, (-)-ECG, (-)-EGCG from the tea infusates with a "catechin index" varying from 17,204 to 30,749 (Table 8.2). For comparison, one of the tea brand "catechin index" was 82% higher than U.K. Twinings–Lapsang black tea and 73% higher than Yorkshire tea reported by Khokhar and Magnusdottir [82]. The concentrations of procyanidin B1 and B2 dimers approached levels of 5.0 mg/g dry weight in certain tea brands, while the levels of (+)-C and (−)-EGC in the tea infusates were relatively low. Linear regression analyses produced a high correlation coefficient with contents in total proanthocyanidins and total phenols, while flavonoids seem to weakly influence the antioxidant potentials of the tea infusates. The greatest contribution of phenolic compounds to the antioxidant activities of the tea infusates came from (−)-EGCG, (−)-ECG, (−)-EC, and GA. It is noteworthy that in the manufacturing of black tea during the process commonly known as fermentation, the monomeric flavan-3-ols undergo polyphenol oxidase-dependent oxidative polymerization, leading to the formation of bisflavanols, theaflavins, thearubigins, and other oligomers [85], which are also contributive to the antioxidative efficacy of the extracts. Literature data on the antioxidant power of different teas, notably black teas using the FRAP assay, [86] are consistent with the FRAP values measured for the Mauritian teas. It is, however, noteworthy that the FRAP data obtained with the tea brand containing the greatest amount of phenolic derivatives was 59% higher than China Black tea, 67% higher than Pu Li tea, 43% higher than oolong tea, a semi fermented tea, 32% and 24% higher than the Jasmine and Japanese teas.

The numbers of the phenolic hydroxyl groups have been intrinsically linked to the potential of the tea antioxidants (Figure 8.4). So the antioxidant index $(-)$-EGCG (8 groups)$>(-)$-ECG (7 groups)$>$GC (6 groups)$>(-)$-EGC (6 groups)$>$ $(-)$-EC (5 groups) are widely suggested [87,88]. The correlations reported in this study: $(-)$-EGCG$>(-)$-ECG$>(-)$-EC$=$GA$>(-)$-EGC in the tea infusate are consistent with this.

However, the antioxidant capacities have a different order, depending on the mode with which the antioxidant index is determined. Gardner et al. [89] suggested the antioxidant activity profile: $(-)$-EGCG$>(-)$-ECG$>(-)$-EC$>(-)$-EGC$>$ $(+)$-C$>$GA based on the use of the Fremy's radical (potassium nitrosodisulpho- nate) as the oxidant in an aqueous medium. However, use of galvanoxyl as an oxidizing agent in ethanol produced a different order: $(-)$-EGCG$>(-)$-ECG$>$ GA$>(-)$-EC$=(+)$-C$>(-)$-EGC. Thus, ranking of antioxidant activity is system- dependent, as highlighted above. The application and/or the end use of the plant- derived antioxidant should determine the methodology used.

Several cross-sectional surveys in a wide range of population groups have been conducted to estimate the daily intake of antioxidants. Hertog et al. [90] showed that the average daily intake of five flavonoids (quercetin, kaempferol, myricetin, luteolin, and apigenin) in the Dutch diet was 23 mg, while it was only 12.9 mg in the U.S. diet. A similar investigation found that the average flavonoid daily intake in the U.S. diet was 20.1 mg. An earlier study indicated that the total polyphenol daily intake in the U.S.A. was close to 1 g [91] comprising 16% flavonols, flavones and flavanones, 17% anthocyanins; 20% catechins and 45% biflavones. For comparison, using our data from the total phenolic and total flavonoid assays, the consumption of 180 and 30 g of red Chinese guava, respectively, would give the above phenolic and flavonoid values.

Likewise, the potential for all types of Mauritian teas to contribute significantly to the dietary intake of antioxidant power is high. Assuming average tea consumption in Mauritius to be 4 cups (1.5% w/v) per day, the calculated total phenols intake from the tea infusates will be approximately 0.85 g/day, which is very close to the 1 g intake from the U.S.A. diet [91].

During the last decade natural antioxidants, particularly phenolics, have been under very close scrutiny as potential therapeutic agents against a wide range of ailments. The free radical scavenging and antioxidant activities of phenolics are dependent upon the arrangement of functional groups about the nuclear structure. Both the number and configuration of H-donating hydroxyl groups are the main structural features influencing the antioxidant capacity of phenolics [92,93]. An investigation of antioxidant properties of phenolic reference compounds conducted by our group [94] shows that the greatest antioxidant activity was observed for dimeric procyanidins like B1 and B2 occuring in Mauritian fruits and teas. The high antioxidant activity of the procyanidin dimers is attributed to their hydroxyl functions that are potent hydrogen donors (Figure 8.4). Furthermore, the conjugated double bonds allow electron delocalisation across the molecule, thus stabilizing the phenoxyl radical [66]. Several studies have highlighted the cardioprotective effect of proanthocyanidins. Grape seed proanthocyanidin extract

(GSPE), a mixture of 75%–80% oligomeric proanthocyanidin and 3%–5% monomeric proanthocyanidin, has proved its efficacy against the incidence of ischemia-reperfusion injury, apoptosis of cardiomyocytes and reduction of foam cell development. The mechanistic pathways of cardioprotection exerted by the proanthocyanidin rich extract included (1) potent hydroxyl and other free radical scavenging abilities; (2) anti-apoptotic, anti-necrotic, and anti-endonucleolytic potentials; (3) modulatory effect on apoptotic regulatory bcl-X_L, p53, and c-myc genes; (4) cytochrome P450 2E1 inhibitory activity; (5) inhibitory effects on proapoptotic, cardioregulatory genes c-JUN and JNK-1 and (6) inhibition of vasoconstriction of vascular smooth muscle and endothelium [95,96].

Among the monomeric flavan-3-ols widely distributed in Mauritian teas, the hierarchy of antioxidant activities has been reported to decrease from (−)-EGCG >(−)-ECG>(−)-EGC>(−)-EC>(+)-C [94].

Flavanols are also potent antioxidants in lipid systems, where they reduce oxidative modifications of membranes by restricting the access of oxidants to the bilayer and the propagation of lipid oxidation in the hydrophobic membrane matrix [97]. Several studies have shown that catechins have a protective role in neurodegeneration. Pretreatment with (-) EC attenuated neurotoxicity induced by oxidized LDL in mouse-derived striatal neurons, as evidenced by apoptotic DNA fragmentation and caspase-3 activation [98]. The (-)-EGCG enhanced the activity of superoxide dismutase (SOD) and catalase in mouse striatum, thus suggesting that flavan-3-ols can also exhibit their neuroprotective effect via regulation of gene expression [99]. Reference [100] reported that (-) EGCG decreased the expression of proapoptotic genes Bax, Bad, caspase-1 and -6, cyclin-dependent kinase inhibitor p21, cell-cycle inhibitor gadd45, fas-ligand, and TNF-related apoptosis-inducing ligand TRAIL, in SH-SY5Y neuronal cells. The decline in bax expression by (−) EGCG may favor the increase in the ratio of bcl-2/bcl-xL to bax/bad proteins, thereby contributing to mitochondrial stability and regulation of the mitochondrial permeability transition pore (mPTP) [101].

Flavonol derivatives exist mainly in fruits, vegetables, and teas as sugar conjugates, the main aglycones being quercetin, myricetin, and kaempferol. The qualitative and quantitative determination of individual flavonoid glycosides in foods is quite difficult, as many of the reference compounds are not commercially available. The hydrolysis of flavonoid glycosides to aglycones is a practical approach for the quantitative determination of flavonoids in plant extracts. The antioxidant capacity in this particular class of flavonoid is also related to the number of hydroxyl groups present in the following order: quercetin>myricetin >kaempferol. The presence of a third hydroxyl group in the B ring at C-5 position does not enhance the effectiveness of myricetin compared to quercetin with simply an o-diphenolic structure. This is supported by the findings that the presence of three hydroxyl groups on the aromatic nucleus does not improve antioxidant efficiency. The low antioxidant activity of kaempferol among the flavonol aglycones could be attributed to the presence of a single hydroxyl group in the B ring, which apparently makes little contribution even in the presence of the conjugated double bond system and the 3-OH group. The free radical scavenging by flavonoids is highly dependent on the presence of a free 3-OH. Both flavonol and flavone have identical hydroxyl configuration, but the absence of the 3-OH functional group in the flavone explains the lower antioxidant activity of the latter. Thus blocking the 3-OH group in the C ring of quercetin as a glycoside while retaining the $3',4'$-dihydroxystructure in the B ring, as in hyperoside and quercitrin can significantly decrease the antioxidant activity [94]. Burda and Olsezek also reported that substitution of 3-OH by a methyl or glycosyl group decreases the activity of quercetin and kaempferol against β-carotene oxidation in linoleic acid [102]. In addition to a reduction in the number of OH group, removal of the 3-OH affects the conformation of the molecule. The torsion angle of the B ring with respect to the rest of the molecule strongly influences free radical scavenging ability. Flavonols and flavanols with a 3-OH are planar, while the flavones and flavanones, lacking this feature, are slightly twisted. Planarity permits conjugation, electron dislocation, and a corresponding increase in flavonoid phenoxyl radical stability [103]. Eliminating this hydrogen bond effects a minor twist of the B ring, abrogates coplanarity, conjugation and electron delocalisation capacity, thereby compromising scavenging ability. Due to this intramolecular hydrogen bonding, the influence of a 3-OH is potentiated by the presence of $3'4'$-catechol, explaining the potent antioxidant activity of flavan-3-ols and flavonols [14].

Ishige et al. reported that three structural features of flavonol were crucial for the neuroprotective effect: the presence of the hydroxyl group on the C3 position, an unsaturated C ring, and hydrophobicity [104]. The importance of the C3 hydroxyl group for antioxidant and free radical scavenging activities has also been observed in the present study. The neuroprotective effect of flavonol is also substantially increased by the presence of the catechol structure. The unsaturation of the C ring in flavonoid is essential for protection from glutamate toxicity. Quercetin with the unsaturated C ring showed protective action while flavanones and flavan-3-ols are totally ineffective [104]. Studying the neuroprotective effects of flavonoids in a PC 12 model, [105] reported that the three aforementioned structural features were critical for cytoprotective activities. Furthermore the implication of increased COX-2 levels in several inflammation-mediated pathologies have suggested that downregulation of COX-2 expression can be a suitable target for prophylaxis [106]. Several flavonoids, genistein, and epigallocatechin gallate in particular [107,108], have been found to downregulate COX-2 expression mainly by modulating upstream enzymes (protein kinase C, ERK, p38 MAPK) and the transcription factor NF-κB known to regulate COX-2 [107,109].

In line with literature data, it is suggested that the following structural features of flavonoids are important for antioxidant and free radical scavenging activities: (a) an o-diphenolic group (in ring b), (b) a 2–3 double bond conjugated with 4-oxo function, and (c) hydroxyl groups in positions 3 and 5 (Figure 8.6).

8.1.3 PROPHYLACTIC PHENOLICS: OVERVIEW OF THEIR METABOLISM AND BIOAVAILABILITY

It is critical to bear in mind that the reported antioxidant activities are based on phenolic derivatives as present in plants using in vitro models. However, several studies have shown that phenolics are extensively metabolized in vivo, mainly

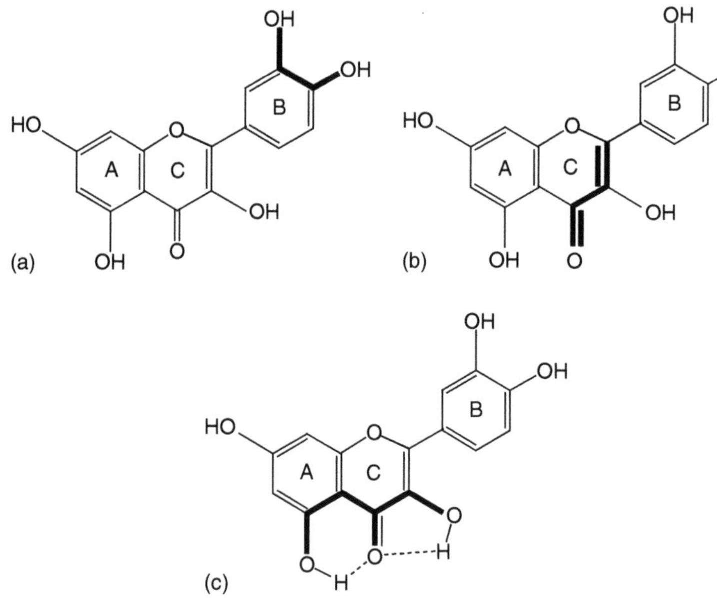

FIGURE 8.6 Structural groups for radical scavenging.

during transfer across the small intestine, by colonic microflora and in the liver, resulting in significant alteration in their redox potential [110,111]. Thus, to delineate the prophylactic potential of phenolic compounds, it is essential to evaluate their efficacy within target tissues.

Phenolic compounds had been considered as antinutrients, due to their adverse effects on digestion in particular protein, and it was common thinking that phenolics were poorly assimilated. However, during the last decade, researchers, and food manufacturers have become increasingly interested in phenolics, due to their protective role in the development of various chronic and degenerative disorders. Extensive knowledge of the bioavailabilty of phenolics is thus essential if their health effects are to be fully understood.

Phenolics, except flavan-3-ol derivatives, are usually present as glycosides in food plants. Phenolic glycosides are stable to cooking, microwaving, and boiling, but are partially hydrolyzed during fermentation process [112]. Furthermore, the glycosides and proanthocyanidin are stable to stomach pH and secreted gastric enzymes. Thus, most of the phenolic compounds consumed reach the gut intact, where their metabolism is initiated (Figure 8.7) [113].

It was long believed that because phytophenolics are fairly large and hydrophilic molecules they could not be absorbed directly, and that their hydrolysis by bacterial enzymes was a prerequisite for their absorption [114]. Pioneering work by [115] shows that phenolic glucoside can be absorbed directly in the small intestine via the sugar transporter sodium-dependent glucose transporter (SGLT1). However, the efficiency of such absorption is limited by efflux of phenolic glycosides by the apical transporter multidrug resistance-associated protein 2 (MRP2) or the *P*-glycoprotein. Inside the enterocytes, the glycosides are then hydrolyzed by a broad-specific β-glucosidase enzyme (BSβG) [116]. Alternatively the glycosides are deglycosylated by lactase phlorizin hydrolase (LPH), an enzyme located in the brush border of the small intestine and responsible for lactose hydrolysis. The enzyme is outside the epithelial cells, so molecules can be deglycosylated in the lumen and the free aglycone can then diffuse into the epithelial cell either passively or by facilitated diffusion [117]. Both pathways give rise to intracellular aglycone that undergoes conjugation by Phase II enzymes in the enterocytes. This process includes three main types of conjugation: methylation, sulfation, and glucuronidation. The conjugation mechanisms are highly efficient, and aglycones are generally either absent in blood or present in very low concentrations [118]. Phenolic metabolites are transported bound to plasma protein. Albumin is the primary protein responsible for the binding, however, affinity of phenolics for albumin varies according to their chemical structure [119].

Phenolic fractions that are not absorbed in the small intestine reach the colon. The colonic microflora, especially anaerobic species *Bacteroides distasonis, B. uniformis,* and *B. ovatus,* hydrolyze glycosides into aglycones and extensively metabolize the aglycones into various aromatic acids [114]. Aglycones are split by the opening of the heterocycle at different points, depending on their chemical structure: flavonols mainly produce hydroxyphenylacetic acids, flavones, and flavanones mainly produce hydroxyphenylpropionic acids, and flavanols mainly produce phenylvalerolactones and hydroxyphenylpropionic acids. These microbial

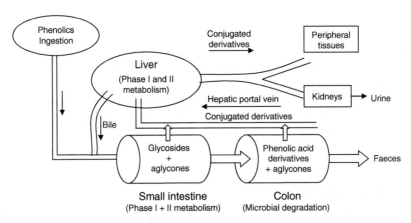

FIGURE 8.7 Schematic representation of phenolic metabolism following oral ingestion. In the small intestine the phenolic glucosides are either absorbed directly via the sodium-dependent glucose transporter (SGLT1) or are hydrolyzed in the intestinal lumen by lactase phlorizin hydrolase (LPH) and subsequently undergo phase I and II metabolism in the enterocytes. Phenolic compenents reaching the large intestine are metabolized by colonic microflora producing mainly phenolic acid derivatives. The phenolic conjugates are transported to the liver via the hepatic portal vein where further metabolism takes place. From the liver, the phenolic conjugates are excreted into bile (resulting in enterohepatic cycle) or enter the systemic circulation gaining access to peripheral tissues and also excreted via the kidneys.

metabolites are absorbed and conjugated with glycine, glucuronic acid, or sulfate [119].

Most of the phenolic metabolites absorbed are transported—bound to albumin—to the liver via the portal vein. In the liver, the phenolics may undergo further modification by Phase II enzymes; however, the conjugation is much complex than that occurring in the intestine and it involves a complex set of conjugating enzymes and carrier systems involved in the regulation of uptake and the production and release of various phenolic metabolites by the hepatocytes [120]. The activity of these enzymes and carrier systems depends on the nature of the phenolics and may also be influenced by genetic polymorphisms. The conjugates are then exported, probably by MRP2 into the bile or systemic circulation [121]. Excretion into the bile results in re-entry into the small intestine and passage of the excreted conjugate into the colon. This is followed by degluronidation by microbes in the ileum or colon and reabsorption of the phenolic further down the GI tract. This is termed enterohepatic cycling.

Part of the phenolic metabolites present in the systemic circulation will be excreted by the kidney while the rest will be delivered to tissues throughout the body. Phenolics have been detected in a wide range of tissues, including brain, endothelial cells, heart, kidney, spleen, pancreas, prostate, uterus, ovary, mammary gland, testes, bladder, bone, and skin [119], and the concentration obtained in these tissues ranged from 30 to 3000 ng aglycone equivalents/g tissue. Conjugation of phenolic compounds usually leads to a decrease in their specific activities. However, in vivo phenolic conjugates may exert their biological activities after deconjugation at the cellular level. Many cells possess β-glucuronidase activity, found both in the lysosome and endeoplasmic reticulum. Sulfatase activity is also present, and has been demonstrated to act on steroid and other sulfates inside in the cell [119,120].

Knowledge on the metabolism and bioavailability of phenolic compounds has been increasing for the past years. These data are primordial for a reasonable evaluation of their potential health effects and for a proper understanding of the biological activities at the tissue and cellular level, leading to a comprehensive view on how phenolics influence metabolic pathways and consequently affect human health.

8.2   CONCLUSION

Strategies for the intervention and prevention of diseases require an understanding of the basic molecular mechanism(s) by prophylactic dietary antioxidant factors from food plants that may potentially prevent or reverse their promotion or progression. Plant food-derived antioxidants may be important in protecting against a number of diseases and reducing oxidation processes in food systems. In order to establish this, it is imperative to measure the markers of baseline oxidative stress, particularly in human health and disease, and examine how they are affected by use of the pure compounds or complex plant foods. It has become important to evaluate how the bioactive components in plant foods affect cellular signaling processes and modulate oxidative stress-mediated responses. These are properties distinct from the classical antioxidant actions. Along this line a clinical trial using Mauritian teas has been initiated and it is strongly expected that the outcome of the study will significantly contribute to the debate on the benefit of tea drinking for prophylactic purposes and, more particularly, on its contribution to cardiovascular health. The implication that tea may reduce the risk for cardiovascular diseases is significant, with their estimated impact directly on medical costs and indirectly through productivity losses [121]. Although the quantitative roles of antioxidants are not precisely known in relation to their health benefits there are grounds for encouraging the use of foods rich in phenolics. Mauritian fruits, vegetables and teas are potent examples. Seminal research reports are now beginning to delineate the absorption of phenolics, enabling an understanding of their bioefficacy and neuroefficacy. The knowledge acquired so far should be used and integrated into applied nutrition programs with optimal health benefits.

REFERENCES

1. Kris-Etherton, P. M., Hecker, K. D., Bonanome, A., Coval, S. M., Binkoski, A. E., Hilpert, K. F., Griel, A. E., and Etherton, T. D., Bioactive compounds in foods: their role in the prevention of cardiovascular disease and cancer, *Am. J. Med.*, 113, 71S–88S, 2002.

2. Vinson, J. A. and Zhang, J., Black and green teas equally inhibit diabetic cataracts in a streptozotocin-induced rat model of diabetes, *J. Agric. Food. Chem.*, 53, 3710–3713, 2005.

3. Bazzano, L. A., He, J., and Ogden, L. G., Legume consumption and risk of coronary heart disease in U.S. men and women: NHANES I epidemiologic follow-up study, *Arch. Int. Med.*, 161, 2573–2578, 2001.

4. Nakachi, K., Imai, K., and Suga, K., Epidemiological evidence for prevention of cancer and cardiovascular disease by drinking green tea, In *Food Factors for Cancer Prevention*, Ohigashi, H., Osawa, T., Watanabe, J., and Yoshikawa, T. Eds., Springer, Tokyo, pp. 105–108, 1996.

5. Singh, R. D., Dubnov, G., Niaz, M. A., Ghosh, S., Singh, R., Rastogi, S. S., Manor, O., Pella, D., and Berry, E. M., Effect of an Indo-Mediterranean diet on progression of coronary artery disease in high risk patients (Indo-Mediterranean Diet Heart Study): a randomized single-blind trial, *Lancet*, 360, 1455–1461, 2002.

6. Dauchet, L., Amouyel, P., Hercberg, S., and Dallongeville, J., Fruit and vegetable consumption and risk of coronary heart disease: A meta-analysis of cohort stuides, *J. Nutr.*, 136, 2588–2593, 2006.

7. Bahorun, T., Luximon-Ramma, A., Crozier, A., and Aruoma, O. I., Total phenol, flavonoid, proanthocyanidin and vitamin C levels and antioxidant activities of Mauritian vegetables, *J. Sci. Food Agric.*, 84, 1553–1561, 2004.

8. Bordoni, A., Hrelia, S., Angeloni, C., Giordano, E., Guarnieri, C., Caldarera, C. M., and Biagi, P. L., Green tea protection of hypoxia/reoxygenation injury in cultured cardiac cells, *J. Nutr. Biochem.*, 13, 103–111, 2002.

9. Luximon-Ramma, A., Bahorun, T., and Crozier, A., Antioxidant actions and phenolic and vitamin C contents of Common Mauritan exotic fruits, *J. Sci. Food Agric.*, 83, 496–502, 2003.

10. Vági, E., Rapavi, E., Hadolin, M., Vásárhelyiné Pérédi, K., Balázs, A., Blázovics, A., and Simándi, B., Phenolic and triterpenoid antioxidants from Origanum majorana L. Herb and extracts obtained with different solvents, *J. Agric. Food Chem.*, 53, 17–21, 2005.

11. Middleton, E. Jr., Kandaswami, C., and Theoharides, T. C., The effects of plant flavonoids on mammalian cells: implications for inflammation, heart disease and cancer, *Pharmacol. Rev.*, 52, 673–839, 2000.

12. Noguchi, N. and Niki, E., Phenolic antioxidants: a rationale for design and evaluation of novel antioxidant drug for atherosclerosis, *Free Radic. Biol. Med.*, 28, 1538–1546, 2000.

13. Mandel, S. and Youdim, M. B. H., Catechin polyphenols: neurodegeneration and neuroprotection in neurodegenerative diseases, *Free Radic. Biol. Med.*, 37, 304, 2004.

14. Bors, W., Heller, W., Michel, C., and Saran, M., Flavonoids as antioxidants: determination of radical scavenging efficiencies, *Methods Enzymol.*, 186, 343–355, 1990.

15. Hou, Z., Lambert, J. D., Chin, K. V., and Yang, C. S., Effects of tea polyphenols on signal transduction pathways related to cancer chemoprevention, *Mutat. Res.*, 555, 3–19, 2003.

16. Kundu, J. K. and Surh, Y. J., Breaking the relay in deregulated cellular signal transduction as a rationale for chemoprevention with anti-inflammatory phytochemicals, *Mutat. Res.*, 591, 123–146, 2005.

17. Mira, L., Feranadez, M. T., Santos, M., Rocha, R., Florencio, M. H., and Jennings, K. R., Interactions of flavonoids with iron and copper ions: a mechanism for their antioxidant activity, *Free Radic. Res.*, 36, 1199–1208, 2002.

18. Toyokuni, S., Tanaka, T., Kawaguchi, W., Lai Fang, N. R., Ozeki, M., Akatsuka, S., Hiai, H., Aruoma, O. I., and Bahorun, T., Effects of the phenolic contents of Mauritian endemic plant extracts on promoter activities of antioxidant enzymes, *Free Radic. Res.*, 37, 1215–1224, 2003.

19. Mauremootoo, J. R., Watt, I., and Florens, F. B. V., The biodiversity of Mauritius and Rodrigues, Edmond, R., Langrand, O. L., Galindo-Leal, C., State of the Hotspots, Madagascar and Indian Ocean Islands. *Conservation International*, Washington D.C., submitted.

20. Kueffer, C. and Lavergne, C., Case studies on the status of invasive woody plant species in the western Indian Ocean, Reunion, Forest Health and Biosecurity working papers, Food and Agricultural Organization, p. 37, 2004.

21. Lavergne, R., Plantes du jardin d'acclimatation de Saint Denis de l'Ile de la Reunion, St Denis, Reunion, Museum d'Histoire Naturelle, 2001.

22. Heeroo, R., Documentation of the weed status of exotic plants in Mauritius based on the records of the Mauritius Herbarium, Bsc Thesis, University of Mauritius, 2000.

23. http://statsmauritius.gov.mu/report/natacc/agri04/sect3.pdf (accessed on November 2005).

24. http://statsmauritius.gov.mu/report/natacc/anudig04/popu.pdf (accessed on November 2005).

25. Halliwell, B., Free radicals and antioxidants: a personal view, *Nutr. Rev.*, 52, 253–265, 1994.

26. Aruoma, O. I., Kang, K. S., Bahorun, T., Sung, M. K., and Rahman, I., Oxidative damage and chronic inflammation induced by smoking: Potential antioxidant and peripheral biomarker considerations, *Cancer Prevention Res.*, 10, 149–158, 2005.

27. Ursini, F., Maiorino, M., Brigelius-Flohé, R., Aumann, K. D., Roveri, A., Schomburg, D., and Flohé, L., The diversity of glutathione peroxidase, *Methods Enzymol.*, 252, 38–63, 1995.

28. Kagan, V. E., Serbinova, E. A., and Packer, L., Antioxidant effects of ubiquinones in microsomes and mitochondria are mediated by tocopherol recycling, *Biochem. Biophys. Res. Commun.*, 169, 851–857, 1990.

29. Aruoma, O. I., Nutrition and health aspects of free radicals and antioxidants, *Food Chem. Toxicol.*, 32, 671–683, 1994.

30. Aruoma, O. I., Assessment of potential pro-oxidant and antioxidant actions, *J. Am. Oil Chem. Soc.*, 73, 1617–1625, 1996.

31. Rice-Evans, C., Miller, N. J., and Paganga, G., Antioxidant properties of phenolic compounds, *Trends Plant Sci.*, 2, 152–159, 1997.

32. Chen, C. and Kong, A. N. T., Dietary cancer-chemopreventive compounds: from signaling and gene expression to pharmacological effects, *Trends Pharmacol. Sci.*, 26, 318–326, 2005.

33. Burton, G. W. and Traber, M. G., Vitamin E: antioxidant activity biokinetics and bioavailability, *Annu. Rev. Nutr.*, 10, 357–382, 1990.

34. Bendich, A., Machlin, L. J., Scandurra, O., Burton, G. W., and Wayner, D. D. M., The antioxidant role of vitamin C, *Free Radic. Biol. Med.*, 2, 419–444, 1986.

35. Regoli, F. and Winston, G. W., Quantification of total oxidant scavenging capacity of antioxidants for peroxynitrite, peroxyl radicals, and hydroxyl radicals, *Toxicol. Appl. Pharmacol.*, 156, 96–105, 1999.

36. Held, K. D., Models for thiol protection of DNA in cells, *Pharmacol. Ther.*, 39, 123–313, 1988.

37. Halliwell, B. and Gutteridge, J. M. C., Role of free radicals and catalytic metal ions in human disease: an overview, *Methods Enzymol.*, 186, 1–85, 1990.

38. Aruoma, O. I., Halliwell, B., Gajewski, E., and Dizdaroglu, M., Copper-ion dependent damage to the bases in DNA in the presence of hydrogen peroxide, *Biochem. J.*, 273, 601–604, 1991.

39. Aruoma, O. I. and Halliwell, B., Action of hypochlorous acid on the antioxidant enzymes SOD, catalase and glutathione peroxidase, *Biochem. J.*, 241, 273–278, 1987.

40. Aruoma, O. I., Whiteman, M., England, T. G., and Halliwell, B., Antioxidant action of ergothioneine: assessment of its ability to scavenge peroxynitrite, *Biochem. Biophys. Res. Commun.*, 231, 389–391, 1997.

41. Asmus, K. D., Bensasson, R. V., Bernier, J. L., Houssin, R., and Land, E. J., One-electron oxidation of ergothioneine and analogues investigated by pulse radiolysis: redox reaction involving ergothioneine and vitamin C, *Biochem. J.*, 315, 625–629, 1996.

42. Moran, J. F., Klucas, R. V., Grayer, R. J., Abian, J., and Becana, M., Complexes of iron with phenolic compounds from soybean nodules and other legume tissues: preoxidant and antioxidant properties, *Free Radic. Biol. Med.*, 22, 861–870, 1997.

43. Yardim-Akaydin, S., Sepici, A., Ozkan, Y., Torun, M., Simsek, B., and Sepici, V., Oxidation of uric acid in rheumatoid arthritis: is allantoin a marker of oxidative stress?, *Free Radic. Res.*, 38, 623–628, 2004.

44. Wink, D. A. and Mitchell, J. B., Nitric oxide and cancer: an introduction, *Free Radic. Biol. Med.*, 34, 951–954, 2003.

45. Mhatre, M., Floyd, R. A., and Hensley, K., Oxidative stress and neuroinflammation in Alzheimer's disease and amyotrophic lateral sclerosis: Common links and potential therapeutic targets, *J. Alzheimer's Diseases*, 6, 147–157, 2004.

46. Lefer, J. and Granger, D. N., Oxidative stress and cardiac disease, *Am. J. Med.*, 109, 315–323, 2000.

47. Crack, P. J. and Taylor, J. M., Reactive oxygen species and the modulation of stroke, *Free Radic. Biol. Med.*, 38, 1433–1444, 2005.

48. Dennery, P. A., Introduction to serial review on the role of oxidative stress in diabetes mellitus, *Free Radic. Biol. Chem.*, 40, 1–2, 2006.

49. Junqueira, V. B. C., Barros, S. B. M., Chan, S. S., Rodrigues, L., Giavarotti, L., Abud, R. L., and Deucher, G. P., Aging and oxidative stress, *Mol. Aspects Med.*, 25, 5–16, 2004.

50. Barnes, P. J. and Karin, M., Nuclear factor-kappa B: a pivotal transcription factor in chronic inflammatory diseases, *N. Engl. J. Med.*, 336, 1066–1071, 1997.

51. Brownlee, M., Biochemistry and molecular cell biology of diabetic complications, *Nature*, 414, 813–820, 2001.

52. DeFronzo, R. A., Pathogenesis of type 2 diabetes: metabolic and molecular implications for identifying diabetes genes, *Diabetes Rev.*, 5, 177–269, 1997.

53. Evans, J. L., Goldfine, I. D., Maddux, B. A., and Grodsky, G. M., Oxidative stress and stress-activated signaling pathways: a unifying hypothesis of type 2 diabetes, *Endocr. Rev.*, 23, 599–622, 2002.

54. Kyriakis, J. M. and Avruch, J., Sounding the alarm: protein kinase cascades activated by stress and inflammation, *J. Biol. Chem.*, 271, 24313–24316, 1996.

55. Weinbrenner, T., Cladellas, M., Covas, M. I., Fito, M., Toma's, M., Sentı, M., and Marrugat, J., High oxidative stress in patients with stable coronary heart disease, *Atherosclerosis*, 168, 99–106, 2003.

56. Gao, W. D., Liu, Y., and Marban, E., Selective effects of oxygen free radicals on excitation-contraction coupling in ventricular muscle. Implications for the mechanism of stunned myocardium, *Circ. Res.*, 94, 2597–2604, 2006.

57. Finkel, T., Signal transduction by reactive oxygen species in non-phagocytic cells, *J. Leukoc. Biol.*, 65, 337–340, 1999.

58. Shah, A. M. and Channon, K. M., Free radicals and redox signalling in cardiovascular disease, *Heart*, 90, 486–487, 2004.

59. Liochev, S. F. and Fridovich, I., Mutant Cu, Zn superoxide dismutases and familial amyotrophic lateral sclerosis: evaluation of oxidative hypotheses, *Free Radic. Biol. Med.*, 34, 1383–1389, 2003.

60. Mhatre, M., Floyd, R. A., and Hensleya, K., Oxidative stress and neuroinflammation in Alzheimer's disease and amyotrophic lateral sclerosis: common links and potential therapeutic targets, *J. Alzheimer's Dis.*, 6, 147–157, 2004.

61. Ishikawa, Y., Kusaka, E., Enokido, Y., Ikeuchi, T., and Hatanaka, H., Regulation of Bax translocation through phosphorylation at Ser-70 of Bcl-2 by MAP kinase in NO-induced neuronal apoptosis, *Mol. Cell. Neurosci.*, 24, 451–459, 2003.

62. Hsu, T. C., Young, M. R., Cmarik, J., and Colburn, N. H., Activator protein 1 (AP-1) and nuclear factor κB (NF-κB)-dependent transcriptional events in carcinogenesis, *Free Radic. Biol. Med.*, 28, 1338–1348, 2000.

63. Bravo, L., Polyphenols: chemistry, dietary sources, metabolism, and nutritional significance, *Nutr. Rev.*, 56, 317–333, 1998.

64. Luximon-Ramma, A., Bahorun, T., Crozier, A., Zbarsky, V., Datla, K. K., Dexter, D. T., and Aruoma, O. I., Characteristion of the antioxidant functions of flavonoids and proanthocyanidins in Mauritian black teas, *Food Res. Int.*, 38, 357–367, 2005.

65. Hollman, P. C. and Katan, M. B., Bioavailability and health effects of dietary flavonols in man, *Arch. Toxicol. Suppl.*, 20, 237–248, 1998.

66. Ariga, T., Koshiyama, I., and Fukushima, M., Antioxidative properties of procyanidin B1 and B3 from azuki beans in aqueous system, *J. Agric. Biol. Chem.*, 52, 2717–2722, 1988.

67. Arteel, G. E. and Sies, H., Protection against peroxynitrite by cocoa polyphenol oligomers, *FEBS Lett.*, 462, 167–170, 1999.

68. Bors, W. and Michel, C., Antioxidant capacity of flavanols and gallate esters: pulse radiolysis studies, *Free Radic. Biol. Med.*, 27, 1413–1426, 1999.

69. Kaur, C. K. and Kapoor, H. C., Antioxidants in fruits and vegetables-the millenium's health, *Int. J. Food Sci. Technol.*, 36, 703–725, 2001.

70. Deighton, N., Brennan, R., Finn, C., and Davies, H. V., Antioxidant properties of domesticated and wild Rubus species, *J. Agric. Food. Chem.*, 80, 1307–1313, 2000.

71. Li, Y., Ou-Lee, T. M., Raba, R., Amundson, R. G., and Last, R. L., Arabidopsis flavonoid mutants are hypersensitive to UV-B irradiation, *Plant Cell*, 5, 171–175, 1993.

72. Cao, G., Sofic, E., and Prior, R. L., Antioxidant capacity of tea and common vegetables, *J. Agric. Food Chem.*, 4, 3426–3431, 1996.

73. Chu, Y. F., Sun, J., Wu, X., and Liu, R. H., Antioxidant and antiproliferative activities of common vegetables, *J. Agric. Food. Chem.*, 50, 6910–6916, 2000.

74. Proteggente, A. R., Pannala, A. S., Paganga, G., Van Buren, L., Wagner, E., Wiseman, S., Van de Puy, F., Dacombe, C., and Rice-Evans, C., The antioxidant activity of regularly consumed fruit and vegetables reflects their phenolic and vitamin C composition, *Free Radic. Res.*, 36, 217–233, 2002.

75. Kaur, C. K. and Kapoor, H. C., Antioxidant activity and total phenolic content of come Asian vegetables, *Int. J. Food Sci. Technol.*, 36, 217–233, 2002.

76. Aruoma, O. I., Methodological considerations for characterizing potential antioxidant actions of bioactive components in plant foods, *Mutat. Res.*, 523–524, 9–20, 2003.

77. Schlesier, K., Harwat, M., Bóhm, V., and Bitsch, R., Assessment of antioxidant activity by using different in vitro methods, *Free Radic. Res.*, 36, 177–187, 2002.

78. Johnson, I. T., Glucosinolates in the human diet. Bioavailability and implications for health, *Phytochem. Rev.*, 1, 183–188, 2002.

79. Zhang, Y. and Talalay, P., Anticarcinogenic activities of organic isothiocyanates: chemistry and mechanisms, *Cancer Res.*, 54, 1976S–1981S, 1994.

80. Cohen, J., Krital, R., and Stanford, J., Fruit and vegetable intakes and prostate cancer, *J. Natl Cancer Inst.*, 9, 61–68, 2000.

81. Hoefler, A. C. and Coggon, P., Reversed phase high performance liquid chromatography of tea constituents, *J. Chromatogr.*, 129, 460–463, 1976.

82. Khokhar, S. and Magnusdottir, S. G. M., Total phenol, catechin, and caffeine contents of teas commonly consumed in the United Kingdom, *J. Agric. Food Chem.*, 50, 565–570, 2002.

83. Treutter, D., Chemical reaction detection of catechins and proanthocyanidins with 4-dimethylaminocinnamaldehyde, *J. Chromatogr.*, 467, 185–193, 1989.

84. Shao, W., Powell, C., and Clifford, M. N., The analysis by HPLC of green, black, and Pu'er teas produced in Yunnan, *J. Sci. Food Agric.*, 69, 535–540, 1995.

85. Lin, J. K., Juan, I. M., Chen, Y. C., Liang, Y. C., and Lin, J. K., Composition of polyphenols in fresh tea leaves and associations of their oxygen-radical absorbing capacity with antiproliferative actions in fibroblast cells, *J. Agric. Food Chem.*, 44, 1387–1394, 1996.

86. Benzie, I. F. F. and Szeto, Y. T., Total antioxidant capacity of teas by the ferric reducing antioxidant power assay, *J. Agric. Food Chem.*, 47, 633–636, 1999.

87. Wiseman, S., Balentine, D. A., and Frei, B., Antioxidants in tea, *Crit. Rev. Food Sci. Nutr.*, 37, 705–718, 1997.

88. Lien, E. J., Ren, S., Bui, H., and Wang, R., Quantitative structure–activity relationship analysis of phenolic antioxidants, *Free Radic. Biol. Med.*, 26, 285–294, 1999.

89. Gardner, P. T., Mc Phail, D. B., and Duthie, G. G. D., Electron spin resonance spectroscopic assessment of the antioxidant potential of teas in aqueous and organic media, *J. Sci. Food Agric.*, 76, 257–262, 1998.

90. Hertog, M. G. L., Fesken, E., Hollman, P., Katan, M., and Kromhout, D., Dietary antioxidant flavonoids and risk of coronary heart disease: the Zutphen elderly study, *Lancet*, 342, 1007–1010, 1993.

91. Kühnau, J., The flavonoids. A class of semi-essential food compounds: their role in human nutrition, *World Rev. Nutr. Cancer*, 24, 117–191, 1976.

92. Cao, G., Sofic, E., and Prior, R. L., Antioxidant and prooxidant behaviour of flavonoids: structure–activity relationships, *Free Radic. Biol. Med.*, 22, 749–760, 1997.

93. Shekher Pannala, A., Chan, T. S., O'Brien, P. J., and Rice-Evans, C. A., Flavonoids B-ring chemistry and antioxidant activity: fast reaction kinetics, *Biochem. Biophys. Res. Commun.*, 282, 1161–1168, 2001.

94. Soobrattee, M. A., Neergheen, V. S., Luximon-Ramma, A., Aruoma, O. I., and Bahorun, T., Phenolics as potential antioxidant therapeutic agents: mechanism and actions, *Mutat. Res.*, 579, 200–213, 2005.

95. Bagchi, D., Sen, C. K., Ray, S. D., Das, D. K., Bagchi, M., Preus, H. G., and Vinson, J. A., Molecular mechanisms of cardioprotection by a novel grape seed proanthocyanidin extract, *Mutat. Res.*, 9462, 1–11, 2003.

96. Pernow, J. and Wang, Q. D., Endothelin in myocardial ischaemia and reperfusion, *Cardiovasc. Res.*, 33, 518–526, 1997.

97. Verstraeten, S. V., Keen, C. L., Schmitz, H. H., Fraga, C. G., and Oteiza, P. I., Flavan-3-ols and procyanidins protect liposomes against lipid oxidation and disruption of the bilayer structure, *Free Radic. Biol. Med.*, 34, 84–92, 2003.

98. Schroeter, H., Spencer, J. P., Rice-Evans, C., and Williams, R. J., Flavonoids protect neurons from oxidized low-density lipoprotein-induced apoptosis involving c-Jun *N*-terminal kinase (JNK), c-Jun and caspase-3, *Biochem. J.*, 358, 547–557, 2001.

99. Levites, Y., Weinreb, O., Maor, G., Youdim, M. B. H., and Mandel, S., Green tea polyphenol (-)-epigallocatechin 3-gallate prevents *N*-methyl-4-phenyl-1,2,3,6-tetrahydropyridine-induced dopaminergic neurodegenration, *J. Neurochem.*, 78, 1073–1082, 2001.

100. Levites, Y., Amit, Y., Mandel, S., and Youdim, M. B. H., Neuroprotection and neurorescue against amyloid beta toxicity and PKC dependent release of non-amyloidogenic soluble precursor protein by green tea polyphenol (−)-epigallocatechin 3-gallate, *FASEB J.*, 17, 952–954, 2003.

101. Merry, D. E. and Korsmeyer, S. J., Bcl-2 gene family in the nervous system, *Ann. Rev. Neurosci.*, 20, 245–267, 1997.

102. Burda, S. and Oleszek, W., Antioxidant and antiradical activities of flavonoids, *J. Agric. Food Chem.*, 49, 2774–2779, 2001.

103. van Acker, S. A. B. E., de Groot, M. J., van de Berg, D. J., Tromp, M. N. J. L., den Kelder, G. D. O., van der Vijgh, W. J. F., and Bast, A., A quantum chemical explanation of the antioxidant activity of flavonoid, *Chem. Res. Toxicol.*, 9, 1305–1312, 1996.

104. Ishige, K., Schubert, D., and Sagara, Y., Flavonoids protect neuronal cells from oxidative stress by three distinct mechanisms, *Free Radic. Biol. Med.*, 30, 433–446, 2001.
105. Dajas, F., Rivera-Megret, F., Blasina, F., Arredondo, F., Abin- Carriquiry, J. A., Costa, G., Echeverry, C. et al., Neuroprotection by flavonoids, *Braz. J. Med. Biol. Res.*, 36, 1613–1620, 2003.
106. Wu, T., Cyclooxygenase-2 and prostaglandin signaling in cholangiocarcinoma, *Biochim. Biophys. Acta*, 1755, 135–150, 2005.
107. Kundu, J. K., Na, H. K., Chun, K. S., Kim, Y. K., Lee, S. J., Lee, S. S., Lee, O. S., Sim, Y. C., and Surh, Y. J., Inhibition of phorbol ester-induced COX-2 expression by epigallocatechin gallate in mouse skin and cultured human mammary epithelial cells, *Am. Soc. Nutr. Sci.*, S3805–S3810, 2003.
108. Ye, F., Wu, J., Dunn, T., Yi, J., Tong, X., and Zhang, D., Inhibition of cyclooxygenase-2 activity in head and neck cancer cells by genistein, *Cancer Lett.*, 211, 39–46, 2004.
109. Surh, Y. J., Chun, K. S., Cha, H. O., Han, S. H., Keum, Y. S., Park, K. K., and Lee, S. S., Molecular mechanisms underlying chemopreventive activities of anti-inflammatory phytochemicals: down-regulation of COX-2 and iNOS through suppression of NF-κB activation, *Mutat. Res.*, 480–481, 243–268, 2001.
110. Day, A. J. and Williamson, G., Absorption of quercetin glycosides, In *Flavonoids in Health and Disease*, Rice-Evans, C. and Packer, L. Eds., Marcel Dekker, New York, pp. 31–412, 2003.
111. Seeram, N. P., Henning, S. M., Zhang, Y., Suchard, M., Li, Z., and Heber, D., Pomegranate juice ellagitannin metabolites are present in human plasma and some persist in urine for up to 48 hours, *J. Nutr.*, 136, 2481–2485, 2006.
112. Price, K. R., Bacon, J. R., and Rhodes, M. J. C., Effect of storage and domestic processing on the content and composition of flavonol glucosides in onion (Allium cepa), *J. Agric. Food Chem.*, 45, 938–942, 1997.
113. Williamsom, G., The use of flavonoid aglycones in in vitro systems to test biological activities: based on bioavailability data, is this a valid approach?, *Phytochem. Rev.*, 1, 215–222, 2002.
114. Manach, C., Regerat, F., Texier, O., Agullo, G., Demigne, C., and Remesy, C., Bioavailability, metabolism and physiological impact of 4-oxo-flavonoids, *Nutr. Res.*, 16, 517–544, 1996.
115. Hollman, P. C. H., de Vries, J. H. M., van Leeuwen, S. D., Mengelers, M. J. B., and Katan, M. B., Absorption of dietary quercetin glycosides and quercetin in healthy ileostomy volunteers, *Am. J. Clin. Nutr.*, 62, 1276–1282, 1995.
116. Day, A. J., Dupont, M. S., Ridley, S., Rhodes, M., Rhodes, M. J. C., Morgan, M. R. A., and Williamson, G., Deglycosylation of flavonoid and isoflavonoid glycosides by human small intestine and liver β-glucosidase activity, *FEBS Lett.*, 436, 71–75, 1998.
117. Day, A. J., Canada, F. J., Diaz, J. C., Kroon, P. A., Mclauchlan, R., Faulds, C. B., Plumb, G. W., Morgan, M. R. A., and Williamson, G., Dietary flavonoid and isoflavone glycosides are hydrolysed by the lactase site of lactase phlorizin hydrolase, *FEBS Lett.*, 468, 166–170, 2000.
118. Scalbert, A. and Williamson, G., Dietary intake and bioavailability of polyphenols, *J. Nutr.*, 130, 2073S–2085S, 2000.
119. Manach, C., Scalbert, A., Morand, C., Remesy, C., and Jimenez, L., Polyphenols: food sources and bioavailability, *Am. J. Clin. Nutr.*, 79, 727–747, 2004.

120. Manach, C. and Donovan, J., Pharmacokinetics and metabolism of dietary flavonoids in humans, *Free Radic. Res.*, 38, 771–785, 2004.
121. Walle, U. K., Galijatovic, A., and Walle, T., Transport of the flavonoid chrysin and its conjugated metabolites by the human intestinal cell line Caco-2, *Biochem. Pharmocol.*, 58, 431–438, 1999.
122. Blumberg, J., Introduction to the proceedings of the third scientific symposium on tea and human health, *J. Nutr.*, 133, S3244–S3246, 2003.
123. Luximon-Ramma, A., Neergheen, V. S., Bahorun, T., Crozier, A., Zbarsky, V., Datla, K. P., Dexter, D. T., and Aruoma, O. I., Assessment of the polyphenolic composition of the organic extracts of Mauritian black teas: A potenital contributor to their antioxidant functions, *BioFactors.*, 27, 79–91, 2006.

9 Functional Foods in Mediterranean and Middle Eastern Countries: History, Scope and Dietary Habits

Fatih Yildiz, Paterna Kotzekidou, Alexandra-Maria Michaelidou, and Giuseppe Nocella

CONTENTS

9.1 INTRODUCTION

The Mediterranean Diet (MD) has been evolving for thousands of years hand-in-hand with the economic development, as well as the scientific and technological progress of the countries which form this region, and has been continually shaping both choice of foods and methods of cooking. Although the MD is common to a large geographical area, identifying a single food which is common to the multifarious cultures it contains, is certainly not easy. Ferro-Luzzi and Sette [1] attempted to identify a common denominator amongst the many MDs and recognized them all to be high in cereals (more than 60% of total energy), low in total fats (less than 30%), with a predominant use of olive oil which represents more than 70% of total lipids, and relatively rich in a variety of fruits and vegetables, which provide at least half of the total amount of dietary fiber (30 g/day).

As is well known, this type of diet has been under the scrutiny of scholars from all over the world, due to the fact that it would appear that its ingredients can reduce death caused by coronary heart disease (CHD) and cancers. In fact, several studies demonstrate that the MD can have a positive effect on the well being of individuals and in particular, in the prevention of heart disease [2,3]. In a study conducted in seven countries, Keys found that CHD was closely related to dietary lifestyles. Groups of people that followed a MD with olive oil as the primary source of fat had the longest lifespan and the lowest incidence of CHD and cancers. These findings are of paramount importance, considering that CHD accounts for 7.2 million deaths a year and is the leading cause of death in industrialized nations. Additional studies among North American and European populations have clearly shown that diets which, according to the International Consensus Statement [4], are "rich in complex carbohydrates and fiber and whose fat source is primarily monounsaturated fatty acids, as found in the olive oil-rich Mediterranean-style diet, lowers low-density lipoprotein (LDL) cholesterol and are associated with a low incidence of CHD." One thriving example of this is the island of Crete, the hub of the MD, where the population has an incidence of coronary disease 37 times lower than Americans.

9.1.1 "FUNCTIONALITY" AND PRINCIPLES OF THE MEDITERRANEAN DIET

The health of the individual and the population in general is the result of interactions between genetics and a number of environmental factors. Nutrition appears to be an environmental factor of major importance [5] as epidemiological, pre-clinical, and clinical studies continue to provide fundamental insights into the dynamic relationships between nutrients and health [6]. Although, much of the attention on nutrients during the past decade has focused on their antioxidant activity, new, and emerging genomic and proteonomic approaches and technologies offer exciting opportunities for identifying molecular targets for dietary components and, thus, determining mechanisms by which they influence the quality of life [6]. In fact, genomic technologies are new weapons in the scientific arsenal that arm nutrition scientists with the ability to leave behind the reductionist method of investigating single nutrient effects on a biological system for a more holistic approach of exploring the molecular details of food nutrient effects on an entire biological organism [7].

Hence, the advances in understanding the relationship between nutrition and health, often at the molecular level, led to the concept of "functional foods," i.e., foods that should have a relevant effect on well-being and health or result in a reduction in disease risk. The functional component of a functional food can be an essential macronutrient or micronutrient, a nutrient that is not considered essential or a non-nutritive component [8,9]. In this respect, fruit, vegetables, and whole grains, delivering "packages" of constituents that may promote health (e.g., vitamins, fiber, and plant chemicals) could be designated as functional foods [10,11]. However, nutrition experts emphasize that the degree of healthy eating needs to be judged on the level of the whole diet [12]. Thus, assessing the relationship of health with overall diet rather than with single nutrients, foods or food groups has intuitive appeal. Free-living people eat combinations of foods containing a mix of nutrients and non-nutrients. The food combinations consumed reflect individual food preferences modulated by a mix of genetic, cultural, social, health, environmental, lifestyle, and economic determinants [13]. For instance, the MD, i.e., the dietary pattern of the geographical region that borders the Mediterranean Sea is certainly shaped by climatic conditions. This part of the world is blessed with a remarkably mild climate that supports great biodiversity. In addition to mainland produce, the Mediterranean generously offers its people a wealth of fish and seafood [14].

Consequently, the MD is generally described as providing an abundance of plant foods (fruits, vegetables, bread, cereal products, legumes, nuts and seeds), favoring the consumption of locally grown, seasonally fresh and minimally processed foods [15]. Fish consumption is also increased, depending on proximity to the sea. Additionally, since olive cultivation is a symbolic feature of the traditional Mediterranean agriculture [16], olive oil is the principal source of fat. Usually, more of some foods means that less of other foods is being consumed [13]. In this respect, the Mediterranean food culture comprises a low proportion of animal fat, little meat and a moderate intake of milk and dairy products [15]. This dietary pattern corresponds to the period of the early 1960s, as contemporary lifestyle has changed food habits introducing more meat and processed food than earlier. Nevertheless, the MD, as described above, remains a target-pattern for a health-promoting diet.

Given the broad geographical region, including at least 16 countries that border the Mediterranean Sea [17], a regional variability should be expected. Thus, the Italian variant of the MD is characterized by higher consumption of pasta, whereas in Spain, fish consumption is particularly high. The Greek version is dominated by the consumption of olive oil and by a high consumption of vegetables and fruits [15].

However, despite regional variability and disparities, the philosophy of the food culture is a common denominator in the Mediterranean region. The principles governing cooking are the same in every place and encompass the use of olive oil, herbs, and fresh vegetables. Thus, the functional benefits of the individual ingredients are maximized through effective combinations improving the capacity to resist disease and enhancing health.

On the road to optimum nutrition, which is an ambitious and long-term objective [9], the Mediterranean-style diet is an interesting and stimulating concept that could serve as a general nutritional suggestion with personalized dietary interventions when necessary. As recent literature is characterized by burgeoning interest in examining the health benefits of this dietary pattern, it appears that the current momentum towards the MD has solid biological foundation and does not represent a transient fashion [18]. Besides, the MD bears a centuries-old tradition that provides a sense of pleasure and well being, and forms a vital part of the world's collective cultural heritage [19].

9.1.2 Similarities and Distinctions Between Mediterranean and Middle Eastern Functional Foods

Mediterranean and Middle Eastern Dietary habits and food choices differ in some respects. More vegetable oils (other than olive oil) and more red meat is used in the Middle Eastern Cookery. Food choices in the Middle East include hummus, falafel, tahini, couscous, and bulgur. Mediterranean food choices include, more olive oil dishes, fish, pasta, paella (rice cooked with fish), wine, vinegar, and a variety of green salads.

9.2 ENVIRONMENTAL EFFECTS ON FUNCTIONAL PROPERTIES

9.2.1 Agriculture, Climate, Soil and Crops. Four Climatic Zones: Arid, Semi-Arid, Sub-Humid, Humid

Growing season, location, the amount of sun exposure, rain, and the type of soil will affect the amount and variety of functional ingredients in plant and animal functional foods

The Mediterranean climate, one of the most distinctive in the world, is characterized by mild, wet winters, and hot, long dry summers. According to the historical, archeological, and geological evidence; until the end of 3000 B.C. there was much greater humidity and rainfall throughout the dry zones of the region; this enabled a rich fauna and flora to thrive. Before 2000 B.C. conditions were slightly more arid than at present; since 1000 B.C. climatic condition began to improve, and since that time, there have been only short-term variations. Mediterranean or dry-summer subtropical climate also occurs in central and coastal southern California, Central Chili, western Cape province in South Africa, and portions of Southern Australia. The large size and the mountains of the Mediterranean produce a variety of local modifications that range from desert to humid mountain Climates. Precipitation generally decreases towards the south and east and ranges from more than 2540 mm (100 in.) per year near Dalmatian to less than 255 mm (10 in.) annually in parts of the North Africa.

Typical summer-month average temperatures are 25°C (77°F); the coolest-month average is 8°C–10°C (46°F–50°F). Centuries of the use of the natural resources of the soil and the vegetation induced erosion, whereby the surface of the land is gradually removed by the action of wind or water. The key to protect the soil erosion is reforestation and terracing at lower elevations, where cultivation can be practiced. Soil and water conservation is also needed in the lowlands.

Four main climatic zones have been identified as arid, semi-arid, humid, and sub-humid regions. The main forms of the land use in the four zones are shown, together with the most important crops in rain-fed farming. Temperature and rainfall are illustrated by data from selected meteorological stations. Within each of the major zones there is a multitude of variations in physical conditions which arise from differences in topography and from the influence of mountain and the sea.

The topographical map shows how the high mountains dominate in the coastal landscape in the Mediterranean. These physical characteristics, together with soil and geological formations, have largely determined the composition and distribution of the native vegetation, including forests. The generally steep and rocky coasts are often deeply indented and interrupted by small scattered plains. Most rivers entering into the Mediterranean carry sediments; however the only large deltas are those of the Nile, Rhone, and the Po rivers. Other important rivers—the Ebro in Spain, the Arno, Tiber, Isonzo in Italy, Vardar, Striman, and Nestos in Greece, and Gediz, Menderes, Seyhan, and Ceyhan in Turkey—form the fertile plains.

The entire Mediterranean basin is tectonically active with frequent earthquakes, particularly in Greece and Turkey. Many of the islands are the peaks of the volcanoes, some of which are still active.

The arid zone includes parts of North Africa which extend to the coast in Libya and Egypt and the deserts of Syria and Iraq. Vegetation consists of scattered trees, shrubs, chaparrals, shrublets various succulent and perennial grasses and herbs including such genera as Stipa (grass) Artemissia (woody shrubs), and Acacia, Pistachio (trees). Following heavy early rains the farmers sow barley and hard wheat (*Triticum durum*) in valley bottoms where water collects.

The semi-arid zone embraces North African Coast, eastern Mediterranean, parts of Spain, Central Anatolia, intermediate areas between the coast and desert in Egypt, Syria, and Israel.

Typical Mediterranean flora is seen in this zone. Evergreen trees, figs, olives, carobs, cork oak, valonia oak, and Aleppo pine have their habitat here, as well as Eucalyptus. While rainfall is still the most important limiting factor in the semi-arid zone, as in the arid zone, it is heavier and less erratic, and fairly continuous arable farming becomes a possibility. Summer farming is rarely possible in this zone without irrigation.

The sub-humid zone covers parts of the Mediterranean littoral and islands, parts of the highlands, much of Yugoslavia, Northeastern Turkey, and lower mountain areas. The evergreen oak forests and cork forests are found in North Africa, Spain, France, and Italy. Livestock rearing is important, particularly pig feeding on the acorns in Italy and France.

Soft wheat (*Triticum vulgare*), oats, pulse, and flax are found in addition to the hard wheat and barley. Arable cultivation is mixed, with scattered groves of olives, carobs (*Ceratonia siliqua*) and almonds. In the foothills and mountains, vines become of increasing importance, while a variety of summer crops such as sorghum, cotton, and maize, are also grown.

The humid zone includes the higher mountain areas, particularly the northern and western slopes in Yugoslavia, Italy, France, Spain, and Portugal and parts of southern and northeastern Asia Minor. Tree crops are grown without irrigation, including chestnut, hazelnut, plums, walnuts, and tea. The elevated areas of the zone have chestnut, beech, pine, cedar, and poplars constituting the forests of the region.

In the Arid and semi-arid zones, the slopes of the mountains are covered by shrubs and chaparrals, which are woody perennial plants that have more than one main branch or stem. Shrubs are composed of several hundred species of plants,

TABLE 9.1
Rich Bioactive Compounds Containing Locally Utilized Mediterranean Crops (Fruits, Vegetables and Plant Species)

The fig (*Ficus carica*)
The pomegranate (*Punica granatum*)
The persimmon (*Diospyros kaki*)
The loquat (*Eriobotrya japonica*)
The cactus pear (*Opuntia* spp.)
The quince (*Cydonia oblonga* Mill.)
The chestnut (*Castanea sativa* Mill.)
The pistachio (*Pistacia vera*)
The tree strawberry (*Arbutus unedo* L.)
The cornelian cherry (*Cornus mas* L.)
The medlar (*Mespilus germanica* L.)
The jujube (*Zizyphus vulgaris* L.)
Oleaster fruit (*Elaeagus angustofolia*)
The azarole (*Crataegus azarolus* L.)
The service tree (*Sorbus domestica* L.)
The mullbery tree (*Morus* spp. and other *Moraceae*)
The carob tree (*Cerotonia silique* L.)
Rose hips (*Rosa* spp. *Rosaceae*)
Gum mastic tree (*Pistacia atlantica*)

most of which are evergreen belonging to families such as rose, oak, heath, capers, and sumac.

9.2.2 Underutilized and Locally-Known Functional Crops in the Mediterranean Basin

Mediterranean countries consume large quantities of fruits and vegetables in their diet, during the growing seasons. These fruits and vegetables include crops that are rare or underutilized in other parts of the world. Table 9.1 give a short listing of these crops. Most of these crops are known to be a very rich source of phenolic compounds.

9.3 BIOACTIVE COMPOUNDS IN PLANT AND ANIMAL FOOD PRODUCTS

The Mediterranean Diet, as previously mentioned, is largely vegetarian in nature. Scientific evidence indicates, that a high consumption of plant-based foods, such as fruit, vegetables, nuts, whole grains and legumes, is associated with a significantly lower risk of chronic diseases. The protective effects of these foods are probably mediated through multiple beneficial nutrients contained in these foods [20].

Along with macronutrients, plant foods contain appreciable amounts of some vitamins, other organic compounds and minerals, as well as dietary fiber.

Dietary fiber encompasses very diverse macromolecules exhibiting a large variety of physicochemical properties. The properties that are nutritionally relevant are mainly particle size and bulk volume, surface area characteristics, hydration, and rheological properties, and the adsorption or entrapment of minerals and organic molecules. Amongst these properties, the viscosity and ion exchange capacity are the main contributors to metabolic effects (glucose and lipid metabolisms), whereas fermentation pattern, bulking effect and particle size are strongly involved in effects on colonic function [21].

There is a substantial body of literature showing that dietary fibers, in particular soluble fibers, decrease blood cholesterol concentrations and may thereby modify the risk of coronary artery disease. Additionally, they may affect risk of large bowel cancers through mechanisms such as altering bile acid metabolism, increasing fecal bulk, or decreasing gut transit time [22]. Thus, the high consumption of fruit and vegetables as introduced by the MD leads to an increased intake of dietary fiber with all the potential health benefits.

Furthermore, the Mediterranean eating pattern includes the consumption of legumes and cereals, foods rich in dietary fiber as well. The relatively high soluble-fiber content of peas and beans was shown to lower blood cholesterol concentrations in feeding studies [22]. Cereals form the basis of the Mediterranean food pyramid. Oats, rye, and barley contain about one-third soluble fiber and two-thirds insoluble fiber, but wheat is lower in soluble fiber [23]. Wheat and its products can be consumed in many ways around the Mediterranean countries. Bread, pitas, and pizzas are the major wheat products. Traditional recipes using local ingredients, without the addition of substantial amounts of sugar and shortenings, promote even more the healthy aspects of these products. In this respect traditional whole-grain products are even better, as refined grains are low in total dietary fiber. Beside fiber, grains, and legumes contain oligosaccharides [23,24]. Oligosaccharides are thought to have effects similar to those of soluble dietary fibers in the human gut. In addition, studies consistently showed that they are able to alter the human fecal flora. Many human studies found that consumption of fructooligosaccharides increased bifidobacteria in the gut while decreasing concentrations of *Escherichia coli*, clostridia, and bacteroides [25]. As far as legumes are concerned, their low glycemic index should also be mentioned; consumption of such foods could be beneficial for diabetics to help control of the level of plasma glucose [26].

Products of plant origin are good sources of minerals as well. In fact, diets rich in potassium, magnesium, and cereal fiber reduce the risk of stroke, particularly among hypertensive men [27]. Furthermore, adequate intake of minerals such as calcium and potassium—specifically derived from foods, where they coexist with other essential nutrients—contributes to cardiovascular as well as overall health [28]. However, in some cases, mostly in legumes and nuts, the presence of phytic acid lowers the bioavailability of these nutrients [29]. Nevertheless, many antinutrients have recently been found to exert positive health effects. Phytic acid (inositol hexaphosphate), for example, is nowadays being promoted as an anticancer agent.

Its proposed mechanism of action includes gene alteration, enhanced immunity and antioxidant properties [30].

Undoubtedly, plant products offer high amounts of vitamins. Whole cereals are rich in vitamins of the B group. Fruit and vegetables are good sources particularly of vitamin A in the form of β-carotene and vitamin C [31]. Vitamin E is present in whole grains and nuts; it is a potent intracellular antioxidant, that protects polyunsaturated fatty acids in cell membranes from oxidative damage [23]. Oxidative stress is involved in the pathology of cancer, atherosclerosis, malaria, and rheumatoid arthritis and could play a role in neurodegenerative diseases and aging processes [32]. Thus, dietary antioxidants, which inactivate reactive oxygen species and provide protection from oxidative damage, are being considered as important preventive strategic molecules. From the point of view of antioxidant activity, vitamin C and carotenoids are not as potent antioxidants as vitamin E. Vitamin C could act synergistically with vitamin E to protect LDL from the attack of free radicals [33]. Furthermore, they could contribute to good health through other mechanisms, such as being co-factors for certain enzymes involved in redox reactions in the body [34]. It has been suggested that vitamin E has other effects on cellular functions that may reduce the risk of CHD, as (a) the inhibition of interleukin-1 secretion from monocytes, which is involved in smooth muscle cell proliferation and monocyte differentiation, by inhibiting protein kinase C, (b) the reduction of platelet reactivity, (c) the maintenance of endothelial membrane integrity, and (d) the modulation of the release of prostacyclin [33].

Green leafy vegetables are good sources of folate, a nutrient with apparent significance in the cancer process. Its essential role in the de novo biosynthesis of purines and pyrimidines, and thus in DNA replication and cell division, and for the synthesis of S-adenosylmethionine, a methyl donor for more than 100 biochemical reactions including methylation of DNA, places it in a unique position relative to DNA stability [6]. Furthermore, folate, B_6 and B_{12} are factors known to influence homocysteine metabolism. Since an elevated level of plasma homocysteine is a risk factor for developing cardiovascular disease [35,36], increase in folate intake would be beneficial. Traditional Mediterranean dishes contain fresh vegetables and salads to accompany meat and fish. Thus, the consumption of seasonably fresh or minimally processed foods enhances the intake of vitamins. Apart the cultivated leafy vegetables, wild plants endemic of the Mediterranean environment are of importance as food ingredients. For example, purslane, a plant commonly grown in Greece is rich in α-linolenic acid [LNA, 18:3 (n-3)] thus improving the ratio of (n-6) to (n-3) fatty acids [5].

Walnuts are a good source of α-linolenic acid ($\sim 7\%$), too. Generally, nuts may protect against coronary artery disease through other mechanisms. Most nuts are rich in arginine, which is the precursor of endothelium-derived relaxing factor, nitric oxide (NO). Nitric oxide is a potent vasodilator that can inhibit platelet adhesion and aggregation [20,37]. Nuts are also a good source of fat, mainly monounsaturated and polyunsaturated.

Thus, plant-derived foods, although they vary in their nutritional composition profiles, are generally good sources of important nutrients (i.e., fiber, vitamins, minerals). Recently, the health effects of these "daily" consumed natural products have been attributed to extranutritional constituents that typically occur in small quantities in plant products [38]. Plants need these secondary metabolites for pigmentation, growth, reproduction, resistance to pathogens and many other functions. Consequently, a plant-based diet, such as the MD, grants an elevated intake of such "non-nutrients" that may transpose their biological activity from the fruit, in which they have developed, to the human body [39], acting as antioxidants, activating liver detoxification enzymes, blocking the activity of bacterial or viral toxins, inhibiting cholesterol absorption, decreasing platelet aggregation or destroying harmful gastrointestinal bacteria [7]. Among them, the most important classes are flavonoids and more complex phenolics, lycopene, phytosterols, and the glucosinolates [40].

Flavonoids constitute a broad class of low molecular weight, secondary plant phenolics. The propensity of a flavonoid to inhibit free-radical mediated events is governed by its chemical structure. Since these compounds are based on the flavan nucleus, the number, positions and types of substitutions influence radical scavenging and chelating activity [41]. Flavonols, flavanones, flavones, flavanols, anthocyanins and isoflavones are particularly common in the diet. Flavonols are the most abundant flavonoids in foods (onions, cherries, apples, kale, tomato, etc.) with quercetin, kaempferol and myricetin being the most studied members of this group. Flavanones are mainly found in citrus fruit, and flavones in celery, parsley and thyme [42]. Catechins, belonging to flavanols, are present in large amounts in green and black tea and in red wine, whereas anthocyanins are found in strawberries and other berries. However, the flavonoid content of foods often varies by variety, climatic and storage conditions, and by part of the plant [43]. The fact that the accumulation of plant flavonoids is enhanced in response to increased light exposure, especially ultraviolet-B ray, implies that the Mediterranean products are favored because of the sunny climate. For example, vegetables grown in Spain were shown to contain four- to five-fold more flavonols than those in the United Kingdom, where greenhouses are used for plant cultivation [44,45]. In addition, there is a trend toward higher flavonol levels in wines from grapes grown in warmer, sunnier regions than in cooler ones. Moreover, the time when fruits are picked has a bearing on flavonol content. In this respect, Mediterranean wines have a relatively higher flavonol content, since grapes grown there are allowed to ripen on the vine, in contrast to grapes in cool, damp regions, which are often picked as soon as they reach a certain sugar level to minimize the risk of rain damaging the crop [45,46]. Thus, the flavonol content of Mediterranean plant products could be assumed to be greater than that of products cultivated in other regions.

In the Mediterranean region, the daily intake of these compounds is expected to be greater, not only because of the higher content of the plant tissues but also because of the high consumption of fruit and vegetables, which is one of the desirable key features of the traditional common diet. Besides, aromatic herbs and spices, used in the preparation of local specialties in the different Mediterranean cuisines (like oregano, sage, rosemary, basil, parsley, thyme), could also serve as sources of flavonoids, as well as other phenolic compounds [47–50]. Dietary habits and preferences could also affect the level of intake. For example, wild edible greens frequently eaten in rural Greece in the form of salads and pies have a considerably higher flavonoid content than an equal quantity of red wine or black tea, which are considered principle sources of flavonoids for North European countries [18].

Flavonoids are generally considered as an important category of antioxidants in the human diet [51]. There are suggestions that flavonoids such as quercetin and cyanidin are better antioxidants than vitamin C or E in reducing the free radical generation in vitro, on a mole to mole basis [52,53]. For [54], they may have beneficial health effects because of their antioxidant properties and their inhibitory role in various stages of tumor development in animal studies. Suppression of abnormal angiogenesis may provide therapeutic strategies in the treatment of angiogenesis-dependent disorders. A substantial body of recent literature suggests that these natural compounds could make up an important group of angiogenesis inhibitors [55,56]. However, there is still much to be learnt about absorption and metabolism of flavonoids [57]. If they have to be absorbed in order to exert their functional activity, it is worth saying that the bioavailability of certain flavonoids differs markedly depending on the food source. For example, the absorption of quercetin from onions has been shown to be four-fold greater than from apples or tea [42,58]. Alternatively, some health effects of polyphenols may not require their absorption through the gut barrier. They may have a direct impact on the gut mucosa and protect it against oxidative stress or the action of carcinogens [59]. Moreover, the low levels detected in the circulation, in combination with modulation of the antioxidant activities and polarities on conjugation and metabolism, suggest that the ability of certain flavonoids to provide health benefits may not necessarily depend on the ability of the native compound to act as a scavenger of free radicals or reactive nitrogen species. It may depend on the influence of the in vivo forms on gene expression, as well as on the ability of the metabolites to interact with cell signaling cascades, ultimately influencing the cell at the transcriptional level and downregulating pathways leading to cell death [60]. As already mentioned, there is a paucity of information on dose response and toxicity of these compounds. Ongoing research will help elucidate their role in the significant benefits of the traditional MD.

Phytoestrogens, plant-derived compounds that structurally or functionally mimic mammalian estrogens, are considered to play an important role in the prevention of cancers, heart disease, menopausal symptoms and osteoporosis [61–63]. They can be classified as selective estrogen receptor modulators (SERMs) [64]. SERMs are non-steroidal chemicals with a similar structure to mammalian estrogen 17β-estradiol (E_2) and an affinity toward estrogen receptors. They are unique in that they can function as agonists or antagonists depending on tissue, estrogen receptor (ER) and concentration of circulating endogenous estrogens [65]. Both genomic and nongenomic mechanisms have been proposed to explain phytoestrogenic effects on human health [66].

There are several classes of phytoestrogens. The more ubiquitous are the phenolic estrogens, isoflavones, coumestans and lignans. For the Mediterranean population, food sources of isoflavones are oilseeds, such as the sunflower seed and nuts from different botanical families. Lignans are widespread in foodstuffs such as cereals, fruit and vegetables [67]. The most well-known phytoestrogenic lignans are secoisolariciresinol and matairesinol, which are converted by bacterial action in the gut into enterodiol and enterolactone, mammalian lignans not found in plants. Pumpkin seeds, also consumed in some Mediterranean countries, contain secoisolariciresinol and trace amounts of lariciresinol [68]. Secoisolariciresinol is found in many food and beverage categories, while matairesinol is found in smaller amounts and only trace amounts in food legumes [60].

Another compound that has shown estrogenic activity is resveratrol trihydroxystilbene [3–5]. It is found in a variety of plants and functions as a phytoalexin to protect against fungal infections [69]. Major dietary sources of interest for the Mediterraneans are grapes and wine. Seeds and skin of the grapes are enriched in resveratrol, which has relatively poor solubility in water. However, it is soluble in low percentages of alcohol and is thus effectively absorbed by the body from wine [51].

The heart-protective effects of resveratrol are at least in part related to the anti-inflammatory and antioxidant activities, including reduction of LDL oxidation, platelet aggregation and inhibition of cyclooxygenase enzymes [51]. Furthermore, this therapeutic molecule has recently been reported to be an angiogenesis inhibitor [51,66,70].

Lycopene, from red fruit and vegetables, is also a natural antioxidant but not a polyphenol. It is an open hydrocarbonic chain of 40 carbon atoms, containing 11 conjugated and two non-conjugated double bonds arranged in a linear array. Since lycopene lacks β-ionone ring structure, it lacks provitamin A activity [71]. Tomatoes and tomato-based products are the main source of lycopene for the Mediterraneans. However, its level varies widely among different varieties and stages of fruit ripening [36].

Bioavailability is enhanced by cooking food sources of lycopene, particularly in the presence of oil or fats. For that reason, the Mediterranean way of preparing many dishes, combining cooked tomatoes with olive oil, favors maximal absorption of this antioxidant [36]. Although the antioxidant properties of lycopene are thought to be primarily responsible for its beneficial properties, evidence is accumulating to suggest that other mechanisms, such as intercellular gap junction communication, hormonal and immune system modulation and metabolic pathways, may also be involved [71]. However, discussing, the health benefits of tomatoes, one must consider that they are also rich in other nutrients such as β-carotene, vitamin C, folate and potassium [72]. Thus, it is clear that isolating one single active nutrient from the complex interplay of substances contained in foods can be fraught with difficulty [73].

Phytosterols form another important subclass contained in plant foods. Two sterol molecules that are synthesized by plants are β-sitosterol and its glycoside. In animals, these two molecules exhibit anti-inflammatory, anti-neoplastic, anti-pyretic and immune-modulating activity. In the body, phytosterols can compete with cholesterol in the intestine for uptake, and aid in the elimination of cholesterol from the body. Competition with cholesterol for absorption from the intestine is not unexpected, as the structure of plant sterols is similar to that of cholesterol [74,75].

Sulfur-containing phytochemicals like glucosinolates and sulfides may have a number of beneficial properties, as well. Glucosinolates, which are present in cruciferous vegetables, are activators of liver detoxification enzymes. Consumption of cruciferous vegetables (cabbage, cauliflower, broccoli, etc.) offers a phytochemical strategy for providing protection against carcinogenesis, mutagenesis and other forms of toxicity of electrophiles and reactive forms of oxygen [74]. Sulfides are found in large quantities in garlic and other bulbous plants. The main component, which is believed to be the active one, is allicin, formed from its precursor, alliin, by the enzyme allinase. Garlic is widely used as a flavoring agent in the Mediterranean cuisine. The proposal that garlic consumption can have a significant antioxidative effect is almost unquestioned because of the wealth of scientific literature supporting these effects, which include both animal and human studies. Far less certain is the identity of the specific compounds from garlic or garlic products responsible for its most antioxidant effects, and how to use them most effectively in various pathophysiological conditions [76].

The functionality of the MD, in other words, its healthy and preventive dimensions, has so far been attributed to the increased consumption of fruit, vegetables, cereals and legumes. However, the cardinal characteristic of this diet is that olive oil serves as the principal source of dietary fat [46]. The health promoting properties of olive oil are conferred by the monounsaturated oleic acid, the high intake of squalene and the unique profile of the phenolic fraction [77]. In addition, a positive health contribution comes from the content of α-tocopherol, the tocopherol with the highest vitamin E activity [78].

Historically, the MD has been associated with a lower rate of cardiovascular disease in those populations that consume it [79], as this dietary model, enriched in monounsaturated fat (MUFA), decreases the level of LDL-cholesterol in the plasma when replacing a saturated fat (SFA) enriched diet. Current dietary guidelines suggest replacing SFA with complex carbohydrates. Substantial evidence exists that high-density lipoprotein cholesterol levels are higher and triglycerides are lower on a high-MUFA than a low-fat/carbohydrate-rich, blood cholesterol-lowering diet [80]. Polyunsaturated fatty acid (PUFA) is another dietary alternative for SFA, but, in that case, qualitative changes could affect the risk for CHD. A high intake of PUFA could increase their concentration in LDL particles and favor oxidation, fundamental to the onset of arteriosclerosis [81]. Furthermore, the protective effect of dietary MUFA on arteriosclerosis goes beyond cholesterol [82], as MUFA diets could protect endothelial cells [83]. and might decrease prothrombotic environment, modifying different hemostatic components.

Extra virgin olive oil, or "liquid gold" as the ancient poet Homer used to call it, is obtained from the whole fruit by means of physical pressure without the use of chemicals. During this procedure, all of the components of the drupe are transferred to the oil as opposed to seed-oils that are deprived of this important group during the various stages of refining [20]. Thus, extra virgin olive oil has a higher total phenol content compared to refined virgin olive oil [84]. The absolute concentration of phenols in olive oils is the result of a complex interaction between several factors, including cultivar, degree of maturation, and climate. It usually decreases with over-maturation of olives, although there are some exceptions to this rule. For instance, olives grown in warmer climates, despite a more rapid maturation, yield oils that are richer in phenols [85]. Olive oil is a source of at least 30 phenolic compounds [86]. Hydroxytyrosol (3,4-dihydroxyphenylethanol; DOPET) is the major o-diphenol of virgin olive oil, present in both a free and esterified form (oleuropein aglycon) [87].

In vitro, *ortho*-diphenolic (catecholic) compounds such as oleuropein and hydroxytyrosol exert potent antioxidant activities, such as inhibition of low density lipoprotein oxidation and free radical scavenging [88]. It is possible to speculate that the lower rate of CHD related to the consumption of virgin olive oil could be partly attributed to the antioxidant properties of its phenolic components [89]. One of the mechanisms that might improve antioxidative cellular defenses may involve a direct effect of the biophenols on the DNA transcription of glutathione-related enzyme [90]. Furthermore, the ability of DOPET to arrest cell proliferation and induce apoptosis in cultured human cells indicates its possible chemopreventive nature [91]. Despite the great numbers of published references in recent decades on the beneficial effects of olive oil, it is only recently that the role in the body of the phenolic compounds present in olive oil has been investigated. Several studies with hydroxytyrosol in Caco-2 cells and in rats have shown that hydroxytyrosol is metabolized to homovanillic acid; in rats it is metabolized to at least five metabolites. Two of the metabolites, homovanillic acid and homovanillic alcohol, are almost as strong radical scavengers as hydroxytyrosol [90,92].

Olive oil also contains significantly higher amounts of squalene than seed oils. Since squalene is to a large extent transferred to the skin, its major protective effect is thought to be against skin cancer. This is supported by studies showing inhibition of this neoplasm in rodents and low incidence within Mediterranean populations. The mechanism is probably by scavenging singlet oxygen generated by ultraviolet light [77].

Besides olive oil, another liquid element that the Mediterranean nature provides to the inhabitants of the region is wine. Its production takes place throughout the area, although its chemical composition is profoundly influenced by enological techniques, the grape cultivar from which it originates and climatic factors [93]. The interest in compounds present in grapevines was stimulated when epidemiological studies showed an inverse correlation between red wine consumption and incidence of cardiovascular diseases. Wine contains ethanol, which acts through a haemostatic mechanism and an increase in circulating high-density lipoproteins [94]. However, part of the protective effect of red wine against atherosclerotic cardiovascular disease should be attributed to the presence of phenolic compounds, like phenolic acids, resveratrol and flavonoids. Mechanisms by which these compounds have been proposed to provide health benefits, in addition to being direct chemical protectants, involve modulatory effects on a variety of metabolic and signaling enzymes [68,95]. Another health aspect may, however, be related to the cultural importance of wine in the daily life of Mediterranean people. Traditionally, they enjoy a glass of wine during the course of a meal. Thus, wine drinking is often associated with pleasure and conviviality. Whether a happy lifestyle may protect against CHD is an open question that warrants further investigation [16].

Being the cornerstones of the traditional MD, products of plant origin are allies for a healthy lifestyle. Their role in disease prevention is beyond dispute, but discussing the health effects of each constituent separately is an oversimplified hypothesis. Considering the great variety of bioactive substances, it appears extremely unlikely that any one substance is responsible for all of the associations seen between plant foods and disease prevention [96]. In this respect, it is considered that consumption of tomato products with olive oil but not sunflower oil increases the antioxidant activity of plasma [97]. Furthermore, a broader spectrum of chemoprevention with less adverse effects can be obtained through effective combination of functional foods [98].

Thus, far from being a nutritional magic bullet, the traditional MD is a dietary approach, which, by encouraging a general increase in the consumption of plant foods, ensures the built-in redundancy of multiple agents with independent overlapping and perhaps interactive mechanisms [25]. Moreover, the abundant consumption of plant foods provides palatability and promotes satiety, resulting in a dietary pattern that incorporates animal foods in small amounts.

Animal foods are important sources of proteins, vitamins (such as vitamin B_{12}) and minerals, but they also contain saturated fatty acids and cholesterol. In the Mediterranean countries, consumption of beef, pork, lamb and goat meat has traditionally been low. This low consumption of red meat limits the intake of elements, whose over-consumption could have negative effect on human health. However, the major benefit from eating less meat, seems to be the concomitant increase in plant foods intake, as the positive association between meat and chronic diseases might not be directly linked to meat components [35]. It would rather be related to a limited intake of antioxidants present mainly in products of plant origin. Moreover, according to traditional recipes in Mediterranean countries, meat is cooked in olive oil based sauces with onion, garlic, herbs or wine. This way of cooking reduces the quantity of salt added and favors the low production of harmful oxidants, thus contributing to the impressive antioxidant potential of the MD [99].

As recently reported [100–102], the lipid fraction of animal products (meat and dairy products) contains minor constituents with health-promoting effects. Thus, the conjugated linoleic acid (CLA) isomers, for example, may be a novel therapeutic nutrient. Moreover, in the Mediterranean region, this lipid fraction could be a good source of α-linolenic acid (LNA) 18:3 (n-3), in the case of grass fed animals [103]. n-3 fatty acids are said to contribute to the good functioning of the cardiovascular system, on the basis of various physiological effects [104]. As already mentioned, locally grown wild plants, like purslane, are good sources of LNA. In this respect and based on the food chain, chickens that wander on farms, eating grass, purslane, dried figs and other good sources of (n-3) fatty acids, are expected to produce eggs richer in this nutrient. As expected, this also holds true for milk products such as cheese and yoghurt, coming from milk-giving animals that graze rather than being fed grain [5]. Fermented dairy products are closely related to the tradition of the Mediterranean basin. These dairy products, initially developed as a means to preserve milk, are consumed in low to moderate amounts. As currently nutrition interest moves beyond the study of essential nutrients, it is important to recognize that fermented milk products could deliver nutrition beyond that provided by essential nutrients. Thus, apart from being a good source of protein and calcium, they also contain bioactive peptides with multifunctional activities, as reviewed by Smacchi [105]. Moreover, the presence of live lactic acid bacteria, especially abundant in yoghurt, is thought to be essential to exert immunostimulatory effects. Clinical reports have suggested that dietary consumption of fermented foods, such as yogurt, can alleviate some of the symptoms of atopy and might also reduce the development of allergies, possibly via a mechanism of immune regulation. Whey proteins and CLA are believed to contribute to yoghurt's beneficial effects, as well [106].

Marine products are also popular in the Mediterranean food culture, although not in the same way everywhere. As it is expected, fish consumption is a function of proximity to the sea. Seafood is a good source of selenium. Octopus, for example, a typical Greek appetizer, contains 90 µg selenium/100 g [5] and can, thus, be regarded as a cardioprotective food, since increased consumption of selenium-rich foods by cardiac patients has been shown to result in a reduced cardiac mortality rate in several trials [107].

Furthermore, the flesh of oil-rich fish such as mackerel and sardines is rich in the very-long-chain derivatives of α-linolenic acid, eicosapentaenoic acid (EPA) and docosahexaenoic acid (DHA) [102]. The consumption of foods rich in these cardioprotective nutrients would also favor a lower ratio of (n-6) to (n-3) fatty acids. The most important aspect of essential fatty acids in the prevention of mammary cancer is the ratio of (n-6) to (n-3), rather than the absolute concentration of either [108]. Research data indicate that a ratio of 1:1 or 2:1 protects most against the development and growth of mammary cancer [5].

In conclusion, the MD is rich in bioprotective nutrients such as oleic acid, omega-3 fatty acids, fiber, vitamins of the B group and various antioxidants, and low in saturated and trans fatty acids. Therefore, the expected benefits for the prevention of CHD go far beyond an antioxidant effect and include lipid and blood pressure lowering effects, anti-inflammatory effects and the prevention of arterial plaque rupture and thrombosis, as well as protection against malignant ventricular arrhythmia and heart failure [109]. Furthermore, many of the components have been shown, when studied separately, to potentially prevent cancer initiation or metastasis, prevent angiogenesis and induce apoptosis [5].

To simplify the picture, think of a classical Mediterranean meal consisting of fish just caught from the Mediterranean Sea, wild greens with drops of pure olive oil and yoghurt with nuts, honey and succulent fresh fruit as a dessert! Imagine how health-promoting this would be.

9.4 HEALTH ATTRIBUTES OF MEDITERRANEAN DIET AND EPIDEMIOLOGICAL EVIDENCE

Mediterranean foods contain a large number of health-Protective Phytochemicals, and these compounds are kept intact during cooking and serving. The following a list of these compounds and their functions:

1-Flavonoids: Yellow pigments of the plant tissues. Present in berries, citrus, cucumbers, beans, onions, and yams, include such chemicals, Flavones, flavanones, phytoestrogens, isoflavones (genistein), quercetin, present in most fruits and vegetables. These compounds keep cancer-causing hormones from attaching to the surface of the cells. Genistein present in beans prevents tumor formation, quercetin present in citrus prevents growth of tumor cells in anticarcinogenic effects. Both of these phytochemicals prevent, breast and prostate cancer by reducing the tumor growth and migration. Flavonoids also function as antiinflamatory agents in humans.

2-Isothiocyanates: Phenethyl isothiocyanate (PEITC), and sulforaphane are found in cabbage, turnips, broccoli, cauliflowers, turnips, kale, brussels sprouts. Several other isothiocyanates are found in physiologically significant amounts in these vegetables. They give chemoprotection against tumorogenesis and act as anti-initiation factor and detoxification agent, but not as an antiprogression factor, especially in lung and breast cancers.

3-Monoterpenes: These, found only in essential oils of citrus, cherries, spearmint, dill and caraway function as chemorepellants. Perillyl alcohol and D-limonene, (monoterpenes of orange oil) inhibit hepatic tumors and their proliferation thus preventing liver, pancreatic and breast cancers.

4-Organosulfides: Another class of bioactive, non-nutritive dietary phytochemicals. Allium species (onions, scallions, leeks, garlics, and chives) are released when the plant is chewed. Allyl mercaptan, diallyl sulfide, and diallyl disulfide (DAS) are effective inhibitors of tumor initiation and promotion. Allium vegetables are associated with reduced risk of skin, mammary glands, oral mucosa, esophagus, stomach, and colon cancers. Diallyl sulfide (DAS) reduces the formation of DNA adducts and tumors in all cancers. Selenium containing diallyl selenide is even more potent in the prevention of chemically induced tumors.

5-Polyphenols: Plant phenolic compounds include vanillin, gallic acid, ellagic acid, p-coumaric acid, chlorogenic acid (beans), rutin (orange peel), curcumin (turmeric) and tangeritin (tangerines). Plant phenolics are potent scavengers of free electrons, and are effective converters of phenols to phenoxy radicals. Plant phenolics also inhibit the nitrosation of amines and amides to carcinogen nitrosamines and nitrosamides. Polyphenols act as chemical nucleophiles and deactivators of electron-poor radicals, thus inhibiting both tumor initiation and progression. Epicatechin derivatives (present in tea) and silybin and silimarin (found in artichokes) have delayed and retarded the progression of tumors of the skin, lungs, stomach, duodenum, esophagus, prostate, and colon. Tea polyphenols are potent antioxidants, preventing oxidative damage to DNA and strong inhibitors of the tumor-promoting enzymes. Tea polyphenols also inhibit the nitrosation of organic compounds.

6-Natural antioxidant vitamins: Includes vitamin A, E, C, and carotenoids, anthocyanins. Carotenoids may quench free radicals thus prevents tumor initiation. Anthocyanins stimulate immune function, thus preventing diseases.

7-Other compounds: Capsaicin, abundant in hot peppers, keeps toxic molecules from attaching to DNA at the molecular level thus prevents tumors and cancers.

Indole-3-carbinol: Present in cauliflower and cabbage, reduces breast cancer risk.

Lignans: Present in flax seed and linseed, act as phytoestrogen (also present in pomegranate) These compounds are weak oestrogens. They can imitate the protective action of oestrogen on the cardiovascular system and on bone, thus preventing heart diseases and osteoporosis in woman. It is suggested that to get enough of these phytochemicals one should consume at least 400 g of fruits, vegetables (including 30 g of pulses, nuts, and seed), daily, which make five servings of the plant foods each day.

9.4.1 MICROBIAL FUNCTIONAL ACTIVITIES

Microbial cultures occurring in functional foods contribute to food safety and/or offer one or more organoleptic, technological, nutritional, or health advantages. During fermentation processes, functional starter cultures or co-cultures contribute to the production of natural and healthy products, whereas probiotic[*] strains although also classified as functional starters or co-cultures, have inherent health-promoting properties. Common microorganisms used in probiotic preparations are predominantly *Lactobacillus* species, such as *Lactobacillus acidophilus, Lactobacillus casei, Lactobacillus reuteri, Lactobacillus rhamnosus, Lactobacillus johnsonii*, and *Lactobacillus plantarum* and *Bifidobacterium* species, such as *Bifidobacterium longum, Bifidobacterium breve, Bifidobacterium lactis* [110]. The implementation of carefully selected strains can help to achieve in situ expression of the desired properties, i.e., lactic acid bacteria are able to produce antimicrobial substances, sugar polymers, sweeteners, aromatic compounds, vitamins, useful enzymes, nutraceuticals or have probiotic properties. So, they enhance shelf life and microbial safety, improve texture, and contribute to the pleasant sensory profile of the end product, as well as contributing to the replacement of chemical additives by natural compounds (including organic acids, carbon dioxide, hydrogen peroxide, diacetyl, ethanol, bacteriocins, reuterin, and reutericyclin) providing the consumer with new, attractive food products [111].

Since the early belief of Elie Metchnikoff that the "friendly" microbes present in fermented foods could contribute to human health and well-being, a wealth of experiments have described the use of selected microorganisms, mainly belonging to the lactic acid bacteria, for the prevention or treatment of pathological conditions [112]. New research technologies have supported earlier suggestions of health-promoting properties of lactobacilli and bifidobacteria, including stabilisation of the intestinal microflora by competition against pathogens, reduction of lactose intolerance, prevention of antibiotic-induced diarrhea, prevention of colon cancer, and stimulation of the immune system [113]. An important employment for the use of immunomodulatory probiotics in health care is in the control of microbial pathogens obtained by several potential mechanisms, including: that localised lactic acid production by probiotics in the gastrointestinal tract can limit pathogen growth; that anti-pathogen substances secreted by the probiotics (i.e., bacteriocins) are directly microbicidal; that seeding the gut mucosa with de novo "friendly" bacteria can limit pathogen attachment (i.e., competitive exclusion); or that immunomodulatory signals generated by probiotics can stimulate host immunity sufficiently to afford a degree of increased protection against pathogens [114]. By probiotics, a safe and non-pharmaceutical combating of microbial pathogens at the gastrointestinal tract, in the respiratory tract tissues [115] and the urogenital tract [116] can be obtained.

---

* The word "probiotic" is derived from the Greek and means "for life."

Probiotic products appear in three main categories: conventional foods such as fermented products with addition of probiotic bacteria, consumed primarily for nutritional purpose; food supplements or fermented milks with food formulations used as a delivery vehicle for probiotic bacteria; and dietary supplements in the form of capsules and other formulations consumed for health effects [117]. The most commonly used probiotics are lactic acid bacteria and nonpathogenic, antibiotic-resistant, ascospore yeasts, principally *Saccharomyces boulardii* [118]. Milk-based functional foods are particularly interesting from the scientific and the applicable point of view. As a daily intake of 10^9–10^{10} CFU viable cells is considered the minimum dose shown to have positive effects on host health [117], probiotic yogurts in which bacteria belonging to *Lactobacillus* and *Bifidobacterium* genera have been incorporated, are capable of exerting their beneficial effect by balancing the intestinal flora and eventually competing with pathogens for gut colonization [118].

The most common use of probiotic microorganisms is in fermented dairy products, but the use of probiotics as starter cultures for fermented meat products is possible. Combining probiotic potential and technological performances of *Lactobacillus* strains would lead to interesting probiotic starters for novel dry fermented sausage manufacture [119,120]. Particularly, the strains of *L. plantarum*-group isolated from traditional dry fermented sausages in Southern Italy displayed good probiotic properties [121]. A large majority of *Lactobacillus sakei*, *Lactobacillus curvatus* and *L. plantarum* strains isolated from Greek-style fermented sausage also have potential probiotic properties [121].

Among species of bifidobacteria and lactobacilli isolated from faecal samples of healthy elderly Italian subjects, *Lactobacillus fermentum* and *B. longum* were the most represented species. Elderly people have a decreasing number of bifidobacteria in the intestinal microflora, which is associated with the high incidence of and susceptibility to degenerative and infectious diseases [122]. The above confirmation opens up the possibility of reversing such trends by an optimal nutrition regimen containing a combination thereof called a synbiotic to be used as functional food ingredients, or prebiotics that selectively encourage the growth of lactic acid producing bacteria in the intestine. Thus, *L. fermentum* and *B. longum* are considered good candidates to be utilized for the design of functional foods to fortify the intestinal microflora of the elderly [122].

Nutraceuticals from bacterial origin through specific physiological action contribute to the health of the consumer [123]. In fermented dairy products, lactic acid bacteria strains producing high amounts of low-calorie polyols contribute to the reduction of the sugar content. Mannitol-producing lactic acid bacteria can be applied in fermented foods, leading to products with an extra nutritional value, as mannitol is assumed as an antioxidant and a non-metabolizable sweetener [124]. On the other hand, oligosaccharides produced by lactic acid bacteria, due to their low-calorie character, their fibre-like nature and their bifidogenic effect, give to the fermented products health-promoting properties [125]. Cereals can be used as sources of nondigestible carbohydrates that can selectively stimulate the growth of lactobacilli and bifidobacteria present in the colon and act as prebiotics [126].

Vitamins, such as folate, produced by *Lactobacillus delbrueckii* subsp. *bulgaricus* and *Streptococcus thermophilus* increase the folate content of yoghurt [127].

Pectins and pectic-oligosaccharides, fibres found commonly in plants and vegetables, have the ability to inhibit *E. coli* O157:H7 Shiga toxin. New food developments that exploit this interaction could have important consequences for the prevention and/or alleviation of gastrointestinal symptoms caused by this toxin, taking into account the therapeutic and prophylactic potential of pectic-oligosaccharides [128].

Lactic acid bacteria due to their proteolytic activitites, liberate health-enhancing bioactive peptides from milk. The most probable liberation of bioactive peptides during food processing occurs via proteolysis and microbial enzymes, especially of lactic acid bacteria; or secondary starters may be the major producers [129]. The bioactive peptides stimulate the immune system, improve absorption in the intestinal tract, exert antihypertensive or antithrombotic effects, display anti-microbial activity, and function as carriers for minerals, especially calcium [10]. The ability of lactic acid bacteria to liberate de novo immunoregulatory peptides from proteins via enzymatic hydrolysis contributes to immune regulation. Peptides inhibitory to angiotensin-I-converting enzyme and endo- and aminopeptidases of lactic acid bacteria and *Pseudomonas fluorescens* were isolated from Italian cheeses such as Mozzarella, Italico, Crescenza, and Gorgonzola [130]. On the other hand, microbial strains with high proteolytic enzyme activity potentially increase the risk for histamine and tyramine formation in food systems, by increasing the availability of free histidine and tyrosine, respectively, [131]. Foods likely to contain elevated levels of biogenic amines include fish and fish products, dairy products, meat and meat products, fermented vegetables and soy products, and alcoholic beverages such as wine and beer [132]. Thus, all starter culture candidates (especially strains of *Lactobacillus, Pediococcus*) should be carefully checked for their potential of biogenic amine formation under the appropriate processing conditions [131]. Although during cheese ripening the decarboxylation of free amino acids results in formation of amines, in Feta cheese the low pH and the high salt content keep the total amount of biogenic amines relatively low [133]. Between biogenic amines, tyramine has been predominantly found in Idiazabal cheese [134], Feta cheese [133] Beyaz cheese [135], as well as in Azeitao [136] and different cheeses in Egypt [137].

As probiotic therapy uses bacterial interference and immunomodulation in the control of several infectious, inflammatory, and immunologic conditions [138] the search for new disease-specific probiotics seems to be warranted for the future [139] and the Mediterranean region is a promising biotop.

9.4.2 Anti-Aging Properties of Mediterranean Foods

The MD and its anti-aging properties have been outlined in Table 9.2, which is compiled from world population census (U.S. Census Bureau, International Data Base). Non-Mediterranean countries included for comparison.

The Mediterranean countries which had the highest consumption levels of olive oil, fruits and vegetables also had the highest longevity.

TABLE 9.2
The Number of People Over the Age of 80 in the Following Selected Mediterranean and Middle Eastern Countries

Country or Area in 2000	Age	Total No.	Male (M)	Female (F)	% of Population Over 80 (M+F)
Albania	80+	39,725	13,357	26,368	1.1
Algeria	80+	167,133	75,714	91,419	0.5
Armenia	80+	37,432	11,891	25,541	1.1
Azerbaijan	80+	73,235	19,146	54,089	0.9
Bosnia	80+	32,372	10,688	21,684	0.8
Bulgaria	80+	175,782	67,582	108,200	2.2
Croatia	80+	96,647	27,199	69,448	2.3
Cyprus	80+	18,733	7,492	11,241	2.5
Egypt	80+	281,524	96,673	184,851	0.4
France	80+	2,217,564	699,165	1,518,399	3.7
Georgia	80+	83,986	21,371	62,615	1.7
Greece	80+	378,960	153,443	225,517	3.6
Israel	80+	138,217	57,013	81,204	2.4
Italy	80+	2,315,793	773,785	1,542,008	4.0
Lebanon	80+	34,140	14,472	19,668	1.0
Libya	80+	25,348	11,497	13,851	0.5
Macedonia	80+	23,921	10,218	13,703	1.2
Jordan	80+	19,940	8,667	11,273	0.4
Malta	80+	9,363	3,338	6,025	2.4
Monaco	80+	1,897	629	1,268	6.0
Portugal	80+	285,043	93,396	191,647	2.8
Romania	80+	398,420	140,106	258,314	1.8
Slovenia	80+	47,525	13,620	33,905	2.5
Syria	80+	63,506	30,089	33,417	0.4
Turkey	80+	597,397	229,322	368,075	0.9
USA	80+	9,251,968	3,093,305	6,158,663	3.3
China	80+	11,778,971	4,317,168	7,461,803	0.9
India	80+	6,107,200	3,148,660	2,958,540	0.6
Sweeden	80+	454,560	158,059	296,501	5.1
Afganistan	80+	65,658	37,212	28,486	0.3
Angola	80+	16,919	7,430	9,489	0.2

Source: U.S. Census Bureau, International Data Base-2000.

9.5 MEDITERRANEAN VS. MIDDLE EASTERN COOKERY AND DIETARY PATTERNS

In the light of the positive influences of the MD on heart diseases and, as a result, on longevity, a number of questions arise regarding this diet. For example, which foods characterize this diet? What is the daily intake of calories for people in these countries? Furthermore, seeing that the Mediterranean is geographically such a far reaching area, can there be such a thing as a single MD? Is there a unifying food in such diverse cultures? And if we accept that some Mediterranean countries are also Middle Eastern countries, what dietary differences and similarities can be observed in these two culturally dissimilar blocs?

In order to answer these questions, trends in the average calories daily intake of food consumption in the different countries which make up the Mediterranean region, including those which are considered belonging to the Middle East, will be examined. The comparison is not simple, due to the sheer size and cultural, social and economic complexity of an area shared by several continents. In fact, while in Northern Mediterranean countries food consumption trends have been related to economic development, food availability, cultural values, socio-demographic characteristics, preferences and lifestyles, in most of the Southern Mediterranean and Middle East countries the relationship is different. In the latter countries it would appear that little attention has been paid to consumer preferences and this has led to the consumption of a restricted variety of low quality food products. Here the main objective of food policies has been to cover basic food needs for most of the population. In fact, some food products have been heavily subsidized [140]. Thus, in order to simplify this task, trends between 1961 and 2001 will be examined for the following three groups of countries:[*]

1. Mediterranean Countries belonging to the EU (MedEUc): Cyprus, France Greece, Italy, Malta, Portugal and Spain;
2. Mediterranean countries outside the EU (MedexEUc): Albania, Algeria, Egypt, Jordan, Israel, Lebanon, Libya, Morocco, Syria, Tunisia, and Turkey;
3. Middle East countries (MEc): United Arab Emirates, Iran, Kuwait, Saudi Arabia and Yemen.

In order to compare these countries, we will refer to data provided by the FAO's *Food Balance Sheet* (FBS) despite its shortcomings [141]. In fact, although these data in some cases overestimate apparent consumption[†] and are likely to contain errors regarding registration, they still remain the only source of food consumption data pro capita over such a long time span and for most of the countries involved in the present study.[‡] What follows in this section is an attempt to trace the evolution of the daily calorie intake from vegetal sources and from animal sources. Section 9.5 will examine how the consumption of cereals, vegetables oils, pulses, fruit, and vegetables, meat, animal fats and milk has evolved in Mediterranean countries.

---

[*] Organizations such as UNESCO classify some of these countries both as "Mediterranean" and as "Middle East" countries too. In this study it has been decided to consider all countries bordering on the Mediterranean Sea as being "Mediterranean."

[†] The consumption in question is apparent and is thus indirectly estimates with the following equation: Apparent food consumption = Production + Imports + Initial stocks − Exports − Non food industrial uses − Feed − Seed − Waste (post harvest to retail) − Final stocks.

[‡] For some Middle East and Mediterranean countries data were not available.

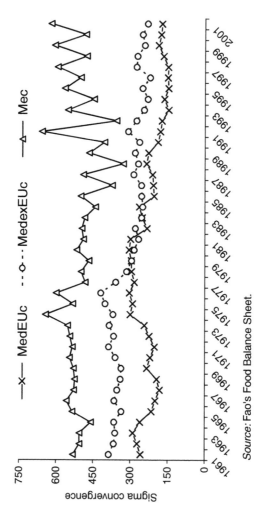

Source: Fao's Food Balance Sheet.

FIGURE 9.1 Sigma convergence of daily calorie intake in MedEUc, MedexEUc and MEc (1961–2001).

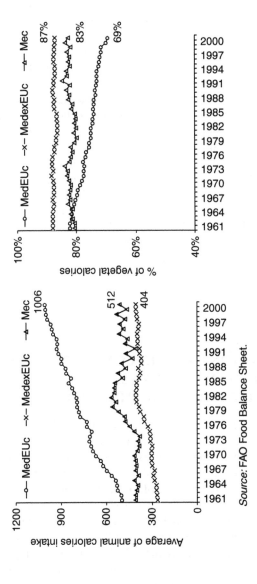

Source: FAO Food Balance Sheet.

FIGURE 9.2 Evolution of animal and vegetal calories intake (1961–2001).

In 2001 the average total calorie intake per person in MedEUc, MedExEUc, and MEc was, respectively, 3576, 3163, and 2951. Firstly, the analyses of data from 1961 to 2001 has shown that all three figures are the final result of an evident trend of continual growth. This growth has been faster in countries outside the EU and in fact, since 1961 the three groups have registered the following growths: 29% (MedEUc), 43% (MedexEUc) and 38% (MEc). This is typical of what was postulated in neoclassical growth models in which the further below its steady state an economy starts out, the faster it tends to grow thus in this case, the lower the calorie intake, theoretically the faster growth should be. Figure 9.1 illustrates the trend in the sigma convergence[*] of the daily intake of calories for the three groups of countries. This helps us understand whether within each group differences in calorie intake are diminishing. Countries within the Mediterranean Basin are converging towards the same amount of daily calorie intake, while in MEc, after a period in which the disparities in calories seemed to ebb (i.e., 1975–1990), the gap between countries began to widen again.

In terms of total calories consumed which are of vegetal origin and those of animal origin, the three groups are markedly different (Figure 9.2). In MedEUc e MEc, vegetal foods still make up the major source of calories consumed. In fact 85% of calorie intake in these two groups is of vegetal origin, while the percentage was lower, 69%, in MedEUc in 2001. Thus, while for MedexEUc and MEc vegetal commodities make up a significant portion of people's daily calorie intake, for MedEUc, calories deriving from foods of animal origin are growing faster than in the other two groups: over the period considered, consumption of calories from animal sources increased in MedEUc, MedexEUc, and MEc by respectively, 103%, 53%, and 25%.

9.6 THE EVOLUTION OF FOOD CONSUMPTION IN MEDITERRANEAN COUNTRIES: A TRADE-OFF BETWEEN TRADITION AND INNOVATION

During the period under examination, people's daily calorie intake deriving from the consumption of cereals in both groups of countries has diminished. However, foods which have always differentiated the MD from other type of diets, i.e., vegetal products, still make up almost three-fifth of calorie intake in MedexEUc, while in MedEUc, these foods make up two-fifth of the total intake (Figure 9.3). The two groups do not particularly differ with regards to the consumption of fruit and vegetables while, with regard to vegetable oils, even though their use is increasing in both groups, it has been observed that in MedEUc they show a greater variety of vegetal source calories consumed (almost 20% in the last twenty years).

The analysis of this group of foods is significant because it contains the one food which distinguishes the MD from all other dietary styles: olive oil. Our data analysis

---

[*] In the literature on economic growth, sigma convergence can be represented by standard deviation and it occurs if the dispersion in income declines over time (Barro and Sala-i-Martin 1995).

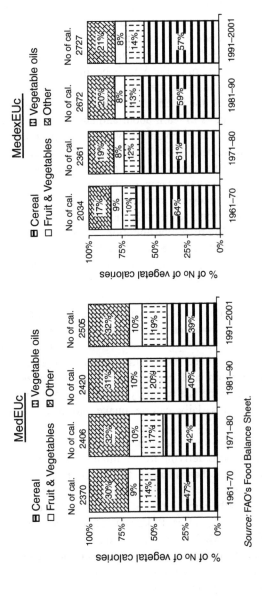

Source: FAO's Food Balance Sheet.

FIGURE 9.3 Share and comparison of selected vegetal foods in the diet of MedEU and MedexEU countries (1961–2001).

shows that in the period examined, consumption of olive oil in MedEUc has risen in all countries but remains the most important source of calories obtained from vegetable oils only in Greece, Italy, and Spain. Significantly, except France, which has never been a great consumer of olive oil, two other varieties of vegetable oils are emerging: soya bean oil in Cyprus, Malta, and Italy and sunflower oil in all the remaining countries. In MedexEUc the calorie intake from olive oil is diminishing. In fact, at present it only remains the major source of calories obtained from vegetable oils in Libya. MedexEUc emerge as consumers of sunflower, soya bean and in some cases cottonseed, palm, rape, and mustard oil. However, why is it that olive oil, the traditional symbol of the MD, is beginning to be substituted by other oils? Surely one of the causes is the high cost of olive oil compared to other vegetable oils. Furthermore, the internationalization of markets over the last twenty years has allowed consumers to learn about and subsequently purchase surrogates for olive oil. It is likely that in countries where the consumption of olive oil is diminishing, people are starting to cook and season with other types of oil. Perhaps housewives use extra-virgin olive oil as a dressing and other oils for frying, a cooking technique which wastes a large amount of oil.

With regards to the category labeled "other," the most important foods it contains are sugar, starch roots (especially the consumption of potatoes and sweet potatoes) and pulses. In MedEUc consumption of "other" foods remains stable at 31% while in MedexEUc it has risen from 17% between 1961 and 1970 to 21% between 1991 and 2001. Sugar is the food which contributes most significantly to the daily calorie intake in MedEUc.

Looking at foods deriving from animals (Figure 9.4), meat and dairy foods (apart from butter) make up roughly 3/4 of the total daily calorie intake. However, the daily calorie intake of foods deriving from animals is higher in MedEUc, even if in percentage terms in the last 30 years it has remained stable at around 40%. There are two reasons for this trend. Firstly, this might be linked to the relationship between food consumption and economic and social development which has characterized MedEUc in the last twenty years. In fact, higher earnings per capita and the fact that people's daily calorie intake has nearly reached its highest threshold, together with factors such as low population growth, small nuclear families, aging populations and changing lifestyles, have triggered off a process which tends to favours foods deriving from animal origin to foods of vegetal origin. Secondly, it is important to bear in mind that MedexEUc are Arab countries and therefore for religious reasons the consumption of pork, a food which is generally rich in calories, is basically zero. In fact, if Israel is excluded from MedexEUc one person consumes on average circa 19 kg/year of meat against almost 70 kg/year consumed in MedEUc. In MedexEUc, dairy products represent the most significant quota of calorie intake from animal sources. These are increasing and on average in the 1991–2001 period, they made up roughly 44% of people's daily calorie intake, Turkey was the only country in which the average kg/year per capita consumption of these products diminished, passing from 173 kg/year in 1961–1970 to 133 kg/year. Despite this, Turkey, immediately after Israel, is the MedexEUc with the highest annual per capita consumption of dairy foods. Dairy foods are also increasing in MedEUc, thus showing such products to be more income-elastic than meat products,

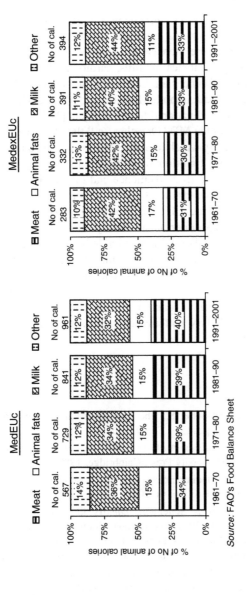

FIGURE 9.4 Share and comparison of selected animal foods in the diet of MedEU and MedexEU countries (1961–2001).

whose consumption is unlikely to increase significantly over the next few years [140].

So far this analysis seems to point to the fact that in the majority of MedEUc, the MD is very different from the prototypical diet of the sixties in which the principal foods were of vegetal origin. Furthermore, the growing calorie intake in MedexEuc is leading these countries to consumption levels which are typical of industrialized countries. These patterns, together with changes in lifestyle, largely associated with rapid urbanization, which demands less physical energy (more sedentary work and leisure) are responsible for the increase in so-called Non-Communicable Diseases [142]. As a result obesity, diabetes mellitus, cardiovascular disease, hypertension, strokes, and some types of cancer are becoming increasingly significant causes of disability and premature death in both MedEUc and MedexEUc, placing additional burdens on already overtaxed national health budgets [143].

So what is the future of the MD? Or rather, will the socio-economic changes and stressful rhythms of a society which in many Mediterranean countries no longer give women the time which they traditionally dedicated to the preparation of food, push families more and more towards convenience and fast food? Surely there will be a further thrust in MedEUc towards the lifestyles of Northern European countries, and this is likely to occur in MedexEUc vs. MedEUc too. Hopefully this change will come about in respect of a Mediterranean dietary tradition which will allow the old model to blend with the new. This will be possible if, on the one hand, technological innovation and scientific research is able to find the means of transferring the MD into the new dietary model, and on the other hand, if the older generations are able to pass on these traditions. In a scenario in which young consumers are concerned with product information and quality, functional foods could well become a benchmark for new Mediterranean dietary patterns, as well as a challenge for the agro-food industry to create new products [144].

REFERENCES

1. Ferro-Luzzi, A. and Sette, S., The Mediterranean diet: an attempt to define its present and post composition, *Eur. J. Clin. Nutr.*, 43 (2), 13–29, 1989.

2. Keys, A., *Seven Countries: A Multivariate Analysis of Death and Coronary Heart Disease*, Harvard University Press, Cambridge, 1980.

3. Trichopoulou, A., Mediterannean diet: the past and the present, *Nutr. Metab. Cardiovasc.*, 11(suppl. to No. 4), 2001.

4. International Consensus Statement, *Dietary fat, the Mediterranean Diet, and Lifelong Good Health 2000*, International Conference on the Mediterranean Diet, Royal College of Physicians London January, 13–14, 2000. (http://europa.eu.int/comm/agriculture/prom/olive/medinfo/uk_ie/consensus/index.htm).

5. Simopoulos, A. P., The Mediterranean diets: what is so special about the diet of Greece? The scientific evidence, *J. Nutr.*, 131, 3065S–3073S, 2001.

6. Milner, J. A., Strategies for cancer prevention: the role of diet, *Br. J. Nutr.*, 87 (suppl. 2), S265–S272, 2002.

7. Go, V. L. W., Butrum, R. R., and Wong, D. A., Diet, nutrition, and cancer prevention: the postgenomic era, *J. Nutr.*, 133, 3830S–3836S, 2003.

8. Mollet, B. and Rowland, I., Functional foods: at the frontier between food and pharma, *Curr. Opin. Biotechnol.*, 13, 483–485, 2002.

9. Roberfroid, M. B., Concepts in functional foods: the case of inulin and oligofructose, *J. Nutr.*, 129 (suppl. 7), 1398S–1401S, 1999.

10. Richardson, D. P., The grain, the wholegrain and nothing but the grain: the science behind wholegrain and the reduced risk of heart disease and cancer, *Nutr. Bull.*, 25 (4), 353–360, 2000.

11. Pennington, J. A. T., Food composition databases for bioactive food components, *J. Food. Comp. Anal.*, 15, 419–434, 2002.

12. Lähteenmäki, L., In *Consumers and Functional Foods, Functional Dairy Products*, Mattila-Sandholm, T. and Saarela, M. Eds., CRC Press, Boca Raton, FL, pp. 346–358, 2003.

13. Kant, A. K., Dietary patterns and health outcomes, *J. Am. Diet Assoc.*, 104, 615–635, 2004.

14. Georgakis, S. A., Meat and meat products, In *The Mediterranean Diet: Constituents and Health Production*, Matalas, A. L., Zampelas, A., Stavrinos, V., and Wolinsky, I. Eds., CRC Press, Boca Raton, FL, pp. 157–179, 2001.

15. Trichopoulou, A. and Lagiou, P., The Mediterranean diet: definition, epidimiological aspects, and current patterns, In *The Mediterranean Diet: Constituents and Health Production*, Matalas, A. L., Zampelas, A., Stavrinos, V., and Wolinsky, I. Eds., CRC Press, Boca Raton, FL, pp. 53–76, 2001.

16. de Lorgeril, M., Salen, P., Paillard, F., Laporte, F., Boucher, F., and de Leiris, J., Mediterranean diet and the French paradox: two distinct biogeographic concepts for one consolidated scientific theory on the role of nutrition in coronary heart disease, *Cardiovasc. Res.*, 54, 503–515, 2002.

17. Kris-Etherton, P., Eckel, R. H., Howard, B. V., St. Jeor, S., and Bazzarre, T. L., Lyon Diet Heart Study. Benefits of a Mediterranean-style, *National Cholesterol Education Program/American Heart Association Step I Dietary Pattern on Cardiovascular Disease*, 103, 1823–1825, 2001.

18. Trichopoulou, A. and Vasilopoulou, E., Mediterranean diet and longevity, *Br. J. Nutr.*, 84 (suppl. 2), S205–S209, 2000.

19. Anonymous, Dietary fat consensus statements, *Am. J. Med.*, 113 (suppl. 2), 5S–8S, 2002.

20. Hu, F. B., Plant-based foods and prevention of cardiovascular disease: an overview, *Am. J. Clin. Nutr.*, 78 (suppl.), 544S–551S, 2003.

21. Guillon, F. and Champ, M., Structural and physical properties of dietary fibres, and consequences of processing on human physiology, *Food Res. Int.*, 33, 233–245, 2000.

22. Kushi, L. H., Meyer, K. A., and Jacobs, D. R. Jr., Cereals, legumes, and chronic disease risk reduction: evidence from epidemiologic studies, *Am. J. Clin. Nutr.*, 70 (suppl.), 451S–458S, 1999.

23. Slavin, J. L., Martini, M. C., Jacobs, D. R. Jr., and Marquart, L., Plausible mechanisms for the protectiveness of whole grains, *Am. J. Clin. Nutr.*, 70 (suppl.), 459S–463S, 1999.

24. Mathers, J. C., Pulses and carcinogenesis: potential for the prevention of colon, breast and other cancers, *Br. J. Nutr.*, 88 (suppl. 3), S273–S279, 2002.

25. Buddington, R. K., Williams, C. H., Chen, S. C., and Witherly, S. A., Dietary supplement of neosugar alters the fecal flora and decreases activities of some reductive enzymes in human subjects, *Am. J. Clin. Nutr.*, 63, 709–716, 1996.

26. Katsilambros, N. L. and Zampelas, A., Diabetes mellitus, obesity, and the Mediterranean diet, In *The Mediterranean Diet: Constituents and Health Production*, Matalas, A. L., Zampelas, A., Stavrinos, V., and Wolinsky, I. Eds., CRC Press, Boca Raton, FL, pp. 225–242, 2001.

27. Ascherio, A., Rimm, E. B., Hernan, M. A., Giovannucci, E. L., Kawachi, I., Stampfer, M. J., and Willett, W. C., Intake of potassium, magnesium, calcium, and fiber and risk of stroke among U.S. men, *Circulation*, 98 (12), 1198–1204, 1998.

28. McCarron, D. A. and Reusser, M. E., Are low intakes of calcium and potassium important causes of cardiovascular disease? *Am. J. Hypertens.*, 14, 206S–212S, 2001.

29. Sandberg, A. S., Bioavailability of minerals in legumes, *Br. J. Nutr.*, 88 (suppl. 3), S281–S285, 2002.

30. Fox, C. H. and Eberl, M., Phytic acid (IP6), novel broad spectrum anti-neoplastic agent: a systematic review, *Complement. Ther. Med.*, 10, 229–234, 2002.

31. Nanos, G. D. and Gerasopoulos, D. G., Fruits, vegetables, legumes and grains, In *The Mediterranean Diet: Constituents and Health Production*, Matalas, A. L., Zampelas, A., Stavrinos, V., and Wolinsky, I. Eds., CRC Press, Boca Raton, FL, pp. 97–126, 2001.

32. Briante, R., Febbraio, F., and Nucci, R., Antioxidant properties of low molecular weight phenols present in the Mediterranean diet, *J. Agric. Food Chem.*, 51, 6975–6981, 2003.

33. Zampelas, A., Hourdakis, M., and Yiannakouris, N., The Mediterranean diet and coronary heart disease, In *The Mediterranean Diet: Constituents and Health Production*, Matalas, A. L., Zampelas, A., Stavrinos, V., and Wolinsky, I. Eds., CRC Press, Boca Raton, FL, pp. 243–291, 2001.

34. Weisburger, J. H., Mechanisms of action of antioxidants as exemplified in vegetables, tomatoes and tea, *Food Chem. Toxicol.*, 37, 943–948, 1999.

35. Brouwer, I. A., van Dusseldorp, M., West, C. E., Meyboom, S., Thomas, C. M. G., Duran, M., van het Hof, K. H., Eskes, T. K. A. B., Hautvast, J. G. A. J., and Steegers-Theunissen, R. P. M., Dietary folate from vegetables and citrus fruit decreases plasma homocysteine concentrations in humans in a dietary controlled trial, *J. Nutr.*, 129, 1135–1139, 1999.

36. Appel, L. J., Miller, E. R., Jee, S. H., Stolzenberg-Solomon, R., Lin, P. H., Erlinger, T., Nadeau, M. R., and Selhub, J., Effect of dietary patterns on serum homocysteine: results of a randomized, controlled feeding study, *Circulation*, 102 (8), 852–857, 2000.

37. Wu, G. and Meininger, C. J., Arginine nutrition and cardiovascular function, *J. Nutr.*, 130, 2626–2629, 2000.

38. Kris-Etherton, P. M., Hecker, K. D., Bonanome, A., Coval, S. M., Binkoski, A. E., Hilpert, K. F., Griel, A. E., and Etherton, T. D., Bioactive compounds in foods: their role in the prevention of cardiovascular disease and cancer, *Am. J. Med.*, 113 (suppl. 2), 71S–88S, 2002.

39. Visioli, F., Bellomo, G., and Galli, C., Free radical-scavenging properties of olive oil polyphenols, *Biochem. Biophys. Res. Commun.*, 247, 60–64, 1998.

40. Cassidy, A. and Dalais, F. S., Phytochemicals, In *Nutrition & Metabolism*, Gibney, M. J., Macdonald, I. A., and Roche, H. M. Eds., Blackwell Publisher, UK, pp. 307–317, 2003.

41. Heim, K. E., Tagliaferro, A. R., and Bobilya, D. J., Flavonoid antioxidants: chemistry, metabolism and structure-activity relationship, *J. Nutr. Biochem.*, 13, 572–584, 2002.

42. Ren, W., Qiao, Z., Wang, H., Zhu, L., and Zhang, L., Flavonoids: promising anticancer agents, *Med. Res. Rev.*, 23 (4), 519–534, 2003.

43. Le Marchand, L., Cancer preventive effects of flavonoids—a review, *Biomed. Pharmacother.*, 56, 296–301, 2002.

44. Stewart, A. J., Bozonnet, S., Mullen, W., Jenkins, G. I., Lean, M. E. J., and Crozier, A., Occurrence of flavonols in tomatoes and tomato-based products, *J. Agric. Food Chem*, 48, 2663–2669, 2000.

45. McDonald, M. S., Hughes, M., Burns, J., Lean, M. E. J., Matthews, D., and Crozier, A., Survey of the free and conjugated myricetin and quercetin content of red wines of different geographical origins, *J. Agric. Food Chem.*, 46, 368–375, 1998.

46. Aherne, S. A. and O'Brien, N. M., Dietary flavonols: chemistry, food content, and metabolism, *Nutrition*, 18, 75–81, 2002.

47. Zheng, W. and Wang, S. Y., Antioxidant activity and phenolic compounds in selected herbs, *J. Agric. Food Chem.*, 49, 5165–5170, 2001.

48. Exarchou, V., Nenadis, N., Tsimidou, M., Gerothanassis, I. P., Troganis, A., and Boskou, D., Antioxidant activities and phenolic composition of extracts from greek oregano, greek sage, and summer savory, *J. Agric. Food Chem.*, 50, 5294–5299, 2002.

49. Mamalakis, G. and Kafatos, A., Mediterranean diet and longevity, In *The Mediterranean Diet: Constituents and Health Production*, Matalas, A. L., Zampelas, A., Stavrinos, V., and Wolinsky, I. Eds., CRC Press, Boca Raton, FL, pp. 205–223, 2001.

50. Trichopoulou, A. and Vasilopoulou, E., Mediterranean diet and longevity, *Br. J. Nutr.*, 84 (suppl. 2), S205–S209, 2000.

51. Trichopoulou, A., Vasilopoulou, E., Hollman, P., Chamalides, Ch., Foufa, E., Kaloudis, T. r., Kromhout, D., Miskaki, Ph., Petrochilou, I., Poulima, E., Stafilakis, K., and Theophilou, D., Nutritional composition and flavonoid content of edible wild greens and green pies: a potential rich source of antioxidant nutrients in the Mediterranean diet, *Food Chem.*, 70, 319–323, 2000.

52. Datla, K. P., Christidou, M., Widmer, W. W., Rooprai, H. K., and Dexter, D. T., Tissue distribution and neuroprotective effects of citrus flavonoid tangeretin in a rat model of Parkinson's disease, *NeuroReport*, 12, 3871–3875, 2001.

53. Rice-Evans, C. A., Miller, N. J., and Paganga, G., Antioxidant properties of phenolic compounds, *Trends Plant Sci.*, 2 (4), 152–159, 1997.

54. Hollman, P. C. H. and Katan, M. B., Dietary flavonoids: intake, health effects and bioavailability, *Food Chem. Toxicol.*, 37, 937–942, 1999.

55. Cao, Y., Cao, R., and Bråkenhielm, E., Antiangiogenic mechanisms of diet-derived polyphenols, *J. Nutr. Biochem.*, 13, 380–390, 2002.

56. Losso, J. N., Targeting excessive angiogenesis with functional foods and nutraceuticals, *Trends Food Sci. Technol.*, 14, 455–468, 2003.

57. Walle, T., Absorption and metabolism of flavonoids, *Free Radic. Biol. Med.*, 36 (7), 829–837, 2004.
58. Hollman, P. C. H., van Trijp, J. M. P., Buysman, M. N. C. P., v.d. Gaag, M. S., Mengelers, M. J. B., de Vries, J. H. M., and Katan, M. B., Relative bioavailability of the antioxidant flavonoid quercetin from various foods in man, *FEBS Lett.*, 418, 152–156, 1997.
59. Scalbert, A., Morand, C., Manach, C., and Rémésy, C., Absorption and metabolism of polyphenols in the gut and impact on health, *Biomed. Pharmacother.*, 56, 276–282, 2002.
60. Rice-Evans, C., Flavonoids and isoflavones: absorption, metabolism, and bioactivity, *Free Radic. Biol. Med*, 36 (7), 827–828, 2004.
61. Setchell, K. D. R., Phytoestrogens: the biochemistry, physiology, and implications for human health of soy isoflavones, *Am. J. Clin. Nutr.*, 68 (suppl.), 1333S–1346S, 1998.
62. Adlercreutz, H., Phyto-oestrogens and cancer, *Lancet Oncol.*, 3, 364–373, 2002.
63. Kronenberg, F. and Fugh-Berman, A., Complementary and alternative medicine for menopausal symptoms: a review of randomized, controlled trials, *Ann. Intern. Med.*, 137, 805–813, 2002.
64. Brzezinski, A. and Debi, A., Phytoestrogens: the "natural" selective estrogen receptor modulators? *Eur. J. Obstet. Gynecol. Reprod. Biol.*, 85, 47–51, 1999.
65. Ososki, A. L. and Kennelly, E. J., Phytoestrogens: a review of the present state of research, *Phytother. Res.*, 17, 845–869, 2003.
66. Anderson, J. J. B., Anthony, M., Messina, M., and Garner, S. C., Effects of phyto-oestrogens on tissues, *Nutr. Res. Rev.*, 12 (1), 75–116, 1999.
67. Ibarreta, D., Daxenberger, A., and Meyer, H. H. D., Possible health impact of phytoestrogens and xenoestrogens in food, *APMIS*, 109, 161–184, 2001.
68. Sicilia, T., Niemeyer, H. B., Honig, D. M., and Metzler, M., Identification and stereochemical characterization of lignans in flaxseed and pumpkin seeds, *J. Agric. Food Chem.*, 51, 1181–1188, 2003.
69. Frémont, L., Biological effects of resveratrol, *Life Sci.*, 66 (8), 663–673, 2000.
70. Reddy, L., Odhav, B., and Bhoola, K. D., Natural products for cancer prevention: a global perspective, *Pharmacol. Ther.*, 99, 1–13, 2003.
71. Rao, A. V. and Agarwal, S., Role of lycopene as antioxidant carotenoid in the prevention of chronic diseases: a review, *Nutr. Res.*, 19 (2), 305–323, 1999.
72. Arab, L. and Steck, S., Lycopene and cardiovascular disease, *Am. J. Clin. Nutr.*, 71 (suppl.), 1691S–1695S, 2000.
73. de Lorgeril, M., Mediterranean diet in the prevention of coronary heart disease, *Nutrition*, 14 (1), 55–57, 1998.
74. Dillard, C. J. and German, J. B., Phytochemicals: nutraceuticals and human health, *J. Sci. Food Agric.*, 80, 1744–1756, 2000.
75. Orzechowski, A., Ostaszewski, P., Jank, M., and Berwid, S. J., Bioactive substances of plant origin in food—impact on genomics, *Reprod. Nutr. Dev.*, 42, 461–477, 2002.
76. Banerjee, S. K., Mukherjee, P. K., and Maulik, S. K., Garlic as an antioxidant: the good, the bad and the ugly, *Phytother. Res.*, 17, 97–106, 2003.
77. Owen, R. W., Giacosa, A., Hull, W. E., Haubner, R., Würtele, G., Spiegelhalder, B., and Bartsch, H., Olive-oil consumption and health: the possible role of antioxidants, *Lancet Oncol.*, 1, 107–112, 2000.

78. Wahrburg, U., Kratz, M., and Cullen, P., Mediterranean diet, olive oil and health, *Eur. J. Lipid Sci. Technol.*, 104, 698–705, 2002.

79. Willett, W. C., Sacks, F., Trichopoulou, A., Drescher, G., Ferro-Luzzi, A., Helsing, E., and Trichopoulos, D., Mediterranean diet pyramid: a cultural model for healthy eating, *Am. J. Clin. Nutr.*, 61, 1402S–1406S, 1995.

80. Kris-Etherton, P. M., Monounsaturated fatty acids and risk of cardiovascular disease, *Circulation*, 100 (11), 1253–1258, 1999.

81. Mata, P., Alonso, R., Lopez-Farre, A., Ordovas, J. M., Lahoz, C., Garces, C., Caramelo, C., Codoceo, R., Blazquez, E., and de Oya, M., Effect of dietary fat saturation on LDL oxidation and monocyte adhesion to human endothelial cells in vitro, *Arterioscler. Thromb. Vasc. Biol.*, 16, 1347–1355, 1996.

82. Pérez-Jiménez, F., López-Miranda, J., and Mata, P., Protective effect of dietary monounsaturated fat on arteriosclerosis: beyond cholesterol, *Atherosclerosis*, 163, 385–398, 2002.

83. Pérez-Jiménez, F., Castro, P., López-Miranda, J., Paz-Rojas, E., Blanco, A., López-Segura, F., Velasco, F., Marín, C., Fuentes, F., and Ordovás, J. M., Circulating levels of endothelial function are modulated by dietary monounsaturated fat, *Atherosclerosis*, 145, 351–358, 1999.

84. Owen, R. W., Giacosa, A., Hull, W. E., Haubner, R., Spiegelhalder, B., and Bartsch, H., The antioxidant/anticancer potential of phenolic compounds isolated from olive oil, *Eur. J. Cancer*, 36, 1235–1247, 2000.

85. Visioli, F. and Galli, C., Olive oil phenols and their potential effects on human health, *J. Agric. Food Chem.*, 46, 4292–4296, 1998.

86. Tuck, K. L. and Hayball, P. J., Major phenolic compounds in olive oil: metabolism and health effects, *J. Nutr. Biochem.*, 13, 636–644, 2002.

87. Manna, C., D'Angelo, S., Migliardi, V., Loffredi, E., Mazzoni, O., Morrica, P., Galletti, P., and Zappia, V., Protective effect of the phenolic fraction from virgin olive oils against oxidative stress in human cells, *J. Agric. Food Chem.*, 50, 6521–6526, 2002.

88. Visioli, F., Galli, C., Galli, G., and Caruso, D., Biological activities and metabolic fate of olive oil phenols, *Eur. J. Lipid Sci. Technol.*, 104, 677–684, 2002.

89. Visioli, F. and Galli, C., Antiatherogenic components of olive oil, *Curr. Atheroscler. Rep.*, 3 (1), 64–67, 2001.

90. Masella, R., Varì, R., D'Archivio, M., Di Benedetto, R., Matarrese, P., Malorni, W., Scazzocchio, B., and Giovannini, C., Extra virgin olive oil biophenols inhibit cell-mediated oxidation of LDL by increasing the mRNA transcription of glutathione-related enzymes, *J. Nutr.*, 134, 785–791, 2004.

91. D'Angelo, S., Manna, C., Migliardi, V., Mazzoni, O., Morrica, P., Capasso, G., Pontoni, G., Galletti, P., and Zappia, V., Pharmacokinetics and metabolism of hydroxytyrosol, a natural antioxidant from olive oil, *Drug Metab. Dispos.*, 29, 1492–1498, 2001.

92. Tuck, K. L., Hayball, P. J., and Stupans, I., Structural characterization of the metabolites of hydroxytyrosol, the principal phenolic component in olive oil, in rats, *J. Agric. Food Chem.*, 50, 2404–2409, 2002.

93. Soleas, G. J., Diamandis, E. P., and Goldberg, D. M., Wine as a biological fluid: history, production, and role in disease prevention, *J. Clin. Lab. Anal.*, 11, 287–313, 1997.

94. Rimm, E. B., Klatsky, A., Grobbee, D., and Stampfer, M. J., Review of moderate alcohol consumption and reduced risk of coronary heart disease: is the effect due to beer, wine, or spirits? *Br. Med. J.*, 312, 731–736, 1996.

95. Birt, D. F., Hendrich, S., and Wang, W., Dietary agents in cancer prevention: flavonoids and isoflavonoids, *Pharmacol. Ther.*, 90, 157–177, 2001.

96. Lee, A., Thurnham, D. I., and Chopra, M., Consumption of tomato products with olive oil but not sunflower oil increases the antioxidant activity of plasma, *Free Radic. Biol. Med.*, 29 (10), 1051–1055, 2000.

97. Bhuvaneswari, V., Chandra Mohan, K. V. P., and Nagini, S., Combination chemoprevention by tomato and garlic in the hamster buccal pouch carcinogenesis model, *Nutr. Res.*, 24, 133–146, 2004.

98. Papoutsakis-Tsarouhas, C. and Wolinsky, I., Cancer and the Mediterranean diet, In *The Mediterranean Diet: Constituents and Health Production*, Matalas, A. L., Zampelas, A., Stavrinos, V., and Wolinsky, I. Eds., CRC Press, Boca Raton, FL, pp. 293–340, 2001.

99. Roche, H. M., Noone, E., Nugent, A., and Gibney, M. J., Conjugated linoleic acid: a novel therapeutic nutrient?, *Nutr. Res. Rev.*, 14, 173–187, 2001.

100. Gnädig, S., Xue, Y., Berdeaux, O., Chardigny, J. M., and Sebedio, J. L., Conjugated linoleic acid (CLA) as a functional ingredient, In *Functional Dairy Products*, Mattila-Sandholm, T. and Saarela, M. Eds., CRC Press, Boca Raton, FL, pp. 263–298, 2003.

101. Masso-Welch, P. A., Zangani, D., Ip, C., Vaughan, M. M., Shoemaker, S., Ramirez, R. A., and Ip, M. M., Inhibition of angiogenesis by the cancer chemopreventive agent conjugated linoleic acid, *Cancer Res.*, 62, 4383–4389, 2002.

102. Stanner, S., n-3 Fatty acids and health, *Nutr. Bull.*, 25, 81–84, 2000.

103. Connor, W. E., Importance of n-3 fatty acids in health and disease, *Am. J. Clin. Nutr.*, 71 (suppl.), 171S–175S, 2000.

104. Smacchi, E. and Gobbetti, M., Bioactive peptides in dairy products: synthesis and interaction with proteolytic enzymes, *Food Microbiol.*, 17, 129–141, 2000.

105. Meydani, S. N. and Ha, W. K., Immunologic effects of yogurt, *Am. J. Clin. Nutr.*, 71, 861–872, 2000.

106. de Lorgeril, M., Salen, P., Accominotti, M., Cadau, M., Steghens, J. P., Boucher, F., and de Leiris, J., Dietary and blood antioxidants in patients with chronic heart failure. Insights into the potential importance of selenium in heart failure, *Eur. J. Heart Fail.*, 3, 661–669, 2001.

107. Cowing, B. E. and Saker, K. E., Polyunsaturated fatty acids and epidermal growth factor receptor/mitogen-activated protein kinase signaling in mammary cancer, *J. Nutr.*, 131, 1125–1128, 2001.

108. Shortt, C., The probiotic century: historical and current perspectives, *Trends Food Sci. Technol.*, 10, 411–417, 1999.

109. Leroy, F. and De Vuyst, L., Lactic acid bacteria as functional starter cultures for the food fermentation industry, *Trends Food Sci. Technol.*, 15, 67–78, 2004.

110. Metchnikoff, E., *Prolongation of Life: Optimistic Studies*, Putnam's Sons, New York, 1998.

111. Temmerman, R., Pot, B., Huys, G., and Swings, J., Identification and antibiotic susceptibility of bacterial isolates from probiotic products, *Int. J. Food Microbiol.*, 81, 1–10, 2002.

112. Cross, M. L., Microbes versus microbes: immune signals generated by probiotic lactobacilli and their role in protection against microbial pathogens, *FEMS Immunol. Med. Microbiol.*, 34, 245–253, 2002.

113. Alvarez, S., Herrero, C., Bru, E., and Perdigon, G., Effect of *Lactobacillus casei* and yogurt administration on prevention of *Pseudomonas aeruginosa* infection in young mice, *J. Food Prot.*, 64, 1768–1774, 2001.

114. Reid, G., Beuerman, D., Heinemann, C., and Bruce, A. W., Probiotic Lactobacillus dose required to restore and maintain a normal vaginal flora, *FEMS Immunol. Med. Microbiol.*, 32, 37–41, 2001.

115. Sanders, M. E. and Veld, J. H., Bringing a probiotic-containing functional food to the market: microbiological, product regulatory and labelling issues, *Antonie van Leeuwenhoek*, 76, 293–315, 1999.

116. Alvarez-Olmos, M. I. and Oberhelman, R. A., Probiotic agents and infectious diseases: a modern perspective on a traditional therapy, *Clin. Infect. Dis.*, 32, 1567–1576, 2001.

117. Fasoli, S., Marzotto, M., Rizzotti, L., Rossi, F., Dellaglio, F., and Torriani, S., Bacterial composition of commercial probiotic products as evaluated by PCR-DGGE analysis, *Int. J. Food Microbiol.*, 82, 59–70, 2003.

118. Silvi, S., Verdenelli, M. C., Orpianesi, C., and Cresci, A., EU project Crownalife: functional foods, gut microflora and healthy ageing. Isolation and identification of *Lactobacillus* and *Bifidobacterium* strains from faecal samples of elderly subjects for a possible probiotic use in functional foods, *J. Food Eng.*, 56, 195–200, 2003.

119. Andlauer, W. and Fürst, P., Nutraceuticals: a piece of history, present status and outlook, *Food Res. Int.*, 35, 171–176, 2002.

120. Wisselink, H. W., Weusthuis, R. A., Eggink, G., Hugenholtz, J., and Grobben, G. J., Mannitol production by lactic acid bacteria: a review, *Int. Dairy J.*, 12, 151–161, 2002.

121. Charalampopoulos, D., Wang, R., Pandiella, S. S., and Webb, C., Application of cereals and cereal components in functional foods: a review, *Int. J. Food Microbiol.*, 79, 131–141, 2002.

122. Ruas-Madiedo, P., Hugenholtz, J., and Zoon, P., An overview of the functionality of exopolysaccharides produced by lactic acid bacteria, *Int. Dairy J.*, 12, 163–171, 2002.

123. Crittenden, R. G., Martinez, N. R., and Playne, M. J., Synthesis and utilisation of folate by yoghurt starter cultures and probiotic bacteria, *Int. J. Food Microbiol.*, 80, 217–222, 2002.

124. Olano-Martin, E., Williams, M. R., Gibson, G. R., and Rastall, R. A., Pectins and pectic-oligosaccharides inhibit *Escherichia coli* O157:H7 Shiga toxin as directed towards the human colonic cell line HT29, *FEMS Microbiol. Lett.*, 218, 101–105, 2003.

125. Gobbetti, M., Stepaniak, L., De Angelis, M., and Di Cagno, R., Latent bioactive peptides in milk proteins: activation and significance in dairy processing, *Crit. Rev. Food Sci. Nutr.*, 42, 223–239, 2002.

126. Papamanoli, E., Tzanetakis, N., Litopoulou-Tzanetaki, E., and Kotzekidou, P., Characterization of lactic acid bacteria isolated from a Greek dry-fermented sausage in respect of their technological and probiotic properties, *Meat Sci.*, 65, 859–867, 2003.

127. Pennacchia, C., Ercolini, D., Blaiotta, G., Pepe, O., Mauriello, G., and Villani, F., Selection of Lactobacillus strains from fermented sausages for their potential use as probiotics, *Meat Sci.*, 67, 309–317, 2004.

128. Smacchi, E. and Gobetti, M., Peptides from several Italian cheeses inhibitory to proteolytic enzymes of lactic acid bacteria, *Pseudomonas fluorescens* ATCC 948 and to the angiotensin I-converting enzyme, *Enz. Microbiol. Technol.*, 22, 687–694, 1998.

129. Bodmer, S., Imark, C., and Kneubühl, M., Biogenic amines in foods: histamine and food processing, *Inflamm. Res.*, 48, 296–300, 1999.

130. Shalaby, A. R., Significance of biogenic amines to food safety and human health, *Food Res. Int.*, 29, 675–690, 1996.

131. Valsamaki, A., Michaelidou, A., and Polychroniadou, A., Biogenic amine production in Feta cheese, *Food Chem.*, 71, 259–266, 2000.

132. Ordonez, J. A., Ibanez, F. C., Torre, P., and Barcina, Y., Formation of biogenic amines in Idiazabal ewe's milk cheese: effect of ripening, pasteurisation and starter, *J. Food Prot.*, 60, 1371–1375, 1997.

133. Durlu-Oezkaya, F., Alichanidis, E., Litopoulou-Tzanetaki, E., and Tunail, N., Determination of biogenic amines content of Beyaz cheese and biogenic amine production ability of some lactic acid bacteria, *Michwissenschaft*, 54, 680–682, 1999.

134. Pinho, O., Ferreira, I. M., Mendes, E., Oliveira, B. M., and Ferreira, M., Effect of temperature on evolution of free amino acid and biogenic amine contents during storage of Azeitao cheese, *Food Chem.*, 75, 287–291, 2001.

135. El-Sayed, M. M., Biogenic amines in processed cheese available in Egypt, *Int. Dairy J.*, 6, 1079–1086, 1996.

136. Salminen, S. and Arvilommi, H., Probiotics demonstrating efficacy in clinical settings, *Clin. Infect. Dis.*, 32, 1577–1578, 2001.

137. Gil, J.M. Gracia, A. and Angula, A.M., Trends in the consumption of animal food products in Mediterranean countries, *Options Mediferranéennes*, 11, 31, 1993.

138. Brunisma, J., *World Agriculture: Towards 2015/30*, an FAO Perspective. Earthscan (London) and FAO (Rome), 2003.

139. Barro, R. and Sala-i-Martin, X., *Economic Growth*, McGraw-Hill, New York, 1995.

140. Alexandratos, Nikos, *The Mediterranean Diet in a World Context*, Paper presented at the V Barcelona Congress on the Mediterranean Diet, FAO, Rome, 2004.

141. WHO, *Diet, nutrition and the prevention of chronic disease: Report of a joint WHO/FAO expert consultation*, Geneva, 2003.

142. Trichopoulou, A., Mediterannean diet: the past and the present, *Nutr. Metab. Cardiovasc.*, 11 (suppl. to No. 4), 2001.

143. Trichopoulou, A. and Lagiou, P., Healthy traditional Mediterranean Diet: an expression of culture, history, and lifestyle, *Nutr. Rev.*, 55, 383–389, 1997.

144. Yildiz, F., *Phytoestrogens in Functional Foods*, CRC Press, Taylor and Francis Group, 2006.

10 Functional Foods in the European Union

Andreu Palou, Paula Oliver, Ana Mª Rodríguez, and Antoni Caimari

CONTENTS

10.1   INTRODUCTION

The relationship between nutrition and health is gaining public acceptance. The traditional idea of an adequate diet in the mere sense of providing enough nutrients to ensure the survival of an individual, satisfying the traditionally considered metabolic needs, and pleasantly gratifying the sensation of hunger has become insufficient to cover the present situation in Europe. The emphasis is now placed on the food potential to promote health, improve well-being, and reduce the risk of illness [1–3]. Thus, the trend is to replace the concept of "adequate nutrition" with "optimum nutrition." The notion of "Functional Foods"—"Foods for Specified Health Use," "Health Promoting Foods" or "Foods for particular nutritional uses"—has appeared to indicate that some foods can exert beneficial actions on certain functions in the organism which go beyond their traditionally considered nutritional effects [4–6].

When defining functional foods, there is no universally accepted agreement. The concept of Functional Food emerged in Japan during the eighties and triggered the introduction of a health claim system for "Foods for Specified Health Uses" (FOSHU) in 1991 (see [7]). In Japan, foods identified as FOSHU must be approved by the Minister of Health and Welfare after the submission of comprehensive science-based evidence to support the claims for the foods when they are consumed as part of an ordinary diet. In the U.S.A., "reduction of risk of disease" claims have been permitted since 1993 for certain foods. Health claims are authorized by the U.S. Food and Drug Administration (FDA) on the basis of "the totality of publicly available scientific evidence and where there is significant scientific agreement amongst qualified experts that the claims are supported by the evidence."

In the European Union (EU) still in 2006 there is a general prohibition about attributing to any foodstuff the property of preventing, treating, or curing a human disease—or referring to such properties. In the absence of specific provisions at the European level, each EU member state has applied its own interpretation. Thus, different EU countries have adopted different legislation or measures to regulate the use of such claims, and a claim that is allowed in one country may be prohibited in another. In addition, producers have to deal with a variety of regulations and practices. With the recent initiative for a regulation of the European Parliament and the Council on the nutrition and health claims made on foods [8], the situation is going to change considerably.

10.2   THE FUFOSE AND PASSCLAIM CONSENSUS IN EUROPE

In 1998–1999 a European consensus document was achieved within the FUFOSE project supported by the European Commission (EC) describing that the positive effect of a functional element may be both its contribution to maintaining and promoting a healthy state of well-being as well as reducing the risk of suffering a certain illness or disorder [3,9,10].

The FUFOSE project looked at six main areas of science and health: growth, development, and differentiation; substrate metabolism; defense against reactive oxidative species; the cardiovascular system; gastrointestinal physiology and function; and the effects of foods on behavior and psychological performance. The report takes the position that functional foods should be in the form of normal foods and must demonstrate their effects in amounts that can normally be expected to be consumed in the diet [10]. FUFOSE considered the functional food concept, in the context of nutrition, as being quite distinct from other approaches such as food fortification or supplementation; thus, functional foods are considered different from nutraceuticals, pharmafoods, or vitafoods. Functional foods are food products to be taken as part of the usual diet in order to have beneficial effects that go beyond what are known as traditional nutritional effects. Moreover, these beneficial effects have to be demonstrated scientifically to justify any health claim [10,11].

Later, the European Concerted Action PASSCLAIM group was established with the following objectives [12]: (a) to produce a generic tool with principles for assessing the scientific support for health-related claims for foods and food components; (b) to critically evaluate the existing schemes that assess the scientific substantiation of claims; and (c) to select common criteria for how markers should be identified, validated, and used in well-designed studies to explore links between diet and health. In its first phase [13], the focus was on three physiological areas: (a) diet-related cardiovascular disease (CVD), (b) bone health and osteoporosis, and (c) physical performance and fitness. In the second phase, the project continued with the following areas [14]: (d) body weight regulation, insulin sensitivity and risk of diabetes, (e) diet-related cancer, (f) mental state and performance, and (g) gut health and immunity.

Functional foods in Europe have gained a lot of attention among several social partners: the food industry, the scientific community, consumer organizations, and European and local health authorities. It is perceived that the European perspective differs from the American perspective. In Europe functional foods are more clearly considered food, and thus exert their beneficial effects in quantities that are normally consumed in a balanced diet, excluding pills, capsules, and tablets [15]. Thus, the development of safe and effective functional foods that can positively affect consumer health and well-being appears as a major opportunity to contribute to improving public health and the quality of life.

As revised by [3,11], "functional food" has unique features: (1) it is a conventional or everyday food; (2) it is consumed as part of the normal or usual diet, (3) it is composed of naturally occurring components, perhaps in unnatural concentrations, or which are present in foods that would not normally supply them; (4) it has a positive effect on target function(s) beyond nutritive value or basic nutrition; (5) it may enhance well-being and health, reduce the risk of disease, or provide a health benefit so as to improve the quality of life including physical, psychological, and behavioral performance; and (6) it has authorized and scientifically-based claims.

A functional food must remain a food and must demonstrate its effects in amounts that normally can be expected to be consumed in the diet; thus, it is not a pill or a capsule, but part of the normal food pattern [10].

10.3 DESIGNING FUNCTIONAL FOODS

The first step in the research and development of a functional food is to identify a(some) functional factor(s), condition(s) or compound(s) that may produce a specific effect that is potentially beneficial for health, and the potential functional effect must be identified by sufficient scientific evidence. It is necessary to elucidate the interaction of the specific compound responsible for the beneficial effect with other elements in the diet and find out its function within the organism at different levels (gene, biochemical, molecular, cellular, and physiologic), including elucidation or substantial knowledge of the mechanism of action [16,17]. To cover this knowledge in depth requires basic fundamental research to gain enough knowledge on the mechanism of action.

Then, it must be proven in relevant animal models and in humans by means of well-designed epidemiological and intervention studies that could show a valid statistical relationship between the intake of the compound in question and the specific benefit it provides [2,3,15]. It must be taken into account that a foodstuff may be functional for practically the whole population or only most of the people within a particular subgroup (e.g., people with cardiovascular risk factors, overweight or obesity, etc.). Functionality may also depend on one or more individual particularities [5,18]. Also of great importance is the evaluation of the security margins of the effective doses to produce the functional effects, and these margins must be safe and applicable to all main population groups, including those who can be expected to consume them most [19,20].

Identification and validation of good, quantifiable biomarkers that are sensitive to dietary modulation and reflect the steps in the process between food intake and functional effects are of great help [3,10,21]. By way of example, a list of putative markers for key functions related to the cardiovascular system is described in Table 10.1 [10]. Biological markers need to meet a number of requirements before their use can become widespread: they should be measurable in easily accessible material, be linked to the outcome involved in the biological process being studied, represent relatively immediate outcomes that can be used to assess interventions in a reasonable time, and be rigorously validated.

The "molecule-based" approach that deals with enriched common foods such as cereals, bread, dairy products, margarine, and beverages has the advantage of clarity of communication, provided that the molecules that are used to enrich the food have been proven to elicit the desired functional effects. The "whole food" approach, which is based on the large body of epidemiological knowledge

TABLE 10.1
Examples of Potential Markers for Key Target Functions Related to the Cardiovascular System and Candidate Food Components for Modulation of these Functions

Target Functions	Potential Biomarkers	Candidate Food Components
Lipoprotein homeostasis	Lipoprotein profile	SFA ($\downarrow$)
	LDL-cholesterol	MUFA, PUFA
	HDL-cholesterol	Tocotrienols
	Triacylglycerols	Plant sterols and stanols
		Soluble fiber
		Soy protein
		Fat replacers
		β-glucan
		Trans-fatty acids ($\downarrow$)
Endothelial and arterial integrity	Growth factors	Certain antioxidants
	Adhesion molecules	Vitamin E
	Cytokines	n-3 PUFA
Thrombogenic potential	Platelet function	n-3 PUFA
	Clotting factors	Linoleic acid
		Certain antioxidants
Control of hypertension	Systolic and diastolic blood pressure	Total energy intake ($\downarrow$)
		Sodium chloride ($\downarrow$)
		n-3 PUFA
		Peptides from milk (VPP and IPP)
Control of homocysteine	Plasma homocysteine levels	Folic acid
		Vitamin B6
		Vitamin B12

LDL, low-density lipoprotein; HDL, high-density lipoprotein; $\downarrow$, reduced intake; SFA, saturated fatty acids; MUFA, monounsaturated fatty acids; n-3 PUFA, n-3 series of long-chain fatty acids; VPP, Valine-Proline-Proline; IPP, Isoleucine-Proline-Proline.

Source: Adapted from Diplock, A. et al., Scientific concepts of functional foods in Europe: consensus document, *Br. J. Nutr.* 81 (suppl. 1) S1–S27, 1999.

showing that some diets (e.g., rich in vegetables and fruits or fish) are associated with long term health benefits, has the advantage that there is a cultural and practical heritage of use; both approaches can be clearly connected. The debate of nutritional profiles in Europe (see below in 10.7.2) with respect to the new legislation on health claims can also be referred to here: there is the view that products that do not have a "desirable" nutritional profile, such as sugar-based sweets, high salt and high fat snacks, or high fat and sugar biscuits and cakes should not be allowed to carry claims [8]. In this way it is highly possible that traditionally considered healthy diets such as the Mediterranean diet [22] will join newly designed functional modifications in the future.

10.4 NUTRIGENOMICS

Post-genomic technologies (transcriptomics, proteomics, metabolomics, and epigenomics) within Nutrigenomics will help to better define molecular composition, and provide solid evidence for cause-and-effect relationships between food composition and health outcomes. Nutrigenomics is the study of how food and genes interact, and the consequences. It links genome research, biotechnology, and molecular nutrition research, thus providing new developments in the field of nutrition and health. In Europe, the NuGO research network of excellence has been established [23].

Nutrigenomics will allow a greater than ever understanding of food influences in our homeostatic systems, allowing approaches to estimate potential adverse or beneficial effects of foods in precocious phases before a disease is committed to occurring. Additionally, it will help explaining to what extent the influence of diet on health depends on an individual's genetic makeup. Difficulties are expected, both due to the complexity of food and food practices, and due to the very complexity of our metabolic systems—and the fact that they are finely interrelated. Most foods are a vastly complex and synergistic or non-synergistic mix of several nutrients and many other components. Besides, food intake is not independent of several other factors: physical activity, feeling, and emotion, social, economic, and other environmental factors.

Much more knowledge is needed on the molecular mechanisms of the processes involved in the action of functional foods. Not just of individual molecules in isolation, but also of complex mixtures from food matrices. A goal for the future is to translate the studies of the in vivo response of our organisms to specific receptors/cells/tissues/organs in appropriate in vitro assays [24].

10.5 SOME EXAMPLES OF FUNCTIONAL FOODS IN EUROPE

In Europe the concept of functional foods emerged approximately 15 years ago, and we can say that they have been successfully and naturally accepted and tolerated by the population—and continue to grow in importance. Either way, successful brands are often more the result of commercial strategies than of innovations in science, and are more about broad consumer education than narrow health claims. It is also important whether the consumer, who is increasingly more concerned about his or her health, understands the benefit offered by a functional food.

As we know when talking of functional foods, we can take into account different kinds of products providing a health benefit due to the presence of an intrinsic component of the food. In the nineties, thanks to the progress of nutrition science in discovering the intrinsic health properties of components present naturally in different foodstuffs, food companies began to praise the beneficial inherent properties of their products, without needing to add specially isolated bioactive compounds. For example, since 1994, the potential benefits of the intake of tomato in reducing the risk of prostate cancer have been actively communicated, due to their content in carotenoids (lycopene); and since 1998, the lycopene content has been mentioned on the labels of processed tomato products. The same has happened with other products, such as tea, olive oil, etc., and as nutrition science continues to reveal more about the intrinsic health benefits of everyday foods, the strategy of marketing intrinsic healthy food components will become ever more common.

Apart from these "natural" functional foods, there is increasing scientific knowledge about the health properties or disadvantages of different food components and, thanks to food technology, there have also appeared functional foods that are rich or deprived in specific compounds. In the success of all these products, not only is the choice of the ingredient and the health benefit important, but so are the other commercial strategies such as innovations in packaging and marketing.

The experience of Europe's probiotic daily-dose market shows what is possible when you bring together packaging innovation and marketing based on simple wellness messages. It is a market that was practically nonexistent in 1994 but, in 2005, it has tremendous sales throughout Europe. Europe's probiotic daily-dose market can be illustrated by the 100 mL Actimel (Danone) drink. These small bottles were an innovation back in the mid-1990s, and have totally succeeded in the European market.

Another important example of functional foods with great success in Europe are the plant sterol-enriched products, such as Pro-Activ. This brand began with margarine (yellow fat spreads) enriched with plant sterol and stanol esters and later added milk-like products (Benecol, Unilever). More recently, the cholesterol-lowering Pro-Activ brand has extended to the launch of a 100 mL dairy drink.

In conclusion, Europe is part of a world in which health is at last becoming the basis of everything the food industry does. Nowadays there is a wide list of foods that can potentially be considered "functional" and we can guess that in the near future most food will be functional. Next, we will present a few examples of functional foods that are already commercialized in Europe. The three main examples are those referring to probiotics and foods enriched with some plant sterols and with n-3 fatty acids (Table 10.2). Apart from functional foods, we also will consider some supplements that may become functional food ingredients in the future.

10.5.1 FOODS WITH PLANT STEROLS

In order to preserve cardiovascular health it is appropriate to control circulating lipid levels. High cholesterol levels (mainly as low-density lipoprotein (LDL)-cholesterol) are a well-known risk factor for development of atherosclerosis, so a reduction in these levels could be associated with a reduction in the risk of CVD [25]. Plant sterols or phytosterols (stanols and sterols) are vegetal-derived substances (mainly present in vegetal oils), structurally and functionally analogous to cholesterol, that have the capacity of decreasing blood cholesterol levels, thus playing an important role in protection against atherosclerosis and related CVD in hypercholesterolemic people [26–28].

The presence of plant sterols in the diet (between 1 and 3 g/day [29]) have been shown to be effective in lowering total cholesterol and LDL-cholesterol circulating levels by affecting mainly cholesterol absorption, without negatively affecting other

TABLE 10.2
Examples of Functional Foods

Functional Food	Active Food Component	Target Function
Dairy products	Calcium	Prevention of osteoporosis
Yogurts	Cereals	Control of glycemia
Milks		Reduce risk of cancer
	Folic acid	Reduce risk of neural tube defects in the newborn
	Oleic acid	Reduce risk of cancer and cardiovascular disease (CVD)
		Decrease low-density lipoprotein (LDL) and total cholesterol
		Increase high-density lipoprotein (HDL)-cholesterol
		Decrease risk of CVD
	Omega-3 fatty acids	Control of lipid profile and hypertension
		Decrease risk of CVD
	Plant sterols	Decrease LDL and total cholesterol
		Decrease risk of CVD
	Probiotics (*Lactobacillus* sp., *Bifidobacteria* sp.)	Optimal intestinal function and intestinal microbial balance
	Prebiotics (non-digestible oligosaccharides)	
	Soy and isoflavones	Reduce risk of breast and prostate cancer
		Improve the symptomatology associated to menopause
Fats	Calcium	Prevent osteoporosis
Margarines	Plant sterols	Decreased LDL and total cholesterol
		Decreased risk of CVD
Edible oils	Diacylglycerols	Reduce serum triacylglycerol
		Increase degradation of free fatty acids and reduce fat deposition
Eggs	Omega-3 fatty acids	Control of lipid profile and hypertension
		Decrease risk of CVD
Meat and derivates	Less saturated fat	Reduce risk of CVD

Food	Component	Claimed effect
Cereals Breakfast enriched cereals Cereal bars	Fiber	Beneficial effects in the digestive process Reduce risk of developing cancer Reduce serum lipid levels and control of insulinemia
	Folic acid	Reduce risk of neural tube defects in the newborn Reduce risk of cancer and CVD
	Soy and isoflavones	Reduce risk of breast and prostate cancer Improve the symptomatology associated to menopause
Drinks	Calcium	Prevention of osteoporosis
	Fiber	Beneficial effects in the digestive process Reduce risk of developing cancer Reduce serum lipid levels and control of insulinemia
Juices and enriched drinks	Omega-3 fatty acids	Control of lipidic profile and hypertension Decrease risk of CVD
	Probiotics (*Lactobacillus* sp., *Bifidobacteria* sp.) Prebiotics (non-digestible oligosaccharides)	Optimal intestinal function and intestinal microbial balance
	Soy and isoflavones	Reduce risk of breast and prostate cancer Improve the symptomatology associated to menopause
	Vitamin C	Antioxidant effects and cancer prevention Enhancement of the immune system
Tahitian Noni juice	Juice of *Morinda citrifolia*	Beneficial effects in different disorders
Bread and bakery products	Cereals	Control of glycemia Reduce risk of cancer
	Fiber	Beneficial effects in the digestive process Reduce risk of developing cancer Reduce serum lipid levels and control of insulinemia
	Folic acid	Reduce risk of neural tube defects in the newborn Reduce risk of cancer and CVD
	Omega-3 fatty acids	Control of lipid profile and hypertension Decrease risk of CVD

(continued)

TABLE 10.2 (Continued)

Functional Food	Active Food Component	Target Function
	Prebiotics	Intestinal function and intestinal microbial balance
	Selenium	Increase antioxidant protection
		Reduce risk of cancer
Salt	Salt with less sodium and supplemented with potassium and magnesium	Reduce hypertension and risk of cardiovascular diseases
Special dietary products (low-lactose products, gluten-free foods, etc.)	Low levels or absence of a harmful food ingredient	Reduce risk of food allergies, food intolerances and absorption disorders

processes except the absorption of certain liposoluble nutrients, particularly β-carotene [29,30]. Plant sterols have a chemical structure similar to that of cholesterol and compete with it in the formation of lipid micelles in the gut, thus decreasing cholesterol absorption [31]. Moreover, it has been shown that they also increase cholesterol removal processes from the gut epithelial cells to the intestinal lumen [32].

The introduction of functional foods based on plant sterols was initiated in Finland (stanol enrichment) before the existence of the novel food EU regulation [33]. Yellow fat spreads or margarines enriched with plant sterols (8%) were authorized as a Novel Food for commercialization in the EU in 2000 [34], Since then, a number of other food products have been enriched (ranging from bakery products to drinks) [8,35].

10.5.2   FISH AND OMEGA-3 ($n-3$) POLYUNSATURATED FATTY ACIDS

There is good scientific evidence that dietary fatty acid composition is involved in the etiology of many diseases (insulin-related problems, obesity, development of atherosclerotic plaques, and CVD) [36]. For many years it has been known that polyunsaturated fatty acids (PUFA) have a certain capacity for lowering adiposity and triacylglycerol levels, which is due to the effect of activating fatty acid hepatic catabolism, activating thermogenesis, and inhibiting liver lipogenic capacity [37]. All these factors may play a role in reducing the risk of CVD but, apart from the improved plasma lipid and lipoprotein concentrations [38], PUFA has been reported to exhibit anti-thrombotic [39], anti-arrhythmic, and anti-inflammatory [40] effects that also work in favor of a lower risk of cardiovascular illness.

Polyunsaturated fatty acids may be divided into two categories, the omega-3 ($n-3$) and the omega-6 ($n-6$) families, depending on whether the double bond closest to the non-carboxyl end is located at C-3 or C-6 respectively. Humans are unable to synthesize $n-3$ and $n-6$ PUFA, therefore these fatty acids are essential and must be supplied by diet. The main unsaturated fatty acids are linolenic acid ($18:3n-3$) and linoleic acid ($18:2n-6$). These fatty acids can be elongated and desaturated into their longer-chain derivatives arachidonic acid ($20:4n-6$), eicosapentaenoic acid (EPA; $20:5n-3$), and docosahexaenoic acid (DHA; $22:6n-3$). Health benefits observed with the intake of PUFA are nearly exclusive to EPA and DHA [41], thus, both acids are considered "conditionally essential." Omega-3 PUFA exert a consistent hypotriacylglycerolemic effect that is dose-dependent and persistent. In terms of disease status, epidemiological studies have indicated that the incidence of CVD is inversely associated with consumption of these two fatty acids (EPA and DHA).

Health authorities, therefore, recommend an increase in foods containing omega-3 (mainly EPA and DHA) fatty acids in order to reduce the risk of CVD and, apart from increasing fish intake, a good way can be the supplementation of quotidian foods. Modern food technology makes it possible for a wide array of foods to be enriched in omega-3 and, in fact, a high variety of enriched products exists all over Europe. Some examples are: dairy products (mainly milk), bread and bakery products, eggs, and derivates, pasta, sauces, juices, non-alcoholic drinks, etc. [42].

When considering enrichment with omega-3 fatty acids, it must be taken into account that they get oxidized rapidly and, for that reason, vitamin E and/or other antioxidants have to be used in order to prevent oxidation that could alter organoleptic properties.

Docosahexaenoic acid is naturally found in breast milk and has been reported to support visual and cognitive development in infants [43]. Several infant formulas now contain DHA to more closely mimic breast milk [43].

10.5.3 OLIVE OIL, OLEIC ACID

The known "Mediterranean paradox" expresses the reduced cardiovascular pathology incidence in the Mediterranean countries, despite the high proportion of fat in the diet. It is partially related to the consumption of olive oil and, moderately, red wine [44], together with other foods (bread and cereals, fish, fruits, and vegetables, legumes, etc.). In these countries, an important part of the dietary fat comes from olive oil, whose fundamental fatty acid is oleic acid. Traditionally, virgin olive oil was consumed [45], which has different phytochemical compounds with antioxidant activity [46]. Oleic acid (octadecaenoic; 18:1n-9) is the main dietary acid representing the monounsaturated fatty acids (MUFA) [47]. Besides exerting relatively minor effects on the quantitative and qualitative regulation of cholesterol levels, oleic acid also appears to interfere directly with the inflammatory response that characterizes early atherogenesis [48]. Moreover, possibly in concert with other more highly unsaturated fatty acids, oleic acid could contribute to preventing atherosclerosis by its modulation of gene expression in endothelial leukocyte adhesion molecules [48]. On the other hand, some eicosanoids with vasodilatation and antiagregant activity are derived from oleic acid [48]. This oil reduces the serum levels of triacylglycerol, total and LDL-cholesterol and its oxidation, and is one of the few substances that appears to be able to increase the serum levels of high-density lipoprotein (HDL)-cholesterol [49]. Thus, due to its interesting health promoting properties, oleic acid has been used to enrich different dietary products such as milk and fermented milk. There is also high-oleic sunflower oil, etc.

10.5.4 DIACYLGLYCEROL OILS

Diacylglycerol is an intermediate product of triacylglycerol hydrolysis that comprises up to 10% of glycerides in plant-derived edible fats and oils. Although it is absorbed in the same way as triacylglycerols, it is metabolized differently: more of it seems to be burned and less of it is stored as fat [50].

It has been reported that the intake of an oil containing 70% of unusual 1,3-diacylglycerol species has metabolic characteristics that are distinct from those of triacylglycerol of similar fatty acid composition [50]. The main reported effect of the ingestion of diacylglycerols (primarily in the form of 1,3-diacylglycerols) is a reduction in the levels of serum triacylglycerols, a decrease in postprandial hyperlipidemia, a modest decrease in body weight in higher body weight individuals, and, in some cases, a moderate reduction in body fat, both in humans and experimental models [50,51]. Reacylation to triacylglycerols in small intestinal cells was, in animal models, found to be slower with diacylglycerol feeding than with triacylglycerol feeding [51]. Diacylglycerol consumption has also been shown to decrease the activities of enzymes of fatty acid synthesis and increase the expression and activities of enzymes involved in the β-oxidation pathway, favoring in this way the degradation in the liver of free fatty acids rather than their deposition as body fat [51].

An edible oil (Enova oil) containing about at least 80% of diacylglycerol molecules esterified at positions 1 and 3 has been recently evaluated by the dietetic products, nutrition and allergies (NDA) Scientific Panel of the European Food Safety Authority (EFSA) as a first step in the process of introducing this product into the EU market. Enova oil has been given GRAS (generally recognized as safe) status by the Food and Drug Administration. It can be used for home cooking, as well as an ingredient in salad dressings, mayonnaise, spreads, and other processed foods. But although Enova oil has been available in the U.S.A. since January 2005, it has been sold in Japan since 1999 and is used widely as cooking oil.

10.5.5   LESS SATURATED FATS

The link between saturated fats and the risk of CVD has been known since the early 1900s because their intake raises serum cholesterol levels [52]. For this reason, the manufacture of foods in which saturated fats are replaced by cholesterol-lowering unsaturated fatty acids, e.g., soluble omega-3 and omega-6 fatty acids that increase the proportion of polyunsaturated fats in the end product, appears of great interest. In Europe (Finland) a technology has been developed which allows this process by substituting saturated with unsaturated fat. This concept has been applied in the manufacture of processed meat products where its benefits are particularly evident, but in the future it can also be applied to other food groups.

10.5.6   PROBIOTICS AND PREBIOTICS

The intestinal flora that colonizes the distal part of the small and large intestine, in normal conditions, is made up of a great variety of microorganisms, an important portion of which are beneficial for our health. Other intestinal microorganisms could be bad for the organism, although they coexist in the gut in equilibrium [53]. When this equilibrium is lost (e.g., with the use of antibiotics) some intestinal disorders can appear, reflecting the importance of maintaining the activity of beneficial bacteria in the gut. Lactic acid bacteria have interesting potential health effects that first arose in relation to the prevention and cure of intestinal disorders to protect the intestines from harmful microbes and toxins and to reduce allergic reactions.

Dietary modulation of the human gut flora has been carried out for many years and, in humans, there are positive aspects to gut fermentation [53]. For instance, the bifidobacteria and lactobacilli may help to reduce the risk of disease in different ways. Lactobacilli may aid digestion of lactose for intolerant people, reduce constipation, and infantile diarrhea, help to resist some infections, prevent traveler's diarrhea and help to relieve irritable bowel syndrome [54]. On the other hand, bifidobacteria are thought to stimulate the immune system, inhibit pathogen growth, produce B vitamins, and help to restore the normal flora after antibiotic therapy [55]. In general, due to the numerous potential health promoting effects, there is currently much interest in increasing the numbers and activities of these bacteria in the large gut [53]. In this way, we should differentiate prebiotics and probiotics. The premise behind the former is to stimulate certain indigenous bacteria resident in the gut rather than to introduce exogenous species, as is the case with probiotics.

10.5.6.1 Probiotics

We can define probiotics as living microorganisms which, when taken in certain amounts, can provide the organism with beneficial effects [56], such as species of *Lactobacillus* and *Bifidobacterium*. The lactobacilli and bifidobacteria are mainly found in fermented dairy products. Nevertheless, not all lactobacilli and bifidobacteria are considered probiotic; it is necessary to demonstrate that they have a beneficial effect for the organism which is distinct to the purely nutritional aspect. Apart from intestinal disorders, certain probiotics could act in other disorders, although more studies are needed to establish which probiotics are the best for every potential situation [9]. The product family includes milk, fermented milk, yogurt, juice, daily-dose drink, and capsules.

With respect to the regulation of probiotics in the EU, there is no special one about the criteria determining that a kind of bacterium can be considered probiotic. Nor is there a regulation for its use in elaborating food, or one that applies to their nutritional claims [20].

10.5.6.2 Prebiotics

Another approach to promote commensal bifidobacteria or lactobacilli is by the intake of certain nonviable substrates that are known as prebiotics [53]. The first description of a prebiotic, by Gibson and Roberfroid [55], was as a "non-digestible food ingredient that beneficially affects the host by selectively stimulating the growth and/or activity of one or a limited number of bacteria in the colon, and thus improves health." So, we can understand a prebiotic as a nonviable food ingredient selectively metabolized by beneficial intestinal bacteria.

At least three criteria are required for a substrate to be considered a prebiotic. The first is that the substrate must not be substantially hydrolyzed or absorbed in the stomach or small intestine. The second is that it must be selective for beneficial commensal bacteria in the colon, such as the bifidobacteria. And the third is that fermentation of the substrate should induce beneficial luminal/systemic effects within the host [53].

Although any dietary component that reaches the colon intact is a potential prebiotic, the main interest is non-digestible oligosaccharides. We can find some prebiotics naturally in the diet, in fruits and vegetables that contain prebiotic oligosaccharides, but the levels in these foods can be too low to exert significant effects. Thus, there exists much value in the approach of fortification of commonly ingested foodstuffs with prebiotics [53]; and, as mentioned, current prebiotics seem to be mainly the oligosaccharides that are non-digestible in the upper gut.

Oligosaccharides (short-chain polysaccharides) are sugars of approximately 3–20 monosaccharide units. They can be extracted from fruits and vegetables, but some of them can be commercially produced by the hydrolysis of polysaccharides or through enzymatic generation [53]. Among the most studied prebiotics are inulin and galactooligosaccharides. Oligosaccharides such as lactulose, fructo-, galacto-, and soybean oligosaccharides, lactosucrose, isomalto-, gluco, and xylo-oligosaccharides, and palatinose have been suggested as having prebiotic potential [57].

Different foods can be supplemented with prebiotics (virtually any carbohydrate containing food); these include beverages and fermented milks, health drinks, bakery products, table spreads, sauces, infant formulae and weaning foods, cereals, biscuits, dairy products, etc. [53].

10.5.7 DIETARY FIBER

Dietary fiber is the denomination used for those vegetal substances (mainly carbohydrates) that cannot be digested by human enzymes although they are partially fermented by the colonic bacteria [44]. There are insoluble and soluble types of fiber. Insoluble fiber includes cellulose, hemicelluloses, and lignin, with interesting properties such as feces volume increase, promotion of intestinal motility, antioxidant properties, reduction of cholesterol intestinal absorption, etc. [58–60]. Soluble fiber includes pectins, gums, mucilages, and some hemicelluloses, and their main characteristic is that they can take up water and form viscous gels (with laxative properties). These kinds of fibers may have different beneficial effects in the digestive process, as well as effects on the reduction of cholesterol, triacylglycerols, and postprandial insulinemia [59,61,62]. Soluble fiber has other potentially interesting actions, such as the promotion of antiproliferative effects by the butyrate produced by the colonic bacteria using these fibers as a substrate [63].

Both types of fiber are found in varying proportions in food, although insoluble fiber is predominant in whole cereals while the soluble one is abundant in fruits, vegetables, and tubers. There are different products in the food industry that are enriched with fiber, such as some cereals, bread, buns, drinks, pâtés, and cold meat [44].

10.5.8 CEREALS (WHOLE-GRAIN)

Cereals contain many compounds with health promoting effects (fiber, minerals, B-group vitamins, and various phytochemicals such as phytoestrogens and antioxidants) and thus provide another alternative for the production of functional foods [64]. Whole-grain consumption has been associated to a reduction in the risk of developing certain kinds of cancers (especially gastrointestinal cancers such as gastric and colonic, and hormonally-dependent cancers including breast and prostate), CVD, diabetes, obesity, and other chronic diseases [64]. Components in the whole grains that can be protective against cancer include compounds that affect the gut environment, such as dietary fiber, resistant starch, and oligosaccharides. Fermentation of fiber in the colon results in the production of short chain fatty acids that have been shown to be antineoplasic and to reduce serum cholesterol [65]. Moreover, whole grains are rich in antioxidants, including vitamins, trace minerals, phenolic compounds, and phytoestrogens with potential hormonal effects that have been proposed to be important in cancer and CVD [64]. Whole grains may also be protective against diseases by modulation of the glycemic response: refining grains tends to increase the glycemic response, and thus whole grains should slow glycemic response [66], helping to control insulin liberation to the bloodstream. This in turn may help to control diabetes and obesity [67].

In addition, cereals can be used as fermentable substrates for the growth of probiotic microorganisms; and as sources of non-digestible carbohydrates that besides promoting several beneficial physiological effects can also selectively stimulate the growth of bacteria present in the colon and thus act as prebiotics [68]. Finally, cereal constituents, such as starch, can be used as encapsulation materials for probiotics in order to improve their stability during storage and enhance their viability during their passage through the adverse conditions of the gastrointestinal tract [68].

Traditionally, cereals have been used for baking, so we can naturally find them in bread and bakery products. However, due to their beneficial effects, they are also used in other foodstuffs such as snacks and cereal bars, yogurts, etc.

10.5.9 CAROTENOIDS AND LYCOPENE

Much interest has arisen about diets rich in fruits and vegetables containing carotenoids, due to their potential health benefit against prevalent chronic diseases such as cancer and CVD. In fact, there are numerous reviews about the relationship of diet and lung cancer and CVD in humans and wide research has been carried out in this field over the past 30 years (see [69–71]). Epidemiological studies have suggested that there is a reduced risk for people with a high dietary intake of vegetables and fruits rich in carotenoids, and there is an important association between high serum β-carotene levels and the reduced incidence of lung cancer in prospective studies.

Nevertheless, when two important intervention studies were carried out with high doses (20 mg/day or more) of synthetic pure β-carotene or with specific formulations containing β-carotene, the results indicated a higher incidence of lung cancer in risk populations of smokers rather than a protective effect [71–73]. In fact, the role of other carotenoids or other compounds from vegetables and fruits, and associated dietary or life style patterns, has not been adequately explored in the epidemiological studies [71].

Some hypotheses have been made to explain these contradictory results. One is that higher dietary intakes of tomatoes and tomato-based products, resulting in higher blood levels of lycopene, may help reduce the risk of certain cancers, and that β-carotene is not the only important bioactive compound in a fruit and vegetable-rich diet. A review has been made grouping 72 epidemiological studies including cohort, case–control, diet-based, and biomarker-based studies investigating the consumption of tomatoes and related products, blood lycopene levels, and cancer incidence at various anatomical sites [74]. The data from this review indicate that high consumers of tomatoes and tomato products are at a substantially decreased risk of numerous cancers, with the strongest associations found with cancers of the prostate gland, lung, and stomach—all across numerous diverse populations. Prospective long-term clinical trial data are needed to go further into the role of lycopene in disease prevention, although the present data are suggestive of a protective effect.

Apart from its possible healthy effects protecting from cancer, an increased lycopene intake may also play a protective role in prevention of CVD, thus contributing to cardiovascular health [75]. Due to their ingredients, a lot of natural and common foods could be—or could become—considered as functional foods, and among those that contain β-carotene or lycopene, vegetables such as broccoli, cauliflower, cabbage, and tomato are of importance.

10.5.10 FOLIC ACID

Folic acid is important in the formation of the unborn baby's spinal cord and additional supplies in the expectant mother's diet can reduce the risk of neural tube defects in the newborn [76]. Thus, in women planning a pregnancy or who are pregnant, it is recommended that they take an additional folic acid supplement and that they use fortified foods as well as food naturally containing folate. In addition, folic acid deficiency with age is not rare, and a folate deficit has been associated with the loss of mental brightness related to ageing [77,78]. B vitamins, particularly folate, may give considerable protection against serious diseases such as cancer and heart disease [79].

In this way, the intake of fortified foods can also be used as a nutritional strategy for old people. In the market there exist fortified-foods enriched in folic acid, such as flours, for preventing disease and increasing health [80]. Cereal flours are enriched or fortified with vitamins, minerals, and folic acid to provide specific health benefits.

A clear dose–response relationship has not been clearly defined for folic acid, as well as for many bioactive compounds. The threshold between beneficial and adverse effects appears to be a critical point to be addressed.

10.5.11 VITAMIN C

At the beginning of the '70s, the Nobel laureate Linus Pauling advised that daily intakes of 1 g or more of vitamin C can protect against the common cold (see [81]) and his hypothesis has been very popular though the years. Moreover, this advice was followed by other claims of beneficial effects on a variety of conditions and it is generally thought that high amounts of vitamin C are necessary to maintain good health.

Because of the media attention given to these claims and the apparently low toxicity of vitamin C there has been extensive human exposure to intakes of up to 10 g/day [81]. Nevertheless, some doubts and disillusionment arose with the results from several clinical trials questioning the beneficial effect of vitamin C [82], but new evidence leads to increasing confidence in the properties of vitamin C, with the establishment of a new theory stating that aging and many diseases originate from the accumulation of free radicals [83].

Free-radical scavengers are usually known as antioxidants and vitamin C is one of the most effective and widespread antioxidants in biological systems [84,85], although scavenging of reactive oxygen species and organic radicals is not the only function of vitamin C in the metabolism. Moreover, new perspectives could result from the analysis of the role of vitamin C in many enzyme-mediated functions [86].

One important function of vitamin C is preventing scurvy, which results from its deficiency. The best understood function of vitamin C is in the synthesis of collagen, which promotes the formation of hydroxyproline [87]. Non-hydroxylated collagen is unstable, and is unable to form the triple helix required for normal structure in tissues such as subcutaneous, cartilage, bone, and teeth [88].

Vitamin C has also been related to the prevention of common diseases in our society like cancer [89] and ischemic heart disease [90–93]. Moreover, vitamin C intake improves impaired endothelial vasodilatation in patients with essential hypertension [94] and improves congestive heart failure [95]. Other possible roles have been suggested in preventing vitamin C and iron deficiency anemia [96–98], preventing some adverse effects of diabetes mellitus [99], improving some disorders associated to renal transplantation [95], fighting against oxidation processes in Alzheimer's disease [100–102], reducing the clinical signs of recurrent herpes labialis [103] and reducing (together with vitamin E) the oxidative stress and the viral load in HIV infected patients [104].

With respect to the famous putative preventing action of vitamin C in the common cold, this is still controversial. A review of different studies suggests a reduction of at least 80% in the incidence of pneumonia in supplemented vitamin C groups, and it provided an important benefit in elderly hospital patients with bronchitis or pneumonia [105,106]. Nevertheless, the investigations carried out have shown that long-term daily supplementation with high doses (1 g daily during winter months) of vitamin C does not appear to prevent colds, although there could be a modest beneficial effect in reducing the duration of cold symptoms [107–109].

Recent reviews of vitamin C by the Food and Nutrition Board in the U.S.A. and the Expert Group on Vitamins and Minerals in the U.K. have recommended a guidance level of 1 g/day, as a supplemental intake. In this sense, it is important to consider that fruits and vegetables are rich in vitamin C and that vitamin C is used in a number of foodstuffs as an additive.

10.5.12 VITAMIN E

There has been intense research about vitamin E action over the past decades, but there are still controversial data about its usefulness as a dietary supplement. The term "vitamin E" groups a family of molecules of related structure: four tocopherols and four tocotrienols. All these molecules possess antioxidant activity, although α-tocopherol is the most active [110]. The other natural forms (β-, γ-, and δ-tocopherols and the four tocotrienols) are not converted into α-tocopherol by humans, and they are poorly recognized by the α-tocopherol transfer protein in the liver [110].

α-Tocopherol has antioxidant activity, which has led to the expectation that dietary supplementation with vitamin E could present health promoting effects in disorders such as the oxidation of LDL and the pathogenesis of atherosclerosis [111,112]. Nevertheless, the large clinical trials carried out so far provide little supporting evidence [113]. On the other hand, recently there has been emerging evidence showing that vitamin E could have other physiological activities apart from the antioxidant actions. Some of these actions may be involved in the prevention of atherosclerosis by a way distinct to the oxidant role. Other actions may be at the level of gene transcription regulation and vitamin E has also been implicated in the prevention of cancer [110].

10.5.13 CALCIUM

Osteoporosis is quite a common problem in old people. In order to avoid osteoporosis complications, prevention is of great importance: to obtain the maximum bone mass before or around the age of thirty and preserve it until the most dangerous age, and, particularly, until the menopause in women [114]. For these reasons, it is of great importance to ingest a proper quantity of calcium daily by eating foods that naturally have it or to eat supplemented foods [115].

Moreover, calcium has been shown to reduce body weight, probably by binding fat in the intestine and increasing its excretion from the body. The mechanism by which calcium increases fat excretion is probably an interaction between calcium and fatty acids, resulting in the formation of insoluble calcium fatty acid soaps and hence in reduced fat absorption. These facts would mean that an increased calcium intake could also favor weight loss [116].

There is a wide range of examples of calcium-enriched foods, mainly directed to preventing osteoporosis, such as margarines and dairy products (milk, yogurts, cheese, etc.). The form in which calcium is present in foods appears of particular importance as it determines big differences in calcium absorption.

10.5.14 SELENIUM

Selenium is essential for a number of enzymes that perform important metabolic functions that are necessary for good health, such as glutathione peroxidase, an enzyme that participates in the antioxidant protection of cells. Epidemiological research has suggested a link between higher levels of selenium and the reduced risk of a number of cancers, including breast, prostate, and colon [117].

Selenium enters the food system from soils, and there is wide evidence to indicate that the world's soils vary considerably with respect to their content of biologically available selenium. As a consequence, levels in the European diet are lower than in American or Canadian diets, due to the lower levels of the mineral in Europe, influencing the amount absorbed by foods such as cereals. As a result, people in many countries do not appear to consume adequate amounts of selenium to support the maximal expression of the selenium enzymes [117]. For these reasons, it seems of potential interest to have foods enriched in this mineral. It has been shown that the intake of wheat naturally enriched with selenium and its products increases glutathione peroxidase activity in blood and decreases lipid parameters and glucose in blood [118]. Selenium-fortified bread has been launched in the U.K.

10.5.15 Soy and Isoflavones

Phytoestrogens are vegetal molecules with an estrogen-like structure; in fact, they behave as partial agonists for estrogen receptors. Thus, different healthy actions have been postulated for these molecules, such as a reduction of osteoporosis [119], a reduction of the incidence of breast and prostate cancer [120], improvement in the symptomatology associated with menopause, and positive effects on the cardiovascular system [121]. The main natural source for phytoestrogens—isoflavones [119]—are legumes, especially soy (which has about 25–40 mg per serving). Nevertheless, the presence of isoflavones is not the only functional feature of legumes such as soy. Apart from the important content of soluble and insoluble fiber, legumes have important proportions of different bioactive molecules and micronutrients, such as riboflavin, folic acid, and minerals with a high bioavailability such as zinc, copper, selenium, iron, and calcium (this last one, especially in soy) [122,123]. Due to the mentioned health-promoting potential action, different commercially available foods have been supplemented with soy, such as milks, fermented milks, snack bars, etc.

10.5.16 Flavonoids

Another phytochemical group that appears of interest for use as supplements or to enrich different kinds of foods, and that is also important in non-supplemented food, is made up of phenolic compounds. There are over 5000 different molecules of this type, and of particular importance are the flavonoids [124].

Over 4000 flavonoids have been identified in plants [125]. These phenolic compounds can be classified as isoflavones (mentioned above); flavonols, such as quercetin, kaempterol, myricetin, and isorhammetin (mainly found in onions, apples, teas, berries, olives, bananas, lettuce, plums, and red wine); flavones, such as luteolin and apigenin (mainly found in apples, celery, celeriac, lemons, parsley, oregano, lettuce, and beets); flavanones, such as hesperetin, naringenin, and eriodictyol (mainly found in oranges, grapefruits, and lemons); anthocyanidins, such as cyandin, delphinidin, malvidin, etc. (mainly found in berries); flavan-3-ols, such as catechin, gallocatechin, theaflavin, etc. (mainly found in green and black tea, plums, apples, and cranberries); and procyanidins, such as polymeric catechins and epicatechins (in cocoa, chocolate, cinnamon, cranberries, pinto beans, kidney beans, hazelnuts, and pecans) [125].

Flavonoids act as antioxidants by chelating redox-active metals and by scavenging free radicals. The quelation of iron and copper by flavonoids can prevent peroxyl radical and lipid peroxidation [126–129]. These compounds can also act as terminators of free radicals by the donation of electrons to form stable products, and they are good scavengers of hydroxyl and peroxyl radicals, as well as quenching superoxide radicals and singlet oxygen [126,130,131]. Moreover, some flavonoids present anti-inflammatory activity; e.g., quercetin inhibits Tumor necrosis factor-alpha (TNF-α), as well as nitric oxide production by lipopoly-saccharide-activated macrophages [132]. Due to their different properties, flavonoids are related to cancer prevention by dietary agents, and different potential mechanisms for flavonoid and isoflavonoid substances have been postulated, such as estrogenic and antiestrogenic activity, antiproliferation, cell cycle arrest and apoptosis, antioxidation, induction of detoxification enzymes, regulation of host immune function, and others [133].

Thus, apart from isoflavones, several flavonoids can be very interesting in the field of functional foods. They are present in numerous natural foods (as mentioned above) and they can also be considered for food supplementation.

10.5.17 Less Salt

Along with cancer, CVD are today the most common cause of premature death. Besides cholesterol, a high salt content in the diet has been shown to present a significant cardiovascular risk because of the increase in blood pressure [134]. The mechanisms by which dietary salt increases arterial pressure are not fully understood, but they seem to be related to the inability of the kidneys to excrete large amounts of salt [134]. In Finland, a mineral salt product has been developed containing a balanced mineral composition—reduced in sodium and with added potassium and magnesium—that is currently being exported to several countries around the world (PanSalt) and that is claimed to help balance mineral intake which affects blood pressure.

10.5.18 Special Dietary Products

The problem of food allergies and absorption disorders has been taken seriously for years in developed countries and, thus, there is an important variety of products on the market for those following a special diet. Here, we will review only some representative examples. For instance, lactose intolerance is relatively common and different low-lactose products have been developed and manufactured, taking into account that complete elimination of lactose is not necessary to ensure tolerance by most lactose maldigesters. It has been shown that they may be able to tolerate foods containing 6 g of lactose or less per serving [135]. Other special products are gluten-free foods for celiac people. Celiac disease is triggered by the cereal protein gluten and elimination of gluten is the effective therapy; thus, to increase food safety for celiac patients, gluten has been included in food regulations introducing obligatory labeling [136].

Another disorder where dietary treatment is important is phenylketonuria, which is a congenital absence of phenylalanine hydroxylase (an enzyme that converts phenylalanine into tyrosine), thus allowing phenylalanine to accumulate in blood and seriously impair early neuronal development. This can produce severe mental retardation if not correctly treated. The defect in phenylketonuria can be controlled by diet and, in this disorder, protein substitutes are an essential component in its management, thus highlighting the importance of research into creating better protein substitutes and amino acid tablet preparations for patients [137]. A number of other foods for particular nutritional purposes could be added here, as this field will be particularly active in the area of food innovations in the near future.

Other products, like sucrose substituting products, can be mentioned. In 1994, the European Directive 94/35/CE authorized the use of five intense sweeteners for which acceptable daily intakes were established [138]. Nevertheless, it was shown a long time ago that sweeteners such as fructose and sorbitol have no advantages over sucrose on the effect on blood glucose in well-regulated adult diabetics, and it could be unnecessary to have specially sweetened foods designed for diabetics [139].

10.5.19 TAHITIAN NONI JUICE

The plant *Morinda citrifolia* L., more commonly known as "noni" in the Hawaian islands, as well as its fruit, has been used for a long time as a medicinal staple of the islands of the South Pacific in the treatment of different diseases such as diabetes, heart disease, high blood pressure, kidney, and bladder disorders [140]. Its fruit, leaves, and bark have also been used as a poultice for sores, cuts, and boils [141]. Tahitian Noni juice is a fruit juice mixture of about 90% Noni fruit (*Morinda citrifolia* L.) and 10% common grape and blueberry juice concentrates and natural flavors.

Both the history of the juice and different studies carried out have suggested potential health benefits for this product [142]. In general, a consumption of 30 mL/day has been recommended. Tahitian Noni juice has been marketed in the U.S.A. since 1996, and in Canada, Japan, Australia, Mexico, Norway, and Hong Kong. It has been recently introduced into Europe.

10.6 FOOD SAFETY AND FUNCTIONAL FOODS

Functional foods are a recent phenomenon in Europe, and are as yet not covered as such by any specific legislation. Thus, at present, the development of functional foods in Europe has evolved by taking into account the safety standards derived from the various general frameworks concerning food legislation. Safety aspects have been emphasized in Europe during the last decade in a context where there has been a process of integration in the EU of new member states, and a general European food law was not in force until 2002 [143]. A reference to some European legislation on food safety (see http://europa.eu.int/comm/food/fs/sfp/sfp_index_en.html) and to the recent initiative to regulate nutrition and health claims made on foods set up by the EC [8] has to be made.

10.6.1 THE FOOD SAFETY FRAMEWORK IN THE EUROPEAN UNION: EFSA

In 1996–1997, a profound reform of the system in Europe was put into practice. Until then, the EU had an assessment structure linked to general guidance that was more related to legislation and commerce; the connection with the interest in consumer health was not very evident. Controversy arose above all as a result of the epidemic crisis of Bovine Spongiform Encephalopathy (BSE), otherwise known as "mad cow disease."

The received criticism for their insufficient action rendered in the relationship between BSE in cattle and the new variety of the Creutzfeldt–Jakob disease in humans. A perception of excessive dominance or influence of industrial interests within the committees and administration was detected among consumers and other social sectors, as well. The European Parliament widely expressed its preoccupation on the scientific assessment of the EU in terms of food safety and the need for a greater link of this assessment with the public interest in consumer health. In short, the European treaty of 1997 signified the commitment of the EU member states to a joint policy of food safety.

In April, 1997, the Commission dissolved the old scientific committees and nine new scientific committees were created, including the Scientific Committee on Food (SCF) (for scientific assessment of all aspects of food safety and in the different steps of the food chain), and were put under the auspices of the then DG XXIV, which was in charge of defending consumer interests and health. The reform meant a great qualitative change in the focus on food safety problems by the EC. The procedures for ensuring principles of excellence, independence, and transparency were renewed at all levels. For example, although the SCF used a majority system for decision-making, minority opinions were guaranteed to be taken in every detail, which meant a guarantee of transparency. The decision that opinions (and also the minutes of the sessions) are immediately to be made public on the Internet Web page is also key. In this way, because the management decision depends on previous scientific assessments, all individual, business, or sectorial interests have the same access to information, which, in turn, makes it easier to manage problems in a participative way.

In 2002, the creation of the EFSA as an independent entity, with its headquarters presently in Parma (Italy), represented the logical solution of continuity of its precursor scientific committees, and fulfils the overall characteristics of an era (1996–2002) that has been decisive for food safety in Europe. In essence, EFSA (initially structured as eight scientific panels, listed in Table 10.3, and a coordinator committee) incorporated the main rules of procedure of the SCF and related committees—and its independence from other General Directorates of the EC has been stressed. The key function of this Authority (http://europa.eu.int/comm/food/fs/efa/index_en.html) is the expert analysis and scientific evaluation of risks, and to communicate these risks to all interested sectors and the public in general. The task of "risk management," that is, the process by which different alternatives (detected by expert, e.g., EFSA, analysis) are pondered and decisions are made to correct problems (new legislation, controls, studies, sanctions, etc.) remains in the hands of the state members and of the EC.

10.6.2 Safety of Functional Foods: Legislation and Scientific Basis

The two key-aspects in potential food evaluation are safety and efficacy. Safety is guaranteed in Europe by the application of various regulations that are in charge of controlling any novelty in food components and the obtaining processes. As mentioned above, there is no specific European regulation for functional foods. Depending on the type of food novelty—whether or not it can be considered to be functional—that is trying to be introduced, the evaluation of its safety will correspond to one type of specific legislation or another. In some cases, it will

TABLE 10.3
Scientific Panels of The European Food Safety Authority (EFSA)

Panels	Description
Scientific panel on food additives, flavorings, processing aids and materials in contact with food [AFC]	The Panel on food additives, flavorings, processing aids, and materials in contact with food deals with questions of safety in the use of food additives, flavorings, processing aids, and materials in contact with food; with associated subjects concerning the safety of other deliberately added substances to food; and with questions related to the safety of processes
Scientific Panel on additives and products or substances used in animal feed [FEEDAP]	The Panel on additives and products or substances used in animal feed deals with questions on safety for the animal, the user/worker, the consumer of products of animal origin, the environment, and with the efficacy of biological and chemical products/substances intended for deliberate addition/use in animal feed
Scientific panel on plant health, plant protection products and their residues [PPR]	The Panel on plant health, plant protection products, and their residues deals with questions on the safety of plant protection products for the user/worker, the consumer of treated products, and the environment, as well as plant health
Scientific panel on genetically modified organisms [GMO]	The Panel on genetically modified organisms deals with questions on genetically modified organisms as defined in Directive 2001/18/EC, such as micro-organisms, plants, and animals relating to deliberate release into the environment and genetically modified food and feed, including their derived products
Scientific panel on dietetic products, nutrition, and allergies [NDA]	The Panel on dietetic products, nutrition, and allergies deals with questions on dietetic products, human nutrition and food allergy, and other associated subjects such as novel foods
Scientific panel on biological hazards [BIOHAZ]	The Panel on biological hazards deals with questions on biological hazards relating to food safety and food-born disease, including food-born zoonoses and transmissible spongiform encephalopathies, microbiology, food hygiene, and associated waste management
Scientific panel on contaminants in the food chain [CONTAM]	The Panel on contaminants in the food chain deals with questions on contaminants in food and feed, associated areas, and undesirable substances such as natural toxicants, mycotoxins, and residues or non-authorized substances not covered by another Panel
Scientific panel on animal health and welfare [AHAW]	The Panel on animal health and welfare deals with questions on all aspects of animal health and animal welfare, primarily relating to food producing animals including fish

correspond to the area covered by the legislation of the so-called Novel Foods [33], the regulations on nutritional supplements [143], or those referring to foods for particular nutritional uses (PARNUTS) [144].

In addition to these, other European legislation and directives cover aspects on food safety (see http://europa.eu.int/comm/food/fs/sfp/sfp_index_en.html) such as those related to pollutants, microbiological risks, hygiene, irradiation, dietary foods, supplements, and fortified foods, additives, and flavorings, food contact materials, natural mineral waters, and questions pertaining to official controls and rapid warning systems.

In particular, the evaluation of "novel foods" involves new ground in the evaluation of food issues [33]. Let us not forget, for instance, that foods (the traditional ones) have never been systematically evaluated for their safety; in fact, quite the opposite, we have always accepted their need, taking the net beneficial advantages of their effects for granted.

As far as efficacy is concerned, both the guidelines as to the scientific evidence required to validate the efficacy of new components and associated foods, and the question of so-called claims, allegations, or health claims that may accompany the eventual processes of commercialization are yet to be defined.

10.6.3   REGULATION OF NOVEL FOODS

Any novel food or food ingredient that has not been significantly consumed in the EU before May 15, 1997, must be evaluated with respect to its safety according to the European legislation on Novel Foods (EC Regulation No. 258/97) [145]. This regulation covers a very wide range of foods and ingredients. It included transgenic foods; foods and ingredients that have a new molecular structure; those consisting of or isolated from microorganisms, fungi, and algae; those from animals and plants reproduced using non-traditional methods; and those foods and food ingredients obtained with a new production process that gives rise to significant changes in the composition or structure of the foods or ingredients in a way that affects their nutritional value, metabolism, or levels of undesirable substances. At present, the assessment process of genetically modified organisms and their products is being considered separately [35,146].

In short, the Novel Foods regulation establishes that a novel food must fulfill three criteria in order to be accepted as such: (a) it must not be dangerous to the consumer, (b) it must not be deceiving for the consumer, and (c) it must not differ from the traditional food it could displace, so long as this is not advantageous to the consumer. To help in the implementation, an EC Recommendation has been published concerning the scientific aspects and the presentation of information necessary to support applications for the placing on the European market of novel foods and novel food ingredients, and the preparation of initial assessment reports under the Regulation of the Novel Foods [147].

All in all, to have complete, extremely rigorous information is necessary for the evaluation of novel foods or novel food ingredients. This information should be structured in roughly eleven sections (see Table 10.4) depending on the specific characteristics of the food type [147].

The marketing of novel foods or novel food ingredients can follow a more simplified procedure when they have already been considered to be substantially equivalent to existing foods or ingredients with respect to their composition, nutritional value, metabolism, foreseeable habitual consumption, and levels of undesirable substances by the committees authorized by the EU member states.

TABLE 10.4
Summary of the Essential Information Required in the Evaluation of Non-Genetically Modified Novel Foods

Specifications on the origin and composition of the novel food, ensuring its identification
Effects that can be produced by the production processes of the novel food
Uses and characteristics of the organisms (and their products) used as a source of novel foods
Effects of the modification introduced on the properties of food. Consideration of the intentional effects
    and the non-intentional ones
Stability of the modification introduced
Nutritional impacts. Evaluation of the impact of the introduction of the novel food on the consumption
    patterns and habits in the population
Nutritional analysis: composition and foreseeable impact in the population's diet
Study of the allergenic potential and effects in sensitive populations
Toxicological information
Microbiological information
Any information as to the effects of total or partial exposure of human populations to the novel food

Note: Depending on each case, additional scientific information may be required.

A very useful tool in the assessment of novel foods is based on the concept of "substantial equivalence," which was first introduced by the Organization for Economic Cooperation and Development [148] and was later extended by FAO/ WHO [149] with particular reference to foods produced by modern biotechnology. This concept presents the idea that existing organisms used as food or food sources can serve as a basis for comparison when assessing the safety of a novel food. Substantially equivalent novel foods are considered comparable, with respect to safety, to their conventional counterparts. Establishment of substantial equivalence is not a safety or nutritional assessment in itself, but an approach to compare a potential novel food with its appropriate counterpart or related traditional food. It constitutes a starting point to orientate further analysis or research into more concrete safety aspects.

Conversely, if a novel food does not show substantial equivalence to an existing one, this does not imply that it is unsafe: it merely indicates the need to evaluate it on the basis of its particular composition or properties. The concept of substantial equivalence can also be extended to the assessment of novel sources and new processes. It was derived with transgenic foods in mind but is currently used as a tool to apply to non-genetically modified novel foods.

10.6.4 Fortified Foods, Food Supplements—Upper Levels of Nutrients

Functional foods are generally produced by using different approaches: by eliminating a component known to cause a deleterious effect to the consumer, by adding a component that is not normally present, by increasing the concentration of a component naturally present in foods to produce beneficial effects, and by replacing a component (potentially deleterious) with another for which beneficial effects have been demonstrated. Taking into account that functional foods are usually made by adding or removing some functional substances, the present control strategy for food supplements and enriched or fortified foods has a significant implication for functional foods [9,150].

The fortification or enrichment of foods has been conducted for the purpose of a nutritional need or for a commercial purpose. Fortification means the addition of one or more essential nutrients to a food—whether or not they are normally contained in the food—for preventing or correcting a demonstrated deficiency of one or more nutrients in the population or in specific population groups. Supplementation consists of introducing nutrients or other ingredients which consumers believe to have particular beneficial properties and can be sold in the form of capsules, tablets, or powders.

Directive 2002/46/EC of the European Parliament and of the Council of June 10, 2002 on the approximation of the laws of the Member States relating to food supplements establishes harmonized rules for the labeling of food supplements and introduces specific rules on vitamins and minerals in food supplements [143]. At present, there are an increasing number of products marketed in the Community as foods containing concentrated sources of nutrients and presented to supplement the intake of those nutrients in the normal diet. The aim of this new legislation is to ensure that these products are safe and appropriately labeled so that consumers can make informed choices.

As a first stage, this directive deals essentially with vitamins and minerals used as ingredients in food supplements. It is envisaged that rules concerning other substances with a nutritional or physiological effect used as ingredients in food supplements should be laid down at a later stage, provided that appropriate scientific data about them become available.

Important work in regard to supplements was recently developed in the EU concerning the establishment of maximum tolerable levels for the intake of vitamins and minerals [151]. Minimum intake needs of these nutrients have been established as limits in order to avoid the appearance of alterations or illnesses due to their deficiency. In contrast, maximum tolerable levels of nutrients were defined to adequately control the maximum quantities that can be eaten, e.g., in relation to supplemented foods. From the point of view of nutritional security, legislation does not require minimal nutritional values for common foods; this is only required for nutritional allegations. Nevertheless, the European Council has raised some important questions that have yet to be dealt with, such as the risk involved in modifying feeding habits due to the negative perception of those foods that do not carry associated allegations of health, rising risk of allergies, and the possible appearance of new diseases related to feeding.

The definition of clear guidelines to establish scientific evidence should be laid down, and would probably include human studies with appropriate animal models. These guidelines should also identify markers, as mentioned above [3]. Studies tailored on a case by case basis will also be needed.

10.7 TOWARDS FUNCTIONAL FOOD HEALTH CLAIMS AND NUTRITIONAL CLAIMS IN EUROPE

The interest in health claims is rising in Europe. Discussing food claims means addressing the complex issue of communication of health benefits of foodstuffs between consumers and producers (see [19,152–154]). The relationship between optimal nutrition and a healthy life is gaining public acceptance and consumers are increasingly health-conscious and want to obtain more information about the food they buy. At the same time, the industry wants to take advantage of the developments in food science and is investing heavily in innovative projects. But only a food that successfully conveys its health benefits in a meaningful way to the consumer is one that eventually becomes an investment incentive for the industry.

A lot of effort has been made internationally and in the EU to demonstrate that scientific substantiation of claims is possible and to establish valid criteria for this process. The fundamental principle defining food allegations or claims is that they must be scientifically proven, must not be ambiguous, and must be clear to the consumer. Council Directive 90/496/EEC on nutrition labeling includes a definition of "nutrition claim" as any representation and any advertising message that states, suggests, or implies that a foodstuff has particular nutritional properties due to the energy or nutrient content [155]. They are more precisely defined in the coming European legislation (see below).

It is evident that the evolution and development of functional foods must be accompanied by the application of suitable controls and follow-up to guarantee that the claims are used to obtain their best effect in promoting public health and protecting the consumer against information that could be false or erroneous.

10.7.1 NUTRITIONAL AND HEALTH CLAIMS IN EUROPE

The European Community has adopted detailed rules on labeling and nutrition labeling of foods (see references in 8). With regard to claims, it must be remembered that there is the basic provision that claims should not mislead the consumer. Furthermore, Article 2.1.B of Directive 2000/13/EC [156] on the labeling, presentation, and advertising of foods prohibits the attribution of preventing, treating, and curing properties to foods. However, the EC Proposal for a Regulation of the European Parliament and of the Council on Nutrition and Health Claims made on Foods, 2003/0165 (COD), was released in July 2003 [8]. Nutrition and health claims used in the labeling, presentation, and advertising of foods have been approved by the European parliament in 2006.

A "claim" is defined as any message or representation that is not mandatory under Community or national legislation, including pictorial, graphic, or symbolic representation that states, suggests, or implies that a food has particular characteristics. "Nutrition claim" covers any claim which states, suggests or implies that a food has particular nutrition properties due to: the energy (calorific value) it provides; whether it provides at a reduced or increased rate, or does not provide; the nutrients or other substances it contains, contains in reduced or increased proportions, or does not contain. "Health claim" means any claim that states, suggests, or implies that a relationship exists between a food category, a food, or one of its constituents and health. "Reduction of disease risk claim" means any health claim that states, suggests, or implies that the consumption of a food category, a food, or one of its constituents significantly reduces a risk factor in the development of a human disease.

The situation is now changing from a prohibition of claims related to "prevent, treat, or cure a human disease" to allowing these claims in a certain way. For instance, it can be realized that new legislation does not refer directly to human disease but to "… reduce a risk factor in the development of a human disease."

There has been a voluntary code of practice in place since 1999. It was adopted by the Confederation of the Food and Drink Industries of the EU (CIAA) on the use of health claims to help companies prepare the documentation necessary for the substantiation of health claims, a procedure that did not use to be shared by consumer organizations.

10.7.2 NUTRIENT PROFILES AND OTHER CONTENTIOUS ISSUES

The EC Proposal for the regulation of Health Claims made on foods involved considerable debate in the European Parliament. In general, it establishes better protection of consumers when nutrition and health claims are made, facilitates the free movement of health foods within the community, provides better legal security to manufacturers who make health claims, and promotes food scientific research, development, and innovation.

The new legislation requires claims to be based on and be substantiated by generally accepted scientific data. Claims may be made only in cases where the average consumer can be expected to understand the beneficial effects expressed in the claim. All claims must also relate to the food as ready-to-eat in accordance with the manufacturer's instructions.

The initial proposal [8] prohibits the use of nutrition or health claims for foods that do not meet established nutrient profiles, e.g., candies, high salt, high saturated fat, and snack foods, and limits the claims that may be made about alcoholic beverages. In the definitive text a compromise was attained that foods that do not respect the nutrient profiling would still be able to make health claims as long as only one food component exceeds the limit and this is clearly indicated in the label.

The impact of this new legislation will be complemented by a number of voluntary initiatives by the different sectors in the food chain, particularly consumers and the food distribution sector. For instance, the introduction of a new "traffic light" labeling system that will highlight the total fat, saturated fat, salt, and sugar in a product by red, amber, and green color coding next to the calorie and carbohydrate information.

Nutrient profiles will be established that will be based on scientific knowledge about diet and nutrition, and their relationship to health and, in particular, the role of nutrients and other substances with a nutritional or physiological effect on chronic diseases. They are to be established by reference to the composition of fat, saturated fatty acids, trans-fatty acids, sugars, and salt/sodium. In setting the nutritional profiles, the Commission will seek the advice of the EFSA and carry out consultations with interested parties, in particular, food business operators and consumer groups.

It is expected that to prepare its opinion on a proposed claim, the EFSA shall verify: (a) that the proposed wording of the health claim is substantiated by scientific data; (b) that the wording of the heath claim complies with the criteria laid down in the regulation; and (c) that the proposed wording of the health claim is understandable and meaningful to the consumer. In the event of an opinion in favor of approving the health claim, the opinion shall include: (a) the name and address of the applicant; (b) the designation of the food or category of food in respect of which a claim is to be used and its particular characteristics; (c) the recommended wording, in all Community languages, of the proposed health claim; and (d) where necessary, conditions of use of the food or an additional statement or warning that should accompany the health claim on the label.

REFERENCES

1. Palou, A., Bonet, M. L., and Serra, F., Eds., *Obesity and Functional Foods in Europe*, European Commission, Luxembourg, pp. 1–409, 2002.
2. Bellisle, F., Functional Foods in Europe: designing foods for the prevention and treatment of obesity, In Obesity and Functional Foods in Europe, Palou, A., Bonet, M.L., and Serra, F., Eds., Luxembourg, pp. 293–301, 2002.
3. Bellisle, F., Blundell, J. E., Dye, L., Fantino, M., Fern, E., Fletcher, R. J., Lambert, J. et al., Functional food science, behavior and psychological functions, *Br. J. Nutr.*, 80, 173–193, 1998.
4. Arai, S., Morinaga, Y., Yoshikawa, T., Ichiishi, E., Kiso, Y., Yamazaki, M., Morotomi, M., Shimizu, M., Kuwata, T., and Kaminogawa, S., Recent trends in functional food science and the industry in Japan, *Biosci. Biotechnol. Biochem.*, 66, 2017–2029, 2002.
5. Eckhardt, R. B., Genetic research and nutritional individuality, *J. Nutr.*, 131, 336–339, 2001.
6. Contor, L., Functional food science in Europe, *Nutr. Metab. Cardiovasc. Dis.*, 11, 20–23, 2001.
7. Hirahara, T., Key factors for the success of functional foods, *Biofactors*, 22, 289–294, 2004.
8. EC, Proposal for a regulation of the European Parliament and the council on the nutrition and health claims made on foods, *COM* (2003) 424 final, 2003.
9. Roberfroid, M. B., Prebiotics and probiotics: are they functional foods? *Am. J. Clin. Nutr.*, 71, 1682S–1687S, 2000 (discussion 1688S–1690S).
10. Diplock, A., Aggett, P., Ashwell, M., Bornet, F., Fern, E. B., and Roberfroid, M. B., Scientific concepts of functional foods in Europe: consensus document, *Br. J. Nutr.*, 81 (suppl. 1), S1–S27, 1999.
11. Roberfroid, M. B., Global view on functional foods: European perspectives, *Br. J. Nutr.*, 88 (suppl. 2), S133–S138, 2002.
12. Richardson, D. P., Affertsholt, T., Asp, N. G., Bruce, A., Grossklaus, R., Howlett, J., Pannemans, D., Ross, R., Verhagen, H., and Viechtbauer, V., PASSCLAIM—synthesis and review of existing processes, *Eur. J. Nutr.*, 42 (suppl. 1), I96–I111, 2003.

13. Asp, N. G. and Contor, L., Process for assessment of scientific support for claims on food (PASSCLAIM): overall introduction, *Eur. J. Nutr.*, 42 (suppl. 1), I3–I5, 2003.
14. Contor, L. and Asp, N. G., Process for the assessment of scientific support for claims on food (PASSCLAIM) phase two: moving forward, *Eur. J. Nutr.*, 43 (suppl. 2), II3–II6, 2004.
15. Hasler, C. M., Functional foods: benefits, concerns and challenges—a position paper from the American council on science and health, *J. Nutr.*, 132, 3772–3781, 2002.
16. Daniel, H., Genomics and proteomics: importance for the future of nutrition research, *Br. J. Nutr.*, 87 (suppl. 2), S305–S311, 2002.
17. Woychik, R. P., Klebig, M. L., Justice, M. J., Magnuson, T. R., and Avner, E. D., Functional genomics in the post-genome era, *Mutat. Res.*, 400, 3–14, 1998.
18. Milner, J. A., Functional foods and health promotion, *J. Nutr.*, 129, 1395S–1397S, 1999.
19. Palou, A., Serra, F., and Pico, C., General aspects on the assessment of functional foods in the European Union, *Eur. J. Clin. Nutr.*, 57 (suppl. 1), S12–S17, 2003.
20. Palou, A., Picó, C., and Bonet, M. L., La seguridad de los nuevos alimentos en Europa: alimentos funcionales, In *Tendencias en alimentación funcional*, Serrano Ríos, M., Sastre Gallego, A., and Cobo Sanz, J. M. Eds., You & Us, SA, Madrid, pp. 123–140, 2005.
21. Hasler, C. M., Kundrat, S., and Wool, D., Functional foods and cardiovascular disease, *Curr. Atheroscler. Rep.*, 2, 467–475, 2000.
22. Trichopoulou, A. and Vasilopoulou, E., Mediterranean diet and longevity, *Br. J. Nutr.*, 84 (suppl. 2), S205–S209, 2000.
23. NuGO, European Nutrigenomics Organization, Network of excellence: linking genomics, nutrition and health research, FOOD-CT-2004-506360, 2004, http://www.nugo.org/everyone
24. Gidley, M. J., Naturally functional foods: challenges and opportunities, *Asia Pac. J. Clin. Nutr.*, 13, S31, 2004.
25. Kiechl, S. and Willeit, J., The natural course of atherosclerosis. Part I: incidence and progression, *Arterioscler. Thromb. Vasc. Biol.*, 19, 1484–1490, 1999.
26. Amundsen, A. L., Ose, L., Nenseter, M. S., and Ntanios, F. Y., Plant sterol ester-enriched spread lowers plasma total and LDL cholesterol in children with familial hypercholesterolemia, *Am. J. Clin. Nutr.*, 76, 338–344, 2002.
27. Hallikainen, M. A., Sarkkinen, E. S., and Uusitupa, M. I., Plant stanol esters affect serum cholesterol concentrations of hypercholesterolemic men and women in a dose-dependent manner, *J. Nutr.*, 130, 767–776, 2000.
28. Plat, J., Kerckhoffs, D. A., and Mensink, R. P., Therapeutic potential of plant sterols and stanols, *Curr. Opin. Lipidol.*, 11, 571–576, 2000.
29. SCF, Scientific-Committee-on-Food, General view of the Scientific Committee on Food on the long-term effects of the intake of elevated levels of phytosterols from multiple dietary sources, with particular attention to the effects on beta-carotene, *SCF/CS/NF/DOS/20 ADD 1 Final*, 2002.
30. Noakes, M., Clifton, P., Ntanios, F., Shrapnel, W., Record, I., and McInerney, J., An increase in dietary carotenoids when consuming plant sterols or stanols is effective in maintaining plasma carotenoid concentrations, *Am. J. Clin. Nutr.*, 75, 79–86, 2002.
31. Moghadasian, M. H., Pharmacological properties of plant sterols in vivo and in vitro observations, *Life Sci.*, 67, 605–615, 2000.
32. Plat, J. and Mensink, R. P., Increased intestinal ABCA1 expression contributes to the decrease in cholesterol absorption after plant stanol consumption, *FASEB J.*, 16, 1248–1253, 2002.

33. EC, Regulation (EC) No. 258/97 of the European Parliament and of the Council of 27 January concerning novels foods and novel foods ingredients, *Off. J. Eur. Commun.*, L43, 1–7, 1997.

34. SCF, Opinion of the SCF on a request for the safety assessment of the use of phytosterol esters in yellow fat spreads, *SCF/CS/NF/DOS/1 Final*, 2000.

35. EC, Regulation 1830/2003 of the European Parliament and of the Council of 22 September 2003 concerning the traceability and labeling of genetically modified organisms and the traceability of food and feed products produced from genetically modified organisms and amending Directive 2001/18/EC, *Off. J. Eur. Commun.*, L268, 24–28, 2003, http://europa.eu.int/eurlex/pri/en/oj/dat/2003/l_268/l_26820031018en00240028.pdf.

36. Roche, H. M., Dietary carbohydrates and triacylglycerol metabolism, *Proc. Nutr. Soc.*, 58, 201–207, 1999.

37. Duplus, E., Glorian, M., and Forest, C., Fatty acid regulation of gene transcription, *J. Biol. Chem.*, 275, 30749–30752, 2000.

38. Roche, H. M. and Gibney, M. J., Postprandial triacylglycerolaemia: the effect of low-fat dietary treatment with and without fish oil supplementation, *Eur. J. Clin. Nutr.*, 50, 617–624, 1996.

39. Yamaguchi, K., Mizota, M., Hashizume, H., and Kumagai, A., Antiatherogenic action of eicosapentaenoic acid (EPA) in multiple oral doses, *Prostaglandins Leukot. Med.*, 28, 35–43, 1987.

40. Calder, P. C., Immunoregulatory and anti-inflammatory effects of $n-3$ polyunsaturated fatty acids, *Braz. J. Med. Biol. Res.*, 31, 467–490, 1998.

41. Freese, R. and Mutanen, M., Alpha-linolenic acid and marine long-chain $n-3$ fatty acids differ only slightly in their effects on hemostatic factors in healthy subjects, *Am. J. Clin. Nutr.*, 66, 591–598, 1997.

42. Trautwein, E. A., $n-3$ Fatty acids—physiological and technical aspects for their use in food, *Eur. J. Lipid. Sci. Technol.*, 103, 45–55, 2001.

43. Auestad, N., Scott, D. T., Janowsky, J. S., Jacobsen, C., Caroll, R. E., Montalto, M. B., Halter, R. et al., Visual, cognitive, and language assessments at 39 months: a follow-up study of children fed formulas containing long-chain polyunsaturated fatty acids to 1 year of age, *Pediatrics*, 112, e177–e183, 2003.

44. Silveira Rodriguez, M. B., Monereo Megias, S., and Molina Baena, B., Functional nutrition and optimal nutrition: near or far? *Rev. Esp. Salud Publica*, 77, 317–331, 2003.

45. Ramirez-Tortosa, M. C., Suarez, A., Gomez, M. C., Mir, A., Ros, E., Mataix, J., and Gil, A., Effect of extra-virgin olive oil and fish-oil supplementation on plasma lipids and susceptibility of low-density lipoprotein to oxidative alteration in free-living spanish male patients with peripheral vascular disease, *Clin. Nutr.*, 18, 167–174, 1999.

46. Owen, R. W., Giacosa, A., Hull, W. E, Haubner, R., Wartele, G., Spiegelhalder, B., and Bartsch, H., Olive-oil consumption and health: the possible role of antioxidants, *Lancet Oncol.*, 1, 107–112, 2000.

47. Ortiz Leyba, C., Jiménez Jiménez, F. J., Garnacho Montero, J., and García Garmendia, J. L., In *Nuevos Sustratos Lipídicos En Nutrición Artificial, Nutrición Clínica: Implicaciones Del Estrés Oxidativo Y De Los Alimentos Funcionales*, Gil Hernández, A., Ruiz López, M. D., Sastre Gallego, A., and Schwartz Riera, S. Eds., McGraw-Hill, Madrid, pp. 44–47, 2001.

48. Massaro, M., Carluccio, M. A., and De Caterina, R., Direct vascular antiatherogenic effects of oleic acid: a clue to the cardioprotective effects of the Mediterranean diet, *Cardiologia*, 44, 507–513, 1999.

49. Mataix, J., *Nutrición y alimentación humana*, Ergon, Madrid, 2001.

50. Tada, N. and Yoshida, H., Diacylglycerol on lipid metabolism, *Curr. Opin. Lipidol.*, 14, 29–33, 2003.

51. Tada, N., Physiological actions of diacylglycerol outcome, *Curr. Opin. Clin. Nutr. Metab. Care*, 7, 145–149, 2004.

52. Mensink, R. P. and Katan, M. B., Effect of dietary trans fatty acids on high-density and low-density lipoprotein cholesterol levels in healthy subjects, *N. Engl. J. Med.*, 323, 439–445, 1990.

53. Manning, T. S. and Gibson, G. R., Microbial-gut interactions in health and disease: prebiotics, *Best Pract. Res. Clin. Gastroenterol.*, 18, 287–298, 2004.

54. Salminen, S., Ramos, P., and Fonden, R., Substrates and lactic acid bacteria, *Lactic Acid Bacteria*, Salminen, S., and Von Wright, A., Eds., Marcel Dekker, New York, pp. 1–442, 2003.

55. Gibson, G. R. and Roberfroid, M. B., Dietary modulation of the human colonic microbiota: introducing the concept of prebiotics, *J. Nutr.*, 125, 1401–1412, 1995.

56. Salminen, S., Bouley, C., Boutron-Ruault, M. C., Cunnings, J. H., Franck, A., Gibson, G. R., Isolauri, E., Moreau, M. C., Robertfroid, M., and Rowland, I., Functional food science and gastrointestinal physiology and function, *Br. J. Nutr.*, 80 (suppl. 1), S147–S171, 1998.

57. Gibson, G. R., Berry Ottaway, P., and Rastall, R. A., *Prebiotics: New Developments in Functional Foods*, Chandos Publishing Limited, Oxford, 2000.

58. Delargy, H. J., O'Sullivan, K. R., Fletcher, R. J., and Blundell, J. E., Effects of amount and type of dietary fiber (soluble and insoluble) on short-term control of appetite, *Int. J. Food Sci. Nutr.*, 48, 67–77, 1997.

59. Kay, R. M. and Truswell, A. S., Dietary fiber: effects on plasma and biliary acids in man, In *Medical Aspects of Dietary Fiber*, Spiller, G. A. and Kay, R. M. Eds., Plenum Medical Book Co., New York, pp. 153–173, 1980.

60. Howarth, N. C., Saltzman, E., and Roberts, S. B., Dietary fiber and weight regulation, *Nutr. Rev.*, 59, 129–139, 2001.

61. Jenkins, D. J., Axelsen, M., Kendall, C. W., Augustin, L. S., Vuksan, V., and Smith, U., Dietary fiber, lente carbohydrates and the insulin-resistant diseases, *Br. J. Nutr.*, 83 (suppl. 1), S157–S163, 2000.

62. Ramirez-Tortosa, C., Lopez-Pedrosa, J. M., Suarez, A., Ros, E., Mataix, J., and Gil, A., Olive oil- and fish oil-enriched diets modify plasma lipids and susceptibility of LDL to oxidative modification in free-living male patients with peripheral vascular disease: Spanish Nutrition Study, *Br. J. Nutr.*, 82, 31–39, 1999.

63. Cummings, J. H., Pomare, E. W., Branch, W. J., Naylor, C. P., and Macfarlane, G. T., Short chain fatty acids in human large intestine, portal, hepatic and venous blood, *Gut*, 28, 1221–1227, 1987.

64. Slavin, J. L., Martini, M. C., Jacobs, D. R. Jr., and Marquart, L., Plausible mechanisms for the protectiveness of whole grains, *Am. J. Clin. Nutr.*, 70, 459S–463S, 1999.

65. McIntyre, A., Gibson, P. R., and Young, G. P., Butyrate production from dietary fibre and protection against large bowel cancer in a rat model, *Gut*, 34, 386–391, 1993.

66. Jenkins, D. J., Wolever, T M. , Jenkins, A. L., Giordano, C., Giudici, S., Thompson, L. U., Kalmusky, J., Josse, R. G., and Wong, G. S., Low glycemic response to traditionally processed wheat and rye products: bulgur and pumpernickel bread, *Am. J. Clin. Nutr.*, 43, 516–520, 1986.

67. Slavin, J. L., Mechanisms for the impact of whole grain foods on cancer risk, *J. Am. Coll. Nutr.*, 19, 300S–307S, 2000.
68. Charalampopoulos, D., Wang, R., Pandiella, S. S., and Webb, C., Application of cereals and cereal components in functional foods: a review, *Int. J. Food Microbiol.*, 79, 131–141, 2002.
69. IARC, International Agency for Research on Cancer, *IARC Handbook of Cancer Prevention*, Vol. 2, *Carotenoids*, 1998.
70. Ziegler, R. G., Mayne, S. T., and Swanson, C. A., Nutrition and lung cancer, *Cancer Causes Control*, 7, 157–177, 1996.
71. SCF, Opinion of the Scientific Committee on Food on the safety of use of beta carotene from all dietary sources, *SCF/CS/ADD/COL/159 Final*, 2000.
72. ATBC-Study-Group, The effects of vitamin E and β carotene on the incidence of lung cancer and other cancers in male smokers, *N. Engl. J. Med.*, 330, 1029–1356, 1994.
73. Omenn, G. S., Goodman, C. E, Thornquist, M. D., Balmes, J., Cullen, M. R., Glass, A., Keogh, J. P. et al., Effects of a combination of beta carotene and vitamin A on lung cancer and cardiovascular disease, *N. Engl. J. Med.*, 334, 1150–1155, 1996.
74. Giovannucci, E., Tomatoes, tomato-based products, lycopene, and cancer: review of the epidemiologic literature, *J. Natl Cancer Inst.*, 91, 317–331, 1999.
75. Rissanen, T., Voutilainen, S., Nyyssonen, K., and Salonen, J. T., Lycopene, atherosclerosis, and coronary heart disease, *Exp. Biol. Med. (Maywood)*, 227, 900–907, 2002.
76. Czeizel, A. E. and Dudas, I., Prevention of the first occurrence of neural-tube defects by periconceptional vitamin supplementation, *N. Engl. J. Med.*, 327, 1832–1835, 1992.
77. Moretti, R., Torre, P., Antonello, R. M., Cattaruzza, T., Cazzato, G., and Bava, A., Vitamin B12 and folate depletion in cognition: a review, *Neurol. India*, 52, 310–318, 2004.
78. Rampersaud, G. C., Kauwell, G. P., and Bailey, L. B., Folate: a key to optimizing health and reducing disease risk in the elderly, *J. Am. Coll. Nutr.*, 22, 1–8, 2003.
79. Lucock, M. D. and Yates, Z., A differential role for folate in development disorders, vascular disease and other clinical conditions: the importance of folate status and genotype, In Folate and Human Development, Massaro, E. J. and Rosers, J. M., Eds., Humana Press, New Jersey, pp. 263–298, 2002.
80. Molloy, A. M. and Scott, J. M., Folates and prevention of disease, *Public Health Nutr.*, 4, 601–609, 2001.
81. Miller, D., Vitamin excess and toxicity, *Nutr. Toxicol.*, 1, 81–133, 1982.
82. Moertel, C. G., Fleming, T. R., Creagan, E. T., Rubin, J., O'connell, M. J., and Ames, M. M., High-dose vitamin C versus placebo in the treatment of patients with advanced cancer who have had no prior chemotherapy: a randomized double-blind comparison, *N. Engl. J. Med.*, 312, 137–141, 1985.
83. Marx, J. L., Oxygen free radicals linked to many diseases, *Science*, 235, 529–531, 1987.
84. Frei, B., England, L., and Ames, B. N., Ascorbate is an outstanding antioxidant in human blood plasma, *Proc. Natl Acad. Sci. U.S.A.*, 86, 6377–6381, 1989.
85. Rose, R. C. and Bode, A. M., Biology of free radical scavengers: an evaluation of ascorbate, *FASEB J.*, 7, 1135–1142, 1993.
86. Arrigoni, O. and De Tullio, M. C., Ascorbic acid: much more than just an antioxidant, *Biochim. Biophys. Acta*, 1569, 1–9, 2002.
87. Peterkofsky, B., Ascorbate requirement for hydroxylation and secretion of procollagen: relationship to inhibition of collagen synthesis in scurvy, *Am. J. Clin. Nutr.*, 54, 1135S–1140S, 1991.

88. Bsoul, S. A. and Terezhalmy, G. T., Vitamin C in health and disease, *J. Contemp. Dent. Pract.*, 5, 1–13, 2004.

89. Rock, C. L., Jacob, R. A., and Bowen, P. E., Update on the biological characteristics of the antioxidant micronutrients: vitamin C, vitamin E, and the carotenoids, *J. Am. Diet Assoc.*, 96, 693–702, 1996 (quiz 703–704).

90. Ting, H. H., Timimi, F. K., Haley, E. A., Roddy, M.-A., Ganz, P., and Creager, M. A., Vitamin C improves endothelium-dependent vasodilation in forearm resistance vessels of humans with hypercholesterolemia, *Circulation*, 95, 2617–2622, 1997.

91. Hamabe, A., Takase, B., Vehata, A., Kurita, A., Ohsuzu, F., and Tamai, S., Impaired endothelium-dependent vasodilation in the brachial artery in variant angina pectoris and the effect of intravenous administration of vitamin C, *Am. J. Cardiol.*, 87, 1154–1159, 2001.

92. Enstrom, J. E., Kanim, L. E., and Klein, M. A., Vitamin C intake and mortality among a sample of the United States population, *Epidemiology*, 3, 194–202, 1992.

93. Sahyoun, N. R., Jacques, P. F., and Russell, R. M., Carotenoids, vitamins C and E, and mortality in an elderly population, *Am. J. Epidemiol.*, 144, 501–511, 1996.

94. Taddei, S., Virdis, A., Ghiadoni, L., Magagna, A., and Salvetti, A., Vitamin C improves endothelium-dependent vasodilation by restoring nitric oxide activity in essential hypertension, *Circulation*, 97, 2222–2229, 1998.

95. Williams, M. J., Sutherland, W. H., McCormick, M. P., De Jong, S. A., McDonald, J. R., and Walker, R. J., Vitamin C improves endothelial dysfunction in renal allograft recipients, *Nephrol. Dial. Transplant.*, 16, 1251–1255, 2001.

96. Hallberg, L., Iron and vitamins, *Bibl. Nutr. Dieta*, 20–29, 1995.

97. Sharma, D. C. and Mathur, R., Correction of anemia and iron deficiency in vegetarians by administration of ascorbic acid, *Indian J. Physiol. Pharmacol.*, 39, 403–406, 1995.

98. Mao, X. and Yao, G., Effect of vitamin C supplementations on iron deficiency anemia in Chinese children, *Biomed. Environ. Sci.*, 5, 125–129, 1992.

99. Cunningham, J. J., The glucose/insulin system and vitamin C: implications in insulin-dependent diabetes mellitus, *J. Am. Coll. Nutr.*, 17, 105–108, 1998.

100. Kontush, A., Mann, U., Arlt, S., Ujeyl, A., Luhrs, C., Muller-Thomsen, T., and Beisiegel, U., Influence of vitamin E and C supplementation on lipoprotein oxidation in patients with Alzheimer's disease, *Free Radic. Biol. Med.*, 31, 345–354, 2001.

101. Riviere, S., Birlouez-Aragon, I., Nourhashemi, F., and Vellas, B., Low plasma vitamin C in Alzheimer patients despite an adequate diet, *Int. J. Geriatr. Psychiatry*, 13, 749–754, 1998.

102. Morris, M. C., Beckett, L. A., Scherr, P. A., Hebert, L. E., Bennett, D. A., Field, T. S., and Evans, D. A., Vitamin E and vitamin C supplement use and risk of incident Alzheimer disease, *Alzheimer Dis. Assoc. Disord.*, 12, 121–126, 1998.

103. Hovi, T., Hirvimies, A., Stenvik, M., Vuola, E., and Pippuri, R., Topical treatment of recurrent mucocutaneous herpes with ascorbic acid-containing solution, *Antiviral Res.*, 27, 263–270, 1995.

104. Allard, J. P., Aghdassi, E., Chau, J., Tam, C., Kovacs, C. M., Salit, I. E., and Walmsley, S. L., Effects of vitamin E and C supplementation on oxidative stress and viral load in HIV-infected subjects, *AIDS*, 12, 1653–1659, 1998.

105. Hemila, H. and Douglas, R. M., Vitamin C and acute respiratory infections, *Int. J. Tuberc. Lung Dis.*, 3, 756–761, 1999.

106. Hunt, C., Chakravorty, N. K., Annan, G., Habibzadeh, N., and Schorah, C. J., The clinical effects of vitamin C supplementation in elderly hospitalised patients with acute respiratory infections, *Int. J. Vitam. Nutr. Res.*, 64, 212–219, 1994.

107. Douglas, R. M., Hemila, H., D'Souza, R., Chalker, E. B., and Treacy, B., Vitamin C for preventing and treating the common cold, *Cochrane Database Syst. Rev.*, CD000980, 2004.

108. Douglas, R. M., Chalker, E. B., and Treacy, B., Vitamin C for preventing and treating the common cold, *Cochrane Database Syst. Rev.*, CD000980, 2000.

109. Van Straten, M. and Josling, P., Preventing the common cold with a vitamin C supplement: a double-blind, placebo-controlled survey, *Adv. Ther.*, 19, 151–159, 2002.

110. Schneider, C., Chemistry and biology of vitamin E, *Mol. Nutr. Food Res.*, 49, 7–30, 2005.

111. Esterbauer, H., Dieber-Rotheneder, M., Striegl, G., and Waeg, G., Role of vitamin E in preventing the oxidation of low-density lipoprotein, *Am. J. Clin. Nutr.*, 53, 314S–321S, 1991.

112. Jessup, W., Kritharides, L., and Stocker, R., Lipid oxidation in atherogenesis: an overview, *Biochem. Soc. Trans.*, 32, 134–138, 2004.

113. Upston, J. M., Terentis, A. C., and Stocker, R., Tocopherol-mediated peroxidation of lipoproteins: implications for vitamin E as a potential antiatherogenic supplement, *FASEB J.*, 13, 977–994, 1999.

114. Heaney, R. P., Calcium in the prevention and treatment of osteoporosis, *J. Intern. Med.*, 231, 169–180, 1992.

115. Heaney, R. P., Calcium, dairy products and osteoporosis, *J. Am. Coll. Nutr.*, 19, 83S–99S, 2000.

116. Zemel, M. B., Role of dietary calcium and dairy products in modulating adiposity, *Lipids*, 38, 139–146, 2003.

117. Combs, G. F. Jr., Selenium in global food systems, *Br. J. Nutr.*, 85, 517–547, 2001.

118. Djujic, I. S., Jozanov-Stankov, O. N., Milovac, M., Jankovic, V., and Djermanovic, V., Bioavailability and possible benefits of wheat intake naturally enriched with selenium and its products, *Biol. Trace Elem. Res.*, 77, 273–285, 2000.

119. Messina, M. J., Legumes and soybeans: overview of their nutritional profiles and health effects, *Am. J. Clin. Nutr.*, 70, 439S–450S, 1999.

120. Messina, M. J., Persky, V., Setchell, K. D., and Barnes, S., Soy intake and cancer risk: a review of the in vitro and in vivo data, *Nutr. Cancer*, 21, 113–131, 1994.

121. Lichtenstein, A. H., Soy protein, isoflavones and cardiovascular disease risk, *J. Nutr.*, 128, 1589–1592, 1998.

122. Lynch, S. R., Beard, J. L., Dassenko, S. A., and Cook, J. D., Iron absorption from legumes in humans, *Am. J. Clin. Nutr.*, 40, 42–47, 1984.

123. Weaver, C. M. and Plawecki, K. L., Dietary calcium: adequacy of a vegetarian diet, *Am. J. Clin. Nutr.*, 59, 1238S–1241S, 1994.

124. Nijveldt, R. J., van Nood, E., van Hoorn, D. E., Boelens, P. G., van Norren, K., and van Leeuwen, P. A., Flavonoids: a review of probable mechanisms of action and potential applications, *Am. J. Clin. Nutr.*, 74, 418–425, 2001.

125. Kris-Etherton, P. M., Leferre, M., Beecher, G. R., Gross, M. D., Keen, C. L., and Etherton, T. D., Bioactive compounds in nutrition and health-research methodologies for establishing biological function: the antioxidant and anti-inflammatory effects of flavonoids on atherosclerosis, *Annu. Rev. Nutr.*, 24, 511–538, 2004.

126. Bravo, L., Polyphenols: chemistry, dietary sources, metabolism, and nutritional significance, *Nutr. Rev.*, 56, 317–333, 1998.

127. Brown, A. S., Moro, M. A., Masse, J. M., Cramer, E. M., Radomski, M., and Darley-Usmar, V., Nitric oxide-dependent and independent effects on human platelets treated with peroxynitrite, *Cardiovasc. Res.*, 40, 380–388, 1998.

128. Afanas'ev, I. B., Dorozhko, A. I., Brodskii, A. V., Kostyuk, V. A., and Potapovitch, A. I., Chelating and free radical scavenging mechanisms of inhibitory action of rutin and quercetin in lipid peroxidation, *Biochem. Pharmacol.*, 38, 1763–1769, 1989.

129. Sugihara, N., Arakawa, T., Ohnishi, M., and Furuno, K., Anti- and pro-oxidative effects of flavonoids on metal-induced lipid hydroperoxide-dependent lipid peroxidation in cultured hepatocytes loaded with alpha-linolenic acid, *Free Radic. Biol. Med.*, 27, 1313–1323, 1999.

130. Jovanovic, S. V. and Simic, M. G., Antioxidants in nutrition, *Ann. NY Acad. Sci.*, 899, 326–334, 2000.

131. Simic, M. G. and Jovanovic, S. V., Mechanisms of inactivation of oxygen radicals by dietary antioxidants and their models, *Basic Life Sci.*, 52, 127–137, 1990.

132. Manjeet, K. R. and Ghosh, B., Quercetin inhibits LPS-induced nitric oxide and tumor necrosis factor-alpha production in murine macrophages, *Int. J. Immunopharmacol.*, 21, 435–443, 1999.

133. Birt, D. F., Hendrich, S., and Wang, W., Dietary agents in cancer prevention: flavonoids and isoflavonoids, *Pharmacol. Ther.*, 90, 157–177, 2001.

134. Meneton, P., Jeunemaitre, X., de Wardener, H. E., and MacGregor, G. A., Links between dietary salt intake, renal salt handling, blood pressure, and cardiovascular diseases, *Physiol. Rev.*, 85, 679–715, 2005.

135. Hertzler, S. R., Huynh, B. C., and Savaiano, D. A., How much lactose is low lactose?, *J. Am. Diet. Assoc.*, 96, 243–246, 1996.

136. Stern, M., A major step towards a practical and meaningful gluten analysis, *Eur. J. Gastroenterol. Hepatol.*, 17, 523–524, 2005.

137. Lenner, R. A., Specially designed sweeteners and food for diabetics–a real need? *Am. J. Clin. Nutr.*, 29, 726–766, 1976.

138. Garnier-Sagne, I., Leblanc, J. C., and Verger, P., Calculation of the intake of three intense sweeteners in young insulin-dependent diabetics, *Food Chem. Toxicol.*, 39, 745–749, 2001.

139. MacDonald, A., Daly, A., Davies, P., Asplin, D., Hall, S. K., Rylance, G., and Chakrapani, A., Protein substitutes for PKU: what's new? *J. Inherit. Metab. Dis.*, 27, 363–371, 2004.

140. Abbot, I. A. and Shimazu, C., The geographical origin of the plants most commonly used for medicine by Hawaiians, *J. Ethnopharm.*, 14, 213–222, 1985.

141. Krauss, B., *Plants in Hawaiian Culture*, University of Hawaii Press, Honolulu, 1993.

142. Hirazumi, A., Furusawa, E., Chou, S. C., and Hokama, Y., Anticancer activity of *Morinda citrifolia* (noni) on intraperitoneally implanted Lewis lung carcinoma in syngeneic mice, *Proc. West Pharmacol. Soc.*, 37, 145–146, 1994.

143. EC, Regulation No. 178/2002 of the European Parliament and of the Council of 28 January 2002 laying down the general principles and requirements of food law, establishing the European Food Safety Authority and laying down procedures in matters of food safety, *Off. J. Eur. Communities*, L31, 1–24, 1989.

144. EC, Council Directive 89/398/EEC of 3 May 1989 on the approximation of the laws of Member States relating to foodstuffs intended for particular nutritional uses, *Off. J. Eur. Communities*, L186, 27–32, 1989.

145. EC, Regulation No 258/97 of the European Parliament and of the Council of 27 January 1997 concerning novel foods and novel food ingredients, *Off. J. Eur. Communities*, L43, 1–7, 1997.

146. EFSA, Guidance document of the Scientific Panel on Genetically Modified Organisms for the risk assessment of genetically modified plants and derived food and feed, *EFSA J.*, 99, 1–93, 2004.

147. EC, 97/618/EC: Commission Recommendation of 29 July 1997 concerning the scientific aspects and the presentation of information necessary to support applications for the placing on the market of novel foods and novel food ingredients and the preparation of initial assessment reports under Regulation (EC) No 258/97 of the European Parliament and of the Council, *Off. J. Eur. Communities*, L253, 1–36, 1997.

148. OECD, Safety evaluation of foods derived by modern biotechnology: concept and principles, 1993.

149. WHO/FAO, Safety aspects of genetically modified foods of plant origin, *Report of a joint FAO/WHO expert consultation on foods derived from biotechnology, 29 May–2 June*, 2000.

150. Kwak, N. S. and Jukes, D. J., Functional foods. Part 2: the impact on current regulatory terminology, *Food Control*, 12, 109–117, 2001.

151. EFSA, "Tolerable Upper Intake Levels for Vitamins and Minerals", EU Publication Office ISBN 92-9199-014-0, 2006.

152. Palou, A., Pico, C., and Bonet, M. L., Food safety and functional foods in the European Union: obesity as a paradigmatic example for novel food development, *Nutr. Rev.*, 62, S169–S181, 2004.

153. Palou, A. and Serra, F., Perspectivas europeas sobre los alimentos funcionales, *Alimentación, Nutrición y Salud*, 7 (3), 76–90, 2000.

154. Palou, A., Bonet, M. L., and Serra, F., *Obesity and Functional Foods in Europe*, European Commission, Luxembourg, 2002.

155. EEC, Council Directive 90/496/EEC of 24 September 1990 on nutrition labeling for foodstuffs, *Off. J. Eur. Communities*, L276, 40–44, 1990.

156. EC, Directive 2000/13/EC of the European Parliament and of the Council of 20 March 2000 on the approximation of the laws of the Member States relating to the labelling, presentation and advertising of foodstuffs, *Off. J. Eur. Communities*, L109, 29–42, 2000.

11 Functional Foods in the European Union: Main Issues and Impact on the Food Industry

*Maurizio Canavari, Alessandra Castellini,
Giuseppe Nocella, and Carlo Pirazzoli*

CONTENTS

Abstract

In Europe the demand for functional foods varies remarkably from country to country, on the basis of alimentary traditions, enforced legislation and different cultural heritage that people have acquired. The opportunities for expansion on the market seem to be quite favorable and the interest of consumers is rather high.

However, several obstacles are slowing down the diffusion of these products in the European common market area. A major one stems from the lack of an officially recognized

251

definition for these terms, necessary in order to assign clearly these products to the food sector rather than to the pharmaceutical one.

It is virtually impossible to carry out a complete survey of this field, due to the lack of homogeneous and trustworthy statistical data and to the hodgepodge of definitions used for the key terms. In fact, every country adopts its own national legislation and includes different products in the class of functional foods.

The different meanings that the term "functional food" has assumed in the EU member Countries could also hinder free trade, even within EU boundaries. In fact, in the name of safeguarding human health, each partner can block the admission of a product, even if it comes from another EU member country.

This lack of clarity at the production phase is reflected in stricter controls at the consumption phase. Incorrect assumptions made by vendors and, consequently, by consumers, can involve some risks due not only to nutritional opportunities lost by not using functional foods, but also possible damage to health. Ambiguous definitions and gaps in knowledge regarding the composition and the effects of these products, in fact, can interfere with proper consumer awareness and also hamper the organized development of the industry.

11.1   INTRODUCTION

The presence of products enriched with nutrients and foods conducive to good health is not a new phenomenon in the European market; however, a recent development is the identification of genuine categories, such as nutraceuticals and functional foods, into which they have been grouped. Consequently, relevant research is in its initial stages.

The diffusion of these products and growing consumer success have both been encouraged by the social–economic trend typical of industrialized countries in recent years. Rises in income and an increase in disposable funds have indeed made it possible for people to look after their own well-being, having satisfied their basic needs. Obviously, this has influenced every aspect of consumer lifestyle, including eating habits. The choice of a particular food is no longer connected with the need to demonstrate a certain social or economic status (as was the case with the increase in sales of red meat about 40 years ago). It now depends more on the desire to be in good health both physically and mentally and to avert the risk of illness. This concept sometimes broadens into a vision of caring for one's own body, beyond the restricted medical sense of the word. "Taking care of yourself" is seen as the replenishment of certain substances and restoration of functions which people feel have been lost due to time, work and stress.

Recent population trends have seen a substantial increase in the proportion of "old" people, a natural result of a rise in the average life expectancy, while the birth rate has fallen. Accompanying the increase in age there has been greater recourse to products for maintaining and improving one's health that are not necessarily pharmaceuticals. On the contrary, the preference shown for non-chemical therapies, self-medication, and medicines alternative to conventional treatments has, without doubt, stimulated the health food sector—a phenomenon doubtless also influenced by the increasing prices of prescription drugs.

From the standpoint of consumers, then, a real change has been experienced with regard to their relationship with food. An additionional variable the health aspect, has been added to the factors influencing a buying decision. This chapter will therefore focus on the functional foods sector of the European union (EU).

11.2 OBJECTIVES

This chapter sets out to examine the functional foods sector of the EU, considering both legal and marketing aspects. The legislation in force will be summarized, in order to clarify what is meant by the term "functional foods" and which foods belong or do not belong to this category. At the same time, gaps in the regulations will be brought to light. Since the EU is the subject of research, it is necessary to analyze primarily the legislative measures at the European level, which prevail over any national legislation. In this case, the EU regulations consist mainly of Directives, which are not directly applicable as laws in the member countries, but must be enforced through a national regulation respecting the guidelines given by the Directive.

Once the parameters of the existing laws have been clarified, an attempt will be made to define functional foods and to understand what special features distinguish them from other food products. In this regard, numerous references will be made to "nutraceuticals" and we will compare the two terms, which are often confused in practice. In connection with this, it should first be said that no official documents defining one or other category exist, so we rely on the opinions and studies of scholars, operators, and experts. They are not only from Europe but also from other countries where the use of these terms has already produced excellent results in sales.

Finally, the EU market will be examined from the standpoint of a qualitative analysis of demand and supply. The main research findings regarding consumer behavior will be highlighted in order to understand both how marketers can better target their products and how scholars can push the frontiers of research, the fruits of which can help to stimulate the demand.

11.3 DEFINITIONS AND LEGISLATION

11.3.1 DEFINITION OF THE PRODUCT AND THE SECTOR

No official definition exists for the terms "nutraceutical" and "functional food" but they are now commonly accepted and associated with enriched foods that are connected with the health and well-being of the individual. This connection is very often rather vague and knowledge of more detailed aspects of the effects involved is often very sketchy. Thus, in order to clarify this issue we will discuss the legislative aspects of providing a definition for nutraceuticals.

Nutraceuticals is a term that was coined by the U.S. market, the first country (together with Japan) where the use of these products has become widespread [30]. The word derives from combining elements of the words "nutrients" and "pharmaceuticals" and the intended meaning is quite evident, even if these terms can encompass a wide variety of product categories. This chapter, although focusing on the EU, will use a definition of nutraceuticals taken from the Nutrition Business Journal [1], one of the most authoritative American magazines in the field; the NBJ

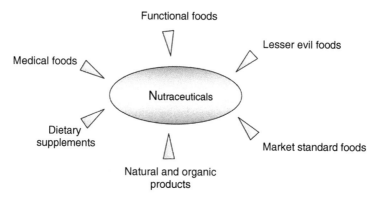

FIGURE 11.1 Nutraceutical categories. (Authors' elaboration from *Nutrition Business Journal*.)

describes the nutraceuticals as all consumable food, beverage, and supplement products purchased predominantly for health reasons, including the prevention and cure of disease.

Thus we are dealing with a category containing an extremely wide range of products (see Figure 11.1) lacking precise boundaries, and which includes natural and organic foods, supplements, functional foods, some lesser evil foods (foods with unhealthy ingredients taken out such as fat, sugar, caffeine, salt, etc.) and some market standard foods (foods consumed predominantly for health reasons such as cranberry and prune juice, a certain percentage of foods frequently purchased for health reasons like orange juice, yogurt, fish, fruit, and herbal tea and former functional and lesser evil foods that have become the market standard like enriched flour, iodized salt and low-fat milk).

Consequently, it is not always easy to see the difference between a nutraceutical in general and a conventional food when each is displayed on the shop counter.

According to this definition, functional foods represent a narrower category of products, which includes all foods enriched with ingredients that can provide benefits to health. Depending on the author, the definition of functional foods may cover a greater or lesser range of meaning. In the first case the class includes, besides enriched foods, food supplements and other nutrients administered in concentrated form (in capsules, for example). In the second case we are limited to considering only genuine foods, excluding pills, powders or tablets which in themselves do not constitute a meal.

According to the FLEP[*] Working Group on Functional Foods, "functional foods are everyday foods that contain added ingredient such as probiotics in bio-yogurts and fermented milk drinks… and other ingredients specifically designed to give food a positive health benefit" [2].

---

[*] European Food Law Enforcement Practitioner's Forum, an organization of food control in the EU, that established a Working Group about the case of functional foods.

In theory, it could be said that every food capable of providing vitamins, fibres or any nutrient might belong to this group. What makes the difference, however, is that the composition of the functional product must be *enriched* and the positive effect must be superior (or different) from that of the original ingredients. The working group mentioned is carrying out studies in order to consider a more, precise definition of the sector but several experts maintain that perhaps it is not necessary to create an ad hoc category for functional foods: it could be sufficient to enact clear laws to regulate what may or may not be shown on labels (especially any claimed connection between human health and the effects of the food) and the product would automatically fall into a class.

The absence of a genuine commodity class for functional foods in Europe may also be explained by the difficulty of precisely identifying a single type of these products, which often belong to different classes and food types. Thus it is a question, more than anything, of an "across the classes" group of nutritional products, brought together by some common features but often with other, very different properties of their own (for example, functional foods are present in both the beverage and solid food sectors, there are functional food snacks and sauces and so on, for every moment of the day). Furthermore, consideration must be given to the fact that many of the foods which companies would like to market now as functional foods are often no more than products already on the market to which an ingredient has simply been added in order to enhance certain nutritive qualities.

11.3.2 Regulation

An analysis of the legislation underlying the functional foods sector or, more generally, related to the category of nutraceuticals, proves to be rather difficult. In the EU there is in fact no specific body of legislation for these products so far. On the other hand, its construction would be somewhat complex. The primary need is to establish a definition of such references unequivocally.

The spread of this food category is increasing significantly and its regulation has now become a necessity. For this reason a study and development of norms for these products has been promoted: the EU made the first intervention, and it represents the point of departure for the common legislation of these "particular" foods.

11.3.2.1 Commodity Definition

Functional foods are products that are removed from the term "food" as understood in the more traditional sense and in many aspects are more akin to the field of pharmaceuticals and/or cosmetics; thus they maintain a certain ambiguity in their nature. Let us imagine the functions of a cosmetic product, which aims, for example, at improving a person's appearance or protecting the condition of a person's skin. Such features are decidedly similar to the advantages claimed for foods containing antioxidant substances. The difference lies mainly in the way in which they are used and the benefit gained, but the underlying active ingredient could be derived from vegetables.

According to European law, however, functional foods are simply food products and cannot for any reason claim therapeutic properties. Indeed, the latter is reserved exclusively for medicines. Food products must not be promoted in any way as a treatment for the prevention of pathologies or for the restoration or the modification of a physiological function of the human body. This is a principle that the EU has also underscored in several documents on the subject, and it is a rule that all the laws governing this sector have in common in the different countries where functional foods are to be found, especially in Western Europe, Japan, and the U.S. [3].

But the absence of common EU legislation forces European members to rely on their own national laws. This, however, has led to substantial differences in the way products are regarded when they cross borders. Furthermore, competitive and commercial conditions on the market are different in each country; sometimes legislation assigns a product to a certain marketable category in one country but to a different category in other European partners, thus creating considerable difficulties in trade.

Furthermore, it must be pointed out that on May 1, 2004, ten countries joined the EU, bringing their own their laws and rules, including those about food issues. Those Countries have also implemented Directive 2002/46/EC by the 1st of August 2005, together with the other food laws. It is expected be a complex and probably slow process [4].

11.3.2.2 Composition

The composition of functional foods and the legislation concerning the ingredients permitted vary somewhat from country to country in terms of dosage, origin and source of raw materials, treatments or processes to which they may or may not be subjected, etc. In this regard, two examples can be mentioned: the first concerns the legislation covering probiotics (lactic acid bacteria, in particular) and the second deals with the regulations for fatty acids [5]. For most European partners the addition of lactic acid bacteria (like Lactobacillus and Bifidobacteria strains) to a substance is freely permitted. In Denmark however, approval must be requested each time they are used in a new product, while in Italy the strains come under the law for foods intended for special diets* and, consequently must respect the laws in force for that particular sector, which is very close to medical foods.†

The use of fatty acids is also allowed by many EU members, but in Belgium their use must always be followed by a notification to the national health authorities. Denmark and Finland permit fatty acids only if naturally occurring, and in France they can be added only to dietetic foods. European Community legislation should attempt to overcome these inconsistencies.

---

* Legislative Decree no. 111 of the 27th of January 1992 "Implementation of the directive 89/398/CEE relating to foodstuffs intended for particular nutritional use."

† Medical foods are products prescribed by medical doctors when a patient has special nutritional needs for the prevention or treatment of a particular health condition [6,7]. In Italy, they must be sold exclusively in pharmacies, prepared on demand by a dispensing chemist on the basis of the patient's requirements.

When constituents (such as antioxidants, carotenoids, isoflavonoids, etc.) of a functional food are derived from medicinal plants or other botanical species, the absence of appropriate legislation becomes even more acute. In most EU countries there is no pertinent legislation, and each case has to be judged on its own merits. Belgium and the U.K. are exceptions: in Belgium the use of these substances is allowed in accordance with a list (in which the species and extracts that may or may not be used are catalogued); in the U.K. the only constraint is that their "safety" is guaranteed.

The legislation in force for this group of products in Italy is quite similar to that of the other European partners, in that it does not refer to functional foods or nutraceuticals with precision. Nevertheless two general laws can be applied. One is the basic law for the food sector, no. 283 of 1962 (updated and amended several times), regarding "Hygiene regulations in the production and sale of foods and drinks." It regulates the production and marketing of substances intended as foodstuffs and requires the inspection of the various phases by competent health authorities.

The second law, briefly mentioned earlier, is Legislative decree no. 111 of 1992, "Implementation of Directive 1989/398/EEC, Relating to Foodstuffs Intended for Particular Nutritional Uses." This category includes, among others, foods enriched with nutrients such as carbohydrates, proteins or vitamins or with other kinds of substances, including those derived from plants. These foods, however, are intended to produce particular metabolic alterations and the public must be informed by means of special labels, which display the word "dietetic" or a similar term. These laws, then, do not have a precise effect on the issue, but until EU legislation is passed and implemented they will comprise one of the main reference points for the national sector.

In April, 2002, the Decree of the President of the Italian Republic no. 2002/57 implemented "Directive 1999/21/EC Regarding Dietary Foodstuffs Intended for Special Medicinal Use." It specifies the classification and labeling regulations for these kinds of products.[*]

As already mentioned, the EU considers Community intervention in this sector as absolutely necessary. Firstly, it has regulated certain food supplements, which are not functional foods in the strict sense. However, it is important to mention this rule because it represents not only one of the most recent and innovative European interventions, but also the starting phase of a new regulation process. Directive 2002/46/EC of the European Parliament and Council of June 10, 2002 deals with the approximation of the laws of the Member states relating to food supplements. It came fully into effect on 1st August 2005, laying down the essential requirements

---

* Dietary foodstuff intended for special medical use are divided in three classes:

  - complete food from a nutritional point of view with a standard formulation of the nutrients;
  - complete food from a nutritional point of view with a formulation of the nutrients right for a specific disease;
  - incomplete food from a nutritional point of view with a standard formulation of the nutrients or appropriate for a specific disease.

and principles that must be considered in the field of food supplements. The basic intent and the principal aims that could be reached through an approximation of the rules are:

- To establish univocal rules able to favor the free circulation of these products without bureaucratic barriers that "... create unequal condition of competition[*]";
- To prepare understandable labels and claims in order to "ensure a high level of protection for consumers and facilitate their choice";
- To safeguard Public Health [8].

The push towards such a norm has come from the awareness of the increasing importance of food supplements that are bought by consumers in order to supplement their normal diets, but who are often unaware of safe levels of intake. This Directive only marks the beginning of Community intervention in this sector; in fact it only deals with vitamin and mineral supplements, but in future the regulation will be extended to other oligo-elements.

Also, in a second phase the Directive will be extended to functional foods that present a difficult and complicated situation, which requires precise rules able to supercede the regulations of each partner.

Firstly, this Directive establishes that *food supplements* are "... foodstuffs the purpose of which is to supplement the normal diet and which are concentrated sources of nutrients or other substances with a nutritional or physiological effect, alone or in combination, marketed in dose form, namely forms such as capsules, pastilles, tablets, pills and other similar forms, sachets of powder, ampoules of liquids, drop dispensing bottles and other similar forms of liquids and powders designed to be taken in measured small unit quantities"(art. 2).

Furthermore this Directive also includes two explicative annexes: the first contains the list of vitamins and minerals which may be used in the manufacture of food supplements with their measure units (e.g., Vitamin A (μg RE), Vitamin B1 (mg), Vitamin B12 (μg), Calcium (mg), Iron (mg), Iodine (μg) etc.); the second annex refers to vitamin and mineral *substances* permitted as food ingredients (e.g., for Vitamin A: retinol, beta-carotene, etc.; for Vitamin C: L-ascorbic acid, etc., and for Minerals: calcium carbonate, calcium chloride, etc.).

This Directive is valid for all EU member Countries, which must adapt it to their own national realities, respecting its main guidelines. In this way, apart from some differences, it will be possible to have a common guideline in functional foods regulation.

In Italy, Directive 2002/46/EC has been implemented by the Legislative Decree no. 169 of May 21, 2004. Before this, another document has been promoted regarding these issues: the Ministry of Health Circular letter [29] no. 3 of July 18, 2002, followed by several explanatory notes. This document deals with the labeling

---

[*] Directive 2002/46/EC of the European Parliament and of the Council of 10 June 2002 on the approximation of the laws of the Member states relating to food supplements.

of therapeutic products made with plant derivatives and will be examined below. In order to outline its action range, this document focuses on "… vitamin and mineral supplements (like the foods enriched with these substances) and all the products which bring a certain amount of other elements involved in nutrition and in quantities of a nutritive meaning and consistent with food placement." This definition is not very clear and, although following some guide-lines of dir. 2002/46/EC, it leaves some ambiguous areas. The Circular letter explains that these products, whatever their therapeutic purposes, must be placed in the food category: in fact, it does not refer to therapeutic functions, but rather to optimizing nutritional intake favoring health conditions and helping physiological functions. For this document, all these effects are compatible with the food classification of the substances.

In this way, food supplements acquire a physiological effect in addition to "simple" nutritional functions.

Some operators uphold that this Circular letter has helped to create a particular herbal supplement class for the Italian market: it has classified supplements made with herbal derivates (from officinal plants) like food supplements. The Legislative Decree 169 has clarified some critical issues, but it still refers to further intervention by the Eu legislator.

11.3.2.3   Communication of Health Benefit

Another burning issue which is still poorly regulated, concerns claims. The claim is defined by the BEUC (*Bureau Européen des Union de Consommateurs*) as

> …any message, reference or presentation, whatever the means of transmission (including trademarks) stating, implying or suggesting that a foodstuff has special characteristics, properties or effects linked to its nature, composition, nutritional value, method of production or/and processing or any other quality…[9]

.   Various types of claims are recognized (nutrient, health, etc.) here, but none of them is actually well-defined. Moreover, the meaning attributed to each of them tends to vary from one institution to another.[*] The need for clarity in this regard is not of particular interest to consumers, for whom recognition of the various categories of claims is of secondary concern, but it is necessary to define their terminology legally in order to coordinate the procedures of different countries so that the consumer can trust the reliability of a claim when choosing what to buy.

There are two pertinent laws in this regard in the EU: Directive 1990/496/EC, which is the basis for regulating nutritional labeling, currently under revision, and Directive 2000/13/EC, regarding the presentation and advertising of foodstuffs. Neither of the two however specifically deals with claims for foodstuffs [3].

---

* E.g., The definition of "nutrient function claim" in the Codex Alimentarius is quite similar to the definition of "health claim" in the U.K.'s Joint Health Claims Initiative [9].

The attention given to claims is considerable, as these represent the main channel for conveying information between the producer and the food and health consumer, but it is precisely this relationship that must be defined in a way that is not misleading in the description of its effects. Thus, a major problem for which the EU is trying to find an answer is consistent legislation by the various partners regulating these matters. The activities of the European Commission (EC) also arise from the fact that, following a consumer survey, it became apparent that almost all consumers wished to be informed completely and clearly concerning the effects, whether positive or negative, of a given product in order to be able to make informed choices. Some European countries—the Netherlands, Spain, Finland, U.K., Belgium, and Sweden—have already formed their own sets of regulations in this regard, while France and Denmark are in the process of doing so [5]. Other member countries, including Italy, have not yet started this updating process.

It is therefore a matter of great importance as it will affect the advertising and all other contact with the public. The question that has yet to be answered is: what terms can be used to communicate the references to a possible influence of a food component on health? The answer is still not forthcoming, and this leaves consumers exposed to possible risks. One thing is certain: medical claims are prohibited in the case of foodstuffs and references to treatment, cure or prevention of diseases (understood in the medical sense of the word) must be avoided and the National Consumer Council maintains that this prohibition must be upheld and respected [3].

The interest shown by the EC is quite substantial and has given rise to numerous studies and projects. Furthermore, a discussion featuring public health, including the question of functional foods, was started in 1998 within the European Council. Of the various opinions expressed by the partners, the prevailing viewpoint was that of Sweden, which proposed that consumers must be informed especially regarding two main aspects in order to be able to exercise an informed choice:

- The real relationship between the food (and its intake over a period of time) and health and
- The nutritional composition of the product

The group was finally asked to produce a series of guidelines clarifying the main topics inherent in this sector.

The need for common legislation within the territory of the EU and the penetration of these foods on the market, also in European countries, have spurred Community institutions to give greater consideration to the subject: the need has become apparent for consumers to be protected and for the sector to be reorganized. The fundamental question is whether all foods are suitable for the addition of other nutrients. In particular, the attention of the EU Institutions is focused on vitamins and minerals as ingredients allowed to enrich foods. These were examined in 1992 in a report about their energetic and nutritional contribution. On this subject a task force has been created with the aim of developing research, collecting data and, above all, establishing safety standards for the consumption of these supplements [5].

The Council Directive 1990/496/EEC on nutrition labeling concerns foods to be delivered directly to the consumer but "shall not be apply to diet integrators/food supplements.*" Furthermore, the application of this Directive must follow the terms established in Directive 1989/398/EEC on the approximation of the laws of the Member States relating to foodstuffs intended for particular nutritional uses, and other specific Directives.

As a result, until 2002 both functional foods and food supplements had been regarded as standard foods, without a specific law regulating its labeling.

The strong position of BEUC at that time was that it may be useful to make distinctions between different types of claims for technical reasons, but it is important that legislative proposals focus on how consumers perceive health-related claims (rather than making detailed distinctions between different types of claims which will be irrelevant to most people). In 2002, the EU began to draft a proposal on these themes. The approach adopted in this document, which is in favor of BEUC, is that usage of health claims like "reduction of disease risk factor" must be combined with scientifically and clinically tested evidence. This authorization must be released by an independent institution that must test and confirm the claim's declaration.

Without doubt, the EU is lagging behind the U.S. and Japan as regards legislation for functional foods and health claims. The US Food and Drug Administration has established how the latter are to be expressed and which of them need or to be subjected to controls [31, 37–39]. Japan is in the forefront, with a law that is very clear about this and which defines the category of functional foods in the greatest detail [30, 33].

In 2002, after several months work, Italy issued the Ministry of Health Circular Letter No. 3 that takes into consideration some topics treated in Directive 2002/46/EC and anticipates, for some themes, its legal application. The Directive refers to the "application of the notification procedure of label treated in the art. 7 of the D.L. n. 111 of 1992 on the products made up with ingredients or derivates from plants having health purposes." This procedure of label authorization by the Ministry of Health allows the government to determine whether a product is suitable for sale and consumption.† The importance of this document lies in the fact that this procedure of notification controls the commercialization of the reference, too.

In the EU, the difficulty of arriving at a single law for functional foods which can regulate the market (in the broadest accepted meaning of the term) is certainly daunting. It is a process that will have to deal with the desires of individual countries, their eating habits and their respective prerogatives for safeguarding the health of the population. In general, we are dealing with matters in which each partner would like to maintain some degree of autonomy. However, the confused situation that exists at present is surely counterproductive, not only for the

---

* Directive 1990/496/EC does not apply either to natural mineral waters or other waters intended for human consumption.

† This procedure cannot be applied to vegetable ingredients that are traditionally used for nutrition, like camomile or tea.

consumers who are inadequately protected, but also for the sector itself, which is not able to develop in a well-defined direction.

It has been recognized that two basic objectives of the EU Directive 2002/46/EC—informing and ensuring a high level of protection for consumers—may be inconsistent. In fact, informing consumers does not necessarily mean that they will make the best choice for their health but simply that they will be able to make this choice. Furthermore, in different countries, the hope of holding back several pathologies and diseases, spread by unwise dietary habits (e.g., obesity) or physiological alteration linked to consumption, depends only on the consumer's willingness to make the right choice spontaneously.

The priority at the basis of the actions of the European legislator is to allow consumers to make an informed choice.

11.4   THE MARKET

Despite the fact that forecasts from the most prominent market research companies indicate that the growth rate of functional foods is continually dropping in industrialized countries (compared to the 2%–3% growth rate of the entire food industry),the figures are nonetheless extremely impressive [10]. In fact, these forecasts show rises which swing between 5 and 15% and possibilities of development which no other agri-food industry segment appears able to offer. For example, Euromonitor has determined that between 1988 and 2003 the French functional food market has more than doubled, to settle around 1.5 billion euro. Furthermore, forecasts for 2008 estimate that retail sales will reach 2.8 billion euro, while in the U.K., "functional" dairy, bakery, and snacks alone will obtain a revenue of almost 80 million pounds. However, it must be underscored that functional foods still make up a very small part of what is on offer in the agri-food market [11].

These estimates often do not match because there are still no guidelines available which unequivocally identify the borderline between conventional and functional foods. All this generates confusion, not only in consumers who must choose between similar products containing unclear information, but also in those who collect and analyze market information. For example, the economic dimension of a certain category of products may change according to the way in which this food is categorized as being "functional." If we add to this the fact that legislation within the EU is not uniform, it is easy to see how hard it is to compare countries. Nevertheless, despite the fact that we are unable to give a precise dimension to the functional food market, significant opportunities undoubtedly exist within this market for companies ready to invest in innovative processes and products. In fact, the market potential and the possibility of creating added value seems to have been perceived by the majority of operators in the EU.

Despite these difficulties, the main categories of functional food on the market may be identified as follows:

- Beverages (e.g., juices with added vitamin A, C, E, energy drinks, etc.)
- Bread, bakery products, snacks, and cereal (e.g., breakfast products with added calcium, iron, etc.)
- Confectionery (e.g., chewing gum which helps prevent tooth decay)

- Dairy products (e.g., milk with extra calcium, yogurt with special microbial cultures, Omega-3 fatty acids, etc.);
- Oils and fats (e.g., products which reduce the level of "bad cholesterol");
- Vegetables (e.g., potatoes with added selenium).

The development of this market both in Europe and in the rest of the world requires higher scientific standards and more complex technological features than have been adopted so far in the preparation of traditional foods [12]. Companies which are investing in this market can follow two roads to success [13]: they can either attempt to understand consumer demand (pull) or exploit superiority at the supply end (science and technology push).

11.4.1 THE DEMAND

In a market which is ever more driven by demand, the introduction of innovative products such as functional foods must be based upon an understanding of consumer behavior. In fact, various studies carried out in Europe, the U.S. and other industrialized countries have aimed to examine how social-demographic factors and cognitive and attitudinal components can influence consumer choices with regards to functional foods.

A survey conducted in Northern Ireland [14] to test consumer understanding of functional dairy foods showed that women respondents between 18 and 24 years of age and the more highly educated were the two groups most knowledgable about such products. In general, the description of the functional food consumer type is rather variable depending on age and the specific functions the product must fulfill. For example, functional foods which act on the physical state of the person are very popular among the youngest consumer classes, as they improve their physical capabilities; but at the same time they are also bought by persons over 70 to give energy to a physique that is beginning to suffer tiredness and debilitation. Products having the effect of disease prevention enjoy some success among the middle age class, between 40 and 65 years old [15]. The awareness of the term "functional food" among consumers has increased over time, but in different ways within diverse areas of the EU. A recent survey [16] reports that the share of respondents who are aware of functional products substantially differs across nations: 38% in Germany, 26% in France, and 14% in the U.K.

But what are the factors would encourage consumers to choose functional foods? One answer was supplied by a study carried out in Finland [17], where by means of laddering interview techniques an attempt was made to identify the reasons which led consumers to choose between certain functional foods, including yogurt, spreads, juices, carbonated soft drinks, sweets, and ice cream. Results confirm the findings of other studies in which European consumers considered healthiness an important choice attribute [17–21]. However, in this study it emerged that healthiness in functional food can be considered a multidimensional choice factor; in fact, in some cases, it can be linked to a general state of well being, such as the consumption of ice cream, while in others it is linked to the prevention of disease, such as the case of spreads. Taste was seen as an important factor in choice too, even if in some products it was not considered at all. This was explained by the authors, who suggested that the functional element was considered more important than reasons of choice.

Such studies thus highlight that the level of acceptance, and consequently the development of demand, of these products is influenced by beliefs and attitudes that consumers have towards functional foods [28]. Beliefs linked to consumption of functional foods can be identified in terms of the benefits that can be reaped in terms of health, as well as the possibility of preventing certain diseases, and, more generally, the importance given to well-being and diet. A recent study carried out in Belgium [10] set out to examine how constructs such as "health benefit beliefs" and the "perceived role of food for health" could influence the level of acceptance of functional foods, using a multivariate Probit model. The results confirmed the importance of health benefit beliefs and how these correlate positively to levels of both consumer acceptance and consumer knowledge of such foods. However, demographic and social–economic variables were not statistically significant, probably owing to the small sample; the perceived role of food health was correlated negatively.

The results synthesized so far seem to indicate that European consumers are becoming more sensitive towards the link between food and health. Nevertheless, we must bear in mind that within the EU the consideration of this relationship varies from one country to another, on the basis of widespread eating habits, public health policies, differences in cultural heritage and the socio-economic level reached. For example, consumer choices vary according to the commonest pathologies in a given area: functional foods having the effect of alleviating or helping to prevent certain problems will be sold mainly in countries where these situations arise more frequently, and vice versa. For example, in the United Kingdom diseases related to the circulatory system and to stress are the most widespread [5] and consequently it is the functional foods claiming to regulate blood pressure, keep cholesterol levels low, etc., that enjoy the greatest success.

Socioeconomic research and studies regarding demand should be expanded in order to understand how the choice process of consumers can be influenced by technology used to enrich foods (traditional or GMO), by health claims (physiological or prevention) and, obviously, by price [22]. It is likely that these factors have a different impact not only on consumers in different European countries, but also on consumers in the U.S. and other industrialized countries. For example, we could hypothesize that most European consumers could be hostile to functional foods if they were produced using genetic engineering, while North American consumers might be indifferent to the use of gene gun technology. Furthermore, the enrichment of such products could be looked upon with suspicion by some consumers. So, to what point can trust in health claims influence consumer choice? And, if science and technology could unequivocally demonstrate that these foods could have a positive impact on the longevity of the human beings, what price would consumers be willing to pay? Finally, in terms of social justice, should not all citizens be able to buy these foods and definitively substitute them for traditional foods?

It would appear that despite the fact that various studies conducted in the EU strongly indicate a potential market among particular groups of consumers, it is necessary to investigate the questions raised above in order to guide the food industry in the right direction. At present both public and private sectors should invest more in communication and consumer education if they do not want to miss the opportunity to create added value to the food industry.

11.4.2 MARKET SUPPLY AND COMPANIES

Companies have seen enormous potential for success in the functional foods market mainly on account of the message that these products convey: "eat something that you like and which, besides just nourishing you, will also promote your health simply by helping to maintain wellness." This message fits in well with consumption growth registered by the market during the last decade. Obviously, the impact of this phenomenon on company strategies, e.g., food product advertising, sometimes emphasizes health-related aspects more than other sensory features.

Attracted by the positive performance and favorable prospects for development which this sector seems to offer, the number of companies showing interest in breaking into this market is increasing.

According to Menrad [23], the types of actors operating in the Functional Food segment in the EU consumer market, are the following:

– Multinational food companies, managing a broad range of products
– Pharmaceutical industries and/or producers of dietary specialties
– Category leaders at national level
– Small and medium sized food companies
– Large retail companies managing their own brands
– Suppliers of functional ingredients (business to business)

However, there are not only food producers involved, but also pharmaceutical companies, sometimes in collaboration with companies involved in agricultural chemistry. The majority belong to multinationals.[*] The professional categories that operate together are therefore different, and this is mainly due to the heterogeneous nature of the products. Each type of actor is coping with the market exploiting advantages but also facing specific problems.

On one hand, food sector companies may experience severe difficulties crossing over into areas of a pharmaceutical nature, which they are not capable of handling. On the other, when the product is managed by a pharma brand, the risk of losing the main characteristics of a certain food, with its related attributes of taste, desirability, and digestibility, to become a kind of pharmaceutical product is also important. On the basis of information about consumer experiences, this development is to be avoided: the product should aim to improve (or maintain) a general state of good

---

[*] Several multinationals, in the latest years, have open a specific department for the research and the development of functional foods and nutraceuticals: Nestlè, Kraft, Kellogg, etc.

health, instead of being stressed as a curative means, because it seems that consumers reject food advertised to resemble a medical preparation. It seems that healthy eating is acceptable but the concept of food must be kept in mind. Furthermore, references to particular pathologies that are too explicit may frighten away the healthy consumer, who believes that the food is suitable only for those suffering from a specific pathology [24]. Considering these problems, some multinational food enterprises have developed specific brands for marketing functional food products.

It is also clear that, from the standpoint of production, the absence of legislation and lack of clarity that are typical of this sector constitute an obstacle to expansion. Despite this fact, enterprises willing to develop new segments and introduce innovative products have tested the European markets for functional foods since the mid 1990s. Most of them have been multinational food companies and pharmaceutical industries [23].

The advantages for multinational food companies mainly involve the availability of resources to invest in product development and marketing, the presence of a R&D department with the necessary expertise, and the limited impact on the company as a whole that may be caused by a failure of a product that is just a small part of a wide portfolio.

Pharma industries also entered the market quite early, exploiting their specific competence in organizing clinical trials to support health claims associated with a specific product. The lack of experience in food marketing, the different competition level in the food industry, and the high rate of new product failures typical of this market represent the other side of the coin. However, the development of this market was dramatically influenced by their ability to introduce and substantiate medical notions in the food industry, establishing the basis for success of fortified soft drinks, as well as functional dairy products.

Category leaders, as well as small and medium-sized food enterprises, have had the opportunity to start from a favorable position in specific markets, either a specific market area or a niche market, developing a "me-too" functional foodstuff, sometimes as a functional version of a well-established product. The lack of resources to develop a pioneering approach is the main constraint for these companies.

In this regard, in Japan and in the U.S., where this market is more mature than in the EU, these companies may refer to several service agencies which have the main task of assisting companies wishing to enter this market, depending on the type of product and the nature of the company itself, in order to deal with bureaucratic regulations or legislative constraints [30]. In Europe this kind of service is still rarely found, and generally very expensive.

Other two types of actors played a role in the development of the market: large food retail companies and suppliers of food ingredients. The former, considering the increasing role of products labeled with the retailer's brand in sales related to a wider variety of categories, motivated these firms to differentiate their products and follow recent trends. The latter, acting as a "push" operator for the market, actively developing innovations related to ingredients linked to natural sources, like animal foodstuff and agricultural produce, e.g., vitamins, flavonoids, procyanidins, Omega-3 fatty acids, herbal extracts, etc.

However, further developments in the functional food market are expected in the near future, and new types of actors may be interested in entering the market. In this regard, an interesting case study involves an Italian consortium of farmers, marketers, and manufactures operating in the potato industry, that developed a selenium-enriched fresh potato. This product gained an interesting market quota and a relevant price premium in the first two years after its introduction, and since then the selenium-enriched onion has also been developed.

A key role in Europe may be played by public research bodies, both on the side of providing scientific evidence for food claims, and on the side of raising interest for new ingredients or components of food that may be taken into consideration for their functional characteristics.

Interesting perspectives may be gained by agricultural biotechnology, once clear information on the chance to achieve safe, ethical, and efficacious development of agricultural biotechnology products is available.

11.4.3 MARKETING STRATEGIES

The development of some products in the European market provides material for reflection on the market strategies adopted by companies.

With regard to products, some of them have preferred to operate in the mass market, with these goods typified by a general or at least not too specific functionality, assuming a position which is not very different from that of conventional foodstuffs. Other companies have opted instead for niche markets: in this case, however, the product must have properties aimed at resolving specific problems and must be able to satisfy the better-informed consumer demand that also requires more detailed information on the functional foods that are chosen for purchase.

European consumers are increasingly seeking health-enhancing foods with the aim of improving their quality of life, as well as mental and physical well-being, to help prevent disease and promote longevity [25,32]. This need has become apparent, and both product innovation and communication strategies have been affected. The food industry in Europe, as well as in other developed countries, tries to respond to these trends by placing increasing emphasis on positive features, rather than insisting on providing solutions aimed at diminishing some negative effects linked to food, like fat, cholesterol, and salt content [14]. From a consumer perspective, the development of health-enhancing products seems to be more appealing than just providing a "something-free" food.

Many experts have highlighted a method (pointed out as the "big idea in functional foods") which is relatively common to all companies to market a product. Once the consumer's primary needs of the moment have been identified and an ingredient has been studied which has a certain success, they tend to introduce it in the greatest possible number of references, letting a food become "functional overnight" [26]. This is how, in a very short time, a group of new functional foods comes on the market with a functional ingredient common to all of them. According to Johnson [16], from 2003 to 2004, over 100 new products have been launched on the market, but it is difficult to discover how many of them succeeded.

The ingredient's health benefit concerns what is believed to be the momentary objective of the consumers. At the present time, a common health fears is heart attack, and it is believed that the risks can be reduced by lowering cholesterol levels; thus the market has seen an increase in the supply of products performing actions involving cholesterol levels. Data regarding the U.K. (Table 11.1) shows the relative importance (on a qualitative scale) of some food characteristics; these are also increasingly taken into consideration in Mediterranean countries, where the incidence of heart disease is growing, due to changing life-styles and despite healthier diets.

Functional foods can therefore represent an important opportunity for food production in general. In fact, a saturation of traditional markets can already be seen, and consequently new possibilities for earnings are being sought by creating different kinds of products. The enrichment of the products makes it possible to start from an already known base on which to build by changing its composition and future market positioning.

The various businesses that operate in the functional foods market, however, also have a need to distinguish their product from those of direct competitors. To achieve this aim, there is the tendency to emphasize the scientific aspect and content of the product. By suggesting similarity to medical foods, the differentiation may lead to advantages—but it means entering a commercial field which is very different from that of functional foods. Continuing to market the products as functional foods, one forgets that the consumer does not look at the scientific background when purchasing a food product but pays much more attention to the brand which explains why a food is functional, also the result of thorough (and costly) studies and laboratory research, find themselves in the position of competing with products that are much less effective but which, for example, have a much better known brand name. This is a contradiction that exists in the market.

Although the element of benefits expected from consumption is obviously central, other aspects linked to the product profile must not be neglected. The European consumer is certainly strongly diversified (both between and across countries), but sensory characteristics, convenience, variety, and emotional, intangible aspects remain prominent factors, which are developed and incorporated

TABLE 11.1
UK Consumers: Choice in Buying Products for Health

	Most Important (%)	Important (%)
Lower in fat	33	55
Lower in sugar	26	56
Lower in cholesterol	19	38
Lower in calories	15	30

Source: The Henley Centre on New Nutrition Business, June, 2000. With permission.

in sound solutions regarding packaging, labeling, and branding, the latter considered of paramount importance for consumer acceptance [14].

The importance of the brand name thus remains crucial. A company will have a great deal of difficulty, as has been the experience of some English companies, breaking into a food market in which their name was not hitherto known [27]. The process of entering a market is lengthy, but made easier when the brand name of functional foods is already known for that sector. A company producing enriched yogurt will probably be more successful if it is already known for dairy products than if it had been operating in the confectionery sector up to that point. Of course this argument does not imply that the opposite cannot happen.

Closely tied to this aspect, communication and promotion activities are devoted to allow the product to achieve the more favorable position of capturing target market demands. For the consumer's sake, the benefits obtained from the functional product are generally shown on the label. The terms used may be scientific to varying degrees, depending on the interest shown by the consumers in the functions explained by the reference and thus the target market. Some companies have promoted a system of stepwise differentiated information, based on the personal desire of the consumer to be informed [34]. This system begins with the information on the product label (the essentials), progresses to the leaflet inside the packaging (more descriptive), then to the leaflets available at the sales outlet and finally the most detailed information provided by the vendor or possibly an Internet site, depending on the method of purchase. In this way everyone is free to obtain information to the desired extent before choosing a product for purchase.

Information supplied to consumers of functional foods are of vital importance for two main reasons:

- these products are somehow linked to the health of the consumer, who must be given the possibility of making a careful purchase;
- the functions performed by functional foods are relatively few, compared to the number of available products; consequently competition is intense: market shares will be conquered only through an adequate education of the consumer, who should be able to value the efficacy of the different products.

These aspects call for well-prepared salespersons and for widespread distribution of scientifically-based, independent, and reliable information. For this reason, it is generally thought that functional foods cannot be considered equal to conventional foods; their sale requires special attention, especially for those products suitable for particular health conditions.

As discussed in previous sections, the importance of claims is crucial, and the likelihood of purchase would be greater if product health benefits were clinically proven, possibly with the endorsement of public research bodies or at governmental level [14].

Observing the behavior of companies in terms of pricing strategies, it has become apparent that products sold at higher prices, with the intention of suggesting superiority over conventional foods, have not been successful in many cases.

In contrast, functional foods sold at prices which are more or less similar to those of the non-enriched foods from which they are derived have been well received. This fact is probably connected to the concept that the product is still considered a food in spite of its special properties. The stronger constraints are experienced by products that have been developed just by adding specific ingredients, without technological innovation described as the basis of the product innovation. In all these cases, it is generally difficult to protect the distinctive characteristics, so a truly competitive advantage may not be achieved and the market is rapidly invaded by followers and me-too producers. In other cases, the benefits provided by the functional food can be easily substituted by conventional foods, or the trade-off between perceived value of the benefits is not high enough to justify a premium price.

As regards the channel of trade used for selling enriched foods, it is necessary to make a distinction between a broad definition of the term "functional foods" (which also includes food supplements) and a more limited meaning, which is the case so far, to genuine functional foods. In both cases they can be regularly found in the small and large sales outlets unless they are foods which are designated for specific medical purposes, in which case they are sold exclusively at the chemist's. Furthermore, they are sold in specialized shops and sometimes even by herbalists.

However, if we also consider supplements, the legislation is rather different from country to country allowing the division of the members into two groups: the more restrictive group (to which Italy belongs) in which the supplements are sold mainly in specialized shops or chemist's, and the more "liberal" countries in which these products, besides their normal appearance in mass markets, can easily be bought even by mail order, internet or television.

There are important factors that affect the choice of the marketing channel. Selling a product through a regular food retail channel implies large volumes and a competition with other products for the available space on the shelves. Moreover, consumers are facilitated in comparing prices with direct competitors (similar products or different products with similar capacity to fill their needs) or indirect competitors (regular foods to which similar healthy properties are attributed). This distribution strategy may result in relatively low price margins, and high total margins.

If the chosen distribution channels are mainly limited to specialty food stores, or more specifically to pharmacies or health food stores, the competition is differently deployed, and a higher price margin, together with lower volumes are expected. Across the different EU member states the choice of the marketing channel may also be affected by the different taxation systems applied to regular food products, to supplements and to medicines [36].

11.5   FINAL REMARKS

According to some experts one of the success factors of functional foods is their ability to shift across categories: they are different from normal foods but do not go as far as to constitute a separate category aimed only at a target of consumers who need them as a kind of "therapy." Their health features are suited to all types of consumer for the fact that they are quite "general" in their functions and have small quantities of active ingredients. In fact these products must be suitable for repeated spontaneous consumption over a period of time, sometimes with daily frequency and not regulated from the quantitative standpoint just as the case is for any food. The interest in this category has been encouraging so far with significant growth rates in both industrialized countries, even if at an initial stage, and developing countries.

There is a need then to find satisfying answers to the problems discussed in order to permit the sector to develop in a rational way. Above all, in the matter of their legislative "treatment": the official definition of what constitutes a functional food, could bring about the specification of a single regulation to which reference can be made, ensuring greater safety for the consumer and avoiding the need for production companies to resort to self-regulation methods in order to guarantee a product in the absence of a legislative and institutional action. In general, the entire class of nutraceuticals show the same problems and issues. A solution to the problem of the structure and content of claims would also probably result from this.

Observing the example of Japan which has regulated the sector with precise legislation, one can understand how they encourage opportunities for success in the market: with the support of the Government they have a decisively higher rate of consumption and they can act without fear of repercussion in promotional activities since there is clarity concerning what may or may not be done and/or said.

In the same way, the definitive clarification of the nature of these foods could allow the use of one kind of commercial channel, thus making the situation more homogeneous among the various European members. Moreover, with a description of content and cataloguing, there would no longer be a basis for raising non-tariff barriers against free trade. In fact in many instances, a different definition between one member and another of the EU has obstructed free trade that should be taking place in the Community area.

A need which is particularly felt by companies operating in the functional foods sector is for studies and research into the question. At the moment studies are carried out mostly at a private level by the companies themselves. At the institutional level, interest is still rather limited, even if the matter probably deserves greater attention in view of market growth experienced by functional foods. Other advantages are to be gained: an informed consumer, who, having adequate material available, has a greater choice among the various brands and various products. Recent surveys, among other things, have shown that the proportion of consumers with a clear idea of these issues is rather small [13,15,16], especially if compared with the proportion of those who would be interested in knowing more.

REFERENCES

1. http://www.nutritionbusiness.com/research/definitions.cfm (accessed on November 10, 2006).
2. FLEP Working Group, Discussion Paper on Nutritional claims and Functional claims—Draft FLEP response, Commission of Consumer Affairs, Bruxelles, 2001, www.europa.eu.int/.

3. National Consumer Council, Nutrition and functional claims, A response from the National Consumer Council to consultation on the European Commission's Discussion Paper (SANCO/1341/2001), Bruxelles, 2001.

4. Directive 2002/46/EC of the European Parliament and of the Council of 10 June 2002 on the approximation of the laws of the member States relating to food supplements.

5. Geiser, S., Marketing functional foods in Europe—Health Product Information, Vitafoods International Conference Proceedings, April 13–15, Geneve, 1999.

6. Aarts, T., How long will the medical food window of opportunity remain open? *J. Nutra. Funct. Med. Foods*, 1 (3), 45–57, 1997.

7. Center for Food Safety and Applied Nutrition, Medical Foods, US Food and Drug Administration, Office of Special Nutritionals, 1997, http://vm.cfsan.fda.gov/

8. Directive 2000/13/EC of the European Parliament and of the Council of 20 March 2000 on the approximation of the laws of the Member States relating to the labeling, presentation and advertising of foodstuffs.

9. BEUC, Discussion paper on Nutrition and Functional claims. BEUC comments, Bruxelles, 2001, http://europa.eu.int/comm/food/food/labellingnutrition/resources/fl_com9150.pdf (retrieved April 16, 2004.)

10. Verbeke, W. Consumer acceptance of functional foods: socio-demographic, cognitive, and attitudinal determinants, *Food Qual. Pref.*, 16 (1), 45–57, 2005.

11. I do not find details for this source (Euromonitor, 2004) anywhere among the listed references.

12. Diplock, A. T., Aggett, P. J., Ashwell, M., Bornet, F., Fern, E. B., and Roberfroid, M. B., Scientific concepts of functional foods in Europe: consensus document, *Br. J. Nutr.*, 81 (1), S1–S27, 1999.

13. van Kleef, E., van Trijp, H. C. M., Luning, P., and Jongen, W. M. F., Consumer-oriented functional food development: how well do functional disciplines reflect the voice of the consumer? *Trends Food Sci. Technol.*, 13 (3), 93–101, 2002.

14. Gray, J., Consumer perception of the functional dairy food market in the Northern Ireland, *Int. J. Con. Stu.*, 26 (2), 154–158, 2002.

15. Gilbert, L., The consumer market for functional foods, *J. Nutraceuticals, Funct. Med. Foods*, 1 (3), 5–21, 1997.

16. Johnson, S., Consumer attitudes to functional foods, Vitafoods International Conference 2004, Delegate Documentation, Geneve, 2004.

17. Urala, N. and Lätheenmäki, L., Reasons behind consumers' functional food choices, *Nutr. Food Sci.*, 33 (4), 148–158, 2003.

18. Jonas, M. S. and Beckmann, S. C., Functional foods: consumer perceptions in Denmark and England, MAPP Working Paper No. 55, Aarhus School of Business, Aarhus, 1998.

19. Lappalainen, R., Kearney, J., and Gibney, M., A pan European survey of consumer attitudes to food, nutrition, and health: an overview, *Food Qual. Pref.*, 9 (6), 467–478, 1998.

20. Nielsen, N. A., Bech-Larsen, T., and Grunert, K. G., Consumer purchase motives and product perception: a laddering study on vegetable oil in three countries, *Food Qual. Pref.*, 9 (6), 455–466, 1998.

21. Zanoli, R. and Naspetti, S., Consumer motivations in the purchase of organic food. A means-end approach, *Br. Food J.*, 8 (8), 643–653, 2002.

22. Larsen, T. B. and Grunert, K. G., The perceived healthiness of functional foods. A conjoint study of Danish, Finnish and American consumers' perception of functional foods., *Appetite*, 40, 9–14, 2003.

23. Menrad, K., Market and Marketing of functional foods in Europe, *J. Food Eng.*, 56, 181–188, 2003.
24. Mellentin, J. and Heasman, M., Heart health become an everyday marketing message, 2001, www.new-nutrition.com
25. Sloan, A. E., The new market: foods for the not-so-healthy, *Food Technol.*, 53 (2), 63, 1999.
26. Mellentin, J. and Heasman, M., Life and death marketing, *New Nutrition Business*, February 2001, www.new-nutrition.com
27. Heasman, M. and Mellentin, J., *The Functional Foods Revolution*, James & James/Earthscan, London, 2001.
28. Childs, N. M. and Poryzees, G. H., Foods that help prevent disease: consumer attitudes and public policy implications, *J. Consumer Marketing*, 14 (6), 433–447, 1997.
29. Circular letter n. 3 of July 18, 2002 of the Italian Health Ministry about the Application of the notification procedure of claims by art. N. 7 of Legislative Decree n, 111/1992 to the health products containing plants and vegetables derivates.
30. Dennin, R. J., Overview of the U.S. and Japan natural product markets, Vitafoods International Conference Proceedings, 1999, April 13–15, Geneve.
31. Food and Drug Administration, Dietary Supplement Health and Educational Act, US Food and Drug Administration, Center for Food Safety and Applied Nutrition, 1995, December 1.
32. Gilbert, L., Defining the nutraceuticals/functional foods customer, *Natural Foods Merchandiser*, 10, 1999.
33. Heasman, M., The regulation and marketing of functional foods in Japan, *New Nutrition Business*, May, 1999.
34. Klompenhouwer, T. and Van Den Belt, H., Regulating functional foods in the (EU): informed choice versus informed consumer protection? *J. Agric. Environ. Ethics*, 16 (6), 545–556, 2003.
35. Legislative Decree no. 111 of the 27th of January 1992. Implementation of the directive 89/398/CEE relating to foodstuffs intended for particular nutritional use.
36. Mark-Herbert, C., Development and marketing strategies for functional foods, *AgBioForum*, 6 (1&2), 75–78, 2003.
37. U.S. Government, Dietary Supplements, Title 21, Chapter 1, Part 190 Code of Federal Regulations of the U.S., U.S. Government Printing Office via GPO, (Access on 1999).
38. U.S. Food and Drug Administration, Staking a claim to good health – FDA and Science stand behind health claims on foods, FDA Consumer, November–December, 1998.
39. U.S. Food and Drug Administration, An FDA guide to dietary supplements, FDA Consumer, September–October, 1999.

12 Functional Foods Legislation in Brazil

Franco M. Lajolo and Eliane Miyazaki

CONTENTS

12.1 AN INTRODUCTION TO BRAZILIAN FOOD REGULATIONS

Food legislation has important functions—it protects and sets the standard for the market, and guarantees the safety of consumers. However, through the years, those factors have changed due to the development of new foods, new habits, new technologies and new discoveries. It is important that these new foods are in accordance with some appropriate standard but it is also important to have flexible legislation that is able to keep up with those changes.

In Brazil, foods must have the approval of either the Ministry of Health or the Ministry of Agriculture if the products are dairy, meat, fish or beverage [1–3] before they are ready for commercialization, and for that they must follow the specification of identity and quality standards for the product. This specification indicates microbiological and microscopic limits, and authorized additives.

Inspected products have their specifications and labeling analyzed in order to verify their conformity with the law. Besides security, labeling has become an important item for inspection since Brazilian authorities are increasingly concerned about misleading labeling.

The food labeling laws in Brazil may be classified as:

a. General food labeling
b. Nutritional food labeling
c. Food label claim

The following information must appear on the label of prepackaged foods according to general food labeling regulation: food name, ingredients list, net contents, manufacturer, packer, distributor or importer name and address, country of origin, lot identification, shelf life and instructions for storage and use.

Due to the importance of nutrition and nutrition information for the consumer, and the raising levels of obesity, nutritional food labeling has been mandatory in Brazil since 2000. In December, 2003, other Mercosur countries—Argentina, Paraguay, and Uruguay—also adopted mandatory rules. The deadline for labels' conformity is July, 2006.

In Mercosur, the nutrients that must appear on any nutritional labeling are: energetic value, carbohydrates, proteins, total fat, saturated fat, trans-fat, dietary fiber and sodium (Figure 12.1).

The Brazilian label claim legislation is more recent and covers content claims (nutrients and non-nutrients), functional claims and also health claims.

Nutrient content claims were authorized many years ago in Brazil. Minerals and vitamins were freely added into food and beverages till 1998, when the addition of those nutrients was restricted to the limit of 100% of RDA. The products can present the content claims "source" or "rich" as for example "Food X is a source of calcium."

Due to market evolution, new rules have been developed and a new category of food has been created and identified as food for special dietary uses. In 1988 the first sugar-free beverages and foods were legally released as "diet" food—with sugar replaced by a sweetener [4].

In 1998, ten years later, fat-free and sodium-free foods were released and diet food was classified as "food for special dietary uses" [5,6].

Nutritional information portion of ... g (serving size)	
Quantity per portion %	%VD (*)
Energetic value ...kcal = ... kJ	...
Carbohydrates ... g	...
Protein ... g	...
Total fat ... g	...
Saturated fat ... g	...
Trans fat ... g	VD not specified
Dietary fiber ... g	...
Sodium ... mg	...
* "%Dietary Reference Intakes for a 2000 kcal/ 8400 kJ diet. Daily intakes may vary according to personal energetic needs."	

FIGURE 12.1 Mercosul nutritional labeling model.

Since then, foods for special dietary uses are classified as follows:

a. Foods with nutrient restriction (sugar, fat, protein, and sodium);
b. Foods with controlled nutrient (e.g., weight control food, sport food);
c. Foods for specific population (e.g., infant formula).

In some ways, foods for special dietary uses and fortified with vitamins and minerals were considered functional food's precursors, since they have been formulated for a special function.

Vitamins and minerals, herbs, amino acids, fatty acids, carotenoids added to foods or beverages, or in pharmaceuticals form (supplements in capsules or tablets) as well as food designed for athletes, diabetic persons and those seeking weight control were first commercialized in the Brazilian market with or without functional claims, and sometimes even with functional claims for the cure, prevention and treatment of illness.

There were a large number of products in Brazil until 1999, many advertised as "natural" products. They were being sold as foods or food supplements, bearing unproven preventive and curative claims. Some of these products consisted of complex mixtures of herbs extracts, containing very active pharmacological compounds such as amphetamines (ephedrine) and alkaloids. At the same time, foods with claims on their labels related to added ingredients like omega-3 fatty acids, phytosterols, fibers, oligossacharides, etc., were displayed on supermarket shelves as well. The claims were associated with intestinal function and health, osteoporosis, heart diseases, cholesterol levels, cancer, diabetes, stress, menopause, immune function, etc.

In order to organize the production, import and commerce of those supplements and the new functional foods, ANVISA (National Sanitary Surveillance Agency) a kind of FDA, linked to the Ministry of Health [7], set up a special work group to propose specific legislation. That work group had representatives from academia, industry, the International Life Sciences Institute (ILSI) and ANVISA and studied the existing regulations in several countries all over the world, seeking a balanced position [8]. The suggested legislation for Brazil was issued on these previously established concepts and assumptions:

a. Functional foods were not to be legally defined as a new or different category but considered as food with health claims;
b. They should be useful to improve the well-being of at least some of the population groups and should not oppose the nutrition and health policy of the country;
c. They should have their efficacy and safety proved through a science-based process and this should be conducted by an expert specially nominated by a scientific technical committee;
d. Claims should be true and not misleading and written in language understandable by consumers.

12.2 ORGANIZING THE MARKET: FOOD AND DRUGS

The basic idea was that a clear distinction between *conventional* or *traditional foods*, as opposed to *drugs* (including phytoterapeutics), should be maintained. *Drugs* have a distinctive purpose (cure, treat, mitigate pain, prevent disease), some known and accepted risk, and are used by individuals for a short period of time, rather than by the general population. There are only two categories under which to classify consumable products in Brazil: as a food or a drug. There is not a third classification, such as supplement or natural products. And it is not permissable to sell drugs in supermarkets in Brazil.

Even if not legally included as a category and officially defined, the common understanding was that functional foods are foods, in most cases similar to conventional foods, consumed as part of the usual diet, able to produce physiological or metabolic effects useful to maintain good health and reduce the risk of non-transmissible chronic disorders, beyond their basic nutritional properties.

The idea was also that being a functional *food* could not be presented in pharmaceutical form (capsules or tablets). A *nutraceutical*—a term reserved for bioactive compound, presented in pharmaceutical forms, usually in a concentration higher than the one in foods—is named officially in Brazil as *bioactive*.

Based on the legislation issued at the end of 2001, the safety, efficacy, and adequacy of claims for approximately 2000 existing products were assessed, but less than 400 were approved and are still being produced by several manufacturers or importers (today, fewer than 50 can be really called functional foods).

In the process of reorganization of the market, hundreds of products (many imported), which would be or were actually classified in other countries as *dietary supplements*, were not approved as foods. Some of them were prohibited. Others were registered as *drugs,* mostly as phytoterapeutics.

Following the same line, for registration purposes, mixtures of foods with isolated compounds or with herbals and herbal extracts having pharmacological activity were considered drugs, not foods, and as such had to be analyzed; they were thus being submitted to pharmaceutical—not food—regulations.

12.3   REGULATION OF FUNCTIONAL AND HEALTH CLAIMS

Brazil was the first country in Latin American to issue legislation for functional food claims. As we have seen, functional foods have not been officially defined as a food classification in Brazil and the legislation basically asks for demonstration of safety of products and efficacy of claims. There are now four regulations which were issued in 1999 and a more recent one issued in 2002. These are:

    a. Resolution no. 16, of April 30, 1999 (republished on 03/12/1999):
        To approve the technical regulation on procedures for food registration and/or new ingredients [9].
    b. Resolution no. 17, of April 30, 1999 (republished on 03/12/1999):
        To approve the technical regulation establishing the basic guidelines for evaluation of risk and safety of foods [10].
    c. Resolution no. 18, of April 30, 1999 (republished on 03/12/1999):
        To approve the technical regulation establishing the basic guidelines for analysis and proof of functional and/or health claims on food labels [11].

d. Resolution no. 19, of April 30, 1999 (republished on 10/12/1999):
    To approve the technical regulation on procedures for food registration
    with functional and/or health claims on their labels [12].

The Technical Scientific Advisory Commission on Functional Foods and New
Foods (CTCAF) was instituted by the Brazilian Sanitary Surveillance Agency
(ANVISA), which had been in charge of applying the former regulations [13].
Professional scientific members from academia and research institutes with
expertise in different areas of food science, nutrition, and toxicology form this
Commission.
The principles that guide CCTCAF evaluation are:

- Safety evaluation and scientific risk analysis;
- Scientific efficacy evaluation;
- Not to define functional food but to approve food claims;
- Evaluate each proposal based on current scientific facts;
- A company/producer is responsible for security probation and claim
  efficacy;
- Products and claims in accordance with Health Policies;
- Previous decisions may be reevaluate based on new scientific evidences;
- Allegations to referring to disease prevention, treatment or cure; are
  not allowed.
- Allegations must be easily understandable for consumers.

Recently, on January 2002, a new regulation (Resolution RDC #2) was issued,
covering "Bioactive Substances and Probiotics" with functional or health claims. Its
aim was to make as clear as possible a distinction between *functional foods* and
bioactive compounds (*nutraceuticals*), usually sold as pills, capsules, etc., and *OTC
drugs,* mostly herbs and other phytoterapeutics [14].
Altogether the resolution rules on demonstration of safety and efficacy of novel
foods, food ingredients and bioactive compounds and food with functional and
health claims (Figure 12.2). All these products must be registered and approved by
the health authority (ANVISA).
Novel foods or ingredients are those without a history of use in Brazil (e.g.,
Olestra, Trehalose. Genetically modified foods are not covered here) or those
containing substances added in higher concentrations than those usually used or
presented in foods (like, for instance, oligossacharides or carotenoids).
Foods with claims on their labels are regulated by the legislation that allows both
functional and health claims. Functional claims are defined as those referring to the
metabolic or physiological role of a nutrient or a non-nutrient in normal body
functions. Health claims are defined as those that imply, suggest or state the
existence of a relationship between the food or food component and a disease or a
condition related to health.
Based on the existing regulation in Brazil, functional foods can be considered as
foods bearing health claims, approved by the health authority; a pre-market approval
is, therefore, mandatory.

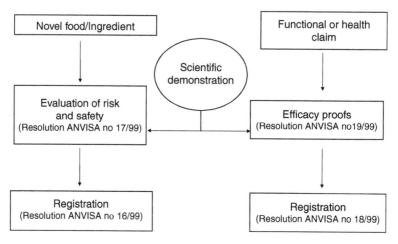

FIGURE 12.2 Safety of products and efficacy of claims. All regulations are available on the website: http://anvisa.gov.br/ing/legis/index.htm.

To register a product with the Ministry of Health (or the Ministry of Agriculture when the product is a dairy, meat, fish or beverage) the industry must submit a "Technical Report" containing the following information: the name of the product; the recommended uses and intakes; a complete description of the origin and technological process used for production, the chemical composition and molecular characterization (when needed, with description of the analytical methodology); a copy of the label with the proposed claim and the scientific support for it, plus evidence in support of the claim. The report should also contain information about the status of the product in other countries, if it was considered safe, if the claim was approved, and possible uses and restrictions. The report is then submitted to the assessment committee for evaluation of safety and efficacy based on the evidence presented.

Besides these documents, the Technical Report must present scientific studies that prove the product safety and efficacy.

The scientific safety and efficacy evidence that should be presented in the dossier of the product to get it registered within ANVISA has to include all or some of the following data: chemical composition and molecular characterization (when it applies); biochemical assays (in vitro); nutritional, physiological, metabolic, and toxicological essays with animals; cellular, and clinical and epidemiological evidence. Demonstration of historical or traditional use of the products by indigenous populations with benefits and no reported risks may be accepted in some cases as supporting evidence, provided they are well documented.

12.4   USING FUNCTIONAL AND HEALTH CLAIMS

Functional claims are allowed provided they are scientifically proven. Health properties and health claims based on these properties have more strict limits in legislation. Those claiming health maintenance are allowed and have been relatively common; those professing health risk reduction, although possible, are very scarce, and claims on mitigation of symptoms, prevention or cure of diseases are prohibited.

The statement on the label should be scientifically proven by research that may be done by the industry itself or by published literature available in peer-reviewed journals. The extent of the demonstration expected is proportional to the type and importance of the claim made. For instance, a nutrient or bioactive component content claim may be based only on the chemical analysis of the compound and the description of its chemical structure. But claims on "bioavailability" due to the influence of the matrix of the food, may depend on this also or maybe only on an animal assay. A risk reduction statement would hardly be approved without mechanistic, clinical and/or epidemiological studies using suitable biomarkers.

Besides demonstrating safety and efficacy, claims must inform the target population (if any) that the claimed effect is to be observed with normal consumption of the food, and to be obtained in the context of the normal diet, with all possible interactions. The claim should not conflict with the nutrition and health policy of the country; and it should be true, not misleading and reflect scientific agreement. The extent of demonstration, type and number of studies, extent of agreement or acquisition of a real consensus, are a matter of debate in Brazil (as in other countries) among regulators, academia and industry.

The Commission (CTCAF) makes the final assessment. The information provided by the industry throughout a "dossier" on the product is analyzed by the commission according to scientific criteria and based on the quantity and quality of the literature provided and otherwise available. A final decision on the approval or prohibition of the product is based on a consensus Dialog with the industry and requests for additional information before a decision may also be part of the process.

12.5   EXAMPLES OF CLAIMS APPROVED

The claims allowed in Brazil are mostly functional claims as can be seen in some examples below:

   a. Spreads with phytosterols:
      Help to maintain healthy levels of cholesterol when associated with healthy diet and life style.
   b. Milk containing long chain n-3 fatty acids (EPA and DHA):
      Helps to control triglyceride levels, blood fluidity, inflammatory and immune response when associated with a healthy diet and life style. Enhances immune defense.
      Allegations not approved: reduces the risk of heart diseases and reduces blood cholesterol.
   c. Milk with n-3 and n-6 fatty acids from vegetable oils:
      Low in saturated fats and enriched with n-3 and n-6 fatty acids and needed to keep cholesterol low and a healthy heart.

    d. Milk and other products with prebiotics:
       Contribute to a healthy intestinal flora and helps maintaining the intestinal flora balance.
       Allegation not approved: assures to your son a healthy digestive system promoting his optimal development and well being and has the capacity to adjust intestinal flora and help digestion.

    e. Probiotics (Lactobacilli and Bifidobacteria):
       Help in maintaining the intestinal flora balance.
       Help in reducing harmful bacteria.
       Help in increasing the beneficial flora.
       Allegations not approved: increases antibodies and also strengthen natural defenses against daily aggression and stress.

    f. Chitosan:
       Helps reducing fat and cholesterol absorption.
       Helps in weight control and cholesterol reduction when associated with a hypocaloric diet.

    g. Oat, soy flakes, wheat germ, pectin, Psyllium:
       Fibers help intestinal motility.

    h. Oat bran:
       Helps cholesterol reduction when associated a low fat and cholesterol diet.
       Allegation not approved: reduces the risk of cardiovascular diseases and osteoporosis (this was made only for soy).

    i. Chewing gums with xylitol:
       Help lowering pH thus reducing the bacterial tooth plaque, it is not fermentable by mouth bacteria, neutralize acids that harm teeth.
       Allegation not allowed: helps reducing cavities.

12.5.1 Recent Changes in Regulation of Health Claim—Horizontal Claims (2007)

After five years experience of applying the legislation claims, CTCAF (in 2006) reviewed the claims previously approved, and a new list was be issued and made available by Internet. It is important to mention that due to misuse, mostly extension of the meaning of several claims as originally allowed, the ANVISA proposed horizontal claims to be used by all the products with similar properties. These claims have now defined wording that will have to be used as such. These horizontal claims are presented in Table 12.1.

In some situations, risk management and risk communication measures were adopted. For instance: post-marketing surveillance (for olestra and chitosan products) and a suggestion for research in the country (fat spreads with phytosterols). Products in pharmaceutical forms should have the following statement: "This product is not intended to treat or cure any disease" on their label, and those intended for elderly, children or pregnant women must have the special precautions clearly stated.

Dietary fibers have been discussed since the ANVISA Commission was created. In Brazil there has been a reduction of fiber consumption in the last few decades, due

TABLE 1
Function and Health Claims Allowed Presently in Brazil (From 2007)
FATTY ACIDS

FATTY ACIDS OF THE OMEGA 3 FAMILY
Claim
"The consumption of Omega 3 fatty acids helps maintaining healthy levels of triglycerides, provided this is associated with a balanced diet and healthy living habits"
Observations
1. This claim must only be used for long-chain Omega 3 fatty acids coming from fish oils (EPA - Eicosapentaenoic Acid and DHA – Docosahexaenoic Acid).   2. The following warning phrase, highlighted and in bold letters, must be included on the label of products isolated in capsules, tablets, pills, powders and the like: "People who have some form of illness or physiological alteration, particularly blood clotting alterations, pregnant women, nursing mothers and children should consult their doctor before using this product".   3. The amount of omega 3 fatty acid must be stated in the nutritional information table. The model with the full declaration of total fats, saturated fats, mono-unsaturated and poly-unsaturated fats must be used. Below the poly-unsaturated fats data the label should declare the type of Omega 3 fatty acids, specifying whether EPA and/or DHA and/or Linolenic acid, as the case may be.   4. The fatty acids of the Omega 3 families content may only be declared on the label when, in the daily dose of the product ready for consumption, they represent at least: 0.1 g of EPA and DHA   0.2 g of alpha-linolenic acid

CAROTENOIDS

LYCOPENE
Claim
"Lycopene has an antioxidating action that protects cells against free radicals. It should be consumed in association with a balanced diet and healthy living habits"
Observations
The amount of lycopene contained in the daily dose of the product ready for consumption must be declared on the label, next to the claim.

LUTEIN
Alegação
"Lutein has an antioxidating action that protects cells against free radicals. It should be consumed in association with a balanced diet and healthy living habits"
Observations
The amount of lutein contained in the daily dose of the product ready for consumption must be declared on the label, next to the claim.

DIETARY FIBERS

DIETARY FIBERS
Claim
"Dietary fibers help the intestines to function normally. They should be consumed in association with a balanced diet and healthy living habits"
Observations
1. This claim must be used for soluble and insoluble fibers provided that the daily dose of the product ready for consumption meets at least the "source" attribute of dietary fibers established by <u>Ministerial Directive SVS/MS 27/98.</u>   2. The amount of soluble or insoluble fiber, as the case may be, must be declared in the nutritional information table, below the dietary fibers.   3. The following warning phrase, highlighted and in bold letters, must be included on the label of products isolated in capsules, tablets, pills, powders and the like: **"This product must be consumed along with an intake of liquids".**

BETA GLUCANE
Claim
"Beta-glucan (dietary fiber) helps reduce the absorption of cholesterol. It should be consumed in association with a balanced diet and healthy living habits"

(Continued)

TABLE 12.1 *(Continued)*

Observations
1. The daily dose of the product ready for consumption must at least meet the "source" attribute for dietary fibers established by Ministerial Directive SVS/MS 27/98.
2. The amount of beta glucane, in the form of dietary fiber, must be declared in the nutritional information table, below the dietary fibers.
3. The following warning phrase, highlighted and in bold letters, must be included on the label of products isolated in capsules, tablets, pills, powders and the like: "**This product must be consumed along with an intake of liquids**".

FRUITOLIGOSACCHARIDES

Claim
Fruitoligosaccharides (FOS) help to balance intestinal flora. They should be consumed in association with a balanced diet and healthy living habits"

Observations
1. The daily dose of the product ready for consumption must at least meet the "source" attribute for dietary fibers established by Ministerial Directive SVS/MS 27/98
2. The amount of fruit oligosaccharides must be declared in the nutritional information table, below the dietary fibers.

INULIN

Claim
Inulin helps to balance intestinal flora. It should be consumed in association with a balanced diet and healthy living habits"

Observations
1. The daily dose of the product ready for consumption must at least meet the "source" attribute for dietary fibers established by Ministerial Directive SVS/MS 27/98.
2. The amount of inulin must be declared in the nutritional information table, below the dietary fibers.

LACTULOSE

Claim
Lactulose helps the intestines to function normally. It should be consumed in association with a balanced diet and healthy living habits"

Observations
1. The daily dose of the product ready for consumption must at least meet the "source" attribute for dietary fibers established by Ministerial Directive SVS/MS 27/98.
2. The amount of lactulose as dietary fibers must be declared in the nutritional information table.
3. The following warning phrase, highlighted and in bold letters, must be included on the label of products isolated in capsules, tablets, pills, powders and the like: "**This product must be consumed along with an intake of liquids**".

PSYILLINUM OR PSYLLIUM

Claim
Psyllium (dietary fiber) helps reduce the absorption of fat. It should be consumed in association with a balanced diet and healthy living habits"

Observations
1. The product must be called either Psillium or Psyllium.
2. The only type already evaluated comes from *Plantago ovata*. Any other species must be evaluated as to the safety of its use.
3. The daily dose of the product ready for consumption must at least meet the "source" attribute for dietary fibers established by Ministerial Directive SVS/MS 27/98.
4. The amount of Psyllium fiber must be declared in the nutritional information table, below the dietary fibers.
5. The following warning phrase, highlighted and in bold letters, must be included on the label of products isolated in capsules, tablets, pills, powders and the like: "**This product must be consumed along with an intake of liquids**".

QUITOSANE

Claim
"Quitosane helps reducing the absorption of fat and cholesterol. It should be consumed in association with a balanced diet and healthy living habits"

Observations
1. The daily dose of the product ready for consumption must at least meet the "source" attribute for dietary fibers established by Ministerial Directive SVS/MS 27/98.
2. The amount of quitosane must be declared in the nutritional information table, below the dietary fibers.
3. The following warning phrase, highlighted and in bold letters, must be included on the label: "People allergic to fish and shellfish must avoid eating this product".

TABLE 12.1 *(Continued)*

4. The following warning phrase, highlighted and in bold letters, must be included on the label of products isolated in capsules, tablets, pills, powders and the like: "**This product must be consumed along with an intake of liquids**".
5. The company must present an analysis showing the inorganic contamination levels for arsenic, cadmium, lead and mercury, separately. Use as a reference Decree Law 55871, of March 26, 1965 (DOU 09/04/65) - category: other foodstuffs.

PHYSTEROLS

PHYTOSTEROLS
Claim
"Phytosterols help reducing the absorption of cholesterol. They should be consumed in association with a balanced diet and healthy living habits"
Observations
1. The company may include in the claim, in brackets, the type of phytosterol used in the product.   2. The amount of phyrosterol contained in the daily dose of the product ready for consumption must be stated on the label, next to the claim.   3. The following warning phrase, highlighted and in bold letters, must be included on the label: "**People with high levels of cholesterol should seek medical guidance**".

PROBIOTICS

LACTOBACILLUS ACIDOPHILUS   **LACTOBACILLUS CASEI SHIROTA**   **LACTOBACILLUS CASEI, VARIETY RHAMMOSUS**   **LACTOBACILLUS CASEI, VARIETY DEFENSIS**   **LACTOBACILLUS DELBRUECKII SUB-SPECIES BULGARICUS**   **BIFIDOBACTERIUM BIFIDUM**   **BIFIDOBACTERIUM LACTIS**   **BIFIDOBACTERIUM LONGUM**   **STREPTOCOCCUS SALIVARIUS SUB-SPECIES THERMOPHILLUS**
Claim
The (Probiotics) helps in maintaining the intestinal flora balanced. It should be consumed in association with a balanced diet and healthy living habits"
Observations
1. The amount of probiotic in colony forming units (CFU), contained in the daily dose of the product ready for consumption, must be stated on the label, next to the claim.   2. The company must present an analysis of the product proving the amount of live micro-organism until the product's use-by date.

Bifidobacterium Animalis
Claim
Bifidobacterium animallis helps the intestines to function normally. It should be consumed in association with a balanced diet and healthy living habits"
Observations
1. The amount of probiotic in colony forming units (CFU), contained in the daily dose of the product ready for consumption, must be stated on the label, close to the claim.   2. The company must present an analysis of the product proving the amount of live micro-organism until the product's use-by date.

Soy Protein
Claim
"The daily consumption of at least 25 gr of soy protein may help to reduce cholesterol. It should be consumed in association with a balanced diet and healthy living habits"
Observations
The amount of soy protein contained in the daily dose of the product ready for consumption must be stated on the label, next to the claim.

Note: The exact phrase suggested must be used on food labels—Horizontal claims

Source: From BRASIL. Ministério da Saúde. Agência Nacional de Vigilância Sanitária. **Alimentos com Alegações de Propriedades Funcionais e ou de Saúde. Novos Alimentos/Ingredientes, Substâncias Bioativas e Probióticos,** 2007. Available in: http: http://www.anvisa.gov.br/alimentos/comissoes/tecno.htm

to changes in eating habits and lowered ingestion of traditional staple foods, such as beans. This fact, and the recognized importance of fiber for gut health, has been a motivation for the production of fibers and fiber—added foods carrying several types of claims, including a single "content" claim. This has caused some discussion and led to legal problems, mostly due to the lack of a clear consensus on an adopted definition, a problem that seems to be occurring all over the world.

The official Brazilian definition for *fiber* is "any eatable material not hydrolyzed by the endogenous enzymes of human digestive tract" [15] which was adopted after the discussion in the Mercosur. Considering the well-known functional aspects of fibers (see below), this is a very limited definition.

In June, 2000 AACC issued this definition: "Dietary fiber is the edible parts of plants or analogous carbohydrates that are resistant to digestion and absorption in the human small intestine with complete and partial fermentation in the large intestine. Dietary fibers include polysaccharides, oligossacharides, lignin, and associated plant substance. Dietary fibers promote beneficial physiological effects including laxation, and/or blood cholesterol attenuation and/or blood glucose attenuation" [16].

In 2002, the U.S. Institute of Medicine/National Academy of Sciences proposed these definitions: "Dietary fiber consists of non digestible carbohydrates and lignin that are intrinsic and intact in plants. Functional fiber consists of isolated, non-digestible carbohydrates that have beneficial physiological effects in humans. Total fiber is the sum if Dietary Fiber and Functional Fiber" [17].

The Codex Committee on Nutrition and Food for Special Dietary Uses proposed in 2004 that dietary fiber be defined as an edible and not digestible material composed of carbohydrate polymers with a degree of polymerization (DP) not lower than 3 or of carbohydrate polymers (DP>3) processed (by physical, enzymatic or chemical means) or synthetic. Dietary fiber is neither digested nor absorbed in the small intestine and has at least one of the following properties: increases stool production; stimulates colonic formulation; reduces fasting cholesterol levels; reduces postprandial blood sugar and/or insulin levels [18].

Even if fiber is not well defined in Brazil, labeling may include information on it when a claim is presented. Since not all types of fibers have the capacity for lowering cholesterol, reducing blood glucose levels and being fermented to produce short chain fatty acids, the position adopted in Brazil has been to consider that each product needs a proper characterization of the fiber it contains, according to the claim made. Even a statement that a product is "rich" or a "good source of fiber" depends on a definition of what is considered fiber. Several products have, for instance, undigested poly-saccharides that behave physiologically like fiber but legally are not defined as such.

The recent definition of the U.S. National Academy of Science that included dietary fiber and also "added fiber" may help to introduce more defined claims. Clearly there is a need for an international harmonization and a single definition of dietary fiber.

12.6   BIOACTIVE COMPOUNDS IN BRAZIL

Dietary supplements or nutraceuticals are common terms in other countries but do not exist in food or pharmaceutical regulations in Brazil. Only recently, in 2002, a specific norm—ANVISA Resolution no. 2—(14) was issued to cover what was called "Bioactive Compounds." Brazilian health authorities consider it important to keep a clear distinction between these products and foods.

It was important to consider the safety aspect because many of these substances contained *concentrated* forms of components. Even if normally contained in foods, an increased concentration as found in pills, capsules, etc., could have adverse effects after a long-term usage, or could interact with drugs, reducing their efficiency.

Bioactive compounds have been defined in the regulation as "natural or synthetic substances having a demonstrated metabolic or physiologic activity." They were classified in seven different chemical groups: carotenoids, phytosterols, flavonoids, phospholipids, organosulfur compounds, polyphenols, prebiotics, and probiotics. Probiotics were included due to their appearance in non-food forms and were defined as "live microorganisms having the ability to improve the microbial intestinal equilibrium and thus to produce beneficial effects to the individuals health."

Bioactive compound products have to be approved and registered before going into the market, in a process identical to that previously described for functional foods. They should be proven safe and must have an allegation (functional or health claim) explaining what their actions and benefits are. In case of probiotics, they must warrant that the physiological property alleged is maintained during the shelf life indicated. It is important to stress that a bioactive compound product is not approved if it has no well-defined claims on its label.

Products with amino acids and peptides like creatine, taurine, arginine, and carnitine, which are common in U.S.A., are not included in the bioactive category. These products are used to increase muscular mass and to improve performance. Doubts were raised concerning the safety of some of these products, specifically because of the lack of data on the effects of long-term ingestion. Considering their uncontrolled consumption by some population groups, and also the unproven or very limited efficacy and the doubts on their safety, these products were not allowed to be sold in supermarkets. Amino acids can be used in the enteral and parenteral nutrition products.

Foods and beverages made from soya bean or that have soya bean proteins in different forms or obtained through different technologies are widely used in several products in Brazil and have become more popular due to the publicity on the health properties of isoflavones.

There are also isoflavones in capsules advertised as having several health benefits that are becoming popular among women. But isoflavones in capsule form have been recently considered as drugs by the health authority. This was based on the fact that the efficacy for stated properties like cancer prevention and treatment, osteoporosis reduction, use for hormone therapy replacement and heart protection, were not considered proved and could not be indicated to general population. Although they were considered capable of reducing some of the symptoms of menopause, care should be taken and they should be always used under medical supervision.

With this regulation there is a clear classification of products containing isolated compounds from foods. It should be noted that plant parts, plant extracts, botanicals or herbals are covered under the pharmaceutical regulation and as such may not be sold in supermarkets but only in pharmacies, which is an important difference between Brazil and other countries.

12.7 CONCLUSION

The evolution of science, technology, and longevity, and the search for improved quality of life led to the discovery of substances that can be used either isolated or in foods. Foods fortified with vitamins and minerals and those designed for special dietary uses had already presented that tendency. This generated the need of specific and current regulations in Brazil as in other countries.

Regulation in Brazil is focused on safety and efficacy and both functional and health claims are allowed, provided they are scientifically proven. Issuing regulations for food claims allowed the production of functional foods within the country and promoted or renewed research efforts in important areas of food science and nutrition.

The big challenge is to define what criteria—as, for example, biomakers—must be used as the scientific evidence for functional or health claims that introduce benefits besides those of basic nutrition, and how to inform the consumer using a message that is not misleading and employs understandable language.

In Brazil the label claim legislation managed to organize the market and is always updated in order to keep up with scientific developments and national health policies.

Health authorities, academia, and the industry sector have been gathering to discuss functional foods science's state of the art in scientific meetings. However, it is important that international forums are created in order to exchange scientific information and to set the standard for legislation around the world.

REFERENCES

1. Brasil, *Regulamento da Inspeção Industrial e Sanitária de Produtos de Origem Animal (RIISPOA)*, Decreto no. 30691, de 29 de março de 1952; Ministério da Agricultura: Brasília, Diário Oficial da União, Poder Executivo, de 7 de julho de 1952.
2. Brasil, *Institui normas básicas sobre alimentos*, Decreto-Lei no. 986, de 21 de outubro de 1969; Ministério da Marinha de Guerra, Ministério do Exército e Ministério da Aeronáutica Militar: Brasília, Diário Oficial da União, 1969.
3. Brasil, *Dispõe sobre a Padronização, a Classificação, o Registro, a Inspeção, a Produção e a Fiscalização de Bebidas, Autoriza a Criação da Comissão Intersetorial de Bebidas e dá outras Providências*, Lei no. 8918, de 14 de julho de 1994; Congresso Nacional: Brasília, Diário Oficial da União, Poder Executivo, de 15 de julho de 1994.
4. Brasil, *Aprova normas sobre Alimentos Dietéticos*, Portaria no. 23 de 1988; Ministério da Saúde: Brasília, 1988.
5. Brasil, *Regulamento Técnico de Alimentos para Fins Especiais*, Portaria no. 29, de 13 de janeiro de 1998; Ministério da Saúde, Secretaria da Vigilância Sanitária: Brasília, Diário Oficial da União, Poder Executivo, de 15 de janeiro de 1998.

6. Brasil, *Regulamento Técnico de Alimentos Adicionados de Nutrientes*, Portaria no. 31, de 13 de janeiro de 1998; Ministério da Saúde, Secretaria da Vigilância Sanitária: Brasília, Diário Oficial da União, Poder Executivo, de 16 de janeiro de 1998.

7. Brasil, *Define o Sistema Nacional de Vigilância Sanitária, cria a Agência Nacional de Vigilância Sanitária e dá outras providências*, Lei no. 9.782, de 26 de janeiro de 1999; Congresso Nacional: Brasília, Diário Oficial da União, Poder Executivo, de 27 de janeiro de 1999.

8. Agência Nacional de Vigilância Sanitária, Alimentos com alegações de propriedades funcionais, II Seminário, Brasília, p.95, 2002.

9. Brasil, *Regulamento Técnico de Procedimentos para o Registro de Alimentos e ou Novos Ingredientes*, Resolução no. 16, de 30 de abril de 1999; Agência Nacional de Vigilância Sanitária: Brasília; Diário Oficial da União, Poder Executivo, de 03 de dezembro de 1999.

10. Brasil, *Regulamento Técnico que Estabelece as Diretrizes Básicas para Avaliação de Risco e Segurança dos Alimentos*, Resolução no. 17, de 30 de abril de 1999; Agência Nacional de Vigilância Sanitária: Brasília, Diário Oficial da União, Poder Executivo, de 03 de dezembro de 1999.

11. Brasil, *Regulamento Técnico que Estabelece as Diretrizes Básicas para Análise e Comprovação de Propriedades Funcionais e ou de Saúde Alegadas em Rotulagem de Alimentos*, Resolução no. 18, de 30 de abril de 1999; Agência Nacional de Vigilância Sanitária: Brasília, Diário Oficial da União, Poder Executivo, de 03 de dezembro de 1999.

12. Brasil, *Regulamento Técnico para Procedimento de Registro de Alimento com Alegações de Propriedades Funcionais e ou de Saúde em Sua Rotulagem*, Resolução no. 19, de 30 de abril de 1999; Agência Nacional de Vigilância Sanitária: Brasília, Diário Oficial da União, Poder Executivo, de 10 de dezembro de 1999.

13. Brasil, *Institui a Comissão de Assessoramento Técnico-científico em Alimentos Funcionais e Novos Alimentos*, Portaria no. 15, de 30 de abril de 1999; Agência Nacional de Vigilância Sanitária: Brasília, Diário Oficial da União, Poder Executivo, de 03 de maio de 1999.

14. Brasil, *Regulamento Técnico de Substâncias Bioativas e Probióticos Isolados com Alegação de Propriedade Funcional e ou de Saúde*, Resolução RDC no. 2, de 7 de janeiro de 2002; Agência Nacional de Vigilância Sanitária: Brasília, Diário Oficial da União, Poder Executivo, de 09 de janeiro de 2002.

15. Brasil, *Regulamento Técnico sobre Rotulagem Nutricional de Alimentos Embalados*, Resolução RDC no. 360, de 23 de dezembro de 2003; Agência Nacional de Vigilância Sanitária: Brasília, Diário Oficial da União, Pode Executivo, de 26 de dezembro de 2003.

16. American Association of Cereal Chemists (AACC), *The definition of dietary fiber*, Report of the Dietary Definition Committee to the Board of Directors of the American Association of Cereal Chemists, 46 (3), 112-126, 2001.

17. Institute of Medicine/National Academy of Sciences, *Dietary Reference Intakes for Energy, Carbohydrates, Fiber, Fat, Protein and Amino Acids*, National Academic Press, Washington, DC, 2002.

18. Codex Alimentarius, *Proposals for a Definition, Method of Analysis and Conditions for Dietary Fibre Content*, Codex Committee on Nutrition and Food for Special Dietary Uses, CX/NFSDU 04/3-Add.1, 2004.

Part II

The "Ohmics" Technologies and Functional Foods

13 Principles of Proteomics

Lianji Jin, Lenore Polo Rodicio, and
Patrick A. Limbach

CONTENTS

13.1 INTRODUCTION

The term proteome was derived from genome to describe the set of proteins encoded by the genome. Proteomics can be defined as the study of proteomes much like genomics refers to the study of genomes. In practice, proteomics includes not only the characterization and study of proteins along with their isoforms and post-translational modifications (PTMs), but can also involve the generation of protein expression profiles from external stimulation to the system of interest and the study of protein–protein interactions and protein function.

The field of proteomics has emerged to directly characterize proteins at a global level. With the advances arising in genomics, the identified functional genes can provide a "blueprint" for the possible gene products that are the focal point of proteomics [1]. A genome is a static entity, whereas a proteome is dynamic and must be studied individually. Proteomes vary due to splicing of mRNA during transcription and due to PTM. Moment-by-moment snapshots of a proteome can reflect the up- and down-regulation of proteins, their modification status, and their interacting partners at a given cell state.

Proteomics covers a broad range of subjects aimed at understanding complex cellular functions in a systematic manner. It is generally categorized in three areas, which include: (1) large-scale protein identifications including isoforms and PTMs; (2) quantitative proteomics or global analysis of protein expression; and (3) the characterization of protein–protein interactions.

The identification of all of the proteins present in a cell provides the basis for understanding cell regulation and function. A single gene can produce multiple gene products as a result of alternative splicing generating a protein having many isoforms, each of which must be characterized. Post-translational modifications are covalent modifications of protein amino acid residues. Post-translational modifications such as phosphorylation and glycosylation can be involved in important cell processes such as the regulation of enzyme activity, cell signaling and modulation of molecular interactions.

In addition to identification of proteins present in a proteome, it is often useful to generate quantitative information relating to these proteins. Most commonly, relative information about changes in protein abundance is determined. In these experiments, one measures the relative abundance ratios of proteins present in two systems (e.g., cells or biological fluids) to evaluate perturbation-induced protein expression. A technically more challenging measurement is the determination of absolute protein amounts within a given proteome.

The third area of proteomics involves characterizing protein–protein interactions. The success in obtaining genomic information has increased interest in studying the encoded protein networks which govern cellular function. Elucidating protein–protein interactions on a proteome-wide scale often requires combining the techniques associated with protein identification and the techniques required for protein quantification with additional biochemical techniques that probe protein–protein interactions.

This chapter will provide an overview of the instrumentation and techniques common to proteomics. Initially, descriptions of methods used in isolating proteins prior to their analysis are presented. Next, the mass spectrometry techniques used for protein identification are discussed. Finally, relevant examples of proteomics are provided to illustrate the breadth and depth of this emerging field.

13.2 SAMPLE PREPARATION

Success in proteomics often is dependent upon appropriate isolation and purification of the protein(s) of interest. In this section, basic approaches for cell lysis and tissue extraction, protein separation and proteolytic digestion are presented.

13.2.1 CELL LYSIS AND TISSUE EXTRACTION

Reliable proteome analysis starts with the quantitative transfer of cytosolic and membrane-bound proteins into solution phase for subsequent separation, purification and analysis. A standard protocol for cell lysis and tissue extraction uses a cocktail containing 2 M thiourea, 5 M urea, 0.25% 3-(3-cholamidopropyl) diethyl-ammonio-1-propane sulfonate (CHAPS), 0.25% Tween 20, 0.25% N-decyl-N, N-dimethyl-3-ammonio-1-propane sulfonate (SB 3–10), 100 mM dithiothreitol (DTT), 0.25% carrier ampholytes (Bio-Lyte 3–10 (Bio-Rad), Servalyte 3–10 (Serva), Ampholine 3.5–9.5 (Amersham Pharmacia Biotech) and Resolyte 4–8 (BDH, Poole, Dorset, UK), 1:1:1:1), 10% isopropanol, 12.5% water, saturated isobutanol, 5% glycerol, 1 mM sodium vanadate and 1× Complete protease inhibitor cocktail. Cells should be lysed by the addition of lysis/extraction solution to yield a concentration of approximately 20,000 cells/L. After scraping, further extraction with the cocktail described above for 0.5–1 h with constant mixing on a Nutator is required [2]. This protocol allows for the solubilization of proteins in a single step without the need to shear nucleic acids, and this approach is compatible with two-dimensional (2-D) gel electrophoresis.

13.2.2 POLYACRYLAMIDE GEL ELECTROPHORESIS

1-D or 2-D polyacrylamide gel electrophoresis (PAGE) has been a primary tool for protein separations and studies for decades, and is one of the methods of choice for proteome fractionation before mass spectrometry characterization in proteomics. Isoelectric focusing (IEF), the first dimension in 2-D PAGE, separates proteins and peptides based on their isoelectric points. Sodium dodecyl sulfate polyacrylamide gel electrophoresis (SDS-PAGE), the second dimension in 2-D PAGE and typically used for 1-D PAGE, nominally separates proteins based on their molecular weight. Numerous commercial kits and protocols are available for 1-D or 2-D PAGE.

After 1-D or 2-D PAGE separation of proteins, the proteins must be visualized by staining. Currently, a selection of stains is available for gel staining including Coomassie blue, silver, copper, zinc or fluorescent dyes, such as the SYPRO-based fluorescent dyes. Table 13.1 lists current detection limits for various staining approaches as well as their compatibility with subsequent mass spectrometric analysis.

13.2.3 ENZYMATIC DIGESTION

In the bottom-up proteomics approach, peptides derived from enzymatic digestion of proteins are characterized by mass spectrometry. In-gel digestion refers to the

TABLE 13.1

Detection Limits for Protein Staining Techniques and Recommended Compatibility with Mass Spectrometry

	Organic Dyes	
Coomassie Blue	500 ng/mm^2	Not recommended
Modified colloidal Coomassie G-250	1–10 ng/mm^2	Recommended
	Silver Stains	
Silver nitrate	0.25 ng/mm^2	Not recommended
	Negative Stains	
Copper chloride	5 ng protein/mm	Not recommended
Zinc chloride	10–12 ng	Not recommended
	Fluorescent Dyes	
SYPRO Red	0.5–10 ng protein/mm^2	Recommended
SYPRO Orange	0.5–10 ng protein/mm^2	Recommended
SYPRO Tangerine	4–10 ng protein/mm^2	Not recommended
SYPRO Ruby	0.25–1 ng protein/mm^2	Highly recommended

proteolytic digestion of proteins separated by 1-D or 2-D PAGE. With in-gel digestions, after visualization of the proteins on the gel, gel spots or bands are excised and the protein is digested within the gel matrix. In-solution digestion refers to the proteolytic digestion of proteins isolated by solution-phase approaches such as immunoprecipitation or High-performance liquid chromatography (HPLC).

Table 13.2 provides a convenient protocol for in-gel digestion. The in-gel trypsin digestion protocol described here is applicable to 2-D gels visualized by Coomassie blue, Daiichi silver, SYPRO Orange, SYPRO Red, SYPRO Ruby, or SYPRO Tangerine [3]. In-solution digestion is much simpler than in-gel digestion. For in-solution tryptic digests, the sample of interest is suspended in several hundred microliters of 50 mM ammonium bicarbonate (pH 8). Trypsin is added so that the final protein-to-enzyme mass ratio is around 50:1. After overnight incubation at 37°C, the reaction is halted by the addition of 4 μL of 10% trifluoroacetic acid. The sample is then ready for further analysis.

Besides trypsin, which cleaves proteins at lysine and arginine residues, a variety of other proteases are available. Table 13.3 lists some of the more commonly used proteases in proteomics. A standard methodology for generating sufficient proteolytic fragments that cover nearly the entire sequence of a protein is the use of multiple enzymes. For example, Lys-C cleaves at lysine residues and Asp-N at aspartic acid residues. The use of trypsin and Lys-C will generate sequence information which is overlapping and confirmatory. Asp-N provides sequence information from regions that are not covered by either trypsin or Lys-C. Thus, a combination of these three enzymes can result in enhanced sequence coverage that would otherwise not be possible using each individual enzyme.

TABLE 13.2
Typical In-Gel Digestion Protocol

1 Excise the protein band of interest from the gel and cut into small pieces
2 Wash the gel pieces with 100 μL of deionized water
3 Add 50 μL of acetonitrile and incubate for 5 min
4 Destain the gel pieces by discarding the supernatant and replacing with 50 μL of acetonitrile/50 mM NH₄HCO₃, pH 8.5, until the pieces are completely destained
5 Lyophilize the gel pieces
6 Add 10 mM dithiothreitol (DTT) in 100 mM NH₄HCO₃ to cover the gel pieces. Incubate at 56°C for 1 h to reduce the protein
7 After cooling to room temperature, replace the DTT solution with an equal volume of 55 mM iodoacetamide in 100 mM NH₄HCO₃. Incubate at room temperature in the dark for 45 min
8 Wash the gel pieces with 100 μL of 50 mM NH₄HCO₃
9 Remove the supernatant and incubate the pieces with 100 μL of acetonitrile for 5 min. Repeat this step until the gel pieces are white
10 Lyophilize the gel pieces, rehydrate, and add 20 μL of 10 ng/μL trypsin in 50 mM NH₄HCO₃, 5 mM CaCl₂ in an ice bath
11 After 45 min, remove the enzyme solution and replace with 50 mM NH₄HCO₃, 5 mM CaCl₂
12 Allow the gel pieces to incubate overnight at 37°C
13 Add 50 μL of acetonitrile to the solution, and allow to incubate for 10 min
14 Remove the supernatant to a clean tube and incubate the gel pieces with 30 μL of water, 50 μL of acetonitrile for 10 min more
15 Remove the supernatant and combine with the supernatant resulting from step 11
16 Lyophilize and redissolve in 20 μL of 50% acetonitrile/0.1% formic acid for mass spectrometric analysis

Source: From Borchers, C., Peter, J. F., Hall, M. C., Kunkel, T. A., and Tomer, K. B., *Anal. Chem.*, 72, 1163–1168, 2000; Shevchenko, A., Wilm, M., Vorm, O., and Mann, M., *Anal. Chem.*, 68, 850–858, 1996.

TABLE 13.3
Common Proteases Used in Proteomics

Proteases	Amino Acid Sequence Specificity
Trypsin	X-Lys/-X and X-Arg/-X
Lys-C	X-Lys/-X
Clostripain	X-Arg/-X
Protease V8	X-Glu/-X and X-Asp/-X
Asp-N	X-/Asn-X
Pepsin	X-/Phe-X, X-/Trp-X, and X-/Tyr-X
Proteinase K	Non specific
Subtilisin	Non specific

The location of hydrolysis is denoted by a slash (/) before or after the amino acid responsible for specificity. X stands for an arbitrary amino acid.

13.3 INSTRUMENTATION IN PROTEOMICS

The elucidation of protein properties and function requires knowledge of the sequence and structure of that protein. It can be argued that mass spectrometric techniques have been best developed for protein and peptide analysis. The first peptides analyzed by mass spectrometry were characterized in the early 1980s. Since that time, matrix-assisted laser desorption/ionization (MALDI) and electrospray ionization (ESI) techniques have expanded the capabilities for analyzing these molecules, as have mass analyzers with greater accuracy and higher resolution.

13.3.1 MATRIX-ASSISTED LASER DESORPTION/IONIZATION TIME-OF-FLIGHT MASS SPECTROMETRY

Matrix-assisted laser desorption/ionization time-of-flight mass spectrometry (MALDI-TOFMS) is a versatile instrumental approach for biomolecule analysis and is particularly well-suited for the structural characterization of peptides or proteins. Attractive features of MALDI-TOFMS include its high sensitivity (with detection limits in the low femtomole range routinely achieved), rapid analysis time, high duty cycle and ease of automation.

In MALDI-MS, a low molecular weight organic acid (i.e., the matrix) is co-crystallized with the analyte of interest prior to spotting this mixture onto an instrument-specific sample target. Upon irradiation with a UV laser, the matrix preferentially absorbs laser energy and is brought into the gas phase as a plume during which the analyte is lifted as well. The matrix and analyte undergo a series of gas-phase reactions leading to positively charged peptide or protein ions [4]. These ions enter a field-free drift tube and travel at a velocity that is inversely proportional to their m/z ratio. The time for these ions to travel the length of the field-free region is measured and used to calculate the velocity and ultimately m/z of each ion. MALDI-TOFMS is particularly amenable to mixture analysis.

13.3.2 ELECTROSPRAY IONIZATION-MASS SPECTROMETRY

In ESI, the analyte of interest is placed into solution and this solution is then nebulized into an electric field. A combination of solvent evaporation and droplet fission lead to the final desolvated analyte ions [5]. In proteomics, one finds ESI most often coupled to three types of mass analyzers: quadrupole time-of-flight (Q-TOF), ion traps (either quadrupole or linear ion traps) or Fourier transform ion cyclotron resonance (FTICR) mass analyzers. The latter two types are both tandem-in-time mass analyzers.

The Q-TOF has a quadrupole mass filter that passes the ions of interest into a collision cell for fragmentation. The fragment ions are then separated and detected using a time-of-flight mass analyzer. The quadrupole ion trap (QIT) consists of three electrodes (one ring and two endcap electrodes) on which overlapping DC and RF potentials are applied to confine ions. Ions are then sequentially ejected from the trap and detected. The linear ion trap (LIT), a more recent development, can simply be thought of as a quadrupole mass filter with endcap electrodes. Advantages of the LIT over the QIT include greater ion storage capacity, higher duty cycles, faster scanning times and a greater number of scan modes. Fourier transform ion cyclotron resonance mass analyzers operate by trapping ions in an ICR cell in a way similar to QIT mass analyzers. FTICRMS is capable of resolving and identifying up to thousands of components of a complex mixture without prior wet chemical separation, offering 10–100 times higher mass resolution, high resolving power, and high mass accuracy than any other mass analysis technique to date [6].

13.3.3 COLLISION-INDUCED DISSOCIATION

The most common tandem mass spectrometry experiment in proteomics is the product ion scan in combination with collision-induced dissociation (CID). The product ion scan consists of three steps. First, user-specified sample ions, usually the molecular ion $(M+H)^+$ arising from a particular component, are selected or isolated. These ions then undergo collisions with a nonreactive gas such as N_2 or Ar, which deposit internal energy into the ion leading to bond cleavage and the generation of fragment ions. Finally, the fragment ions are separated and detected. A common approach used in peptide sequencing is data-directed or data-dependant analysis (DDA), in which a user-defined number (usually three) of the most abundant peptide ions are selected for MS/MS.

A scheme for peptide fragmentation is shown in Figure 13.1. If the charge resides on the N-terminal fragment of the peptide, then fragments of type a, b, and c are produced. If the charge is retained on the C-terminal fragment, then fragments of type x, y, and z are produced. A subscript indicates the position of the cleavage counting from the N-terminus (for a, b, c fragments) or the C-terminus (x, y, z fragments). In addition, a " after the letter/subscript designation indicates that the fragment was produced by the addition of two rather than one hydrogen. For peptides, typically b- or y-type ions are generated and used for sequence interpretation.

A unique dissociation approach used in combination with FTICRMS is electron capture dissociation (ECD). Electron capture dissociation is a radical site dissociation method, which yields a different fragment ion series than conventional CID [7]. Electron capture is postulated to occur initially at a protonated site releasing an energetic H˙ radical that deposits internal energy locally at the site of cleavage, leading to dissociation along the peptide backbone. Unlike CID methods which mainly produce b- and y-ion series, ECD typically produces c- and z-ion series. A side-by-side comparison between ECD and CID find that posttranslational modifications such as carboxylation, glycosylation, and sulfation are less easily lost in ECD than in CID.

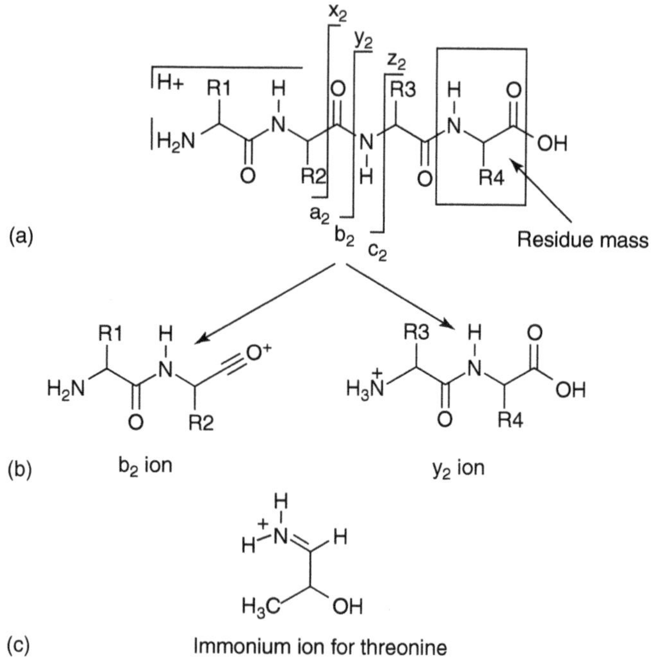

(a)

(b)   b₂ ion   y₂ ion

(c)   Immonium ion for threonine

FIGURE 13.1 Peptide fragmentation nomenclature as proposed by Roepstorff. (From Roepstorff, P., *Biomed. Mass Spectrom.*, 11, 601, 1984. With permission.)

13.3.4 HIGH-PERFORMANCE LIQUID CHROMATOGRAPHY

High-performance liquid chromatography is a versatile instrumental technique for separating peptides derived from enzymatic digestion of proteins. Separation of a mixture of analytes, such as peptides, is effected through interaction between the stationary phase and one or more physical properties of the analyte. Physical properties commonly exploited include size, charge or hydrophobicity. One-dimensional reversed-phase HPLC using hydrophobic stationary phases is most commonly used to separate simple mixtures of peptides or proteins.

Multidimensional chromatography separates peptides with increased resolution and peak capacity. These gains arise because multidimensional separation relies on two or more properties for fractionation. Since peak capacity is defined as the number of individual components that can be resolved by a particular separation method, multidimensional approaches yield increased peak capacity due to a multiplicative effect. Two-dimensional chromatography in the form of strong cation exchange chromatography as the first dimension followed by reversed-phase HPLC as the second dimension is most commonly used in proteomics. Multidimensional protein identification technology (MudPIT) is an online pseudo two-dimensional approach using a biphasic microcapillary column packed with strong cation exchange (SCX) and reversed phase (RP) stationary phase material (Figure 13.2).

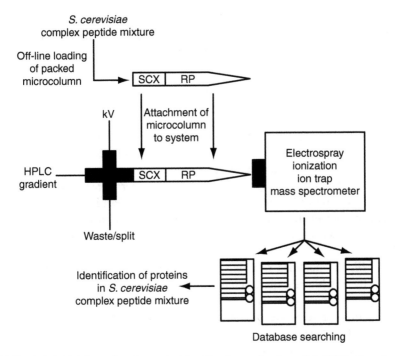

FIGURE 13.2 Multidimensional protein identification technology (MudPIT). Based on the method of Link et al., complex peptide mixtures from different fractions of a *Saccharomyces cerevisiae* whole-cell lysate were loaded separately onto a biphasic microcapillary chromatography column packed with strong cation exchange (SCX) and reversed-phase (RP) packing materials, respectively. After loading the complex peptide mixture into the microcapillary column, the column was inserted into the instrumental setup. Xcalibur software, high-performance liquid chromatography (HPLC), and mass spectrometer were controlled simultaneously by means of the user interface of the mass spectrometer. Peptides directly eluted into the tandem mass spectrometer because a voltage (kV) supply is directly interfaced with the microcapillary column. As described in the experimental protocol, peptides were first displaced from the SCX to the RP by a salt gradient and eluted off the RP into the MS/MS. In an iterative process, the microcolumn was re-equilibrated and an additional salt step of higher concentration displaced peptides from the SCX to the RP. Peptides were again eluted by an RP gradient into the MS/MS, and the process was repeated. The tandem mass spectra generated were correlated to theoretical mass spectra generated from protein or DNA databases by the SEQUEST algorithm. (From Washburn, M. P., Wolters, D., and Yates, J. R. I., *Nat. Biotechnol.*, 19, 242–247, 2001. With permission.) Copyright 2001 Nature Publishing Group.

13.4   PROTEOMIC APPROACHES

Current proteomics approaches are generally categorized in two categories: bottom-up and top-down. In the bottom-up approach, peptides generated from the proteolytic digestion of purified protein(s) are analyzed by mass spectrometric approaches. In the top-down approach, intact proteins are initially characterized by mass spectrometric approaches. Both approaches, independently or collaboratively, are implemented in current proteomic studies.

13.4.1   PEPTIDE MASS FINGERPRINTING

Peptide mass fingerprinting (PMF) or peptide mass mapping (PMM), pioneered by Henzel and colleagues for the rapid identification of proteins separated on a two-dimensional polyacrylamide gel [8], is the process of identifying an unknown protein by mass determination of its peptide fragments generated by specific enzymatic or chemical cleavage. Peptide mapping experiments have been routinely carried out by non-MS techniques such as Edman N-terminal sequencing. Mass spectrometry, however, offers a faster method for peptide mapping with resolution and mass accuracies that are three to four orders of magnitude higher than in SDS-PAGE.

In PMF, the peptides resulting from cleavage of the unknown protein are identified by matching the mass of the peptide to peptide masses in a database. Three or more peptide mass values are needed for a positive match of the parent protein. When multiple proteins are present in a single spot, a second search may yield positive identification of a second protein with excluded peptide mass values from the first protein. Most protein databases, as well as the software required to search them, are available on-line. Table 13.4 lists the Internet websites for some of these databases and search programs.

The instrumentation most often used for PMF is MALDI-MS. Peptide mass fingerprinting using MALDI-MS has found its greatest utility in identifying proteins

---

TABLE 13.4
Protein Databases and Search Algorithms

	Internet Address
Protein Databases	
MSDB	ftp://ftp.ncbi.nih.gov/repository/MSDB/MSDB.nam
NCBInr	http://www.ncbi.nlm.nih.gov
SWISS-PROT	http://www.expasy.org/sprot
DbEST	http://www.ncbi.nlm.nih.gov/dbEST/index.html
Database Search Algorithms	
SEQUEST	http://fields.scripps.edu/sequest/
MASCOT	http://www.matrixscience.com
Protein prospector	http://prospector.ucsf.edu
Genomic solutions	http://bioinformatics.genomicsolutions.com/ProteinId.html
De Novo Sequencing	
PEAKS	http://www.bioinformaticssolutions.com/products/peaksoverview.php
SEQUIT	http://www.sequit.org/sequit.htm

separated by 2-D gel electrophoresis (Figure 13.3a and Figure 13.3b). Matrix-assisted laser desorption/ionization time-of-flight mass spectrometry is particularly well suited for these types of analyses because it generates spectra containing primarily singly-charged ion peaks. The resulting spectrum is essentially a "fingerprint" for the protein of interest (Figure 13.3c).

13.4.2 SEQUENCE TAG ANALYSIS

An alternative and often more effective approach is to sequence peptides by tandem mass spectrometry to generate sequence tags (i.e., partial sequence information), which can be searched against the database to identify the protein of interest. The protocol begins with the enzymatic digestion of the protein as before. In the LC-ESI-MS/MS approach, the resulting peptides are applied to a RP-HPLC column (typically C18 columns are used). Peptides are retained on the RP column and contaminating salts and buffer components are washed away. The peptides are then eluted from the RP column into the mass spectrometer using an acetonitrile gradient. As peptides are eluted from the RP column into the ionization source of the mass spectrometer, the instrument acquires tandem (MS/MS) mass spectra of the eluting peptides (Figure 13.3c and Figure 13.3d). The acquired MS/MS spectra of peptides are correlated with predicted amino acid sequences in translated genomic databases and the resulting list of peptide sequences are used to identify the protein. Alternatively, sequence tags can be generated by use of MALDI TOF/TOF mass spectrometry, wherein particular peptides generated in the standard PMF experiment are selected for gas-phase fragmentation instrumentally.

As an alternative to 2-D gel electrophoresis separation of proteins, protein mixtures (e.g., complexes or whole cell lysates) can be enzymatically digested and analyzed by 2-D chromatography coupled with tandem mass spectrometry (Figure 13.4) [9]. This method is particularly useful for identifying proteins, such as hydrophobic membrane proteins or proteins with extreme pI values, which are not normally well-represented on a 2-D polyacrylamide gel. Identification is done by the sequence tag technique with significant bioinformatics support.

13.4.2.1 Top–Down Protein Characterization

While the bottom-up approach is the most widely applied and successful protocol for protein identification, this approach is not without limitations. DNA sequence errors can occur which result in protein sequence errors limiting accurate assignments during database searches.

Additionally, the correlation between the DNA and protein sequence can be low due to gene splicing during mRNA transcription or PTMs. In such cases, a bottom-up approach may yield inaccurate results or results which do not reflect splicing or posttranslational modification.

The top-down approach has been developed as an alternative to the bottom-up approach [10]. This approach is based upon the accurate molecular mass determination of the intact protein followed by gas-phase fragmentation. This

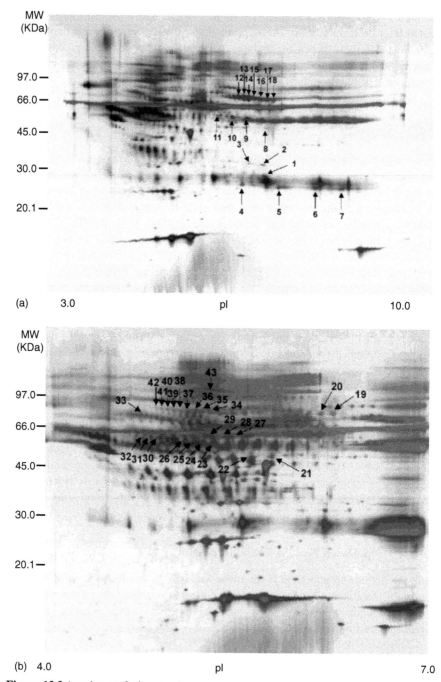

Figure 13.3 (caption on facing page)

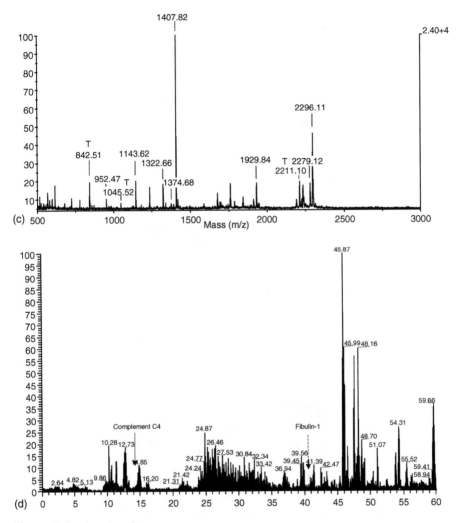

Figure 13.3 (*continued*)

FIGURE 13.3 Silver-stained 2-D gels of CSF with (a) p*I* 3–10 linear; and (b) p*I* 4–7 linear. (c) Matrix-assisted laser desorption/ionization (MALDI) mass spectrum of the tryptic digest derived from spot 1. Each matched peptide *m/z* was labeled. Trypsin autolysis products were labeled with T. (d) LCMS identification of protein spot 35. Base-peak ion chromatogram of the tryptic peptide mixture. The arrows indicate the retention time of a MS/MS spectrum of a peptide derived from complement C4, and an MS/MS spectrum of a peptide derived from fibulin-1. (From Yuan, X. and Desiderio, D. M., *J. Proteome Res.*, 2, 476–487, 2003. With permission.) Copyright 2003 American Chemical Society.

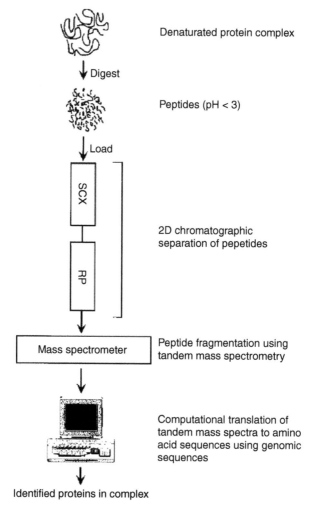

Denaturated protein complex

Digest

Peptides (pH < 3)

Load

SCX

2D chromatographic
separation of pepetides

RP

Mass spectrometer

Peptide fragmentation using
tandem mass spectrometry

Computational translation of
tandem mass spectra to amino
acid sequences using genomic
sequences

Identified proteins in complex

FIGURE 13.4 Direct analysis of large protein complexes (DALPC). In the flow diagram, the rectangles represent a strong cation exchange (SCX) and a reversed-phase (RP) liquid chromatography column. Typically, a denatured protein complex is digested with trypsin. The acidified peptide mixture is loaded onto the SCX column. A discrete fraction of peptides is displaced from the SCX column to the RP column. This fraction is eluted from the RP column into the mass spectrometer. This iterative process is repeated, obtaining the fragmentation patterns of peptides in the original peptide mixture. The program SEQUEST is used to correlate the tandem mass spectra of fragmented peptides to amino acid sequences using nucleotide databases. The filtered outputs from the program are used to identify the proteins in the original protein complex. (From Link, A. J., Eng, J., Schieltz, D. M., Carmack, E., Mize, G. J., Morris, D. R., Garvik, B. M., and Yates, J. R. I., *Nat. Biotechnol.*, 17, 676–682, 1999. With permission.) Copyright 1999 Nature Publishing Group.

approach is most commonly implemented using FTICRMS and ECD. Initially, an accurate molecular weight of the protein is obtained and used to search against the protein database to find a potential match. When an exact match is not obtained, sequencing via ECD allows one to determine differences between the experimental and theoretical sequences.

Top-down and bottom-up approaches can be integrated to fully exploit the advantages of both methods. An example of such integration is seen in Figure 13.5, which illustrates the proteomic analysis of *Shewanella oneidensis* [11]. The bottom-up approach identified a total of 868 proteins from *S. oneidensis* which corresponds to approximately 17% of the theoretical proteome. The top-down approach identified 22 proteins out of 70 detected. Where the bottom-up approach excels in the sheer numbers of identified proteins, the top-down approach provides a holistic view of the intact proteins and such complementary data greatly increase the confidence level of positive identifications.

13.4.2.2   Data Processing and Database Searching

Bioinformatics is of paramount importance in proteomics. In PMF, the peptides produced from MALDI spectra are searched directly against a database (Table 13.4). The essence of this approach is based on a probability matching between experimentally measured peptide masses and those theoretically predicted from proteins within a particular database. The search algorithms typically consider the number of matches between experimental and theoretical masses, the uniqueness of the peptide mass and the molecular weight of the peptide. Because matches are recorded solely on the basis of probabilistic fit to database entries, PMF is best limited to prokaryotic organisms with sequenced genomes. Protein identifications for higher-order organisms are better done using the sequence-tag approach.

There are two methods used to examine tandem mass spectral data generated in the sequence-tag technique for protein ID. The most common involves a probability based matching between the experimental fragment ion data and that predicted from peptide sequences generated from proteins within a particular database. SEQUEST is one of the more widely used algorithms for such searches [12]. In SEQUEST a cross-correlation function, Xcorr, is used to provide a measurement of similarity between the m/z values for the fragment ions predicted from amino acid sequences obtained from the database and the fragment ions observed in the tandem mass spectrum. In general, a difference greater than 0.1 between the normalized cross-correlation functions of the first- and second-ranked search results indicates a successful match between sequence and spectrum.

The other method for determining a peptide sequence from tandem mass spectral data is de novo sequencing. De novo sequencing interprets the tandem mass spectral data with little or no assistance from database information. As with the methods already discussed, the most probable sequence is derived by comparing experimental data to theoretical data and then evaluating the goodness of fit between the two. A common strategy implemented with the sequence-tag

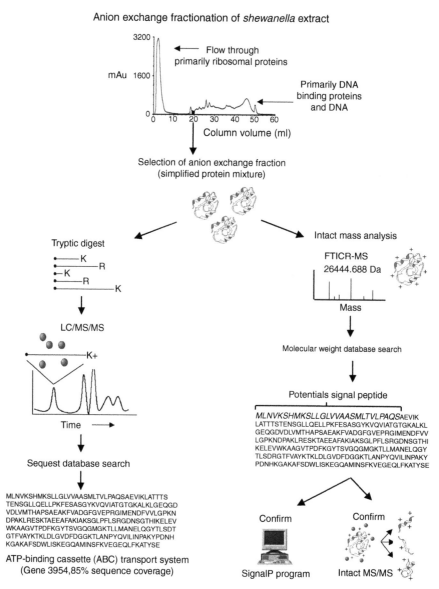

FIGURE 13.5 Combination of the "bottom-up" and "top-down" approaches to proteomic analysis. Extracts are first fractionated via anion exchange and are then split for proteolytic digestion and molecular weight determination. Analysis of the proteolytic digestion products is performed using 1-D or 2-D LC/MS/MS techniques followed by database searching. The undigested fractions undergo a dialysis step followed by analysis using Fourier transform ion cyclotron resonance (FTICR). By simplifying the fractions via anion exchange and using a 9.4 T ES-FTICR, increased dynamic range is obtained for intact mass analysis. In this example, a putative periplasmic protein (ABC transporter system) is identified using the "bottom-up" approach and its subsequent signal peptide confirmed via intact mass analysis, use of the SignalP program, and MS/MS of the intact protein. (From VerBerkmoes, N. C.,

technique is to initially search the tandem MS data by a database search approach. Then, any fragment ion spectra remaining unassigned are subjected to de novo sequencing.

13.5 POST-TRANSLATIONAL MODIFICATIONS

Post-translational modifications, although important for biological function, are difficult to characterize accurately on a large scale compatible with proteomics. The primary limitation arises from the incongruity between approaches developed for protein identification and those required for mapping PTMs. In protein identification, only a subset of the total protein sequence need be accurately characterized to provide confident assignments. In PTM mapping, the entire protein sequence must be characterized to identify differences between that predicted from genomic information and that obtained at the protein level.

One of the simplest approaches has been based on conventional bottom-up approaches wherein a larger and more comprehensive peptide mass fingerprinting analysis is done (Figure 13.6). Protease A is usually trypsin which cleaves C-terminal to the arginine and lysine and protease B can be Asp-N or Glu-C which cleave N-terminal aspartic acid and lysine. Even if using two proteases, 100% sequence coverage may not be resulted and a modified peptide may still escape detection. This problem can be addressed by selectively capturing modified peptides by affinity methods if a specific modification is of interest, hence greatly reduce the peptide complexity and identify the modified peptide.

In some instances, particular PTM's can be selectively enriched by chemical, biochemical or immunological approaches. The most successful example of selective enrichment has been the use of immobilized metal (Fe^{3+}) affinity chromatography (IMAC) for phosphopeptides [13]. Alternatively, immunoprecipitation of tyrosine-phosphorylated proteins using antiphosphotyrosine antibodies can be used to enrich this subset. The general advantage of selective enrichment is that only modified proteins or peptides must be characterized, which overcomes a limitation of the general mapping strategy described above.

13.6 QUANTITATIVE PROTEOMICS

13.6.1 DIFFERENCE GEL ELECTROPHORESIS

Protein expression profiling can be done using 2-D-PAGE by labeling two (or more) cell states with different fluorescent dyes. This technique, commonly referred to as 2-D difference gel electrophoresis (2-D-DIGE), overcomes many of the prior

Bundy, J. L., Hauser, L., Asano, K. G., Razumovskaya, J., Larimer, F., Hettich, R. L., and Stephenson, J. L. Jr., *J. Proteome Res.*, 1, 239–252, 2002. With permission.) Copyright 2002 American Chemical Society.

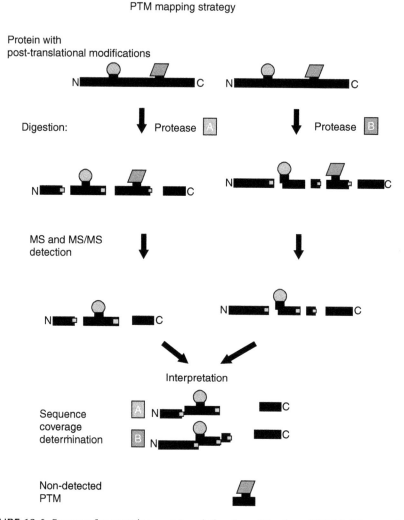

PTM mapping strategy

FIGURE 13.6 Strategy for mapping post-translational modifications (PTMs). The protein is shown as a line, and modifications are indicated by the symbols. The two columns show enzymatic digestion by two different enzymes to cover as much as possible of the protein sequence with peptides in the preferred mass range for MS analysis (500–3000 Da). (From Mann, M. and Jensen, O.N., *Nat. Biotechnol.*, 21, 255–261, 2003. With permission.) Copyright 2003 Nature Publishing Group.

limitations of gel-based quantification approaches. Because the various populations are labeled first and then separated on the same 2-D gel, higher precision in such analyses are possible. As typically implemented, DIGE is done separately from any protein identification using mass spectrometry due to loss of spot information once the gel is removed from the fluorescence imager and due to interferences arising from the fluorescent dyes.

13.6.2 STABLE ISOTOPE LABELING

A significant number of experimental methods are based upon the use of stable isotopes as a means of differentiating control from experimental samples. The two strategies used to analyze protein expression are global coding and targeted coding. In global coding, all peptides in a protein are isotopically labeled, either before, during or after proteolysis, and then a sub-fraction is selected for analysis. In targeted coding, the isotopic labeling and selection occur essentially in a single-step as only a sub-fraction of peptides are labeled.

Stable isotope labeling in the form of metabolic coding is one of the most common and successful global coding protocols. In metabolic coding, stable isotope labeled nutrients in growth media or isotopically labeled amino acids are used to differentially label control and experimental cells. Stable isotope tagging via proteolytic ^{18}O labeling is another global coding protocol [14]. Two ^{18}O atoms are incorporated into the carboxyl termini of all tryptic peptides during proteolytic cleavage. In general, these global coding methods are easy to implement and provide representative proteome coverage in the labeling step yet are subject to reduced precision as the labeling step is dependent on chemical or biochemical reactions.

The most recognized targeted coding protocol is the isotope-coded affinity tag (ICAT) protocol developed for large-scale quantitative profiling of proteins [15]. Figure 13.7 is a schematic of the quantitation process using the ICAT technique. The protocol targets cysteine-containing proteins from two different cell states which are labeled with isotopically light (d0) and heavy (d8) biotin affinity tags. Enzymatic digestion followed by a three-step chromatography fractionation of tryptic peptides and tandem mass spectrometry yield light and heavy isotope tagged peptides appearing as doublets in the mass spectra, and the peak area ratios are then calculated from which quantitative protein information is obtained. Solid-phase isotope tagging is an improved ICAT protocol for quantitative proteome profiling [16]. A side-by-side comparison with ICAT showed that solid-phase stable isotope labeling is simpler, more efficient, and more sensitive.

13.7   BIOMARKER DISCOVERY

Mass Spectrometry has become a reliable screening tool of high speed and throughput for disease diagnosis and biomarker discovery in clinical proteomics. Surface-enhanced laser desorption/ionization mass spectrometry (SELDI-MS) is similar to MALDI-MS and is used to detect peptides and proteins absorbed onto derivatized sample targets. This technique was investigated for profiling human serum proteome to identify ovarian cancer from normal subject with the assistance of a new bioinformatics tool [17]. Proteomic patterns from control samples were studied to obtain a statistical profile. The subsequent mass spectra of diseased and normal samples were then contrasted to this statistical profile, and identification of ovarian cancer samples was postulated to be possible. The unknown species from the mass spectra are under investigation to find their origin and full identity, which are hydrophobic and of low molecular mass, could be derived from the host organ, the cancer, or the constitute metabolic fragments (Figure 13.8).

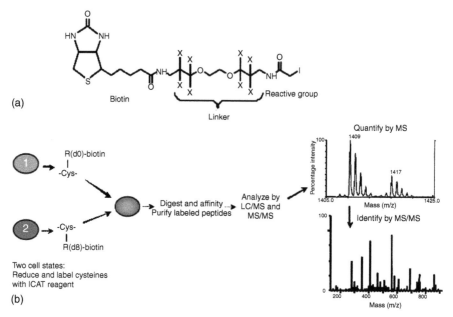

FIGURE 13.7 (a) Structure of isotope-coded affinity tag (ICAT) reagent. The reagent has three elements: an affinity tag (biotin) that is used to isolate ICAT labeled peptides, a linker that can incorporate stable isotopes, and a reactive group with specificity toward thiol groups (cysteines [Cys]). The reagent exists in two forms: heavy (contains eight deuteriums [d8]) and light (no deuteriums [d0]). (b) ICAT strategy for quantitating differential protein expression. Two protein mixtures representing two different cell states are treated with d0 and d8 ICAT reagents, respectively; an ICAT reagent is covalently attached to each cysteinyl residue in every protein. Protein mixtures are combined and proteolyzed to peptides; ICAT-labeled peptides are isolated by use of the biotin tag. These peptides are separated by microcapillary high-performance liquid chromatography (LC). A pair of ICAT-labeled peptides is chemically identical and easily visualized because it essentially coelutes. An 8-Da difference was measured by scanning mass spectrometer (4 m/z [mass-to-charge] U difference for a doubly charged ion). Ratios of the original amounts of proteins from the two cell states are strictly maintained in the peptide fragments. Relative quantification is determined by the ratio of the ion currents for the d0-and d8-tagged peptide pairs. Every other scan is devoted to fragmenting and then recording sequence information for the eluting peptide (tandem mass spectrum [MS/MS]). The protein is identified by computer search of the tandem mass spectrum against large protein databases. In theory, every peptide pair in the mixture is, in turn, measured and fragmented, resulting in the relative quantitation and identification of each protein in the mixture during a single analysis. MS, mass spectrum. (From Aebersold, R., *J. Infect. Dis.*, 187 (suppl. 2), S315–S320, 2003. With permission.) Copyright 2003 The University of Chicago Press.

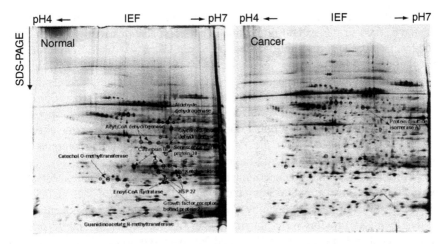

FIGURE 13.8 Analysis of differential expression of liver tissue proteins in hepatocellular carcinoma (HCC) patients. Soluble proteins were prepared and 150 μg proteins were separated on pH 4–7 IPG strips and on 7.5%–17.5% gradient sodium dodecyl sulfate polyacrylamide gel electrophoresis (SDS-PAGE). Staining was by silver nitrate. 2-DE gel images of soluble-fraction proteins from normal liver and cancer tissues of a HCC patient. The differentially expressed proteins are marked with circles. The circles on the "normal" gel show the spots that decrease in intensity in cancer tissue; those on the "cancer" gel show the spots increasing in cancer tissue. Differentially expressed protein spots identified by peptide mass fingerprinting are labeled with protein name. (From Kim, J., Kim, S. H., Lee, S. U., Ha, G. H., Kang, D. G., Ha, N. Y., Ahn, J. S. et al., *Electrophoresis*, 23, 4142–4156, 2002. With permission.) Copyright 2002 Wiley-VCH.

13.8 LARGE-SCALE PROTEOMICS

13.8.1 PROTEIN–PROTEIN INTERACTIONS

Elucidating protein–protein interactions on a proteome-wide scale is feasible using current mass spectrometry based proteomics technology [18]. Using budding yeast, *Saccharomyces cerevisiae*, as a test model 725 bait proteins representing different functional classes detected 3617 associated proteins covering 25% of the yeast proteome. Numerous protein complexes were identified including many new interactions involved in various signaling pathways and in DNA damage response. The global yeast protein interaction map can shed light on and provide a partial framework for understanding more complex proteome networks.

13.8.2 THE *ORYZA SATIVA* MITOCHONDRIAL PROTEOME

The rice (*Oryza sativa*) mitochondrial proteome has been comprehensively studied using proteomic approaches [19]. A modified two-Percoll gradient density separation technique was used for isolation of mitochondria largely free of contamination from other cell compartments. The purified rice mitochondrial proteins were separated by 2-D PAGE or BN (blue-native) SDS-PAGE. The BN

SDS-PAGE separates proteins based on native mass in the first dimension and denatured mass in the second dimension. This approach was used for membrane proteins, which are insoluble under IEF conditions. A wide range of rice mitochondrial proteins were identified (6.7–252 kDa) including hydrophilic/hydrophobic proteins (hydropathicity −1.27 to +0.84) and acidic and basic proteins (p*I*4.0–12.5). A total of 232 protein spots from the PAGE separations were excised, trypsin digested and analyzed by tandem MS. From those spots, 149 proteins were identified by searching open reading frames and six-frame expressed sequence tags translated from the *O. sativa* genome. Eight-five proteins were assigned functions expected within the mitochondrial organelle.

Additional studies involving complete digestion of mitochondrial proteins with trypsin followed by LC/MS/MS produced 170 MS/MS spectra that matched 72 sequence entries from rice open reading frames and expressed sequence tags. Among them, 45 proteins were obtained using LC/MS/MS alone and 28 proteins were detected in both the LC/MS/MS and PAGE-based approaches. Overall 136 non-redundant rice proteins were identified including 23 proteins of unknown function.

REFERENCES

1. Tyers, M. and Mann, M., *Nature*, 422, 193–197, 2003.
2. Leimgruber, R. M., Malone, J. P., Radabaugh, M. R., LaPorte, M. L., Violand, B. N., and Monahan, J. B., *Proteomics*, 2, 135–144, 2002.
3. Lauber, W. M., Carroll, J. A., Dufield, D. R., Kiesel, J. R., Radabaugh, M. R., and Malone, J. P., *Electrophoresis*, 22, 906–918, 2001.
4. Zenobi, R. and Knochenmuss, R., *Mass Spectrom. Rev.*, 17, 337–366, 1998.
5. Fenn, J. B., *Angew. Chem. Int. Ed.*, 42, 3871–3894, 2003.
6. Marshall, A. G., *Int. Mass Spectrom.*, 200, 331–356, 2000.
7. Zubarev, R. A., Horn, D. M., Fridriksson, E. K., Kelleher, N. L., Kruger, N. A., Lewis, M. A., Carpenter, B. K., and McLafferty, F. W., *Anal. Chem.*, 72, 563–573, 2000.
8. Henzel, W. J., Billeci, T. M., Stults, J. T., Wong, S. C., Grimley, C., and Watanabe, C., *Proc. Natl Acad. Sci. U.S.A.*, 90, 5011–5015, 1993.
9. Link, A. J., Eng, J., Schieltz, D. M., Carmack, E., Mize, G. J., Morris, D. R., Garvik, B. M., and Yates, J. R. I., *Nat. Biotechnol.*, 17, 676–682, 1999.
10. Kelleher, N. L., Lin, H. Y., Valaskovic, G. A., Aaserud, D. J., Fridriksson, E. K., and McLafferty, F. W., *J. Am. Chem. Soc.*, 121, 806–882, 1999.
11. VerBerkmoes, N. C., Bundy, J. L., Hauser, L., Asano, K. G., Razumovskaya, J., Larimer, F., Hettich, R. L., and Stephenson, J. L. Jr., *J. Proteome Res.*, 1, 239–252, 2002.
12. Eng, J. K., McCormack, A. L., and Yates, J. R. I., *J. Am. Soc. Mass Spectrom.*, 5, 976–989, 1994.
13. Andersson, L. and Porath, J., *Anal. Biochem.*, 154, 250–254, 1986.
14. Yao, X., Freas, A., Ramirez, J., Demirev, P. A., and Fenselau, C., *Anal. Chem.*, 73, 2836–2842, 2001.
15. Aebersold, R., *J. Infect. Dis.*, 187 (suppl. 2), S315–S320, 2003.
16. Zhou, H., Ranish, J. A., Watts, J. D., and Aebersold, R., *Nat. Biotechnol.*, 19, 512–515, 2002.

17. Petricoin, E. F., Ardekani, A. M., Hitt, B. A., Levine, P. J., Fusaro, V. A., Steinberg, S. M., Mills, G. B. et al., *Lancet*, 359, 572–577, 2002.

18. Ho, Y., Gruhler, A., Heilbut, A., Bader, G. D., Moore, L., Adams, S. L., Millar, A. et al., *Nature*, 415, 180–183, 2002.

19. Heazlewood, J. L., Howell, K. A., Whelan, J., and Millar, A. H., *Plant Physiol.*, 132, 230–242, 2002.

20. Roepstorff, P., *Biomed. Mass Spectrom.*, 11, 601, 1984.

21. Washburn, M. P., Wolters, D., and Yates, J. R. I., *Nat. Biotechnol.*, 19, 242–247, 2001.

22. Yuan, X. and Desiderio, D. M., *J. Proteome Res.*, 2, 476–487, 2003.

23. Mann, M. and Jensen, O. N., *Nat. Biotechnol.*, 21, 255–261, 2003.

24. Kim, J., Kim, S. H., Lee, S. U., Ha, G. H., Kang, D. G., Ha, N. Y., Ahn, J. S. et al., *Electrophoresis*, 23, 4142–4156, 2002.

14 Toxicogenomics and Systems Toxicology

Michael D. Waters and Jennifer M. Fostel

CONTENTS

Abstract

Toxicogenomics combines transcript, protein, and metabolite profiling with conventional toxicology to investigate the interaction between genes and environmental stress in disease causation. The patterns of altered molecular expression that are caused by specific exposures or disease outcomes have revealed how toxicants act and cause disease. Despite numerous success stories, the field faces significant challenges in discriminating the molecular basis of toxicity. In this chapter we argue that toxicology is gradually evolving into a systems toxicology. This will ultimately allow us to describe all the toxicological interactions that occur within a living system under stress and use our knowledge of toxicogenomic responses in one species to predict the modes-of-action of similar agents in other species.

14.1 INTRODUCTION

The ability to discern mechanisms of toxicity as related to health issues is an important challenge facing scientists, public health decision-makers, and regulatory authorities whose aim is to protect humans and the environment from exposures to hazardous drugs, chemicals, and environmental stressors (such as global warming or non-ionizing radiation) [1–5]. In addition, the problems of identifying environmental factors involved in the etiology of human disease and of performing safety and risk assessments for drugs and chemicals, and even dietary constituents, have long been formidable issues.

Toxicology, the study of poisons, is focused on the substances and exposures that cause adverse effects in living organisms. A critical part of this study is the empirical and contextual characterization of adverse effects at the level of the organism, the tissue, the cell, and intracellular molecular systems. Thus, studies in toxicology measure the effects of an agent on an organism's food consumption and digestion, its body and organ weights, on microscopic histopathology, and on cell viability, immortalization, necrosis, and apoptosis [6].

The rapid accumulation of genomic sequence data and associated gene and protein annotation has catalyzed the application of gene expression analysis to understand the modes-of-action of chemicals and other environmental stressors on biological systems (see Figure 14.1). These developments have facilitated the emergence of the field of toxicogenomics, which aims to study the response of a whole genome to toxicants or environmental stressors [7–18]. The related field of toxicoproteomics [19–21] is similarly defined with respect to the proteome, the protein subset of the genome. Global technologies, such as cDNA and oligonucleotide microarrays, protein chips, and nuclear magnetic resonance (NMR)-based molecular profiling, for example, can simultaneously measure the expression of numerous genes, proteins, and metabolites, respectively, thus providing the potential to accelerate the discovery of toxicant pathways, modes-of-action, and specific chemical and drug targets. Toxicogenomics therefore combines toxicology with genetics, global -omics technologies (see Box 14.1) and appropriate pharmacological and toxicological models (Figure 14.1) to provide a comprehensive view of the function of the genetic and biochemical machinery of the cell.

This review explores the new field of toxicogenomics, delineates some of its research approaches and success stories, and describes the challenges it faces. It discusses how integrating data derived from transcriptomic, proteomic, and metabonomic studies can contribute to the development of a toxicogenomics knowledgebase (Figure 14.2, Box 14.1) and to the evolution of systems toxicology as it relates to molecular expression profiling. In many ways, current gene, protein and metabolite expression profiles are simple "snapshots"; by contrast, systems toxicology, like systems biology [22,23], attempts to define the interactions of all the elements in a given biological system, under stress or toxicant perturbation, to achieve a mechanistic understanding of the toxicological response.

BOX 14.1   DESCRIPTIONS OF SELECTED -OMICS TECHNOLOGIES

The terms Transciptomics, Proteomics, and Metabonomics or Metabolomics refer to highly parallel analytical technologies wherein simultaneous measurements are made of expressed genes, proteins, or metabolites, respectively. These technologies are used to ascertain the function of the genome. Toxicogenomics makes use of all of these functional genomics technologies in the study of toxicology. The terms toxicoproteomics and toxicometabolomics are sometimes used in a technology-centric sense to discuss the response of the proteome or metabolome to toxicants.

Transcriptomics—cDNA Microarray Hybridization and Analysis

Early gene-expression profiling experiments that were carried out for toxicogenomics studies employed cDNA microarrays [59]. Although this cDNA technology is rapidly being supplanted by synthetic—short and long—oligonucleotide microarrays, the technological concepts underlying the two approaches are largely analogous: cDNAs are derived from sequence-verified clones representing the 3 ends of the genes, which are either spotted onto glass slides using a robotic arrayer or synthesized in situ. Each RNA sample is labeled with dye-conjugated dUTP by reverse transcription from an oligo dT primer. The fluorescently labeled cDNAs are then hybridized to the microarray and the microarray is scanned using laser excitation of the fluorophores [24]. Raw pixel intensity images derived from the scanner are analyzed to locate targets on the array, measure local background for each target, and subtract it from the target intensity value.

Proteomics

An established proteomics strategy [110] uses global protein stratification systems, such as polyacrylamide gel electrophoresis (PAGE), followed by protein identification by mass spectrometry (MS). Two-dimensional PAGE separation, by charge and by mass, can resolve thousands of proteins to near homogeneity. This separation is a necessary prerequisite to enzymatic digestion and MS identification, which requires unique peptide fingerprint masses or amino-acid sequence tags. Where proteins are separated by liquid chromatography (LC) instead of PAGE, a new and promising platform involving multidimensional LC can be used to fractionate and reduce the complexity of the protein mixture before peptide sequencing by MS or Tandem MS (LC/MS/MS). This approach is being augmented by SELDI (Surface enhanced laser desorption/ionization time-of-flight mass spectrometry; SELDI-TOF MS), a method that results in the isolation of tens- to hundreds-of-thousands of low molecular weight fragments representing a proteome.

Metabolomics and Metabonomics

Quantitative analytical methods have been developed to identify metabolites in pathways or classes of compounds. This collective directed approach has been called metabolite profiling or metabolomics. Semi-quantitative, NMR-based metabolic fingerprinting has also been applied to high abundance metabolites and has been termed "metabonomics" [111]. Peaks detected in NMR spectra carry information regarding the structure of the metabolites, whereas peaks detected by MS have associated molecular weights. In addition, specific MS methods can be established to fragment the parent molecule, allowing metabolites to be identified through the investigation of fragmentation patterns.

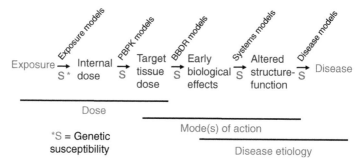

FIGURE 14.1 The role of genetic susceptibility and computational model on the continuum from exposure to disease outcome. The sequence of events between initial exposure and final disease outcome are shown from left to right. Following exposure, the body's "ADME" (absorption, distribution, metabolism, and excretion) systems control local concentrations of a chemical stressor in various body compartments. The impact of genetics is felt in specific alleles encoding various transporters, xenobiotic metabolizing enzymes, etc. Mathematical models such as exposure models, PB/PK, and BBDR models can be used to approximate these processes. PB/PK models are a set of differential equations structured to provide a time course of a chemical's mass-balance disposition (wherein all inputs, outputs, and changes in total mass of the chemical are accounted for) in pre-selected anatomical compartments. BBDR models are dose response models based on underlying biological processes. Once the target tissue is exposed to a local stressor the cells respond and adapt or undergo a toxic response; this process can be modeled with systems toxicology approaches. Finally, the disease outcome itself can be mimicked by genetic or chemically-induced models of particular diseases, for instance in the Zucker rat model of diabetes or the streptozotocin-treated rat model.

14.2 TOXICOGENOMICS: AIMS AND METHODS

Toxicogenomics has three principal goals: to understand the relationship between environmental stress and human disease susceptibility (Figure 14.1), to identify useful biomarkers of disease and exposure to toxic substances, and to elucidate the molecular mechanisms of toxicity. A typical toxicogenomics study might involve an animal experiment with three treatment groups: high- and low-dose treatment groups, and a vehicle control group that has received only the solvent used with the test agent. These groups will be observed at two to three points in time, with three to five animal subjects per group. In this respect, a toxicogenomics investigation resembles a simple, acute toxicity study. Where the two approaches differ is in the scope of the response they each aim to detect, and in the methods used. The highest dose regimen is intended to produce an overtly toxic response, which in a toxicogenomics study can be detected using the global measurement techniques described below (see also Box 14.2).

In a typical toxicogenomics experiment, lists of significantly differentially expressed genes are created for each biological sample [24]. Alternatively, profile analysis methods can be applied to dose- and time-course studies [24] to identify genes and gene profiles of interest [25]. Then, with the aid of the relevant knowledge that is systematically extracted and assembled [26] through literature mining, comparative analysis, and iterative biological modeling of molecular expression

---

BOX 14.2   DATABASES AND STANDARDS FOR EXCHANGE OF DATA

Databases

Public databases allow the scientific community to publish, share, and compare the data obtained from toxicology and toxicogenomics experiments. They are a resource for data mining, and for the discovery of novel genes/proteins through their co-expression with known molecules. They also help to identify and minimize the use of experimental practices that introduce undesirable variability into toxicogenomics datasets.

Guidelines

Public data repositories promote international database and data exchange standards [112–115] through guidelines developed by specific regulatory agencies. For example, the Clinical Data Interchange Standards Consortium (CDISC) develops guidelines for the electronic submission of clinical data, whereas the Standards for Exchange of Nonclinical (SEND) Data Consortium) addresses the submission of toxicology study data (see online links box). Minimum Information About a Microarray Experiment (MIAME) guidelines [116] specify sufficient and structured information to be recorded to correctly interpret and replicate microarray experiments or to retrieve and analyze the data from a public microarray database (such as ArrayExpress (Europe) [114], GEO (U.S.) [112], or CIBEX (Japan) [117]). Similar guidelines describing what information should be included in a published set of toxicogenomics data are under development by the Microarray Gene Expression Data Society (MGED). The MGED Toxicogenomics Working Group has recently broadened its scope to include environmental genomics and nutrigenomics and has changed its name to Reporting Structure for Biological Investigations (RSBI). The RSBI has proposed a tiered checklist to describe a biological investigation; such a checklist should enhance harmonization of related disciplines and reduce overlap in data collection. Tier I includes checklists (such as the "Investigation Design Description") and describes the study design, including the role of each experimental subject, the treatment given, and the times of tests or other events in the study. Tier II is comprised of checklists for studies of toxicology (MINTox), environmental genomics (MIAME/Env), and nutrigenomics (MIAME/Nut). Tier III includes technology modules for transcriptomics (MIAME), proteomics (MIAPE [118]), and metabonomics (SMRS). Modules in Tier II and III aim to collect the data necessary to interpret an investigation in a specific domain.

---

datasets, it is possible to differentiate adaptive responses of biological systems from those changes (or biomarkers) associated with or precedent to clinical or visible adverse effects [27–29]. Over the past five years the field of toxicogenomics has validated the concept of gene expression profiles as "signatures" of toxicant classes, disease subtypes or other biological endpoints. These signatures have effectively directed the analytical search for predictive biomarkers of toxicant effects and contributed to the understanding of the dynamic alterations in molecular mechanisms associated with toxic and adaptive responses.

The experimental work involved in a toxicogenomics study and the amount of gene expression data generated are vast. To examine only one tissue per animal requires 18–45 microarrays (more if technical replicates are used) and the attendant measurement of as many as 20,000 or more transcripts per array. In addition, each

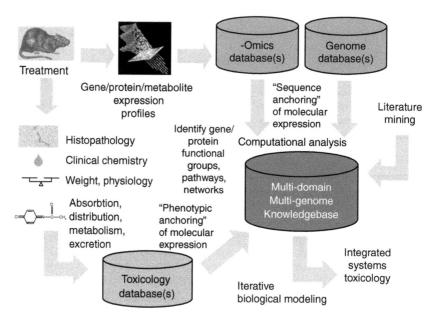

FIGURE 14.2 A framework for systems toxicology. The figure indicates the paths from the initial observation (rat in upper left) to an integrated toxicogenomics knowledgebase and thence to systems toxicology. The -omics data stream is shown by the clockwise path from rat to knowledgebase, and the "traditional" toxicology approach is shown in the counter-clockwise path. The knowledgebase will integrate both data streams, along with literature knowledge, and by virtue of iterative modeling lead to a systems toxicology understanding. The framework involves "phenotypic anchoring" (to toxicological endpoints and study design information) and "sequence anchoring" (to genomes) of multi-domain molecular expression datasets in the context of conventional indices of toxicology and the iterative biological modeling of resulting data.

animal will typically have treatment-associated data on total body and organ weight measurements, clinical chemistry measurements (often up to 25 parameters), and microscopic histopathology findings for several tissues [6]. The careful collection, management, and integration of these data, in the context of the experimental protocol, is essential for interpreting toxicological outcomes. Thus, all data must be recorded in terms of dose, time, and severity of the toxicological or histopathological phenotype(s). The compilation of such experimental data, together with toxicoinformatics tools and computational modeling will be important in deriving a new understanding of toxicant-related disease [12].

Toxicogenomics is just beginning to integrate the multiple data streams derived from transcriptomics, proteomics, and metabonomics with traditional toxicological and histopathological endpoint evaluation (Figure 14.2). This integration has the potential to synergize our understanding of the relationship between toxicological outcomes and molecular genetics. Furthermore, toxicology and toxicogenomics are progressively developing from studies done predominantly on individual chemicals and stressors into a knowledge-based science [16]. However, the evolution of a truly "predictive toxicology"—wherein knowledge of toxicogenomic responses of a prototypic agent in one species and strain is used to predict the mode-of-action of a similar agent in a related strain or another species—will require that the results of numerous toxicogenomics investigations across genotypes and species be assimilated into a multi-domain, multi-genome knowledgebase (Figure 14.2). This knowledgebase must be searchable by chemical formula/stressor-type, by gene/protein/metabolite molecular signature, or by phenotypic outcome, among other entities, to find results analogous to those observed with a newly tested agent. Toxicology will then have become an information science, and public health and risk assessment will be the beneficiaries.

14.3   THE EVOLUTION OF THE FIELD OF TOXICOGENOMICS

Toxicogenomics has evolved from early gene expression studies, which described the response of a biological system to a particular toxicant or panel of reference agents, towards more mature investigations that integrate several omics domains with toxicology and pathology data (see Table 14.1). Exposure- and outcome-specific patterns of gene, protein, and metabolite profiles have been used to identify molecular changes that serve as biomarkers of toxicity [13,30–37] and provide insights into mechanisms of toxicity [38–48] and disease causation [49–53]. Critical to this evolution were extensive and ongoing genome sequencing and annotation efforts [54,55] and the ability to describe response profiles in genetically and toxicologically important species such as mouse, rat, dog, and human. Another important contribution to toxicogenomics has been the formation of collaborative research consortia [56–58] that bring together scientists from regulatory agencies, industrial laboratories, academia, and the government to identify and address important issues for the field.

14.3.1   Profiles of Response to Toxicants

Nuwaysir et al. popularized the term "toxicogenomics" to describe the use of microarrays to measure the responses of toxicologically relevant genes, and to identify selective, sensitive biomarkers of toxicity [59]. The first published toxicogenomics study compared the gene expression profiles of human cells responding to the inflammatory agent lipopolysaccharide (LPS) or to mitogenic activation by phorbol myristate acetate (PMA) [60]. RNA samples isolated at various times following exposure showed the expected increases in cytokine, chemokine, and matrix metalloproteinase transcripts. Similar expression profiles were seen insynoviocytes and chondrocytes from a patient with rheumatoid arthritis, confirming the ability of the system to mimic the biological changes that occur during inflammatory disease. Subsequent studies extended this type of observation in other tissues and for a wide variety of toxicants, enabling the association of specific molecular profiles with specific toxicities.

TABLE 14.1
The Scope and Evolution of Toxicogenomics

Study Aim	Key References
Toxicogenomics tools and model systems	Toxicogenomics began with "toxicology-specific" cDNA microarrays designed to measure the levels of acute phase and xenobiotic-metabolizing enzymes such as cytochrome P450s [59,126]. These were superseded as commercial platforms were developed for toxicologically important species such as rats. The armamentarium of pre-clinical gene expression platforms was completed with the canine microarray [55]. It is now possible to use commercial oligonucleotide microarrays (see Online Links Box) to measure expression responses in species ranging from nematode (Caenorhabditis elegans), to frog (Xenopus laevis) and zebrafish (Danio rerio) to rodents (rat and mouse), to non-human primates and man. Toxicogenomics tools for sentinel aquatic species have been developed as well [74]. Later experiments began to focus on more challenging subjects such as subcellular organelles [127], non-standard tissue such as saliva [128], less well-characterized species [129], genetic models of diseases [130], and integration of data from different -omics disciplines [39,47,52,74,75,131]. Additionally, comprehensive studies of yeast have become increasingly important [99,100,132]
Some tissues used in toxicogenomics studies	Most toxicogenomics studies to date have involved hepatotoxicants [13,24,30–36,40,42,43,45–48,50,52,53,120,131,133,134], as the liver is the primary source of xenobiotic metabolism and detoxification, and because liver injury is the principal reason for withdrawal of new drugs from the market [135]. However, toxicogenomics studies have also addressed, nephrotoxicity [36,38,44], neurotoxicity [136,137] and reproductive toxicity [41], for example, as well as lung toxicity [31,49], skin toxicity [138] and cardiotoxicity [139]
Phenotypic anchoring	Phenotypic anchoring relates expression profiles to specific adverse effects defined by conventional measures of toxicity such as histopathology or clinical chemistry [12,16,63]
	Experiments have been designed to correlate expression patterns with disease pathologies such as necrosis, apoptosis, fibrosis, or inflammation [24,30,49,55,140]. Additionally, phenotypic anchoring can be used to provide biological context for toxicogenomics observations made at subtoxic doses [33,46]
Some classes of toxicants characterized	Studies have examined responses to toxicants with established mechanisms of toxicity [30,35,36,42,43,45,53,141], environmental toxicants [50,74,138,142,143], or exposures to suprapharmacological levels of drugs [31,33,39,40,46,47,52,120,139]

Examples of toxicant or stressor mechanisms	Acetaminophen [33,39,47,120,144] (see Box 14.3) Estrogenic agents [41,145] Oxidant stress [132,146] Peroxisome proliferators [13,34,36,43,45]
Importance of reporting husbandry and other technical details	Expression profiles are altered by experimental conditions including harvest method, in vitro culture method, vehicle used to deliver an agent, time of day of sacrifice, and diet. Up to 9% of the transcripts in mouse liver fluctuated with circadian cycling [147]. These included genes controlling glucose metabolism and vesicle trafficking or cytoskeleton, as might be anticipated from changes in the diet of animals during the day and night. In addition, however, transcript levels of Cyp17 and Cyp2a4, which are important for steroid synthesis, and Cyp2e1, which is important for detoxification of xenobiotics, also fluctuated. These changes might be expected to impact the response to test agents, and reflect a requirement to report the time of day of dosing and sacrifice, along with the diet, vehicle, and harvest and culture methods when summarizing or publishing results of toxicogenomics studies
Commercial database resources for toxicogenomics profiles	Toxicogenomics studies for the purpose of developing commercial databases have been performed by both GeneLogic and Iconix (http://www.genelogic.com/, http://www.iconixpharm.com/). These companies have each gathered data from several hundreds of samples produced from short-term exposures of agents at pharmacological and toxicological dose levels. Customers of either company can access the respective databases to classify the mode-of-action of novel agents of interest
Integration	*Of Toxicogenomics Efforts:* ILSI Committee on the Application of Toxicogenomics to Risk Assessment [56,57], Toxicogenomics Research Consortium (TRC), Consortium on Metabonomics and Toxicology (COMET) [58]. Through such Consortia the technical factors affecting data can be identified and overcome, approaches to data analysis and interpretation can be agreed upon, and high-quality public datasets prepared. The field of toxicoproteomics is currently not represented by a consortium (however, see Human Proteome Organization HUPO in the Online Links Box), although the ILSI Genomics Committee and the TRC are working toxicogenomics consortia in transcriptomics, and COMET is a working toxicogenomics consortium in metabonomics *Of Data Domains:* [39,47,74,75,134]. Integration of data can provide a more complete picture of the expression profiles associated with a particular treatment, shedding light not only on what the cell is planning (transcriptomics) but what occurred in the proteome and metabonome

Note: Peroxisome proliferators are compounds that induce increased numbers of peroxisomes—single-membrane cytoplasmic organelles that metabolize long-chain fatty acids.

14.3.2 Phenotypic Anchoring

Conventional toxicology has employed surrogate markers correlated with toxic responses to monitor adverse outcomes in inaccessible tissues [61]. For example, liver enzymes alanine aminotransferase (ALT) and aspartate aminotransferase (AST) are released following hepatic damage, and levels of these enzymes in serum correlate with histopathological changes in the liver [61,62]. These serum enzyme markers, in conjunction with histopathology, facilitate the "phenotypic anchoring" of molecular expression data [12,16,63]. "Phenotypic anchoring" is the process of determining the relationship between a particular expression profile and the pharmacological or toxicological phenotype of the organism for a particular exposure or dose at a particular time [12]. The dose and time alone are often insufficient to define the toxicity experienced by an individual animal, thus another measure of toxicity is needed for full interpretation of the data obtained during a toxicogenomics study. Conversely, the phenotype alone may be insufficient to anchor the molecular profile, since an elevated value for serum ALT can be observed both before peak toxicity (as it rises) and after peak toxicity (as it returns to baseline). Thus, anchoring the molecular expression profile in phenotype, dose, and time helps to define the sequence of key molecular events in the mode-of-action of a toxicant.

Phenotypic anchoring can also be used in conjunction with lower doses to classify agents and to explore the mechanisms of toxicity that occur before histopathological changes are seen. For example, transcriptional changes that occur following both low- and high-dose exposures of acetaminophen were identified, indicating that biological responses can be detected using transcriptome measurements before histopathological changes are easily detected [33]. Additionally, phenotypic anchoring can help to elucidate a toxicant's mechanism of action; for example, the transcriptional responses in a rat model to superpharmaceutical doses of WAY-144122 (a negative regulator of insulin) were observed before histopathological changes were seen in either the liver or ovary, and reflected different mechanisms of toxicity in the two organs [46].

14.3.3 Biomarkers

Some toxicities lack conventional biomarkers, increasing risk in clinical trials and motivating the search for novel pre-clinical biomarkers to support drug development. A class of lead compounds identified in a discovery program based on gamma secretase inhibition as therapy for Alzheimer's disease also had an undesirable effect of inhibiting cleavage by Notch1 of the Hes1 gene product, a process important for the differentiation of intestinal epithelial cells. Through the use of gene expression profiling and subsequent protein analysis, Searfoss et al. [64] identified adipsin as a novel biomarker for this toxicity.

Carcinogenic potential is conventionally measured using a two-year study, incurring significant expense in both animals and human resources. It is therefore of great interest to identify biomarkers of carcinogenicity that can be detected in acute, short-term studies, and efforts towards this have been reported [24,32,51–53,65,66]. Biomarkers with clinical relevance have also been found using toxicogenomic approaches. For example, Petricoin et al. [67] found a set of protein markers that distinguished patients with high levels of prostate specific antigen (PSA), a clinical marker correlated with prostate cancer, from those with low PSA levels and thus presumed to be healthy. In addition, the marker set also correctly predicted 71% of patients with intermediate PSA levels.

14.3.4 CONSORTIA

The issues facing toxicogenomics are larger than can be solved by scientists independently, and the rapid advancement of the field requires common efforts towards data collection and comparison. Three main collaborative research consortia have been formed principally to standardize measurements and to guide the interpretation of toxicogenomics experiments. These groups of scientists from industry, government, and academic laboratories, as well as from regulatory agencies, were organized by research institutions around a relevant scientific question.

The ILSI/HESI Genomics Committee, the first of these groups, began its work in 1999 and reported its main findings in 2004. These findings included the mechanisms of toxicity of several agents (hepatotoxicants clofibrate and methapyrilene [56], and the nephrotoxicants cisplatin [68], gentamicin and puromycin [69]), the successful applications of toxicogenomics to genotoxicity [65], and the establishment of a collaboration with the European Bioinformatics Institute (EBI) to develop a public toxicogenomics database [6]. The Toxicogenomics Research Consortium (TRC) of the NIEHS National Center for Toxicogenomics (NCT) is engaged in a project to standardize toxicogenomics investigations and to analyze environmental stress responses. In 2003, Consortium for Metabonomics Technology (COMET) reported interim progress towards producing a metabonomics database containing studies of 80 agents [48]. Member laboratories reported data free of inter-laboratory bias, suggesting that the COMET standardized method was robust, and that findings obtained in different laboratories could be subjected to longitudinal data mining for patterns associated with various toxicity endpoints.

The ILSI Genomics Committee also found that microarray results from different laboratories and different platforms were comparable in identifying a common biological response profile, although the responses of individual genes contributing to the pattern differed between platforms [68,70–72]. This, together with the metabonomics reproducibility reported by COMET, is a critically important finding that supports the use of public toxicogenomics databases for meaningful meta-analysis of results obtained in different laboratories. Although some researchers [10,73] are concerned that the capacity to assemble data on drug and toxicant effects using these technologies could result in inappropriate safety and risk decisions, collective efforts such as these will do much to help develop scientific consensus on the appropriate uses of gene expression data.

14.4 INTEGRATION OF DATA

A key objective in toxicogenomics is to integrate data from different studies and analytical platforms to produce a richer and biologically more refined understanding of the toxicological response of a cell, organ, or organism (see Box 14.3). For example, one would like to describe the interplay between protein function and gene expression, or between the activity of certain metabolizing enzymes and the excretion into serum or urine of populations of small metabolites. The integration of data from different domains—such as proteomics and transcriptomics [47,74,75] or transcriptomics and metabonomics [39]—has been reported. In these experiments, tissue samples using different technologies derived from the same individual animals or from comparably treated animals were analyzed in parallel using different technologies. However, the data from different studies were integrated only after a short list of differentially responsive transcripts or protein spots had been derived.

The experience gained from integrating global proteomics or metabonomics data, such as spot intensities from 2-D gels or metabonomics fingerprint data from NMR, tells us that cluster or principle component analysis may be performed to derive global signatures of molecular expression in much the same way as in transcriptomics analyses. If biological samples segregate into unique clusters that show similar expression characteristics, additional efforts can be undertaken to discern the novel proteins or metabolites that are expressed in these samples. Further steps can also be taken to evaluate these proteins or metabolites as potential biomarkers and as a means to determine the underlying toxicological response.

Although software is plentiful for managing expression profiling data at the laboratory level there is a compelling need for public databases that combine profile data with associated biological, chemical and toxicological endpoints [6,76]. Comparisons of gene, protein, and metabolite data within public databases will be valuable for promoting a global understanding of how biological systems function and respond to environmental stressors [58,77]. As these repositories are developed, experiments will be deposited from disparate sources, using different experimental designs, yet targeting the same toxicity endpoint or a similar class of toxicant [78,79]. In these cases, it will be important that the databases integrate data from related studies before performing data mining.

To maximize the value of deposited datasets, the repositories must also be able to integrate data from different technological domains (see Box 14.1 and Box 14.3). Furthermore, a standard representation of data types within each domain is a prerequisite for efficient and accurate storage, access, analysis, comparison, and data exchange. International standards that reach across technological and biological domains are under development by the MGED RSBI Working Groups (see Box 14.2). Additionally, members of regulatory bodies are working with scientists from industry, academic, and government laboratories participating in the ILSI Genomics Committee and CDISC/SEND Consortia to develop standards for the exchange, analysis, and interpretation of transcriptomics data.

A proposal has been made to extend toxicogenomics and combine it with computational approaches such as physiologically-based pharmacokinetic (PB/PK) and pharmacodynamic modelling [16]. PB/PK modeling can be used to derive quantitative estimates of the dose of the test agent or its metabolites that are present in the target tissue at any time after treatment, thereby allowing molecular expression profiles to be anchored to internal dose, as well as to the time of exposure

BOX 14.3 INTEGRATION OF ACETAMINOPHEN (APAP) TOXICOGENOMICS PROFILES

Acetaminophen (APAP, paracetamol) overdose is a leading cause of hospitalization for acute liver failure in the U.S., and its mechanism of toxicity is well-characterized [119]. Data from six toxicogenomics studies are compared as an example of the power of integrating data derived under different conditions and for different purposes. In the figure, genes with altered expression in any of the studies were organized into functional categories. Circles in (a) represent the number of genes in a given category of biological activity identified in the referenced study. The size of a circle is proportional to the number of expressed genes (smallest circles, one gene; largest, 23 genes). The total number of genes represented is 228. The matrix in (b) compares the referenced studies by experimental subject and the technology employed.

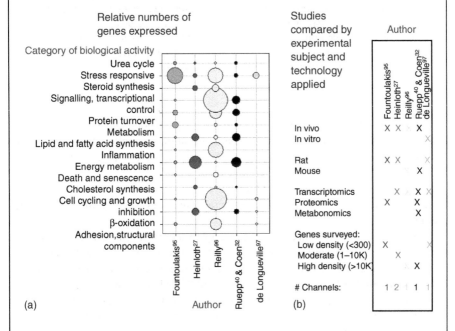

(a)

(b)

1. Fountoulakis et al. [120] performed a proteomics study of a C57BL/6 mouse liver following exposure to either APAP or its non-toxic isomer AMAP. The work was based on a database of 256 identified proteins. Changes in histopathology and the levels of 35 proteins were observed at 8 h post-exposure. Proteins identified were those expected based on knowledge of the mechanism (free-radical based protein adducts). AMAP-related changes were generally a subset of those seen following treatment with APAP.

2. Heinloth et al. [33] used a high-density (6000 gene) two-color cDNA array to observe expression changes in F-344/N rats exposed to one of three APAP doses, at pre-toxicity (6 h), peak toxicity (24 h) or recovery (48 h), as judged by conventional histopathology. Even at low doses, cellular energy loss and occasional mitochondrial damage was observed.

3. Reilly et al. [121] used high-density DNA single-channel oligo arrays (> 11,000 genes) and RNA from livers of C57Bl/6 X 129/Ola hybrid mice 6 h following APAP exposure. Significant alterations in nearly 100 genes from seven major categories of biological activity were detected.

4. Reupp et al. [47] exposed CD-1 mice to subtoxic and toxic doses of APAP, sampled at 15 min–4 h, and then performed microarray (mouse ToxBlots of 450 genes) and 2D-PAGE proteomics characterizations of liver (mitochondrial subfractions). They reported changes in glutathione S-transferase, inflammatory signaling molecules, and mitochondrial proteins within 15 min of exposure, before changes in transcript levels were observed.

5. Coen et al. [59] performed the metabonomics characterization of liver and plasma from AP-1 mice exposed to APAP (from 15 min to 4 h), and then integrated the findings with parallel microarray analysis (high-density oligonucleotide array) of the liver. They reported an increased rate of hepatic glycolysis and changes in lipid and energy metabolism.

6. A study by de Longueville et al. [122] used low density arrays (59 genes) to identify changes in expression seen in rat hepatocytes in vitro after exposure to a variety of toxicants, including APAP.

and the toxicant-induced phenotype. Relationships among gene and metabolite expression can then be described both as a function of the applied dose of an agent, and the ensuing kinetic and dynamic dose–response behaviors that occur in various tissue compartments [80]. Such models also must take into account the fact that the transcriptome, proteome, and metabolome are themselves dynamic systems, and are therefore subject to significant environmental influences, such as time of day and diet [81–83].

Despite the numerous successes of toxicogenomics in the context of toxicology, a poorly addressed but confounding issue pertinent to drug safety and human risk assessment is the impact of the individual genetic background on the response of the individual animal or human patient. The PharmGKB pharmacogenetics knowledgebase [84] catalogues the relationship between different human genetic backgrounds and susceptibility to drug therapy. In addition, the NIEHS Environmental Genome Project (EGP) [14] is identifying SNPs in genes that are important in environmental disease, detoxification, and repair. Linking toxicogenomics knowledgebases with those containing information about SNPs and human susceptibility will gradually lead to a more complete picture of the relevance of the responses and genotypes of surrogate animal species to human risk assessment.

14.5   CHALLENGES AND TECHNICAL CONSIDERATIONS

Predicting potential human health risks from chemical stressors raises three general challenges. These are: the diverse properties of thousands of chemicals and other stressors that are present in the environment, the time and dose parameters that define the relationship between exposure to a chemical and disease, and the genetic and experiential diversity of human populations and of organisms used as surrogates to determine the adverse effects of a toxicant. Figure 14.1 illustrates the effect of genetic susceptibility on the continuum from toxic exposure to disease outcome. Knowledge of this continuum, and the role that genetics has in it, can help us to understand environmentally induced diseases, assess risk, and make public health decisions. Associated with these challenges are others of a more technical nature; these pertain specifically to toxicogenomics studies, and are described below.

14.5.1 Technical Issues

Although genome-wide alterations in mRNA, protein, or metabolite levels in tissue extracts clearly are useful in identifying "signature" gene changes, verifying that one or more gene products are involved in a toxic process depends on knowing the specific cell types in which the target-gene transcripts and products are located. Northern or western blotting, or real-time PCR, are typically used to verify the expression profile of a gene or to selectively analyze its expression as a function of toxicant dose or time of exposure. In situ hybridization, immunohistochemistry, and other techniques must be used to identify the specific cell types that express the gene(s).

The ability to focus molecular expression analysis on only a limited number of cell types depends upon cell separation methods that minimize the opportunity for other cell types to contribute to gene expression in situ. Even the most carefully gathered biological samples contain many cell types, especially if the sample is from inflamed or necrotic tissue. More homogeneous samples are provided by laser capture microdissection (LCM), a method that isolates individual cells or sections of tissue from a fixed sample [85–89]. The use of LCM minimizes contributions by non-target cell populations in comparisons of diseased and normal tissues, but also introduces handling and preparation steps that can affect detection accuracy.

Concomitant with new technology that selectively samples cell populations must come the ability to reliably detect signals from increasingly smaller samples. For example, it will frequently be necessary to amplify mRNA from the same biological sample that was used for transcriptomics analysis. The need to detect weak signals or small but biologically important changes in expression levels remains, as toxicologists explore the initial steps in biological signaling cascades and compensatory processes. At present, cDNA microarray hybridization can detect strong signals within a mixed cell population in samples that are diluted by up to 20-fold [90]. Thus, this technology is likely able to detect a strong signal from a population comprising 5%–10% of the total tissue, but might miss more subtle changes associated with signaling or other initial responses to a stressor. With LCM, a relatively pure cell population could be sampled; thus, the technology would be expected to detect much more subtle changes. For instance, responses seen only in a subpopulation, or asynchronous responses, occurring in 10% of the cells at the time of sampling. The ultimate goal would be the ability to quantify genomic changes occurring in a single cell.

Although mRNA analysis is a powerful tool for recognizing toxicant-induced effects, analysis of protein structure and modification and, more importantly, of global protein expression provides distinct advantages for understanding the functional state of the cell or tissue. Promising new methods are emerging, including the capacity to profile proteins with antibody arrays [91] and surface-enhanced laser desorption mass spectrometry (SELDI, see Box 14.1) [92,93]. Alterations in patterns of mRNA and protein expression in accessible tissues such as serum [20,80,94] may offer new insights into the function of genes in the context of toxicity and guide the search for protein biomarkers of toxicant exposure or predictive toxicity.

14.5.2 BIOINFORMATICS CHALLENGES

Full realization of the potential of molecular profiling in toxicogenomics requires a very substantial investment in bioinformatics in order to extract biological sense from the myriad of interrelated numerical molecular identifiers and their associated annotations. Advances in bioinformatics and mathematical modeling provide powerful approaches for identifying the patterns of biological response that are imbedded in genomic datasets. However, facile interpretation of global molecular datasets derived from -omics technologies is currently constrained by the "bioinformatics bottleneck." Bioinformatics must improve in gene, protein, and metabolite identification and annotation to open the field of toxicogenomics to high-throughput applications in drug development and toxicant evaluation. Several useful resources address the annotation problem by linking identifiers used in genomic databases at the NCBI, EMBL, and DDBJ to other annotation resources (see Online Links Box). Critical to resolving annotation inconsistencies is the knowledge of the sequence of the actual nucleotide or protein that is used to query the genome.

The use of advanced bioinformatics tools to extract information from microarray results [95] is valuable only if the data employed by these tools have a high degree of internal specificity and accuracy [6]. Additionally, the interpretation of molecular expression profiles must emphasize both biological coherence as well as statistical validity when deriving knowledge from toxicogenomics experiments. This means that once a set of genes with altered expression is identified, their biological functions must be ascertained. Mechanistic interpretation of transcript changes may be impeded by the non-standard or imprecise annotation of a sequence element (i.e., gene). Without appropriate synonyms for gene names, the effectiveness of a literature search may be limited. Differences in annotation within and among different microarray platforms may hamper the comparison of results. Such inconsistency frequently arises from annotation resources using different lexicons, or from annotation information being compiled at different times.

Additional bioinformatics and interpretive challenges arise at many levels of biological organization (Figure 14.3). Our current focus and level of understanding of the global molecular landscape encompasses only the lower levels of complexity (genes/proteins, gene/protein groups, functional pathways). The resolution of this knowledge might be termed linear toxicoinformatics, which is the description of environmental stimuli and responses over dose and time following a toxicological stress.

Toxicologists and risk assessors typically define a sequence of key events and linear modes-of-action for environmental chemicals and drugs [96–98]. By contrast,

---

ONLINE LINKS BOX

Some Annotation Resources

Gene Ontology (GO): http://www.geneontology.org

Source: http://source.stanford.edu

EASE: http://david.niaid.nih.gov/david/

MatchMiner: http://discover.nci.nih.gov/matchminer/html/MatchMinerInteractiveLookup.jsp.

Commercial array-specific databases such as Affymetrix's NetAffx: http://www.affymetrix.com/analysis/index.affx.

Multi-species Commercial Microarray Platforms

Affymetrix: http://www.affymetrix.com/index.affx

Agilent: http://www.chem.agilent.com/Scripts/PCol.asp?lPage=494

Groups Working towards Standardization and Exchange Standards for Toxicogenomics Data:

- National Institute of Environmental Health Sciences NCT (http://www.nih.niehs.gov/nct).
- TRC http://www.niehs.nih.gov/nct/trc.htm
- Food and Drug Administration (FDA), CDISC/SEND Consortium (http://www.cdisc.org/)
- EMBL-EBI European Bioinformatics Institute (http://www.ebi.ac.uk/microarray/Projects/ilsi/index.html),
- International Life Sciences Institute's Health and Environmental Sciences Institute (http://hesi.ilsi.org/ and http://hesi.ilsi.org/index.cfm?pubentityid=1) Technical Committee on the Application of Genomics to Mechanism Based Risk Assessment
- Microarray Gene Expression Data (MGED) Society (www.mged.org/).
- MGED Toxicogenomics Working Group (MGED, http://www.mged.org/).
- Human Proteome Organization (http://www.hupo.org/).
- Protein Standard Initiative (PSI) (http://psidev.sourceforge.net/).
- Protein interaction format [148] (http://www.nature.com/cgi-taf/DynaPage.taf?file=/nbt/journal/v22/n2/abs/nbt926.html).
- PEDRo prescribes standards for proteomic data for databases [149] (http://psidev.sourceforge.net/).
- MIAPE, represents Minimum Information About a Proteomics Experiment [118] (http://psidev.sourceforge.net/).
- Standard Metabonomics Reporting Structure (SMRS) Group is developing metabonomics exchange standards (www.smrsgroup.org).

---

the networks and systems level of biological organization may demonstrate highly non-linear cellular expression state changes in response to environmental stimuli [99,100]. Thus, the statistical and bioinformatics-based separation of the complex adaptive, pharmacological, and toxicological responses of drugs, chemicals, and

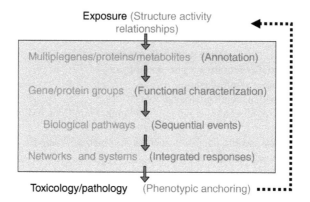

FIGURE 14.3 Bioinformatics challenges and biological complexity. The focus of bioinformatics (red) in interpreting molecular expression data depends on the level of biological complexity (blue)—here shown progressing from genes/proteins/metabolites to networks and systems. For toxicology/pathology, the focus is on phenotypic anchors—observed biological responses that can be related to the chemical structure of the test agent or exposure.

even dietary constituents will probably very much be a matter of degree—reflecting the kinetic and dynamic responses of specific tissues to toxicants as directed by the genome, the genetic heritage of the individual, and that individual's current and prior exposures.

14.6   SYSTEMS TOXICOLOGY

Ideker et al. [22] used the phrase "systems biology" to describe the integrated study of biological systems at the molecular level—involving perturbation of systems, monitoring molecular expression, integrating response data, and modeling the systems' molecular structure and network function. Here we similarly use the phrase "systems toxicology" to describe the toxicogenomics evaluation of biological systems, involving perturbation by toxicants and stressors, monitoring molecular expression and conventional toxicological parameters, and iteratively integrating response data to model the toxicological system [16].

A number of approaches are being developed to model network behavior, with different assumptions, data requirements, and goals. However, it is not likely that toxicogenomics and systems toxicology models will be assembled exclusively from knowledge of cellular components, without equivalent knowledge of the response of these components to toxicants [99]. Thus the "stress testing" of the structural biology of the system and the capture of that data in the context of the functioning organism adapting, surviving, or succumbing to the stress will be required.

Development of a knowledgebase to accurately reflect network-level molecular expression and to facilitate a systems-level biological interpretation requires a new paradigm of data management, data integration, and computational modeling.

BOX 14.4 THE CHEMICAL EFFECTS IN BIOLOGICAL SYSTEMS (CEBS) KNOWLEDGEBASE

To promote a systems biology approach to understanding the biological effects of environmental chemicals and stressors, the CEBS knowledgebase is being developed to house data from many complex data streams in a manner that will allow extensive and complex queries from users. Unified data representation will occur through a systems biology object model (a system for managing diverse -omics and toxicology/pathology data formats) that incorporates current standards for data capture and exchange (CEBS SysBio-OM) [123]. Data streams will include gene expression, protein expression, interaction, and changes in low molecular weight metabolite levels on agents studied, in addition to associated toxicology, histopathology, and pertinent literature [108].

The conceptual design framework for CEBS (see Figure 14.4) is based on functional genomics approaches that have been used successfully for analyzing yeast gene expression datasets [15,16,100]. Because CEBS will contain data on molecular expression, and associated chemical/stressor induced effects in multiple species (e.g., from yeast to humans), it will be possible to derive functional pathway and network information based on cross-species homology. Genomic homology can be tapped within a knowledgebase such as CEBS to gain new understanding in toxicology, as well as in basic biology and genetics.

CEBS will index and sequence-align to the respective genomes all datasets known to the knowledgebase. Thus, changes or differences in the expression patterns of entire genomes at the levels of mRNA, protein, and metabolism can be determined. It will be possible to query CEBS globally, i.e., to "BLAST" [124] the knowledgebase with a profile of interest and have it return information on similar profiles observed under defined experimental conditions of dose, time, and phenotype. CEBS will provide dynamic links to relevant sites such as genome browsers, animal model databases, genetic quantitative trait (QTL) and SNP susceptibility data, and PB/PK and BBDR modeling. Using search routines optimized for parsing known gene/protein groups onto toxicologically relevant pathways and networks, CEBS will automatically survey the literature and integrate this new knowledge with existing knowledgebase annotations. The current status of the CEBS infrastructure and that of other toxicogenomics databases is described in a recent review [6]. These repositories offer the regulatory community reference resources for comparison with toxicogenomics data submitted in the compound registration process [125]. Progress in the development of CEBS can be monitored at http://cebs.niehs.nih.gov/ and http://www.niehs.nih.gov/cebs-df/.

A knowledge base that fully embraces systems toxicology (see Box 14.4) will use precise sequence data to define macromolecules, interaction data based experimentally on co-localization, co-expression and analyses of protein–protein interactions, and functional and phenotypic data based on gene knockouts, knockins, and RNA-interference studies, in addition to studies of responses to chemical, physical, and biological stressors. These data will allow specific molecules to be accurately related to biological phenomena that reflect the normal as well as the stressed cell, tissue, organ, and organism. In the best of circumstances, a systems toxicology approach will build a toxicogenomics understanding from global molecular expression changes that are informed by PBPK/PD modeling and biologically based dose–

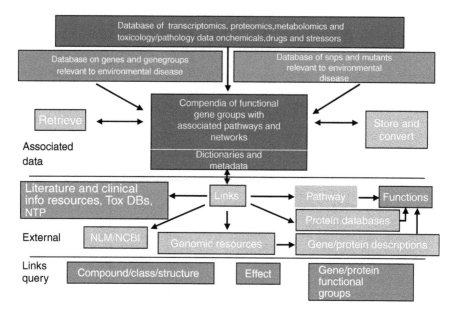

FIGURE 14.4 Conceptual framework for the development of the chemical effects in biological systems (CEBS) knowledgebase. CEBS knowledgebase is a cross-species reference toxicogenomics information system on chemicals/stressors and their effects. The Figure has three sections—the upper section, indicating data associated in CEBS; the center section, indicating external links from CEBS; and the lower section, indicating sample query types that CEBS will support. The boxes in the upper section include primary data (blue), important genetic loci (red), and genetic markers such as SNPs (green). The tasks CEBS will carry out are shown with gray boxes. In the central section, the links to databases are shown in gray, and the links to unstructured data are in green. NTP, National Toxicology Program; NLM, National Library of Medicine; NCBI, National Center for Biotechnology Information.

response (BBDR) modeling. The challenge in constructing a robust systems toxicology knowledgebase is formidable.

14.7 THE FUTURE OF TOXICOGENOMICS

New toxicogenomics methods have the power and potential to revolutionize toxicology. Technological innovations that are already in use permit RNA profiling of formalin-fixed tissues [101], potentially making archived tissues from generations of toxicological studies accessible to gene expression analysis. Methods to array hundreds of toxicologically relevant protein antibodies and to profile hundreds of small molecules in high throughput mode using GC/LC/MS are in development.

Toxicoproteomics research is anticipated to lead to the identification, measurement, and evaluation of proteins and other biomarkers that are more accurate, sensitive, and specific than those available now—and which might be targeted to particular human genetic subpopulations [19,94,102]. Metabonomics research will help to identify small endogenous molecules as important changes in a sequence of key metabolic event; such "metabolite fingerprints" might then help to diagnose and define the ways in which specific chemicals, environmental exposures, or stressors cause disease. This, coupled with the ability to detect damage to particular organs by observing alterations in serum and urine components, is expected to lead to the more sensitive detection of exposure or risk factors [102]. Additional considerations in assessing the toxicogenomic response to environmental exposures are the individual genotype, lifestyle, age, and exposure history [82]. Toxicogenomics and the related field of Nutrigenomics [82,83,103–106] will help to ascertain the degree to which these factors influence the balance between healthy and diseased states.

Toxicogenomics will increase the relevance of toxicology through the global observation of genomic responses with therapeutically and environmentally realistic dose-regimens. It will help to delineate the mode-of-action of various classes of agents and the unique genetic attributes of certain species and population subgroups that render them susceptible to toxicants [15,97]. Studies on strains within a species that are sensitive or resistant to the chemical induction of specific disease phenotypes will be particularly valuable. Extending this thinking to the phylogenetic analysis of both core conserved biological processes [107] and the toxicological responses seen in different species will provide additional comparative insight on genetic susceptibility and probable disease outcomes.

The combined application of the -omics technologies will improve our overall understanding of mechanisms of toxicity and disease etiology as integrated toxicogenomics databases are developed more fully [15]. Data on gene/protein/metabolite changes collected in the context of dose, time, target tissue, and phenotypic severity across species from yeast to nematode to man will provide the comparative information needed to assess the genetic and molecular basis of gene-environment interactions. Toxicology will emerge as an information science that will facilitate scientific discovery across biological species, chemical classes, and disease outcomes [12]. Although there are large challenges in developing public toxicogenomic data repositories, the nucleotide sequence databases—GenBank, EMBL, and DDBJ—provide an excellent example of the benefit of sharing data to the larger scientific and medical community.

Concomitant with development of toxicogenomics databases must be the evolution of bioinformatics methods and data mining tools—and individuals trained to apply them [95]. We believe that a predictive systems toxicology will gradually evolve, aided by knowledge that is systematically generated [26] through literature mining [108,109], comparative analysis, and iterative biological modeling of molecular expression datasets over time. Given the vast numbers and diversity of drugs, chemicals, and environmental agents, and the diversity of species in which they act, we believe, however, that it is only through the development of a comprehensive and public knowledgebase that toxicology and environmental health can rapidly advance. The ultimate goal of the NCT is to create the Chemical Effects in Biological Systems (CEBS) knowledgebase, a public resource (see Box 14.4) that will enable health scientists and practitioners to understand and mitigate or prevent adverse environmental exposures and related diseases in the 21st century.

14.8   ABBREVIATIONS

ALP	alkaline phosphatase
ALT	alanine aminotransferase
AMAP	3-acetamidophenol
AST	aspartate aminotransferase
APAP	acetaminophen, paracetamol
BLAST	Basic Local Alignment Search Tool
CDISC	Clinical Data Interchange Standards Consortium
CEBS	Chemical Effects in Biological Systems
DDBJ	DNA Data Bank of Japan
EBI	European Bioinformatics Institute
EMBL	European Molecular Biology Laboratory
GO	Gene Ontology
ILSI/HESI	Health and Environmental Sciences Institute of the International Life Sciences Institute
LC	liquid chromatography
LCM	laser capture microdissection
LPS	lipopolysaccharide
MAGE-ML	Microarray gene expression-markup language
MAGE-OM	MicroArray Gene Expression-Object Model
MGED	Microarray Gene Expression Database Society
MIAME	Minimum Information About a Microarray Experiment
MS	mass spectrometry
NIEHS	National Institute of Environmental Health Sciences
NCBI	National Center for Biotechnology Information
NCT	National Center for Toxicogenomics
NLM	National Library of Medicine
NMR	nuclear magnetic resonance
NTP	National Toxicology Program
PAGE or 2-D PAGE	2-dimensional polyacrylamide gel electrophoresis
PMA	phorbol myristate acetate
QTL	quantitative trait loci
RT-PCR	real time polymerase chain reaction
SEND	Standards for Exchange of Nonclinical Data
SELDI	surface-enhanced laser desorption ionization

14.9   GLOSSARY

Biologically based dose response (BBDR) modeling   The science of establishing dose–response models based on underlying biological processes.

Biomarker   A pharmacological or physiological measurement that is used to predict a toxic event in an animal.

Chondrocytes   Cartilage cells that produce the structural components of cartilage.

Functional genomics   The development and application of global (genome-wide or system-wide) experimental approaches to assess gene function by making use of information and reagents provided by physical mapping and sequencing of genomes.

Information science   The systematic study and analysis of the sources, development, collection, organization, dissemination, evaluation, use, and management of information in all its forms, including the channels (formal and informal) and technology used in its communication.

Knowledgebase   An archival and computational system that uses data, information, and knowledge captured from experts to carry out tasks that create new information and new understanding.

Lead compounds   Chemicals or drugs that show promise for commercialization.

Longitudinal data mining   The process of locating previously unknown patterns and relationships within data that result from multiple observations of a population of genes, animals, or patients.

Metabolomics or metabolite profiling   The directed use of quantitative analytical methods for analyzing the entire metabolic content of a cell or organism at a given time (the metabolome).

Metabonomics   Techniques that detect changes in the concentration of low molecular weight metabolites present in a cell or organism at a given time (the metabonome) by using nuclear magnetic resonance or mass spectrometry coupled to gas or liquid chromatography.

Mode-of-action   The sequence of events from absorption of a compound into an organism to a toxic outcome or death.

Necrosis   The localized death of living cells.

Nuclear magnetic resonance (NMR)   An analytical chemistry technique that is used to study molecular structure and dynamics; it explores spectrum differences that are caused by the differential alignment of atomic spins in the presence of a strong magnetic field.

Nutrigenomics   The study of the nutritional environment and related cellular/genetic processes at the level of the genome.

Parsing   The process of determining the syntactic structure of a sentence or string of symbols in a language.

Pharmacodynamic (PD) modeling   Involves the development of a mathematical description of a toxicological or disease outcome after therapy.

Physiologically-based pharmacokinetic (PB/PK) modeling   Involves deriving a set of mathematical (differential) equations that are structured to provide a time course of a chemical's mass-balance disposition (wherein all inputs, outputs, and changes in total mass of the chemical are accounted for) in pre-selected anatomical compartments.

Principle component analysis A statistical method that seeks to reduce the dimensionality of a data set by projecting the data onto new axes that align with the variability within the data.

Protein chip A genomic set of proteins that are arrayed on a solid surface without denaturation.

Proteomics A collection of techniques used to measure the structural and functional properties of proteins through the use of 2-dimensional gel electrophoresis or liquid chromatography, typically followed by protein identification using some form of mass spectrometry.

RNA interference An ancient natural antiviral mechanism that directs silencing of gene expression in a sequence-specific manner and can be exploited artificially to inhibit the expression of any gene of interest.

Synoviocytes Cells believed to be responsible for the production of synovial fluid components in joints for absorption from the joint cavity, and for blood/synovial fluid exchanges.

Systems biology The integrated study of biological systems (cells, tissues, organs, or entire organisms) at the molecular level. It involves perturbing systems, monitoring molecular expression, integrating response data, and modeling the molecular structure and network function of the system. See Gene Regulatory Network.

Systems toxicology The study of the perturbation of biological systems by chemicals and stressors, monitoring changes in molecular expression and conventional toxicological parameters, and iteratively integrating response data to describe the functioning organism.

Tandem mass spectrometry The use of two mass spectrometers in series to detect and identify substances based on mass and charge.

Target tissue The tissue or tissues that are damaged as a result of exposure to a toxicant or stressor.

Toxicoinformatics The description of a toxicological stress and the annotation of the dose-dependent molecular responses that are elicited over time.

Transcriptomics Techniques that measure the full complement of activated genes, mRNAs, or transcripts in a particular tissue at a particular time, typically through the use of cDNA or oligonucleotide microarrays.

REFERENCES

1. Aubrecht, J. and Caba, E., Gene expression profile analysis: an emerging approach to investigate mechanisms of genotoxicity, *Pharmacogenomics*, 6, 419–428, 2005.

2. Ekins, S., Systems-ADME/Tox: resources and network approaches, *J. Pharmacol. Toxicol. Methods*, 53, 38–66, 2006.

3. Leighton, J. K., Application of emerging technologies in toxicology and safety assessment: regulatory perspectives, *Int. J. Toxicol.*, 24, 153–155, 2005.

4. Oberemm, A., Onyon, L., and Gundert-Remy, U., How can toxicogenomics inform risk assessment? *Toxicol. Appl. Pharmacol.*, 207(suppl.2), 592–598, 2005.

5. Reynolds, V. L., Applications of emerging technologies in toxicology and safety assessment, *Int. J. Toxicol.*, 24, 135–137, 2005.

6. Mattes, W. B., Pettit, S. D., Sansone, S. A., Bushel, P. R., and Waters, M. D., Database development in toxicogenomics: issues and efforts, *Environ. Health Perspect.*, 112, 495–505, 2004.

7. Aardema, M. J. and MacGregor, J. T., Toxicology and genetic toxicology in the new era of "toxicogenomics": impact of "-omics" technologies, *Mutat. Res.*, 499, 13–25, 2002.

8. Afshari, C. A., Perspective: microarray technology, seeing more than spots, *Endocrinology*, 143, 1983–1989, 2002.

9. Ulrich, R. and Friend, S. H., Toxicogenomics and drug discovery: will new technologies help us produce better drugs?, *Nat. Rev. Drug Discov.*, 1, 84–88, 2002.

10. Fielden, M. R. and Zacharewski, T. R., Challenges and limitations of gene expression profiling in mechanistic and predictive toxicology, *Toxicol. Sci.*, 60, 6–10, 2001.

11. Hamadeh, H. K., Amin, R. P., Paules, R. S., and Afshari, C. A., An overview of toxicogenomics, *Curr. Issues Mol. Biol.*, 4, 45–56, 2002.

12. Tennant, R. W., The national center for toxicogenomics: using new technologies to inform mechanistic toxicology, *Environ. Health Perspect.*, 110, A8–A10, 2002.

13. Thomas, R. S., Rank, D. R., Penn, S. G., Zastrow, G. M., Hayes, K. R., Pande, K., Glover, E. et al., Identification of toxicologically predictive gene sets using cDNA microarrays, *Mol. Pharmacol.*, 60, 1189–1194, 2001.

14. Olden, K. and Guthrie, J., Genomics: implications for toxicology, *Mutat. Res.*, 473, 3–10, 2001.

15. Waters, M. D., Olden, K., and Tennant, R. W., Toxicogenomic approach for assessing toxicant-related disease, *Mutat. Res.*, 544, 415–424, 2003.

16. Waters, M. D., Boorman, G., Bushel, P., Cunningham, M., Irwin, R., Merrick, A., Olden, K. et al., Systems toxicology and the chemical effects in biological systems knowledge base, *Environ. Health Perspect.*, 111, 811–824, 2003.

17. Lobenhofer, E. K., Bushel, P. R., Afshari, C. A., and Hamadeh, H. K., Progress in the application of DNA microarrays, *Environ. Health Perspect.*, 109, 881–891, 2001.

18. Burchiel, S. W. et al., Analysis of genetic and epigenetic mechanisms of toxicity: potential roles of toxicogenomics and proteomics in toxicology, *Toxicol. Sci.*, 59, 193–195, 2001.

19. Merrick, B. A. and Tomer, K. B., Toxicoproteomics: a parallel approach to identifying biomarkers, *Environ. Health Perspect.*, 111, A578–A579, 2003.

20. Petricoin, E. F., Rajapaske, V., Herman, E. H., Arekani, A. M., Ross, S., Johann, D., Knapton, A. et al., Toxicoproteomics: serum proteomic pattern diagnostics for early detection of drug induced cardiac toxicities and cardioprotection, *Toxicol. Pathol.*, 32(suppl.1), 122–130, 2004.

21. Wilkins, M. R., Pasquali, C., Appel, R. D., Ou, K., Golaz, O., Sanchez, J. C., Yan, J. X. et al., From proteins to proteomes: large scale protein identification by two-dimensional electrophoresis and amino acid analysis, *Biotechnology (N Y)*, 14, 61–65, 1996.

22. Ideker, T., Galitski, T., and Hood, L., A new approach to decoding life: systems biology, *Annu. Rev. Genomics Hum. Genet.*, 2, 343–372, 2001.

23. Nurse, P., Understanding cells, *Nat. Biotechnol.*, 424, 883, 2003.

24. Hamadeh, H. K., Knight, B. L., Haugen, A. C., Sieber, S., Amin, R. P., Bushel, P. R., Stoll, R. et al., Methapyrilene toxicity: anchorage of pathologic observations to gene expression alterations, *Toxicol. Pathol.*, 30, 470–482, 2002.

25. Gant, T. W. and Zhang, S. D., In pursuit of effective toxicogenomics, *Mutat. Res.*, 575, 4–16, 2005.

26. Zweiger, G., Knowledge discovery in gene-expression–microarray data: mining the information output of the genome, *Trends Biotechnol.*, 17, 429–436, 1999.

27. Ruepp, S. et al., Assessment of hepatotoxic liabilities by transcript profiling, *Toxicol. Appl. Pharmacol.*, 207(suppl.2), 161–170, 2005.

28. Watson, W. P. and Mutti, A., Role of biomarkers in monitoring exposures to chemicals: present position, future prospects, *Biomarkers*, 9, 211–242, 2004.

29. Thukral, S. K., Nordone, P. J., Hu, R., Sullivan, L., Galambos, E., Fitzpatrick, V. D., Healy, L., Bass, M. B., Cosenza, M. E., and Afshari, C. A., Prediction of nephrotoxicant action and identification of candidate toxicity-related biomarkers, *Toxicol. Pathol.*, 33, 343–355, 2005.

30. Waring, J. F., Jolly, R. A., Ciurlionis, R., Lum, P. Y., Praestgaard, J. T., Morfitt, D. C., Buratto, B., Roberts, C., Schadt, E., and Ulrich, R. G., Clustering of hepatotoxins based on mechanism of toxicity using gene expression profiles, *Toxicol. Appl. Pharmacol.*, 175, 28–42, 2001.

31. Mortuza, G. B., Neville, W. A., Delaney, J., Waterfield, C. J., and Camilleri, P., Characterisation of a potential biomarker of phospholipidosis from amiodarone-treated rats, *Biochim. Biophys. Acta*, 1631, 136–146, 2003.

32. Kramer, J. A., Curtiss, S. W., Kolaja, K. L., Alden, C. L., Blomme, E. A. G., Curtiss, W. C., Davila, J. C., Jackson, C. J., and Bunch, R. T., Acute molecular markers of rodent hepatic carcinogenesis identified by transcription profiling, *Chem. Res. Toxicol.*, 17, 463–470, 2004.

33. Heinloth, A. N., Irwin, R. D., Boorman, G. A., Nettesheim, P., Fannin, R. D., Sieber, S. O., Snell, M. L. et al., Gene expression profiling of rat livers reveals indicators of potential adverse effects, *Toxicol. Sci.*, 80, 193–202, 2004.

34. Hamadeh, H. K., Bushel, P. R., Jayadev, S., DiSorbo, O., Bennett, L., Li, L., Tennant, R. et al., Prediction of compound signature using high density gene expression profiling, *Toxicol. Sci.*, 67, 232–240, 2002.

35. Bulera, S. J., Eddy, S. M., Ferguson, E., Jatkoe, T. A., Reindel, J. F., Bleavins, M. R., and De La Iglesia, F. A., RNA expression in the early characterization of hepatotoxicants in Wistar rats by high-density DNA microarrays, *Hepatology*, 33, 1239–1258, 2001.

36. Bartosiewicz, M. J., Jenkins, D., Penn, S., Emery, J., and Buckpitt, A., Unique gene expression patterns in liver and kidney associated with exposure to chemical toxicants, *J. Pharmacol. Exp. Ther.*, 297, 895–905, 2001.

37. Verhoeckx, K. C., Bijlsma, S., de Groene, E. M., Witkamp, R. F., van der Greef, J., and Rodenburg, R. J., A combination of proteomics, principal component analysis and transcriptomics is a powerful tool for the identification of biomarkers for macrophage maturation in the U937 cell line, *Proteomics*, 4, 1014–1028, 2004.

38. Cutler, P., Bell, D. J., Birrell, H. C., Connelly, J. C., Connor, S. C., Holmes, E., Mitchell, B. C. et al., An integrated proteomic approach to studying glomerular nephrotoxicity, *Electrophoresis*, 20, 3647–3658, 1999.

39. Coen, M., Ruepp, S. U., Lindon, J. C., Nicholson, J. K., Pognan, F., Lenz, E. M., and Wilson, I. D., Integrated application of transcriptomics and metabonomics yields new insight into the toxicity due to paracetamol in the mouse, *J. Pharm. Biomed. Anal.*, 35, 93–105, 2004.

40. Donald, S., Verschoyle, R. D., Edwards, R., Judah, D. J., Davies, R., Riley, J., Dinsdale, D. et al., Hepatobiliary damage and changes in hepatic gene expression caused by the antitumor drug ecteinascidin-743 (ET-743) in the female rat, *Cancer Res.*, 62, 4256–4262, 2002.

41. Fertuck, K. C., Eckel, J. E., Gennings, C., and Zacharewski, T. R., Identification of temporal patterns of gene expression in the uteri of immature, ovariectomized mice following exposure to ethynylestradiol, *Physiol. Genomics*, 15, 127–141, 2003.

42. Fountoulakis, M., de Vera, M. C., Crameri, F., Boess, F., Gasser, R., Albertini, S., and Suter, L., Modulation of gene and protein expression by carbon tetrachloride in the rat liver, *Toxicol. Appl. Pharmacol.*, 183, 71–80, 2002.

43. Hamadeh, H. K., Bushel, P. R., Jayadev, S., Martin, K., DiSorbo, O., Sieber, S., Bennett, L. et al., Gene expression analysis reveals chemical-specific profiles, *Toxicol. Sci.*, 67, 219–231, 2002.

44. Huang, Q., Dunn, R. T., 2nd, Jayadev, S., DiSorbo, O., Pack, F. D., Farr, S. B., Stoll, R. E., and Blanchard, K. T., Assessment of cisplatin-induced nephrotoxicity by microarray technology, *Toxicol. Sci.*, 63, 196–207, 2001.

45. Kramer, J. A., Blomme, E. A., Bunch, R. T., Davila, J. C., Jackson, C. J., Jones, P. F., Kolaja, K. L., and Curtiss, S. W., Transcription profiling distinguishes dose-dependent effects in the livers of rats treated with clofibrate, *Toxicol. Pathol.*, 31, 417–431, 2003.

46. Peterson, R. L., Casciotti, L., Block, L., Goad, M. E., Tong, Z., Meehan, J. T., Jordan, R. A. et al., Mechanistic toxicogenomic analysis of WAY-144122 administration in Sprague-Dawley rats, *Toxicol. Appl. Pharmacol.*, 196, 80–94, 2004.

47. Ruepp, S. U., Tonge, R. P., Shaw, J., Wallis, N., and Pognan, F., Genomics and proteomics analysis of acetaminophen toxicity in mouse liver, *Toxicol. Sci.*, 65, 135–150, 2002.

48. Waring, J. F., Gum, R., Morfitt, D., Jolly, R. A., Ciurlionis, R., Heindel, M., Gallenberg, L., Buratto, B., and Ulrich, R. G., Identifying toxic mechanisms using DNA microarrays: evidence that an experimental inhibitor of cell adhesion molecule expression signals through the aryl hydrocarbon nuclear receptor, *Toxicology*, 181–182, 537–550, 2002.

49. Wagenaar, G. T., ter Horst, S. A., van Gastelen, M. A., Leijser, L. M., Mauad, T., van der Velden, P. A., de Heer, E., Hiemstra, P. S., Poorthuis, B. J., and Walther, F. J., Gene expression profile and histopathology of experimental bronchopulmonary dysplasia induced by prolonged oxidative stress, *Free Radic. Biol. Med.*, 36, 782–801, 2004.

50. Lu, T., Liu, J., LeCluyse, E. L., Zhou, Y. S., Cheng, M. L., and Waalkes, M. P., Application of cDNA microarray to the study of arsenic-induced liver diseases in the population of Guizhou, China, *Toxicol. Sci.*, 59, 185–192, 2001.

51. Hamadeh, H. K., Jayadev, S., Gaillard, E. T., Huang, Q., Stoll, R., Blanchard, K., Chou, J. et al., Integration of clinical and gene expression endpoints to explore furan-mediated hepatotoxicity, *Mutat. Res.*, 549, 169–183, 2004.

52. Iida, M., Anna, C. H., Hartis, J., Bruno, M., Wetmore, B., Dubin, J. R., Sieber, S. et al., Changes in global gene and protein expression during early mouse liver carcinogenesis induced by non-genotoxic model carcinogens oxazepam and Wyeth-14,643, *Carcinogenesis*, 24, 757–770, 2003.

53. Ellinger-Ziegelbauer, H., Stuart, B., Wahle, B., Bomann, W., and Ahr, H. J., Characteristic expression profiles induced by genotoxic carcinogens in rat liver, *Toxicol. Sci.*, 77, 19–34, 2004.

54. Twigger, S., Lu, J., Shimoyama, M., Chen, D., Pasko, D., Long, H., Ginster, J. et al., Rat Genome Database (RGD): mapping disease onto the genome, *Nucleic Acids Res.*, 30, 125–128, 2002.

55. Higgins, M. A., Berridge, B. R., Mills, B. J., Schultze, A. E., Gao, H., Searfoss, G. H., Baker, T. K., and Ryan, T. P., Gene expression analysis of the acute phase response using a canine microarray, *Toxicol. Sci.*, 74, 470–484, 2003.

56. Ulrich, R. G., Rockett, J. C., Gibson, G. G., and Pettit, S. D., Overview of an interlaboratory collaboration on evaluating the effects of model hepatotoxicants on hepatic gene expression, *Environ. Health Perspect.*, 112, 423–427, 2004.

57. Pennie, W., Pettit, S. D., and Lord, P. G., Toxicogenomics in risk assessment: an overview of an HESI collaborative research program, *Environ. Health Perspect.*, 112, 417–419, 2004.

58. Lindon, J. C., Nicholson, J. K., Holmes, E., Antti, H., Bollard, M. E., Keun, H., Beckonert, O. et al., Contemporary issues in toxicology the role of metabonomics in toxicology and its evaluation by the COMET project, *Toxicol. Appl. Pharmacol.*, 187, 137–146, 2003.

59. Nuwaysir, E. F., Bittner, M., Trent, J., Barrett, J. C., and Afshari, C. A., Microarrays and toxicology: the advent of toxicogenomics, *Mol. Carcinog.*, 24, 153–159, 1999.

60. Heller, R. A., Schena, M., Chai, A., Shalon, D., Bedilion, T., Gilmore, J., Woolley, D. E., and Davis, R. W., Discovery and analysis of inflammatory disease-related genes using cDNA microarrays, *Proc. Natl. Acad. Sci. U.S.A.*, 94, 2150–2155, 1997.

61. Loeb, W. F. and Quimby, F. W., Eds., *The Clinical Chemistry of Laboratory Animals*, Taylor and Francis, Philadelphia, PA, 1999.

62. Travlos, G. S., Morris, R. W., Elwell, M. R., Duke, A., Rosenblum, S., and Thompson, M. B., Frequency and relationships of clinical chemistry and liver and kidney histopathology findings in 13-week toxicity studies in rats, *Toxicology*, 107, 17–29, 1996.

63. Paules, R., Phenotypic anchoring: linking cause and effect, *Environ. Health Perspect.*, 111, A338–A339, 2003.

64. Searfoss, G. H. et al., Adipsin, a biomarker of gastrointestinal toxicity mediated by a functional gamma-secretase inhibitor, *J. Biol. Chem.*, 278, 46107–46116, 2003.

65. Newton, R. K., Aardema, M., and Aubrecht, J., The utility of DNA microarrays for characterizing genotoxicity, *Environ. Health Perspect.*, 112, 420–422, 2004.

66. van Delft, J. H., van Agen, E., van Breda, S. G., Herwijnen, M. H., Staal, Y. C., and Kleinjans, J. C., Comparison of supervised clustering methods to discriminate genotoxic from non-genotoxic carcinogens by gene expression profiling, *Mutat Res.*, 575, 17–33, 2005.

67. Petricoin, E. F., 3rd., Ornstein, D. K., Paweletz, C. P., Ardekani, A., Hackett, P. S., Hitt, B. A., Velassco, A. et al., Serum proteomic patterns for detection of prostate cancer, *J. Natl. Cancer Inst.*, 94, 1576–1578, 2002.

68. Thompson, K. L., Afshari, C. A., Amin, R. P., Bertram, T. A., Car, B., Cunningham, M., Kind, Cl. et al., Identification of platform-independent gene expression markers of cisplatin nephrotoxicity, *Environ. Health Perspect.*, 112, 488–494, 2004.

69. Kramer, J. A., Pettit, S. D., Amin, R. P., Bertram, T. A., Car, B., Cunningham, M., Curtiss, S. W. et al., Overview on the application of transcription profiling using selected nephrotoxicants for toxicology assessment, *Environ. Health Perspect.*, 112, 460–464, 2004.

70. Baker, V. A., Harries, H. M., Waring, J. F., Duggan, C. M., Ni, H. A., Jolly, R. A., Yoon, L. W. et al., Clofibrate-induced gene expression changes in rat liver: a cross-laboratory analysis using membrane cDNA arrays, *Environ. Health Perspect.*, 112, 428–438, 2004.

71. Chu, T. M., Deng, S., Wolfinger, R., Paules, R. S., and Hamadeh, H. K., Cross-site comparison of gene expression data reveals high similarity, *Environ. Health Perspect.*, 112, 449–455, 2004.

72. Waring, J. F., Ulrich, R. G., Flint, N., Morfitt, D., Kalkuhl, A., Staedtler, F., Lawton, M., Beekman, J. M., and Suter, L., Interlaboratory evaluation of rat hepatic gene expression changes induced by methapyrilene, *Environ. Health Perspect.*, 112, 439–448, 2004.

73. Smith, L. L., Key challenges for toxicologists in the 21st century, *Trends Pharmacol. Sci.*, 22, 281–285, 2001.

74. Hogstrand, C., Balesaria, S., and Glover, C. N., Application of genomics and proteomics for study of the integrated response to zinc exposure in a non-model fish species, the rainbow trout, *Comp. Biochem. Physiol. B Biochem. Mol. Biol.*, 133, 523–535, 2002.

75. Juan, H.-F., Lin, J. Y.-C., Chang, W.-H., Wu, C.-Y., Pan, T.-L., Tseng, M.-J., Khoo, K.-H., and Chen, S.-T., Biomic study of human meyloid leukemia cells differentiation to macrophages using DNA array, proteomic, and bioinformatic analytical methods, *Electrophoresis*, 23, 2490–2504, 2002.

76. Hayes, K. R. and Bradfield, C. A., Advances in toxicogenomics, *Chem. Res. Toxicol.*, 18, 403–414, 2005.

77. Amin, R. P., Hamadeh, H. K., Bushel, P. R., Bennett, L., Afshari, C. A., and Paules, R. S., Genomic interrogation of mechanism(s) underlying cellular responses to toxicants, *Toxicology*, 181–182, 555–563, 2002.

78. Raghavan, N., Amaratunga, D., Nie, A. Y., and McMillian, M., Class prediction in toxicogenomics, *J. Biopharm. Stat.*, 15, 327–341, 2005.

79. McMillian, M., Nie, A., Parker, J. B., Leone, A., Kemmerer, M., Bryant, S., Herlich, J. et al., Drug-induced oxidative stress in rat liver from a toxicogenomics perspective, *Toxicol. Appl. Pharmacol.*, 207(suppl.2), 171–178, 2005.

80. Merrick, B. A. and Madenspacher, J. H., Complementary gene and protein expression studies and integrative approaches in toxicogenomics, *Toxicol. Appl. Pharmacol.*, 207(suppl.2), 189–194, 2005.

81. Kita, Y. et al., Implications of circadian gene expression in kidney, liver and the effects of fasting on pharmacogenomic studies, *Pharmacogenetics*, 12, 55–65, 2002.

82. Kaput, J. and Rodriguez, R. L., Nutritional genomics: the next frontier in the postgenomic era, *Physiol. Genomics*, 16, 166–177, 2004.

83. Kaput, J., Diet-disease gene interactions, *Nutrition*, 20, 26–31, 2004.

84. Klein, T. E. and Altman, R. B., PharmGKB: the pharmacogenetics nad pharmacogenomics knowledge base, *Pharmacogenomics J.*, 4, 1, 2004.

85. Wittliff, J. L. and Erlander, M. G., Laser capture microdissection and its applications in genomics and proteomics, *Methods Enzymol.*, 356, 12–25, 2002.

86. Jain, K. K., Application of laser capture microdissection to proteomics, *Methods Enzymol.*, 356, 157–167, 2002.

87. Emmert-Buck, M. R. et al., Laser capture microdissection, *Science*, 274, 998–1001, 1996.

88. Bonner, R. F., Emmert-Buck, M., Cole, K., Pohida, T., Chuaqui, R., Goldstein, S., and Liotta, L. A., Laser capture microdissection: molecular analysis of tissue, *Science*, 278, 1481–1483, 1997.

89. Karsten, S. L., Van Deerlin, V. M., Sabatti, C., Gill, L. H., and Geschwind, D. H., An evaluation of tyramide signal amplification and archived fixed and frozen tissue in microarray gene expression analysis, *Nucleic Acids Res.*, 30, E4, 2002.

90. Hamadeh, H. K. et al., Detection of diluted gene expression alterations using cDNA microarrays, *Biotechniques*, 32, 322–324, 2002, (See also 326–329).

91. Huang, R. P., Detection of multiple proteins in an antibody-based protein microarray system, *J. Immunol. Methods*, 255, 1–13, 2001.

92. Merchant, M. and Weinberger, S. R., Recent advancements in surface-enhanced laser desorption/ionization-time of flight-mass spectrometry, *Electrophoresis*, 21, 1164–1177, 2000.

93. Liotta, L. and Petricoin, E., Molecular profiling of human cancer, *Nat. Rev. Genet.*, 1, 48–56, 2000.

94. Merrick, B. A. and Bruno, M. E., Genomic and proteomic profiling for biomarkers and signature profiles of toxicity, *Curr. Opin. Mol. Ther.*, 6, 600–607, 2004.

95. Quackenbush, J., Computational analysis of microarray data, *Nat. Rev. Genet.*, 2, 418–427, 2001.

96. Farland, W. H., The U.S. environmental protection agency's risk assessment guidelines: current status and future directions, *Toxicol. Ind. Health*, 8, 205–212, 1992.

97. Farland, W. H., Cancer risk assessment: evolution of the process, *Prev. Med.*, 25, 24–25, 1996.

98. Larsen, J. C., Farland, W., and Winters, D., Current risk assessment approaches in different countries, *Food Addit. Contam.*, 17, 359–369, 2000.

99. Begley, T. J., Rosenbach, A. S., Ideker, T., and Samson, L. D., Damage recovery pathways in saccharomyces cerevisiae revealed by genomic phenotyping and interactome mapping, *Mol. Cancer Res.*, 1, 103–112, 2002.

100. Hughes, T. R., Marton, M. J., Jones, A. R., Roberts, C. J., Stoughton, R., Armour, C. D., Bennett, H. A. et al., Functional discovery via a compendium of expression profiles, *Cell*, 102, 109–126, 2000.

101. Lewis, F., Maughan, N. J., Smith, V., Hillan, K., and Quirke, P., Unlocking the archive—gene expression in paraffin-embedded tissue, *J. Pathol.*, 195, 66–71, 2001.

102. Wetmore, B. A. and Merrick, B. A., Toxicoproteomics: proteomics applied to toxicology and pathology, *Toxicol. Pathol.*, 32, 619–642, 2004.

103. German, J. B., Genetic dietetics: nutrigenomics and the future of dietetics practice, *J. Am. Diet. Assoc.*, 105, 530–531, 2005.

104. Ordovas, J. M. and Corella, D., Nutritional genomics, *Annu. Rev. Genomics Hum. Genet.*, 5, 71–118, 2004.

105. van Ommen, B., Nutrigenomics: exploiting systems biology in the nutrition and health arenas, *Nutrition*, 20, 4–8, 2004.

106. Fenech, M., The genome health clinic and genome health nutrigenomics concepts: diagnosis and nutritional treatment of genome and epigenome damage on an individual basis, *Mutagenesis*, 20, 255–269, 2005.

107. Stuart, G. W. and Berry, M. W., A comprehensive whole genome bacterial phylogeny using correlated peptide motifs defined in a high dimensional vector space, *J. Bioinform. Comput. Biol.*, 1, 475–493, 2003.

108. Chaussabel, D. and Sher, A., Mining microarray expression data by literature profiling, *Genome Biol.*, 3, RESEARCH0055, 2002.

109. Sluka, J. P., Extracting knowledge from genomic experiments by incorporating the biomedical literature, In *Methods of Microarray Data Analysis II*, Lin, S. M. and Johnson, K. F. Eds., Kluwer Academic Publishers, Boston, MA, pp. 195–209, 2002.

110. Patterson, S. D. and Aebersold, R. H., Proteomics: the first decade and beyond, *Nat. Genet.*, 33 (suppl.), 311–323, 2003.

111. Nicholson, J. K., Connelly, J., Lindon, J. C., and Holmes, E., Metabonomics: a platform for studying drug toxicity and gene function, *Nat. Rev. Drug Discov.*, 1, 153–161, 2002.

112. Edgar, R., Domrachev, M., and Lash, A. E., Gene expression omnibus: NCBI gene expression and hybridization array data repository, *Nucleic Acids Res.*, 30, 207–210, 2002.

113. Mattingly, C. J., Colby, G. T., Rosenstein, M. C., Forrest, J. N., Jr., and Boyer, J. L., Promoting comparative molecular studies in environmental health research: an overview of the comparative toxicogenomics database (CTD), *Pharmacogenomics J.*, 4, 5–8, 2004.

114. Brazma, A., Parkinson, H., Sarkans, U., Shojatalab, M., Vilo, J., Abeygunawardena, N., Holloway, E. et al., ArrayExpress—a public repository for microarray gene expression data at the EBI, *Nucleic Acids Res.*, 31, 68–71, 2003.

115. Ball, C. A., Sherlock, G., Parkinson, H., Rocca-Sera, P., Brooksbank, C., Causton, H. C., Cavalieri, D. et al., Microarray Gene Expression Data (MGED) Society. Standards for microarray data., Science, 298, 539, 2002.

116. Brazma, A., Hingamp, P., Quackenbush, J., Sherlock, G., Spellman, P., Stoeckert, C., Aach, J. et al., Minimum information about a microarray experiment (MIAME)-toward standards for microarray data, *Nat. Genet.*, 29, 365–371, 2001.

117. Ikeo, K., Ishi-i, J., Tamura, T., Gojobori, T., and Tateno, Y., CIBEX: center for information biology gene expression database, *C R Biol.*, 326, 1079–1082, 2003.

118. Orchard, S., Hermjakob, H., Julian, R. K., Jr., Runte, K., Sherman, D., Wojcik, J., Zhu, W., and Apweiler, R., Common interchange standards for proteomics data: Public availability of tools and schema, *Proteomics*, 4, 490–491, 2004.

119. Bessems, J. G. and Vermeulen, N. P., Paracetamol (acetaminophen)-induced toxicity: molecular and biochemical mechanisms, analogues and protective approaches, *Crit. Rev. Toxicol.*, 31, 55–138, 2001.

120. Fountoulakis, M., Berndt, P., Boelsterli, U. A., Crameri, F., Winter, M., Albertini, S., and Suter, L., Two-dimensional database of mouse liver proteins: changes in hepatic protein levels following treatment with acetaminophen or its nontoxic regioisomer 3-acetamidophenol, *Electrophoresis*, 21, 2148–2161, 2000.

121. Reilly, T. P, Bourdi, M., Brady, J. N., Pise-Masison, C. A., Radonovich, M. F., George, J. W., Pohl, L. R., Expression profiling of acetaminophen liver toxicity in mice using microarray technology, *Biochem. Biophys. Res. Commun.*, 282, 321–328, 2001. Erratum in: *Biochem. Biophys. Res. Commun.*, 283 (2), 536, 2001.

122. de Longueville, F., Atienzar, F. A., Marcq, L., Dufrane, S., Evrard, S., Wouters, L., Leroux, F. et al., Use of a low-density microarray for studying gene expression patterns induced by hepatotoxicants on primary cultures of rat hepatocytes, *Toxicol. Sci.*, 75, 378–392, 2003.

123. Xirasagar, S., Gustafson, S., Merrick, B. A., Tomer, K. B., Stasiewicz, S., Chan, D. D., Yost, K. J., 3rd. et al., CEBS object model for systems biology data, SysBio-OM, *Bioinformatics*, 20, 2004–2015, 2004.

124. Altschul, S. F., Gish, W., Miller, W., Myers, E. W., and Lipman, D. J., Basic local alignment search tool, *J. Mol. Biol.*, 215, 403–410, 1990.

125. Petricoin, E. F., 3rd., Hackett, J. L., Lesko, L. J., Puri, R. K., Gutman, S. I., Chumakov, K., Woodcock, J., Feigal, D. W., Jr., Zoon, K. C., and Sistare, F. D., Medical applications of microarray technologies: a regulatory science perspective, *Nat. Genet.*, 32(suppl.), 474–479, 2002.

126. Bartosiewicz, M., Trounstine, M., Barker, D., Johnston, R., and Buckpitt, A., Development of a toxicological gene array and quantitative assessment of this technology, *Arch. Biochem. Biophys.*, 376, 66–73, 2000.

127. Jiang, X. S., Zhou, H., Zhang, L., Sheng, Q. H., Li, S. J., Li, L., Hao, P. et al., A high-throughput approach for subcellular proteome: identification of rat liver proteins using subcellular fractionation coupled with two-dimensional liquid chromatography tandem mass spectrometry and bioinformatic analysis, *Mol. Cell Proteomics*, 3, 441–455, 2004.

128. Vitorino, R., Lobo, M. J., Ferrer-Correira, A. J., Dubin, J. R., Tomer, K. B., Domingues, P. M., and Amado, F. M., Identification of human whole saliva protein components using proteomics, *Proteomics*, 4, 1109–1115, 2004.

129. Talamo, F., D'Ambrosio, C., Arena, S., Del Vecchio, P., Ledda, L., Zehender, G., Ferrara, L., and Scaloni, A., Proteins from bovine tissues and biological fluids: defining a reference electrophoresis map for liver, kidney, muscle, plasma and red blood cells, *Proteomics*, 3, 440–460, 2003.

130. Reddy, P. H., McWeeney, S., Park, B. S., Manczak, M., Gutala, R. V., Partovi, D., Jung, Y. et al., Gene expression profiles of transcripts in amyloid precursor protein transgenic mice: up-regulation of mitochondrial metabolism and apoptotic genes is an early cellular change in Alzheimer's disease, *Hum. Mol. Genet.*, 13, 1225–1240, 2004.

131. Heijne, W. H., Stierum, R. H., Slijper, M., van Bladeren, P. J., and van Ommen, B., Toxicogenomics of bromobenzene hepatotoxicity: a combined transcriptomics and proteomics approach, *Biochem. Pharmacol.*, 65, 857–875, 2003.

132. Weiss, A., Delproposto, J., and Giroux, C. N., High-throughput phenotypic profiling of gene-environment interactions by quantitative growth curve analysis in Saccharomyces cerevisiae, *Anal. Biochem.*, 327, 23–34, 2004.

133. Boess, F., Kamber, M., Romer, S., Gasser, R., Muller, D., Albertini, S., and Suter, L., Gene expression in two hepatic cell lines, cultured primary hepatocytes, and liver slices compared to the in vivo liver gene expression in rats: possible implications for toxicogenomics use of in vitro systems, *Toxicol. Sci.*, 73, 386–402, 2003.

134. Heijne, W. H., Slitt, A. L., van Bladeren, P. J., Groten, J. P., Klaassen, C. D., Stierum, R. H., and van Ommen, B., Bromobenzene-induced hepatotoxicity at the transcriptome level, *Toxicol. Sci.*, 79, 411–422, 2004.

135. Lee, W. M., Drug-induced hepatotoxicity, *N. Engl. J. Med.*, 349, 474–485, 2003.

136. Xie, T., Tong, L., Barrett, T., Yuan, J., Hatzidimitriou, G., McCann, U. D., Becker, K. G., Donovan, D. M., and Ricaurte, G. A., Changes in gene expression linked to methamphetamine-induced dopaminergic neurotoxicity, *J. Neurosci.*, 22, 274–283, 2002.

137. Dam, K., Seidler, F. J., and Slotkin, T. A., Transcriptional biomarkers distinguish between vulnerable periods for developmental neurotoxicity of chlorpyrifos: implications for toxicogenomics, *Brain Res. Bull.*, 59, 261–265, 2003.

138. Hamadeh, H. K., Trouba, K. J., Amin, R. P., Afshari, C. A., and Germolec, D., Coordination of altered DNA repair and damage pathways in arsenite-exposed keratinocytes, *Toxicol. Sci.*, 69, 306–316, 2002.

139. Hu, D., Cao, K., Peterson-Wakeman, R., and Wang, R., Altered profile of gene expression in rat hearts induced by chronic nicotine consumption, *Biochem. Biophys. Res. Commun.*, 297, 729–736, 2002.

140. Waring, J. F., Ciurlionis, R., Jolly, R. A., Heindel, M., and Ulrich, R. G., Microarray analysis of hepatotoxins in vitro reveals a correlation between gene expression profiles and mechanisms of toxicity, *Toxicol. Lett.*, 120, 359–368, 2001.

141. Hamadeh, H. K., Bushel, P., Paules, R., and Afshari, C. A., Discovery in toxicology: mediation by gene expression array technology, *J. Biochem. Mol. Toxicol.*, 15, 231–242, 2001.

142. Witzmann, F. A., Bobb, A., Briggs, G. B., Coppage, H. N., Hess, R. A., Li, J., Pedrick, N. M., Ritchie, G. D., Iii, J. R., and Still, K. R., Analysis of rat testicular protein expression following 91-day exposure to JP-8 jet fuel vapor, *Proteomics*, 3, 1016–1027, 2003.

143. Heijne, W. H., Jonker, D., Stierum, R. H., van Ommen, B., and Groten, J. P., Toxicogenomic analysis of gene expression changes in rat liver after a 28-day oral benzene exposure, *Mutat. Res.*, 575, 85–101, 2005.

144. Huang, Q., Jin, X., Gaillard, E. T., Knight, B. L., Pack, F. D., Stoltz, J. H., Jayadev, S., and Blanchard, K. T., Gene expression profiling reveals multiple toxicity endpoints induced by hepatotoxicants, *Mutat. Res.*, 549, 147–167, 2004.

145. Adachi, T., Koh, K. B., Tainaka, H., Matsuno, Y., Ono, Y., Sakurai, K., Fukata, H., Iguchi, T., Komiyama, M., and Mori, C., Toxicogenomic difference between diethylstilbestrol and 17beta-estradiol in mouse testicular gene expression by neonatal exposure, *Mol. Reprod. Dev.*, 67, 19–25, 2004.

146. Nadadur, S. S., Schladweiler, M. C., and Kodavanti, U. P., A pulmonary rat gene array for screening altered expression profiles in air pollutant-induced lung injury, *Inhal. Toxicol.*, 12, 1239–1254, 2000.

147. Akhtar, R. A., Reddy, A. B., Maywood, E. S., Clayton, J. D., King, V. M., Smith, A. G., Gant, T. W., Hastings, M. H., and Kyriacou, C. P., Circadian cycling of the mouse liver transcriptome, as revealed by cDNA microarray, is driven by the suprachiasmatic nucleus, *Curr. Biol.*, 12, 540–550, 2002.

148. Hermjakob, H., Montecchi-Palazzi, L., Bader, G., Wojcik, J., Salwinski, L., Ceol, A., Moore, S. et al., The HUPO PSI's molecular interaction format—a community standard for the representation of protein interaction data, *Nat. Biotechnol.*, 22, 177–183, 2004.

149. Jones, A., Hunt, E., Wastling, J., Pizarro, A., and Stoeckert, C. J., Jr., An object model and database for functional genomics, *Bioinformatics*, 20, 1583–1590, 2004.

Part III

Angiogenesis and Chronic Degenerative Diseases

15 Endothelial Cell Responses to Physiological and Pathophysiological Environments

R. W. Siggins and C. A. Hornick

CONTENTS

15.1 INTRODUCTION

In vivo cells are never static. They are constantly altering the expression and sensitivity of their constituent proteins in maintaining functional, homeostatically stable internal and external environments. In this regard, changes in the extracellular milieu can have profound effects on cell function. Under basal physiological conditions, minor changes in the surrounding environment are quickly sensed and appropriate homeostatic responses are evoked. However, sometimes the insult to cells is greater than their ability to respond, leading to abnormal protein expression profiles. Some pathologies can cause major changes in cellular microenvironments, and the reaction to these perturbations often invokes a coordinated, interactive response from many cell types.

Endothelial cells that take part in responses to most environmental stressors are found in every tissue of the body. Viewed as a whole, the endothelium is one of the largest organs in humans, with an approximate mass of 1 kg containing about ten trillion (10^{13}) cells [1,2]. Despite its considerable mass and ubiquitous presence in tissues, the endothelium has only recently begun to be considered more than a "layer of nucleated cellophane," or a passive barrier between the tissues and the blood [1,3]. The first research to attribute direct functionality to the endothelium was

349

reported in 1980 by Furchgott and Zawadzki. They described the role of the endothelium in effecting changes in vascular smooth muscle tone in a dose-dependent manner to acetylcholine administration through a molecule they named Endothelial-Derived Relaxing Factor (EDRF). Their research ultimately led to the elucidation of EDRF as nitric oxide (NO) [4].

Although the endothelium can be thought of as a single organ, a heterogeneity of ECs exists along the vascular tree. For example, NO is the major endothelium-derived vasorelaxing factor for the large conduit arteries, but at the level of arteriolar resistance (vessels 100 μm), prostaglandins are the main source of vasorelaxation. However, the differences seen in endothelial cell (EC) responses are not limited to those based on vessel structure alone. Heterogeneity in ECs is also seen in expression patterns of proteins, mRNA, and signaling pathways, as well as major endothelial functions such as nutrient trafficking, barrier function, and angiogenesis [3]. These differences underlie some of the difficulty in studying EC physiology, as making generalizations across tissue types is problematic. Despite this hetero-geneity, ongoing research is finding that a number of EC responses to stimuli, such as those seen in pathological situations, are similar along the vascular tree, and current research is finding new therapeutic targets for disease intervention based on these similarities [5–7].

Endothelial cells have been implicated in many disease processes including thrombus formation, atherogenesis, hypertension, arthritis, autoimmunity, inflam-mation, and cancer. In healthy organisms, endothelial cells have been shown to have a vasorelaxing, antithrombotic, anti-inflammatory, and quiescent phenotype. However, this normal physiological expression profile is altered in all of the above mentioned disease states through a variety of mechanisms. Changes in the cellular environment are typically first sensed through receptors expressed on the cell surface. Cytokines, growth factors, extracellular matrix adhesion, and sheer stress can stimulate ECs through many receptor complexes, eliciting functional arousal from the quiescent state.

15.2   ENDOTHELIAL CELL HETEROGENEITY AND MICROENVIRONMENTS

Endothelial cells show remarkable heterogeneity, which is dependent upon both genetics and environmental cues [8]. During embryonic development, the cardiovascular system is the first to form through the processes of vasculogenesis and angiogenesis [9,10]. Vasculogenesis is marked by the budding of the hemangioblast into an immature network of tube-like structures. These structures differentiate into arteries, veins, and lymphatics very early in this process. The "specialized" ECs in the arterial and venous systems begin to secrete matrix proteins and specific paracrine factors to influence the cells in the surrounding tissues to migrate, proliferate, and differentiate.

Mesenchymal, stromal, and mural cells migrate toward the rudimentary vasculature to begin the process of organogenesis. In the embryo as well as the adult, tissues are known to secrete many cytokines, growth factors, and matrix proteins that, in turn, promote structural organization of the newly forming vessels.

This crosstalk between ECs and the surrounding cells of the developing tissues leads to the specialization of vessels that are capable of meeting the metabolic and physiologic needs of the vessel-promoting tissues [11]. Heterogeneity among ECs is easily seen morphologically. Differences in the histology of ECs are dependent upon vessel size, organ localization, intra-organ localization, and even age [8,11]. EC phenotypes are distinguishable through cell size, shape, nuclear localization with regards to blood flow, level of fenestration, organization of EC–EC junctions, number of plasmalemmal bodies, and the specific matrix proteins that comprise the vessel wall [8,11]. For example, hepatocyte–EC interactions in the liver lead to a highly fenestrated open microvasculature, while astrocyte-EC communication in the brain engenders a microvasculature with tightly joined endothelia cells. These tissue-specific adaptations of endothelial junctional bonding regulate the permissive or restrictive transendothelial transport of substances from the blood to the parenchyma in the liver or brain respectively.

The microenvironment in which a vessel is found regulates the EC phenotype. The largest vessels are organized into three discernable layers: the intima, the media, and the adventitia. ECs in the intima attach directly to a layer of extracellular matrix (ECM) that separates ECs from the elastic lamina (EL) of the media. The EL is marked by concentric layers of vascular smooth muscle cells (VSMCs) in between matrices containing primarily collagens, elastin, and fibrillins. The adventitia surrounds the EL and is composed mainly of fibroblasts and the collagens these cells produce [12]. The vascular cells that are resident in the vessel wall and their associated ECM correspond to the size of the vessel. As the arterial tree branches from muscular arteries to the level of arterioles, fibroblasts disappear. The ECM of arterioles has been shown to contain largely elastin, collagens I and III, and fibrillin, while the larger arteries have matrices composed of collagens IV, V, VI, fibronectin, and proteoglycans, as well as the ECM components of arterioles. Veins between 1 and 5 mm in diameter differ from muscular arteries by also containing collagens XII and XIV while lacking fibroblasts. Venules (-20 to 100 J.lm) contain ECs, VSMCs, and perivascular cells that secrete a matrix of laminin, collagen IV, and fibronectin. The smallest vessels of the body (capillaries 20 µm) are comprised of a similar ECM as that found in venules, but with two distinct differences. Collagen IV is the ECM protein in greatest abundance, as opposed to laminin in the venules, and the ECM of capillaries contains a quantifiable amount of heparin sulfate proteoglycan. The level of heterogeneity among ECs is also increased by the presence of the tissues specific stromal cells that also contribute to vessel homeostasis [13,14]. In the microenvironment of capillaries, a codependant relationship exists between the only two vascular cell types present: ECs and pericytes.

Pericytes are solitary, perivascular, smooth muscle-like mural cells that invest the walls of the microvasculature, being most abundant on venules but also common on capillaries [15,16]. Pericytes are highly plastic in nature with the ability to differentiate into other mesenchymal cells such as osteoblasts, smooth muscle cells, and even back to fibroblast-like cells that produce collagen type-1 during wound healing and inflammation [15,16]. Pericytes also contain myofibrils and have been shown to be contractile in vitro, although whether or not this ability translates to control of capillary blood flow in vivo has not been established [17]. Coverage of capillaries by pericytes varies in different tissues and organs with more abundant pericytes typically present in tissues with higher hydrostatic pressure such as the lower extremities [15]. Pericyte distribution on capillary surfaces often reflects the specialized function of the tissue they serve. For example vascular surfaces engaged in nutrient or gas exchange are typically pericyte-free as in the lungs, kidney, and placenta [18]. In neural tissue and in the retina, in which pericyte density on microvessels is very high, their functions remain largely unknown, although it has been suggested that they contribute to the blood–brain and blood–retinal barriers [18]. Pericytes are typically found adjacent to or bridging endothelial junctions and covering endothelial gaps during inflammatory reactions, although their multi-functional interactions with the endothelium demonstrate that their role is much more substantial than simple mechanical support of microvessel integrity. Pericytes extend long finger-like processes over the surface of the ECs with which they share a common basement membrane containing holes with numerous points of cell to cell membrane contact in which gap junctions, communicating junctions, tight junctions, and adhesion plaques are present [17]. This intimate morphological contact underscores the crosstalk between the two cell types.

This is particularly well documented in the process of neovascularization or angiogenesis. Physiologically, angiogenesis is rare in the adult, occurring mainly during endometrial growth during the female reproductive cycle. However, in the altered microenvironment of many disease states including: cancer, diabetic retinopathy, rheumatoid arthritis, psoriasis, inflammatory bowel disease, aids dementia complex, and many others, angiogenesis is regarded as either causative or contributing to disease progression [18–20].

It has been believed for some time that the initial endothelial vessels of an angiogenic sprouting reaction formed without pericytes, and that subsequent investment of the vessels with pericytes led to remodeling, maturation, and stabilization of the vascular network, based mostly on data from retinal models [21]. However, newer data has clearly shown pericytes present at very early stages of angiogenesis; and, in some cancer models, the new vessel wall in a small percentage of cases can be composed of pericytes alone with no endothelium present [18,22,23]. A number of investigators have suggested that the targeting of both pericytes and endothelial cells may be the most effective approach to tumor treatment [24]. Pericytes can be a major source of VEGF-A (Vascular Endothelial Growth Factor-A) which is secreted in response to endothelial-derived NO [25]. A graded distribution of VEGF-A in turn, controls both the migration and orientation of endothelial cells as well as EC proliferation within the growing capillary stalk [26]. The spreading of pericytes along the growing sprout is controlled by the secretion of the pericyte mitogen Platelet Derived Growth Factor-B (PDGF-B) from the capillary tip-cell [16]. Platelet Derived Growth Factor-B is recognized by the pericytes that express the PGDF-B receptor (PGDFR-β). This PDGF-B and PGDFR-β based crosstalk underlies the co-recruitment of pericytes and endothelial cells during vesicular angiogenesis [27,28]. In a microenvironment lacking pericytes, as in PDGF-B- and PGDF receptor-β deficient mice, endothelial hyperplasia, increased capillary diameter, abnormal EC shape and number, and increased transendothelial permeability result in perinatal lethality [29].

The commonality among resident constituent cell types between different vessels underlies similarities that can be seen when examining vessels from one organ system to another. Vessels throughout the vasculature share many of the same ECM components; however, the individual matrix components are found in differing amounts between ECMs. Interactions between these insoluble factors are facilitated by many organizational proteins that are capable of altering the lattice structure of the matrix [30,31]. In addition to these findings, ECMs have been shown to act as a reservoir for growth factors and proenzymes. Vascular Endothelial Growth factor and fibroblast growth factor (FGF), for example, are sequestered in the ECM [32].

Maintenance of the basement membrane is tightly regulated through the activation of matrix metalloproteinases (MMP2, MMP3, MMP9, and urokinase plasminogen activator) and their inhibitors [33]. These proteases cleave both ECM proteins and plasma proteins, releasing sequestered proangiogenic growth factors as noted above. The ECM and plasma protein breakdown products themselves also contain a variety of anti-angiogenic substances, including angiostatin from the plasma protein plasminogen, endostatin from collagen XVIII, and tumstatin from collagen IV [31,32,34]. Under normal physiological conditions, ECM proteins have a very long halflife, reported to be 60–70 days in the arterial wall. Despite this longevity, physiological levels of ECM breakdown products acting as endogenous endothelium-specific angiogenic inhibitors, including tumstatin and endostatin, appear to be sufficient to maintain a quiescent phenotype under normal conditions [34,35]. For a switch from a quiescent to an angiogenic phenotype to occur, the normal balance between pro- and anti-angiogenic factors must be altered in favor of the former. Many disease processes are capable of upsetting this balance by causing an alteration in the expression of EC proteins [36,37].

ECs and other vascular cells interact with the ECM primarily through a single family of transmembrane receptors—the integrins. Integrins are a heterodimeric family of proteins consisting of an alpha and a beta subunit. At least ten different heterodimers are expressed by endothelial cells in a spatio-temporal manner with certain integrins present during the quiescent state and others appearing when ECs become activated during the migratory and proliferative events of angiogenesis, enabling cells to respond to alterations in the microenvironment [36,38]. Several of the integrins have as few as one or two ligands; however some integrins bind promiscuously to many ECM proteins or ECM degradation products. For example, Integrin as1ß1 binds only fibronectin, and integrins a1ß1 and a2ß1 are collagen I receptors, but avß3 and avßs are capable of binding fibronectin, vitronectin, fibrinogen, osteopontin thrombospondin, endostatin, and von Willebrand factor (vWF) [31,39]. Antagonists to the ß1 integrins are capable of blocking angiogenesis; however, gene knock out studies reveal normal angiogenesis in mice lacking avß3 and avß3s. This suggests a considerable level of redundancy among integrin activity. The integrins are crucial in providing survival signals to the ECs during angiogenesis on appropriate matrices [38,40,41]. Most integrins activate focal adhesion kinases (FAKs) or Shc, which leads to activation of downstream intracellular targets [42]. Ras, Raf-1, Mek 1, and Erk 1 and 2 play an important role in promoting angiogenesis by activating ECs to divide and migrate [43].

These pathways are also activated by proangiogenic growth factors such as VEGF and FGF.

Several studies have shown crosstalk between specific integrins and growth factors. For example, when angiogenesis is stimulated by FGF, inhibiting integrin av133 or as131 will activate apoptosis [44]. It has also been demonstrated that the ability of VEGF to promote angiogenesis is disrupted in the absence of integrin avl3s ligation to the matrix [39]. Soluble ligands from a proteolyzed matrix often elicit opposite effects from integrins than an intact matrix. For example, some matrix fragments are capable of initiating apoptosis rather than migration and proliferation [32]. Integrins, therefore, appear to serve a regulatory role by maintaining EC viability in the face of fluctuations in soluble signals like growth factor withdrawal.

The importance of a vessel's microenvironment in affecting EC phenotype can be observed utilizing in vitro co-culture systems and in vivo transplant experiments. ECs grown in cell culture are separated from their natural microenvironment; therefore their phenotype "dedifferentiates" and the cells become exhausted after only a few passages. ECs from young adult donors can be cultured through five passages, but the ability to proliferate diminishes with the age of the donor [1]. The finding that ECs manifest an age-dependent phenotype may be related to the fact that cardiovascular diseases are rare in humans before the age of thirty, the proposed lifespan of ECs [45]. Cell culture systems typically lack appropriate ECMs, hemodynamic forces, cytokines, growth factors, and cell–cell interactions. Much of the in vivo phenotype can be restored by co-culturing EC with EGM from the native vessels, non-endothelial cells from the tissue of origin—or a combination of both. EC isolated from one tissue can also be driven to express the phenotype of ECs from other tissues by growing them under the appropriate conditions. Aortic ECs, for example, can be transformed into endothelium that resembles the pulmonary microvascular by growing the aortic ECs on EGM from the lung [46]. In other experiments, astroctyes inserted into the chick chorioallantoic membrane (CAM) model will engender a continuous endothelium, similar to that seen in the brain [47]. These findings underscore the importance of both perivascular cells and ECMs in forming the microenvironment.

Similar results can be obtained in vivo. Tissue grafts containing vascular networks transplanted between the brain and periphery will assume the morphology of the new tissue [48]. Similarly, when neonatal heart tissue is transplanted into the ears of mice, the expanding auricular vessels begin expressing vWF. This protein is commonly expressed in the microvasculature of the heart, but not in the ears. These studies offer further evidence that the EC phenotype is largely dependent upon its microenvironment, and ECs, like pericytes maintain a certain amount of phenotypic plasticity [49].

In addition to matrix proteins and integrin expression profiles, growth factors also contribute to the balance between pro- and anti-angiogenic molecules in the microenvironment. In a recently published study, Ozawa transfected mice myoblasts with the gene encoding the VEGF165 splice variant [50]. Transformed cells were transplanted into mouse ear myofibrils. Before transplantation, VEGF-expressing myoblasts were isolated according to the amount of VEGF that they produced. An overabundance of this cytokine leads to aberrant angiogenesis, while in transplanted myoblasts that showed modest VEGF expression, the angiogenesis was morphologically and physiologically normal. One might conclude that the total level of proangiogenic growth factor is not as important as the microenvironmental level of VEGF present [50]. Interestingly, VEGF165 is one of the isoforms that is known to be sequestered by fibronectin, but this binding is both pH and heparin dependent [51].

Disease states elicit a wide repertoire of responses from the endothelium that are unique to the microenvironment of specific ECs. The EC phenotype determines the response, with changes in the expression of soluble proteins, cell surface molecules, and components that specifically alter the ECM. The ways ECs respond to changing microenvironments during a pathological process may elicit immune activation, angiogenesis, and even apoptosis.

15.3 INFLAMMATION

Local and systemic inflammation share similar effectors in activating the endothelium. For example, during sepsis, the proinflammatory cytokines tumor necrosis factor (TNFu), IL-1, and IL-6 are the first to appear in the blood. These molecules directly act on the endothelium by increasing permeability through diminished junctional integrity [52,53]. ECs respond by producing tissue factor and upregulating cell adhesion molecules (CAMs). CAMs, such as P- and E-selectin, intercellular adhesion molecule (ICAM), and vascular cell adhesion molecule (VCAM) initiate tissue acquisition of circulating immune cells and platelets [54]. In accordance with acting as a barrier between the blood and tissue, ECs express very low levels of these molecules under basal physiological conditions.

Lipopolysaccharide (LPS) is a predominant immune-activating molecule found on the surface of gram-negative bacteria, such as *E. coli*. Its receptor, toll-like receptor 4 (TLR4), is constitutively expressed on all endothelia. Lipopolysaccharide upregulates the expression of TLR4, inducing ECs to initiate a change in their expression profile [55]. However, LPS, TNFu, and IFNy all lead to the presentation of a different TLR on the surface of the activated endothelium, TLR2. ECs do not normally express TLR2.

Some of the most compelling evidence for microenvironmental determination of EC heterogeneity comes from transgenic mice studies. Constructs containing a LacZ reporter gene under control of a vWF promoter segment were incorporated into the endothelium of mice. Assaying for LacZ gave positive results in the brain, heart, and skeletal muscle microcirculation. Reporter gene expression was found only in the capillaries of the heart, and significantly, no expression was found in the endothelium of the coronary arteries or veins, penetrating arteries, or endocardium. However, RT-PCR analysis confirmed expression of wild-type vWF mRNA in the liver, spleen, lung, kidney, aorta, and the megakaryocyte lineage [49]. These results, in conjunction with the heart transplantation studies cited above, indicate that ECs are heterogeneous in their transcriptional control, and that this heterogeneity is due to the microenvironment [49].

Microenvironments are formed through the unique association of ECs, perivascular cells, and stromal cells in close proximity to each other. These cells each contribute to the architecture and composition of the ECM as well as to the soluble factors found in the surrounding milieu. Interactions between these factors and the cells involved determine their unique phenotypes, and therefore functions. These phenotypes and functions can be profoundly affected by disease processes that produce perturbations in the expression of these cell types and the factors they secrete.

15.4 PATHOLOGIES WITH ENDOTHELIAL CELL INVOLVEMENT

The specificity of TLR2 accounts for the ability of ECs to respond to a much wider array of pathogen classes including gram-positive bacteria, spirochetes, mycobacteria, and fungi [2].

In addition to the first-response cytokines observed during inflammation, changes in growth factors (such as VEGF), chemokines, and enzymes have also been measured. Matrix metalloproteinases and nitric oxide (NO) from nitric oxide synthase (NOS) are present at increased levels. In addition to degrading the matrix and causing vasodilation, these molecules also stimulate ECs to produce cytokines that contribute to the microenvironmental changes that occur [56].

Intravenous LPS, when present in amounts seen during sepsis, can denude the endothelium [57]. High levels of ECs in the blood strongly correlate to the mortality of patients with sepsis, as patients with the lowest number of circulating ECs have a much better prognosis [58]. Assuming the patient does survive, the phenotype of the new endothelium is always different from that of the native ECs. Cell densities of renewed endothelium can be twice as high as the original ECs, and they typically show a more vasoconstricting and prothrombogenic phenotype [45].

15.4.1 ATHEROSCLEROSIS

During the process of atherogenesis, much of the microenvironment becomes disturbed. Atherosclerotic plaques form in the vessel walls and are capable of altering the phenotype of mural cells and ECs. These phenotypic transformations lead directly to changes in the ECM components [59]. From our perspective on the endothelium, it is important to note that atherosclerosis does not affect all vessels equally. Given this, atherosclerosis is an archetypal disease process that elucidates the differences between the functioning of a healthy physiological microenvironment and one that has been altered in the face of pathology.

The endothelium of an atherosclerotic vessel will mostly maintain its phenotype distal to the site of injury. However, segments that are directly affected by plaques lose their normal physiologic functioning. For instance, acetylcholine (Ach) is a known vasodilator, but in vessels with plaques, Ach has a vasoconstricting effect [60]. This response is entirely dependent on the endothelium, because when treated with nitroglycerine, all segments of a diseased vessel are capable of dilating as measured by increases in blood flow [60]. In this context, the NO generated is known to act directly on smooth muscle cells in the artery wall; therefore, enough phenotypic stability exists in the VSMGs to maintain a semi-functional artery.

In addition to functional changes, EGs can undergo remarkable morphological alterations in the area of an atherosclerotic plaque. Giant, multinucleated cells of EG origin have been observed in advanced lesions [61]. The exact function of these mega-EGs remains to be elucidated. Nevertheless, alterations in the secreted factors from EGs localized to lesions have shown an increase in endothelin-1 (ET-1) formation, while the production of prostacyclin is remarkably decreased. The changes in these vasoactive substances are directly correlated to the severity of the lesion [62,63]. It is interesting to note that ET-1 production in the healthy endothelium is considered to be inversely proportional to NOS activity [64]. Atherosclerotic EGs also begin to express MHC II molecules, which are generally only expressed on the surface of putative antigen-presenting cells [65]. All these data taken together demonstrate that ECs in atherosclerotic plaques express a proinflammatory and prothrombotic phenotype.

In conjunction with the phenotypic switch seen in the ECs, infiltration of atherosclerotic plaques with immune cells, such as monocytes and macrophages, is typical. The homing of these immune cells to the lesion comes as a result of the proinflammatory phenotype expressed by the ECs [66]. Most of the same molecules present during a typical inflammatory response are also expressed by ECs localized in plaques, including the CAMs. Once macrophages infiltrate the vessel wall, they begin producing cytokines and chemokines, which in turn, alter the ECM. As plaque macrophages engulf excess lipid, their phenotype is also subject to change. They become foam cells, named for their large vacuolar morphology, which are central in the production of many of the proinflammatory cytokines that have been shown to modify EC function [67,2]. The phenotypic change from macrophages to foam cells can be blocked by antibodies to $\alpha_V\beta_3$ integrin [68].

Researchers have observed that some VSMCs also acquire an altered phenotype at the site of plaque formation. As described above, VSMCs align themselves in concentric rings throughout the EL of vessels. The phenotype, for at least some VSMCs, during atherosclerosis is marked by a dendritic appearance [69]. This secretory phenotype leads the VSMCs to remodel the ECM of the vessel wall. In normal vessels, VSMCs and fibroblasts are the principal producers of the ECM; however, the secretory VSMCs express a different set of ECM proteins. Plaques typically contain an abundance of collagens, accounting for 60% of the total protein. Collagens I and III are supplemented by the addition of types IV, V, and VI. The total amount of collagen, the number of resident VSMCs, and the presence of a thick versus thin fibrous cap all determine plaque integrity [14].

Atherosclerosis is a disease of radical microenvironmental alteration. ECs, VSMCs, and immune cells undergo transformations from their normal physiological phenotypes, and all contribute to the gross remodeling of the entire microenvironment. Ongoing research into the distinct contribution that each cell type makes to the altered milieu offers the potential for a more integrated understanding as well as new preventive strategies and therapies.

15.4.2 Cancer

In the study of EC physiology, no other disease state presents as many challenges to understanding the functioning of ECs in a pathological setting as cancer. There are many reasons for this, the most significant being the numerous etiologies underlying the myriad forms of cancers that have been described to date. The EC phenotype in cancer varies with the host tumor [70,71]. This is largely due to the nature of the microenvironmental interaction between the endothelium and the specific tumor type [19,72,73]. The universal requirement of tumors for a rich vasculature to fuel their metabolic needs, without which they cannot grow beyond 1 mm in diameter, has pointed toward endothelial angiogenesis as a likely therapeutic target [19,72,74]. Thus, one goal of many research projects has been to define common characteristics of the many EC phenotypes with the hope of finding inhibitors of angiogenesis.

Vessel morphology in growing tumors is reminiscent of that seen during embryological vasculogenesis and angiogenesis. The vessels that form during tumor growth often lack the hierarchy observed in the adult vasculature in terms of vessel size and branching [75]. Coordination between tumors and ECs is much less structured, in that tissue metabolic need does not necessarily determine vessel growth. Distinct regions within a tumor might be starved for nutrients, while other locations in the same tumor will have an abundance of vessels supplying them [76]. Because tumors grow into occupied spaces, the pressure exerted on the surrounding tissue and inside the tumor creates a situation in which not all vessels are even functional—due to vascular compression [77,78].

ECs lining the vessels of tumors are not always continuous. In fact, whole segments of microvessels might lack endothelium entirely, as noted in our discussion of pericytes above—although other vascular segments may contain ECs growing in thick layers [72]. The ECs often exhibit a much more columnar morphology as opposed to the typical elongated ECs that are seen in healthy vasculature. This strange morphology represents potential sites for tumor metastasis [79].

ECs from tumor vasculature, while displaying a wide range of histological differences, function similarly to healthy endothelium in distinct and important ways. The endothelium retains both the intrinsic and extrinsic apoptotic pathways that are witnessed in normal ECs. Anchorage-dependent growth is a requisite for ECs, both in normal and tumor vasculature. Interference with this interaction leads to apoptosis in both microenvironments [80].

The microenvironment of tumors may contain a variety of matrix proteins that can be assembled in a variety of ways. The coordination in creating the microenvironment is lost as the tumor cells disproportionately drive the functioning of the microenvironment. Unregulated secretion of growth factors, cytokines, and ECM proteins select for an endothelium that is capable of growing in this environment. Interrupting one or even a few of these signals does not guarantee the destruction of the entire tumor vasculature. Other locations in the tumor may secrete different factors, and create a distinct microenvironment. This secondary site may promote the growth of new vessels that do not respond to the same antagonists. Despite these difficulties, inhibiting tumor angiogenesis and promoting tumor vessel regression remains a viable treatment option for eliminating cancer because the altered phenotypic expression of molecular markers activating both tumor ECs and pericytes yields a distinct signature for therapeutic targeting [70,81–83].

REFERENCES

1. Thorin, E. and Shreeve, S. M., Heterogeneity of vascular endothelial cells in normal and disease states, *Pharmacol. Ther.*, 78, 155–166, 1998.
2. Galley, H. F. and Webster, N. R., Physiology of the endothelium, *Br. J. Anaesth.*, 93, 105–113, 2004.
3. Aird, W. C., Endothelium as an organ system, *Crit. Care Med.*, 32, 8271–8279, 2004.
4. Furchgott, R. F. and Zawadzki, J. V., The obligatory role of endothelial cells in the relaxation of arterial smooth muscle by acetylcholine, *Nature*, 288, 373–376, 1980.
5. Rosen, L., Antiangiogenic strategies and agents in clinical trials, *Oncologist*, 5 (suppl.), 20–27, 2000.
6. Sanz, L. and Alvarez-Vallina, L., The extracellular matrix: a new turn-of-the-screw for antiangiogenic strategies, *Trends Mol. Med.*, 9, 256–262, 2003.
7. Nanda, A. and St. Croix, B., Tumor endothelial markers: new targets for cancer therapy, *Curr. Opin. Oncol.*, 16, 44–49, 2004.
8. Aird, W. C., Endothelial cell heterogeneity, *Crit. Care Med.*, 31, 8221–8230, 2003.
9. Drake, C. J., Hungerford, J. E., and Little, C. D., Morphogenesis of the first blood vessels, *Ann. N.Y. Acad. Sci.*, 857, 155–179, 1998.
10. Nikolova, G. and Lammert, E., Interdependent development of blood vessels and organs, *Cell Tissue Res.*, 314, 33–42, 2003.
11. Cleaver, O. and Melton, D. A., Endothelial signaling during development, *Nat. Med.*, 9, 661–668, 2003.
12. Brooke, B. S., Kamik, S. K., and Li, D. Y., Extracellular matrix in vascular morphogenesis and disease: structure versus signal, *Trends Cell Biol.*, 13, 51–56, 2003.
13. Karnik, S. K., Brooke, B. S., Bayes-Genis, A., Sorensen, L., Wythe, L. D., Schwartz, R. S., Keating, M. T., and Li, D. Y., A critical role for elastin signaling in vascular morphogenesis and disease, *Development*, 130, 411–423, 2003.
14. Bou-Gharios, G., Ponticos, M., Rajkumar, V., and Abraham, D., Extra-cellular matrix in vascular networks, *Cell Prolif.*, 37, 207–220, 2004.
15. Sims, D. E., Diversity within pericytes, *Clin. Exp. Pharmacol. Physiol.*, 27, 842–846, 2000.
16. Betsholtz, C., Lindblom, P., and Gerhardt, H., Role of pericytes in vascular morphogenesis, *EXS*, 115–125, 2005.
17. Allt, G. and Lawrenson, L. G., Pericytes: cell biology and pathology, *Cells Tissues Organs*, 169, 1–11, 2001.
18. Gerhardt, H. and Betsholtz, C., Endothelial-pericyte interactions in angiogenesis, *Cell Tissue Res.*, 314, 15–23, 2003.
19. Folkman, J., Tumor angiogenesis: theraputic implications, *N. Engl. J. Med.*, 285, 1182–1186, 1971.
20. Risau, W., Mechanisms of angiogenesis, *Nature*, 386, 671–674, 1997.
21. Benjamin, L. E., Hemo, L., and Keshet, E., A plasticity window for blood vessel remodelling is defined by pericyte coverage of the preformed endothelial network and is regulated by PDGF-B and VEGF, *Development*, 125, 1591–1598, 1998.

22. Ozerdem, U. and Stallcup, W. B., Early contribution of pericytes to angiogenic sprouting and tube formation, *Angiogenesis*, 6, 241–249, 2003.

23. Sun, J. F., Phung, T., Shiojima, I., Felske, T., Upalakalin, J. N., Feng, D., Kornaga, T. et al., Microvascular patterning is controlled by fine-tuning the Akt signal, *Proc. Natl Acad. Sci. U.S.A*, 102, 128–133, 2005.

24. Bergers, G., Song, S., Meyer-Morse, N., Bergsland, E., and Hanahan, D., Benefits of targeting both pericytes and endothelial cells in the tumor vasculature with kinase inhibitors, *J. Clin. Investig.*, 111, 1287–1295, 2003.

25. Reynolds, L. P., Grazul-Bilska, A. T., and Redmer, D. A., Angiogenesis in the corpus luteum, *Endocrine*, 12, 1–9, 2000.

26. Gerhardt, H. and Betsholtz, C., How do endothelial cells orientate?, *EXS*, 3–15, 2005.

27. Hellstrom, M., Kalen, M., Lindahl, P., Abramsson, A., and Betsholtz, C., Role of PDGFB and PDGFR-beta in recruitment of vascular smooth muscle cells and pericytes during embryonic blood vessel formation in the mouse, *Development*, 126, 3047–3055, 1999.

28. Lindblom, P., Gerhardt, H., Liebner, S., Abramsson, A., Enge, M., Hellstrom, M., Backstrom, G. et al., Endothelial PDGF-B retention is required for proper investment of pericytes in the microvessel wall, *Genes Dev.*, 17, 1835–1840, 2003.

29. Hellstrom, M., Gerhardt, H., Kalen, M., Li, X., Eriksson, U., Wolburg, H., and Betsholtz, C., Lack of pericytes leads to endothelial hyperplasia and abnormal vascular morphogenesis, *J. Cell Biol.*, 153, 543–553, 2001.

30. Brewer, C. F., Miceli, M. C., and Baum, L. G., Clusters, bundles, arrays and lattices: novel mechanisms for lectirr-saccharide-mediated cellular interactions, *Curr. Opin. Struct. Biol.*, 12, 616–623, 2002.

31. Jain, R. K., Molecular regulation of vessel maturation, *Nat. Med.*, 9, 685–693, 2003.

32. Bix, G. and Iozzo, R. V., Matrix revolutions: 'tails' of basement-membrane components with angiostatic functions, *Trends Cell Biol.*, 15, 52–60, 2005.

33. Heissig, B., Hattori, K., Friedrich, M., Rafii, S., and Werb, Z., Angiogenesis: vascular remodeling of the extracellular matrix involves metalloproteinases, *Curr. Opin. Hematol.*, 10, 136–141, 2003.

34. Sund, M., Hamano, Y., Sugimoto, H., Sudhakar, A., Soubasakos, M., Yerramalla, U., Benjamin, L. E. et al., Function of endogenous inhibitors of angiogenesis as endothelium-specific tumor suppressors, *Proc. Natl Acad. Sci. U.S.A.*, 102, 2934–2939, 2005.

35. Nissen, R., Cardinale, G. J., and Udenfriend, S., Increased turnover of arterial collagen in hypertensive rats, *Proc. Natl Acad. Sci. U.S.A.*, 75, 451–453, 1978.

36. Stupack, D. G. and Cheresh, D. A., Integrins and angiogenesis, *Curr. Top. Dev. Biol.*, 64, 207–238, 2004.

37. Chavakis, E. and Dimmeler, S., Regulation of endothelial cell survival and apoptosis during angiogenesis, *Arterioscler. Thromb. Vasc. Biol.*, 22, 887–893, 2002.

38. Stupack, D. G., Integrins as a distinct subtype of dependence receptors, *Cell Death Differ.*, 2005.

39. Stupack, D. G. and Cheresh, D. A., Apoptotic cues from the extracellular matrix: regulators of angiogenesis, *Oncogene*, 22, 9022–9029, 2003.

40. Stupack, D. G. and Cheresh, D. A., Get a ligand, get a life: integrins, signaling and cell survival, *J. Cell Sci.*, 115, 3729–3738, 2002.

41. Stupack, D. G., Integrins as a distinct subtype of dependence receptors, *Cell Death Differ.*, 2005.

42. Eliceiri, B. P., Puente, X. S., Hood, J. D., Stupack, D. G., Schlaepfer, D. D., Huang, X. Z., Sheppard, D., and Cheresh, D. A., Src-mediated coupling of focal adhesion kinase to integrin alpha(v)beta5 in vascular endothelial growth factor signaling, *J. Cell Biol.*, 157, 149–160, 2002.

43. Howe, A. K., Aplin, A. E., and Juliano, R. L., Anchorage-dependent ERK signaling-mechanisms and consequences, *Curr. Opin. Genet. Dev.*, 12, 30–35, 2002.

44. Friedlander, M., Brooks, P. C., Shaffer, R. W., Kincaid, C. M., Vamer, J. A., and Cheresh, D. A., Definition of two angiogenic pathways by distinct alpha v integrins, *Science*, 270, 1500–1502, 1995.

45. Vanhoutte, P. M. and Scott-Burden, T., The endothelium in health and disease, *Tex. Heart Inst. J.*, 21, 62–67, 1994.

46. Augustin-Voss, H. G., Johnson, R. C., and Pauli, B. U., Modulation of endothelial cell surface glycoconjugate expression by organ-derived biomatrices, *Exp. Cell Res.*, 192, 346–351, 1991.

47. Janzer, R. C. and Raff, M. C., Astrocytes induce blood–brain barrier properties in endothelial cells, *Nature*, 325, 253–257, 1987.

48. Stewart, P. A. and Wiley, M. J., Developing nervous tissue induces formation of blood–brain barrier characteristics in invading endothelial cells: a study using quail—chick transplantation chimeras, *Dev. Biol.*, 84, 183–192, 1981.

49. Aird, W. C., Edelberg, L. M., Weiler-Guettler, H., Simmons, W. W., Smith, T. W., and Rosenberg, R. D., Vascular bed-specific expression of an endothelial cell gene is programmed by the tissue microenvironment, *J. Cell Biol.*, 138, 1117–1124, 1997.

50. Ozawa, C. R., Banfi, A., Glazer, N. L., Thurston, G., Springer, M. L., Kraft, P. E., McDonald, D. M., and Blau, H. M., Microenvironmental VEGF concentration, not total dose, determines a threshold between normal and aberrant angiogenesis, *J. Clin. Investig.*, 113, 516–527, 2004.

51. Goerges, A. L. and Nugent, M. A., pH regulates vascular endothelial growth factor binding to fibronectin: a mechanism for control of extracellular matrix storage and release, *J. Biol. Chem.*, 279, 2307–2315, 2004.

52. Tamm, I., Cardinale, I., Krueger, J., Murphy, J. S., May, L. T., and Sehgal, P. B., Interleukin 6 decreases cell–cell association and increases motility of ductal breast carcinoma cells, *J. Exp. Med.*, 170, 1649–1669, 1989.

53. McLaughlin, F., Hayes, B. P., Horgan, C. M., Beesley, J. E., Campbell, C. J., and Randi, A. M., Tumor necrosis factor (TNF)-alpha and interleukin (IL)-1 beta down-regulate intercellular adhesion molecule (ICAM)-2 expression on the endothelium, *Cell Adhes. Commun.*, 6, 381–400, 1998.

54. Verma, S. and Anderson, T. J., Fundamentals of endothelial function for the clinical cardiologist, *Circulation*, 105, 546–549, 2002.

55. Faure, E., Thomas, L., Xu, H., Medvedev, A., Equils, O., and Arditi, M., Bacterial lipopolysaccharide and IFN-gamma induce Toll-like receptor 2 and Toll-like receptor 4 expression in human endothelial cells: role of NF-kappa B activation, *J. Immunol.*, 166, 2018–2024, 2001.

56. Webster, N. R., Sepsis and the systemic inflammatory response syndrome, *J.R. Coll. Surg. Edinb.*, 45, 345, 2000.

57. Reidy, M. A. and Bowyer, D. E., Scanning electron microscopy of aortic endothelium following injury by endotoxin and during subsequent repair, *Prog. Biochem. Pharmacol.*, 13, 175–181, 1977.

58. Mutunga, M., Fulton, B., Bullock, R., Batchelor, A., Gascoigne, A., Gillespie, J. I., and Baudouin, S. V., Circulating endothelial cells in patients with septic shock, *Am. J. Respir. Crit. Care Med.*, 163, 195–200, 2001.

59. Spyridopoulos, I. and Andres, V., Control of vascular smooth muscle and endothelial cell proliferation and its implication in cardiovascular disease, *Front Biosci.*, 3, d269–d287, 1998.

60. El Tamimi, H., Mansour, M., Wargovich, T. J., Hill, J. A., Kerensky, R. A., Conti, C. R., and Pepine, C. J., Constrictor and dilator responses to intracoronary acetylcholine in adjacent segments of the same coronary artery in patients with coronary artery disease. Endothelial function revisited, *Circulation*, 89, 45–51, 1994.

61. Antonov, A. S., Nikolaeva, M. A., Klueva, T. S., Romanov, Y., Babaev, V. R., Bystrevskaya, V. B., Perov, N. A., Repin, V. S., and Smirnov, V. N., Primary culture of endothelial cells from atherosclerotic human aorta. Part 1. Identification, morphological and ultrastructural characteristics of two endothelial cell sub-populations, *Atherosclerosis*, 59, 1–19, 1986.

62. Bacon, C. R., Cary, N. R., and Davenport, A. P., Endothelin peptide and receptors in human atherosclerotic coronary artery and aorta, *Circ. Res.*, 79, 794–801, 1996.

63. Jones, G. T., van Rij, A. M., Solomon, C., Thomson, I. A., and Packer, S. G., Endothelin-1 is increased overlying atherosclerotic plaques in human arteries, *Atherosclerosis*, 124, 2535, 1996.

64. Lavallee, M., Takamura, M., Parent, R., and Thorin, E., Crosstalk between endothelin and nitric oxide in the control of vascular tone, *Heart Fail. Rev.*, 6, 265–276, 2001.

65. Page, C., Rose, M., Yacoub, M., and Pigott, R., Antigenic heterogeneity of vascular endothelium, *Am. J. Pathol.*, 141, 673–683, 1992.

66. Lalor, P. F. and Adams, D. H., Lymphocyte homing to allografts, *Transplantation*, 70, 1131–1139, 2000.

67. Gerrity, R. G., The role of the monocyte in atherogenesis: II. Migration of foam cells from atherosclerotic lesions, *Am. J. Pathol.*, 103, 191–200, 1981.

68. Antonov, A. S., Kolodgie, F. D., Munn, D. H., and Gerrity, R. G., Regulation of macrophage foam cell formation by alpha Vbeta3 integrin: potential role in human atherosclerosis, *Am. J. Pathol.*, 165, 247–258, 2004.

69. Dey, N. B., Foley, K. F., Lincoln, T. M., and Dostmann, W. R., Inhibition of cGMP dependent protein kinase reverses phenotypic modulation of vascular smooth muscle cells, *J. Cardiovasc. Pharmacol.*, 45, 404–413, 2005.

70. Bergers, G. and Benjamin, L. E., Tumorigenesis and the angiogenic switch, *Nat. Rev. Cancer*, 3, 401–410, 2003.

71. Folberg, R. and Maniotis, A. J., Vasculogenic mimicry, *APMIS*, 112, 508–525, 2004.

72. Folkman, J., Watson, K., Ingber, D., and Hanahan, D., Induction of angiogenesis during the transition from hyperplasia to neoplasia, *Nature*, 339, 58–61, 1989.

73. Blouw, B., Song, H., Tihan, T., Bosze, J., Ferrara, N., Gerber, H. P., Johnson, R. S., and Bergers, G., The hypoxic response of tumors is dependent on their microenvironment, *Cancer Cell*, 4, 133–146, 2003.

74. Vajkoczy, P., Farhadi, M., Gaumann, A., Heidenreich, R., Erber, R., Wunder, A., Tonn, J. C., Menger, M. D., and Breier, G., Microtumor growth initiates angiogenic sprouting with simultaneous expression of VEGF, VEGF receptor-2, and angiopoietin-2, *J. Clin. Investig.*, 109, 777–785, 2002.

75. Dewhirst, M. W., Tso, C. Y., Oliver, R., Gustafson, C. S., Secomb, T. W., and Gross, J. F., Morphologic and hemodynamic comparison of tumor and healing normal tissue microvasculature, *Int. J. Radiat. Oncol. Biol. Phys.*, 17, 91–99, 1989.

76. Moeller, B. J., Cao, Y., Vujaskovic, Z., Li, C. Y., Haroon, Z. A., and Dewhirst, M. W., The relationship between hypoxia and angiogenesis, *Semin. Radiat. Oncol.*, 14, 215–221, 2004.

77. Milosevic, M. F., Fyles, A. W., and Hill, R. P., The relationship between elevated interstitial fluid pressure and blood flow in tumors: a bioengineering analysis, *Int. J. Radiat. Oncol. Biol. Phys.*, 43, 1111–1123, 1999.

78. Milosevic, M., Fyles, A., Hedley, D., and Hill, R., The human tumor microenvironment: invasive (needle) measurement of oxygen and interstitial fluid pressure, *Semin. Radiat. Oncol.*, 14, 249–258, 2004.

79. Hida, K. and Klagsbrun, M., A new perspective on tumor endothelial cells: unexpected chromosome and centrosome abnormalities, *Cancer Res.*, 65, 2507–2510, 2005.

80. McEwen, A., Emmanuel, C., Medbury, H., Leick, A., Walker, D. M., and Zoellner, H., Induction of contact-dependent endothelial apoptosis by osteosarcoma cells suggests a role for endothelial cell apoptosis in blood-borne metastasis, *J. Pathol.*, 201, 395–403, 2003.

81. Molema, G., Meijer, D. K., and de Leij, L. F., Tumor vasculature targeted therapies: getting the players organized, *Biochem. Pharmacol.*, 55, 1939–1945, 1998.

82. Berger, M., Bergers, G., Arnold, B., Hammerling, G. J., and Ganss, R., Regulator of G protein signaling-S induction in pericytes coincides with active vessel remodeling during neovascularization, *Blood*, 105, 1094–1101, 2005.

83. Kuldo, J. M., Ogawara, K. I., Werner, N., Asgeirsdottir, S. A., Kamps, J. A., Kok, R. J., and Molema, G., Molecular pathways of endothelial cell activation for (targeted) pharmacological intervention of chronic inflammatory diseases, *Curr. Vasc. Pharmacol.*, 3, 11–39, 2005.

16 Angiogenic Switch: Roles of Estrogenic Compounds

Sushanta K. Banerjee, Gibanananda Ray, Peter Van Veldhuizen, and Snigdha Banerjee

CONTENTS

16.1   INTRODUCTION

Cancer spreads from its primary sites through a metastatic process by which cancer cells penetrate into lymphatic and blood vessels, circulating through the bloodstream, ultimately invading and growing in distant sites of the body. It is a complex and multi-step life-threatening event. Multiple studies have been conducted in the last several years and various laboratories are still actively engaged in uncovering the molecular mechanisms that make this deadly event possible. Despite massive effort, scientists are still unable to discover weapons that have the potency to wipe out the metastatic process. Yet, studies point out that new blood vessel formation from existing blood vessels, known as angiogenesis, is crucial for cancer development and metastasis. Therefore, demolition of angiogenesis may help in blocking the progression of cancer metastasis.

Formation of new blood vessel/angiogenesis from existing blood vessels is a normal physiological event. It is a multi-step process, involving cellular interactions with the components of the extracellular matrix, endothelial cell proliferation, migration, and differentiation into capillaries [1]. In normal conditions, angiogenesis is virtually absent in adult organisms. However, angiogenesis is required for growth and development in the human body [2]. During the menstrual cycle in women, angiogenesis occurs as new blood vessels form in the lining of the uterus [3]. It is also required for the repair and regeneration of cells during wound healing in adults [4]. Angiogenesis is implicated in the pathogenesis of a variety of disorders including proliferative retinopathies, age-related macular degeneration, rheumatoid arthritis, and tumors [5,6]. In tumor angiogenesis, the blood vessel network infiltrates into cancerous tissue, which send signals to surrounding normal host tissue to activate a cascade of genes to precipitate the growth of new blood vessels [6]. Architecturally, each tumor blood vessel is different from its normal counterpart. The tumor blood vessels are irregular in shape, dilated, and tortuous. They are unable to organize into definitive venules, arterioles and capillary-like structures [1,7,8]. Tumor angiogenesis is often leaky and hemorrhagic—partly due to the overproduction of angiogenic factors, like vascular endothelial growth factor (VEGF)/vascular permeability factor (VPF).

Several positive and negative regulators of angiogenesis are associated with this multi-step event. An imbalanced expression of positive- and negative-angiogenic factors results in an "angiogenic switch" (see Figure 16.1). Cancer cells secrete several positive angiogenic factors, such as VEGF, and VEGF receptors (VEGFR), acidic fibroblast growth factor (aFGF), basic fibroblast growth factor (bFGF) [9], interleukin-8 (IL-8) [10], angiogenin [11], angiotropin [12], epidermal growth factor (EGF), platelet derived endothelial cell growth factor (PDGF) [13], transforming growth factor α (TGF-α), transforming growth factor β (TGF-β), hepatocyte growth factor (HGF) [14] and others that play critical roles in growing endothelial cells. There are some anti-angiogenic molecules, such as thrombospondin [15], angiostatin [16] and endostatin [17] that prevent angiogenesis, which are also secreted by tumor cells. But the level of production is very low to undetectable.

Estrogens have been associated with several types of human cancers [18–20], and play critical roles in the regulation of angiogenesis [21,22]. In our laboratory, anti-apoptotic activity and cell proliferation by estrogen has been attributed to the increment of the most potent cell cycle regulatory protein, cyclin B1 in MCF-7 cells [23]. We and others reported the involvement of estrogen in many carcinogenic processes through angiogenesis [24–27], and, in addition, antiestrogens have been shown to inhibit tumor angiogenesis [28]. 2-Methoxyestradiol (2-ME$_2$), an estrogen metabolite, has shown potential as an anti-angiogenic agent [29,30]. Despite the present understanding, the precise molecular mechanisms by which estrogen regulates angiogenesis have not been defined. In this review, we discuss the role of estrogen and its related compounds and their possible molecular mechanisms that modulate many pro-angiogenic factors. We will discuss how these pro-angiogenic factors eventually turn on the angiogenic switch leading to the development of cancer, and the anti-angiogenic factors that may prevent this angiogenic switch.

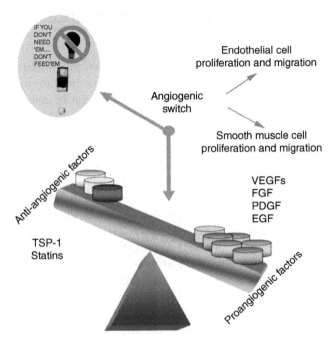

FIGURE 16.1 "Balance Hypothesis" of tumor angiogenic switch. Tumor angiogenesis is a multi-step process that is regulated by imbalance expression of several positive and negative angiogenic factors.

16.2 ANGIOGENIC SWITCH: A UNIQUE AND CRUCIAL EVENT IN CARCINOGENESIS

Tumor angiogenic switch is the hallmark of cancer. It can occur at various stages of tumor growth and progression, depending on the tumor type and its environments [7,31]. Tumor angiogenic switch essentially starts with tumor cells releasing positive factors, which transmit molecular signals to surrounding normal host tissue. This molecular signaling ultimately triggers multiple downstream genes in the host tissue that, in turn, potentiate the formation of new blood vessels surrounding tumors [7]. Activation of this angiogenic switch is a prerequisite for tumor growth and progression. It occurs in both benign and malignant cancers except for astrocytomas, where the activation of angiogenic switch is a late event and occurs after the regression of the existing blood vessels [7,31,32]. Although, the angiogenic switch is reported to occur during the progression of the tumor beyond the microscopic size, angiogenesis can be switched on at an early stage of this disease [33,34]. This activation can be potentiated by mutational activation of an oncogene(s). For example, K-ras mutations in chronic pancreatitis and pancreatic adenocarcinoma may be associated with the events that increase angiogenesis and it may potentiate or promote tumor angiogenesis in both pre-and post neoplastic lesions of the pancreas [33].

16.3 ESTROGENS: ACTIVATORS OF THE ANGIOGENIC SWITCH

Estrogens are a group of steroid hormones in women that play fundamental roles in the development of the sex organs and breasts, the regulation of the menstrual cycle and during pregnancy [35,36]. They also play crucial roles in maintaining normal fetal development, differentiation, and reproductive physiology [37]. Estrogens also uphold normal physiological balance in maintaining skeletal, cardiovascular, and central nervous systems [38]. They are biosynthesized from cholesterol through a cascade of biochemical reactions with the involvement of a series of enzymes.

The biochemical structure of estrogens is an 18-carbon steroid with four rings, namely A, B, C, and D. There is a phenolic hydroxyl group at C-3 position of the aromatic A ring, and a methyl group at position C-13 in D ring (Figure 16.2). Out of all naturally occurring estrogens, 17β-estradiol (E2) is the most important estrogen, followed by its metabolite estrone (E1). Estriol (E3) is another important estrogen metabolite having potential physiological roles. Cytochrome P450 enzymes play a crucial role in estrogen synthesis and subsequent oxidative metabolism. 17β-Hydroxysteroid dehydrogenase 1 (17β-HD1) is another key enzyme in estrogen metabolism as it exclusively catalyzes the reversible reaction between E2 and E1 [39] (Figure 16.3). E2 is hydroxylized by the help of cytochrome P450 producing

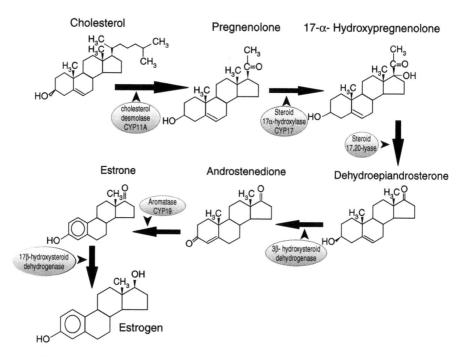

FIGURE 16.2 Sequential events estrogen biosynthesis. Estrogen is an 18 carbon steroid hormone that is produced from cholesterol through various mechanisms. During this multi-step process the key component produced is estrone, which is further converted into estrogen.

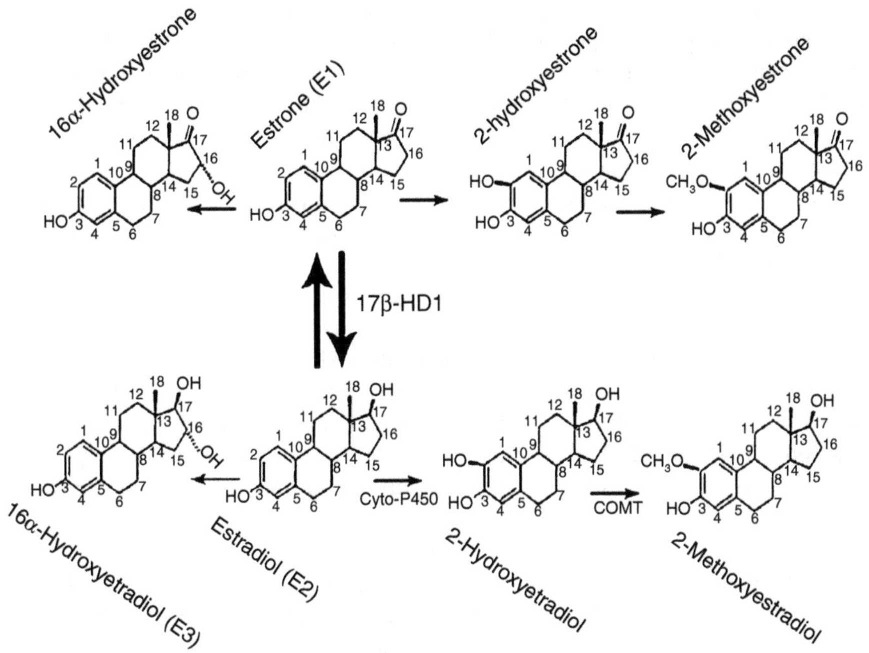

FIGURE 16.3 Sequential events of estrogen metabolism and formation of 2-Methoxyestradiol. Estrogen produced from cholesterol via the formation of estrone (E1), where 17β-hydroxysteroid dehydrogenase 1 (17β-HD1) enzyme plays an important role for the conversion of estrogen (E2) from E1. In one pathway, E2 is converted to 16a-hydroxyestradiol. However, in another pathway E2 is hydroxylized at 2-C position with the help of cytochrome P450, and further methylation at the same position by catechol-*O*-methyltransferase (COMT) and 2-Methoxyestradiol.

2-hydroxy estradiol, an important estrogen metabolite with a carcinogenic effect, showing significant potential to damage DNA [40]. Further enzymatic activation of 2-hydroxy estradiol by catechol-O-methyltransferase (COMT) subsequently produces 2-ME$_2$, which is an important anti-cancer drug (Figure 16.3). E1 is the most prominent estrogen metabolite circulating in postmenopausal women.

In pre-menopausal women, estrogens are produced primarily in the ovaries (and in the placenta during pregnancy). Small amounts are also produced by the adrenals. Follicle-stimulating hormone (FSH) upregulates aromatase expression in granulosa cells. Leutinizing hormone (LH) stimulates production of the substrate for aromatase, whereas FSH increases the amount of the enzyme so that estradiol production can increase by 8- to 10-fold at the time of ovulation [41]. In postmenopausal women, estrogen synthesis takes place nearly exclusively in extra glandular tissues. Androstenedione is produced primarily by the adrenal glands and is converted to estrone by aromatase in the peripheral tissues, such as adipose tissue [42]. The enzyme 17-HD1 then converts estrone to estradiol (Figure 16.3). Through the androstenedione-to-estrone pathway, postmenopausal women produce approximately 100 mg of estrone per day. A substantial fraction of estrone is converted to estradiol to produce circulating concentrations of 10–20 pg/mL [43]. In men, small amounts of estrogens are produced by the adrenal glands and testicles. The actions of estrogens are regulated by their concentrations in the circulation and are taken up by the target tissue depending upon the ER status of that tissue. In pre-menopausal women, the ratio of E2 between breast tissue and plasma is 1:1. However, in postmenopausal women this ratio becomes about 40:1 because the main source of estrogens are the peripheral adipose tissue [40], where aromatase, 17β-HSD1 and estrone sulfatase are actively and locally involved in estrogen production [44,45].

Estrogens bind to the nuclear estrogen receptor (ER) protein and transactives the ER by dimerization [46] and phosphorylation, which effectively activates gene transcription by estrogen response elements (EREs), or acts by nonnuclear mechanisms like rapid activation of kinase signaling cascades [47]. After dimerization and phosphorylation, the ER transactives two transcriptional activation functions, AF-1 and AF-2, depending upon responsive promoter and cell types [48]. ER phosphorylation at serine 118 is important for the transcription activation function of AF-1 [49].

Normally, estrogens persuade the growth, differentiation, and function of tissues of the female reproductive system, such as the ovary, uterus, or breast, and the non-reproductive systems, such as the cardiovascular system and bone, through an ER dependent manner [40]. However, the patho-physiological actions of estrogens are overwhelming. They play a most significant and crucial role in promoting the reproductive organ specific cancers through multiple genomic and non-genomic signaling pathways. E2 significantly induces cell proliferation by the induction G1/S during the cell cycle in breast cancer [50]. E2 inhibits the production of p27 CKD inhibitor, which consequently allows the activation of cyclin E-CDK2 or cyclin D-CDK4 to proliferate the cell through the S phase [51]. Moreover, estrogens have been considered as potent stimulators of tumor angiogenesis and this angiogenic effect is dose dependent [52] and is mediated through the estrogen receptor [22,24,53]. Studies from our laboratory and others have shown that estrogen is able to induce proangiogenic factors including VEGF-A, fibroblast growth factor (FGF), transforming growth factor (TGF), platelet-derived growth factor (PDGF) and others, which have been discussed in subsequent sections [25–27,52]. The concept of "the balance hypothesis" for the angiogenic switch [54] indicates that the modulation of anti-angiogenic factors is also a key step in regulation of this event. Our studies have shown that estrogen not only enhances angiogenic factors, it also inhibits the anti-angiogenic factor, thombospondin-1 (TSP-1) during angiogenesis [55].

16.4   ANGIOGENIC SWITCH BY ESTROGEN: THE ROLE OF VASCULAR ENDOTHELIAL GROWTH FACTOR

Vascular endothelial growth factor is a polypeptide, and is structurally related to PDGF. The gene for VEGF is located on chromosome 6p12. This protein has an apparent molecular mass ∼45 kDa under non-reducing conditions and about

23 kDa under reducing conditions, suggesting that VEGF is a homodimeric glycoprotein [56]. Vascular endothelial growth factor is a predominant inducer of tumor angiogenesis and an important prognostic factor in breast cancer [57]. It plays a pivotal role in angiogenesis, by promoting microvascular endothelial cell proliferation, migration, and assembly into new vessels [58]. Vascular endothelial growth factor is a potent mitogen for micro- and macrovascular endothelial cells [56,59] and induces vasodilatation in vitro in a dose-dependent fashion [60]. Immunohistochemical analysis showed a positive correlation of VEGF with vessel involvement/density and the presence of lymph node metastasis, suggesting that patients with VEGF-positive tumors had a worse prognosis than those with VEGF-negative tumors [61]. Elevations in VEGF levels have also been detected in the serum of cancer patients. Serum levels of VEGF were increased significantly according to stage progression [62,63]. Vascular endothelial growth factor is involved in the development of colorectal cancer, and the measurement of VEGF in the serum may be a useful noninvasive clinical marker for evaluating the disease status [64]. The availability of specific monoclonal antibodies, which are capable of inhibiting VEGF-induced angiogenesis in vivo and in vitro [65,66] indicate a prominent role of VEGF in angiogenesis.

Analysis of the cDNA sequence of VEGF clones predicts five VEGF mRNA species encoding human VEGF isoforms of 121, 145, 165, 189, and 206 amino acids (Figure 16.3). Alternate splicing of mRNA from a single VEGF gene, containing eight exons, generates these isoforms [67]. Vascular endothelial growth factor$_{121}$ and VEGF$_{165}$ are the predominant isoforms secreted by a variety of normal and transformed cells [68–70]. These two isoforms are prime regulators of endothelial cell proliferation, angiogenesis, and vascular permeability [71]. Rat, mouse and bovine VEGF isoforms are predicted to be shorter by one amino acid (i.e., 120, 144, 164, 188, and 205). Recent studies have demonstrated the possible existence of an additional VEGF splice variant in the rat penis tissue, which could encode a protein of 110 amino acid residues [72] (see Figure 16.4).

Vascular endothelial growth factor mediates its mitogenic and vasopermeabilic effects through the two tyrosine kinase family receptors, which are Flt-1 (fms-like tyrosine kinase)/VEGFR1, and KDR/Flk-1 (mouse fetal liver kinase)/VEGFR2 [73]. Both receptors have been shown to be selectively expressed in endothelial cells [74] and sporadically expressed in tumor cells. In addition to these two major receptors, VEGF$_{165}$ binds with a docking receptor, which is known as VEGF$_{165}$R [68,75,76]. This isoform specific docking-receptor is identical to human neuropilin-1(NRP-1), a receptor for collapsin/semaphorin that mediates neural cell guidance [77–79]. NRP-1 expression has been detected in both endothelial and tumor-derived human breast and prostate cells [76]. The expression of this gene can be modulated by tumor necrosis factor-α (TNF-α) in human vascular endothelial cells [80]. The precise role of this co-receptor in regulation of angiogenesis is uncertain. Previous studies suggest that NRP-1 may enhance the bioactivity of VEGF$_{165}$ and also augments the binding of VEGF-A$_{165}$ to KDR [76]. Gene disruption studies imply that NRP-1 may be a crucial regulator of angiogenesis, as mouse embryos, lacking a functional neuropilin gene, decease due to an inappropriate development of their cardiovascular system [81].

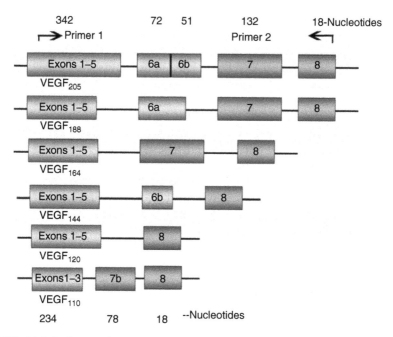

FIGURE 16.4 Isoforms of Vascular endothelial growth factor (VEGF). There are five isoforms of VEGF with 120, 144, 164, 188, and 205 amino acid in rodent and bovine.

Multiple studies from our laboratory demonstrated the importance of VEGF and its receptor, KDR/Flk-1/VEGFR2, as well as NRP-1, in the regulation of tumor angiogenesis during the development of an estrogen-induced pituitary tumor in Fisher 344 rats, which is one of the most intensively studied experimental models of hormonal carcinogenesis and angiogenesis [52]. Vascular endothelial growth factor protein, VEGFR2, and docking/co-receptor NRP-1 were identified using immuno-histochemical analysis, in both endothelial and non-endothelial cells, in rat pituitaries and suggested that VEGF-A$_{165}$, VEGFR2, and NRP-1 may actively participate in the regulation of estrogen-induced tumor angiogenesis. This perception was further established by our subsequent in vivo studies, which indicated that the anti-angiogenic compound, 2-ME$_2$, is able to block estrogen induced rat pituitary tumor angiogenesis through the inhibition of VEGF synthesis [82].

The amplitude expression of VEGF by estrogen has been identified in both endothelial cell and non-endothelial tumor cells [22,25–27,83,84] and this expression is mediated by both estrogen receptor-alpha (ER-α) dependent and independent pathways [25,85]. However, the possible functional roles of estrogen-induced VEGF in tumor cells have been elusive. It is anticipated that upon secretion from tumor cells, this growth factor may contribute to tumor growth stimulation as well as tumor angiogenic stimulation by autocrine–paracrine loops through the interaction with cell surface receptors (Figure 16.4). This anticipation is logical because recent studies from our laboratory and others have shown that estrogen, by regulating expression and secretion of VEGF by non-endothelial cells, regulates the angiogenic switch [30,86].

16.5 ANGIOGENIC SWITCH BY ESTROGEN: THE ROLE OF FIBROBLAST GROWTH FACTOR

There are at least ten distinct members of the FGF family of growth factors. The two originally characterized FGFs identified were termed as FGF1 (acidic-FGF, aFGF) and FGF2 (basic-FGF, bFGF). Acidic fibroblast growth factor and bFGF are polypeptides with multiple biological activities in vivo and in vitro, including roles in mitogenesis, cellular differentiation, and repair of tissue injury and angiogenesis [87,88]. The effect and the characteristic of bFGF have been extensively studied in many cell types. Basic fibroblast growth factor is an important mitogenic and angiogenic factor that stimulates endothelial cell growth and migration. The bFGF expression is complex because at least four (18, 22, 22.5, and 24 kDa) bFGF isoforms in humans are synthesized through the alternative use of translation initiation codons [89,90]. Basic fibroblast growth factor or FGF2 promotes angiogenesis in vitro and in vivo [91]. The expression of bFGF has been described in a wide variety of animal and human tumors [54]. The elevated levels of bFGF, with its clinical implications, have been identified in the urine or serum of a substantial number of patients with malignancies, including renal cell carcinoma [92,93], breast cancer [94,95], lymphoblastic leukemia [96], and non-small cell lung cancer [97]. Several studies have also added the value of this laboratory parameter as a prognostic indicators [98,99].

The effect of estrogen and its compounds on the expression of bFGF showed significant results in normal tissue, several cancer cell types, and animal model systems. Estradiol significantly enhances the angiogenic effect of bFGF in mice [100] and enhances vascular development in $Fgf2$ gene ($Fgf2^{-/-}$) knockout mice through the induction of bFGF [101]. Moreover, the migration and cell proliferation of cultured endothelial cells obtained from the subcutaneous Matrigel plugs of $Fgf2^{+/+}$ mice can be enhanced by estradiol. Western blot analysis also showed that untreated $Fgf2^{+/+}$ endothelial cells expressed mainly the FGF2^{lmw} isoform, whereas estrogen treatment significantly induced the FGF2^{hmw} protein expression. FGF2^{hmw} was found more abundant in the nuclei from estrogen-treated endothelial cells than in untreated controls. The studies thus indicate that the amplitude expression of bFGF by estrogen is a critical event in estrogen-induced tumor angiogenesis.

16.6 ANGIOGENIC SWITCH BY ESTROGEN: THE ROLE OF TRANSFORMING GROWTH FACTOR

Transforming growth factor (TGF) is one of several proteins secreted by transformed cells that can stimulate the growth of normal cells. There are two prominent transforming growth factors, namely TGF-α and TGF-β. TGF-α binds the

EGF receptor (EGFR) and stimulates the growth of various cell types. TGF-α is a small 50 amino acid residue long and is a mitogenic protein that contains three disulfide bridges. It shares about 30% sequence identity with EGF and competes with EGF for the same membrane-bound receptor sites [102], stimulation of which induces its dimerization, autophosphorylation on tyrosine residues, and triggers several signal transduction pathways, such as phosphatidylinositol 3-kinase/protein kinase B (PI 3-kinase/Akt) and the p44/p42 MAPK [103]. Higher TGF-α expression has been found in breast cancer patients [104]. TGF-α and other FGF-like proteins are thought to play a role in angiogenesis [105]. TGF-β is an ubiquitous cytokine that affects various biological processes including angiogenesis [106]. TGF-β family consists of three isoforms: TGF-β1, TGF-β2, and TGF-β [107].

TGF-β plays a biphasic role in cancer. In the earlier stages of carcinogenesis, it is known to act as a tumor suppressor, whereas in the later stages it is known to support tumor progression [108]. It is a potent growth inhibitor in epithelial tissues and has a tumor suppression function that is lost in many tumor-derived cell lines [109,110]. Mutation in TGF-β has been found in nearly all pancreatic cancers [111,112] and colon cancers [113]. It has been found that there is a marked increase in the expression of TGF-β mRNA and protein in human cancers (in vivo), including those of the pancreas, colon, stomach, lung, endometrium, prostate, breast, brain, and bone [110]. TGF-β receptors are known to act as tumor suppressors at early stages of carcinoma development. In tumor cells, TGF-β receptors are often down-regulated or the TGF-β receptor availability at the cell surface is impaired, and these defects are considered to help the cells to escape from the growth inhibitory properties of TGF-β [114,115].

TGF-β1 stimulated VEGF production by lung fibroblasts from Smad2 deficient animals and wild-type animals. In contrast, TGF-β1 did not affect VEGF production by fibroblasts from Samd3 deficient mice [116]. On the other hand, VEGF attenuates TGF-β action in the human endothelial cell, specifically at the level of transcription of the PAI-1 gene and Smad2/3 phosphorylation [117]. Matrix GLA protein treatment in endothelial cells showed increased proliferation, migration, tube formation, and increased release of VEGF-A and bFGF through the induction of TGF-β [118]. Studies indicate that TGF-β1 treatment may regulate angiogenesis in pituitary cells by initially increasing levels of pro-angiogenic VEGF-A and then stimulating the anti-angiogenic molecules TSP-1 and TSP-2 levels [119]. TGF-β1 also induces VEGF-A and TSP-1 in Rat proximal tubular cells NRK52E cells and has been associated with the activation of pathway-restricted receptor-activated Smads (Smad2 and 3) [120], which indicates both the anti- and proangiogenic potential of TGF-β.

TGF-α activation by estrogen has been associated with higher angiogenic rates in invasive breast cancer cases [121]. Estrogen also stimulates the growth of prostatic stromal cells and increases smooth muscle cell markers through an involvement of a TGF- β pathway [122]. Co-treatment with estrogen and 3,3′-Diindolylmethane (DIM) significantly increased the TGF- α protein expression through an ER-dependent manner in Ishikawa cells [123]. Studies suggest that TGF-β2 is regulated in a tissue specific manner and the secretion of TGF-β2 is tightly regulated by hormones. These studies also show that estrogen and prolactin are critical factors in the tissue-specific regulation of the local production of TGF-β2 in the mammary gland and female reproductive tract [124].

16.7 ANGIOGENIC SWITCH BY ESTROGEN: THE ROLE OF PLATELET-DERIVED GROWTH FACTOR

Platelet-derived growth factor plays an important role in tumor growth and angiogenesis [125]. Platelet-derived growth factor was originally identified in platelets and in serum as a mitogen for fibroblasts, smooth muscle cells (SMC), and glia cells in culture [126]. The PDGF is mitogenic for many cells [127,128]. It consists of two monomers: an A-chain (PDGF-AA) and a B-chain (PDGF-BB). In its active form, PDGF is a disulfide-bound dimer of two monomers, and all dimeric combinations are as PDGF-AA, PDGF-AB, and PDGF-BB [129,130].

Platelet-derived growth factor-AA and PDGF-BB have been considered as autocrine and paracrine factors in various types of tumors [131–133]. Platelet-derived growth factor-B also increases the expression of several angiogenic factors that include increased VEGF expression in fibroblasts [134] and ECs [135]. It has been demonstrated that PDGF-B stimulates vascular ECs that express PDGF-Rβ to form tube-like networks in vitro [136]. In fibroblasts and vascular smooth muscle cells, PDGF induces VEGF expression [137,138], and this VEGF induction was in a PKC dependent manner [137]. Platelet-derived growth factor-B is a paracrine factor in U87MG gliomas, and enhances glioma angiogenesis by stimulating VEGF expression in tumor endothelia and by recruiting pericytes to the neovessels [139].

There are two receptor subunits for PDGF, the PDGFα receptor (PDGFRα) and the PDGFβ receptor (PDGFRβ). The binding of PDGF ligands with their receptors induces the dimerization of the subunits [140]. The PDGFRα binds to either the A- or B-chain, whereas the PDGFRβ only binds to the B-chain, thus forming homo or heterodimers [130], and the ligand induces the receptor autophosphorylation [141]. Stimulation of the PDGF receptor activates an enzyme cascade that includes various phosphorylating enzymes, i.e., protein kinase C (PKC), Ras, Raf, and mitogen-activated protein kinase (MAPK), and ultimately triggers cell division [142]. Increased PDGF and PDGFR expression have also been implicated in the progression of human cancers [143,144]. In human glioblastomas, PDGFR expression is increased in tumor cells, whereas PDGFRβ expression is elicited in neovascular ECs, indicating both autocrine and paracrine actions of PDGF in tumor growth and angiogenesis [125]. PDGFRα and PDGFRβ are expressed in a high percentage of epithelial ovarian cancers [145]. Mice deficient in either PDGF-B or PDGFRβ developed hemorrhages or edemas during the later stages of embryogenesis [146]. Exogenously expressed PDGF-B increased proliferation of stromal cells and enhanced angiogenesis in tumors [147].

Estrogen regulates PDGF-A, PDGF-B, and PDGF-Rα, which may be related to the vasculoprotective effect of estrogen [148]. One review demonstrates that MCF-7 breast cancer cells secret PDGF along with some other growth factors, and this activity is regulated by estradiol [149]. A PDGF-like factor also has been found to be secreted by MCF-7 cells, and this activity is enhanced by the induction of estradiol [150].

16.8   ANTI-ANGIOGENIC FACTORS AND THE ROLE OF ESTROGEN

TSP-1, a homotrimeric 450 kDa glycoprotein, is a well studied anti-angiogenesis regulator. In endothelial cells, TSP-1 inhibits cell proliferation [151] and angiogenesis [15] and has been considered an effector molecule for the tumor suppressor gene p53 [152]. A suppressive activity of TSP-I in cancer cell proliferation and metastasis has been shown in various experimental models such as murine melanoma, lung and breast human cell lines [153,154], and in endothelial cells [55]. Decreased expression of TSP-1 has been observed in tumor progression, such as prostate [155] and bladder [156]. p53 has been shown to increase the expression of the major TSP-1 [152,157] and to decrease the expression of the major angiogenic stimulator VEGF [158,159]. The additive effects of TSP- and VEGF-siRNA delayed the onset of tumors in nude mice [160]. TSP-1 inhibits angiogenesis in vivo and in vitro and induces apoptotic cell death by utilizing a transmembrane receptor CD36 and a sequential activation of Src family kinase p59fyn and p38 mitogen-activated protein kinases [161]. Moreover, reduced expression of TSP-1 is the cause of mammary and kidney tumor angiogenesis, and is achieved through the sequential activation of Ras, phosphatidylinositol-3-kinase (PI3K) and Myc phosphorylation [162]. TSP-1 has been found to suppress tumor growth and angiogenesis in mice by inhibiting matrix metalloproteinase-9 and VEGF expression [162]. TSP- mediates endothelial cell apoptosis and inhibits angiogenesis in association with the an increased expression of Bax and a decreased expression of Bcl-2 [163].

A recent study from our laboratory explored the potential role of TSP-1 in inhibiting cell proliferation and invasion in endothelial cells treated with estrogen. The study demonstrated that estrogen inhibits both mRNA and protein expression of TSP-1, and this down-regulation was mediated through the non-genomic ER/MAPKK/ERK1/2 and JNK signaling pathways [164]. The studies have also shown that the exogenous addition of TSP-1 diminishes estrogen-induced endothelial cell proliferation and migration [164]. Therefore, our studies suggest that TSP-1 can be considered an important negative regulator in our understanding of the increased angiogenesis in response to estrogens.

16.9   ROLE OF THE ESTROGEN METABOLITE, 2-METHOXYES-TRADIOL AS AN ANTI-ANGIOGENIC AGENT

2-Methoxyestradiol (2-ME$_2$) is a physiological metabolite of estrogen generated from the sequential hydroxylation and methylation at 2-C position of 17β-estradiol by cytochrome P450 and catechol-O-methyl transferase, respectively [165]. In recent years, 2-ME$_2$ has become a promising anticancer drug [166]. 2-ME$_2$ has been demonstrated to inhibit both tumor growth and angiogenesis in vivo and in vitro [167–169]. A study from our laboratory showed that 2-ME$_2$ inhibited estrogen-induced lactotroph growth and tumor angiogenesis in the female Fischer 344 rat [26]. Further studies in our laboratory reported a novel biphasic role of 2-ME$_2$ in angiogenesis in ER$^+$ tumor cell lines, such as GH3 rat pituitary tumor cells and MCF-7 human breast cancer cells, and ER$^-$ PaCa-2 pancreatic adenocarcinoma cells.

2-ME$_2$ at low doses (1 μM) induced higher VEGF-A mRNA expression in GH3 and MCF-7 cells; however, at higher doses (5 and 10 μM), the VEGF expression was downregulated in both GH3 and PaCa-2 cells. This 2-ME$_2$-induced modulation of VEGF expression was ER mediated because the pure anti-estrogen ICI,182780 altered the VEGF expression [30]. 2-ME$_2$ has been shown to inhibit angiogenesis in C57BL/6 mice by inhibiting proangiogenic factors bFGF and VEGF [167]. Moreover, 2-ME$_2$ downregulated HIF-1α and eventually inhibited angiogenesis through the inhibition of VEGF expression in breast and prostate cancer cells [170]. Basic fibroblast growth factor-induced angiogenesis was also inhibited by 2-ME$_2$, which also inhibited endothelial cell migration and vitronectin-induced bovine pulmonary artery endothelial cells migration in a concentration-dependent manner [171].

16.10 CONCLUSION

Several recent advances in the field of hormonal carcinogenesis have contributed to understanding of the molecular events involved in estrogen signaling in the induction and promotion of cancer. The most revealing finding is that estrogen is capable of inducing tumor angiogenic switch through the activation or inactivation of pro- and anti-angiogenic factors by modulating multiple signal transduction pathways. Interestingly, a metabolite of estrogen, such as 2-Methoxyestradiol, exhibits a contrasting impact on the development of cancer. Therefore, it has become a promising anticancer drug. However, the molecular mechanism(s) associated with the anticancer property of this agent have not yet been fully elucidated. Ongoing studies will no doubt uncover the mechanism of action of this metabolite on the angiogenic switch.

REFERENCES

1. Folkman, J., The role of angiogenesis in tumor growth, *Semin. Cancer Biol.*, 3, 65–71, 1992.
2. Anasti, J. N., Kalantaridou, S. N., Kimzey, L. M., George, M., and Nelson, L. M., Human follicle fluid vascular endothelial growth factor concentrations are correlated with luteinization in spontaneously developing follicles, *Hum. Reprod.*, 13, 1144–1147, 1998.
3. Maas, J. W., Groothuis, P. G., Dunselman, G. A., de Goeij, A. F., Struyker-Boudier, H. A., and Evers, J. L., Endometrial angiogenesis throughout the human menstrual cycle, *Hum. Reprod.*, 16, 1557–1561, 2001.
4. Arnold, F. and West, D. C., Angiogenesis in wound healing, *Pharmacol. Ther.*, 52, 407–422, 1991.
5. Folkman, J. and Klagsbrun, M., Angiogenic factors, *Science*, 235, 442–447, 1987.
6. Folkman, J., What is the evidence that tumors are angiogenesis dependent? *J. Natl Cancer Inst.*, 82, 4–6, 1990.

7. Bergers, G. and Benjamin, L. E., Tumorigenesis and the angiogenic switch, *Nat. Rev. Cancer*, 3, 401–410, 2003.

8. Folkman, J., Fundamental concepts of the angiogenic process, *Curr. Mol. Med.*, 3, 643–651, 2003.

9. Klein, S., Roghani, M., and Rifkin, D. B., Fibroblast growth factors as angiogenesis factors: new insights into their mechanism of action, *EXS*, 79, 159–192, 1997.

10. Fain, J. N. and Madan, A. K., Insulin enhances vascular endothelial growth factor, interleukin-8, and plasminogen activator inhibitor 1 but not interleukin-6 release by human adipocytes, *Metabolism*, 54, 220–226, 2005.

11. Olson, K. A., Byers, H. R., Key, M. E., and Fett, J. W., Inhibition of prostate carcinoma establishment and metastatic growth in mice by an antiangiogenin monoclonal antibody, *Int. J. Cancer*, 98, 923–929, 2002.

12. Hockel, M., Sasse, J., and Wissler, J. H., Purified monocyte-derived angiogenic substance (angiotropin) stimulates migration, phenotypic changes, and "tube formation" but not proliferation of capillary endothelial cells in vitro, *J. Cell. Physiol.*, 133, 1–13, 1987.

13. Toi, M., Atiqur, R. M., Bando, H., and Chow, L. W., Thymidine phosphorylase (platelet-derived endothelial-cell growth factor) in cancer biology and treatment, *Lancet Oncol.*, 6, 158–166, 2005.

14. Sheen-Chen, S. M., Liu, Y. W., Eng, H. L., and Chou, F. F., Serum levels of hepatocyte growth factor in patients with breast cancer, *Cancer Epidemiol. Biomarkers Prev.*, 14, 715–717, 2005.

15. Good, D. J., Polverini, P. J., Rastinejad, F., Le Beau, M. M., Lemons, R. S., Frazier, W. A., and Bouck, N. P., A tumor suppressor-dependent inhibitor of angiogenesis is immunologically and functionally indistinguishable from a fragment of thrombospondin, *Proc. Natl Acad. Sci. U.S.A.*, 87, 6624–6628, 1990.

16. Matsunaga, T., Weihrauch, D. W., Moniz, M. C., Tessmer, J., Warltier, D. C., and Chilian, W. M., Angiostatin inhibits coronary angiogenesis during impaired production of nitric oxide, *Circulation*, 105, 2185–2191, 2002.

17. O'Reilly, M. S., Boehm, T., Shing, Y., Fukai, N., Vasios, G., Lane, W. S., Flynn, E., Birkhead, J. R., Olsen, B. R., and Folkman, J., Endostatin: an endogenous inhibitor of angiogenesis and tumor growth, *Cell*, 88, 277–285, 1997.

18. Brake, T. and Lambert, P. F., Estrogen contributes to the onset, persistence, and malignant progression of cervical cancer in a human papillomavirus-transgenic mouse model, *Proc. Natl Acad. Sci. U.S.A.*, 102, 2490–2495, 2005.

19. Henderson, B. E., Ross, R., and Bernstein, L., Estrogens as a cause of human cancer: the Richard and Hinda Rosenthal Foundation award lecture, *Cancer Res.*, 48, 246–253, 1988.

20. Feigelson, H. S. and Henderson, B., Estrogens and breast cancer, In *Endocrinology of Breast Cancer*, Manni, A. Ed., Humana Press, New Jersey, pp. 55–67, 1999.

21. Cid, M. C., Schnaper, H. W., and Kleinman, H. K., Estrogens and the vascular endothelium, *Ann. N.Y. Acad. Sci.*, 966, 143–157, 2002.

22. Losordo, D. W. and Isner, J. M., Estrogen and angiogenesis: a review, *Arterioscler. Thromb. Vasc. Biol.*, 21, 6–12, 2001.

23. Zoubine, M. N., Weston, A. P., Johnson, D. C., Campbell, D. R., and Banerjee, S. K., 2-Methoxyestradiol-induced growth suppression and lethality in estrogen-responsive MCF-7 cells may be mediated by downregulation of p34cdc2 and cyclin B1 expression, *Int. J. Oncol.*, 15, 639–646, 1999.

24. Johns, A., Freay, A. D., Fraser, W., Korach, K. S., and Rubanyi, G. M., Disruption of estrogen receptor gene prevents 17 beta estradiol-induced angiogenesis in transgenic mice, *Endocrinology*, 137, 4511–4513, 1996.

25. Banerjee, S., Saxena, N., Sengupta, K., and Banerjee, S. K., 17alpha-estradiol-induced VEGF-A expression in rat pituitary tumor cells is mediated through ER independent but PI3K-Akt dependent signaling pathway, *Biochem. Biophys. Res. Commun.*, 300, 209–215, 2003.

26. Banerjee, S. K., Sarkar, D. K., Weston, A. P., De, A., and Campbell, D. R., Over expression of vascular endothelial growth factor and its receptor during the development of estrogen-induced rat pituitary tumors may mediate estrogen-initiated tumor angiogenesis, *Carcinogenesis*, 18, 1155–1161, 1997.

27. Banerjee, S. K., Zoubine, M. N., Tran, T. M., Weston, A. P., and Campbell, D. R., Overexpression of vascular endothelial growth factor164 and its co-receptor neuropilin-1 in estrogen-induced rat pituitary tumors and GH3 rat pituitary tumor cells, *Int. J. Oncol.*, 16, 253–260, 2000.

28. Gagliardi, A. and Collins, D. C., Inhibition of angiogenesis by antiestrogens, *Cancer Res.*, 53, 533–535, 1993.

29. Ricker, J. L., Chen, Z., Yang, X. P., Pribluda, V. S., Swartz, G. M., and Van, W. C., 2-Methoxyestradiol inhibits hypoxia-inducible factor 1alpha, tumor growth, and angiogenesis and augments paclitaxel efficacy in head and neck squamous cell carcinoma, *Clin. Cancer Res.*, 10, 8665–8673, 2004.

30. Banerjee, S. N., Sengupta, K., Banerjee, S., Saxena, N., and Banerjee, S. K., 2-Methoxyestradiol exhibits a biphasic effect on VEGF-A in tumor cells and upregulation is mediated through ER-α: a possible signaling pathway associated with the impact of 2-ME2 on proliferative cells, *Neoplasia*, 5, 417–426, 2003.

31. Bergers, G. and Benjamin, L. E., Angiogenesis: tumorigenesis and the angiogenic switch, *Nat. Rev. Cancer*, 3, 401–410, 2003.

32. Vajkoczy, P., Farhadi, M., Gaumann, A., Heidenreich, R., Erber, R., Wunder, A., Tonn, J. C., Menger, M. D., and Breier, G., Microtumor growth initiates angiogenic sprouting with simultaneous expression of VEGF, VEGF receptor-2, and angiopoietin-2, *J. Clin. Invest.*, 109, 777–785, 2002.

33. Banerjee, S. K., Zoubine, M. N., Mullick, M., Weston, A. P., Cherian, R., and Campbell, D. R., Tumor angiogenesis in chronic pancreatitis and pancreatic adenocarcinoma: impact of K-ras mutations, *Pancreas*, 20, 248–255, 2000.

34. Kitadai, Y., Onogawa, S., Kuwai, T., Matsumura, S., Hamada, H., Ito, M., Tanaka, S., Yoshihara, M., and Chayama, K., Angiogenic switch occurs during the precancerous stage of human esophageal squamous cell carcinoma, *Oncol. Rep.*, 11, 315–319, 2004.

35. Norman, A. W. and Litwack, G., Estrogens and progestins, In *Anonymous-Hormones*, Academic Press, New York, pp. 361–386, 1997.

36. Travis, R. C. and Key, T. J., Oestrogen exposure and breast cancer risk, *Breast Cancer Res.*, 5, 239–247, 2003.

37. Beato, M., Gene regulation by steroid hormones, *Cell*, 56, 335–344, 1989.

38. Manolagas, S. C. and Kousteni, S., Perspective: nonreproductive sites of action of reproductive hormones, *Endocrinology*, 142, 2200–2204, 2001.

39. Poutanen, M., Isomaa, V., Peltoketo, H., and Vihko, R., Regulation of oestrogen action: role of 17 beta-hydroxysteroid dehydrogenases, *Ann. Med.*, 27, 675–682, 1995.

40. Parl, F. F., Estrogen receptor expression in breast cancer, In *Anonymous Estrogens, Estrogen Receptor and Breast Cancer*, IOS Press, Ohmsha, Washington, DC, pp. 135–204, 2002.

41. Santen, R. J., Samojlik, E., and Wells, S. A., Resistance of the ovary to blockade of aromatization with aminoglutethimide, *J. Clin. Endocrinol. Metab.*, 51, 473–477, 1980.

42. Judd, H. L., Judd, G. E., Lucas, W. E., and Yen, S. S., Endocrine function of the postmenopausal ovary: concentration of androgens and estrogens in ovarian and peripheral vein blood, *J. Clin. Endocrinol. Metab.*, 39, 1020–1024, 1974.

43. Kirschner, M. A., Schneider, G., Ertel, N. H., and Worton, E., Obesity, androgens, estrogens, and cancer risk, *Cancer Res.*, 42, 3281s–3285s, 1982.

44. James, V. H., McNeill, J. M., Beranek, P. A., Bonney, R. C., and Reed, M. J., The role of tissue steroids in regulating aromatase and oestradiol 17 beta-hydroxysteroid dehydrogenase activities in breast and endometrial cancer, *J. Steroid Biochem.*, 25, 787–790, 1986.

45. Pasqualini, J. R., Gelly, C., Nguyen, B. L., and Vella, C., Importance of estrogen sulfates in breast cancer, *J. Steroid Biochem.*, 34, 155–163, 1989.

46. Notides, A. C., Lerner, N., and Hamilton, D. E., Positive cooperativity of the estrogen receptor, *Proc. Natl Acad. Sci. U.S.A.*, 78, 4926–4930, 1981.

47. Martinez, E. and Wahli, W., Cooperative binding of estrogen receptor to imperfect estrogen-responsive DNA elements correlates with their synergistic hormone-dependent enhancer activity, *EMBO J.*, 8, 3781–3791, 1989.

48. Tasset, D., Tora, L., Fromental, C., Scheer, E., and Chambon, P., Distinct classes of transcriptional activating domains function by different mechanisms, *Cell*, 62, 1177–1187, 1990.

49. Ali, S., Metzger, D., Bornert, J. M., and Chambon, P., Modulation of transcriptional activation by ligand-dependent phosphorylation of the human oestrogen receptor A/B region, *EMBO J.*, 12, 1153–1160, 1993.

50. Taylor, I. W., Hodson, P. J., Green, M. D., and Sutherland, R. L., Effects of tamoxifen on cell cycle progression of synchronous MCF-7 human mammary carcinoma cells, *Cancer Res.*, 43, 4007–4010, 1983.

51. Foster, J. S. and Wimalasena, J., Estrogen regulates activity of cyclin-dependent kinases and retinoblastoma protein phosphorylation in breast cancer cells, *Mol. Endocrinol.*, 10, 488–498, 1996.

52. Banerjee, S. K., Campbell, D. R., Weston, A. P., and Banerjee, D. K., Biphasic estrogen response on bovine adrenal medulla capillary endothelial cell adhesion, proliferation and tube formation, *Mol. Cell. Biochem.*, 177, 97–105, 1997.

53. Krasinski, K., Spyridopoulos, I., Asahara, T., van der, Z. R., Isner, J. M., and Losordo, D. W., Estradiol accelerates functional endothelial recovery after arterial injury, *Circulation*, 95, 1768–1772, 1997.

54. Hanahan, D. and Folkman, J., Patterns and emerging mechanisms of the angiogenic switch during tumorigenesis, *Cell*, 86, 353–364, 1996.

55. Sengupta, K., Banerjee, S., Saxena, N. K., and Banerjee, S. K., Thombospondin-1 disrupts estrogen-induced endothelial cell proliferation and migration and its expression is suppressed by estradiol, *Mol. Cancer Res.*, 2, 150–158, 2004.

56. Ferrara, N. and Henzel, W. J., Pituitary follicular cells secrete a novel heparin-binding growth factor specific for vascular endothelial cells, *Biochem. Biophys. Res. Commun.*, 161, 851–858, 1989.

57. Relf, M., LeJeune, S., Scott, P. A., Fox, S., Smith, K., Leek, R., Moghaddam, A., Whitehouse, R., Bicknell, R., and Harris, A. L., Expression of the angiogenic factors vascular endothelial cell growth factor, acidic and basic fibroblast growth factor,

tumor growth factor beta-1, platelet-derived endothelial cell growth factor, placenta growth factor, and pleiotrophin in human primary breast cancer and its relation to angiogenesis, *Cancer Res.*, 57, 963–969, 1997.

58. Ferrara, N. and Davis-Smyth, T., The biology of vascular endothelial growth factor, *Endocr. Rev.*, 18, 4–25, 1997.

59. Pepper, M. S., Wasi, S., Ferrara, N., Orci, L., and Montesano, R., In vitro angiogenic and proteolytic properties of bovine lymphatic endothelial cells, *Exp. Cell Res.*, 210, 298–305, 1994.

60. Ku, D. D., Zaleski, J. K., Liu, S., and Brock, T. A., Vascular endothelial growth factor induces EDRF-dependent relaxation in coronary arteries, *Am. J. Physiol.*, 265, H586–H592, 1993.

61. Maeda, K., Chung, Y. S., Ogawa, Y., Takatsuka, S., Kang, S. M., Ogawa, M., Sawada, T., and Sowa, M., Prognostic value of vascular endothelial growth factor expression in gastric carcinoma, *Cancer*, 77, 858–863, 1996.

62. Kondo, S., Asano, M., Matsuo, K., Ohmori, I., and Suzuki, H., Vascular endothelial growth factor/vascular permeability factor is detectable in the sera of tumor-bearing mice and cancer patients, *Biochim. Biophys. Acta*, 1221, 211–214, 1994.

63. Matsuyama, W., Hashiguchi, T., Mizoguchi, A., Iwami, F., Kawabata, M., Arimura, K., and Osame, M., Serum levels of vascular endothelial growth factor dependent on the stage progression of lung cancer, *Chest*, 118, 948–951, 2000.

64. Fujisaki, K., Mitsuyama, K., Toyonaga, A., Matsuo, K., and Tanikawa, K., Circulating vascular endothelial growth factor in patients with colorectal cancer, *Am. J. Gastroenterol.*, 93, 249–252, 1998.

65. Kim, K. J., Li, B., Houck, K., Winer, J., and Ferrara, N., The vascular endothelial growth factor proteins: identification of biologically relevant regions by neutralizing monoclonal antibodies, *Growth Factors*, 7, 53–64, 1992.

66. Presta, L. G., Chen, H., O'Connor, S. J., Chisholm, V., Meng, Y. G., Krummen, L., Winkler, M., and Ferrara, N., Humanization of an anti-vascular endothelial growth factor monoclonal antibody for the therapy of solid tumors and other disorders, *Cancer Res.*, 57, 4593–4599, 1997.

67. Klagsbrun, M. and D'Amore, P. A., Regulators of angiogenesis, *Annu. Rev. Physiol.*, 53, 217–239, 1991.

68. Neufeld, G., Cohen, T., Gitay-Goren, H., Poltorak, Z., Tessler, S., Sharon, R., Gengrinovitch, S., and Levi, B. Z., Similarities and differences between the vascular endothelial growth factor (VEGF) splice variants, *Cancer Metastasis Rev.*, 15, 153–158, 1996.

69. Ferrara, N., The role of vascular endothelial growth factor in pathological angiogenesis, *Breast Cancer Res. Treat.*, 36, 127–137, 1995.

70. Neufeld, G., Cohen, T., Gengrinovitch, S., and Poltorak, Z., Vascular endothelial growth factor (VEGF) and its receptors, *FASEB J.*, 13, 9–22, 1999.

71. Cohen, T., Herzog, Y., Brodzky, A., Greenson, J. K., Eldar, S., Gluzman-Poltorak, Z., Neufeld, G., and Resnick, M. B., Neuropilin-2 is a novel marker expressed in pancreatic islet cells and endocrine pancreatic tumours, *J. Pathol.*, 198, 77–82, 2002.

72. Burchardt, M., Burchardt, T., Chen, M. W., Shabsigh, A., de la Taille, A., Buttyan, R., and Shabsigh, R., Expression of messenger ribonucleic acid splice variants for vascular endothelial growth factor in the penis of adult rats and humans, *Biol. Reprod.*, 60, 398–404, 1999.

73. Claffey, K. P. and Robinson, G. S., Regulation of VEGF/VPF expression in tumor cells: consequences for tumor growth and metastasis, *Cancer Metastasis Rev.*, 15, 165–176, 1996.

74. Bression, D., Brandi, A. M., Le, D. M., Cesselin, F., Hamon, M., Martinet, M., Kerdelhue, B., and Peillon, F., Modifications of the high and low affinity pituitary domperidone- binding sites in chronic estrogenized rat, *Endocrinology*, 113, 1799–1805, 1983.

75. Ortega, N., Hutchings, H., and Plouet, J., Signal relays in the VEGF system, *Front. Biosci.*, 4D141-52, D141–D152, 1999.

76. Soker, S., Takashima, S., Miao, H. Q., Neufeld, G., and Klagsbrun, M., Neuropilin-1 is expressed by endothelial and tumor cells as an isoform-specific receptor for vascular endothelial growth factor, *Cell*, 92, 735–745, 1998.

77. Feiner, L., Koppel, A. M., Kobayashi, H., and Raper, J. A., Secreted chick semaphorins bind recombinant neuropilin with similar affinities but bind different subsets of neurons in situ, *Neuron*, 19, 539–545, 1997.

78. Kolodkin, A. L. and Ginty, D. D., Steering clear of semaphorins: neuropilins sound the retreat, *Neuron*, 19, 1159–1162, 1997.

79. Kolodkin, A. L., Levengood, D. V., Rowe, E. G., Tai, Y. T., Giger, R. J., and Ginty, D. D., Neuropilin is a semaphorin III receptor, *Cell*, 90, 753–762, 1997.

80. Giraudo, E., Primo, L., Audero, E., Gerber, H. P., Koolwijk, P., Soker, S., Klagsbrun, M., Ferrara, N., and Bussolino, F., Tumor necrosis factor-alpha regulates expression of vascular endothelial growth factor receptor-2 and of its co-receptor neuropilin-1 in human vascular endothelial cells, *J. Biol. Chem.*, 273, 22128–22135, 1998.

81. Kitsukawa, T., Shimizu, M., Sanbo, M., Hirata, T., Taniguchi, M., Bekku, Y., Yagi, T., and Fujisawa, H., Neuropilin-semaphorin III/D-mediated chemorepulsive signals play a crucial role in peripheral nerve projection in mice, *Neuron*, 19, 995–1005, 1997.

82. Banerjee, S. K., Zoubine, M. N., Sarkar, D. K., Weston, A. P., Shah, J. H., and Campbell, D. R., 2-Methoxyestradiol blocks estrogen-induced rat pituitary tumor growth and tumor angiogenesis: possible role of vascular endothelial growth factor, *Anticancer Res.*, 20, 2641–2645, 2000.

83. Vanveldhuizen, P. J., Zulfiqar, M., Banerjee, S., Cherian, R., Saxena, N. K., Rabe, A., Thrasher, J. B., and Banerjee, S. K., Differential expression of neuropilin-1 in malignant and benign prostatic stromal tissue, *Oncol. Rep.*, 10, 1067–1071, 2003.

84. Zoubine, M. N., Hamilton, J. W., Vanveldhuizen, P., Weston, A. P., Campbell, D. R., and Banerjee, S. K., Overexpression of VEGF165 and its receptor neuropilin-1 in human breast and prostate cancer cell lines: essential cytokine and its receptor for tumor angiogenesis, *Proc. Am. Assoc. Cancer Res.*, 40, 453–454, 1999 (abstract).

85. Sengupta, K., Banerjee, S., Saxena, N., and Banerjee, S. K., Estradiol-induced vascular endothelial growth factor-A expression in breast tumor cells is biphasic and regulated by estrogen receptor-alpha dependent pathway, *Int. J. Oncol.*, 22, 609–614, 2003.

86. Albrecht, E. D., Babischkin, J. S., Lidor, Y., Anderson, L. D., Udoff, L. C., and Pepe, G. J., Effect of estrogen on angiogenesis in co-cultures of human endometrial cells and microvascular endothelial cells, *Hum. Reprod.*, 18, 2039–2047, 2003.

87. Burgess, W. H. and Maciag, T., The heparin-binding (fibroblast) growth factor family of proteins, *Annu. Rev. Biochem.*, 58, 575–606, 1989.

88. Yamasaki, M., Miyake, A., Tagashira, S., and Itoh, N., Structure and expression of the rat mRNA encoding a novel member of the fibroblast growth factor family, *J. Biol. Chem.*, 271, 15918–15921, 1996.

89. Florkiewicz, R. Z. and Sommer, A., Human basic fibroblast growth factor gene encodes four polypeptides: three initiate translation from non-AUG codons, *Proc. Natl Acad. Sci. U.S.A.*, 86, 3978–3981, 1989.

90. Arnaud, E., Touriol, C., Boutonnet, C., Gensac, M. C., Vagner, S., Prats, H., and Prats, A. C., A new 34-kilodalton isoform of human fibroblast growth factor 2 is cap dependently synthesized by using a non-AUG start codon and behaves as a survival factor, *Mol. Cell Biol.*, 19, 505–514, 1999.

91. Walgenbach, K. J., Gratas, C., Shestak, K. C., and Becker, D., Ischaemia-induced expression of bFGF in normal skeletal muscle: a potential paracrine mechanism for mediating angiogenesis in ischaemic skeletal muscle, *Nat. Med.*, 1, 453–459, 1995.

92. Dosquet, C., Coudert, M. C., Lepage, E., Cabane, J., and Richard, F., Are angiogenic factors, cytokines, and soluble adhesion molecules prognostic factors in patients with renal cell carcinoma? *Clin. Cancer Res.*, 3, 2451–2458, 1997.

93. Fujimoto, K., Ichimori, Y., Kakizoe, T., Okajima, E., Sakamoto, H., Sugimura, T., and Terada, M., Increased serum levels of basic fibroblast growth factor in patients with renal cell carcinoma, *Biochem. Biophys. Res. Commun.*, 180, 386–392, 1991.

94. Sliutz, G., Tempfer, C., Obermair, A., Dadak, C., and Kainz, C., Serum evaluation of basic FGF in breast cancer patients, *Anticancer Res.*, 15, 2675–2677, 1995.

95. Granato, A. M., Nanni, O., Falcini, F., Folli, S., Mosconi, G., De, P. F., Medri, L., Amadori, D., and Volpi, A., Basic fibroblast growth factor and vascular endothelial growth factor serum levels in breast cancer patients and healthy women: useful as diagnostic tools? *Breast Cancer Res.*, 6, R38–R45, 2004.

96. Perez-Atayde, A. R., Sallan, S. E., Tedrow, U., Connors, S., Allred, E., and Folkman, J., Spectrum of tumor angiogenesis in the bone marrow of children with acute lymphoblastic leukemia, *Am. J. Pathol.*, 150, 815–821, 1997.

97. Brattstrom, D., Bergqvist, M., Larsson, A., Holmertz, J., Hesselius, P., Rosenberg, L., Brodin, O., and Wagenius, G., Basic fibroblast growth factor and vascular endothelial growth factor in sera from non-small cell lung cancer patients, *Anticancer Res.*, 18, 1123–1127, 1998.

98. Nguyen, M., Watanabe, H., Budson, A. E., Richie, J. P., Hayes, D. F., and Folkman, J., Elevated levels of an angiogenic peptide, basic fibroblast growth factor, in the urine of patients with a wide spectrum of cancers, *J. Natl Cancer Inst.*, 86, 356–361, 1994.

99. Colomer, R., Aparicio, J., Montero, S., Guzman, C., Larrodera, L., and Cortes-Funes, H., Low levels of basic fibroblast growth factor (bFGF) are associated with a poor prognosis in human breast carcinoma, *Br. J. Cancer*, 76, 1215–1220, 1997.

100. Morales, D. E., McGowan, K. A., Grant, D. S., Maheshwari, S., Bhartiya, D., Cid, M. C., Kleinman, H. K., and Schnaper, H. W., Estrogen promotes angiogenic activity in human umbilical vein endothelial cells in vitro and in a murine model, *Circulation*, 91, 755–763, 1995.

101. Zhou, M., Sutliff, R. L., Paul, R. J., Lorenz, J. N., Hoying, J. B., Haudenschild, C. C., Yin, M. et al., Fibroblast growth factor 2 control of vascular tone, *Nat. Med.*, 4, 201–207, 1998.

102. Moy, F. J., Li, Y. C., Rauenbuehler, P., Winkler, M. E., Scheraga, H. A., and Montelione, G. T., Solution structure of human type-alpha transforming growth factor determined by heteronuclear NMR spectroscopy and refined by energy minimization with restraints, *Biochemistry*, 32, 7334–7353, 1993.

103. Petegnief, V., Friguls, B., Sanfeliu, C., Sunol, C., and Planas, A. M., Transforming growth factor-alpha attenuates *N*-methyl-D-aspartic acid toxicity in cortical cultures by preventing protein synthesis inhibition through an Erk1/2-dependent mechanism, *J. Biol. Chem.*, 278, 29552–29559, 2003.

104. Lundy, J., Schuss, A., Stanick, D., McCormack, E. S., Kramer, S., and Sorvillo, J. M., Expression of neu protein, epidermal growth factor receptor, and transforming growth factor alpha in breast cancer. Correlation with clinicopathologic parameters, *Am. J. Pathol.*, 138, 1527–1534, 1991.

105. Okamura, K., Morimoto, A., Hamanaka, R., Ono, M., Kohno, K., Uchida, Y., and Kuwano, M., A model system for tumor angiogenesis: involvement of transforming growth factor-alpha in tube formation of human microvascular endothelial cells induced by esophageal cancer cells, *Biochem. Biophys. Res. Commun.*, 186, 1471–1479, 1992.

106. Narayan, S., Thangasamy, T., and Balusu, R., Transforming growth factor-beta receptor signaling in cancer, *Front. Biosci.*, 10, 1135–1145, 2005.

107. Piek, E., Heldin, C. H., and Ten, D. P., Specificity, diversity, and regulation in TGF-beta superfamily signaling, *FASEB J.*, 13, 2105–2124, 1999.

108. Sun, L., Tumor-suppressive and promoting function of transforming growth factor beta, *Front. Biosci.*, 9, 1925–1935, 2004.

109. Reiss, M., TGF-beta and cancer, *Microbes Infect.*, 1, 1327–1347, 1999.

110. Gold, L. I. and Parekh, T. V., Loss of growth regulation by transforming growth factor-beta (TGF-beta) in human cancers: studies on endometrial carcinoma, *Semin. Reprod. Endocrinol.*, 17, 73–92, 1999.

111. Goggins, M., Shekher, M., Turnacioglu, K., Yeo, C. J., Hruban, R. H., and Kern, S. E., Genetic alterations of the transforming growth factor beta receptor genes in pancreatic and biliary adenocarcinomas, *Cancer Res.*, 58, 5329–5332, 1998.

112. Villanueva, A., Garcia, C., Paules, A. B., Vicente, M., Megias, M., Reyes, G., de, V. P. et al., Disruption of the antiproliferative TGF-beta signaling pathways in human pancreatic cancer cells, *Oncogene*, 17, 1969–1978, 1998.

113. Grady, W. M., Myeroff, L. L., Swinler, S. E., Rajput, A., Thiagalingam, S., Lutterbaugh, J. D., Neumann, A. et al., Mutational inactivation of transforming growth factor beta receptor type II in microsatellite stable colon cancers, *Cancer Res.*, 59, 320–324, 1999.

114. Knaus, P. I., Lindemann, D., DeCoteau, J. F., Perlman, R., Yankelev, H., Hille, M., Kadin, M. E., and Lodish, H. F., A dominant inhibitory mutant of the type II transforming growth factor beta receptor in the malignant progression of a cutaneous T-cell lymphoma, *Mol. Cell Biol.*, 16, 3480–3489, 1996.

115. Kim, S. J., Im, Y. H., Markowitz, S. D., and Bang, Y. J., Molecular mechanisms of inactivation of TGF-beta receptors during carcinogenesis, *Cytokine Growth Factor Rev.*, 11, 159–168, 2000.

116. Kobayashi, T., Liu, X., Wen, F. Q., Fang, Q., Abe, S., Wang, X. Q., and Hashimoto, M., Smad3 mediates TGF-beta1 induction of VEGF production in lung fibroblasts, *Biochem. Biophys. Res. Commun.*, 327, 393–398, 2005.

117. Yamauchi, K., Nishimura, Y., Shigematsu, S., Takeuchi, Y., Nakamura, J., Aizawa, T., and Hashizume, K., Vascular endothelial cell growth factor attenuates actions of transforming growth factor-beta in human endothelial cells, *J. Biol. Chem.*, 279, 55104–55108, 2004.

118. Bostrom, K., Zebboudj, A. F., Yao, Y., Lin, T. S., Torres, A., and Matrix, G. L. A., protein stimulates VEGF expression through increased transforming growth factor-beta1 activity in endothelial cells, *J. Biol. Chem.*, 279, 52904–52913, 2004.

119. Horiguchi, H., Jin, L., Ruebel, K. H., Scheithauer, B. W., and Lloyd, R. V., Regulation of VEGF-A, VEGFR-I, thrombospondin-1, -2, and -3 expression in a human pituitary cell line (HP75) by TGFbeta1, bFGF, and EGF, *Endocrine*, 24, 141–146, 2004.

120. Nakagawa, T., Li, J. H., Garcia, G., Mu, W., Piek, E., Bottinger, E. P., Chen, Y. et al., TGF-beta induces proangiogenic and antiangiogenic factors via parallel but distinct Smad pathways, *Kidney Int.*, 66, 605–613, 2004.

121. Schmitt, F. C. and Soares, R., TGF-alpha and angiogenesis, *Am. J. Surg. Pathol.*, 23, 358–359, 1999.

122. Hong, J. H., Song, C., Shin, Y., Kim, H., Cho, S. P., Kim, W. J., and Ahn, H., Estrogen induction of smooth muscle differentiation of human prostatic stromal cells is mediated by transforming growth factor-beta, *J. Urol.*, 171, 1965–1969, 2004.

123. Leong, H., Firestone, G. L., and Bjeldanes, L. F., Cytostatic effects of 3,3'-diindolylmethane in human endometrial cancer cells result from an estrogen receptor-mediated increase in transforming growth factor-alpha expression, *Carcinogenesis*, 22, 1809–1817, 2001.

124. Schneider, S. L., Gollnick, S. O., Grande, C., Pazik, J. E., and Tomasi, T. B., Differential regulation of TGF-beta 2 by hormones in rat uterus and mammary gland, *J. Reprod. Immunol.*, 32, 125–144, 1996.

125. Hermanson, M., Funa, K., Hartman, M., Claesson-Welsh, L., Heldin, C. H., Westermark, B., and Nister, M., Platelet-derived growth factor and its receptors in human glioma tissue: expression of messenger RNA and protein suggests the presence of autocrine and paracrine loops, *Cancer Res.*, 52, 3213–3219, 1992.

126. Betsholtz, C., Karlsson, L., and Lindahl, P., Developmental roles of platelet-derived growth factors, *Bioessays*, 23, 494–507, 2001.

127. Kohler, N. and Lipton, A., Platelets as a source of fibroblast growth-promoting activity, *Exp. Cell Res.*, 87, 297–301, 1974.

128. Uren, A., Yu, J. C., Gholami, N. S., Pierce, J. H., and Heidaran, M. A., The alpha PDGFR tyrosine kinase mediates locomotion of two different cell types through chemotaxis and chemokinesis, *Biochem. Biophys. Res. Commun.*, 204, 628–634, 1994.

129. Hart, C. E., Bailey, M., Curtis, D. A., Osborn, S., Raines, E., Ross, R., and Forstrom, J. W., Purification of PDGF-AB and PDGF-BB from human platelet extracts and identification of all three PDGF dimers in human platelets, *Biochemistry*, 29, 166–172, 1990.

130. Claesson-Welsh, L., Platelet-derived growth factor receptor signals, *J. Biol. Chem.*, 269, 32023–32026, 1994.

131. Sundberg, C., Branting, M., Gerdin, B., and Rubin, K., Tumor cell and connective tissue cell interactions in human colorectal adenocarcinoma. Transfer of platelet-derived growth factor-AB/BB to stromal cells, *Am. J. Pathol.*, 151, 479–492, 1997.

132. Kawai, T., Hiroi, S., and Torikata, C., Expression in lung carcinomas of platelet-derived growth factor and its receptors, *Lab. Invest.*, 77, 431–436, 1997.

133. Fudge, K., Bostwick, D. G., and Stearns, M. E., Platelet-derived growth factor A and B chains and the alpha and beta receptors in prostatic intraepithelial neoplasia, *Prostate*, 29, 282–286, 1996.

134. Nauck, M., Roth, M., Tamm, M., Eickelberg, O., Wieland, H., Stulz, P., and Perruchoud, A. P., Induction of vascular endothelial growth factor by platelet-activating factor and platelet-derived growth factor is downregulated by corticosteroids, *Am. J. Respir. Cell Mol. Biol.*, 16, 398–406, 1997.

135. Wang, D., Huang, H. J., Kazlauskas, A., and Cavenee, W. K., Induction of vascular endothelial growth factor expression in endothelial cells by platelet-derived growth factor through the activation of phosphatidylinositol 3-kinase, *Cancer Res.*, 59, 1464–1472, 1999.

136. Battegay, E. J., Rupp, J., Iruela-Arispe, L., Sage, E. H., and Pech, M., PDGF-BB modulates endothelial proliferation and angiogenesis in vitro via PDGF beta-receptors, *J. Cell Biol.*, 125, 917–928, 1994.

137. Finkenzeller, G., Marme, D., Weich, H. A., and Hug, H., Platelet-derived growth factor-induced transcription of the vascular endothelial growth factor gene is mediated by protein kinase C, *Cancer Res.*, 52, 4821–4823, 1992.

138. Brogi, E., Wu, T., Namiki, A., and Isner, J. M., Indirect angiogenic cytokines upregulate VEGF and bFGF gene expression in vascular smooth muscle cells, whereas hypoxia upregulates VEGF expression only, *Circulation*, 90, 649–652, 1994.

139. Guo, P., Hu, B., Gu, W., Xu, L., Wang, D., Huang, H. J., Cavenee, W. K., and Cheng, S. Y., Platelet-derived growth factor-B enhances glioma angiogenesis by stimulating vascular endothelial growth factor expression in tumor endothelia and by promoting pericyte recruitment, *Am. J. Pathol.*, 162, 1083–1093, 2003.

140. Ullrich, A. and Schlessinger, J., Signal transduction by receptors with tyrosine kinase activity, *Cell*, 61, 203–212, 1990.

141. Ek, B. and Heldin, C. H., Characterization of a tyrosine-specific kinase activity in human fibroblast membranes stimulated by platelet-derived growth factor, *J. Biol. Chem.*, 257, 10486–10492, 1982.

142. Bornfeldt, K. E., Raines, E. W., Graves, L. M., Skinner, M. P., Krebs, E. G., and Ross, R., Platelet-derived growth factor. Distinct signal transduction pathways associated with migration versus proliferation, *Ann. N.Y. Acad. Sci.*, 766, 416–430, 1995.

143. Westermark, B., Heldin, C. H., and Nister, M., Platelet-derived growth factor in human glioma, *Glia*, 15, 257–263, 1995.

144. Maher, E. A., Furnari, F. B., Bachoo, R. M., Rowitch, D. H., Louis, D. N., Cavenee, W. K., and DePinho, R. A., Malignant glioma: genetics and biology of a grave matter, *Genes Dev.*, 15, 1311–1333, 2001.

145. Wilczynski, S. P., Chen, Y. Y., Chen, W., Howell, S. B., Shively, J. E., and Alberts, D. S., Expression and mutational analysis of tyrosine kinase receptors c-kit, PDGFRalpha, and PDGFRbeta in ovarian cancers, Hum, *Pathol.*, 36, 242–249, 2005.

146. Lindahl, P., Hellstrom, M., Kalen, M., and Betsholtz, C., Endothelial-perivascular cell signaling in vascular development: lessons from knockout mice, *Curr. Opin. Lipidol.*, 9, 407–411, 1998.

147. Skobe, M. and Fusenig, N. E., Tumorigenic conversion of immortal human keratinocytes through stromal cell activation, *Proc. Natl Acad. Sci. U.S.A.*, 95, 1050–1055, 1998.

148. Savolainen-Peltonen, H., Loubtchenkov, M., Petrov, L., Delafontaine, P., and Hayry, P., Estrogen regulates insulin-like growth factor-1, platelet-derived growth factor A and B, and their receptors in the vascular wall, *Transplantation*, 77, 35–42, 2004.

149. Lippman, M. E., Dickson, R. B., Gelmann, E. P., Rosen, N., Knabbe, C., Bates, S., Bronzert, D., Huff, K., and Kasid, A., Growth regulation of human breast carcinoma occurs through regulated growth factor secretion, *J. Cell Biochem.*, 35, 1–16, 1987.

150. Bronzert, D. A., Pantazis, P., Antoniades, H. N., Kasid, A., Davidson, N., Dickson, R. B., and Lippman, M. E., Synthesis and secretion of platelet-derived growth factor by human breast cancer cell lines, *Proc. Natl Acad. Sci. U.S.A.*, 84, 5763–5767, 1987.

151. Bagavandoss, P. and Wilks, J. W., Specific inhibition of endothelial cell proliferation by thrombospondin, *Biochem. Biophys. Res. Commun.*, 170, 867–872, 1990.

152. Dameron, K. M., Volpert, O. V., Tainsky, M. A., and Bouck, N., Control of angiogenesis in fibroblasts by p53 regulation of thrombospondin-1, *Science*, 265, 1582–1584, 1994.

153. Zabrenetzky, V., Harris, C. C., Steeg, P. S., and Roberts, D. D., Expression of the extracellular matrix molecule thrombospondin inversely correlates with malignant progression in melanoma, lung and breast carcinoma cell lines, Int, *J. Cancer*, 59, 191–195, 1994.

154. Weinstat-Saslow, D. L., Zabrenetzky, V. S., VanHoutte, K., Frazier, W. A., Roberts, D. D., and Steeg, P. S., Transfection of thrombospondin-1 complementary DNA into a human breast carcinoma cell line reduces primary tumor growth, metastatic potential, and angiogenesis, *Cancer Res.*, 54, 6504–6511, 1994.

155. Doll, J. A., Reiher, F. K., Crawford, S. E., Pins, M. R., Campbell, S. C., and Bouck, N. P., Thrombospondin-1, vascular endothelial growth factor and fibroblast growth factor-2 are key functional regulators of angiogenesis in the prostate, *Prostate*, 49, 293–305, 2001.

156. Campbell, S. C., Volpert, O. V., Ivanovich, M., and Bouck, N. P., Molecular mediators of angiogenesis in bladder cancer, *Cancer Res.*, 58, 1298–1304, 1998.

157. Grossfeld, G. D., Ginsberg, D. A., Stein, J. P., Bochner, B. H., Esrig, D., Groshen, S., Dunn, M. et al., Thrombospondin-1 expression in bladder cancer: association with p53 alterations, tumor angiogenesis, and tumor progression, *J. Natl Cancer Inst.*, 89, 219–227, 1997.

158. Kieser, A., Weich, H. A., Brandner, G., Marme, D., and Kolch, W., Mutant p53 potentiates protein kinase C induction of vascular endothelial growth factor expression, *Oncogene*, 9, 963–969, 1994.

159. Mukhopadhyay, D., Tsiokas, L., and Sukhatme, V. P., Wild-type p53 and v-Src exert opposing influences on human vascular endothelial growth factor gene expression, *Cancer Res.*, 55, 6161–6165, 1995.

160. Filleur, S., Courtin, A., it-Si-Ali, S., Guglielmi, J., Merle, C., Harel-Bellan, A., Clezardin, P., and Cabon, F., SiRNA-mediated inhibition of vascular endothelial growth factor severely limits tumor resistance to antiangiogenic thrombospondin-1 and slows tumor vascularization and growth, *Cancer Res.*, 63, 3919–3922, 2003.

161. Jimenez, B., Volpert, O. V., Crawford, S. E., Febbraio, M., Silverstein, R. L., and Bouck, N., Signals leading to apoptosis-dependent inhibition of neovascularization by thrombospondin-1, *Nat. Med.*, 6, 41–48, 2000.

162. Rodriguez-Manzaneque, J. C., Lane, T. F., Ortega, M. A., Hynes, R. O., Lawler, J., and Iruela-Arispe, M. L., Thrombospondin-1 suppresses spontaneous tumor growth and inhibits activation of matrix metalloproteinase-9 and mobilization of vascular endothelial growth factor, *Proc. Natl Acad. Sci. U.S.A.*, 98, 12485–12490, 2001.

163. Nor, J. E., Mitra, R. S., Sutorik, M. M., Mooney, D. J., Castle, V. P., and Polverini, P. J., Thrombospondin-1 induces endothelial cell apoptosis and inhibits angiogenesis by activating the caspase death pathway, *J. Vasc. Res.*, 37, 209–218, 2000.

164. Sengupta, K., Banerjee, S., Saxena, N. K., and Banerjee, S. K., Thombospondin-1 disrupts estrogen-induced endothelial cell proliferation and migration and its expression suppressed by estradiol via non-genomic ER-MAPK-JNK signaling pathways, *Mol. Cancer Res.*, 2, 150–158, 2004.

165. Zhu, B. T. and Conney, A. H., Is 2-Methoxyestradiol an endogenous estrogen metabolite that inhibits mammary carcinogenesis? *Cancer Res.*, 58, 2269–2277, 1998.

166. Lakhani, N. J., Sarkar, M. A., Venitz, J., and Figg, W. D., 2-Methoxyestradiol, a promising anticancer agent, *Pharmacotherapy*, 23, 165–172, 2003.

167. Fotsis, T., Zhang, Y., Pepper, M. S., Adlercreutz, H., Montesano, R., Nawroth, P. P., and Schweigerer, L., The endogenous oestrogen metabolite 2-methoxyoestradiol inhibits angiogenesis and suppresses tumor growth, *Nature*, 368, 237–239, 1994.

168. Klauber, N., Parangi, S., Flynn, E., Hamel, E., and D'Amato, R. J., Inhibition of angiogenesis and breast cancer in mice by the microtubule inhibitors 2-Methoxyestradiol and taxol, *Cancer Res.*, 57, 81–86, 1997.

169. Tsukamoto, A., Kaneko, Y., Yoshida, T., Han, K., Ichinose, M., and Kimura, S., 2-Methoxyestradiol, an endogenous metabolite of estrogen, enhances apoptosis and beta-galactosidase expression in vascular endothelial cells, *Biochem. Biophys. Res. Commun.*, 248, 9–12, 1998.

170. Mabjeesh, N. J., Escuin, D., Lavallee, T. M., Pribluda, V. S., Swartz, G. M., Johnson, M. S., Willard, M. T., Zhong, H., Simons, J. W., and Giannakakou, P., 2ME2 inhibits tumor growth and angiogenesis by disrupting microtubules and dysregulating HIF, *Cancer Cell*, 3, 363–375, 2003.

171. Yue, T. L., Wang, X., Louden, C. S., Gupta, S., Pillarisetti, K., Gu, J. L., Hart, T. K., Lysko, P. G., and Feuerstein, G. Z., 2-Methoxyestradiol, an endogenous estrogen metabolite, induces apoptosis in endothelial cells and inhibits angiogenesis: possible role for stress-activated protein kinase signaling pathway and Fas expression, *Mol. Pharmacol.*, 51, 951–962, 1997.

17 Reactive Oxygen Species and Angiogenesis

Shampa Chatterjee

CONTENTS

17.1 INTRODUCTION

Angiogenesis is a process of new blood vessel growth that occurs in the human body at specific times in development and growth. A naturally occurring process, angiogenesis facilitates neovascularization or the generation of new capillaries by recruitment and stimulation of endothelial cells (EC) from various sources [1,2]. Although crucial for embryonic development and wound healing, angiogenesis also contributes to disease, such as in the growth of solid tumors, chronic inflammation, atherosclerosis, ischemia, and diabetic retinopathy.

A number of inducers of angiogenesis have been identified, there is an emerging concept that reactive oxygen species (ROS) serve as necessary components of mitogenic signaling cascades associated with angiogenesis [3–8]. ROS include superoxide anion (O_2^-), hydroxyl radical (OH), lipid radical (LOO$^\cdot$), peroxy radicals (XOO$^\cdot$) and singlet oxygen. H_2O_2 and lipid hydroperoxides (LOOH) are not truly free radicals but are lumped with ROS (Figure 17.1). In general, ROS such as superoxide (O_2^-) and hydrogen peroxide (H_2O_2) are known to cause cell injury and damage. However, at low concentrations, ROS exert a role as signaling molecules that are involved in signal transduction cascades of numerous growth factor-, cytokine-, and hormone-mediated pathways, and regulate biological effects such as apoptosis, cell proliferation, and differentiation.

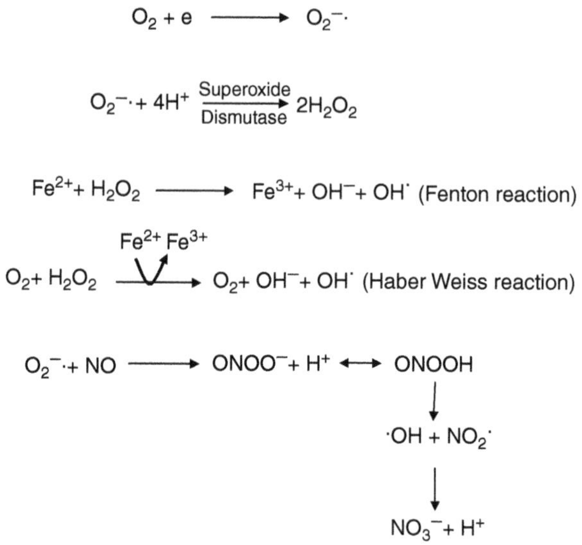

$$O_2 + e \longrightarrow O_2^-\cdot$$

$$O_2^-\cdot + 4H^+ \xrightarrow[\text{Dismutase}]{\text{Superoxide}} 2H_2O_2$$

$$Fe^{2+} + H_2O_2 \longrightarrow Fe^{3+} + OH^- + OH^\cdot \text{ (Fenton reaction)}$$

$$O_2 + H_2O_2 \xrightarrow{Fe^{2+}\ Fe^{3+}} O_2 + OH^- + OH^\cdot \text{ (Haber Weiss reaction)}$$

$$O_2^-\cdot + NO \longrightarrow ONOO^- + H^+ \longleftrightarrow ONOOH$$
$$\downarrow$$
$$\cdot OH + NO_2^\cdot$$
$$\downarrow$$
$$NO_3^- + H^+$$

FIGURE 17.1 Reaction mechanisms for generation of reactive oxygen species (ROS) in the cellular environment.

This review will be focused on the role that ROS signaling plays in angiogenesis. Furthermore, it will be restricted to endothelial proliferation, since generation of blood vessels is the prerequisite of the endothelium. In the EC, ROS generated by either nicotinamide adenine dinucleotide phosphate (NADPH) oxidase-like membrane complexes [9] or through the oxidative metabolism of free arachidonic acid released by ligand dependent phospholipases [10] have a major role in transducing intracellular signals by activated growth factor receptors. Cellular receptors for growth factors and immunological stimuli are, in fact, all linked to ROS-generating systems. Removal of oxygen radicals by chemical or enzyme antioxidants, severely compromises cell response to mitogenic stimulation [11,12]. While the modality of signal transduction by ROS is not well elucidated, key events such as protein tyrosine phosphorylation [13] and activation of transcription factors [14] are general components of the biochemical pathway that lead to cell proliferation.

17.2 ROS IN THE ENDOTHELIUM

There are several enzymatic sources of ROS in mammalian cells, including the mitochondrial electron transport system, xanthine oxidase, the cytochrome p450, the NAD(P)H oxidase(s), and nitric oxide (NO) synthase (NOS) [15]. In ECs, endothelial NAD(P)H oxidase is the major source of ROS. The molecular structure of the endothelial NAD(P)H oxidase is not completely understood, but the subunit composition of the related phagocytic enzyme has been well studied. The neutrophil oxidase consists of a plasma membrane spanning cytochrome b558, comprising gp91phox and p22phox, and cytosolic components p47phox, p67phox, and the small

GTPase Rac1. All the canonical NAD(P)H oxidase subunits characterized in phagocytes are expressed in ECs [16]. On stimulation, the cytosolic components, guided by Rac1, translocate to the plasma membrane where they assemble with the plasma membrane-bound subunits. Electron transfer from NAD(P)H to O_2 generates O_2^-. Unlike neutrophil oxidases that release large amounts of O_2^- in bursts, NAD(P)H oxidases in non-phagocytic cells, such as ECs in a quiescent state, continuously produce low levels of O_2^-. O_2^- dismutates to form H_2O_2, a species that easily diffuses though the cell membrane and is also known to activate transcription factors that may affect cell proliferation. Reactive oxygen species production in EC can be further enhanced by various agonists and growth factors [16]. Agonists such as oxidized low density lipoprotein (LDL) and endothelim-1 increase both gp91phox and O_2^- production in ECs [17,18]. Vascular endothelial growth factor (VEGF) stimulation has been reported to increase O_2^- production in ECs such as human umbilical vein endothelial cells (HUVECs). This was significantly inhibited by the overexpression of dominant negative Rac1 or gp91phox antisense oligonucleotides [19]. In addition to Rac1 and gp91phox, another subunit p47phox was also found to be a functional O_2^- generating component of NADPH oxidase in ECs. Endothelial cells isolated from p47phox$^{-/-}$ mice showed lower O_2^- production upon treatment with agonists such as PMA and TNFα [20].

17.3   ENDOTHELIAL ROS IN SIGNALING

NADPH oxidase derived ROS, the major source of ROS in the endothelium, is implicated in the angiogenic switch in non-tumor tissue [21,22] with the likelihood of similar mechanisms acting in tumors. NAD(P)H oxidases influence tumor cell proliferation via the redox-regulated transcription factor, NF-κB. NF-κB regulates numerous genes involved in apoptosis, cell proliferation, metastasis, and angiogenesis. NF-κB is also constitutively expressed in numerous malignancies [23]. Apart from NF-κB, ROS can affect angiogenesis through VEGF signaling. Reactive oxygen species is involved in the upregulation of VEGF expression at both the protein and mRNA levels [24], and induces angiogenic-related responses such as endothelial cell proliferation [25]. Vascular endothelial growth factor acts via its receptors, the vascular endothelial growth factor receptors (VEGFR), (VEGFR)-1, and -2. Vascular endothelial growth factor binding initiates tyrosine phosphorylation of the VEGFR-2 which results in activation of downstream signaling enzymes including extracellular-regulated kinase-1/2 (ERK1/2), Akt/protein kinase B, and endothelial nitric oxide synthase (eNOS), which contribute to angiogenic phenotype [26]. This phenotype facilitates degradation of extracellular matrix, and migration, proliferation and tube formation of ECs.

Studies have revealed that ROS mediate VEGF-induced angiogenic effects including endothelial cell migration, proliferation, and tube formation [27–29]. In vitro wounding of human keratinocytes has shown that H_2O_2 induces VEGF expression that may support wound healing [29a]. In vivo, there is strong correlation between ROS production with neovascularization and VEGF expression in the eyes of diabetics [30,31] and in balloon injured arteries [32]. Importantly, antioxidants such as pyrrolidine dithiocarbamate [33] and the major components of green tea that have antioxidant properties [34] have been shown to inhibit retinal neovascularization in the mouse model. The thiol antioxidant, N-acetylcysteine (NAC) attenuates endothelial cell invasion and angiogenesis in a tumor model in vivo [35]. Wheeler et al. demonstrated that overexpression of extracellular superoxide dismutase (ecSOD) using adenovirus inhibits tumor vascularization and growth of B16 melanomas in mice [36]. More recently, Roy et al. showed that H_2O_2 supports angiogenesis in vivo. In their wound healing model, low concentrations of H_2O_2 supported the healing process in mice (such as p47^{phox} knockout and MCP-1 deficient) where endogenous H_2O_2 generation is impaired [36a]. Inhibitors of ROS, as well as NAD(P)H oxidase inhibitors, block serum-stimulated endothelial cell proliferation and migration [37]. Hypoxia/reoxygenation, which produces ROS, also elicits capillary tube formation in human microvascular EC [7].

Because NAD(P)H oxidase-derived ROS-dependent signaling in EC are primary features of the process of angiogenesis, its role in endothelial migration and proliferation may provide insight into the components of NAD(P)H oxidase as potential therapeutic targets for treatment of angiogenesis-dependent diseases such as cancer and atherosclerosis and for promoting myocardial angiogenesis in ischemic heart diseases.

17.4 ROS AS MEDIATORS OF ANGIOGENIC SIGNALING

In EC, growth factors such as VEGF and transforming growth factor-β1, as well as cytokines, shear stress, and G-protein-coupled receptor agonists including Angiotensin II (Ang II) have been shown to stimulate the production of ROS via NAD(P)H oxidase [15,38]. Once generated, ROS serve as second messengers to activate multiple signaling pathways leading to proliferation and migration of EC [15,39]. Therefore, molecular targets of ROS in agonist-stimulated signal transduction are important to the understanding of the mechanisms of oxidant signaling ("redox" signaling) in the vasculature. Reactive oxygen species induce VEGF that binds to two tyrosine kinase receptors, VEGFR-1 and -2. In EC the mitogenic and chemotactic effects of VEGF are mediated mainly through VEGFR-2 [40–42]. VEGFR-2 is activated dimerization and transphosphorylation (autophosphorylation) of tyrosine residues in the cytoplasmic kinase domain [43]. This is followed by activation of diverse downstream signaling pathways such as mitogen-activated protein kinases (MAPKs), Akt/protein kinase B and eNOS, which are essential for VEGF-induced EC migration and proliferation [44–46].

Another effect of ROS is the reversible oxidation of proteins. Protein tyrosine phosphatases (PTPs) negatively regulate receptor tyrosine kinase activity and downstream signaling [47]. For instance, PTPs such as SHP-1, SHP-2, and HCPTPA inducibly associate with VEGFR-2 after VEGF stimulation and inhibit VEGF signaling, cell proliferation and cell proliferation [48–50]. Thus, ROS, by inhibiting PTPs, enhance and sustain VEGFR-2 phosphorylation and subsequent angiogenic signaling.

In addition to ROS, it has been well recognized that NO plays an important role in VEGF signaling and postnatal angiogenesis [51]. Exogenous H_2O_2 or Ang II-stimulated increase of ROS potently activates eNOS, which results in NO production in ECs [52,53]. Nitric oxide reacts with superoxide to form peroxynitrite (NOO^-), which may contribute to angiogenic responses [54].

17.4.1 CELL PROLIFERATION

VEGF stimulates DNA synthesis and proliferation via VEGFR-2 and ERK 1/2. Activation of ERK 1/2 is mediated by Ras-Raf-MEK-ERK pathway (Figure 17.2) [55,56]. In addition, VEGF also stimulates the MAPK pathway which is implicated in cell proliferation. There is evidence that the MAPK stimulation occurs via VEGFR-2; pancreatic aortic cells expressing VEGF-2 can activate MAPK in response to VEGF stimulation while cells expressing VEGFR-1 cannot [57]. Vascular endothelial growth factor receptors also facilitate EC migration and invasion [58,59]. Vascular endothelial growth factor induces cell migration by activating factors such as focal adhesion kinase (FAK) and paxillin, and also via the PI3 Kinase/Akt pathway. The p38/MAPK pathway also plays a role in migration as p38 inhibitors decrease cell migration [60]. Experiments with VEGF mutants show that only VEGFR-2 (and not VEGFR-1) is the main mediator of cell migration in ECs [61]. VEGFR-1 activation had no effect on migration of bovine aortic EC [61,62].

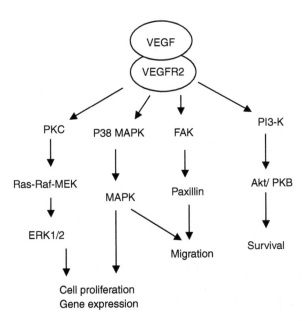

FIGURE 17.2 Vascular endothelial growth factor (VEGF) signaling via VEGFR-2. VEGF binding to VEGFR-2 initiates a signaling cascade that contributes to cell proliferation, migration, and survival. PKC, protein kinase C; ERK, extracellular regulated kinase; MAPK, mitogen activated protein kinase; FAK, focal adhesion kinase; PI3-K, phosphatidyl inositol 3' kinase; Akt/PKB, protein kinase B.

17.4.2 CELL MIGRATION

For a new blood vessel to be formed, ECs must respond to biochemical signaling and begin to migrate toward one another to form microtubules that eventually become new blood vessels. Endothelial cells migration is a mechanically integrated molecular process that involves dynamic, coordinated changes in cell adhesions and cytoskeletal organization. The migration process includes the protrusion of the leading edge, the formation of new adhesions at the front, the contraction of the cell, and the release of adhesions at the rear [63]. Endothelial cells migration occurs by different mechanisms such as chemotaxis (directional migration in response to a concentration gradient of chemoattractants), haptotaxis (directional migration in response to a gradient of immobilized ligands), and mechanotaxis (directional migration induced by mechanical forces). Since VEGF acts as a chemo-attractant, it promotes migration and invasion. In addition to EC, VEGF also stimulates migration of vascular smooth muscle cells, monocytes, mononuclear phagocytes and polymorphonuclear cells, and migration and invasion of tumor cells [64,65]. Molecules such as FAK are activated by the c-terminal region of VEGFR-2; this increases focal adhesion formation and is chemotactic for ECs [66]. Vascular endothelial growth factor stimulates paxillin, a 68 kDa component of the focal adhesion complex; paxillin mediates actin stress fiber formation that in turn controls cytoskeletal reorganization and migration.

17.5 HYPOXIA, REOXYGENATION, AND ANGIOGENESIS

Inadequate blood supply and increasing distance from blood vessels result in areas of low oxygen, i.e., a hypoxic or even anoxic environment in the interior of an artery or blood vessel. Hypoxia triggers the generation of H_2O_2 by mitochondria in low oxygen (1%–2%) [67]. Moreover, hypoxia/reoxygenation also results in ROS, predominantly O_2^- generation. Low oxygen is a stimulus for the accumulation of hypoxia-inducible transcription factors (HIFs) such as HIF-1α ROS increase the DNA binding activity of HIF-1α, a transcription factor responsible for cell adaptation to hypoxia and in charge of the oxygen-dependent expression of the VEGF gene [67]. As previously discussed, VEGF production stimulates cell proliferation through the MAPK and other pathways. The oxygen deprivation hypothesis proposes that neovascularization is an adaptive response to hypoxia resulting from vessel wall thickening. Hypoxia in diseased conditions such as atherosclerosis thus promotes plaque growth in the media and neointima of the blood vessel. This is supported by the observation that atherosclerotic lesions express high levels of VEGF and HIF-1α [68]. Angiogenesis does not initiate plaque formation, but serves as a permissive factor allowing later plaque growth once a critical arterial thickness has been reached.

17.6 TRANSCRIPTION FACTORS AND ANGIOGENIC GENES

ROS produced from hypoxia or hypoxia-reoxygentation activates transcription factors such as the nuclear factor-κB (NF-κB) [69], activator protein-1 (AP-1) [70], hypoxia-inducible factor 1(HIF-1)α [71], p53 [72], and the p21Ras [73]. Low concentrations of H_2O_2 stimulate induction of transcription factor Ets-1, which is required for EC proliferation and tube formation [25]. Exogenous H_2O_2 induces a concentration-dependent increase in monocyte chemoattractant protein (MCP-1) mRNA levels in an AP-1 and NF-κB-dependent manner, which is inhibited by antioxidant NAC in ECs [74].

Ang II-induced ROS derived from NAD(P)H oxidase has been shown to be involved in the increased expression of MCP-1 and NF-κB [75], vascular cell-adhesion molecule-1 (VCAM-1) [76], and STAT [77], all of which regulate cell cycle and cell proliferation. It should be noted that all of these redox-sensitive genes and transcription factors are activated by VEGF, raising the possibility that they are regulated by VEGF-induced ROS. Consistent with this idea, VEGF induced ROS have been shown to be involved in the induction of NFκ-B [78] and the antioxidant enzyme, manganese superoxide dismutase (MnSOD) expression through Rac1-dependent NAD(P)H oxidase in ECs [79].

17.7   ANGIOGENESIS AND MOLECULAR PATHWAYS FOR DISEASE DEVELOPMENT AND PROGRESSION

While angiogenesis is a prerequisite for embryonic development, it also promotes the pathogenesis of diverse chronic human diseases. In diseases where neovascularization is integral to the disease process, such as atherosclerosis, tumors, cancer, ischemic heart disease, diabetic retinopathy, etc., inhibition of angiogenesis is a major goal of therapeutic drug development. Although the link between atherosclerosis and angiogenesis is still inconclusive, an association between neovascularization and atherosclerosis has generally been confirmed by a correlation between the extent of atherosclerosis and plaque neovascularization in human pathological samples [80,81] and in the coronary arteries of hypercholester-olemic primates [82]. Therapeutic approaches thus involve anti-angiogenic drugs such as antibodies against VEGF for the treatment of cancers such as metastatic colorectal carcinoma [83]. Elsewhere, in conditions such as atherosclerosis, VEGF or members of the fibroblast growth factor (FGF) family are administered to stimulate collateral blood vessel formation in the ischemic heart and limb, an approach called therapeutic angiogenesis [84].

However, inhibition of angiogenesis may not be the most suitable therapeutic target in atherosclerotic disease. The available evidence suggests that, although anti-angiogenic therapies may potentially have some effect on the growth of atherosclerotic and neointimal lesions, the benefits may be abrogated by the harmful effects of inhibiting endothelial function and regeneration. The multiplicity of the biological roles of VEGF and the importance of endothelial integrity for vascular function are both strong arguments currently militating against an anti-angiogenic approach to the treatment of cardiovascular disease. The prospects for proangiogenic therapy for ischemic heart disease appear better, but still require unambiguous support from clinical studies.

REFERENCES

1. Folkman, J. and Shing, Y., Angiogenesis, *J. Biol. Chem.*, 267, 10931–10934, 1992.
2. Carmeliet, P. and Jain, R. K., Angiogenesis in cancer and other diseases, *Nature*, 407, 249–257, 2000.
3. Ushio-Fukai, M. and Alexander, R. W., Reactive oxygen species as mediators of angiogenesis signaling: role of NAD(P)H oxidase, *Mol. Cell. Biochem.*, 264, 85–97, 2004.
4. Maulik, N., Redox signaling of angiogenesis, *Antioxid. Redox Signal.*, 4, 805–815, 2002.
5. Cai, H., Griendling, K. K., and Harrison, D. G., The vascular NAD(P)H oxidases as therapeutic targets in cardiovascular diseases, *Trends Pharmacol. Sci.*, 24, 471–478, 2003.
6. Lin, M. T., Yen, M. L., Lin, C. Y., and Kuo, M. L., Inhibition of vascular endothelial growth factor-induced angiogenesis by resveratrol through interruption of Src-dependent vascular endothelial cadherin tyrosine phosphorylation, *Mol. Pharmacol.*, 64, 1029–1036, 2003.
7. Lelkes, P. I., Hahn, K. L., Sukovich, D. A., Karmiol, S., and Schmidt, D. H., On the possible role of reactive oxygen species in angiogenesis, *Adv. Exp. Med. Biol.*, 454, 295–310, 1998.
8. van Wetering, S., van Buul, J. D., Quik, S., Mul, F. P., Anthony, E. C., ten Klooster, J. P., Collard, J. G., and Hordijk, P. L., Reactive oxygen species mediate Rac-induced loss of cell–cell adhesion in primary human endothelial cells, *J. Cell Sci.*, 115, 1837–1846, 2002.
9. Sundaresan, M., Yu, Z. X., Ferrans, V. J., Sulciner, D. J., Gutkind, J. S., Irani, K., Goldschmidt-Clermont, P. J., and Finkel, T., Regulation of reactive-oxygen-species generation in fibroblasts by Rac1, *Biochem. J.*, 318, 379–382, 1996.
10. Woo, C. H., Eom, Y. W., Yoo, M. H., You, H. J., Han, H. J., Song, W. K., Yoo, Y. J., Chun, J. S., and Kim, J. H., Tumor necrosis factor-alpha generates reactive oxygen species via a cytosolic phospholipase A2-linked cascade, *J. Biol. Chem.*, 275, 32357–32362, 2000.
11. Sundaresan, M., Yu, Z. X., Ferrans, V. J., Irani, K., and Finkel, T., Requirement for generation of H_2O_2 for platelet-derived growth factor signal transduction, *Science*, 270, 296–299, 1995.
12. Bae, Y. S., Kang, S. W., Seo, M. S., Baines, I. C., Tekle, E., Chock, P. B., and Rhee, S. G., Epidermal growth factor (EGF)-induced generation of hydrogen peroxide. role in EGF receptor-mediated tyrosine phosphorylation, *J. Biol. Chem.*, 272, 217–221, 1997.
13. Monteiro, H. P. and Stern, A., Redox modulation of tyrosine phosphorylation-dependent signal transduction pathways, *Free Radic. Biol. Med.*, 21, 323–333, 1996.
14. Sun, Y. and Oberley, W., Redox regulation of transcriptional activators, *Free Radic. Biol. Med.*, 21, 335–348, 1996.
15. Griendling, K. K., Sorescu, D., and Ushio-Fukai, M., NAD(P)H oxidase: Role in cardiovascular biology and disease, *Circ. Res.*, 86, 494–501, 2000.
16. Babior, B. M., The NADPH oxidase of endothelial cells, *IUBMB Life*, 50, 267–269, 2000.
17. Rueckschloss, U., Galle, J., Holtz, J., Zerkowski, H. R., and Morawietz, H., Induction of NAD(P)H oxidase by oxidized low-density lipoprotein in human endothelial cells: Antioxidative potential of hydroxymethylglutaryl coenzyme A reductase inhibitor therapy, *Circulation*, 104, 1767–1772, 2001.

18. Duerrschmidt, N., Wippich, N., Goettsch, W., Broemme, H. J., and Morawietz, H., Endothelin-1 induces NAD(P)H oxidase in human endothelial cells, *Biochem. Biophys. Res. Commun.*, 269, 713–717, 2000.

19. Ushio-Fukai, M., Tang, Y., Fukai, T., Dikalov, S., Ma, Y., Fujimoto, M., Quinn, M. T., Pagano, P. J., Johnson, C., and Alexander, R. W., Novel role of gp91phoxcontaining NAD(P)H oxidase in vascular endothelial growth factor induced signaling and angiogenesis, *Circ. Res.*, 91, 1160–1167, 2002.

20. Li, J. M., Mullen, A. M., Yun, S., Wientjes, F., Brouns, G. Y., Thrasher, A. J., and Shah, A. M., Essential role of the NADPH oxidase subunit p47(phox) in endothelial cell superoxide production in response to phorbol ester and tumor necrosis factor-alpha, *Circ. Res.*, 90, 143–150, 2002.

21. Arbiser, J. L., Petros, J., Klafter, R., Govindajaran, B., McLaughlin, E. R., Brown, L. F., Cohen, C. et al., Reactive oxygen generated by Nox1 triggers the angiogenic switch, *Proc. Natl Acad. Sci. U.S.A.*, 99, 715–720, 2002.

22. Gorlach, A., Diebold, I., Schini-Kerth, V. B., Berchner-Pfannschmidt, U., Roth, U., Brandes, R. P., Kietzmann, T., and Busse, R., Thrombin activates the hypoxia-inducible factor-1 signaling pathway in vascular smooth muscle cell: Role of the p22(phox)-containing NADPH oxidase, *Circ. Res.*, 89, 47–54, 2001.

23. Brar, S. S., Kennedy, T. P., Quinn, M., and Hoidal, J. R., Redox signaling of NF-kappaB by membrane NAD(P)H oxidases in normal and malignant cells, *Protoplasma*, 221, 117–127, 2003.

24. Chua, C. C., Hamdy, R. C., and Chua, B. H., Upregulation of vascular endothelial growth factor by H_2O_2 in rat heart endothelial cells, *Free Radic. Biol. Med.*, 25, 891–897, 1998.

25. Yasuda, M., Ohzeki, Y., Shimizu, S., Naito, S., Ohtsuru, A., Yamamoto, T., and Kuroiwa, Y., Stimulation of in vitro angiogenesis by hydrogen peroxide and the relation with ETS-1 in endothelial cells, *Life Sci.*, 64, 249–258, 1999.

26. Dimmeler, S., Dernbach, E., and Zeiher, A. M., Phosphorylation of the endothelial nitric oxide synthase at ser-1177 is required for VEGF-induced endothelial cell migration, *FEBS Lett.*, 477, 258–262, 2000.

27. Colavitti, R., Pani, G., Bedogni, B., Anzevino, R., Borrello, S., Waltenberger, J., and Galeotti, T., Reactive oxygen species as downstream mediators of angiogenic signaling by vascular endothelial growth factor receptor-2/KDR, *J. Biol. Chem.*, 277, 3101–3108, 2002.

28. van Wetering, S., van Buul, J. D., Quik, S., Mul, F. P., Anthony, E. C., ten Klooster, J. P., Collard, J. G., and Hordijk, P. L., Reactive oxygen species mediate Rac-induced loss of cell–cell adhesion in primary human endothelial cells, *J. Cell Sci.*, 115, 1837–1846, 2002.

29. Lin, M. T., Yen, M. L., Lin, C. Y., and Kuo, M. L., Inhibition of vascular endothelial growth factor-induced angiogenesis by resveratrol through interruption of Src-dependent vascular endothelial cadherin tyrosine phosphorylation, *Mol. Pharmacol.*, 64, 1029–1036, 2003.

29a. Sen, C. K., Khanna, S., Babior, B. M., Hunt, T. K., Ellison, E. C., and Roy, S., Oxidant-induced vascular endothelial growth factor expression in human keratinocytes and cutaneous wound healing, *J. Biol. Chem.*, 277, 33284–33290, 2002.

30. Ellis, E. A., Guberski, D. L., Somogyi-Mann, M., and Grant, M. B., Increased H_2O_2, vascular endothelial growth factor and receptors in the retina of the BBZ/Wor diabetic rat, *Free Radic. Biol. Med.*, 28, 91–101, 2000.

31. Ellis, E. A., Grant, M. B., Murray, F. T., Wachowski, M. B., Guberski, D. L., Kubilis, P. S., and Lutty, G. A., Increased NADH oxidase activity in the retina of the BBZ/Wor diabetic rat, *Free Radic. Biol. Med.*, 24, 111–120, 1998.

32. Ruef, J., Hu, Z. Y., Yin, L. Y., Wu, Y., Hanson, S. R., Kelly, A. B., Harker, L. A., Rao, G. N., Runge, M. S., and Patterson, C., Induction of vascular endothelial growth factor in balloon-injured baboon arteries, *Circ. Res.*, 81, 24–33, 1997.

33. Yoshida, A., Yoshida, S., Ishibashi, T., Kuwano, M., and Inomata, H., Suppression of retinal neovascularization by the NF-kappaB inhibitor pyrrolidine dithiocarbamate in mice, *Invest. Ophthalmol. Vis. Sci.*, 40, 1624–1629, 1999.

34. Cao, Y. and Cao, R., Angiogenesis inhibited by drinking tea [Letter], *Nature*, 398, 381, 1999.

35. Cai, T., Fassina, G., Morini, M., Aluigi, M. G., Masiello, L., Fontanini, G., D'Agostini, F., De Flora, S., Noonan, D. M., and Albini, A., *N*-acetylcysteine inhibits endothelial cell invasion and angiogenesis, *Lab. Investig.*, 79, 1151–1159, 1999.

36. Wheeler, M. D., Smutney, O. M., and Samulski, R. J., Secretion of extracellular superoxide dismutase from muscle transduced with recombinant adenovirus inhibits the growth of B16 melanomas in mice, *Mol. Cancer Res.*, 1, 871–881, 2003.

36a. Roy, S., Khanna, S., Nallu, K., Hunt, T. K., and Sen, C. K., Dermal wound healing is subject to redox control, *Mol. Ther.*, 13, 211–220, 2006.

37. Abid, M. R., Kachra, Z., Spokes, K. C., and Aird, W. C., NADPH oxidase activity is required for endothelial cell proliferation and migration, *FEBS Lett.*, 486, 252–256, 2000.

38. Cai, H., Griendling, K. K., and Harrison, D. G., The vascular NAD(P)H oxidases as therapeutic targets in cardiovascular diseases, *Trends Pharmacol. Sci.*, 24, 471–478, 2003.

39. Stone, J. R. and Collins, T., The role of hydrogen peroxide in endothelial proliferative responses, *Endothelium*, 9, 231–238, 2002.

40. Zachary, I. and Gliki, G., Signaling transduction mechanisms mediating biological actions of the vascular endothelial growth factor family, *Cardiovasc. Res.*, 49, 568–581, 2001.

41. Cross, M. J. and Claesson-Welsh, L., FGF nad VEGF function in angiogenesis: Signalling pathways, biological responses and therapeutic inhibition, *Trends Pharmacol. Sci.*, 22, 201–207, 2001.

42. Byrne, A. M., Bouchier-Hayes, D. J., and Harmey, J. H., Angiogenic and cell survival functions of vascular endothelial growth factor (VEGF), *J. Cell. Mol. Med.*, 9, 777–794, 2005.

43. Dougher-Vermazen, M., Hulmes, J. D., Bohlen, P., and Terman, B. I., Biological activity and phosphorylation sites of the bacterially expressed cytosolic domain of the KDR VEGF-receptor, *Biochem. Biophys. Res. Commun.*, 205, 728–738, 1994.

44. Morales-Ruiz, M., Fulton, D., Sowa, G., Languino, L. R., Fujio, Y., Walsh, K., and Sessa, W. C., Vascular endothelial growth factor-stimulated actin reorganization and migration of endothelial cells is regulated via the serine/threonine kinase Akt, *Circ. Res.*, 86, 892–896, 2000.

45. Rousseau, S., Houle, F., Kotanides, H., Witte, L., Waltenberger, J., Landry, J., and Huot, J., Vascular endothelial growth factor (VEGF)-driven actin-based motility is mediated by VEGFR2 and requires concerted activation of stress-activated protein kinase 2 (SAPK2/p38) and geldanamycinsensitive phosphorylation of focal adhesion kinase, *J. Biol. Chem.*, 275, 10661–10672, 2000.

46. Rousseau, S., Houle, F., and Huot, J., Integrating the VEGF signals leading to actin-based motility in vascular endothelial cells, *Trends Cardiovasc. Med.*, 10, 321–327, 2000.

47. Ostman, A. and Bohmer, F. D., Regulation of receptor tyrosine kinase signaling by protein tyrosine phosphatases, *Trends Cell Biol.*, 11, 258–266, 2001.

48. Kroll, J. and Waltenberger, J., The vascular endothelial growth factor receptor KDR activates multiple signal transduction pathways in porcine aortic endothelial cells, *J. Biol. Chem.*, 272, 32521–32527, 1997.

49. Guo, D. Q., Wu, L. W., Dunbar, J. D., Ozes, O. N., Mayo, L. D., Kessler, K. M., Gustin, J. A. et al., Tumor necrosis factor employs a protein-tyrosine phosphatase to inhibit activation of KDR and vascular endothelial cell growth factor induced endothelial cell proliferation, *J. Biol. Chem.*, 275, 11216–11221, 2000.

50. Huang, L., Sankar, S., Lin, C., Kontos, C. D., Schroff, A. D., Cha, E. H., Feng, S. M. et al., HCPTPA, a protein tyrosine phosphatase that regulates vascular endothelial growth factor receptor-mediated signal transduction and biological activity, *J. Biol. Chem.*, 274, 38183–38188, 1999.

51. Murohara, T. and Asahara, T., Nitric oxide and angiogenesis in cardiovascular disease, *Antioxid. Redox Signal.*, 4, 825–831, 2002.

52. Cai, H., Li, Z., Dikalov, S., Hwang, J., Jo, H., Dudley, S. C., and Harrison, D. G., NAD(P)H oxidase derived hydrogen peroxide mediates endothelial nitric oxide production in response to angiotensin II, *J. Biol. Chem.*, 277, 48311–48317, 2002.

53. Cai, H., Li, Z., Davis, M. E., Kanner, W., Harrison, D. G., and Dudley, S. C. Jr., Akt-dependent phosphorylation of serine 1179 and mitogen-activated protein kinase kinase/extracellular signal-regulated kinase 1/2 cooperatively mediate activation of the endothelial nitric-oxide synthase by hydrogen peroxide, *Mol. Pharmacol.*, 63, 325–331, 2003.

54. Marumo, T., Noll, T., Schini-Kerth, V. B., Harley, E. A., Duhault, J., Piper, H. M., and Busse, R., Significance of nitric oxide and peroxynitrite in permeability changes of the retinal microvascular endothelial cell monolayer induced by vascular endothelial growth factor, *J. Vasc. Res.*, 36, 510–515, 1999.

55. Parenti, A., Morbidelli, L., Cui, X. L., Douglas, J. G., Hood, J. D., Granger, H. J., Ledda, F., and Ziche, M., Nitric oxide is an upstream signal of vascular endothelial growth factor induced extracellular signal-regulated kinase1/2 activation in postcapillary endothelium, *J. Biol. Chem.*, 273, 4220–4226, 1998.

56. Pedram, A., Razandi, M., and Levin, E. R., Extracellular signal regulated protein kinase/Jun kinase cross-talk underlies vascular endothelial cell growth factor-induced endothelial cell proliferation, *J. Biol. Chem.*, 273, 26722–26728, 1998.

57. Kroll, J. and Waltenberger, J., The vascular endothelial growth factor receptor KDR activates multiple signal transduction pathways in porcine aortic endothelial cells, *J. Biol. Chem.*, 272, 32521–32527, 1997.

58. Price, D. J., Miralem, T., Jiang, S., Steinberg, R., and Avraham, H., Role of vascular endothelial growth factor in the stimulation of cellular invasion and signaling of breast cancer cells, *Cell Growth Differ.*, 12, 129–135, 2001.

59. Grosskreutz, C. L., Anand-Apte, B., Duplaa, C., Quinn, T. P., Terman, B. I., Zetter, B., and D'Amore, P. A., Vascular endothelial growth factor-induced migration of vascular smooth muscle cells in vitro, *Microvasc. Res.*, 58, 128–136, 1999.

60. Rousseau, S., Houle, F., Landry, J., and Huot, J., p38 MAP kinase activation by vascular endothelial growth factor mediates actin reorganization and cell migration in human endothelial cells, *Oncogene*, 15, 2169–2177, 1997.

61. Gille, H., Kowalski, J., Li, B., LeCouter, J., Moffat, B., Zioncheck, T. F., Pelletier, N., and Ferrara, N., Analysis of biological effects and signaling properties of Flt-1 (VEGFR-1) and KDR (VEGFR-2). A reassessment using novel receptor-specific vascular endothelial growth factor mutants, *J. Biol. Chem.*, 276, 3222–3230, 2001.

62. Bernatchez, P. N., Soker, S., and Sirois, M. G., Vascular endothelial growth factor effect on endothelial cell proliferation, migration, and platelet-activating factor synthesis is Flk-1-dependent, *J. Biol. Chem.*, 274, 31047–31054, 1999.

63. Davis, G. E. and Senger, D. R., Endothelial extracellular matrix: Biosynthesis, remodeling, and functions during vascular morphogenesis and neovessel stabilization, *Circ. Res.*, 97, 1093–1107, 2005.

64. Hoeben, A., Landuyt, B., Highley, M. S., Wildiers, H., Van Oosterom, A. T., and De Bruijn, E. A., Vascular endothelial growth factor and angiogenesis, *Pharmacol. Rev.*, 56, 549–580, 2004.

65. Barleon, B., Sozzani, S., Zhou, D., Weich, H. A., Mantovani, A., and Marme, D., Migration of human monocytes in response to vascular endothelial growth factor (VEGF) is mediated via the VEGF receptor flt-1, *Blood*, 87, 3336–3343, 1996.

66. Qi, J. H. and Claesson-Welsh, L., VEGF-induced activation of phosphoinositide 3-kinase is dependent on focal adhesion kinase, *Exp. Cell Res.*, 263, 173–182, 2001.

67. Chandel, N. S., Maltepe, E., Goldwasser, E., Mathieu, C. E., Simon, M. C., and Schumacker, T. P., Mitochondrial reactive oxygen species trigger hypoxia-induced transcription, *Proc. Natl Acad. Sci. U.S.A.*, 95, 11715–11720, 1998.

68. Yang, P. Y., Rui, Y. C., Lu, L., Li, T. J., Liu, S. Q., Yan, H. X., and Wang, H. Y., Time courses of vascular endothelial growth factor and intercellular adhesion molecule-1 expressions in aortas of atherosclerotic rats, *Life Sci.*, 77, 2529–2539, 2005.

69. Schreck, R., Rieber, P., and Baeuerle, P. A., Reactive oxygen intermediates as apparently widely used messengers in the activation of the NF-kappa B transcription factor and HIV-1, *EMBO J.*, 10, 2247–2258, 1991.

70. Okuno, H., Akahori, A., Sato, H., Xanthoudakis, S., Curran, T., and Iba, H., Escape from redox regulation enhances the transforming activity of Fos, *Oncogene*, 8, 695–701, 1993.

71. Wang, G. L., Jiang, B. H., and Semenza, G. L., Effect of protein kinase and phosphatase inhibitors on expression of hypoxia-inducible factor 1, *Biochem. Biophys. Res. Commun.*, 216, 669–675, 1995.

72. Rainwater, R., Parks, D., Anderson, M. E., Tegtmeyer, P., and Mann, K., Role of cysteine residues in regulation of p53 function, *Mol. Cell. Biol.*, 15, 3892–3903, 1995.

73. Lander, H. M., Ogiste, J. S., Teng, K. K., and Novogrodsky, A., p21Ras as a common signaling target of reactive free radicals and cellular redox stress, *J. Biol. Chem.*, 270, 21195–21198, 1995.

74. Lakshminarayanan, V., Lewallen, M., Frangogiannis, N. G., Evans, A. J., Wedin, K. E., Michael, L. H., and Entman, M. L., Reactive oxygen intermediates induce monocyte chemotactic protein-1 in vascular endothelium after brief ischemia, *Am. J. Pathol.*, 159, 1301–1311, 2001.

75. Tummala, P. E., Chen, X. L., Sundell, C. L., Laursen, J. B., Hammes, C. P., Alexander, R. W., Harrison, D. G., and Medford, R. M., Angiotensin II induces vascular cell adhesion molecule-1 expression in rat vasculature: A potential link between the renin-angiotensin system and atherosclerosis, *Circulation*, 100, 1223–1229, 1999.

76. Chen, X. L., Tummala, P. E., Olbrych, M. T., Alexander, R. W., and Medford, R. M., Angiotensin II induces monocyte chemoattractant protein-1 gene expression in rat vascular smooth muscle cells, *Circ. Res.*, 83, 952–959, 1998.

77. Schieffer, B., Luchtefeld, M., Braun, S., Hilfiker, A., Hilfiker-Kleiner, D., and Drexler, H., Role of NAD(P)H oxidase in angiotensin II-induced JAK/STAT signaling and cytokine induction, *Circ. Res.*, 87, 1195–1201, 2000.

78. Wang, Z., Castresana, M. R., and Newman, W. H., Reactive oxygen and NFkappaB in VEGF-induced migration of human vascular smooth muscle cells, *Biochem. Biophys. Res. Commun.*, 285, 669–674, 2001.
79. Abid, M. R., Tsai, J. C., Spokes, K. C., Deshpande, S. S., Irani, K., and Aird, W. C., Vascular endothelial growth factor induces manganese-superoxide dismutase expression in endothelial cells by a Rac1-regulated NADPH oxidase-dependent mechanism, *FASEB J.*, 15, 2548–2550, 2001.
80. Zamir, M. and Silver, M. D., Vasculature in the walls of human coronary arteries, *Arch. Pathol. Lab. Med.*, 109, 659–662, 1985.
81. Zhang, Y., Cliff, W. J., Schoefl, G. I., and Higgins, G., Immunohistochemical study of intimal microvessels in coronary atherosclerosis, *Am. J. Pathol.*, 143, 164–172, 1993.
82. Williams, J. K., Armstrong, M. L., and Heistad, D. D., Vasa vasorum in atherosclerotic coronary arteries: Responses to vasoactive stimuli and regression of atherosclerosis, *Circ. Res.*, 62, 515–523, 1988.
83. Wood, J. M., Bold, G., Buchdunger, E., Cozens, R., Ferrari, S., Frei, J., Hofmann, F. et al., PTK787/ZK 222584, a novel and potent inhibitor of vascular endothelial growth factor receptor tyrosine kinases, impairs vascular endothelial growth factor-induced responses and tumor growth after oral administration, *Cancer Res.*, 60, 2178–2189, 2000.
84. Patterson, C. and Runge, M. S., Therapeutic angiogenesis. The new electrophysiology? *Circulation*, 99, 2614–2616, 1999.

18 Angiogenesis in Inflammatory Arthritis

Saptarshi Mandal, Smriti Kana kundu-Raychaudhuri, and Siba P. Raychaudhuri

CONTENTS

Abstract

Angiogenesis, the process of new vessel formation requires sequential activation of different growth factors and their cognate receptors. Angiogenesis and an inflammatory reaction are often parallel running processes. Contributing role of angiogenesis in the pathogenesis of various inflammatory diseases is well established. There are several possible mechanisms how angiogenesis influences inflammatory cascades of arthritic diseases. First, neovascularization increases the nutrient supply to tumor like synovium. Second, the expanded endothelial cell surface area in an inflamed joint maximizes the routes of ingress that allow immune and inflammatory cells to adhere and migrate into the synovium. Finally, activated endothelial cells provide a potent source of proinflammatory cytokines, chemokines, and growth factors. Among inflammatory diseases of rheumatologic origin, role of angiogenesis has been most extensively studied in rheumatoid arthritis. Currently there is a major focus to manipulate the regulatory mechanisms involved in neovascularization and endothelial cell activation. Drug development targeting the pathophysiologic processes of angiogenesis is a fast growing discipline in clinical pharmacology. In this chapter we have addressed the role of angiogenesis and its therapeutic implication in rheumatoid arthritis and other inflammatory arthritis.

18.1   INTRODUCTION

Neovascularization or angiogenesis is a multi-step process involving activation, proliferation, and migration of the endothelial cells. Rheumatologist and immunologist have had a special interest in angiogenesis since the early observation of marked prominence of vascular tissue in the pannus of a rheumatoid arthritis synovium. The majority of studies in regard to angiogenesis in inflammatory diseases have been carried out in Rheumatoid arthritis. The role of angiogenesis in psoriasis and psoriatic arthritis is also an active field of research. The other well researched arthritis is osteoarthritis which is classically considered a "degenerative" rather than "inflammatory" condition, and cartilage is the primary site of infliction which is avascular. However, recent data indicate that inflammation and synovitis with angiogenesis may be a part of the initiating events of osteoarthritis [1–3], with synovitis becoming clinically severe in 30% of patients. Data from angiogenesis modulation studies in other systemic inflammatory (autoimmune) diseases, including systemic lupus erythematosus (SLE), systemic sclerosis (SSc), Sjögren's syndrome (SS), mixed connective tissue disease (MCTD), polymyositis/dermato-myositis (PM/DM), and systemic vasculitis, are scarce and inconclusive. Here we will mainly discuss inflammatory angiogenesis in the context of rheumatoid arthritis, and briefly provide available information about other inflammatory rheumatologic diseases.

The outcome of neovascularization in the synovium is highly dependent on the balance or imbalance between angiogenic mediators and inhibitors. There have been several attempts to therapeutically interfere with the cellular and molecular mechanisms underlying RA-associated neovascularization. Most studies have been performed using animal models of arthritis. In addition, a limited number of human clinical trials gave promising results. Chronic inflammatory angiogenesis also increases the pain sensitive sympathetic afferents [2] into an arthritic joint (including osteoarthritis), traveling through vessels. This makes angiogenesis research in arthritis all the more clinically relevant, as pain alleviation is one of the major goals of arthritis treatment. Treating angiogenesis may be able to halt the disease process of arthritis, which makes it an attractive field of research.

18.2 PATHOPHYSIOLOGY OF ANGIOGENESIS: CELLULAR AND MOLECULAR MECHANISMS

Normal endothelial cells (ECs) are quiescent. They do not migrate out of the basement membrane, and rarely proliferate. Their replenishment need is less because of their low apoptosis frequency when blood flow maintains a shearing stress on the endothelium. When an "angiogenic stimulus", which may be hypoxia, soluble angiogenic mediators, metabolic, or mechanical stress, acts on the ECs, they become "activated" and initiate the process of angiogenesis. Angiogenesis generally includes the following events, some of which may temporally overlap:

(1) the release of proteases, (2) degradation of the basement membrane, (3) migration of the ECs into the interstitial space, (4) EC proliferation into solid cordlike outgrowth (sprout), (5) lumen formation in the sprouts, (6) fusion of sprouts to generate connectivity, (7) stabilization (basement membrane deposition, pericyte recruitment, and smooth muscle layer formation; described later), (8) establishment of flow, (9) Regression of disconnected sprouts and remodeling.

This requires turning on a master regulator (transcriptional) switch, which amplifies and maintains expression of the players of the angiogenic cascade. The transcriptional cascade has been, to some extent, worked out in cancer research; other than the Hypoxia Inducible factor (HIF), relatively less information is available about details of other transcription factors in angiogenesis (like Id, etc.). The basic mechanism of the cascade is likely to be stereotyped, possibly with some minor tissue variations.

Compared to the quiescent state, the activated endothelial cell has a different protein expression profile. It synthesizes angiogenic paracrine signals, adhesion molecules (e.g. integrins $\alpha_v\beta_3$, and $\alpha_5\beta_1$), and proteases (each described later in more detail). The feedback from these factors amplifies the angiogenic response and recruits neighboring cells by autocrine and paracrine mechanisms.

An experimental example of a putative angiogenic master regulator transcription switch is HOX B7. It can upregulate proangiogenic mediators FGF2, VEGF, GRO α, IL8, and MMP9 (however, paradoxically downregulates Ang1 and upregulates Ang2) in HOX B7 transfected SKBR3 breast carcinoma cell line—a line that normally lacks HOXB7 (see Table 18.1 for the full names of proangiogenic mediators) [4,5]. Further downstream of proangiogenic growth factors, there is recruitment of secondary activating transcription factors (e.g. HOX D3 downstream of FGF2, AP1 downstream of IL18). These steps further enhance expression mitogens, matrix degradation enzymes, and activated integrins; and cause shedding of proangiogenic soluble cell adhesion molecules (CAM) like CD146, soluble E-selectin, soluble V-CAM1, etc. The activated ECs get loose from the basement membrane following weakening of the basement membrane. Their migration results in the formation of primary capillary sprouts followed by further EC proliferation, migration, synthesis of new basement membrane, and lumen formation within the sprout. The sprouts then anastomose to form capillary loops. Eventually, emigration of ECs out of these sprouts results in the development of second and further generation of new vessels [6]. The proangiogenic mediators are kept from too much action by counterbalancing extracellular anti-angiogenic factors (like endostatin,

TABLE 18.1
Proangiogenic Factors in RA

Category	Name (Short)	Name (Expanded)
1. Growth factors	VEGF	Vascular endothelial growth factor
	bFGF (=FGF2)	Basic fibroblast growth factor
	aFGF (=FGF1)	Acidic fibroblast growth factor
	EGF	Epidermal groth factor
	HGF	Hepatocyte growth factor
	PDGF[a]	Platelet derived growth factor
	IGF-I	Insulin like growth factor-1
	TGF-β[a]	Transforming growth factor-β
	PD-ECGF	Platelet derived epidermal cell growth factor
2. Cytokines	TNF-α[a]	Tumor necrosis factor
	IL-1[a]	Interleukin 1
	IL-6[a]	Interleukin 6
	IL-13[a]	Interleukin 13
	IL-15	Interleukin 15
	IL-18	Interleukin 18
	G-CSF	Granulocyte colony stimulating factor
	GM-CSF	Granulocyte monocyte colony stimulating factor
3. Chemokines	IL-8/CXCL8	Interleukin 8
	ENA-78/CXCL5	Epithelial nutrophil activating protein 78
	Groα/CXCL1	Growth related gene product α
	Groβ/CXCL2	Growth related gene product b
	CTAP-III/CXCL6	Connective tissue activating protein III
	SDF-1/CXCL12	Stromal derived factor 1
4. Extra cellular matrix (ECM) components		Collagen type I
		Fibronectin
		Laminin
		Tenascin
		Heparin and heparan sulfate
		Fibrinogen
5. Proteolytic enzymes	MMPs	Matrix metalloproteases
	uPA	Urokinin type plasminogen activator
6. Cell adhesion molecules (CAMs)		β_1 and β_3 integrins
		E-selectin
	VCAM-1	Vascular cell adhesion molecule-1
	PECAM-1	Platelet-endothelial cell adhesion molecule-1
	CD34	Cluster of differentiation antigen 34
		Sialyl Lewis-X
		Endoglin
7. Other mediators	Ang-1	Angiopoietin-1
	COX-2	Cyclo-oxygenase 2 (inducible cyclo-oxygenase)

(continued)

TABLE 18.1 *(Continued)*

Category	Name (Short)	Name (Expanded)
	PG E$_2$	Prostaglandin E2
	PAF	Platelet activating factor
	SP	Substance P
	Epo	Erythropoetin
	PTN	Pleiotrophin
	Ang	Angiogenin
	MAT	Monocyte angiotropin

[a] Exhibits variable stimulatory and inhibitory acivity (see text).

angiostatin, etc.), some of which are released by longer (or stronger) activation of the same mechanisms that release the proangiogenic factors. If angiogenesis is normally regulated, action of the endothelial activation steps amplified by Hox D3, etc. are temporally succeeded by feedforward inhibition by sequential recruitment of HOX B3 which stabilize vessel morphogenesis [7,8], followed finally by factors like HOX D10 which reinstalls the quiescent state by bringing back the balance of opposing factors.

The vessel stabilization involves secretion of vascular remodeling factors like TGFβ, PDGF, and Ang1; enlargement of the vessel diameter; decrease in the leakiness and thickening of the basement membrane; and recruitment of pericytes and smooth muscles. In pathogenic angiogenesis, the amplification stage continues and the stabilization and remodeling does not adequately set in.

Chronic inflammation and angiogenesis are inter-dependent processes [9]. Inflammation is programmed to involve a component of angiogenesis. Inflammation is the response of tissue to local damage, which initially amplifies the damage (creating a danger signal), and sequentially recruits components of innate and adaptive immunity. The recruitment of immune cells requires endothelial activation and leakiness. To sustain an inflammatory response, many of the inflammatory cells themselves secrete further inflammatory mediators. Acute inflammation, which immediately recruits nutrophils, is often able to destabilize endothelial cells enough to start off an angiogenic response. Adhesion of activated PMNs (Polymorphonuclear cells/Nutrophils) to endothelial cells can cause angiogenesis in vitro and in vivo [10–12]. ICAM1 and E-selectin, two adhesion molecules, are essential for this action of PMNs, and Ets1, a transcription factor acting downstream of ICAM1, acts as a common factor downstream of VEGF and bFGF, and is involved in transcriptional activation of proteases like uPA, MMP 1, 3, and 9.

Macrophage/monocyte lineage cells which are recruited in inflammation with a little delay (compared to polymorphonuclear nutrophils) are capable of producing most growth factors and inflammatory cytokines involved in angiogenesis. But acute inflammation and angiogenesis in healthy tissue generally follows a crescendo-decrescendo pattern, with the new vessels stabilizing once the inflammation is over.

But if it does not recede, ECs may continue to get activated and immature vascular sprouts can continue to form. This in turn can continue to provoke inflammation, as it will recruit more inflammatory cells. Such chronic inflammation associated angiogenesis is a self maintaining deregulated process which is at the root of diseases like rheumatoid arthritis, psoriasis, etc.

Endothelial precursors circulate in systemic blood as a sub-population of $CD34^+$ blood stem cells (discussed in more detail in Section 18.5). Some of these stem cells express vascular endothelial growth factor receptors (VEGFR2). These EC precursors may, under certain circumstances, either be recruited directly to become ECs by the process of "vasculogenesis" (i.e., formation of vessel de novo) or may be deposited in perimuscular zones of the arteries [13–15]. These cells may be important in future therapeutic trials carried out in various vascular disorders including coronary heart disease, obliterative atherosclerosis, stroke, etc.

In vitro models of angiogenesis include EC cultures grown on ECM, such as the Matrigel assay, tissue culture systems, or EC migration, invasion, and chemotaxis assays [6,16,17]. In vivo neovascularization has been studied using the rat, murine, rabbit, and guinea pig corneal micropocket, the chick chorioallantoic membrane, the hamster cheek pouch, the mesenteric assay, the aortic ring, the implanted ECM (sponge/Matrigel plug) assay, and other systems [4–6,18,19]. These models are designed to semi-quantitatively study the angiogenic process, and to test soluble or cell-bound angiogenic or angiostatic agents against suitable controls.

18.3 CLINICAL DESCRIPTION AND PATHOGENESIS OF RHEUMATOID ARTHRITIS

Rheumatoid arthritis (RA) is a chronic multi-organ inflammatory disease, of which a major manifestation is polyarticular arthritis. A vast population throughout the world suffers from this debilitating disease, with a worldwide prevalence of about 1%. In addition to arthritis, inflammatory cascades of RA can cause protean manifestations such as pericarditis, pleural inflammation, interstitial lung disease, systemic vasculitis, and severe inflammatory eye diseases.

According to the American College of Rheumatology guidelines, patients need to have four out of the following seven criteria for a diagnosis of rheumatoid arthritis: (1) Morning stiffness for at least one hour and present for at least six weeks; (2) Swelling of three or more joints for at least six weeks; (3) Swelling of wrist, metacarpophalangeal, or proximal interphalangeal joints for at least six weeks; (4) Symmetric joint swelling; (5) Hand X-ray changes typical of RA that must include erosions or unequivocal bony decalcification; (6) Rheumatoid subcutaneous nodules; (7) Rheumatoid factor.

The pathogenesis of RA is one of the most fascinating fields in immunology research. However, the focus of this chapter is angiogenesis so we will briefly outline the pathologic events. Rheumatoid arthritis is considered an autoimmune disease by the majority of investigators. A key limitation of the autoimmune theory is that despite decades of extensive research in RA the antigen for RA in humans still remains illusive. At best, now we can say that RA is an immune mediated disease with contributing roles of both T and B lymphocytes. Environment, genetic, and immunologic factors all play together in the complex multifactorial disease process of RA.

The first event in RA is probably antigen-dependent activation of T cells. T cell activation subsequently leads to multiple effects, including activation and proliferation of synovial lining and endothelial cells, recruitment and activation of additional proinflammatory cells from the bone marrow and circulation, secretion of cytokines and proteases by macrophages and fibroblast-like synovial cells, and autoantibody production. The synovium in RA organizes itself into an invasive tissue that, if unchecked, can degrade cartilage and bone.

The synovium in RA becomes inflamed and increases greatly in mass because of hyperplasia of the lining cells. This may cause the synovial lining to grow villous projections. The volume of synovial fluid increases, resulting in joint swelling and pain. Blood-derived cells, including T cells, B cells, macrophages, and plasma cells, infiltrate the sublining of the synovium. Although RA shares these histological features (namely, infiltration and hyperplasia) with other inflammatory arthritides, a particularly characteristic feature of RA is the predilection for the synovium to become locally invasive at the synovial interface with cartilage and bone. This invasive, hypervascular and destructive synovium is termed the "pannus". Progressive destruction of the articular cartilage, subchondral bone, and periarticular soft tissues eventually combine to produce the deformities characteristic of longstanding RA. These deformities result in functional deterioration and profound disability in the long term.

18.4 ANGIOGENESIS IN RHEUMATOID ARTHRITIS

In rheumatoid arthritis (RA) there is a spurt of synovial neo-angiogenesis that is considered one of the earliest pathologic features [20]. There is disruption of the vascular arcade and redistribution of more vessels into deeper synovium with a relative lack of vessels near the luminal surface. Increase in vessel density can even be detected non-invasively for example by high resolution ultrasound [21] and can be used as a prognostic marker or treatment response indicator. Inflammation and angiogenesis is generally not uniform all over the synovium and is particularly enhanced at the cartilage-pannus junction. This is true not only in rheumatoid arthritis but also in spondyloarthropathies and even osteoarthritis, as examined by dynamic contrast enhanced MRI (DE-MRI) imaging [22].

The new vessels in the inflamed synovium mostly consist of endothelial sprouts forming leaky capillaries and post capillary venules activated into "high endothelial venule's (HEV)" resembling those in lymphoid tissues [23]. This leads to transudation of fluid, and recruitment of leucocytes: lymphocytes and monocytes mainly into the synovium, and of polymorphonuclear leukocytes into the synovial fluid. Despite increase in the number of microvascular sprouts, there is no improvement in perfusion. Inflammatory cytokines (described in more detail below) lead to an increase in endothelial and stromal proliferation. There is an even higher rate of cell turnover with apoptosis balancing out proliferation. The metabolic demand of capillary and synoviocyte mass outgrows the blood supply in the mature RA synovium. This creates a local ischemia and acidosis [24]. Mean PO2 in rheumatoid synovial fluid is usually below 30 mmHg, and occasionally less than 15 mmHg. The ischemia causes a vicious cycle of increased angiogenesis (with inefficient perfusion) by reduced degradation of the transcription factor hypoxia-inducible factor 1 (HIF-1) that activates transcription of genes of the angiogenic cascade, including those for vascular endothelial growth factor (VEGF) and the VEGF receptor (Figure 18.1).

RA can be considered an "angiogenesis-dependent disease" [25] because without the new blood vessels, there would be no scaffold on which synovitis could grow. The angiogenic cascade activated in RA is more or less similar to angiogenesis observed in tissues, especially tumors.

The invading pannus itself has been considered a tumor like structure which locally invades but does not metastasize. The fibroblast-like type B cells, as opposed to type A or macrophage-like lining cells in the synovium, are suspected to be the main cellular player that show "transformed" invasive behavior in a pannus. Some believe that parallel to HIF1 there is a separate transcription factor AIF (anoxia inducible factor 1) similar to those found in a mouse retrotransposons, and some tumors may also be involved in the invasive nature of the pannus [26], at least in some animal models. And yet another set of evidence suggests that there are mesenchymal precur cells (stem cell like), which may be recruited from blood along with inflammatory cells, that may form some of the synovial "fibroblasts" [27]. The growth factors secreted by them may cause invasiveness and angiogenesis due to the frustrated attempt of these cells to recapitulate limb development and repair. These induce some of the growth factors involved in the synovial angiogenesis.

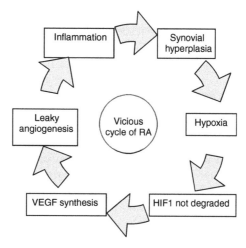

FIGURE 18.1 A simplified representation of the vicious cycle of biological events in respect to inflammation and angiogenesis in rheumatoid arthritis.

The synovitis associated angiogenesis may have a component of "vasculogenesis" as well. In contrast to the embryonic context where vessels can form de novo, here we possibly have a gradual trickling of either circulating endothelial precursors (hemangioblasts) or resident endothelial precurs that may populate a "vasculogenic" zone of arteries at the interface between smooth muscle and advantitia [15]. Of the circulating nucleated cells, bone marrow derived stem cells [$CD34^+$, $CD133^+$, $CD117(c-kit)^+$, $CD90(thy1)^+$] are less than 0.01%. Out of these $<2\%$ are VEGFR2 (Flk1)+. These are the ones considered to be the endothelial precursors, and have been suspected to be recruited to RA synovia, and eventually lose CD133 as they become endothelial lineage committed [28]. Another subset [STRO-1 +, α-SMA + (Smooth Musle Actin)] of multipotent mesenchymal stem cells (MSC) shown to be present in the RA synovia have the potential to differentiate to pericytes and smooth muscles to supplement angiogenesis. However, the role of the MSCs in vessel stabilization is debated. As mentioned earlier, some speculate them to be responsible for exaggerated repair response. This could be caused through recapitulation of developmental signals like Wnt (wingless-Int1) pathway [27]. This has been exemplified by disregulation of this pathway in pseudo-rheumatoid dysplasia—an autosomal recessive defect in Wnt Inducible 3 protein. Reactivation of such developmental signals [29,30] in the RA synovium, may be responsible for induction of matrix metalloproteases (MMP7), and proinflammatory cytokines (IL 6, 8, and 15) in the synovium which, as we shall see below, are involved in angiogenesis.

The angiogenic process, its mediators and inhibitors, and its clinical relevance for diagnostics and therapy have been reviewed earlier by a number of authors [31–34] and in other chapters in this book. Thus, we will review only those angiogenic mediators and inhibitors that have been shown to play a role in RA (Table 18.1 and Table 18.2). Then we will discuss the regulation of the angiogenic process in RA, including important interactions between ECs, angiogenic mediators, and inhibitors. We will also present recent and future therapeutic strategies using

---

TABLE 18.2
Angiogenesis Inhibitors in Rheumatoid Arthritis

1. Growth factors	TGF-β^a, PDGF[a]
2. Cytokines	TNF-α^a, IL-1[a], IL-4, IL-6[a], IL-12, IL-13[a], IFN-α, IFN-γ, LIF
3. Chemokines	CXCs: IP-10 (CXCL10), PF4 (CXCL4), MIG (CXCL9)
	CCs: SLC (CCL21)
4. Protease inhibitors	TIMP-1, TIMP-2, TIMP-3, PAI-1, PAI-2
5. Heparin-binding factors	Thrombospondin-1, PF4
6. ECM fragments	Angiostatin, endostatin
7. Others	SPARC, opioids, retinoids, taxol, troponin I, chondromodulin-1

See text for expansion of acronyms.

[a] Exhibits variable stimulatory and inhibitory activity.

angiogenesis-modulating biological agents, as these treatment modalities may be included in antirheumatic, as well as anti-cancer, therapies.

18.5   ANGIOGENIC MEDIATORS IN THE RHEUMATOID SYNOVIUM

The number of endogenous (soluble and cell surface-bound) and exogenous mediators known to modulate angiogenesis runs into the hundreds (>300 according to some estimates). Mediators shown here are known to be specifically implicated or known to significantly affect RA angiogenesis. These include growth factors, cytokines, chemokines, ECM components, cell adhesion molecules (CAMs), proteolytic enzymes, and other factors. Most of these mediators are produced by ECs, synovial lining cells, fibroblasts and macrophages, lymphocytes, neutrophils, pericytes, smooth muscles, and other cells present in the RA synovium.

18.5.1   Proangiogenic Mediators

18.5.1.1   Growth Factors

One of the best characterized and most potent growth factors expressed in response to hypoxia (transcriptional targets of HIF), that causes endothelial activation and the initial stages of angiogenesis is vascular endothelial growth factor (VEGF). VEGF levels are markedly elevated in synovial fluid aspirate of RA afflicted joints [35,36]. Originally identified as vascular permeability factor (VPF), VEGF causes formation of leaky capillaries [37,38]. VEGF, unlike most other angiogenic growth factors, causes leak by endothelial caveolae to be modified into vesiculovacuolar organelles (VVO) and fenestrations [39]. In addition, it can also loosen up inter-endothelial junctions like many inflammatory mediators [40]. Multiple VEGF isoforms (VEGF121, 145, 165, 189, and 206) are expressed by the RA synovial lining, fibroblasts, smooth muscle cells, neutrophils, and macrophages [35,41,42]. Patients with a high degree of synovitis have more of the VEGHF 165 isoform and the activating receptor KDR (in contrast to Flt1 which is sometimes inhibitoty), and the co-receptor neuropilin1 [41]. VEGF by virtue of its heparin binding property, can be categorized into the heparin binding growth factor family.

The role of receptor tyrosine kinases in angiogenesis has been well established. The high affinity VEGF receptors Flt-1 (Fms like tyrosine kinase) and Flk-1 (fetal liver kinase 1) are receptor tyrosine kinases. Flt-1 has baseline activity even in quiescent endothelial cells. The activating receptor flk-1 is a type III receptor tyrosine kinase, also called VEGFR2 or KDR (kinase insert domain containing receptor). It is involved in embryonic vasculogenesis [43], and is an early marker for circulating EC progenitors [44] which home into the RA synovium [28].

Many articles (especially, older ones) make a distinction between heparin-binding and non-heparin-binding growth factors. "Heparin-binding" ones are bound to negatively charged Glycosamino glycans like heparan sulfate in the ECM (Extracellular matrix). Heparin-sepharose beads can be used to isolate the heparin-binding ones. Heparin-binding growth factors were initially classified into acidic and basic, which were the basic (bFGF/FGF2) and acidic Fibroblast Growth Factors (aFGF/FGF1). Now there are too many growth factors with too many different flavors to allow such simple classifications. Heparin-binding is no longer popular for isolating the HBGFs, as that is likely to pull down a mixture of them. Although the justification to use such nomenclature may not be strong any more, we shall try to salvage these subheadings, as they keep coming up in many rheumatological research publications.

During the initiation or amplification of angiogenic response, heparin-bindings growth factors are released from ECM by EC-derived heparanase and the plasmin/uroplasmin activator(uPA) system [6,16]. Binding to short heparin fragments can enhance the activity of FGF1 and FGF2. Both FGF1 and FGF2 are expressed by macrophages, synovial lining cells, and fibroblasts in the RA synovium in situ [45,46]. FGF2 is among the most potent angiogenic factor secreted by the arthritic synovium [47,48].

Hepatocyte growth factor (HGF)/scatter factor, another heparin-binding angiogenic growth factor is found to be elevated in RA synovium. It is produced by synovial lining fibroblasts and macrophages [49] during synovitis. HGF can directly activate endothelial cells and also indirectly stimulate angiogenesis by tilting the proangiogenic versus anti-angiogenic balance towards the proangiogenic side, for example, by upregulating VEGF and downregulating Thrombospondin1 locally [50].

Most of the other growth factors involved in synovial angiogenesis, for example platelet-derived growth factor (PDGF), transforming growth factor(TGF)-β, platelet-derived endothelial cell growth factor (PD-ECGF)/gliostatin, epidermal growth factor (EGF), and insulin-like growth factor-I (IGF-I) do not bind heparin.

PDGF may act indirectly on angiogenesis by inducing the production of other angiogenic factors including bFGF, VEGF, and type I collagen [51]. It also plays important roles in vessel maturation (especially towards arteriolar forms as opposed to Ang1 which may be involved more in venular maturation) in a normal angiogenic response. PDGF has been found in RA synovial fluids and is expressed by RA synovial tissue macrophages [52]. PDGF also causes synovial fibroblast proliferation.

EGF stimulates most steps of angiogenesis; however, it is less potent than heparin-binding factors [6,16]. EGF has also been detected in RA synovial fluids [53].

IGF-I causes endothelial and smooth muscle proliferation, and Nitric oxide release. The very potent mitogenic and anti-apoptotic effect of IGF-I may add to the local invasive tissue growth and hypoxia that promotes angiogenesis. The amount of IGF-I is increased in RA compared to normal synovial fluids. IGF-I mRNA has also been detected in RA synovial tissues [16,54]. However, whether this is a cause of the pathology or a compensatory response is doubtful, since IGF-I data from coronary angiogenesis studies have indicated IGF-I may have an anti-ischemic (ischemia-preconditioning) and anti-inflammatory role [55]. Thus, IGF-I may play a dual role.

TGF-β has both agonistic as well as antagonistic roles in both inflammation and angiogenesis. TGF-β is expressed by synovial macrophages, and fibroblasts. TGF-β has been speculated to stimulate angiogenesis indirectly by recruiting angiogenic macrophages [56] while it eventually also dampens angiogenic response and stabilizes vessels by pericyte recruitment. Furthermore, TGF-β stimulates the secretion of both the angiogenic urokinase-type plasminogen activator (uPAR) and the angiostatic matrix metalloproteinase (MMP) inhibitors by synovial fibroblasts [57]. In rat models, TGF-β has been shown to have dual effects on synovitis as well: it induces polymorphonuclear cell recruitment into joints, while it suppresses adjuvant-induced arthritis (AIA) [58].

PD-ECGF/gliostatin is structurally homologous to FGFs [6]. PD-ECGF has been detected in the sera and joint effusion aspirate of RA patients. Synovial PD-ECGF levels have been correlated with the release of acute phase response in RA [59].

18.5.1.2   Cytokines

Pro-inflammatory cytokines like tumor necrosis factor-α (TNF-α), interleukin-1 (IL-1), IL-6, IL-8, IL-15, and IL-18 can mediate angiogenesis in RA [6,60]. The effects of IL-6 on angiogenesis may be similar to TGF-ß: agonistic at one dose and antagonistic at another [45]. IL-13, which is considered to be an anti-inflammatory cytokine and generally suppresses angiogenesis under certain circumstances, may also promote EC migration [42,61,62]. Other angiogenic cytokines include granulocyte (G-CSF) and granulocyte-monocyte colony-stimulating factors (GM-CSF), as well as oncostatin M [16,42,45,63]. Most of these cytokines are produced by macrophages, and thus they, especially TNF-α, account for the majority of macrophage-derived angiogenic activity in RA [64–66].

TNF-α is a potent angiogenic cytokine with efficacy comparable to FGFs [66]. Large amounts of TNF-α have been detected in the sera and synovial fluids of RA patients. In the RA synovial tissue, macrophages and ECs express antigenic TNF-α [67–69]. TNF-α in synovial fluid can induce VEGF secretion from the macrophage [70]. However, TNF-α does not induce VEGF in synovial fibroblasts [71,72] and may actually downregulate VEGFR1 and VEGFR2 on endothelial cells [73]. Thus, TNF-α may have a dual role.

The effects of IL-1(α and β) on angiogenesis are controversial, as some groups have shown the angiogenic effects of these cytokines in two different animal models for angiogenesis, while others reported their angiostatic activities in the same systems [16,45]. However, recently IL-1 has been described as a potent angiogenic mediator in a Matrigel assay [74]. Inhibition of IL-1 by anakinra suppressed neovascularization in rats AIA [75]. But caution must be exercised in blocking IL-1, as that would involve a risk of blocking normal tissue healing or remodeling processes. In RA, synovial tissue macrophages and fibroblasts, as well as peripheral blood monocytes, release significant amounts of IL-1 [76–78].

The role of IL-6 in angiogenesis is also controversial. It stimulated EC migration, but inhibited EC proliferation in vitro [79,80]. Therefore, the role of IL-6 in angiogenesis is not fully clear. A significant amount of IL-6 is found in RA synovial fluid [81]. In RA, IL-6 is produced by synovial tissue macrophages and fibroblasts, as well as by lipopolysaccharide-stimulated synovial fluid monocytes [77,82].

IL-13 is generally considered an anti-inflammatory cytokine [83] in RA and other inflammatory situations as it inhibits the production of several other inflammatory mediators. IL-13 gene transfer suppressed neovascularization in rat AIA [84]; therefore IL-13 is considered an angiostatic, rather than an angiogenic cytokine. On the other hand, IL-13 can act as an endothelial chemotaxin, at least in vitro [62]. IL-13 has been detected in the sera, synovial fluids and synovial tissues of RA patients [83].

IL-15 induces blood vessel growth in vivo in nude mice model [85]. High amounts of IL-15 are present in RA synovial fluids and in the lining layer of RA synovial tissues. A lot of recent research has focused on IL-15, as it may be the proinflammatory downstream mediator of IL-1 in RA [86–93].

IL-18 is structurally homologous to IL-1. It induces EC migration, EC tube formation on Matrigel in vitro, and neovascularization in the Matrigel plug or sponge implanted in mice [60]. IL-18 mRNA and protein have been detected in high quantities in RA synovial macrophages and fibroblasts [94]. It has recently been shown that IL-18 may activate VEGF expression via AP1 (Activator Protein 1 complex; a hetero dimmer of the immediate early transcription factors c-fos and c-jun) [95].

Granulocyte and granulocyte-monocyte colony stimulating factors (G-CSF and GM-CSF) are better known as hematopoietic cytokines. They have comparatively less angiogenic activities compared to heparin-binding growth factors [6]. RA synovial tissue macrophages constitutively express GM-CSF. In addition, proinflammatory cytokines stimulate synovial fibroblasts to secrete these CSFs [96–98].

Oncostatin M may act as an angiogenic mediator via bFGF-dependent mechanisms [63]. Nutrophils and macrophages produce this cytokine in the RA synovial tissue. Oncostatin M levels have been found to be high in RA synovial fluids [99]. It has been shown to increase MCP1 and IL1 induced MMP1 and MMP3 production [100], and uPA (urokinase type plasminogen activator) synthesis by synovial fibroblasts [101].

18.5.1.3 Chemokines and Chemikine Receptors

Chemokines comprise a family of small (8–10 kDa) chemotactic cytokines classified into C, CC, CXC, and CX3C family of ligands based on the position of two conserved cysteine residues. C-X-C chemokines containing the ELR (glutamyl-leucyl-arginyl-) amino acid sequence, such as IL-8 (CXCL8), epithelial neutrophil activating protein-78 (ENA-78; CXCL5), growth-related oncogene α (groα; CXCL1), and connective tissue activating protein-III (CTAP-III; CXCL6) have been documented to be angiogenic factors [102–106]. In contrast, most C-X-C chemokines lacking the ELR motif are potent angiostatic factors, except stromal cell-derived factor-1 (SDF-1; CXCL12) which is angiogenic [103]. So far, there is not much information available on the possible angiogenic role of CC chemokines like MIP1 (macrophage inflammatory protein 1), and RANTES (regulated upon activation, normal T cell expressed and secreted) which are known proinflammatory cytokines [103], while MCP1 (monocyte chemoattractant Protein 1), another member of the CC family, has been shown to be angiogenic [107]. Fractalkine (CX3CL1) is the only known C-X3-C chemokine. It is expressed on cytokine-activated EC and promotes angiogenesis. These chemokines play an important role in homing and recruitment of leukocytes at the inflamed endothelium.

Among the ELR containing C-X-C chemokines, the prototype member Interleukin 8 (IL-8 or CXCL8) induces neovascularization both in vivo and in vitro [65,106]. Synovial fibroblasts produce IL-8 mRNA in response to IL-1 and the protein has been found in large quantities in RA joint fluid aspirates. IL-8 is produced by macrophages, and fibroblasts in RA synovium and chondrocytes [108].

Groα (CXCL1), an EC chemoattractant, stimulates angiogenesis. Synovial tissue fibroblasts and synovial fluid mononuclear cells produce Groα upon activation by proinflammatory cytokines and mitogens, respectively. The expression of groα in cultured synovial fibroblasts is induced by IL-1 or TNF-α. In the RA synovial tissue, groα is present in lining cells and interstitial macrophages [108,109].

ENA-78 (CXCL5) a neutrophils potent chemotactic factor, also has angiogenic potency [106]. Significant amounts of ENA-78 are present in RA synovial fluids. Synovial lining fibroblasts, macrophages, and endothelial cells in the pannus produce ENA-78 in large abundance under the influence of TNF-α and IL-1 [110–112].

CTAP-III (CXCL6) is a growth factor derived from human platelet α-granules. CTAP-III is angiogenic and affects many aspects of ECM metabolism [59,65]. RA sera contain high levels of CTAP-III. CTAP-III stimulates the synthesis of various ECM molecules by synovial fibroblasts [113].

SDF-1(CXCL12) is a C-X-C chemokine lacking the ELR motif which specifically binds to the CXCR4 chemokine receptor (fusin) [3,66]. This chemokine induces EC chemotaxis in vitro and dermal angiogenesis in mice in vivo [66]. $CD34^+$/VEGF-2 receptor$^+$ EC precursor stem cells described above express CXCR4 and migrate in response to SDF-1 [10]. Thus, SDF-1 may be the first angiogenic C-X-C chemokine that lacks the ELR motif. SDF-1 is expressed by synovial ECs and is involved in synovial T cell adhesion to intercellular adhesion molecule-1 (ICAM-1) [67]. SDF-1 is involved in CD4+T cell recruitment into the RA synovium [68].

Compared to the CXC chemokines, less data is available on the potential role of C-C chemokines in angiogenesis. Monocyte chemoattractant protein-1 (MCP-1; CCL2) may induce EC chemotaxis in vitro, and is shown to be angiogenic in the chick chorioallantoic membrane assay, Matrigel plug assay and rat aortic sprouting (in absence of leukocytes) assay in vivo. RA synovial fibroblasts produce MCP-1 in response to IL-1 or TNFα. High levels of MCP-1 have been detected in RA synovial fluids. MCP-1-induced neovascularization has been associated with expression of CCR2 on endothelium [107,114].

Fractalkine (CX3CL1) is the only C-X3-C chemokine and is a cell adhesion molecule, unlike the other secreted chemokines. It is chemotactic for monocytes and lymphocytes. It can promote angiogenesis. High levels of fractalkine have been detected in RA synovial fluid samples. In RA synovial tissue, macrophages, fibroblasts, ECs, and dendritic cells express fractalkine [102].

Regarding the possible role of chemokine receptors in angiogenesis, a number of these receptors may be detected on ECs, thus playing a role in chemokine-derived neovascularization. Among the CXC receptors relevant to angiogenesis, CXCR1 binds mainly IL-8; CXCR2 and Duffy antigen mainly bind IL-8, ENA-78, and groα;

and CXCR4 binds only to SDF1 [102,103]. A large body of data indicates CXCR2 may be the most important EC receptor for ELR-containing angiogenic CXC chemokines [33,115]. MCP-1-induced angiogenesis has been associated with increased endothelial expression of CCR2 [107]. Both CXCR2 and CCR2 have been implicated in the pathogenesis of RA [116,117].

18.5.1.4   ECM Components and Matrix Degrading Enzymes

In synovial angiogenesis, endothelial cells need to invade the basement membrane and sprout through the ECM of the synovial tissue containing collagen (mainly type I), fibronectin, laminin, vitronectin, tenascin, thrombospondin, proteoglycans (with glycosaminoglycans), and other matrix components [116–118]. EC migration and chemotaxis are stimulated by type I collagen and, to a lesser extent, types II, III, IV, and V collagen [33]. Fibronectin is chemotactic for ECs and promotes microvessel elongation in explant cultures [119]. Laminin and proteoglycans are involved in EC attachment to the basement membrane and the underlying ECM, and are produced in more than sufficient quantities to support angiogenesis during synovitis [33,120]. Glycosaminoglycans like heparin and heparan-sulphate proteoglycans are imoprtant for the action of heparin-binding angiogenic growth factors [6,33]. Tenascin and Thrombospondin-1 are also an important components of sprouting permissive matrices [121]. When compared to normal synoviocytes, RA synovial fibroblasts show more integrins and more adhesion to collagen, fibronectin, laminin and tenascin [122]. This stickiness may aid the ECs to spread through RA synovium. Type IV collagen, the main component of basement membrane, through which ECs have to invade and which is important in stabilizing vessels in a quiescent form, has been shown to be downregulated in synoviocytes in response to TNF-α and IL1 [123].

Angiogenesis requires proteolytic enzymes for cells to degrade and invade ECM. Furthermore, the cleavage products shed from many large ECM components have angiogenesis modulating properties. RA synovium has been shown to contain matrix metalloproteinases (MMPs), including collagenase (MMP 1), gelatinase A and B (MMP 2 and 9), stromelysin 1 (MMP 3) [but not Stromelysin 2 (MMP 10)], Matrilysin, etc., and urokinase- (uPA) and tissue-type plasminogen activators (tPA) [16,33,124]. Synovial fluid MMP concentration is a marker of synovial inflammation [125]. In the RA synovial tissue, lining cells are the major producers of interstitial collagenase. RA synovial lining cells and subsynovial interstitial macrophages show increased expression of the uPA receptor (CD87) [126] and lining cells express both uPA and tPA [127]. A newer family of metalloprotease molecules termed ADAM (a disintegrin and metalloprotease) and its subset ADAMTS (a disintegrin and metalloprotease with thrombospondin motif) proteinases, which includes aggrecanase-1 and -2 (ADAMTS 4, and ADAMTS5 originally named ADAMTS11) have attained lot of attention in angiogenesis research due to their matrix remodeling properties. Some of the recent members of ADAM family (e.g., ADAM 15 and ADAM 23) can bind αvβ3 integrin through disintegrin domain or through RGD sequence, and thus are thought to mediate similar effects as integrin binding of the matrix ligand. ADAM 15 has been shown to be upregulated in RA under the influence of VEGF [128]. ADAMTS 4 is particularly implicated in the pathogenesis of RA. IL1 is thought to act partly through upregulation of ADAMTS4, and TNF-α blockers are thought to be acting through its downregulation [129]. Both ADAMTS 4 and 5 are constitutively expressed by fibroblast like synoviocytes, but expression of only ADAMTS 4 is upregulated in response to TGF-β [130].

18.5.1.5 Cell Adhesion Molecules

In addition to their dominant role in leukocyte emigration into the RA synovium [131], cell adhesion molecules (CAMs) play very important roles in angiogenesis in RA. Every cell has a characteristic surface expression profile of CAMs which depends on its stage of differentiation (which in turn partly depends on its interaction with surrounding matrix and cells). The CAMs implicated in RA angiogenesis are mostly the ones expressed on activated ECs: most β_1 integrins (especially $\alpha_5\beta_1$), the $\alpha_v\beta_3$ integrin, E-selectin, CD34 (may be a receptor for L-selectin), vascular cell adhesion molecule-1 (VCAM-1), platelet-endothelial cell adhesion molecule-1 (PECAM-1, CD31), endoglin (CD105), VE-cadherin [33,131–134].

Endothelial cells express β_1 integrins (e.g., $\alpha_1\beta_1$, $\alpha_2\beta_1$, $\alpha_3\beta_1$, $\alpha_5\beta_1$, $\alpha_6\beta_1$), and $\alpha_v\beta_5$ integrins, and when activated upregulate $\alpha_5\beta_1$, and switch on $\alpha_v\beta_3$ integrin [135]. These mediate EC-ECM interactions and act as growth factor signaling coreceptors [136–139]. Many β_1 integrins are strongly expressed on RA synovial fluid leukoocytes, synovial tissue lining cells, fibroblasts, and ECs [33,122,136]. VEGF may partly mediate its angiogenic action via β_1 integrin-dependent mechanisms [140]. However, antagonists to $\alpha_v\beta_3$, $\alpha_v\beta_5$ integrins disrupt bFGF and VEGF mediated angiogenesis respectively. The $\alpha_v\beta_3$ integrin has been detected on RA synovial macrophages, lining cells, and fibroblasts [33].

Koch et al. has shown that recombinant soluble VCAM-1 and E-selectin may act as proangiogenic factors based on in vitro and in vivo angiogenesis studies [132]. E-selectin blocking by antibodies inhibited chemotaxis and tube formation of bovine capillary ECs [137]. Soluble ectodomains of VCAM-1 and E-selectin have been found in high levels in the synovial fluids of RA patients [17,141]. VCAM-1 and E-selectin are expressed on RA synovial ECs and synovial lining cells [142]. Soluble P-selectin, also abundant in RA synovial fluids [143], stimulates EC migration in vitro [144].

PECAM-1 (CD31) and endoglin (CD105, one of the TGF-β receptors) have been reported to be angiogenic [133,145]. Both have been detected on most RA synovial ECs, as well as on lining cells and macrophages [146,147]. Lewis^y/H-5-2 glycoconjugate, which is closely related to sialyl Lewis^x (the E-selectin ligand), promotes neovascularization and is abundantly expressed in the RA synovium [148]. Another glycoprotein MUC18 (CD146), a marker for the metastatic potential of tumors (like melanoma), has adhesive and angiogenic properties. Synovial fluid levels of soluble MUC18 in RA correlate with synovial angiogenesis [149].

Thus a number of soluble CAMs, in conjunction with the surface CAMs, may be involved in the angiogenic process. Angiogenic factors, such as proinflammatory cytokines, which are abundantly produced in the RA synovium may upregulate expression of these soluble CAMS and amplify the angiogenic response [131].

18.5.1.6 Other Proangiogenic Molecules

Other angiogenic mediators implicated in RA-associated neovascularization include, cyclooxygenase/prostaglandin system [150], angiogenin [151], angiotropin, angiopoietin-1, pleiotrophin [152], platelet-activating factor (PAF), histamine, substance P, erythropoietin, and adenosine [33].

The cyclooxygenase (COX) pathway [150] and some of its products like prostaglandin E2 [33] may be involved in RA-associated angiogenesis. COX-2 has been implicated in VEGF-dependent neovascularization [153,154].

Angiopoietin-1 (Ang1), the ligand for the Tie1 (Tie=tyrosine kinase with immunoglobulin-like loop and epidermal growth factor homology domains) and Tie2 (also called Tek=Tunica interna endothelial cell kinase) tyrosine kinase exerts no proangiogenic activity on pure endothelial culture in vitro. Instead, it regulates the assembly of non-endothelial vessel wall components such as smooth muscle cells. In contrast to Ang1, the other ligand Angiopoietin 2 (Ang2) is angiostatic when overexpressed [155,156] (Ang2 is further described among the angiostatic factors). Ang1 and Ang2, as well as their receptors Tie1 and Tie2, are also dominantly expressed in sites of active angiogenesis in the RA synovium [157], with Tie2 expressed mainly in the ECs. But the in vivo effects of Ang1 and Ang2 are complex, and involve ECs as well as pericytes and smooth muscles. Ang1 has been shown to also cause an increase in vessel diameter (rather than a number of new sprouts) [158], especially towards venular differentiation, and to stabilize vessels and reduce leakiness by multiple mechanisms [159]. In the context of RA, inhibition of Ang1, for example by Ang2, may have a risk of promoting leakiness [160].

Platelet activating factor (PAF) is also a heparin-dependent proangiogenic factor [161] involved in RA synovial fluid-mediated neovascularization.

The substance P, a member of the tachykinin/neurokinin family, was originally described as the neurotransmitter in the pain sensory afferent nerves. It was also found to be a macrophage-derived proinflammatory and angiogenic factor [6]. Increased substance P levels in RA compared to osteoarthritic synovial fluids and tissues suggest that it may be one of the culprits for the excess synovial angiogenesis in RA [162].

Pleiotrophin is an embryonic growth and differentiation factor with angiogenic properties. It stimulates VEGF synthesis in dermal fibroblasts. In RA, pleiotrophin is expressed by synovial fibroblasts and ECs [152].

Angiogenin is homologous to pancreatic RNAse. It is probably the only member of the RNAse family that is thus far considered an angiogenic factor. It binds heparin and is upregulated in an acute phase response. It is a very potent angiogenic factor (for review see [151]). However, its mode of action is still not fully clear. Angiogenin has been detected in vivo in the RA synovium, and in vitro in synovial fibroblasts culture supernatants [163]. Angiogenin is sometimes confusingly abbreviated as Ang, not to be confused with similar abbreviations of the more famous angiopoietins (see above).

Monocyte AngioTropin (MAT), commonly known as just Angiotropin, is a 4.5 kDa copper binding ribonucleoprotein (ribokine) consisting of a peptide chain of

38 amino acids and an RNA of 43 bases. In several bioassays, angiotropin is found to be a potent angiogenic factor and a strong chemoattractant (but not mitogen) for ECs [164,165]. Success of copper chelation therapy in experimental models of arthritis could potentially stem from their ability to prevent angiotropin action [166].

18.5.2 ANGIOGENESIS INHIBITORS IN RHEUMATOID ARTHRITIS

Our body has evolved a fascinating repertoire of anti-angiogenic factors that can compete with the diverse gamut of proangiogenic mediators. Many of these are present in the RA synovium and can be further supplemented to inhibit angiogenesis in RA, in addition to exogenous pharmaceuticals (Table 18.2) [45]. The table includes some cytokines and growth factors—some of which may be proangiogenic under different circumstances, some angiostatic C-X-C chemokines, heparin-binding factors, corticosteroids, protease inhibitors, antibiotics, tissue-derived inhibitors, anti-cytoskeletal agents, and other compounds.

18.5.2.1 Cytokines and Growth Factors

Some of the growth factors and cytokines, such as IL-1, TNF-α, IL-6, IL-13, and TGF-β and PDGF were already discussed as proangiogenic and can also inhibit neovascularization under various conditions [16].

Interferon-α (IFN-α) and IFN-γ block FGF- and VEGF-independent angiogenesis [6]. Although IFN-γ mRNA is detected in RA synovial tissues and synovial fluid cells, the protein is hardly detectable in RA [77].

IL-4, an anti-inflammatory cytokine, inhibits angiogenesis both in vivo and in vitro [33,167]. It can block synthesis of IL-11 and it antagonizes several effects of IL-1 and TNF-α in RA [168,169].

IL-12 inhibits neovascularization by inducing the production of two angiostatic mediators, IFN-γ and the IP-10 chemokine [170–172]. IL-12 has been detected in RA synovial fluids [173].

Leukemia inhibitory factor (LIF) was originally named for its ability to induce differentiation in murine undifferentiated myeloma (M1) cells, and was later found pleiotropic. Its actions include cell differentiation, acute-phase response induction, calcium release in bone explants, and proteoglycan resorption in arthritis which can potentially be blocked by TGF-β [174]. LIF is angiostatic in vitro [175]. RA synovial fluids contain high levels of LIF. Synovial fibroblasts produce LIF in response to IL-1 or TNF-α stimulation [176].

18.5.2.2 Chemokines and Chemokine Receptors

We mentioned earlier that C-X-C chemokines without the ELR moiety (except SDF1), such as platelet factor-4 (PF4; CXCL4), monokine induced by interferon-γ (MIG; CXCL9), and interferon-γ inducible protein (IP-10; CXCL10) inhibit angiogenesis [33,177]. Large amounts of IP-10 have been detected locally in RA synovial fluids and tissues, but little is detectable in circulation [178]. MIG has also been detected in RA synovial synovium and joint fluid [178]. PF4 is known to inhibit angiogenesis, possibly through direct inhibition of FGF2 and VEGF action.

This action has been localized to a C terminal (47–70 of the 70 amino acids) part of PF4 by in vitro experiments [179], as well as by using an octapeptide (CT112) in vivo inhibition of murine type II collagen-induced arthritis (CIA), a representative model for RA [180]. Secondary lymphoid tissue chemokine (SLC), is a C-C chemokine (CCL21); yet, it binds to the CXCR3 receptor [181] and shows strong antitumor and angiostatic effects.

Chemokine receptor CXCR3 binds to IP-10, MIG, and SLC—the angiostatic cytokines, and thus may play an important role in chemokine-mediated angiogenesis inhibition. CXCR3 exerts abundant expression in the RA synovium [182].

18.5.2.3 Protease Inhibitors

Protease inhibitors, opposite the ECM degrading proteases, are part of forces opposing basement membrane invasion and migration of ECs. They include tissue inhibitors of metalloproteinases (TIMP-1, TIMP-2, and TIMP-3) and plasminogen activator inhibitors (PAI-1 and PAI-2) [45]. TIMPs have been found in abundance in RA synovial fluids and tissues [124]. RA synovia also contains PAIs. Joint aspirate from RA shows higher PAI-1 levels than OA(osteoarthritis) [127]. Synthetic MMP inhibitors have been tried in various angiogenesis models [183]. Adenoviral gene delivery of an uPA/uPAR inhibitor has been tested successfully in blocking angiogenesis in CIA [184].

18.5.2.4 Heparin-Binding Inhibitors

Heparin-binding extracellular molecules can compete with heparin-dependent growth factors (e.g., bFGF) either for activating heparin, or heparan sulfate proteoglycans (coreceptors for the FGF family). One example of such molecules, Thrombospondin-1, is an important ECM component, which also serves as a CAM, while another example PF4, a C-X-C chemokine, was already discussed before. Thrombospondin-1 has been shown to be expressed on RA synovial ECs and macrophages, and to cause a biphasic modulation of synovial blood vessel counts in the rat adjuvant-induced arthritis (AIA) model for RA [185,186].

18.5.2.5 ECM Fragments

Many ECM fragments are known to be anti-angiogenic [187]. The ones most well studied and tested in RA are Endostatin (fragment of type XVII collagen) that binds both α_v as well as α_5 integrins and Angiostatin (fragment of plasminogen) that binds $\alpha_v b_3$ integrin. These are further discussed in the concluding section as potential clinical inhibitors.

18.5.2.6 Other Angiogenesis Inhibitors

Miscellaneous angiostatic factors include Angiopoietin 2 (Ang-2), Secreted Protein Acidic and Rich in Cysteine (SPARC)/osteonectin, opioids, angiostatin, endostatin, retinoids, opioids, troponin I, cartilage-derived natural inhibitors including chondromodulin-1, and others. Many of these agents have been used in cancer therapy trials, and some of them may also be tried in animal models of inflammation, such as arthritis models—and possibly also in humans (see below, or for review see [45]).

Increased SPARC/osteonectin levels have been found in synovial fluids of RA patients. In addition, SPARC secretion in chondrocyte cultures is regulated by angiogenic growth factors and cytokines suggesting the existence of a regulatory feedback mechanism in arthritis [188]. Opioids are natural anti-inflammatory, analgesics and angiostatic agents, and can attenuate progression of rat AIA [189,190]. The cartilage-derived angiogenesis inhibitor chondromodulin-1 can also suppress antigen-induced arthritis [191].

Angiopoietins (Ang-1 and Ang-2) are described in the earlier section. Ang-2 is the endogenous inverse agonist of Tie2 (agonist is the other ligand Ang-1). This inverse agonism inhibits neovascularization [155], but unfortunately may promote leakiness and inflammation [160]. Ang-2, as well as Ang-1, has been detected in the RA synovium [157].

There is a growing amount of data on the inhibitory effects of angiostatin and endostatin on tumor- and inflammation-associated angiogenesis. Angiostatin, a fragment of plasminogen, was purified from the urine of lung tumor-bearing mice [33]. On the other hand, endostatin, a fragment of collagen type XVIII, was produced by murine hemangioendothelioma cells. Both are on trial for angiogenesis inhibition and gene therapy in animal models of arthritis [192–196].

18.6   PSORIATIC ARTHRITIS AND ANGIOGENESIS

Psoriasis is a relatively common chronic inflammatory skin disease, affecting about 2% of the population worldwide [197]. Complaints of joint pain, swelling, morning stiffness, and fatigue in a patient with psoriasis raise a suspicion for concurrent psoriatic arthritis. Psoriatic arthritis is a distinct clinical entity [198]. It is a seronegative inflammatory arthritis associated with psoriasis. Psoriatic arthritis involves small joints of the hands, other peripheral joints, and spondyloarthropathy, including both sacroiliitis and spondylitis. It is not necessary to have the involvement of all these groups of joints together at the same time. The most striking radiologic feature is the coexistence of erosive changes and new bone formation in the distal joints.

The pathogenesis of psoriasis and psoriatic arthritis is unraveling. In the last two decades, extensive work has been done to explore the immunological mechanisms involved in psoriasis. Activated T cells have been found in the affected tissues (both skin and joints) in patients with psoriatic arthritis. Several studies have shown that cytokines secreted from activated T cells and from other mononuclear proinflammatory cells induce the proliferation and activation of synovial and epidermal fibroblasts. An active role of T cells in the pathogenesis is strongly substantiated by the following observations: (i) immunotherapy targeted specifically against CD4+T cells clears active plaques of psoriasis [199], and (ii) in severe combined immunodeficient (SCID) mice, transplanted non-lesional psoriatic skin converts to a psoriatic plaque subsequent to intradermal administration of T cells activated with an antigen cocktail [200]. However, it is equally true that psoriasis treated with agents such as Calcipotriol (Dovonex) and Etritinate (Tegison), which affect the differentiation process of keratinocytes, is very effective in psoriasis. Neither Dovonex nor Tegison are effective in other T cell-mediated cutaneous diseases such as atopic dermatitis or contact dermatitis. No antigen has yet been identified for psoriasis. Using the SCID mouse model, it has been demonstrated that NGF and substance P activated lymphocytes can also convert non-lesional psoriatic skin transplants to psoriasis [201]. Severe combined immunodeficiency mouse-human skin chimeras are a unique animal model for the study of psoriasis and cutaneous inflammation.

These observations suggest that although activated T lymphocytes have an undisputed role in the pathogenesis of psoriasis, there are other regulatory systems that contribute to the inflammatory and proliferative processes of psoriasis. A complete understanding about the pathogenesis of psoriasis and psoriatic arthritis is lacking. Cytokines, chemokines, adhesion molecules, growth factors like NGF, neuropeptides, and T cell receptors act in integrated ways to evolve as unique inflammatory and proliferative processes typical for psoriasis and psoriatic arthritis.

Marked angiogenesis in psoriasis plaques and synovium of psoriatic arthritis is a well recognized event. Specific vascular morphological changes have been described in the psoriasis skin, nailfold capillaries, and, more recently, in PsA synovial membrane—suggesting a common link [202,203]. Angiogenesis is a prominent early event in psoriasis and PsA, elongated and tortuous vessels in the skin and joints, suggest dysregulated angiogenesis resulting in immature vessels [204,205]. Angiogenic growth factors including transforming growth factor β (TGFβ), platelet derived growth factor (PDGF), and vascular endothelial growth factor (VEGF) are markedly increased in psoriasis [206]. VEGF and TGFβ levels are high in the joint fluid in early PsA, and expression of angiopoietins, a novel family of vascular growth factors, colocalise with VEGF protein and mRNA in PsA synovial membrane perivascular areas [207]. Angiopoietin expression is upregulated in perivascular regions in lesional psoriasis skin [208].

18.7   CLINICAL RELEVANCE OF ANGIOGENESIS
RESEARCH IN RHEUMATOLOGY

Objective evaluation of angiogenesis in inflammatory diseases may be helpful in the diagnosing and monitoring of disease activity. In RA, increased synovial vascularity differentiates it from non-inflammatory arthritis, and newly formed blood vessels in biopsy specimens may reflect the progression of the disease [16]. The elevated concentration of the angiogenic soluble CAMs, such as soluble E-selectin and VCAM-1 in the sera and synovial fluid samples of RA patients, may also be a useful marker of increased neovascularization, as well as inflammation [17,141].

Modulating angiogenesis is a novel and a new therapeutic approach to treat rheumatoid arthritis. This has been speculated to have the following advantages:

1. Pannus, as a highly growing and metabolically demanding tissue, is vulnerable (apoptosis-prone) to nutrient deprivation [209].

2. Selective targeting of endothelial cells will reduce the surface area for leukocyte recruitment and reduce the inflammatory cytokine flux produced by the endothelia.

The effect on angiogenesis of various immunomodulators and anti-inflammatory drugs has been evaluated, including NSAIDs, several DMARDs (including methotrexate, leflunomide, gold compounds, and others), and anti-TNF biologicals [16,42,61,103,210–213]. Apart from other modes of action, the angiostatic effects of these drugs need to be taken into consideration during their administration. The selective COX-2 inhibitors celecoxib and rofecoxib suppressed RA-associated MMP production and angiogenesis [153,154]. Infliximab treatment may exert angiostatic effects as it reduced synovial VEGF expression [214] and resulted in increased endostatin production in RA patients [215,216]. The IL-1 receptor antagonist, anakinra, suppressed angiogenesis in rat AIA [75]. A PAF antagonist, BN5730, has been tested in a clinical trial in RA patients, resulting in some clinical improvement in these patients [217,218].

Currently, TNF-α and IL-1 is a major focus for treatment of rheumatoid arthritis [75,214–216,219–221]. Future anti-angiogenic and anti-inflammatory targeting in RA may include the inhibition of cytokine, growth factor, or chemokine production. IL-6 and IL-13 based therapy for RA and other rheumatic diseases has a significant potential [219–222]. In gene transfer studies, IL-13 has been reported to reduce angiogenesis and synovitis in rat AIA [84]. Among angiostatic chemokines, a peptide derived from PF4 (discussed earlier) was able to abrogate murine arthritis [180].

VEGF is the key regulatory growth factor for neoangiogenesis in RA, psoriasis, and psoriatic arthritis, which we have discussed in detail in earlier sections. Antagonizing VEGF is a major area of focus to develop novel therapy for these angiogenesis associated inflammatory conditions. Chimeric antibodies against VEGF and its receptor are now available. A humanized antibody to VEGF is reported to improve neovascularization [214]. A soluble VEGF receptor 1 (VEGFR1) chimeric protein dose-dependently suppressed the proliferation of ECs isolated from arthritic synovial tissues [223]. A number of additional synthetic VEGF and VEGF receptor inhibitors are under development [223,224].

Gold compounds, a well known therapeutic agent, inhibit expression of synovial E-selectin, an angiogenic CAM, in the synovium [225]. Manipulation of the expression of angiogenic CAMs is other possible way to influence angiogenesis and inflammation [16,61,131]. Vitaxin (MEDI-522), a humanized antibody to $\alpha_v\beta_3$ integrin, blocks the interaction of this integrin with its ligands, vitronectin, and osteopontin. Clinical trials with Vitaxin in arthritis are in progress [226].

We have described several angiogenesis inhibitors in the earlier sections such as MMP inhibitors [183], cytoskeleton disrupting agent taxol [212,227], angiostatin, and endostatin [208,228]. Nerve growth factor and its receptor system have been attributed to having a critical role in angiogenesis [229]. Using the SCID mouse model of psoriasis, it has been demonstrated that modulation of the high affinity receptor of NGF(trkA) could be an important therapeutic approach [230]. It is likely that many of these products will undergo further trials in arthritis models and then, possibly, in humans.

However, the long term goal of angiogenesis research in arthritis should not be the mere blocking of angiogenesis, but rather should try to achieve the following:

1. Find a way to remodel the blood vessel architecture by reversing the distribution of vessels in the synovium back to its normal physiologic state, or to manipulate vascular permeability by recruiting the pericytes and smooth muscle.

2. Reinstallation of the correct vasomotor regulation and reduction of pain fibers by modulating the neurovascular functional integration (complete removal of pain fibers is also undesirable as we do not want to create charcot joints).

REFERENCES

1. Walsh, D. A., Angiogenesis in osteoarthritis and spondylosis: successful repair with undesirable outcomes, *Curr. Opin. Rheumatol.*, 16 (5), 609, 2004.
2. Bonnet, C. S. and Walsh, D. A., Osteoarthritis, angiogenesis and inflammation, *Rheumatology (Oxford)*, 44 (1), 7, 2005.
3. Haywood, L., McWilliams, D. F., Pearson, C. I., Gill, S. E., Ganesan, A., Wilson, D., and Walsh, D. A., Inflammation and angiogenesis in osteoarthritis, *Arthritis Rheum.*, 48 (8), 2173, 2003.
4. Meccia, E., Bottero, L., Felicetti, F., Peschle, C., Colombo, M. P., and Care, A., Hoxb7 expression is regulated by the transcription factors nf-y, yy1, sp1 and usf-1, *Biochim. Biophys. Acta*, 1626 (1–3), 1, 2003.
5. Care, A., Felicetti, F., Meccia, E., Bottero, L., Parenza, M., Stoppacciaro, A., Peschle, C., and Colombo, M. P., Hoxb7: a key factor for tumor-associated angiogenic switch, *Cancer Res.*, 61 (17), 6532, 2001.
6. Folkman, J. and Klagsbrun, M., Angiogenic factors, *Science*, 235 (4787), 442, 1987.
7. Gorski, D. H. and Wash, K., The role of homeobox genes in vascular remodeling and angiogenesis, *Circ. Res.*, 87 (10), 865, 2000.
8. Gorski, D. H. and Walsh, K., Control of vascular cell differentiation by homeobox transcription factors, *Trends Cardiovasc. Med.*, 13 (6), 213, 2003.
9. Jackson, J. R., Seed, M. P., Kircher, C. H., Willoughby, D. A., and Winkler, J. D., The codependence of angiogenesis and chronic inflammation, *Faseb. J.*, 11 (6), 457, 1997.
10. Yasuda, M., Shimizu, S., Ohhinata, K., Naito, S., Tokuyama, S., Mori, Y., Kiuchi, Y., and Yamamoto, T., Differential roles of icam-1 and e-selectin in polymorpho-nuclear leukocyte-induced angiogenesis, *Am. J. Physiol. Cell Physiol.*, 282 (4), C917, 2002.
11. Yasuda, M., Shimizu, S., Tokuyama, S., Watanabe, T., Kiuchi, Y., and Yamamoto, T., A novel effect of polymorphonuclear leukocytes in the facilitation of angiogenesis, *Life Sci.*, 66 (21), 2113, 2000.
12. Iba, O., Matsubara, H., Nozawa, Y., Fujiyama, S., Amano, K., Mori, Y., Kojima, H., and Iwasaka, T., Angiogenesis by implantation of peripheral blood mononuclear cells and platelets into ischemic limbs, *Circulation*, 106 (15), 2019, 2002.

13. Peichev, M., Naiyer, A. J., Pereira, D., Zhu, Z., Lane, W. J., Williams, M., Oz, M. C. et al., Expression of vegfr-2 and ac133 by circulating human cd34(+) cells identifies a population of functional endothelial precursors, *Blood*, 95 (3), 952, 2000.

14. Gehling, U. M., Ergun, S., Schumacher, U., Wagener, C., Pantel, K., Otte, M., and Schuch, G., In vitro differentiation of endothelial cells from ac133-positive progenitor cells, *Blood*, 95 (10), 3106, 2000.

15. Zengin, E., Chalajour, F., Gehling, U. M., Ito, W. D., Treede, H., Lauke, H., Weil, J., Reichenspurner, H., Kilic, N., and Ergun, S., Vascular wall resident progenitor cells: a source for postnatal vasculogenesis, *Development*, 2006.

16. Szekanecz, Z., Szegedi, G., and Koch, A. E., Angiogenesis in rheumatoid arthritis: pathogenic and clinical significance, *J. Investig. Med.*, 46 (2), 27, 1998.

17. Koch, A. E., Turkiewicz, W., Harlow, L. A., and Pope, R. M., Soluble e-selectin in arthritis, *Clin. Immunol. Immunopathol.*, 69 (1), 29, 1993.

18. Akhtar, N., Dickerson, E. B., and Auerbach, R., The sponge/Matrigel angiogenesis assay, *Angiogenesis*, 5 (1–2), 75, 2002.

19. Kragh, M., Hjarnaa, P. J., Bramm, E., Kristjansen, P. E., Rygaard, J., and Binderup, L., In vivo chamber angiogenesis assay: an optimized Matrigel plug assay for fast assessment of anti-angiogenic activity, *Int. J. Oncol.*, 22 (2), 305, 2003.

20. Karateev, D. E., Radenska-Lopovok, S. G., Nasonova, V. A., and Ivanova, M. M., Synovial membrane in the early stage of rheumatoid arthritis: clinico-morphological comparisons, *Ter. Arkh.*, 75 (5), 12, 2003.

21. Hau, M., Kneitz, C., Tony, H. P., Keberle, M., Jahns, R., and Jenett, M., High resolution ultrasound detects a decrease in pannus vascularisation of small finger joints in patients with rheumatoid arthritis receiving treatment with soluble tumour necrosis factor alpha receptor (etanercept), *Ann. Rheum. Dis.*, 61 (1), 55, 2002.

22. Rhodes, L. A., Conaghan, P. G., Radjenovic, A., Grainger, A. J., Emery, P., and McGonagle, D., Further evidence that a cartilage-pannus junction synovitis predilection is not a specific feature of rheumatoid arthritis, *Ann. Rheum. Dis.*, 64 (9), 1347, 2005.

23. Middleton, J., Americh, L., Gayon, R., Julien, D., Aguilar, L., Amalric, F., and Girard, J. P., Endothelial cell phenotypes in the rheumatoid synovium: activated, angiogenic, apoptotic and leaky, *Arthritis Res. Ther.*, 6 (2), 60, 2004.

24. Treuhaft, P. S. and McCarty, D. J., Synovial fluid ph, lactate, oxygen and carbon dioxide partial pressure in various joint diseases, *Arthritis Rheum.*, 14 (4), 475, 1971.

25. Colville-Nash, P. R. and Scott, D. L., Angiogenesis and rheumatoid arthritis: pathogenic and therapeutic implications, *Ann. Rheum. Dis.*, 51 (7), 919, 1992.

26. Bodamyali, T., Stevens, C. R., Billingham, M. E., Ohta, S., and Blake, D. R., Influence of hypoxia in inflammatory synovitis, *Ann. Rheum. Dis.*, 57 (12), 703, 1998.

27. Corr, M. and Zvaifler, N. J., Mesenchymal precursor cells, *Ann. Rheum. Dis.*, 61 (1), 3, 2002.

28. Ruger, B., Giurea, A., Wanivenhaus, A. H., Zehetgruber, H., Hollemann, D., Yanagida, G., Groger, M., Petzelbauer, P., Smolen, J. S., Hoecker, P., and Fischer, M. B., Endothelial precursor cells in the synovial tissue of patients with rheumatoid arthritis and osteoarthritis, *Arthritis Rheum.*, 50 (7), 2157, 2004.

29. Sen, M., Wnt signalling in rheumatoid arthritis, *Rheumatology (Oxford)*, 44 (6), 708, 2005.

30. Sen, M., Reifert, J., Lauterbach, K., Wolf, V., Rubin, J. S., Corr, M., and Carson, D. A., Regulation of fibronectin and metalloproteinase expression by Wnt signaling in rheumatoid arthritis synoviocytes, *Arthritis Rheum.*, 46 (11), 2867, 2002.

31. Carmeliet, P., Angiogenesis in life, disease and medicine, *Nature*, 438 (7070), 932, 2005.

32. Carmeliet, P. and Jain, R. K., Angiogenesis in cancer and other diseases, *Nature*, 407 (6801), 249, 2000.

33. Szekanecz, Z., Gaspar, L., and Koch, A. E., Angiogenesis in rheumatoid arthritis, *Front Biosci.*, 10 (1739), 2005.

34. Ferrara, N. and Kerbel, R. S., Angiogenesis as a therapeutic target, *Nature*, 438 (7070), 967, 2005.

35. Fava, R. A., Olsen, N. J., Spencer-Green, G., Yeo, K. T., Yeo, T. K., Berse, B., Jackman, R. W., Senger, D. R., Dvorak, H. F., and Brown, L. F., Vascular permeability factor/endothelial growth factor (vpf/vegf): accumulation and expression in human synovial fluids and rheumatoid synovial tissue, *J. Exp. Med.*, 180 (1), 341, 1994.

36. Koch, A. E., Harlow, L. A., Haines, G. K., Amento, E. P., Unemori, E. N., Wong, W. L., Pope, R. M., and Ferrara, N., Vascular endothelial growth factor. A cytokine modulating endothelial function in rheumatoid arthritis, *J. Immunol.*, 152 (8), 4149, 1994.

37. Keck, P. J., Hauser, S. D., Krivi, G., Sanzo, K., Warren, T., Feder, J., and Connolly, D. T., Vascular permeability factor, an endothelial cell mitogen related to pdgf, *Science*, 246 (4935), 1309, 1989.

38. Senger, D. R., Galli, S. J., Dvorak, A. M., Perruzzi, C. A., Harvey, V. S., and Dvorak, H. F., Tumor cells secrete a vascular permeability factor that promotes accumulation of ascites fluid, *Science*, 219 (4587), 983, 1983.

39. Esser, S., Wolburg, K., Wolburg, H., Breier, G., Kurzchalia, T., and Risau, W., Vascular endothelial growth factor induces endothelial fenestrations in vitro, *J. Cell Biol.*, 140 (4), 947, 1998.

40. Weis, S. M. and Cheresh, D. A., Pathophysiological consequences of vegf-induced vascular permeability, *Nature*, 437 (7058), 497, 2005.

41. Ikeda, M., Hosoda, Y., Hirose, S., Okada, Y., and Ikeda, E., Expression of vascular endothelial growth factor isoforms and their receptors flt-1, kdr, and neuropilin-1 in synovial tissues of rheumatoid arthritis, *J. Pathol.*, 191 (4), 426, 2000.

42. Koch, A. E., Review: Angiogenesis: Implications for rheumatoid arthritis, *Arthritis Rheum.*, 41 (6), 951, 1998.

43. Shalaby, F., Rossant, J., Yamaguchi, T. P., Gertsenstein, M., Wu, X. F., Breitman, M. L., and Schuh, A. C., Failure of blood-island formation and vasculogenesis in flk-1-deficient mice, *Nature*, 376 (6535), 62, 1995.

44. Yamaguchi, T. P., Dumont, D. J., Conlon, R. A., Breitman, M. L., and Rossant, J., Flk-1, an flt-related receptor tyrosine kinase is an early marker for endothelial cell precursors, *Development*, 118 (2), 489, 1993.

45. Auerbach, W. and Auerbach, R., Angiogenesis inhibition: a review, *Pharmacol. Ther.*, 63 (3), 265, 1994.

46. Qu, Z., Huang, X. N., Ahmadi, P., Andresevic, J., Planck, S. R., Hart, C. E., and Rosenbaum, J. T., Expression of basic fibroblast growth factor in synovial tissue from patients with rheumatoid arthritis and degenerative joint disease, *Lab. Invest.*, 73 (3), 339, 1995.

47. Goddard, D. H., Grossman, S. L., Newton, R., Clark, M. A., and Bomalaski, J. S., Regulation of synovial cell growth: basic fibroblast growth factor synergizes with interleukin 1 beta stimulating phospholipase a2 enzyme activity, phospholipase a2 activating protein production and release of prostaglandin e2 by rheumatoid arthritis synovial cells in culture, *Cytokine*, 4 (5), 377, 1992.

48. Goddard, D. H., Grossman, S. L., Williams, W. V., Weiner, D. B., Gross, J. L., Eidsvoog, K., and Dasch, J. R., Regulation of synovial cell growth. Coexpression of transforming growth factor beta and basic fibroblast growth factor by cultured synovial cells, *Arthritis Rheum.*, 35 (11), 1296, 1992.

49. Koch, A. E., Halloran, M. M., Hosaka, S., Shah, M. R., Haskell, C. J., Baker, S. K., Panos, R. J. et al., Hepatocyte growth factor. A cytokine mediating endothelial migration in inflammatory arthritis, *Arthritis Rheum.*, 39 (9), 1566, 1996.

50. Zhang, Y. W., Su, Y., Volpert, O. V., and Vande Woude, G. F., Hepatocyte growth factor/scatter factor mediates angiogenesis through positive vegf and negative thrombospondin 1 regulation, *Proc. Natl Acad. Sci. U.S.A.*, 100 (22), 12718, 2003.

51. Sato, N., Beitz, J. G., Kato, J., Yamamoto, M., Clark, J. W., Calabresi, P., Raymond, A., and Frackelton, A. R. Jr., Platelet-derived growth factor indirectly stimulates angiogenesis in vitro, *Am. J. Pathol.*, 142 (4), 1119, 1993.

52. Remmers, E. F., Sano, H., and Wilder, R. L., Platelet-derived growth factors and heparin-binding (fibroblast) growth factors in the synovial tissue pathology of rheumatoid arthritis, *Semin. Arthritis Rheum.*, 21 (3), 191, 1991.

53. Kusada, J., Otsuka, T., Matsui, N., Hirano, T., Asai, K., and Kato, T., Immuno-reactive human epidermal growth factor (h-egf) in rheumatoid synovial fluids, *Nippon Seikeigeka Gakkai Zasshi*, 67 (9), 859, 1993.

54. Keyszer, G. M., Heer, A. H., Kriegsmann, J., Geiler, T., Keysser, C., Gay, R. E., and Gay, S., Detection of insulin-like growth factor i and ii in synovial tissue specimens of patients with rheumatoid arthritis and osteoarthritis by in situ hybridization, *J. Rheumatol.*, 22 (2), 275, 1995.

55. Conti, E., Carrozza, C., Capoluongo, E., Volpe, M., Crea, F., Zuppi, C., and Andreotti, F., Insulin-like growth factor-1 as a vascular protective factor, *Circulation*, 110 (15), 2260, 2004.

56. Wiseman, D. M., Polverini, P. J., Kamp, D. W., and Leibovich, S. J., Transforming growth factor-beta (tgf beta) is chemotactic for human monocytes and induces their expression of angiogenic activity, *Biochem. Biophys. Res. Commun.*, 157 (2), 793, 1988.

57. Hamilton, J. A., Piccoli, D. S., Leizer, T., Butler, D. M., Croatto, M., and Royston, A. K., Transforming growth factor beta stimulates urokinase-type plasminogen activator and DNA synthesis, but not prostaglandin e2 production, in human synovial fibroblasts, *Proc. Natl Acad. Sci. U.S.A.*, 88 (16), 7180, 1991.

58. Brandes, M. E., Allen, J. B., Ogawa, Y., and Wahl, S. M., Transforming growth factor beta 1 suppresses acute and chronic arthritis in experimental animals, *J. Clin. Invest.*, 87 (3), 1108, 1991.

59. Takeuchi, M., Otsuka, T., Matsui, N., Asai, K., Hirano, T., Moriyama, A., Isobe, I., Eksioglu, Y. Z., Matsukawa, K., and Kato, T., Aberrant production of gliostatin/platelet-derived endothelial cell growth factor in rheumatoid synovium, *Arthritis Rheum.*, 37 (5), 662, 1994.

60. Park, C. C., Morel, J. C., Amin, M. A., Connors, M. A., Harlow, L. A., and Koch, A. E., Evidence of il-18 as a novel angiogenic mediator, *J. Immunol.*, 167 (3), 1644, 2001.

61. Szekanecz, Z. and Koch, A. E., Update on synovitis, *Curr. Rheumatol. Rep.*, 3 (1), 53, 2001.

62. Halloran, M. M., Haskell, C. J., Woods, J. M., Hosaka, S., and Koch, A. E., Interleukin-13 is an endothelial chemotaxin, *Pathobiology*, 65 (6), 287, 1997.

63. Wijelath, E. S., Carlsen, B., Cole, T., Chen, J., Kothari, S., and Hammond, W. P., Oncostatin m induces basic fibroblast growth factor expression in endothelial cells and promotes endothelial cell proliferation, migration and spindle morphology, *J. Cell Sci.*, 110 (Pt 7), 871, 1997.

64. Koch, A. E., Kunkel, S. L., Harlow, L. A., Mazarakis, D. D., Haines, G. K., Burdick, M. D., Pope, R. M., and Strieter, R. M., Macrophage inflammatory protein-1 alpha. A novel chemotactic cytokine for macrophages in rheumatoid arthritis, *J. Clin. Invest.*, 93 (3), 921, 1994.

65. Koch, A. E., Polverini, P. J., Kunkel, S. L., Harlow, L. A., DiPietro, L. A., Elner, V. M., Elner, S. G., and Strieter, R. M., Interleukin-8 as a macrophage-derived mediator of angiogenesis, *Science*, 258 (5089), 1798, 1992.

66. Leibovich, S. J., Polverini, P. J., Shepard, H. M., Wiseman, D. M., Shively, V., and Nuseir, N., Macrophage-induced angiogenesis is mediated by tumour necrosis factor-alpha, *Nature*, 329 (6140), 630, 1987.

67. Koch, A. E., Polverini, P. J., and Leibovich, S. J., Stimulation of neovascularization by human rheumatoid synovial tissue macrophages, *Arthritis Rheum.*, 29 (4), 471, 1986.

68. Lupia, E., Montrucchio, G., Battaglia, E., Modena, V., and Camussi, G., Role of tumor necrosis factor-alpha and platelet-activating factor in neoangiogenesis induced by synovial fluids of patients with rheumatoid arthritis, *Eur. J. Immunol.*, 26 (8), 1690, 1996.

69. Deleuran, B. W., Chu, C. Q., Field, M., Brennan, F. M., Mitchell, T., Feldmann, M., and Maini, R. N., Localization of tumor necrosis factor receptors in the synovial tissue and cartilage-pannus junction in patients with rheumatoid arthritis. Implications for local actions of tumor necrosis factor alpha, *Arthritis Rheum.*, 35 (10), 1170, 1992.

70. Bottomley, M. J., Webb, N. J., Watson, C. J., Holt, P. J., Freemont, A. J., and Brenchley, P. E., Peripheral blood mononuclear cells from patients with rheumatoid arthritis spontaneously secrete vascular endothelial growth factor (vegf): specific upregulation by tumour necrosis factor-alpha (tnf-alpha) in synovial fluid, *Clin. Exp. Immunol.*, 117 (1), 171, 1999.

71. Jackson, J. R., Minton, J. A., Ho, M. L., Wei, N., and Winkler, J. D., Expression of vascular endothelial growth factor in synovial fibroblasts is induced by hypoxia and interleukin 1beta, *J. Rheumatol.*, 24 (7), 1253, 1997.

72. Berse, B., Hunt, J. A., Diegel, R. J., Morganelli, P., Yeo, K., Brown, F., and Fava, R. A., Hypoxia augments cytokine (transforming growth factor-beta (tgf-beta) and il-1)-induced vascular endothelial growth factor secretion by human synovial fibroblasts, *Clin. Exp. Immunol.*, 115 (1), 176, 1999.

73. Patterson, C., Perrella, M. A., Endege, W. O., Yoshizumi, M., Lee, M. E., and Haber, E., Downregulation of vascular endothelial growth factor receptors by tumor necrosis factor-alpha in cultured human vascular endothelial cells, *J. Clin. Invest.*, 98 (2), 490, 1996.

74. Voronov, E., Shouval, D. S., Krelin, Y., Cagnano, E., Benharroch, D., Iwakura, Y., Dinarello, C. A., and Apte, R. N., Il-1 is required for tumor invasiveness and angiogenesis, *Proc. Natl Acad. Sci. U.S.A.*, 100 (5), 2645, 2003.

75. Coxon, A., Bolon, B., Estrada, J., Kaufman, S., Scully, S., Rattan, A., Duryea, D. et al., Inhibition of interleukin-1 but not tumor necrosis factor suppresses neovascularization in rat models of corneal angiogenesis and adjuvant arthritis, *Arthritis Rheum.*, 46 (10), 2604, 2002.

76. Deleuran, B. W., Chu, C. Q., Field, M., Brennan, F. M., Katsikis, P., Feldmann, M., and Maini, R. N., Localization of interleukin-1 alpha, type 1 interleukin-1 receptor and interleukin-1 receptor antagonist in the synovial membrane and cartilage/pannus junction in rheumatoid arthritis, *Br. J. Rheumatol.*, 31 (12), 801, 1992.

77. Brennan, F. M., Field, M., Chu, C. Q., Feldmann, M., and Maini, R. N., Cytokine expression in rheumatoid arthritis, *Br. J. Rheumatol.*, 30 (1), 76, 1991.

78. Shore, A., Jaglal, S., and Keystone, E. C., Enhanced interleukin 1 generation by monocytes in vitro is temporally linked to an early event in the onset or exacerbation of rheumatoid arthritis, *Clin. Exp. Immunol.*, 65 (2), 293, 1986.

79. Rosen, E. M., Liu, D., Setter, E., Bhargava, M., and Goldberg, I. D., Interleukin-6 stimulates motility of vascular endothelium, *Exs*, 59 (194), 1991.

80. Sachs, L., Angiogenesis—cytokines as part of a network, *Exs*, 61 (20), 1992.

81. Houssiau, F. A., Devogelaer, J. P., Van Damme, J., de Deuxchaisnes, C. N., and Van Snick, J., Interleukin-6 in synovial fluid and serum of patients with rheumatoid arthritis and other inflammatory arthritides, *Arthritis Rheum.*, 31 (6), 784, 1988.

82. Field, M., Chu, C., Feldmann, M., and Maini, R. N., Interleukin-6 localization in the synovial membrane in rheumatoid arthritis, *Rheumatol. Int.*, 11 (2), 45, 1991.

83. Isomaki, P., Luukkainen, R., Toivanen, P., and Punnonen, J., The presence of interleukin-13 in rheumatoid synovium and its antiinflammatory effects on synovial fluid macrophages from patients with rheumatoid arthritis, *Arthritis Rheum.*, 39 (10), 1693, 1996.

84. Woods, J. M., Amin, M. A., Katschke, K. J. Jr., Volin, M. V., Ruth, J. H., Connors, M. A., Woodruff, D. C. et al., Interleukin-13 gene therapy reduces inflammation, vascularization, and bony destruction in rat adjuvant-induced arthritis, *Hum. Gene Ther.*, 13 (3), 381, 2002.

85. Angiolillo, A. L., Kanegane, H., Sgadari, C., Reaman, G. H., and Tosato, G., Interleukin-15 promotes angiogenesis in vivo, *Biochem. Biophys. Res. Commun.*, 233 (1), 231, 1997.

86. Baslund, B., Tvede, N., Danneskiold-Samsoe, B., Larsson, P., Panayi, G., Petersen, J., Petersen, L. J. et al., Targeting interleukin-15 in patients with rheumatoid arthritis: a proof-of-concept study, *Arthritis Rheum.*, 52 (9), 2686, 2005.

87. McInnes, I. B. and Gracie, J. A., Interleukin-15: a new cytokine target for the treatment of inflammatory diseases, *Curr. Opin. Pharmacol.*, 4 (4), 392, 2004.

88. McInnes, I. B., Gracie, J. A., Harnett, M., Harnett, W., and Liew, F. Y., New strategies to control inflammatory synovitis: interleukin 15 and beyond, *Ann. Rheum. Dis.*, 62 (suppl. 2), ii51, 2003.

89. Liew, F. Y. and McInnes, I. B., Role of interleukin 15 and interleukin 18 in inflammatory response, *Ann. Rheum. Dis.*, 61 (suppl. 2), ii100, 2002.

90. McInnes, I. B. and Leung, B. P., Innate response cytokines in inflammatory synovitis: a role for interleukin-15, *Curr. Dir. Autoimmun.*, 3 (200), 2001.

91. McInnes, I. B. and Liew, F. Y., Interleukin 15: a proinflammatory role in rheumatoid arthritis synovitis, *Immunol. Today*, 19 (2), 75, 1998.

92. McInnes, I. B., Leung, B. P., Sturrock, R. D., Field, M., and Liew, F. Y., Interleukin-15 mediates t cell-dependent regulation of tumor necrosis factor-alpha production in rheumatoid arthritis, *Nat. Med.*, 3 (2), 189, 1997.

93. McInnes, I. B., al-Mughales, J., Field, M., Leung, B. P., Huang, F. P., Dixon, R., Sturrock, R. D., Wilkinson, P. C., and Liew, F. Y., The role of interleukin-15 in t-cell migration and activation in rheumatoid arthritis, *Nat. Med.*, 2 (2), 175, 1996.

94. Gracie, J. A., Forsey, R. J., Chan, W. L., Gilmour, A., Leung, B. P., Greer, M. R., Kennedy, K. et al., A proinflammatory role for il-18 in rheumatoid arthritis, *J. Clin. Invest.*, 104 (10), 1393, 1999.

95. Cho, M. L., Jung, Y. O., Moon, Y. M., Min, S. Y., Yoon, C. H., Lee, S. H., Park, S. H., Cho, C. S., Jue, D. M., and Kim, H. Y., Interleukin-18 induces the production of vascular endothelial growth factor (vegf) in rheumatoid arthritis synovial fibroblasts via ap-1-dependent pathways, *Immunol. Lett.*, 103 (2), 159, 2006.

96. Alvaro-Gracia, J. M., Zvaifler, N. J., and Firestein, G. S., Cytokines in chronic inflammatory arthritis. Iv. Granulocyte/macrophage colony-stimulating factor-mediated induction of class ii mhc antigen on human monocytes: a possible role in rheumatoid arthritis, *J. Exp. Med.*, 170 (3), 865, 1989.

97. Xu, W. D., Firestein, G. S., Taetle, R., Kaushansky, K., and Zvaifler, N. J., Cytokines in chronic inflammatory arthritis. Ii. Granulocyte-macrophage colony-stimulating factor in rheumatoid synovial effusions, *J. Clin. Invest.*, 83 (3), 876, 1989.

98. Alvaro-Gracia, J. M., Zvaifler, N. J., Brown, C. B., Kaushansky, K., and Firestein, G. S., Cytokines in chronic inflammatory arthritis. Vi. Analysis of the synovial cells involved in granulocyte-macrophage colony-stimulating factor production and gene expression in rheumatoid arthritis and its regulation by il-1 and tumor necrosis factor-alpha, *J. Immunol.*, 146 (10), 3365, 1991.

99. Okamoto, H., Yamamura, M., Morita, Y., Harada, S., Makino, H., and Ota, Z., The synovial expression and serum levels of interleukin-6, interleukin-11, leukemia inhibitory factor, and oncostatin m in rheumatoid arthritis, *Arthritis Rheum.*, 40 (6), 1096, 1997.

100. Langdon, C., Leith, J., Smith, F., and Richards, C. D., Oncostatin m stimulates monocyte chemoattractant protein-1- and interleukin-1-induced matrix metallopro-teinase-1 production by human synovial fibroblasts in vitro, *Arthritis Rheum.*, 40 (12), 2139, 1997.

101. Hamilton, J. A., Leizer, T., Piccoli, D. S., Royston, K. M., Butler, D. M., and Croatto, M., Oncostatin m stimulates urokinase-type plasminogen activator activity in human synovial fibroblasts, *Biochem. Biophys. Res. Commun.*, 180 (2), 652, 1991.

102. Bodolay, E., Koch, A. E., Kim, J., Szegedi, G., and Szekanecz, Z., Angiogenesis and chemokines in rheumatoid arthritis and other systemic inflammatory rheumatic diseases, *J. Cell Mol. Med.*, 6 (3), 357, 2002.

103. Szekanecz, Z. and Koch, A. E., Chemokines and angiogenesis, *Curr. Opin. Rheumatol.*, 13 (3), 202, 2001.

104. Strieter, R. M., Polverini, P. J., Kunkel, S. L., Arenberg, D. A., Burdick, M. D., Kasper, J., Dzuiba, J. et al., The functional role of the elr motif in cxc chemokine-mediated angiogenesis, *J. Biol. Chem.*, 270 (45), 27348, 1995.

105. Strieter, R. M., Polverini, P. J., Arenberg, D. A., and Kunkel, S. L., The role of cxc chemokines as regulators of angiogenesis, *Shock*, 4 (3), 155, 1995.

106. Koch, A. E., Volin, M. V., Woods, J. M., Kunkel, S. L., Connors, M. A., Harlow, L. A., Woodruff, D. C., Burdick, M. D., and Strieter, R. M., Regulation of angiogenesis by the c-x-c chemokines interleukin-8 and epithelial neutrophil activating peptide 78 in the rheumatoid joint, *Arthritis Rheum.*, 44 (1), 31, 2001.

107. Salcedo, R., Ponce, M. L., Young, H. A., Wasserman, K., Ward, J. M., Kleinman, H. K., Oppenheim, J. J., and Murphy, W. J., Human endothelial cells express ccr2 and respond to mcp-1: direct role of mcp-1 in angiogenesis and tumor progression, *Blood*, 96 (1), 34, 2000.

108. Hosaka, S., Akahoshi, T., Wada, C., and Kondo, H., Expression of the chemokine superfamily in rheumatoid arthritis, *Clin. Exp. Immunol.*, 97 (3), 451, 1994.

109. Koch, A. E., Kunkel, S. L., Shah, M. R., Hosaka, S., Halloran, M. M., Haines, G. K., Burdick, M. D., Pope, R. M., and Strieter, R. M., Growth-related gene product alpha. A chemotactic cytokine for neutrophils in rheumatoid arthritis, *J. Immunol.*, 155 (7), 3660, 1995.

110. Hochreiter, W. W., Nadler, R. B., Koch, A. E., Campbell, P. L., Ludwig, M., Weidner, W., and Schaeffer, A. J., Evaluation of the cytokines interleukin 8 and epithelial neutrophil activating peptide 78 as indicators of inflammation in prostatic secretions, *Urology*, 56 (6), 1025, 2000.

111. Halloran, M. M., Woods, J. M., Strieter, R. M., Szekanecz, Z., Volin, M. V., Hosaka, S., Haines, G. K. 3rd et al., The role of an epithelial neutrophil-activating peptide-78-like protein in rat adjuvant-induced arthritis, *J. Immunol.*, 162 (12), 7492, 1999.

112. Koch, A. E., Kunkel, S. L., Harlow, L. A., Mazarakis, D. D., Haines, G. K., Burdick, M. D., Pope, R. M., Walz, A., and Strieter, R. M., Epithelial neutrophil activating peptide-78: a novel chemotactic cytokine for neutrophils in arthritis, *J. Clin. Invest.*, 94 (3), 1012, 1994.

113. Castor, C. W., Smith, E. M., Hossler, P. A., Bignall, M. C., and Aaron, B. P., Connective tissue activation. Xxxv. Detection of connective tissue activating peptide-iii isoforms in synovium from osteoarthritis and rheumatoid arthritis patients: patterns of interaction with other synovial cytokines in cell culture, *Arthritis Rheum.*, 35 (7), 783, 1992.

114. Koch, A. E., Kunkel, S. L., Harlow, L. A., Johnson, B., Evanoff, H. L., Haines, G. K., Burdick, M. D., Pope, R. M., and Strieter, R. M., Enhanced production of monocyte chemoattractant protein-1 in rheumatoid arthritis, *J. Clin. Invest.*, 90 (3), 772, 1992.

115. Addison, C. L., Daniel, T. O., Burdick, M. D., Liu, H., Ehlert, J. E., Xue, Y. Y., Buechi, L., Walz, A., Richmond, A., and Strieter, R. M., The cxc chemokine receptor 2, cxcr2, is the putative receptor for elr+ cxc chemokine-induced angiogenic activity, *J. Immunol.*, 165 (9), 5269, 2000.

116. Ruth, J. H., Rottman, J. B., Katschke, K. J., Qin, S., Wu, L., LaRosa, G., Ponath, P., Pope, R. M., and Koch, A. E., Selective lymphocyte chemokine receptor expression in the rheumatoid joint, *Arthritis Rheum.*, 44 (12), 2750, 2001.

117. Katschke, K. J. Jr, Rottman, J. B., Ruth, J. H., Qin, S., Wu, L., LaRosa, G., Ponath, P., Park, C. C., Pope, R. M., and Koch, A. E., Differential expression of chemokine receptors on peripheral blood, synovial fluid, and synovial tissue monocytes/macrophages in rheumatoid arthritis, *Arthritis Rheum.*, 44 (5), 1022, 2001.

118. Harris, E. D. Jr, Rheumatoid arthritis. Pathophysiology and implications for therapy, *N. Engl. J. Med.*, 322 (18), 1277, 1990.

119. Nicosia, R. F., Bonanno, E., and Smith, M., Fibronectin promotes the elongation of microvessels during angiogenesis in vitro, *J. Cell Physiol.*, 154 (3), 654, 1993.

120. Madri, J. A. and Williams, S. K., Capillary endothelial cell cultures: phenotypic modulation by matrix components, *J. Cell Biol.*, 97 (1), 153, 1983.

121. Canfield, A. E. and Schor, A. M., Evidence that tenascin and thrombospondin-1 modulate sprouting of endothelial cells, *J. Cell Sci.*, 108 (Pt 2), 797, 1995.

122. Rinaldi, N., Schwarz-Eywill, M., Weis, D., Leppelmann-Jansen, P., Lukoschek, M., Keilholz, U., and Barth, T. F., Increased expression of integrins on fibroblast-like synoviocytes from rheumatoid arthritis in vitro correlates with enhanced binding to extracellular matrix proteins, *Ann. Rheum. Dis.*, 56 (1), 45, 1997.

123. Rinaldi, N., Willhauck, M., Weis, D., Brado, B., Kern, P., Lukoschek, M., Schwarz-Eywill, M., and Barth, T. F., Loss of collagen type iv in rheumatoid synovia and cytokine effect on the collagen type-iv gene expression in fibroblast-like synoviocytes from rheumatoid arthritis, *Virchows Arch.*, 439 (5), 675, 2001.

124. Hembry, R. M., Bagga, M. R., Reynolds, J. J., and Hamblen, D. L., Immunolocalization studies on six matrix metalloproteinases and their inhibitors, timp-1 and timp-2, in synovia from patients with osteo- and rheumatoid arthritis, *Ann. Rheum. Dis.*, 54 (1), 25, 1995.

125. Maeda, S. T., Sawai, T., Uzuki, M., Takahashi, Y., Omoto, H., Seki, M., and Sakurai, M., Determination of interstitial collagenase (mmp-1) in patients with rheumatoid arthritis, *Ann. Rheum. Dis.*, 54 (12), 970, 1995.

126. Szekanecz, Z., Haines, G. K., and Koch, A. E., Differential expression of the urokinase receptor (cd87) in arthritic and normal synovial tissues, *J. Clin. Pathol.*, 50 (4), 314, 1997.

127. Belcher, C., Fawthrop, F., Bunning, R., and Doherty, M., Plasminogen activators and their inhibitors in synovial fluids from normal, osteoarthritis, and rheumatoid arthritis knees, *Ann. Rheum. Dis.*, 55 (4), 230, 1996.

128. Komiya, K., Enomoto, H., Inoki, I., Okazaki, S., Fujita, Y., Ikeda, E., Ohuchi, E., Matsumoto, H., Toyama, Y., and Okada, Y., Expression of adam15 in rheumatoid synovium: up-regulation by vascular endothelial growth factor and possible implications for angiogenesis, *Arthritis Res. Ther.*, 7 (6), R1158, 2005.

129. Nicholson, A. C., Malik, S. B., Logsdon, J. M. J.r., and Van Meir, E. G., Functional evolution of adamts genes: evidence from analyses of phylogeny and gene organization, *BMC Evol. Biol.*, 5 (1), 11, 2005.

130. Yamanishi, Y., Boyle, D. L., Clark, M., Maki, R. A., Tortorella, M. D., Arner, E. C., and Firestein, G. S., Expression and regulation of aggrecanase in arthritis: the role of tgf-beta, *J. Immunol.*, 168 (3), 1405, 2002.

131. Szekanecz, Z., Szegedi, G., and Koch, A. E., Cellular adhesion molecules in rheumatoid arthritis: Regulation by cytokines and possible clinical importance, *J. Investig. Med.*, 44 (4), 124, 1996.

132. Koch, A. E., Halloran, M. M., Haskell, C. J., Shah, M. R., and Polverini, P. J., Angiogenesis mediated by soluble forms of e-selectin and vascular cell adhesion molecule-1, *Nature*, 376 (6540), 517, 1995.

133. Horak, E. R., Leek, R., Klenk, N., LeJeune, S., Smith, K., Stuart, N., Greenall, M., Stepniewska, K., and Harris, A. L., Angiogenesis, assessed by platelet/endothelial cell adhesion molecule antibodies, as indicator of node metastases and survival in breast cancer, *Lancet*, 340 (8828), 1120, 1992.

134. Kumar, P., Wang, J. M., and Bernabeu, C., Cd 105 and angiogenesis, *J. Pathol.*, 178 (4), 363, 1996.

135. Stupack, D. G. and Cheresh, D. A., Ecm remodeling regulates angiogenesis: endothelial integrins look for new ligands, *Sci. STKE*, 2002 (119), PE7, 2002.

136. el Gabalawy, H. and Wilkins, J., Beta 1 (cd29) integrin expression in rheumatoid synovial membranes: an immunohistologic study of distribution patterns, *J. Rheumatol.*, 20 (2), 231, 1993.

137. Brooks, P. C., Clark, R. A., and Cheresh, D. A., Requirement of vascular integrin alpha v beta 3 for angiogenesis, *Science*, 264 (5158), 569, 1994.

138. Hood, J. D., Frausto, R., Kiosses, W. B., Schwartz, M. A., and Cheresh, D.A, Differential alphav integrin-mediated ras-erk signaling during two pathways of angiogenesis, *J. Cell Biol.*, 162 (5), 933, 2003.

139. Smyth, S. S. and Patterson, C., Tiny dancers: the integrin-growth factor nexus in angiogenic signaling, *J. Cell Biol.*, 158 (1), 17, 2002.

140. Senger, D. R., Claffey, K. P., Benes, J. E., Perruzzi, C. A., Sergiou, A. P., and Detmar, M., Angiogenesis promoted by vascular endothelial growth factor: regulation through alpha1beta1 and alpha2beta1 integrins, *Proc. Natl Acad. Sci. U.S.A.*, 94 (25), 13612, 1997.

141. Wellicome, S. M., Kapahi, P., Mason, J. C., Lebranchu, Y., Yarwood, H., and Haskard, D. O., Detection of a circulating form of vascular cell adhesion molecule-1: raised levels in rheumatoid arthritis and systemic lupus erythematosus, *Clin. Exp. Immunol.*, 92 (3), 412, 1993.

142. Wilkinson, L. S., Edwards, J. C., Poston, R. N., and Haskard, D. O., Expression of vascular cell adhesion molecule-1 in normal and inflamed synovium, *Lab. Investig.*, 68 (1), 82, 1993.

143. Hosaka, S., Shah, M. R., Pope, R. M., and Koch, A. E., Soluble forms of p-selectin and intercellular adhesion molecule-3 in synovial fluids, *Clin. Immunol. Immunopathol.*, 78 (3), 276, 1996.

144. Morbidelli, L., Brogelli, L., Granger, H. J., and Ziche, M., Endothelial cell migration is induced by soluble p-selectin, *Life Sci.*, 62 (1), PL7, 1998.

145. Maier, J. A., Delia, D., Thorpe, P. E., and Gasparini, G., In vitro inhibition of endothelial cell growth by the antiangiogenic drug agm-1470 (tnp-470) and the anti-endoglin antibody tec-11, *Anticancer Drugs*, 8 (3), 238, 1997.

146. Szekanecz, Z., Haines, G. K., Harlow, L. A., Shah, M. R., Fong, T. W., Fu, R., Lin, S. J., Rayan, G., and Koch, A. E., Increased synovial expression of transforming growth factor (tgf)-beta receptor endoglin and tgf-beta 1 in rheumatoid arthritis: Possible interactions in the pathogenesis of the disease, *Clin. Immunol. Immunopathol.*, 76 (2), 187, 1995.

147. Szekanecz, Z., Haines, G. K., Harlow, L. A., Shah, M. R., Fong, T. W., Fu, R., Lin, S. J., and Koch, A. E., Increased synovial expression of the adhesion molecules cd66a, cd66b, and cd31 in rheumatoid and osteoarthritis, *Clin. Immunol. Immunopathol.*, 76 (2), 180, 1995.

148. Halloran, M. M., Carley, W. W., Polverini, P. J., Haskell, C. J., Phan, S., Anderson, B. J., Woods, J. M. et al., Ley/h: an endothelial-selective, cytokine-inducible, angiogenic mediator, *J. Immunol.*, 164 (9), 4868, 2000.

149. Neidhart, M., Wehrli, R., Bruhlmann, P., Michel, B. A., Gay, R. E., and Gay, S., Synovial fluid cd146 (muc18), a marker for synovial membrane angiogenesis in rheumatoid arthritis, *Arthritis Rheum.*, 42 (4), 622, 1999.

150. Leahy, K. M., Koki, A. T., and Masferrer, J. L., Role of cyclooxygenases in angiogenesis, *Curr. Med. Chem.*, 7 (11), 1163, 2000.

151. Badet, J., Angiogenin, a potent mediator of angiogenesis. Biological, biochemical and structural properties, *Pathol. Biol. (Paris)*, 47 (4), 345, 1999.

152. Pufe, T., Bartscher, M., Petersen, W., Tillmann, B., and Mentlein, R., Expression of pleiotrophin, an embryonic growth and differentiation factor, in rheumatoid arthritis, *Arthritis Rheum.*, 48 (3), 660, 2003.

153. Cha, H. S., Ahn, K. S., Jeon, C. H., Kim, J., and Koh, E. M., Inhibitory effect of cyclo-oxygenase-2 inhibitor on the production of matrix metalloproteinases in rheumatoid fibroblast-like synoviocytes, *Rheumatol. Int.*, 24 (4), 207, 2004.

154. Woods, J. M., Mogollon, A., Amin, M. A., Martinez, R. J., and Koch, A. E., The role of cox-2 in angiogenesis and rheumatoid arthritis, *Exp. Mol. Pathol.*, 74 (3), 282, 2003.

155. Maisonpierre, P. C., Suri, C., Jones, P. F., Bartunkova, S., Wiegand, S. J., Radziejewski, C., Compton, D. et al., Angiopoietin-2, a natural antagonist for tie2 that disrupts in vivo angiogenesis, *Science*, 277 (5322), 55, 1997.

156. Davis, S., Aldrich, T. H., Jones, P. F., Acheson, A., Compton, D. L., Jain, V., Ryan, T. E., Isolation of angiopoietin-1, a ligand for the tie2 receptor, by secretion-trap expression cloning, *Cell*, 87 (7), 1161, 1996.

157. Shahrara, S., Volin, M. V., Connors, M. A., Haines, G. K., and Koch, A. E., Differential expression of the angiogenic tie receptor family in arthritic and normal synovial tissue, *Arthritis Res.*, 4 (3), 201, 2002.

158. Thurston, G., Wang, Q., Baffert, F., Rudge, J., Papadopoulos, N., Jean-Guillaume, D., Wiegand, S., Yancopoulos, G. D., and McDonald, D. M., Angiopoietin 1 causes vessel enlargement, without angiogenic sprouting, during a critical developmental period, *Development*, 132 (14), 3317, 2005.

159. Baffert, F., Le, T., Thurston, G., and McDonald, D. M., Angiopoietin-1 decreases plasma leakage by reducing number and size of endothelial gaps in venules, *Am. J. Physiol. Heart Circ. Physiol.*, 290 (1), H107, 2006.

160. Roviezzo, F., Tsigkos, S., Kotanidou, A., Bucci, M., Brancaleone, V., Cirino, G., and Papapetropoulos, A., Angiopoietin-2 causes inflammation in vivo by promoting vascular leakage, *J. Pharmacol. Exp. Ther.*, 314 (2), 738, 2005.

161. Camussi, G., Montrucchio, G., Lupia, E., De Martino, A., Perona, L., Arese, M., Vercellone, A., Toniolo, A., and Bussolino, F., Platelet-activating factor directly stimulates in vitro migration of endothelial cells and promotes in vivo angiogenesis by a heparin-dependent mechanism, *J. Immunol.*, 154 (12), 6492, 1995.

162. Menkes, C. J., Renoux, M., Laoussadi, S., Mauborgne, A., Bruxelle, J., and Cesselin, F., Substance p levels in the synovium and synovial fluid from patients with rheumatoid arthritis and osteoarthritis, *J. Rheumatol.*, 20 (4), 714, 1993.

163. Liote, F., Champy, R., Moenner, M., Boval-Boizard, B., and Badet, J., Elevated angiogenin levels in synovial fluid from patients with inflammatory arthritis and secretion of angiogenin by cultured synovial fibroblasts, *Clin. Exp. Immunol.*, 132 (1), 163, 2003.

164. Hockel, M., Sasse, J., and Wissler, J. H., Purified monocyte-derived angiogenic substance (angiotropin) stimulates migration, phenotypic changes, and "tube formation" but not proliferation of capillary endothelial cells in vitro, *J. Cell Physiol.*, 133 (1), 1, 1987.

165. Wissler, J. H., Extracellular and circulating redox- and metalloregulated erna and ernp: copper ion-structured rna cytokines (angiotropin ribokines) and bioaptamer targets imparting rna chaperone and novel biofunctions to s100-ef-hand and disease-associated proteins, *Ann. NY Acad. Sci.*, 1022 (163), 2004.

166. Omoto, A., Kawahito, Y., Prudovsky, I., Tubouchi, Y., Kimura, M., Ishino, H., Wada, M., Copper chelation with tetrathiomolybdate suppresses adjuvant-induced arthritis and inflammation-associated cachexia in rats, *Arthritis Res. Ther.*, 7 (6), R1174, 2005.

167. Volpert, O. V., Fong, T., Koch, A. E., Peterson, J. D., Waltenbaugh, C., Tepper, R. I., and Bouck, N. P., Inhibition of angiogenesis by interleukin 4, *J. Exp. Med.*, 188 (6), 1039, 1998.

168. Taki, H., Sugiyama, E., Kuroda, A, Mino, T., and Kobayashi, M., Interleukin-4 inhibits interleukin-11 production by rheumatoid synovial cells, *Rheumatology (Oxford)*, 39 (7), 728, 2000.

169. Sugiyama, E., Kuroda, A., Taki, H., Ikemoto, M., Hori, T., Yamashita, N., Maruyama, M., and Kobayashi, M., Interleukin 10 cooperates with interleukin 4 to suppress inflammatory cytokine production by freshly prepared adherent rheumatoid synovial cells, *J. Rheumatol.*, 22 (11), 2020, 1995.

170. Voest, E. E., Kenyon, B. M., O'Reilly, M. S., Truitt, G., D'Amato, R. J., and Folkman, J., Inhibition of angiogenesis in vivo by interleukin 12, *J. Natl. Cancer Inst.*, 87 (8), 581, 1995.

171. Angiolillo, A. L., Sgadari, C., and Tosato, G., A role for the interferon-inducible protein 10 in inhibition of angiogenesis by interleukin-12, *Ann. NY Acad. Sci.*, 795 (158), 1996.

172. Sgadari, C., Angiolillo, A. L., and Tosato, G., Inhibition of angiogenesis by interleukin-12 is mediated by the interferon-inducible protein 10, *Blood*, 87 (9), 3877, 1996.

173. Schlaak, J. F., Pfers, I., Meyer Zum Buschenfelde, K. H., and Marker-Hermann, E., Different cytokine profiles in the synovial fluid of patients with osteoarthritis, rheumatoid arthritis and seronegative spondylarthropathies, *Clin. Exp. Rheumatol.*, 14 (2), 155, 1996.

174. Carroll, G. J. and Bell, M. C., Leukaemia inhibitory factor stimulates proteoglycan resorption in porcine articular cartilage, *Rheumatol. Int.*, 13 (1), 5, 1993.

175. Pepper, M. S., Ferrara, N., Orci, L., and Montesano, R., Leukemia inhibitory factor (lif) inhibits angiogenesis in vitro, *J. Cell Sci.*, 108 (Pt 1), 73, 1995.

176. Waring, P. M., Carroll, G. J., Kandiah, D. A., Buirski, G., and Metcalf, D., Increased levels of leukemia inhibitory factor in synovial fluid from patients with rheumatoid arthritis and other inflammatory arthritides, *Arthritis Rheum.*, 36 (7), 911, 1993.

177. Strieter, R. M., Kunkel, S. L., Arenberg, D. A., Burdick, M. D., and Polverini, P. J., Interferon gamma-inducible protein 10 (ip-10), a member of the c-x-c chemokine family, is an inhibitor of angiogenesis, *Biochem. Biophys. Res. Commun.*, 210 (1), 51, 1995.

178. Patel, D. D., Zachariah, J. P., and Whichard, L. P., Cxcr3 and ccr5 ligands in rheumatoid arthritis synovium, *Clin. Immunol.*, 98 (1), 39, 2001.

179. Jouan, V., Canron, X., Alemany, M., Caen, J. P., Quentin, G., Plouet, J., and Bikfalvi, A., Inhibition of in vitro angiogenesis by platelet factor-4-derived peptides and mechanism of action, *Blood*, 94 (3), 984, 1999.

180. Wooley, P. H., Schaefer, C., Whalen, J. D., Dutcher, J. A., and Counts, D. F., A peptide sequence from platelet factor 4 (ct-112) is effective in the treatment of type ii collagen induced arthritis in mice, *J. Rheumatol.*, 24 (5), 890, 1997.

181. Vicari, A. P., Ait-Yahia, S., Chemin, K., Mueller, A., Zlotnik, A., and Caux, C., Antitumor effects of the mouse chemokine 6ckine/slc through angiostatic and immunological mechanisms, *J. Immunol.*, 165 (4), 1992, 2000.

182. Qin, S., Rottman, J. B., Myers, P., Kassam, N., Weinblatt, M., Loetscher, M., Koch, A. E., Moser, B., and Mackay, R., The chemokine receptors cxcr3 and ccr5 mark subsets of t cells associated with certain inflammatory reactions, *J. Clin. Invest.*, 101 (4), 746, 1998.

183. Skotnicki, J. S., Zask, A., Nelson, F. C., Albright, J. D., and Levin, J. I., Design and synthetic considerations of matrix metalloproteinase inhibitors, *Ann. NY Acad. Sci.*, 878 (61), 1999.

184. Apparailly, F., Bouquet, C., Millet, V., Noel, D., Jacquet, C., Opolon, P., Perricaudet, M., Sany, J., Yeh, P., and Jorgensen, C., Adenovirus-mediated gene transfer of urokinase plasminogen inhibitor inhibits angiogenesis in experimental arthritis, *Gene Ther.*, 9 (3), 192, 2002.

185. Koch, A. E., Friedman, J., Burrows, J. C., Haines, G. K., and Bouck, N. P., Localization of the angiogenesis inhibitor thrombospondin in human synovial tissues, *Pathobiology*, 61 (1), 1, 1993.

186. Koch, A. E., Szekanecz, Z., Friedman, J., Haines, G. K., Langman, C. B., and Bouck, N. P., Effects of thrombospondin-1 on disease course and angiogenesis in rat adjuvant-induced arthritis, *Clin. Immunol. Immunopathol.*, 86 (2), 199, 1998.

187. Clamp, A. R. and Jayson, G. C., The clinical potential of antiangiogenic fragments of extracellular matrix proteins, *Br. J. Cancer*, 93 (9), 967, 2005.

188. Nakamura, S., Kamihagi, K., Satakeda, H., Katayama, M., Pan, H., Okamoto, H., Noshiro, M. et al., Enhancement of sparc (osteonectin) synthesis in arthritic cartilage. Increased levels in synovial fluids from patients with rheumatoid arthritis and regulation by growth factors and cytokines in chondrocyte cultures, *Arthritis Rheum.*, 39 (4), 539, 1996.

189. Wilson, J. L., Nayanar, V., and Walker, J. S., The site of anti-arthritic action of the kappa-opioid, u-50, 488h, in adjuvant arthritis: importance of local administration, *Br. J. Pharmacol.*, 118 (7), 1754, 1996.

190. Walker, J. S., Chandler, A. K., Wilson, J. L., Binder, W., and Day, O., Effect of mu-opioids morphine and buprenorphine on the development of adjuvant arthritis in rats, *Inflamm. Res.*, 45 (11), 557, 1996.

191. Setoguchi, K., Misaki, Y., Kawahata, K., Shimada, K., Juji, T., Tanaka, S., Oda, H. et al., Suppression of t cell responses by chondromodulin i, a cartilage-derived angiogenesis inhibitory factor: therapeutic potential in rheumatoid arthritis, *Arthritis Rheum.*, 50 (3), 828, 2004.

192. Kurosaka, D., Yoshida, K., Yasuda, J., Yokoyama, T., Kingetsu, I., Yamaguchi, N., Joh, K., Matsushima, M., Saito, S., and Yamada, A., Inhibition of arthritis by systemic administration of endostatin in passive murine collagen induced arthritis, *Ann. Rheum. Dis.*, 62 (7), 677, 2003.

193. Matsumoto, K., Date, K., Ohmichi, H., and Nakamura, T., Hepatocyte growth factor in lung morphogenesis and tumor invasion: role as a mediator in epithelium-mesenchyme and tumor-stroma interactions, *Cancer Chemother. Pharmacol.*, 38 (suppl. (S42)), 1996.

194. Yin, G., Liu, W., An, P., Li, P., Ding, I., Planelles, V., Schwarz, E. M., and Min, W., Endostatin gene transfer inhibits joint angiogenesis and pannus formation in inflammatory arthritis, *Mol. Ther.*, 5 (5 Pt 1), 547, 2002.

195. Kim, J. M., Ho, S. H., Park, E. J., Hahn, W., Cho, H., Jeong, J. G., Lee, Y. W., and Kim, S., Angiostatin gene transfer as an effective treatment strategy in murine collagen-induced arthritis, *Arthritis Rheum.*, 46 (3), 793, 2002.

196. Sumariwalla, P. F., Cao, Y., Wu, H. L., Feldmann, M., and Paleolog, E. M., The angiogenesis inhibitor protease-activated kringles 1-5 reduces the severity of murine collagen-induced arthritis, *Arthritis Res. Ther.*, 5 (1), R32, 2003.

197. Raychaudhuri, S. P. and Farber, E. M., The prevalence of psoriasis in the world, *J. Eur. Acad. Dermatol. Venereol.*, 15 (1), 16, 2001.

198. Moll, J. M. and Wright, V., Psoriatic arthritis, *Semin. Arthritis Rheum.*, 3 (1), 55, 1973.

199. Gottlieb, A. B., Lebwohl, M., Shirin, S., Sherr, A., Gilleaudeau, P., Singer, G., Solodkina, G. et al., Anti-cd4 monoclonal antibody treatment of moderate to severe psoriasis vulgaris: results of a pilot, multicenter, multiple-dose, placebo-controlled study, *J. Am. Acad. Dermatol.*, 43 (4), 595, 2000.

200. Wrone-Smith, T. and Nickoloff, B. J., Dermal injection of immunocytes induces psoriasis, *J. Clin. Invest.*, 98 (8), 1878, 1996.

201. Raychaudhuri, S. P., Dutt, S., Raychaudhuri, S. K., Sanyal, M., and Farber, E. M., Severe combined immunodeficiency mouse-human skin chimeras: a unique animal model for the study of psoriasis and cutaneous inflammation, *Br. J. Dermatol.*, 144 (5), 931, 2001.

202. Braverman, I. M. and Yen, A., Ultrastructure of the capillary loops in the dermal papillae of psoriasis, *J. Invest. Dermatol.*, 68 (1), 53, 1977.

203. Reece, R. J., Canete, J. D., Parsons, W. J., Emery, P., and Veale, D. J., Distinct vascular patterns of early synovitis in psoriatic, reactive, and rheumatoid arthritis, *Arthritis Rheum.*, 42 (7), 1481, 1999.

204. Creamer, D., Sullivan, D., Bicknell, R., and Barker, J., Angiogenesis in psoriasis, *Angiogenesis*, 5 (4), 231, 2002.

205. Fearon, U., Griosios, K., Fraser, A., Reece, R., Emery, P., Jones, P. F., and Veale, D. J., Angiopoietins, growth factors, and vascular morphology in early arthritis, *J. Rheumatol.*, 30 (2), 260, 2003.

206. Creamer, D., Jaggar, R., Allen, M., Bicknell, R., and Barker, J., Overexpression of the angiogenic factor platelet-derived endothelial cell growth factor/thymidine phosphorylase in psoriatic epidermis, *Br. J. Dermatol.*, 137 (6), 851, 1997.

207. Fearon, U., Reece, R., Smith, J., Emery, P., and Veale, D. J., Synovial cytokine and growth factor regulation of mmps/timps: implications for erosions and angiogenesis in early rheumatoid and psoriatic arthritis patients, *Ann. NY Acad. Sci.*, 878 (619), 1999.

208. Kuroda, K., Sapadin, A., Shoji, T., Fleischmajer, R., and Lebwohl, M., Altered expression of angiopoietins and tie2 endothelium receptor in psoriasis, *J. Invest. Dermatol.*, 116 (5), 713, 2001.

209. Firestein, G. S., Starving the synovium: Angiogenesis and inflammation in rheumatoid arthritis, *J. Clin. Invest.*, 103 (1), 3, 1999.

210. Mall, J. W., Myers, J. A., Xu, X., Saclarides, T. J., Philipp, A. W., and Pollmann, C., Leflunomide reduces the angiogenesis score and tumor growth of subcutaneously implanted colon carcinoma cells in the mouse model, *Chirurg*, 73 (7), 716, 2002.

211. Xu, X., Shen, J., Mall, J. W., Myers, J. A., Huang, W., Blinder, L., Saclarides, T. J., Williams, J. W., and Chong, A. S., In vitro and in vivo antitumor activity of a novel immunomodulatory drug, leflunomide: mechanisms of action, *Biochem. Pharmacol.*, 58 (9), 1405, 1999.

212. Oliver, S. J., Cheng, T. P., Banquerigo, M. L., and Brahn, E ., Suppression of collagen-induced arthritis by an angiogenesis inhibitor, agm-1470, in combination with cyclosporin: reduction of vascular endothelial growth factor (vegf), *Cell Immunol.*, 166 (2), 196, 1995.

213. Koch, A. E., Cho, M., Burrows, J., Leibovich, S. J., and Polverini, P. J., Inhibition of production of macrophage-derived angiogenic activity by the anti-rheumatic agents gold sodium thiomalate and auranofin, *Biochem. Biophys. Res. Commun.*, 154 (1), 205, 1988.

214. Lin, Y. S., Nguyen, C., Mendoza, J. L., Escandon, E., Fei, D., Meng, Y. G., and Modi, N. B., Preclinical pharmacokinetics, interspecies scaling, and tissue distribution of a humanized monoclonal antibody against vascular endothelial growth factor, *J. Pharmacol. Exp. Ther.*, 288 (1), 371, 1999.

215. Kucharz, E. J., Gozdzik, J., Kopec, M., Kotulska, A., Lewicki, M., Pieczyrak, R., Widuchowska, M., Zakliczynska, H., Szarzynska-Ruda, M., and Zycinska-Debska, E., A single infusion of infliximab increases the serum endostatin level in patients with rheumatoid arthritis, *Clin. Exp. Rheumatol.*, 21 (2), 273, 2003.

216. Kucharz, E. J., Kotulska, A., Kopec, M., Stawiarska-Pieta, B., and Pieczyrak, R., Serum level of the circulating angiogenesis inhibitor endostatin in patients with hyperthyroidism or hypothyroidism, *Wien Klin Wochenschr*, 115 (5–6), 179, 2003.

217. Hilliquin, P., Guinot, P., Chermat-Izard, V., Puechal, X., and Menkes, C. J., Treatment of rheumatoid arthritis with platelet activating factor antagonist bn 50730, *J. Rheumatol.*, 22 (9), 1651, 1995.

218. Hilliquin, P., Natour, J., Aissa, J., Guinot, P., Laoussadi, S., Benveniste, J., Menkes, C. J., and Arnoux, B., Treatment of carrageenan induced arthritis by the platelet activating factor antagonist bn 50730, *Ann. Rheum. Dis.*, 54 (2), 140, 1995.

219. Kingsley, G., Lanchbury, J., and Panayi, G., Immunotherapy in rheumatic disease: An idea whose time has come—or gone? *Immunol. Today*, 17 (1), 9, 1996.

220. Panayi, G. S., Kingsley, G. H., and Lanchbury, J. S., Immunotherapy of immune-mediated diseases, *Q. J. Med.*, 83 (303), 489, 1992.

221. Kingsley, G., Panayi, G., and Lanchbury, J., Immunotherapy of rheumatic diseases—practice and prospects, *Immunol. Today*, 12 (6), 177, 1991.

222. Saiki, I., Murata, J., Makabe, T., Nishi, N., Tokura, S., and Azuma, I., Inhibition of tumor angiogenesis by a synthetic cell-adhesive polypeptide containing the arg-gly-asp (rgd) sequence of fibronectin, poly(rgd), *Jpn. J. Cancer Res.*, 81 (6–7), 668, 1990.

223. Shibuya, M., Vegf-receptor inhibitors for anti-angiogenesis, *Nippon Yakurigaku Zasshi*, 122 (6), 498, 2003.

224. Manley, P. W., Martiny-Baron, G., Schlaeppi, J. M., and Wood, J. M., Therapies directed at vascular endothelial growth factor, *Expert Opin. Investig. Drugs*, 11 (12), 1715, 2002.

225. Corkill, M. M., Kirkham, B. W., Haskard, D. O., Barbatis, C., Gibson, T., and Panayi, G. S., Gold treatment of rheumatoid arthritis decreases synovial expression of the endothelial leukocyte adhesion receptor elam-1, *J. Rheumatol.*, 18 (10), 1453, 1991.

226. Wilder, R. L., Integrin alpha v beta 3 as a target for treatment of rheumatoid arthritis and related rheumatic diseases, *Ann. Rheum. Dis.*, 61 (suppl. 2), ii96, 2002.

227. Oliver, S. J., Banquerigo, M. L., and Brahn, E., Suppression of collagen-induced arthritis using an angiogenesis inhibitor, agm-1470, and a microtubule stabilizer, taxol, *Cell Immunol.*, 157 (1), 291, 1994.

228. Matsuno, H., Yudoh, K., Uzuki, M., Nakazawa, F., Sawai, T., Yamaguchi, N., Olsen, B. R., and Kimura, T., Treatment with the angiogenesis inhibitor endostatin: a novel therapy in rheumatoid arthritis, *J. Rheumatol.*, 29 (5), 890, 2002.

229. Raychaudhuri, S. K., Raychaudhuri, S. P., Weltman, H., and Farber, E. M., Effect of nerve growth factor on endothelial cell biology: proliferation and adherence molecule expression on human dermal microvascular endothelial cells, *Arch. Dermatol. Res.*, 293 (6), 291, 2001.

230. Raychaudhuri, S. P., Sanyal, M., Weltman, H., and Kundu-Raychaudhuri, S., K252a, a high-affinity nerve growth factor receptor blocker, improves psoriasis: an in vivo study using the severe combined immunodeficient mouse-human skin model, *J. Invest. Dermatol.*, 122 (3), 812, 2004.

19 Angiogenesis and Cardiovascular Diseases

Robert J. Tomanek

CONTENTS

19.1 INTRODUCTION

Since angiogenesis plays a key role in both physiological and pathological processes, its manipulation is an important goal in the control of disease. Currently, angiogenic therapies are being pursued vigorously, since vascular growth is needed for adequate blood perfusion in ischemic diseases. On the other hand, anti-angiogenic therapies are receiving considerable attention because angiogenesis is problematic in other diseases, e.g., cancer, arthritis (Figure 19.1). The goal of this chapter is to provide general insights into the importance of angiogenesis in cardiovascular diseases.

19.2 VASCULOGENESIS, ANGIOGENESIS AND REMODELING

Angioblasts are endothelial cell precursors that originate in the mesoderm (Figure 19.2). During embryonic development, angioblasts assemble into a vascular labyrinth, a process termed vasculogenesis. These vascular channels coalesce and then grow by sprouting or by partitioning (intussusceptive growth). Subsequently, smooth muscle and other adventitial cells are recruited to form arteries and veins. Most vascular growth in the adult occurs via angiogenesis and involves formation of vascular channels from capillaries or venules. Recent findings suggest that neovascularization may, in some cases, involve the incorporation of bone marrow-derived progenitor.[1-3] Thus, circulating angioblasts, characterized by a robust proliferation rate, can participate in

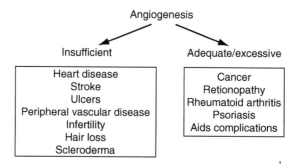

FIGURE 19.1 Angiogenesis facilitates either favorable or unfavorable outcomes in various diseases. Angiogenesis is generally a benefit in cardiovascular diseases, such as narrowing or occlusion of coronary vessels, and in peripheral vascular disease. Current development of therapies for these conditions include growth factors that stimulate angiogenesis. In contrast, tumor metastasis is dependent on neovascularization; thus, therapies are aimed at anti-angiogenic agents.

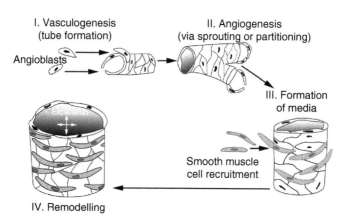

FIGURE 19.2 Vessel formation occurs via: (I) Vasculogenesis (angioblast assembly to form a vascular tube) in embryos and, under some circumstances, in the adult. (II) Subsequent growth in the embryo and most growth in the adult is via angiogenesis (sprouting: i.e., outgrowth of branches or by partitioning the parent vessel). (III) Formation of the media (muscular component) of a precapillary vessel or muscular vein requires recruitment of smooth muscle cells to the preexisting endothelial tube. (IV) Remodeling is a response to altered flow (shear forces). Higher flow states cause an expansion of the vessel (increase in diameter), requiring endothelial and smooth muscle cell proliferation. Lower flow states result in a decrease in vessel diameter.

vessel formation in inflamed, ischemic, or neoplastic tissues. Remodeling involves reorganization of vascular channels in order to adjust for changing perfusion demands. This occurs in collateral vessel enlargement in order to increase flow to an underperfused tissue region.

Vascular growth is a complex phenomenon requiring a host of molecules and signaling events. The initiation of the cascade of events required to grow new vessels necessitates a primary stimulus, i.e., metabolic (ischemia/hypoxia) or mechanical (shear stress, stretch) factors.[4] Angiogenesis necessitates the migration and proliferation of endothelial cells. These events occur when angiogenic factors are favored over angiostatic factors.[5] When endothelial cells are activated, they release proteinases, e.g., metaloproteinases, to dissolve the basement membrane. Pericytes are then recruited to facilitate maturation and maintenance of structural integrity,[6] and the basement membrane is reconstituted. Smooth muscle cells are recruited for the formation of precapillary vessels and veins. These events necessitate the spatial and temporal regulation by growth factors, interactions between cells and the extracellular matrix, and adhesion molecules.[7]

Remodeling plays a major role in the development of collateral vessels (for review see References 8–10). The defining study of Shaper and colleagues[11] revealed that gradual constriction of a coronary artery in dogs stimulates enlargement of collateral vessels via proliferation of endothelial, medial, and adventitial cells. Subsequent studies have also utilized the ameroid occlusion model to create a collateral dependent region and have documented a protective effect attributable to collateral vessel growth.[12–14] Repeated coronary occlusions in dogs also have been shown to stimulate collateralization.[14] Following imposition of the ameroid occluder, the tissue supplied by the vessel becomes ischemic and undergoes a remodeling process. At first, the vessel appears vein-like[11] and then undergoes development of the muscular tunic and restoration of the internal elastic lamina.[15]

If a coronary artery in humans is severely narrowed, collateral flow increases.[16] This finding is consistent with experimental data that indicate that recurrent and severe myocardial ischemia[17,18] stimulates collateral growth. Collateral development, termed arteriogenesis, is largely dependent on shear stress, which occurs in collateral channels as flow is restricted in an epicardial artery. The increase in shear stress activates endothelial cells and the expression of growth factors within the vascular wall.

19.3 GROWTH FACTOR REGULATION OF ANGIOGENESIS

Our understanding of vascular growth, development, and remodeling has been markedly advanced because of newly discovered growth factors and their receptors. Figure 19.3 illustrates two families of growth factors—vascular endothelial growth factors (VEGFs) and angiopoietins that are ligands for endothelial cell specific receptor kinases. These growth factors and their receptors have been shown to play important roles in neovascularization.[19,20]

VEGF is a powerful stimulant for endothelial cell migration, proliferation, and vascular tube formation. The importance of VEGF proteins in vascular development is heightened by the evidence that the loss of a single VEGF allele is lethal in the

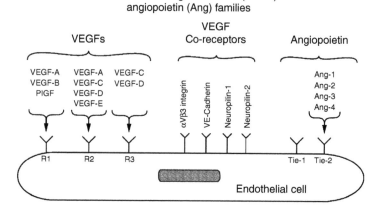

FIGURE 19.3 The VEGF family includes VEGFs A–E and placental growth factor (PlGF). Tyrosine kinase receptors R1 (flt-1), R2 (flk-1), and R3 (flk-4) show some selectivity for the various VEGF family members. VEGF co-receptors are important for activation of the VEGF signaling cascade. Angiopoietins are known to be ligands for the tyrosine kinase receptor Tie-2. Ligands for the Tie-1 receptor have yet to be validated.

embryo. One of the factors that regulates VEGF gene expression is oxygen tension, as demonstrated by its upregulation in cells exposed to low pO_2.[21] This increase also occurs when the myocardium becomes ischemic by occlusion of a coronary artery.[22] The importance of VEGF in vascular growth is further augmented by its monocyte chemotaxis and by its ability to induce serine proteases and promote expression of adhesion molecules (reviewed in Reference 23). As seen in Figure 19.3, optimal VEGF function is dependent upon co-receptors. Some of the known functions of VEGF family members are provided in Table 19.1. Placental growth factor (PlGF) is also a member of the VEGF family since it is a ligand for VEGF-R1. It has been shown to induce angiogenesis in vivo[24] and it appears to facilitate angiogenesis and collateral growth.[25]

Angiopoietins act in a complementary and coordinated fashion to facilitate mature blood vessels. Angiopoietin-1 and its Tie-2 receptor play a key role in cell to cell and cell to matrix interactions, as well as contributing to cell proliferation, migration, and survival.[26] One particularly unique action of angiopoietin-1 is that it counteracts the enhanced vascular leakage associated with VEGF.[27] Angiopoietins are not as well understood as VEGF family members. One underlying factor of our limited understanding of these angiogenic molecules is that their effect on endothelial cells is likely to be cell type and context dependent.

A third family of ligands for thyrosine kinase receptors are the ephrins, which have been shown to mimic the actions of VEGF and angiopoietins in vitro.[28] The ephrins comprise a large family, i.e., 5 Ephrin A's and 3 Ephrin B's, along with 14 Eph receptors.[29] Although the Eph receptors and their ephrin ligands are not well understood, Eph B receptors and Ephrin B ligands are known to be involved in vascular assembly and the acquisition of arteriovenus identity. A recent study has

TABLE 19.1
Some Functions of VEGF's and Angiopoietins

Growth Factor	Function	Reference
VEGF-A (5 splice variants: 121, 145, 165, 189, 206)	Migration, proliferation tube formation	Ferrara et al. (2003)
VEGF-B	Angiogenesis, vascular remodeling	Wanstall et al. (2002) and Silvestre et al. (2003)
VEGF-C	Mitogenic and chemotactic for EC	Witzenblichler et al. (1998) and Shin et al. (2002)
	Enlargement of lymphatic vessels and veins	Saaristo et al. (2002)
VEGF-D	Angiogenic and lymphogenic	Rissamen et al. (2003)
VEGF-E	Similar to VEGF-A	Meyer et al. (1999)
PIGF	Angiogenesis	Ziche et al. (1997)
	Forms hetrodimers with VEGF-A	Cao et al. (1996)
	Angiogenesis and collateral vessel growth	Carmeleit et al. (2001)
Angiopoietin-1	Capillary sprout formation vessel wall stabilization promotes leakage-resistant vessels	Koblizek et al. (1998), Thurston et al. (1999), and Joussen et al. (2002)
Angiopoietin-2	Increases vessel length	Asahara et al. (1998)
	Antagonist for ang-1/Tie-2 pathway	Maisopierre et al. (1997)

See Figure 19.3 for an illustration of these growth factors and their receptors.

documented an increase in endothelial cell migration and tubulogenesis when the Eph B promoter is activated by HoxA9 gene.[30]

As noted in Table 19.1, VEGFs and angiopoietins play key roles in the formation and maintenance of endothelial-lined tubes (capillaries, venules). The recruitment of mural cell progenitors involves angiopoietin-1 and its Tie-2 receptor, as well as PDGF-B (reviewed by Cleaver and Melton[31]). Smooth muscle cell/pericyte differentiation occurs via TGF-β influence. Fibroblast growth factor-2 (FGF-2) is a regulator of numerous cell functions including angiogenesis and vessel remodeling. This growth factor is expressed in cardiomyocytes, fibroblasts, and vascular smooth muscle cells (reviewed in Reference 32). FGF-2 is a powerful angiogenic agent and is known to regulate genes that play a role in angiogenesis, e.g., eNOS, VEGF, ERK 1/2, $\alpha_v\beta_5$ integrin.[33]

19.4 MYOCARDIAL ISCHEMIA AND INFARCTION

Approximately 50% of all deaths in the Western world are directly or indirectly caused by myocardial ischemia and its consequences.[34] Myocardial ischemia is most commonly associated with occlusive arteries, as occurs in atherosclerosis. Current treatments for patients with narrowed coronary arteries include percutaneous transluminal angioplasty or bypass surgery. These patients would also benefit from therapies that stimulate the growth of collateral vessels and growth of the microcirculation, i.e., arterioles and capillaries. Ischemic injury can also occur in hearts with normal coronary arteries if they are enlarged by a pathological stimulus, such as cardiomyopathy or pressure overload due to aortic stenosis or untreated hypertension. Such hearts are more vulnerable to ischemia, as evidenced by the earlier onset of contracture, increased calcium overload during early reperfusion, and a greater accumulation of lactate.[35] Left ventricular hypertrophy (LVH) is well established as a major risk factor. Data from the Framingham Heart Study indicate that LVH is the single most important contribution to cardiovascular morbidity and mortality.[36]

19.5   PERIPHERAL VASCULAR DISEASE

Peripheral artery disease occurs when arteries distal to the aortic arch are significantly narrowed; such narrowing may be due to atherosclerosis, arteritis, local thrombus formation, or embolization. About 12% of the adult U.S. population presents symptoms of the disease,[37,38] which has a strong association with cardiovascular morbidity and mortality. Smoking and diabetes are the two greatest risk factors for peripheral artery disease; other factors include obesity, hyperlipidemia, inactivity, and hypertension.[39] Therapy for peripheral vascular disease is aimed at reducing complications of leg ulcers, ischemia, myocardial infarction, and stroke. Current therapies include antiplatelet drugs, ACE inhibition therapy, lowering LDL cholesterol, antihypertensive drugs, and management of diabetes.[38]

The goal of therapeutic angiogenesis for peripheral vascular disease has received considerable attention during recent years. Gene therapy has been established as a potential method for stimulating angiogenesis in ischemic limb disease.[40] Newer techniques of gene transfer have shown to be effective in preclinical studies and some clinical trials have documented the safety of gene transfer for peripheral vascular disease. Other approaches to stimulate angiogenesis in the ischemic limb include delivery of naked plasmid DNA[41] and electrical stimulation.[42]

The emphasis on gene therapy for peripheral vascular disease is based on some success in animal models in improving perfusion in ischemic limbs. Three areas of improvement have been targeted: (1) prevention of restenosis following baboon angioplasty or stent placement; (2) improving success rate of vascular grafts; and, (3) stimulating vascular growth.[41]

19.6   ATHEROSCLEROSIS AND RESTENOSIS

Angiogenesis occurs in atherosclerotic plaques, an adaptation in response to ischemia in the media of the artery. The ingrowth of small vessels into the media occurs from the vasa vasora, the vascular supply to the artery's outer region. This enhanced vascular supply, intended to prevent tissue hypoxia and necrosis, may in fact enhance plaque formation (reviewed in Reference 43). Hypotheses concerning pathological effects of such neovascularization include vasoconstriction of the vasa vasora or enhanced permeability of the new vessels. The latter would facilitate inflammation, intimal proliferation and matrix deposition.

Arterial injury is also associated with increased wall thickness and consequential intramural ischemia. This is common after therapeutic coronary recanalization procedures. Arterial injury triggers inflammation and the enhancement of key angiogenic growth factors, i.e., VEGF[44] and FGF-2.[45] Balloon dilation also stimulates reactive hyperemia of the arterial wall by promoting growth of the vasa vasora.[46] This response promotes healing, but at the same time may enhance intimal hyperplasia.[47]

19.7　THERAPEUTIC APPROACHES

Therapies for ischemic tissues are aimed at the restoration of blood flow. As noted previously, angioplasty, in combination with stent placement, bypass grafting, and pharmacological therapy are commonly employed. For a significant number of patients, these interventions are not adequate (reviewed in Reference 48). Therefore, the stimulation of neovascularization in ischemic tissues has been the focus of considerable research during the last decade. Some of the research using animal models has encouraged clinical trials.

Much of the research has focused on the delivery of angiogenic factors, e.g., VEGF, FGF. Other candidates for therapeutic angiogenesis in cardiac or limb ischemia include hepatocyte growth/scatter factor (HGF/SF), the monocyte chemo attractant protein-1 (MCP-1), and some chemokines, e.g., interleukin 8 (IL-8).[34,49–52] The goal of therapeutic angiogenesis has been to initiate vascular growth in ischemic tissues by administering growth factor proteins or genes that encode the specific growth factors. There are advantages to both approaches and considerable excitement has been generated by animal studies that have documented angiogenesis. However, a number of problems have been identified that need to be overcome for clinical trials to be successful.

One of the limitations of administering a single growth factor or its gene is that neovascularization may be limited to capillaries rather than encompassing arteriolar and collateral vessel growth. Enhancement of capillary growth alone does not facilitate a significant increase in blood perfusion. The anatomical basis for increased tissue blood flow is an increase in the cross-sectional area of the major resistance vessels. In the case of occlusive vessel disease, growth of collateral vessels is critical. Accordingly, therapeutic neovascularization studies need to focus on arteriogenesis. Tissue ischemia involves an increased expression of hypoxia-inducible factor-1α protein (HIF-1α), which stimulates increased gene expression primarily in VEGF and its Flt-1 receptor and angiopoietin-1. This enhancement of VEGF and angiopoietin-1 mainly affects the growth of capillaries. In order to affect the growth of precapillary vessels (arterioles and arteries), as well as collateral growth, the recruitment of smooth muscle cells is essential. The latter requires additional molecules and signaling. Recent studies concerning improved blood perfusion of the ischemic heart or lower-limb have considered collateral growth as a prime objective of therapy. As noted earlier, such growth is essential when coronary arteries are narrowed by disease, or in the case of peripheral vascular disease that affects primarily the lower limbs. As recently reported,[53] the remodeling of existing collaterals, which increases their diameters, depends on activation of the endothelium via increased shear stress, presence of bone marrow-derived cells, and endothelial and smooth muscle cell proliferation. Monocytes are critical mediators of collateral remodeling and provide proteases necessary for proteolytic activity that permits cellular migration.

Animal models of therapeutic augmentation of collateral growth appear promising. The administration of bFGF and VEGF proteins were shown to increase coronary collateral blood flow in dogs.[54–56] Likewise, bFGF improved perfusion of the ischemic rabbit hind limb.[57] Other animal studies have also demonstrated reason for optimism. However, the road from laboratory to the bedside is long and problematic. First, these earlier studies demonstrated that the growth factors need to be repeatedly administered. In humans, repeated intracoronary artery injection is not a feasible option. Second, studies on humans are less promising (reviewed in References 52 and 53). One major problem is that clinical studies have focused on angiogenesis rather than on arteriogenesis (collateral growth). Another problem is the selection of the patient population. Old patients with end-stage coronary artery disease often have comorbidity and take multiple medications that may modify vascular growth. Still another problem is the choice of the therapeutic agents. Considering the multiple signaling involving vascular growth, strategies that include multiple growth factors and appropriate (optimal) dosing and routes of administration need to be developed.

Studies on therapeutic angiogenesis and arteriogenesis have focused on the administration of recombinant genes or growth factors. An alternative approach is pharmacological treatments that stimulate multiple growth factors and, consequently, angiogenesis or arteriogenesis. Studies in my laboratory have employed a bradycardic drug, alinidine, to stimulate at least both VEGF and bFGF, and the VEGF receptor, FLT-1, in the postinfarcted heart.[58] The significance of this work is that arteriolar and capillary growth was associated with the preservation of coronary reserve. Moreover, treatment was associated by a smaller decline in ejection fraction. We also have shown that diiodothyropropionic acid (a thyroxine analog) stimulates arteriolar growth and modifies myocardial remodeling after infarction.[59] Accordingly, this noninvasive therapeutic avenue provides an alternative to gene therapy.

19.8 SUMMARY

Therapeutic angiogenesis has become the focus of considerable research during the last decade. Studies have shown that proliferation of small vessels and growth of collateral vessels can be induced. These findings are important for preventing tissue ischemic and inadequate oxygenation in a number of conditions, e.g., coronary artery stenosis, cardiac hypertrophy or myopathy, and peripheral vascular disease. There have been important advances in our understanding of the cascade of events that constitute angiogenesis and the growth factors and signaling pathways that regulate it. Studies on animal models have provided considerable information that has encouraged clinical trials. There still remain a number of problems to be solved before the safe and effective delivery of angiogenic agents to patients can be realized. The utility of angiogenic therapy in the treatment of ischemic limb and heart disease will necessitate Phase II and III clinical trials. Considering the major advances in the field, there remains optimism concerning effective treatment protocols in the future.

REFERENCES

1. Asahara, T., Murohara, T., Sullivan, A., Silver, M., van der Zee, R., Li, T., Witzenblichen, B., Schatteman, G., and Isner, J. M., Isolation of putative progenitor endothelial cells for angiogenesis, *Science*, 275, 964–996, 1997.
2. Takahashi, T., Kalka, C., Masuda, H., Chen, D., Silver, M., Kearney, M., Magner, M., Isner, J. M., and Asahara, T., Ischemia- and cytokine-induced mobilization of bone marrow-derived endothelial progenitor cells for neovascularization, *Nat. Med.*, 5, 434–438, 1999.
3. Schatteman, G. C., Hanlon, H. D., Jiao, C., Dodds, S. G., and Christy, B. A., Blood-derived angioblasts accelerate blood-flow restoration in diabetic mice, *J. Clin. Invest.*, 106, 571–578, 2000.
4. Tomanek, R. J., Angiogenesis in nonischemic myocardium, In *Angiogenesis and Cardiovascular Disease*, Ware, J. A. and Simons, M. eds., Oxford University Press, New York, pp. 199–212, 1999.
5. Hanahan, D., Signaling vascular morphogenesis and maintenance, *Science*, 277, 48–50, 1997.
6. Darland, D. C. and D'Amore, P. A., Blood vessel maturation: vascular development comes of age, *J. Clin. Invest.*, 103, 157–158, 1999.
7. Conway, E. M., Collen, D., and Carmeliet, P., Molecular mechanisms of blood vessel growth, *Cardiovasc. Res.*, 49, 507–521, 2001.
8. Koerselman, J., van der Graaf, Y., de Jaegere, P. P., and Grobbee, D. E., Coronary collaterals: an important and underexposed aspect of coronary artery disease, *Circulation*, 107, 2507–2511, 2003.
9. Buschmann, I. and Schaper, W., The pathophysiology of the collateral circulation (arteriogenesis), *J. Pathol.*, 190, 338–342, 2000.
10. Kersten, J. R., Pagel, P. S., Chilian, W. M., and Warltier, D. C., Multifactorial basis for coronary collateralization: a complex adaptive response to ischemia, *Cardiovasc. Res.*, 43, 44–57, 1999.
11. Schaper, W., De Brabander, M., and Lewi, P., DNA synthesis and mitoses in coronary collateral vessels of the dog, *Circ. Res.*, 28, 671–679, 1971.
12. Lamping, K. G., Christensen, L. P., and Tomanek, R. J., Estrogen therapy induces collateral and microvascular remodeling, *Am. J. Physiol. Heart Circ. Physiol.*, 285, H2039–H2044, 2003.
13. Carrol, S. M., Enhancement of coronary collateral development by therapeutic angiogensis, *Cardiovas. Pathobiol.*, 2, 12–24, 1997.

14. Yamamoto, H., Tomoike, H., Shimokawa, H., Nabeyama, S., and Makamura, M., Development of collateral function with repetitive coronary occlusion in a canine model reduces myocardial reactive hyperemia in the absence of significant coronary stenosis, *Circ. Res.*, 55 (5), 623–632, 1984.

15. Schaper, W., Gorge, G., Winkler, B., and Schaper, J., The collateral circulation of the heart, *Prog. Cardiovasc. Dis.*, 31, 57–77, 1988.

16. Cohen, M., Sherman, W., Rentrop, K. P., and Gorlin, R., Determinants of collateral filling observed during sudden controlled coronary artery occlusion in human subjects, *J. Am. Coll. Cardiol.*, 13, 297–303, 1989.

17. Koerselman, J., van der Graaf, Y., de Jaegere, P. P., and Grobbee, D. E., Coronary collaterals: an important and underexposed aspect of coronary artery disease, *Circulation*, 107, 2507–2511, 2003.

18. Tayebjee, M. H., Lip, G. Y., and MacFadyen, R. J., Collateralization and the response to obstruction of epicardial coronary arteries, *QJM*, 97, 259–272, 2004.

19. Yancopoulos, G. D., Davis, S., and Gale, N. W., Vascular-specific growth factors and blood vessel formation, *Nature*, 407, 242–248, 2000.

20. Gale, N. W. and Yancopoulos, G. D., Growth factors acting via endothelial cell-specific receptor tyrosine kinases: VEGFs, angiopoietins, and ephrins in vascular development, *Genes Dev.*, 13 (9), 1055–1066, 1999.

21. Minchenko, A., Bauer, T., Salceda, S., and Caro, J., Hypoxic stimulation of vascular endothelial growth factor expression in vitro and in vivo, *Lab. Invest.*, 71 (3), 374–379, 1994.

22. Hashimoto, E., Ogita, T., Nakaoka, T., Matsuoka, R., Takao, A., and Kira, Y., Rapid induction of vascular endothelial growth factor expression by transient ischemia in rat heart, *Am. J. Physiol.*, 267, H1948–H1954, 1994.

23. Ferrara, N. and Gerber, H. P., The vascular endothelial growth factor family, In *Angiogenesis and Cardiovascular Disease*, Ware, J. A. and Simons, M. eds., Oxford University Press, New York, pp. 101–127, 1999.

24. Ziche, M., Morbidelli, L., Choudhuri, R., Zhang, H. T., Donnini, S., Granger, H. J., and Bicknell, R., Nitric oxide synthase lies downstream from vascular endothelial growth factor-induced but not basic fibroblast growth factor-induced angiogenesis, *J. Clin. Invest.*, 9 (11), 2625–2634, 1997.

25. Carmeliet, P., Moons, L., and Luttun, A., Synergism between vascular endothelial growth factor and placental growth factor contributes to angiogenesis and plasma extravasation in pathological conditions, *Nat. Med.*, 7, 575–583, 2001.

26. Loughna, S. and Sato, T. N., Angiopoietin and Tie signaling pathways in vascular development, *Matrix Biol.*, 20, 319–325, 2001.

27. Thurston, G., Complementary actions of VEGF and Angiopoietin-1 on blood vessel growth and leakage, *J. Anat.*, 200, 575–580, 2002.

28. Adams, R. H., Wilkinson, G. A., Weiss, C., Diella, F., Gale, N. W., Deutsch, U., Risau, W., and Klein, R., Roles of ephrinB ligands and EphB receptors in cardiovascular development: demarcation of arterial/venous domains, vascular morphogenesis, and sprouting angiogenesis, *Genes Dev.*, 13 (3), 295–306, 1999.

29. Augustin, H. G. and Reiss, Y., EphB receptors and ephrinB ligands: regulators of vascular assembly and homeostasis, *Cell Tissue Res.*, 314 (1), 25–31, 2003.

30. Bruhl, T., Urbich, C., Aicher, D., Acker-Palmer, A., Zeiher, A. M., and Dimmeler, S., Homeobox A9 transcriptionally regulates the EphB4 receptor to modulate endothelial cell migration and tube formation, *Circ. Res.*, 94 (6), 743–751, 2004.

31. Cleaver, O. and Melton, D. A., Endothelial signaling during development, *Nat. Med.*, 9 (6), 661–668, 2003.

32. Detillieux, K. A., Sheikh, F., Kardami, E., and Cattini, P. A., Biological activities of fibroblast growth factor2 in the adult myocardium, *Cardiovasc. Res.*, 57 (1), 8–19, 2003.

33. Chen, C. H., Poucher, S. M., Lu, J., and Henry, P. D., Fibroblast growth factor 2: from laboratory evidence to clinical application, *Curr. Vasc. Pharmacol.*, 2 (1), 33–43, 2004.

34. Waltenberger, J. and Hombach, V., Therapeutic angiogenesis for the heart, In *The New Angiotherapy*, Fan, T.-P.Dk. and Hohn, E. C. eds., Humana Press Inc., Totawa, NJ, pp. 279–291, 2002.

35. Friehs, I. and del Nido, P. J., Increased susceptibility of hypertrophied hearts to ischemic injury, *Ann. Thorac. Surg.*, 75, S678–S684, 2003.

36. Ho, D. S., Cooper, M. J., Richards, D. A. B., Uther, J. B., Yip, A. S. B., and Ross, D. L., Comparison of number of extrastimuli versus change in basic cycle length for induction of ventricular tachycardia by programmed ventricular stimulation, *J. Am. Coll. Cardiol.*, 22, 1711–1717, 1993.

37. Criqui, M. H., Fronek, A., Barrett-Connor, E., Klauber, M. R., Gabriel, S., and Goodman, D., The prevalence of peripheral arterial disease in a defined population, *Circulation*, 71 (3), 510–515, 1985.

38. Regensteiner, J. G. and Hiatt, W. R., Current medical therapies for patients with peripheral arterial disease: a critical review, *Am. J. Med.*, 112 (1), 49–57, 2002.

39. Anand, S. S. and Creager, M. A., Peripheral arterial disease, *Am. Fam. Physician*, 65 (11), 2321–2322, 2002.

40. Khan, T. A., Sellke, F. W., and Laham, R. J., Gene therapy progress and prospects: therapeutic angiogenesis for limb and myocardial ischemia, *Gene Ther.*, 10 (4), 285–291, 2003.

41. Manninen, H. I. and Makinen, K., Gene therapy techniques for peripheral arterial disease, *Cardiovasc. Intervent. Radiol.*, 25 (2), 98–108, 2002.

42. Clover, A. J., McCarthy, M. J., Hodgkinson, K., Bell, P. R., and Brindle, N. P., Noninvasive augmentation of microvessel number in patients with peripheral vascular disease, *J. Vasc. Surg.*, 38 (6), 1309–1312, 2003.

43. Post, M., Angiogenesis in atherosclerosis and restenosis, In *Angiogenesis and Cardiovascular Disease*, Ware, J. A. and Simons, M. eds., Oxford University Press, New York, pp. 143–158, 1999.

44. Lindner, V. and Reidy, M. A., Expression of VEGF receptors in arteries after endothelial injury and lack of increased endothelial regrowth in response to VEGF, *Arterioscler. Thromb. Vasc. Biol.*, 16, 1399–1405, 1996.

45. Lindner, V., Lappi, D. A., and Baird, A., Role of basic fibroblast growth factor in vascular lesion formation, *Circ. Res.*, 68, 106–113, 1991.

46. Cragg, A. H., Einzig, S., and Rysavy, J. A., The vasa vasorum and angioplasty, *Radiology*, 148, 75–80, 1983.

47. Edelman, E. R., Nugent, M. A., Smith, L. T., and Karnovsky, M. J., Basic fibroblast growth factor enhances the coupling of intimal hyperplasia and proliferation of vasa vasorum in injured rat arteries, *J. Clin. Invest.*, 89, 465–473, 1992.

48. Abo-Auda, W. and Benza, R. L., Therapeutic angiogenesis: review of current concepts and future directions, *J. Heart Lung Transplant.*, 22, 370–382, 2003.

49. Post, M. J., Laham, R., Sellke, F. W., and Simons, M., Therapeutic angiogenesis in cardiology using protein formulations, *Cardiovasc. Res.*, 49, 522–531, 2001.

50. Simons, M., Bonow, R. O., and Chronos, N. A., Clinical trials in coronary angiogenesis: issues, problems, consensus: an expert panel summary, *Circulation*, 102, E73–E86, 2000.

51. Simons, M. and Ware, J. A., Therapeutic angiogenesis in cardiovascular disease, *Nat. Rev. Drug Discov.*, 2, 863–871, 2003.

52. De Muinck, E. and Simons, M., Re-evaluating therapeutic neovascularization, *J. Mol. Cell. Cardiol.*, 36, 25–32, 2004.

53. Heil, M. and Schaper, W., Influence of mechanical, cellular, and molecular factors on collateral artery growth (arteriogenesis), *Circ. Res.*, 95, 449–458, 2004.

54. Unger, E. F., Banai, S., Shou, M., Lazarous, D. F., Jaklitsch, M. T., Scheinowitz, M., Correa, R., Klingbeil, C., and Epstein, S. E., Basic fibroblast growth factor enhances myocardial collateral flow in a canine model, *Am. J. Physiol.*, 266, H1588–H1595, 1994.

55. Lazarous, D. F., Scheinowitz, M., Shou, M., Hodge, E., Rajanayagam, S., Hunsberger, S., Robison, W. G., Stiber, J. A., Correa, R., and Epstein, S. E., Effects of chronic systemic administration of basic fibroblast growth factor on collateral development in the canine heart, *Circulation*, 91, 145–153, 1995.

56. Banai, S., Jaklitsch, M. T., Shou, M., Lazarous, D. F., Scheinowitz, M., Biro, S., Epstein, S. E., and Unger, E. F., Angiogenic-induced enhancement of collateral blood flow to ischemic myocardium by vascular endothelial growth factor in dogs, *Circulation*, 89, 2183–2189, 1994.

57. Baffour, R., Berman, J., Garb, J. L., Rhee, S. W., Kaufman, J., and Friedmann, P., Enhanced angiogenesis and growth of collaterals by in vivo administration of recombinant basic fibroblast growth factor in a rabbit model of acute lower limb ischemia: dose-response effect of basic fibroblast growth factor, *J. Vasc. Surg.*, 16, 181–191, 1992.

58. Lei, L., Zhou, R., Zheng, W., Christensen, L. P., Weiss, R. M., and Tomanek, R. J., Bradycardia induces angiogenesis, increases coronary reserve, and preserves function of the postinfarcted heart, *Circulation*, 110, 796–802, 2004.

59. Zheng, W., Weiss, R. M., Wang, X., Zhou, R., Arlen, A. M., Lei, L., Lazartigues, E., and Tomanek, R. J., DITPA stimulates arteriolar growth and modifies myocardial postinfarction remodeling, *Am. J. Physiol. Heart Circ. Physiol.*, 286, H1994–H2000, 2004.

20 Angiogenesis and Anti-Angiogenesis in Brain Tumors

Roland H. Goldbrunner

CONTENTS

20.1 SUMMARY

Glioblastomas are common brain tumors bearing a bad prognosis with a mean survival time of about one year despite intense therapeutical efforts. These tumors are characterized by large areas of necrosis and intense neoangiogenesis. A lot of attention has been focused on the basic mechanisms of glioma vascularization, particularly in terms of investigating vascular growth factors and receptors. These studies need an assortment of models, which should allow the investigation of a variety of questions.

Several objectives in basic endothelial cell (EC) biology can adequately be studied in vitro using monolayer assays. Three-dimensional spheroid techniques respect the more complex cell–cell and cell–environment interplay within a three-dimensional culture. To simulate the crucial interaction of human gliomas with host endothelial cells, immunological cells, and extracellular matrix, animal models are mandatory.

Using these models, various approaches for anti-angiogenic treatment strategies have been tested. One of the most promising approaches is interference with the vascular endothelial growth factor (VEGF) family and its receptors. The VEGF/VEGF receptor interaction is the most crucial system for regulating angiogenesis in brain tumors, even though the role of other growth factors like hepatocyte growth factor (HGF)/scatter factor (SF), basic fibroblast growth factor (bFGF), the ephrin family, and the angiopoietins has been established. Synthetic small molecule inhibitors of the VEGF receptor tyrosine kinases have been developed. We investigated the effect of PTK787/ZK222584, which has a high affinity to VEGF receptor 2 (VEGFR2), on vascularization and the growth of VEGF-A sense- and antisense-transfected experimental gliomas in an orthotopic model. Using magnetic resonance imaging (MRI) and histological techniques, we could show that the tumor size of VEGF-A-sense transfected tumors was reduced to the size of VEGF-A-negative gliomas. Additionally, an almost complete down-regulation of neoangiogenesis could be observed. Therefore, it is possible to abrogate the crucial, VEGF-A mediated effects on glioma growth and vascularization using synthetic compounds.

Several flavonoids and epigallocatechin-gallate have been shown to exert anti-tumoral effects in gliomas. However, in contrast to synthetic drugs, currently there are no data about the anti-angiogenic efficacy of micronutritients in gliomas. Due to the poor prognosis of malignant gliomas, evaluation of potential anti-angiogenic effects of micronutrients would be desirable.

20.2   INTRODUCTION

Gliomas are tumors of the glia, which represents the cellular scaffolding of the brain parenchyma. The glia is mainly composed of astrocytes and oligodendrocytes which are the cells of origin for gliomas—the most common intrinsic brain tumor. According to a classification of the World Health Organization (WHO), gliomas are divided into four grades with grade 1 being the most benign form. High grade astrocytomas (WHO grades III and IV) have a poor prognosis: although multimodal therapeutical efforts are employed, the mean survival is approximately 12 months for glioblastoma multiforme (WHO grade IV), and 2–3 years for anaplastic astrocytoma (WHO grade III).

High grade astrocytomas (III and IV) are characterized by rapid invasion of single glioma cells into the surrounding brain tissue and by strong neovascularization. Endothelial cell proliferation and areas of necrosis are histopathological hallmarks for glioblastoma multiforme. Absolute requirements for growth and expansion of normal and neoplastic tissues within the brain, as well as in all other organ systems, are the establishment and maintenance of a sufficient vessel network.

During embryonic development, early vascular progress is performed by mesoderm derived angioblasts, a process called vasculogenesis. The term angiogenesis has been designated to the physiological or pathological formation of new blood vessels by sprouting from preexisting vessels. There is not much known about the regulation of early vasculogenesis in the brain, which is accomplished by angioblasts forming the perineural plexus and capillaries sprouting from the plexus into the cerebrum [1].

Angiogenesis within the brain is a tightly controlled process that is regulated by a variety of growth factors binding to transmembrane tyrosine kinase receptors, which are expressed by EC [2]. Angiogenesis and vasculogenesis also play a crucial role in the growth and progression of primary brain tumors. Therefore, several preclinical attempts have been made to inhibit vascularization of gliomas in order to establish novel therapeutical approaches. This chapter will (a) illuminate the mechanisms of glioma vascularization, (b) focus on the variety of models for assessment of glioma angiogenesis, (c) describe an anti-angiogenic approach using a synthetic compound, and (d) review the literature for anti-angiogenesis by micronutritients.

20.3   MECHANISMS OF ANGIOGENESIS IN BRAIN TUMORS

Malignant gliomas typically present a markedly increased angiogenesis that is crucial for tumor growth and tumor spread [3]. Glioma induced angiogenesis is performed by a combination of two distinct processes: the so-called "classical" angiogenesis with new capillaries sprouting from preexisting host vessels, as well as vasculogenesis through endothelial precursor cells that migrate towards an experimental glioma as individual cells [4]. However, in very early stages, initiation of solid tumor growth is supported by the cooption of pre-existing host vasculature—which could be shown for several tumor types, including gliomas [5].

The levels of proangiogenic growth factors (VEGF-A, bFGF, HGF/SF and angiopoietins) are very low at this early point in time and no angiogenesis occurs. Consequently, angiogenesis-independent tumor growth is associated with increasing tumor cell necrosis. This process leads to the expression of angiogenic growth factors and initiation of angiogenesis. Finally, during further tumor growth, there seems to exist a dynamic balance between vessel regression and growth mediated by angiopoietins and VEGF-A [6].

Recently, a variety of growth factors and corresponding receptors has been identified, and they are required for normal vascular development, as well as pathological glioma vascularization. The VEGF family, which at present comprises six members (VEGF-A, VEGF-B, VEGF-C, VEGF-D, VEGF-E, and placenta growth factor (PlGF)), has been shown to play the central role in physiological and pathological brain angiogenesis [7]. Vascular endothelial growth factors are mitogens for endothelial cells acting on transmembrane tyrosine kinase (TK) receptors, which at present include VEGFR-1 (flt-1), VEGFR-2 (flk-1, KDR), and VEGFR-3 (flt-4). Stimulation of these receptors leads to catalytic activity of the TK which activates the second messenger pathways causing EC activation, increased vascular permeability, and upregulation of neoangiogenesis [1].

Beyond the VEGF family, several other molecules like TNF-α, TGF-β, coagulation factors, or particular integrin receptors have been shown to exert or mediate angiogenic effects. The best investigated proangiogenic agents within the brain are basic fibroblast growth factor (bFGF), hepatocyte growth factor/scatter factor (HGF/SF), the ephrin family, and the angiopoietins. The VEGF/VEGF receptor system and the angiopoietin/tie system [5,8,9] have been considered to be the only EC specific growth factor/receptor systems. However, there is an increasing line of evidence that VEGF receptors are also expressed by tumor cells of different entities including glioma cells [10]. This indicates even more complex tumor host interactions, suggesting possible paracrine stimulation pathways of tumor cells.

Like the VEGF receptors, Tie1 and Tie2 tyrosine kinases are essential for embryological vessel formation. Angiopoietin-1 mediates blood vessel maturation and stability and is a physiological ligand of the receptor tyrosine kinase Tie2. Angiopoietin-2 counteracts the angiopoietin-1 mediated effects leading to widespread vessel discontinuities and is highly expressed under pathological conditions like glioma growth. It seems to mediate pathological angiogenesis by mediating angiogenic effects of other cytokines like VEGF-A, bFGF, angiopoietin-1 and TGF-ss1 [11].

In gliomas, VEGF-A expression levels have been demonstrated to directly correlate with glioma malignancy [12]. VEGF-A is up to 50-fold overexpressed in glioblastoma tissue [13], with a remarkable spatial restriction of VEGF-A expression to perinecrotic palisading cells [3]. Angiopoietin-1 is expressed by tumor cells, whereas Angiopoietin-2 was detected in small tumor vessels with little periendothelial support, suggesting that the angiopoietin/Tie2 system plays a role in the interaction of EC and pericytes or other supportive cells [14].

Altogether, despite increasing knowledge about the basic mechanisms of angiogenesis in gliomas, many questions still are open. Within the next paragraphs, the models that are necessary for finding answers to these questions are discussed— and these models will be used for the development of anti-angiogeneic strategies.

20.4   MODELS FOR ASSESSMENT OF ANGIOGENESIS AND ANTI-ANGIOGENESIS WITHIN THE BRAIN

To investigate the basic objectives of angiogenesis research, in vitro assays are sufficient in most cases. The first step needed for developing an in vitro system for the investigation of angiogenesis is to obtain or establish an EC line or to gain primary cultures of EC. Several animal and human EC lines already have been established and are commercially available. The most widely used EC of human origin is HUVEC—human umbilical vein endothelial cells [15], which represent macrovascular embryonic cells. And yet, are these cells appropriate for the investigation of brain tumor angiogenesis?

20.4.1   SELECTION OF ENDOTHELIAL CELL TYPE

Endothelial cell originate during embryonic development from a line of CD34 negative progenitor cells which differentiate into CD34 positive EC precursor cells

[16] and, finally, into mature endothelial cells. One of the most important determinants in endothelial differentiation is the local environment that mediates the acquisition of organ-specific properties [17]. Because EC are very heterogeneous with regard to cell surface molecules and permeability, the commonly used large vessel EC (like HUVEC) do not adequately reflect the properties of microvascular EC, e.g. in terms of protease expression or reactivity towards angiogenic stimuli [18,19]. Therefore, for the assessment of organ specific EC functions, the use of microvascular EC gained from the organ of interest is absolutely mandatory. However, purification of these microvascular EC is still very difficult, which results in multiple sophisticated methods to isolate these cells in different organ systems [20–23].

Microvascular brain EC represent a highly specialized endothelium, being part of the interface between the blood and the parenchyma of the central nervous system—the so-called blood–brain-barrier (BBB). These EC are characterized by a variety of brain-typical properties like the formation of tight junctions instead of gap junctions, a reduced rate of pinocytosis, and a broad panel of transcellular transporter systems.

With regard to this high specialization of cerebral EC, the availability of microvascular cerebral EC is essential for studying glioma angiogenesis. Cerebral endothelium has been cultured from several species [22,24,25]. However, as EC display species-specificity, isolation and characterization of human cerebral EC is mandatory for the study of human glioma neovascularization. Furthermore, the microvascular endothelium of malignant gliomas differs widely from the resting endothelium of the normal human brain [1]. Among other features, tumor vessels display an irregular shape and increased permeability, as well as the induction of vascular glomeruloid structures, which is mediated by angiopoietin-2 and VEGF overexpression [26]. Additionally, the integrin $\alpha_v\beta_3$ is strongly expressed on glioma blood vessels compared to the resting endothelium of the brain [27]. The same is true for a variety of other adhesion molecules [28]. Therefore, for the study of tumor–host interactions in malignant gliomas, with special regard to neovascularization, the isolation and characterization of intratumoral EC is of particular interest.

For these reasons, the isolation of the normal brain as well as intratumoral EC has been established at our laboratory. Normal brain endothelial cells are isolated from brain specimens gained on the occasion of surgical approaches to deep seated lesions or surgery for epilepsy. Intratumoral EC are isolated from specimens of low grade or high grade gliomas. All tissues are separated from any meninges and large vessels, minced, and homogenized. After digestion, the cells are separated by Percoll gradient centrifugation and the bands containing the endothelial cells are collected. In the case of normal brain, the purity of EC culture usually is above 95%, which has to be confirmed immunohistochemically. In the case of brain tumors, isolation by density gradient alone is not sufficient. For this reason, the purity of EC culture is increased by magnetic bead selection using typical EC markers like CD31, CD105, E-selectin, or VE-cadherin. By the combination of density gradient and bead selection, a purity of over 98% can be achieved [29].

20.4.2   FUNCTIONAL IN VITRO ASSAYS

Most questions about endothelial cell biology can been assessed in two-dimensional systems. Growth-modulating effects of molecules like the VEGF or fibroblast growth factor families can be investigated in monolayer EC assays by direct or indirect cell counting [30]. Two-dimensional transwell assays represent an adequate method for the analysis of the migratory potential of EC, e.g. dependent on protease expression [31]. This assay also allows the investigation of the influence of inhibitors on EC invasion [32]. With respect to these options, monolayer assays also represent the basis for most therapeutic studies for the evaluation of anti-angiogenic strategies. However, two-dimensional models are limited to very basic cell biological questions, as functional EC studies require cell–cell interactions and the interplay of EC with their environment, which are better represented in three-dimensional systems.

Compared to observations in two-dimensional monolayer culture systems, EC behave quite distinctly when cultured and dispersed in 3D systems [33]. Formation of tubular structures is a general key feature of angiogenesis. To assess this phenomenon, in vitro tube formation assays are used: EC are seeded onto three-dimensional collagen-, fibrin- or Matrigel gels. Invasion of cells into the gel and the formation of tube-like structures by EC can be investigated by microscopy or immunostaining [34]. This assay is also suitable for studying the interaction of endothelial and glioma cells in a co-culture system [35].

An additional way to analyze EC behavior in 3D cultures is to mix EC with type I collagen before adding the cell/collagen mixture to tissue culture dishes [36]. There have been many additional 3D models described, like mixing endothelial and glioma cells within collagen plugs [37] or placing blood vessel fragments in fibrin gels [38]. In addition, the spheroid-based confrontation model, which is well established in research on glioma cell invasion [39,40], has been employed in angiogenesis research: spheroids have been prepared from EC and embedded in collagen gels [41]. Endothelial migration can be observed as individual cell or as multicellular cord invasion into the collagen gel. Analogous to the confrontation culture of glioma spheroids with brain aggregates, the interaction of glioblastoma spheroids with human cerebral EC spheroids has been studied to analyze glioma neovascularisation in vitro [42].

20.4.3   IN VIVO MODELS

Since neo-angiogenesis is the typical example for tumor–host interaction, in vivo models are mandatory. Only the in vivo situations allow us to observe the interplay of host EC and brain tumor cells in their more or less natural environment. However, it should be well known that no animal tumor model exactly simulates human brain tumor growth and vascularization. To resemble human gliomas as far as possible, the following principles for brain tumor models should be considered [43]:

- Tumor origin from glial cells,
- Orthotopic implantation and growth,

- Predictability and reproducibility of tumor growth rates,
- Glioma-like characteristics of growth and vascularization within the brain,
- Sufficient duration of tumor growth to permit therapy studies and determination of efficacy, and
- Minimization of model derived immunological interferences.

Several models that are still in use conflict with the demand for an orthotopic technique, the demand for a growth characteristic like human gliomas and, most of all, the demand for respecting the immunological situation. An orthotopic brain tumor model is time consuming and complicated. However, invasive growth patterns of human gliomas are a result of very complex interactions of glioma cells with the brain specific extracellular matrix [44]. Additionally, glioma neovascularization is regulated in a brain specific pattern by the interplay of glioma cells and host vascular cells [45,46], which makes investigation of glioma neoangiogenesis in an orthotopic model necessary.

Human glioma growth is characterized by invasion, perivascular migration, and rapid vascularization, which is best simulated by using *human* glioma cells. This has been demonstrated by studies comparing human glioma growth and rat growth in the rat brain. These studies revealed a slower and more infiltrating growth pattern along white matter tracts in human cell lines, whereas rat gliomas tended to show a rapid and well demarcated growth [47]. In terms of angiogenesis research, U87 and other human glioma cell lines, as well as primary cultures of human brain tumors, have been utilized successfully [48–51].

Partial or complete impairment of the host's immune system is the precondition for the use of human or other xenogeneic glioma cells in an in vivo model. However, it can be shown that interactions of tumor cells and host immunological cells are a crucial part of the complex process of glioma progression [52]. Therefore, the use of immunocompromized animals leads to—in part—artificial growth patterns. A syngeneic model, where tumor and host are derived from the same gene pool, would minimize these immunological problems. Altogether, there are conflicting principles in the selection of the "right" model. Therefore, the final decision for a particular model should depend on the objectives of the intended study.

20.5 ANTI-ANGIOGENESIS IN BRAIN TUMORS

Anti-angiogenesis seems to be one of the most promising approaches in adjuvant glioma therapy. As VEGF-A has evolved to be the most crucial angiogenic growth factor, most treatment strategies developed so far concentrate on effecting the VEGF/VEGF receptor system. Besides the application of antibodies against VEGF-A or the flk-1 receptor, VEGF antisense strategies or the use of small molecules directed against receptor tyrosine kinases have been employed to prevent or slow down vascularization of experimental gliomas.

One of the first studies testing the effects of VEGF inhibition on glioma growth was performed by Kim et al. utilizing a VEGF-A antibody in a subcutaneous nude mouse model. A significant growth inhibition was found in vivo, whereas the antibody did not have any effect in vitro [53]. Millauer et al. used a retrovirus encoding a dominant-negative mutant of the Flk-1/VEGF receptor to infect endothelial target cells in vivo. After endothelial cell transfection, a significant tumor growth inhibition was found in nude mouse models [54]. Additionally, growth of C6 glioma cells could be inhibited in vivo by transfection of antisense-VEGF cDNA [55].

In the years that followed, dozens of papers were published describing novel anti-angiogenic approaches for the treatment of brain tumors. In these papers, inhibitors of metalloproteinases [56], inhibitors of fibroblast growth factor or the corresponding receptor [50,57], interferon beta, tyrosine kinase inhibitors targeted against VEGF receptors [46], novel antibodies against VEGF-A, as well as many other strategies were discussed. Since small molecule tyrosine kinase inhibitors are able to penetrate the blood brain barrier, which usually represents the crucial limitation for any systemic treatment of brain pathologies, our group concentrated on the evaluation of a tyrosine kinase inhibitor that is highly specific for KDR/flk-1, the most important VEGF receptor.

20.5.1 ANTI-ANGIOGENESIS BY A SYNTHETIC COMPOUND

The aim of the following study was to eradicate the VEGF mediated growth effects on an experimental glioma model. For this reason, the C6 rat glioma cell line was transfected with $VEGF_{164}$ cDNA in sense and antisense orientation as described previously [58]. The efficacy of the transfection procedure was controlled by Northern and Western Blot analysis, as well as an enzyme linked immuno sorbent assay (ELISA) for VEGF-A protein. Enzyme linked immuno sorbent assay did not show any VEGF-A expression in antisense-transfected cells and revealed a very high concentration of 36 ng/ml in sense-transfected cells (compared to a mean concentration of about 2 ng/ml in wild-type or mock transfected C6 cells). Stable clones with high VEGF sense (VS) or antisense (AS) expression were used for spheroid production and orthotopic implantation into a rat brain. Wild-type cells (WT) and mock transfected cells (MT) served as controls for in vitro and in vivo experiments. All procedures have been described in detail previously [4,59]. Spheroids were generated from VS, AS, WT and MT monolayer cells by seeding them on soft agar to prevent adhesion. Spheroids with a mean diameter of 300 μm without any signs of central necrosis were selected for orthotopic, subcortical implantation into a rat brain.

To interfere with the VEGF mediated growth effects, PTK787/ZK222584 (co-produced by Novartis, Basle, Switzerland and Schering, Berlin, Germany), an orally bioavailable small molecule tyrosine kinase inhibitor specific for the VEGF-R2, was selected. The animals were divided into five groups with 12 animals in each. Animals in groups 1–3 developed VS gliomas, rats in groups 4 and 5 had AS tumors. Groups 1 and 4 served as controls, groups 3 and 5 were treated with PTK787/ZK222584 orally once a day, beginning the first day after spheroid implantation. Treatment of group 2 (according to the protocol for group 3 and 5) started at day 7 after implantation to allow tumor growth for one week. At day 12 after tumor implantation, an MRI was performed according to a well-established protocol [59]. After the MRI, animals were sacrificed and their brains examined immunohistochemically.

Magnetic resonance imaging morphology revealed the following findings for untreated VS gliomas:

- massive growth with displacement of surrounding structures,
- hyperintense and hypointense signal intensities in T2-w sequences being consistent with intratumoral necrosis and hemorrhage,
- extensive peritumoral edema, and
- strong, but inhomogeneous uptake of contrast medium indicative of the breakdown of the blood–brain barrier and strong vascularization.

PTK787/ZK222584 treatment changed MR morphology completely: tumors became smaller, the mass effect decreased consequently, little perifocal edema was seen, and the intensity of contrast uptake was diminished (Figure 20.1). Antisense gliomas displayed small tumors without significant mass effect or contrast enhancement, a homogeneously hyperintense signal on T2-w images, and barely perifocal edema. There were no morphological differences visible between treated and untreated AS gliomas. The differences in tumor sizes were quantified by MRI volumetry, depicting a reduction of tumor volumes in treated VS gliomas of more than 70% compared to untreated VS gliomas (Figure 20.2). Comparison of early treated VS gliomas (group 3) with untreated AS gliomas (group 4) revealed almost identical tumor volumes indicating that treatment with PTK787/ZK222584 resulted in complete abortion of VEGF mediated growth effects in strongly VEGF expressing tumors compared to primarily VEGF negative tumors.

This striking MR-tomographic finding was further enhanced by the histologic finding that one third of the remaining tumor mass in early treated VS gliomas consisted of necrosis, meaning that an even less viable tumor is left after PTK787/ZK222584 treatment than in any other group. One explanation for this intense effect on tumor growth was provided by the results of vessel staining with anti-CD31: the number of vessels were significantly reduced by treatment with the TK inhibitor compared to the controls (Figure 20.3).

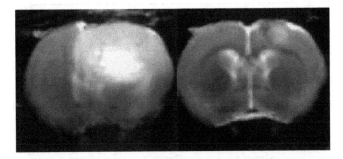

FIGURE 20.1 Magnetic resonance imaging (MRI). T2-weighted MRI of a C6 glioma 12 days after implantation. In comparison to the control tumor (left), treatment with PTK787/222584 (right) led to reduced tumor size, decreased perifocal edema, and reduced mass effect.

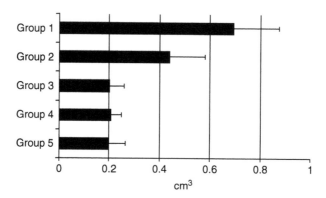

FIGURE 20.2 Tumor volumes. MRI volumetric results. Bars: means of tumor volume in each group, error bars: standard deviation, $n = 12$ per group.

Altogether, this study could confirm that VEGF-A mediated growth effects, which represent a crucial factor for tumor growth, can be completely eliminated by an anti-angiogenic approach using a synthetic VEGF receptor inhibitor [60]. Therefore, clinical trials evaluating this or similar compounds are justified in glioblastoma patients.

20.5.2 ANTI-ANGIOGENESIS IN GLIOMAS: IS THERE A NUTRACEUTICAL APPROACH?

Several naturally occurring agents have been tested for antitumoral effects using glioma cell cultures. Tangeretin and nobiletin, which are both citrus flavonoids and differ from each other by just one methyl group, have been shown to reduce expression of matrix metalloproteinases 2 and 9 (MMP-2 and MMP-9) in glioma

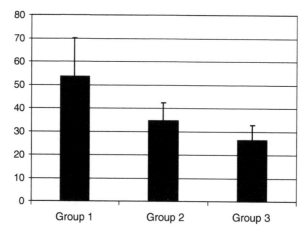

FIGURE 20.3 Vessel counts. Bars: means of vessel counts per high power field. Error bars: standard deviation, $n = 36$ high power fields per group.

cell lines. Matrix metalloproteinases are mandatory for cleavage of extracellular space compounds and, therefore, necessary for glioma cell invasion. Accordingly, tangeretin and nobiletin also could inhibit glioma cell adhesion, migration, and invasion in in vitro assays [61]. However, these results were not found consistently throughout all the glioma cell lines that were used. Tangeretin has been demonstrated to cross the blood–brain barrier; consequently, it should be possible to use this flavonoid clinically [62].

Genistein is a well known isoflavone with established anti-cancer effects [63]. Penar et al. demonstrated that genistein interfered with the epidermal growth factor receptor (EGFR) tyrosine kinase, which is overexpressed or amplified in a high percentage of malignant gliomas. Utilizing this mechanism, glioma cell invasion in a spheroid confrontation model was inhibited by application of genistein [64]. There are also preliminary data that some carotenoids like lycopene reduce motility of glioma cells [65]. Selenium, a trace mineral, is an essential constituent of the anti-oxidant enzyme glutathione peroxidase. Because this enzyme is essential for protection of oxidative damage, selenium seemed to reduce the incidence of—and mortality from—many types of cancer in a large clinical study [66]. However, another large clinical trial could not confirm the efficacy of selenium in preventing skin cancer [67]. In vitro studies support the hypothesis that there might be an anti-glioma effect of selenium by the induction of apoptosis and reduction of glioma cell viability [68,69]. Within the last couple of years, there have been many studies performed in vitro and in vivo investigating the effect of epigallocatechin gallate (EGCG) on various kinds of cancer. Epigallocatechin gallate is the main constituent of green tea polyphenols and was also effective in reducing glioma cell proliferation and inducing apoptosis in glioma cell lines [70].

To summarize, several micronutritients have been shown to exert anti-glioma effects in terms of reducing cell viability, diminishing motility, or inducing apoptosis in glioma cells. Therefore, these substances might be used in future clinical trials. However, there are currently no data about anti-angiogenic effects of any micronutritient in gliomas. Using the model systems described above, it would be of substantial interest to test the effects of substances like flavonoids or EGCG on glioma neoangiogenesis.

20.6 CONCLUSIONS

Increasing knowledge about the mechanisms of glioma neovascularization have led to many promising anti-angiogenic approaches within the last years. These have been tested in well-defined models. Within the large group of synthetic anti-angiogenic drugs, tyrosine kinase inhibitors directed against VEGF receptors appear to be the most promising agents. Despite the existence of some in vitro data about the effects of naturally occurring nutrients against glioma cell viability and motility, there is a complete lack of data concerning the hypothetic anti-angiogenic efficacy of micronutritients in gliomas. As the prognosis of glioblastoma patients is still extremely poor, every effort is justified to develop novel therapeutic approaches against this disease. Therefore, further evaluation of anti-angiogenic or other antitumoral effects of micronutritients, in terms of developing a nutraceutical approach, could contribute to this goal.

REFERENCES

1. Plate, K. H., Mechanisms of angiogenesis in the brain, *J. Neuropathol. Exp. Neurol.*, 58, 313–320, 1999.
2. Carmeliet, P., Mechanisms of angiogenesis and arteriogenesis, *Nat. Med.*, 6, 389–395, 2000.
3. Plate, K. H., Breier, G., Weich, H. A. et al., Vascular endothelial growth factor is a potential tumor angiogenesis factor in human gliomas in vivo, *Nature*, 359, 845–848, 1992.
4. Goldbrunner, R. H., Bernstein, J. J., Plate, K. H. et al., Vascularization of human glioma spheroids implanted into rat cortex is conferred by two distinct mechanisms, *J. Neurosci. Res.*, 55, 486–495, 1999.
5. Holash, J., Maisonpierre, P. C., Compton, D. et al., Vessel cooption, regression, and growth in tumors mediated by angiopoietins and VEGF, *Science*, 284, 1994–1998, 1999.
6. Holash, J., Wiegand, S. J., and Yancopoulos, G. D., New model of tumor angiogenesis: dynamic balance between vessel regression and growth mediated by angiopoietins and VEGF, *Oncogene*, 18, 5356–5362, 1999.
7. Plate, K. H. and Warnke, P. C., Vascular endothelial growth factor, *J. Neurooncol.*, 35, 365–372, 1997.
8. Suri, C., Jones, P. F., Patan, S. et al., Requisite role of angiopoietin-1, a ligand for the Tie2 receptor, during embryonic angiogenesis, *Cell*, 87, 1171–1180, 1996.
9. Davis, S., Aldrich, T. H., Jones, P. F. et al., Isolation of angiopoietin-1, a ligand for the Tie2 receptor, by secretion-trap expression cloning, *Cell*, 87, 1161–1169, 1996.
10. Herold-Mende, C., Steiner, H. H., Andl, T. et al., Expression and functional significance of vascular endothelial growth factor receptors in human tumor cells, *Lab Investig.*, 79, 1573–1582, 1999.
11. Mandriota, S. J. and Pepper, M. S., Regulation of angiopoietin-2 mRNA levels in bovine microvascular endothelial cells by cytokines and hypoxia, *Circ. Res.*, 83, 852–859, 1998.
12. Schmidt, N. O., Westphal, M., Hagel, C. et al., Levels of vascular endothelial growth factor, hepatocyte growth factor/scatter factor and basic fibroblast growth factor in human gliomas and their relation to angiogenesis, *Int. J. Cancer*, 84, 10–18, 1999.
13. Weindel, K., Moringlane, J. R., Marme, D. et al., Detection and quantification of vascular endothelial growth factor/vascular permeability factor in brain tumor tissue and cyst fluid: the key to angiogenesis? *Neurosurgery*, 35, 439–448, 1994.
14. Stratmann, A., Risau, W., and Plate, K. H., Cell type-specific expression of angiopoietin-1 and angiopoietin-2 suggests a role in glioblastoma angiogenesis, *Am. J. Pathol.*, 153, 1459–1466, 1998.
15. Jaffe, E. A., Nachman, R. L., Becker, C. G. et al., Culture of human endothelial cells derived from umbilical veins. Identification by morphologic and immunologic criteria, *J. Clin. Investig.*, 52, 2745–2756, 1973.
16. Reyes, M., Dudek, A., Jahagirdar, B. et al., Origin of endothelial progenitors in human postnatal bone marrow, *J. Clin. Investig.*, 109, 337–346, 2002.
17. Risau, W., Differentiation of endothelium, *FASEB J.*, 9, 926–933, 1995.
18. Hummel, V., Kallmann, B. A., Wagner, S. et al., Production of MMPs in human cerebral endothelial cells and their role in shedding adhesion molecules, *J. Neuropathol. Exp. Neurol.*, 60, 320–327, 2001.

19. Kallmann, B. A., Wagner, S., Hummel, V. et al., Characteristic gene expression profile of primary human cerebral endothelial cells, *FASEB J.*, 16, 589–591, 2002.
20. Kern, P. A., Knedler, A., and Eckel, R. H., Isolation and culture of microvascular endothelium from human adipose tissue, *J. Clin. Investig.*, 71, 1822–1829, 1983.
21. Davison, P. M., Bensch, K., and Karasek, M. A., Isolation and growth of endothelial cells from the microvessels of the newborn human foreskin in cell culture, *J. Investig. Dermatol.*, 75, 316–321, 1980.
22. Bowman, P. D., Betz, A. L., and Goldstein, G. W., Primary culture of microvascular endothelial cells from bovine retina: selective growth using fibronectin coated substrate and plasma derived serum, *In Vitro*, 18, 626–632, 1982.
23. Gossmann, A., Helbich, T. H., Kuriyama, N. et al., Dynamic contrast-enhanced magnetic resonance imaging as a surrogate marker of tumor response to antiangiogenic therapy in a xenograft model of glioblastoma multiforme, *J. Magn. Reson. Imaging*, 15, 233–240, 2002.
24. DeBault, L. E., Kahn, L. E., Frommes, S. P. et al., Cerebral microvessels and derived cells in tissue culture: isolation and preliminary characterization, *In Vitro*, 15, 473–487, 1979.
25. Bowman, P. D., Betz, A. L., Ar, D. et al., Primary culture of capillary endothelium from rat brain, *In Vitro*, 17, 53–362, 1981.
26. Brat, D. J. and Van Meir, E. G., Glomeruloid microvascular proliferation orchestrated by VPF/VEGF: a new world of angiogenesis research, *Am. J. Pathol.*, 158, 789–796, 2001.
27. Gladson, C. L., Expression of integrin alpha v beta 3 in small blood vessels of glioblastoma tumors, *J. Neuropathol. Exp. Neurol.*, 55, 1143–1149, 1996.
28. Gingras, M. C., Roussel, E., Bruner, J. M. et al., Comparison of cell adhesion molecule expression between glioblastoma multiforme and autologous normal brain tissue, *J. Neuroimmunol.*, 57, 143–153, 1995.
29. Miebach, S., Grau, S., Tonn, J. C. et al., Isolation, culture and characterization of microvascular endothelial cells from different intracranial tumors (abstract), *J. Vasc. Res.*, 41, 111, 2004.
30. Maruno, M., Yoshimine, T., Isaka, T. et al., Expression of thrombomodulin in astrocytomas of various malignancy and in gliotic and normal brains, *J. Neurooncol.*, 19, 155–160, 1994.
31. Puyraimond, A., Weitzman, J. B., Babiole, E. et al., Examining the relationship between the gelatinolytic balance and the invasive capacity of endothelial cells, *J. Cell Sci.*, 112, 1283–1290, 1999.
32. Yamaguchi, N., Anand-Apte, B., Lee, M. et al., Endostatin inhibits VEGF-induced endothelial cell migration and tumor growth independently of zinc binding, *EMBO J.*, 18, 4414–4423, 1999.
33. Sankar, S., Mahooti-Brooks, N., Bensen, L. et al., Modulation of transforming growth factor beta receptor levels on microvascular endothelial cells during in vitro angiogenesis, *J. Clin. Investig.*, 97, 1436–1446, 1996.
34. Montesano, R. and Orci, L., Tumor-promoting phorbol esters induce angiogenesis in vitro, *Cell*, 42, 469–477, 1985.
35. Abe, T., Okamura, K., Ono, M. et al., Induction of vascular endothelial tubular morphogenesis by human glioma cells. A model system for tumor angiogenesis, *J. Clin. Investig.*, 92, 54–61, 1993.
36. Goto, F., Goto, K., Weindel, K. et al., Synergistic effects of vascular endothelial growth factor and basic fibroblast growth factor on the proliferation and cord formation of bovine capillary endothelial cells within collagen gels, *Lab Investig.*, 69, 508–517, 1993.

37. Ment, L. R., Stewart, W. B., Scaramuzzino, D. et al., An in vitro three-dimensional coculture model of cerebral microvascular angiogenesis and differentiation, *In Vitro Cell Dev. Biol. Anim.*, 33, 684–691, 1997.
38. Nicosia, R. F., Tchao, R., and Leighton, J., Histotypic angiogenesis in vitro: light microscopic, ultrastructural, and radioautographic studies, *In Vitro*, 18, 538–549, 1982.
39. Tonn, J. C., Haugland, H. K., Saraste, J. et al., Differential effects of vincristine and phenytoin on the proliferation, migration, and invasion of human glioma cell lines, *J. Neurosurg.*, 82, 1035–1043, 1995.
40. Bjerkvig, R., Tonnesen, A., Laerum, O. D. et al., Multicellular tumor spheroids from human gliomas maintained in organ culture, *J. Neurosurg.*, 72, 463–475, 1990.
41. Vernon, R. B. and Sage, E. H., A novel, quantitative model for study of endothelial cell migration and sprout formation within three-dimensional collagen matrices, *Microvasc. Res.*, 57, 118–133, 1999.
42. Wagner, S., Fueller, T., Hummel, V. et al., Influence of VEGF-R2 inhibition on MMP secretion and motility of microvascular human cerebral endothelial cells (HCEC), *J. Neurooncol.*, 62, 221–231, 2003.
43. Goldbrunner, R. H., Wagner, S., Roosen, K. et al., Models for assessment of angiogenesis in gliomas, *J. Neurooncol.*, 50, 53–62, 2000.
44. Goldbrunner, R. H., Bernstein, J. J., and Tonn, J. C., Cell-extracellular matrix interaction in glioma invasion, *Acta Neurochir. (Wien.)*, 141, 295–305, 1999.
45. Vajkoczy, P., Goldbrunner, R., Farhadi, M. et al., Glioma cell migration is associated with glioma-induced angiogenesis in vivo, *Int. J. Dev. Neurosci.*, 17, 557–563, 1999.
46. Vajkoczy, P., Menger, M. D., Goldbrunner, R. et al., Targeting angiogenesis inhibits tumor infiltration and expression of the pro-invasive protein SPARC, *Int. J. Cancer*, 87, 261–268, 2000.
47. Guillamo, J. S., Lisovoski, F., Christov, C. et al., Migration pathways of human glioblastoma cells xenografted into the immunosuppressed rat brain, *J. Neurooncol.*, 52, 205–215, 2001.
48. Teicher, B. A., Menon, K., Alvarez, E. et al., Antiangiogenic and antitumor effects of a protein kinase C beta inhibitor in human T98G glioblastoma multiforme xenografts, *Clin. Cancer Res.*, 7, 634–640, 2001.
49. Yuan, F., Salehi, H. A., Boucher, Y. et al., Vascular permeability and microcirculation of gliomas and mammary carcinomas transplanted in rat and mouse cranial windows, *Cancer Res.*, 54, 4564–4568, 1994.
50. Stan, A. C., Nemati, M. N., Pietsch, T. et al., In vivo inhibition of angiogenesis and growth of the human U-87 malignant glial tumor by treatment with an antibody against basic fibroblast growth factor, *J. Neurosurg.*, 82, 1044–1052, 1995.
51. Ma, J., Fei, Z. L., Klein-Szanto, A. et al., Modulation of angiogenesis by human glioma xenograft models that differentially express vascular endothelial growth factor, *Clin. Exp. Metastasis*, 16, 559–568, 1998.
52. Tada, M. and de Tribolet, N., Recent advances in immunobiology of brain tumors, *J. Neurooncol.*, 17, 261–271, 1993.
53. Kim, K. J., Li, B., Winer, J. et al., Inhibition of vascular endothelial growth factor-induced angiogenesis suppresses tumor growth in vivo, *Nature*, 362, 841–844, 1993.
54. Millauer, B., Shawver, L. K., Plate, K. H. et al., Glioblastoma growth inhibited in vivo by a dominant-negative Flk-1 mutant, *Nature*, 367, 576–579, 1994.
55. Saleh, M., Stacker, S. A., and Wilks, A. F., Inhibition of growth of C6 glioma cells in vivo by expression of antisense vascular endothelial growth factor sequence, *Cancer Res.*, 56, 393–401, 1996.

56. Anand-Apte, B., Pepper, M. S., Voest, E. et al., Inhibition of angiogenesis by tissue inhibitor of metalloproteinase-3, *Investig. Ophthalmol. Vis. Sci.*, 38, 817–823, 1997.

57. Cuevas, P., Carceller, F., Reimers, D. et al., Inhibition of intratumoral angiogenesis and glioma growth by the fibroblast growth factor inhibitor 1,3,6-naphthalene-trisulfonate, *Neurol. Res.*, 21, 481–487, 1999.

58. Sasaki, M., Wizigmann-Voos, S., Risau, W. et al., Retrovirus producer cells encoding antisense VEGF prolong survival of rats with intracranial GS9L gliomas, *Int. J. Dev. Neurosci.*, 17, 579–591, 1999.

59. Goldbrunner, R. H., Bendszus, M., Sasaki, M. et al., Vascular endothelial growth factor-driven glioma growth and vascularization in an orthotopic rat model monitored by magnetic resonance imaging, *Neurosurgery*, 47, 921–929, 2000.

60. Goldbrunner, R. H., Bendszus, M., Wood, J. et al., PTK787/ZK222584, an inhibitor of vascular endothelial growth factor receptor tyrosine kinases, decreases glioma growth and vascularization, *Neurosurgery*, 55, 426–432, 2004.

61. Rooprai, H. K., Kandanearatchi, A., Maidment, S. L. et al., Evaluation of the effects of swainsonine, captopril, tangeretin and nobiletin on the biological behavior of brain tumor cells in vitro, *Neuropathol. Appl. Neurobiol.*, 27, 29–39, 2001.

62. Datla, K. P., Christidou, M., Widmer, W. W. et al., Tissue distribution and neuroprotective effects of citrus flavonoid tangeretin in a rat model of Parkinson's disease, *Neuroreport*, 12, 3871–3875, 2001.

63. Barnes, S., Peterson, T. G., Coward, L. et al., Rationale for the use of genistein-containing soy matrices in chemoprevention trials for breast and prostate cancer, *J. Cell Biochem. (Suppl.)*, 22, 181–187, 1995.

64. Penar, P. L., Khoshyomn, S., Bhushan, A. et al., Inhibition of epidermal growth factor receptor-associated tyrosine kinase blocks glioblastoma invasion of the brain, *Neurosurgery*, 40, 141–151, 1997.

65. Rooprai, H. K., Christidou, M., Pilkington, G. J. et al., The potential for strategies using micronutrients and heterocyclic drugs to treat invasive gliomas, *Acta Neurochir. (Wien.)*, 145, 683–690, 2003.

66. Clark, L. C., Combs, G. F. Jr., Turnbull, B. W. et al., Effects of selenium supplementation for cancer prevention in patients with carcinoma of the skin. A randomized controlled trial, Nutritional Prevention of Cancer Study Group, *JAMA*, 276, 1957–1963, 1996.

67. Duffield-Lillico, A. J., Slate, E. H., Reid, M. E. et al., Selenium supplementation and secondary prevention of nonmelanoma skin cancer in a randomized trial, *J. Natl Cancer Inst.*, 95, 1477–1481, 2003.

68. Sundaram, N., Pahwa, A. K., Ard, M. D. et al., Selenium causes growth inhibition and apoptosis in human brain tumor cell lines, *J. Neurooncol.*, 46, 125–133, 2000.

69. Zhu, Z., Kimura, M., Itokawa, Y. et al., Effect of selenium on malignant tumor cells of brain, *Biol. Trace Elem. Res.*, 49, 1–7, 1995.

70. Yokoyama, S., Hirano, H., Wakimaru, N. et al., Inhibitory effect of epigallocatechin-gallate on brain tumor cell lines in vitro, *Neuro-oncology*, 3, 22–28, 2001.

21 Diabetes and Angiogenesis

Golakoti Trimurtulu, Somepalli Venkateswarlu, and Gottumukkala V. Subbaraju

CONTENTS

21.1 DIABETES AND ANGIOGENESIS

Diabetes mellitus is a chronic condition associated with an abnormally high level of glucose in blood. According to the National Institute of Diabetes and Digestive and Kidney Diseases, 6.3% of the American population have diabetes, and the disease contributes to more than 200,000 deaths each year [1]. The global incidence of diabetes is increasing at an alarming rate and studies indicate that there will be 300 million people with type 2 diabetes world wide by the year 2010 [2]. There are two major types of diabetes mellitus: type 1 diabetes or insulin dependent diabetes mellitus (IDDM) and type 2 diabetes or non-insulin dependent diabetes mellitus (NIDDM).

Type 1 diabetes, also known as juvenile diabetes, is an autoimmune disease. It typically occurs in children or young adults and accounts for 5%–10% of all diabetic cases. People with type 1 diabetes have a primary insulin deficiency, as their

469

pancreas produces little or no insulin, and they must be supplemented with insulin exogenously to enable them to survive.

Type 2 diabetes, the most common form, is a progressive disease that develops when the body does not produce enough insulin or develops insulin resistance. Insulin resistance is considered a metabolic dysfunction that occurs when the body cannot efficiently use the insulin it produces. To compensate, the pancreas releases more and more insulin to try to keep blood sugar levels normal, leading to hyperglycemia (too much glucose in the blood) and hyperinsulinemia (too much insulin in the blood). In the absence of therapeutic intervention, the insulin-producing beta cells in the pancreas become defective and deteriorate gradually causing full-blown diabetes to develop.

Diagnosis and treatment of diabetes have been known to be practiced since ancient times. Plant preparations have widely been used for the treatment of diabetes mellitus in traditional systems of medicine in many cultures across the world [3]. In fact, plant based treatments were the predominant option for diabetic patients until insulin was introduced in 1922. Ancient Indian physicians, Charaka, Shushruta, and Vagbhata had a well-conceived concept of the diabetic condition and described it as madhumeha (honey urine) in Ayurveda [4]. Ancient scholars have mastered the management of madhumeha through an effective balance of Aushada (medicine), Ahar or pathya (diet control), and Vihar (exercise). The aushadic (medical) treatment involved specific foods and medicines of natural origin [5]. In fact, current popular trends for healthy living and the prevention of diabetes and other related diseases seems to have an origin in Ayurveda, which emphasizes health promotion and disease prevention through proper diet, physical exercise, and rejuvenating herbs.

Several types of conventional antidiabetic drugs are currently available in the market [6]. Glipizide and glimepiride are sulfonylureas, which are known to stimulate the pancreas to secret more insulin. Biguanides such as metformin reduce insulin resistance and improves the body's response to insulin. The thiazolidine-dione group of antidiabetic medicines, including rosiglitazone and pioglitazone improve the uptake of glucose by cells and helps the body respond better to insulin. Acarbose and miglitol are alpha-glucosidase inhibitors that slow down the absorption of sugar in the stomach and bowels. Repaglinide and nateglinide, the meglitinide group of drugs, reduce the elevation in blood sugar that generally follows food consumption.

Though allopathic diabetic medicine was driven to prominence by the technological innovations of modern drug discovery, they often are associated with undesirable side effects such as nausea and diarrhea, and some may cause stomach, bowel, and liver problems. Rising healthcare costs have made these medicines inaccessible to many people living under the poverty line. Herbal preparations are still a predominant therapeutic approach in many underprivileged regions of the world. There has been a renewed interest in natural antidiabetic medicines for use as herbal supplements and neutraceuticals. A considerable volume of research has been generated in this area during the past few decades using animal experiments and human clinical studies. There have been 42 randomized and 16 nonrandomized clinical trials on herbal medicine involving individuals with diabetes or impaired glucose tolerance and most of the studies involved patients with type 2 diabetes [7]. There has been positive evidence of improved glucose control in 76% (44 of 58) of the trials and very few studies reported adverse effects.

Hyperglycemia, the metabolic hallmark of diabetes, is responsible for widespread cellular damage. The significant morbidity and mortality of type 1 diabetes and uncontrolled type 2 diabetes result predominantly from diabetes related complications, including blindness, renal failure, amputations, strokes, and cardiac events. Though the main focus of diabetic management has been hyperglycemia, abnormal angiogenesis has been the primary contributing factor for many of the clinical manifestations of diabetes [8]. Metabolic abnormalities of diabetes, as indicated a by a large number of research studies, alter angiogenic processes [9,10]. Diabetes has been associated with the altered expression of several growth factors related to angiogenesis, such as vascular endothelial growth factor (VEGF), fibroblast growth factors (FGF), angiopoietin-1 and nitric oxide (NO) and extracellular matrix (ECM) proteins [11]. The release of these growth factors under appropriate physiologic and pathophysiologic conditions is the critical step in the initiation of angiogenesis. There are multiple mechanisms by which angiogenesis may be vulnerable to the metabolic abnormalities, although the nature of this vulnerability is not completely understood.

As vascularization is a complex mechanism involving angiogenesis and arteriogenesis, diabetes contributes significantly to many vascular complications. A potential link between insulin resistance and microvascular complications appears to be their association with endothelial dysfunction [12]. Endothelial cells, which poorly regulate glucose, may be particularly vulnerable to hyperglycemia. The ability of glucose to scavenge nitric oxide was proposed as a responsible factor for the initiation phase of endothelial dysfunction, and the development of endothelial dysfunction represents a common pathophysiological pathway for many diabetic related complications [13]. NO was reported to be functionally deficient in diabetic animals and humans, although its production from endothelial cells is stimulated by insulin [14–16]. This may explain the observed deficiency of angiogenic responses, which require basal NO production at the sites of intestinal fibrosis.

Hyperglycemia affects vascular structure and function by one or more of the following mechanisms proposed in the literature. They are: (1) enhanced oxidative stress and free radical damage [17]; (2) formation of advanced glycosylation end products [18]; (3) diversion of glucose into the aldose reductase pathway [2]; and (4) activation of one or more isozymes of protein kinase C [19]. Being a metabolic syndrome with hemodynamic abnormalities, diabetes leads to several structural and functional changes in blood vessels. Proliferative vascular disease is the source of most of the diabetic complications. Accordingly, the literature demonstrates a positive correlation between hyperglycemia and the risk of developing vascular abnormalities [20,21]. Furthermore, insulin resistance is associated with increased levels of low-density lipoprotein (LDL or "bad" cholesterol) and triglycerides and low levels of high-density lipoprotein (HDL or "good" cholesterol), which cause atherosclerosis [22]. Thus, people with diabetes have an elevated risk of both developing cardiovascular disease and dying from it—and their prognosis is worse than that for cardiovascular disease alone in people without diabetes [23]. Cardiovascular disease is the leading cause of diabetes-related death.

Diabetes also demonstrates a higher incidence of vascular abnormalities of the retina, kidneys, and fetus, impaired wound healing, increased risk of rejection of transplanted organs, and impaired formation of coronary collaterals when compared to non-diabetic subjects. The abnormalities of angiogenesis can be implicated in the pathogenesis of each of these conditions, and quite possibly in diabetic neuropathy as well [8]. The most clinically apparent effect of abnormal angiogenesis is proliferative diabetic retinopathy. Much of the blindness in diabetes is caused by proliferative vascular disease, an abnormal growth of new blood vessels, or angiogenesis, into the vitreous, the clear gel in the middle of the eye. Selective destruction of pericytes, capillary failure, and hypoxia lead to the release of proangiogenic factors followed by neovascularization. Studies suggest a significant suppression of angiogenesis by the retinal microvasculature during diabetes and implicate advanced glycation end products (AGEs) and AGE receptor interactions in its causations [9].

Impaired wound healing is a common condition in diabetes. Diabetic animal models show reduced angiogenesis in the healing of skin incisions. The circulating dysfunctional endothelial progenitor cells (key elements in angiogenesis) in diabetes may be the cause for this impairment. Studies demonstrate nitric oxide deficiency at the wound site in diabetes, and nitric oxide supplementation exogenously with the nitric oxide donor molsidomine (*N*-ethoxycarbomyl-3-morpholinyl-sidnonimine) was found to reverse the impaired healing in diabetic rats [24]. Diabetic nephropathy is presently the leading cause of chronic renal failure and end stage kidney disease [25,26]. The earliest detectable changes are thickening in the glomerulus and the presence of higher than normal amounts of albumin in urine (microalbuminuria) [26,27]. The risk of diabetic nephropathy is greater in patients with poor hypoglycemic control. Diabetic neuropathy, a nerve disorder caused by diabetes, is characterized by a loss or reduction of sensation in the feet or hands. It is the most common complication of both type 1 and type 2 diabetes mellitus and the cause of greatest morbidity and death among patients with diabetes [28]. The duration and level of hyperglycemia are important determinants of the microvascular complications of diabetes, including neuropathy [29]. Collateral vessel formation, a compensatory mechanism in chronic myocardial and lower limb ischaemia, is significantly reduced in diabetes, probably due to delayed angiogenisis [30].

The hypoglycemic actions of herbal supplements have been reported to be mediated through an insulin secretagogue effect, or through an influence on enzymes involved in glucose metabolism such as hepatic glucokinase, hexokinase, glucose-6-phosphate, fructo-1,6-bisphosphotase, glucose-6-phosphate dehydronase (G6PDH), and phosphofructokinase, etc [31]. Many plant products also improve the diabetic condition by improving the antioxidant status through significantly reducing thiobarbituric acid reactive substances and hydroperoxides, and causing a significant increase in reduced glutathione, superoxide dismutase (SOD), catalase, glutathione peroxidase, and glutathione-*S*-transferase (GST) in the liver and kidney. Recent studies have indicated that plasma concentrations of inflammatory mediators such as tumor necrosis factor alpha (TNF-α) and interleukin-6 (IL-6) are enhanced in the insulin-enhanced states of obesity and diabetes, implicating an underlying role of inflammation in the culmination of type 2 diabetes conditions, especially insulin resistance and impaired β-cell function [32].

The data from clinical trials and epidemiological studies suggest that hyperglycemia is central to the pathogenesis of vascular complications of diabetes. The extent and duration of hyperglycemia may have a bearing on the onset and intensity of diabetes complications. The novel herbal therapies that are being used to treat diabetes such as bitter melon, gymnema, Stevia, jamun, and fenugreek, may also stimulate or inhibit angiogenesis and prevent or treat a few of the complications of diabetes. There are several other natural supplements, such as cinnamon, turmeric, aloe, and guduchi, which effectively control some of the complications of diabetes, but have little or no effect on the control of hyperglycemia. This review will cover a few botanicals with antidiabetic activity and those which are known to ameliorate the complications of diabetes.

21.2 BITTER MELON

Bitter melon is a common name for *Momordica charantia*, whose other popular names include bitter gourd, balsam pear, etc. It is a climbing or scrambling herbaceous vine with tendrils, cultivated throughout India as a vegetable crop up to an altitude of 1500 m. Other known habitats for bitter melon with ethnobotanical significance and a history of folk use include parts of South America, East Africa, the Caribbean, and Asia [33]. Though leaves, seeds, and vines have all been known to possess medicinal properties, fruit is the safest and most potent part of the plant that has wide therapeutic applications. The fruit looks like a warty gourd, usually oblong and resembling a small cucumber. Indigenous tribes in the Amazon and other tropical regions traditionally used bitter melon for an array of conditions, especially as a purported cure for may infectious diseases, cancers, and diabetes [34]. Bitter melon's many therapeutic uses claimed in traditional systems of medicine have now been supported by documented evidence for a wide range of indications [35]. A broad spectrum of chemical constituents including alkaloids, steroids, terpenes, and peptides have been isolated and characterized from bitter melon [36].

The antidiabetic activity of bitter melon was documented as early as the 1960s and it has been authenticated by more than 100 research studies using rabbits, rats, and mouse models—and also human clinical trials. Cerasee, a wild variety of *M. charantia* traditionally used in the form of tea preparations for the treatment of diabetes mellitus in the West Indies and Central America, reduced the level of hyperglycemia by 50% after 5 h in streptozotocin (STZ)-induced diabetic mice. And chronic oral administration to normal mice for 13 days improved glucose tolerance [37]. The bitter melon activity profile seems to have parallels with the actions of both metformin and glibenclamide [38,39]. Whole fresh unripe bitter melon fruit water extract at a dose of 20 mg/kg body weight was reported to reduce the fasting blood glucose by 48%, an effect comparable to that of glibenclamide [40].

Oral administration of bitter melon partially reversed all diabetes-induced effects and normalized the structural abnormalities of peripheral nerves by stimulating glucose uptake into skeletal muscle cells, similar to the response obtained with insulin [41]. The aqueous juice and extracts of bitter melon markedly reduced the diabetes-induced lipid peroxidation (LPO) [42] and reversed the effect of chronic diabetes on the modulation of both P450-dependent monooxygenase activities and Glutathione (GSH)-dependent oxidative stress related LPO, and GST activities [43].

M. charantia significantly retarded diabetic retinopathy and prevented the development of cataractosis in murine alloxan-induced diabetic rats, along with a significant reduction of plasma glucose levels [44–46]. *M. charantia* partially but significantly ($p < 0.05$) prevented renal hypertrophy in diabetic animals as compared to diabetic controls [47]. Bitter melon contains components that show significant effect on lipid profiles and could ameliorate lipid disorders such as hyperlipidemia, especially those triggered by an onset of diabetes [48].

There are several proposed mechanisms by which *M. charantia* may lower blood glucose and improve glucose tolerance, such as decreased uptake of glucose from the intestine, enhanced glycogen synthesis [49], stimulation of insulin secretion [50], increased glucose transporter isoform 4 (GLUT4) protein content [51] in the plasma membrane of the muscle, and insulinomimetic effects as suggested by a number of in vivo studies. The *M. charantia* extracts behave like insulin; their actions are similar to sulfonyl ureas [39]. A steryl glycoside fraction [52] and a lectin fraction [53] with insulinomimetic activities were isolated from the seeds. The alteration in hepatic and skeletal muscle glycogen content and also the key metabolic enzymes involved in carbohydrate metabolism such as hepatic glucokinase, hexokinase, glucose-6-phosphate, fructo-1,6-bisphosphotase, G6PDH, and phosphofructokinase levels in diabetic mice were partially restored by *M. charantia* extracts [31].

Momordica charantia extracts have been evaluated in several human clinical trials and the results showed significant antidiabetic effects. Some early studies (1950–1974) in India and the Caribbean supported the traditional antidiabetic claims attributed to bitter melon [54]. Administration of water-soluble extract of the fruit to nine Asian NIDDM patients significantly reduced blood glucose concentrations during a 50 g oral glucose tolerance test. Daily consumption of dried fruits for 8–11 weeks also produced small but similar effects indicating that karela has the ability to improve glucose tolerance in diabetes [55]. Further evidence in support of these observations came from another trial involving eight NIDDM patients whose glucose tolerance and fasting blood glucose both improved after consuming powdered bitter melon daily for seven weeks [56]. In a later study reported by Srivastava et al. diabetics who were treated with powdered fruit for 3–7 weeks obtained a 25% mean reduction (range 11%–48%) in their post-prandial blood glucose. A group treated with an aqueous extract of the fruit for over 7 weeks in the same study showed a significant fall in their blood and urine sugar levels [57]. A trial for evaluating the effect of bitter melon fruit juice in maturity onset diabetic patients showed significant improvement in glucose tolerance in 73% of the patients, while the rest failed to respond [58]. Consumption of the aqueous homogenized extract of bitter melon fruit pulp by a group of 100 moderate NIDDM patients led to significant reduction ($p < 0.001$) of both fasting and post-prandial serum glucose levels in 86% of the subjects, while another 5% of the cases showed a lowering of only the fasting serum glucose [59].

p-Insulin (polypeptide-p) is a hypoglycemic agent isolated from fruit, seeds, and tissue of bitter melon, whose molecular weight is approximately 11,000 and is composed of 166 amino acid residues. It was found to be effective as a hypoglycemic agent, when administered subcutaneously to gerbils, langurs, and humans [60]. The seed powder of bitter melon was also effective in human trials and it significantly diminished the blood glucose levels in 14 NIDDM and 6 IDDM patients [61]. The reductions of fasting blood glucose, hemoglobin A1c, and postpondrial blood glucose in these trials range between 15 and 45%. Oral administration of the soft extract of bitter melon to fifteen NIDDM patients at a twice daily dose of 200 mg/kg body weight for seven days, along with half of the regular dose of metformin or glibenclamide, or both in combination, caused hypoglycemia greater than that obtained by full doses of the regular medication, suggesting that *M. charantia* acts in synergism with other oral hypoglycemics [62].

21.3   GYMNEMA SYLVESTRE

Gymnema sylvestre, a large and woody climbing plant of the family *Asclepiadaceae*, has been known in the Ayurvedic System of Medicine in India for the treatment of diabetes for many centuries [63–65]. It is found in tropical forests of central and southern India, tropical Africa and Australia. *G. sylvestre* is popularly known as Gurmar, meaning "destroyer of sugar" in Hindi. This vernacular name was coined to depict its remarkable paralyzing effect on taste buds, associated with the temporary suppression of the sweet taste sensation immediately after chewing a couple of *G. sylvestre* leaves. This anti-sweet property has been attributed to some oleanane-type triterpene glycosides, called saponins [66], and probably involves direct interaction of the active ingredients with the sweet taste buds [64]. These anti-sweet principles have been isolated from the leaves and extensively tested in chimpanzees and humans [65,67,68]. Gurmarin is a taste suppressing polypeptide also isolated from *G. sylvestre*.

The leaves of *G. sylvestre* are mostly responsible for the antidiabetic activity. The first documented report on the hypoglycemic action of gymnema leaves dates back to the late 1920s [69] and numerous animal studies have confirmed this initial study. *G. sylvestre* leaf extracts were found to reduce hyperglycemia in diabetic rats, rabbits, and humans [70]. Treatment of STZ-induced diabetic rats [71] with *G. sylvestre* extracts and alloxan-induced diabetic rabbits [72] with dried leaf powder achieved blood glucose homeostasis. The blood glucose homeostasis attained with *Gymnema* treatment was attributed to increase in insulin secretion. The rationale for the enhanced insulin levels appears to be due to repair/regeneration of the endocrine pancreas as manifested by the doubling of islet number and beta cell number in the pancreatic tissue of diabetic rats that were orally administered with *G. sylvestre* extracts [71]. This restorative effect may explain the realization of the blood glucose lowering effect of *G. sylvestre* only in animals with residual pancreatic function, but not in total pancreatectomized animals [73]. Other proposed mechanisms for its antihyperglycemic actions include an increase in glucose uptake and utilization, stimulation of beta cell function, an increase in insulin release due to improved permeability, stimulation of enzymes responsible for uptake and utilization of glucose by insulin dependent pathways, and preventing adrenal hormone from stimulating the liver to produce glucose [70,72].

The chemical ingredients responsible for the hypoglycemic actions were identified as a group of triterpenoid saponins called gymnemic acids [74]. The crude gymnemic acid fraction exhibited a variety of physiological functions. Gymnemic acids inhibit intestinal glucose absorption [75] and reduce plasma glucose [76]. Gymnemic acid IV was found to enhance insulin-releasing action and may contribute to the antihyperglycemic effect of the leaves of G. sylvestre—and it may possibly be an anti-obese and antihyperglycemic pro-drug [74]. Gymnema-genin, obtained from the hydrolysis of gymnemic acid fraction, inhibits glucose absorption.

Some of the in vivo effects of Gymnema were supported by human clinical trials. The leafy parts of G. sylvestre were known to raise the insulin levels in healthy human volunteers [77]. The water-soluble extract (GS4) of the leaves of G. sylvestre was evaluated in two nonrandomized controlled human clinical trials by the same investigators. The results showed positive enhancement of endogenous insulin in both type I and type II diabetic patients. In one clinical trial, 22 type II diabetic patients on conventional oral antihypoglycemic drugs were supplemented with 400 mg of G. sylvestre extract daily for 18–20 months. The serum insulin levels were increased simultaneously with a significant reduction in blood glucose, glycosylated hemoglobin, and glycosylated plasma proteins. The dosage requirement for the conventional drug diminished during treatment and 5 of the 22 patients were able to discontinue their hypoglycemic drugs altogether and maintain blood glucose homeostasis with GS4 alone [78].

The other study demonstrated the effectiveness of water-soluble G. sylvestre extract on insulin-dependent diabetics mellitus (type I). A group of 27 type I diabetic patients on insulin therapy were administered a daily dose of 400 mg extract. The insulin requirements for the patients dropped almost to 50%, along with a significant reduction in fasting blood glucose. The serum lipids returned to near normal levels and the glycosylated hemoglobin (HbA1c) and glycosylated plasma protein level remained higher than the controls, suggesting greater availability of endogenous insulin [79]. A recent preliminary trial [80] in the United States with type I and type II diabetic patients also reported promising results. Sixty five of the participants who completed the trial were supplemented with 800 mg of G. sylvestre extract, standardized to 25% gymnemic acids, for 90 days. In three subsets of participants, the mean daily preprandial plasma glucose concentrations were reduced by 11%–18%, whereas the 2-h postprandial plasma glucose concentrations were lowered by 13%–28%. In addition, 16% of the participants (11 patients) had a decrease in prescription medicine intake.

There are several reports indicating the effect of Gymnema on lipid metabolism [81] and obesity [82] in addition to antidiabetic actions, suggesting its possible use for the treatment of a metabolic condition called syndrome X. A 52 week toxicity study in Wistar rats supplemented by up to 1% of the diet with G. sylvestre extract showed no exposure related changes in body weight, food consumption, hematological examination, serum biochemistry, or the histopathology—establishing its safety [83].

21.4 *STEVIA*

Stevia, scientifically known as *Stevia rebaudiana* is a perennial herbaceous plant native to Brazil, Venezuela, Colombia, and Paraguay. It has been used by the Guarani Indians of Paraguay for centuries, both as a sweetener and a medicine [84]. It was the focus of scientific interest as early as 1887. The leafy part of the plant is estimated to be about 30 times sweeter, with the major component, stevioside, approximately 300 times sweeter, than cane sugar [85]. Steviosides and rebaudiosides are diterpene glucoside constituents in the leaves, which differ in having different sugar molecules. The indigenous tribes in South America knew the therapeutic potential of *Stevia* for diabetes [86]. A few empirical and semi-controlled studies support its traditional use.

The mice pretreated with *Stevia* or stevioside showed lower blood glucose and improved glucose tolerance compared with the control animals, and it protected the test animals from the toxic onslaught of alloxan [87]. Pretreatment with stevioside was more effective in this study. A similar treatment of mice with a combination of stevioside and sodium monoketocholate caused a significant reduction of adrenaline provoked glycemia [88]. Steviol, the aglycone of stevioside, isosteviol, and glucosylsteviol decreased glucose production and oxygen uptake in rat renal tubules [89]. Rebaudioside A, stevioside, and steviol dose dependently enhanced insulin secretion from incubated mouse islets, and the latter two potentiated the same from beta cells INS-1 [90,91]. These insulinotropic effects were critically dependent on the prevailing glucose concentration. Stevioside exerts stimulatory action on hepatic glycogen synthesis under gluconeogenic conditions [92]. Stevioside exerted antihyperglycemic and insulinotropic effects with concomitant suppression of glucagon levels also in diabetic Goto-Kakizaki rats [93].

An early study reported a 35% fall in normal blood sugar levels, 6–8 h following the ingestion of *Stevia* leaf extract [94]. In a randomized control trial, stevioside reduced post-prandial blood glucose levels in type 2 diabetic patients and increased the insulinogenic index by 40%, compared to controls, and decreased glucagon levels [95]. The extracts of *S. rebaudiana* significantly decreased plasma glucose levels during the test and after overnight fasting, and increased glucose tolerance in normal adult human volunteers [96].

A large number of animal experiments with rats and dogs, and also humans, attributed antihypertensive effects to *Stevia* and stevioside. There are conflicting reports regarding the safety of *Stevia* and steviosides, especially related to male fertility and mutagenicity. But other studies found no correlation between *Stevia* and male fertility [97] and also no genotoxic potential was observed for *Stevia* and steviol in a Comet assay [98]. Infact, *Stevia* tea and *Stevia* capsules are approved for sale in Brazil for the treatment of diabetes [99]. Physicians in Paraguay reportedly prescribe *Stevia* leaf tea in the treatment of diabetes [100].

21.5 BANABA

Lagerstroemia speciosa, also called queen crepe myrtle, is an ornamental plant widely cultivated as an avenue tree. The tea from the leaves has traditionally been used in the Philippines as a folk medicine for the treatment and prevention of diabetes. It is popularly known as Banaba [101]. The key chemical ingredients of *L. speciosa* include corosolic acid, maslinic acid, ellagitannin (lagerstroemin), and valoneic acid. Bioassay-guided fractionation of the banaba leaf extract culminated in identifying corosolic acid as the active principle with hypoglycemic activity. It was shown to be a glucose transport activator in Ehrlich ascites tumor cells [102]. Later studies have shown that some other ingredients, with totally unrelated structures, were also responsible for the activity of the whole extract. Lagerstroemin, flosin B, and reginin A are a group of ellagitannins isolated from the leaves of *L. speciosa* that, much like insulin, increased the rate of glucose uptake in rat adipocytes [103]. Lagerstroemin also increased the Extracellular signal-regulated kinases (ERKs) activity in Chinese hamster ovary cells expressing human insulin receptors [104]. These hypoglycemic actions were reported to be caused by activation of the insulin receptors accompanied by increased tyrosine-phosphorylation of the beta-subunit of the insulin receptors, a mechanism different from that employed by insulin [104]. The aqueous extract of banaba, which contains valoneic acid dilactone (VAD) and ellagic acid, may have a dietary use for the prevention and treatment of hyperuricemia. These two ingredients are potential inhibitors of xanthine oxidase (XOD), a key enzyme playing a role in hyperuricemia, catalyzing the oxidation of hypoxanthine to xanthine and then to uric acid [105]. The VAD was reported to be a potent alpha-amylase inhibitor [106].

The elevation of blood plasma glucose level in type II hereditary diabetic mice (KK-AY/Ta Jcl) that were on a cellulose supplemented controlled diet was suppressed almost entirely by swapping the controlled diet with a test diet containing 5% hot water extract or resin purified hot water extract for five weeks [107]. In KK-Ay mice, an animal model of type 2 diabetes, corosolic acid reduced the blood glucose and significantly enhanced the translocation of the muscle facilitative GLUT4 from low-density microsomal membrane to plasma membrane causing a hypoglycemic effect in type 2 diabetes [108]. Insulin and some other drugs upregulate both glucose transport and adipogenic activity (lipid biosynthesis) in adipocytes and this results in weight gain as a serious side effect of diabetes treatment. Banaba extracts, however, showed a desirable combination of glucose uptake stimulatory activity and the absence of adipocyte differtiation activity in addition to effective inhibition of induced adipocyte differtiation in 3T3-L1 cells [109]. Dietary banaba extract exhibited antiobesity effect in obese mice from a genetically diabetic strain (KK-Ay) [110].

The clinical antidiabetic activity of an aqueous ethanolic extract from the leaves of *L. speciosa* standardized to 3% corosolic acid has been demonstrated in a single blind cross over human clinical trial involving twenty two patients with mild non-insulin dependent type II diabetes [111]. These patients were categorized into two groups and treated with a placebo or a formulation containing banaba extract for four weeks followed by four weeks of crossover. A daily dose of 9 tablets containing 1.125 g of standardized extract was administered to the treatment group. The mean blood glucose levels dropped from 169.1 to 132.8 and 128 to 110 mg/dl in the two groups during the banaba treatment period. A placebo controlled clinical study at the Tokyo Jikeikai Medical School in Japan involving oral administration of standardized banaba extract to 24 human subjects with mild cases of type II diabetes also demonstrated significant drop in average blood glucose compared to the placebo group [112]. The recovery of blood glucose to the pre-treatment levels was found to be slow, indicating a memory effect of banaba for blood glucose control. In another clinical study [113] at the South Western Institute of Biomedical Research in Brandenton, Florida, *L. speciosa* standardized to 1% corosolic acid (Glucosol®) was administered successively at the dose levels of 16, 32, and 48 mg/day to type II diabetic patients for 2 weeks at each dose level, with an intermittent 2 week wash out period after each dose level. One group was given the drug in an oil based soft gelatin formulation and the other group was supplemented with a dry powder base in a two-piece hard gelatin capsule. The results confirmed the blood glucose lowering effect found in the earlier studies and suggest that the soft gel formulation has a better bioavailability than a dry-powder formulation. The glucosol® in a soft gel capsule formulation showed a 30% decrease in blood glucose levels compared to a 20% drop seen with the dry-powder filled hard gelatin capsule formulation. The crossover study demonstrated a memory effect for blood glucose control up to four weeks after the termination of the treatment with no signs of adverse effects.

21.6 JAMUN

Eugenia Jambolana is a large evergreen tree on the Indian subcontinent. It is widely cultivated on a large scale on the Indo-Gangetic plains, but it is distributed throughout India and many other parts of the world. *E. Jambolana* is most famously known as Jamun and it produces small purple plums commonly known as Java plum, Indian blackberry, Jambul, etc. The folk healers have used *E. Jambolana* to relieve the symptoms of diabetes, and the seeds of *E. Jambolana* have been used widely by traditional practitioners in India for centuries [114]. A wide spectrum of animal studies conducted on rabbits and rats with various *E. Jambolana* extracts exhibited hypoglycemic actions supporting its folk use as a treatment for diabetes mellitus— and the major activity has been attributed to the seeds [115–118].

Oral administration of the aqueous seed extract of Jamun to alloxan induced diabetic rats at 5 g/kg body weight for six weeks showed significant reduction in blood glucose and elevated total hemoglobin [115]. Similarly, oral administration of Jamun seed alcoholic extract to alloxan-induced diabetic rats at a dose of 100 mg/kg body weight reverted blood glucose, urine sugar, and lipids back to their normal levels in serum and tissue samples [116]. Similar effects found in STZ-induced diabetic rabbits [117] and alloxan-induced diabetic rabbits [118] confirmed the antidiabetic activity of seed extracts. *E. Jambolana* seed extract administered to mild and severe diabetic rabbits at 100 mg/kg body weight for fifteen days achieved a significant fall in fasting blood glucose and glycosylated hemoglobin (GHb) levels with a concomitant increase in serum insulin levels [118]. Seed kernel extract at a concentration of 100 mg/kg body weight reduced blood glucose and cholesterol, and improved glucose tolerance in addition to decreasing the activities of glutamate oxaloacetate transaminase and glutamate pyruvate transaminase in experimental diabetic rats. However, the seed extract showed moderate activity and the seed coat was found to be inactive [119]. The fruit pulp extract of *E. Jambolana* also exhibited hypoglycemic activity and enhanced serum insulin levels in normoglycemic rats and diabetic rats, and stimulated insulin secretion from isolated islets of Langerhans in normal as well as diabetic animals [120].

Seeds contain a glucoside namely, jamboline, and a phenolic substance called ellagic acid in addition to a potent antioxidant substance identified as gallic acid. The seed is also composed of inorganic elements such as zinc, chromium, and vanadium that are known to exhibit hypoglycemic properties. *E. Jambolana* seed kernel extract reverted the compromised antioxidant defense systems of plasma, pancreas, liver, and kidney back to their near normal levels in alloxan and STZ-induced diabetic rats and exhibited protective effects on pancreatic beta-cells [115,121]. The enhanced blood glucose and lipid peroxides, diminished levels of reduced glutathione, and reduced activities of SOD, catalase, and glutathione peroxidase found in the diabetic rats were all restored back to their normal levels after treatment with seed kernel extract [121]. Oral administration of *E. Jambolana* (200 mg/kg) every day for four months reduced the cataract incidence rate to zero, compared to eight out of eight in alloxan-induced diabetic control animals that developed cortical cataract by the end of the trial period [44]. At the same oral dose for 40 days *E. Jambolana* extract also showed a protective effect on kidney function by significantly reducing renal hypertrophy in STZ-induced diabetic rats [47]. The jamun seed extract also demonstrated antihyperlipidaemic effects in alloxan-induced diabetic rats [116] and rabbits [117,118].

In addition to the large number of animal experiments, three non-controlled clinical studies of *E. Jambolana* are known; one of them reported mixed results, but two other studies reported positive outcomes [122,123]. One study showed a statistically significant reduction in mean fasting blood sugar of 51.86 mg/dl at 2 months ($p < 0.001$) but not at 3 months. Additionally, a significant reduction in the response to the glucose tolerance test was observed at both 2 months ($p < 0.001$) and at 3 months ($p < 0.001$) [124]. *E. Jambolana* extracts, powders, and compositions are commercially available. The ripe fruits are eaten throughout India and many other parts of the world and are made into jams, jellies, and health drinks. Madeglucyl® is a commercial product developed from *E. Jambolana* seeds. It was registered as a licensed medicine with new drug application (NDA) in Madagaskar in December 1997. It has widely been used by Malagasy diabetic patients [125].

21.7 FENUGREEK

Fenugreek is a native herb to the regions of the Mediterranean and Western Asia. It is cultivated in bulk quantities mainly from countries like India, China, and Northern Africa. Seeds are the most important components of fenugreek; they are commonly used as a condiment in India and as a medicine among many cultures across the globe. They are widely used for many indications in Ayurveda, Sidda, Unani, and other traditional systems of medicine. The British Herbal Pharmacopoeia listed fenugreek seed as demulcent and hypoglycemic [126].

Fenugreek produced a significant dose related fall in blood glucose levels in the normal as well as alloxan- or STZ-induced diabetic animals. The defatted seeds have reduced basal blood glucose and plasma glucagon levels in normal experimental dogs and reduced hyperglycemia in diabetic and diabetic hypercholestrolemic dogs [127,128]. The diet, supplemented with powdered fenugreek leaves, exhibited significant effect on hyperglycemia, hypoinsulinemia, and glycosylated hemoglobin and improved body weight in STZ-induced diabetes rats, and its effect was found to be similar to that of glibenclamide [129]. A large number of animal studies further supports fenugreek's antidiabetic actions and it is among 10 of the more frequently suggested antidiabetic herbal remedies in countries like Italy. Fenugreek seed contains 4-hydroxyisoleucine (4-OH-Ile), an amino acid that has been identified as one of the active ingredients responsible for blood glucose control [130]. It potentiates insulin secretion in a glucose-dependent manner from isolated NIDDM rat islets and in NIDDM rats, probably through the activation of insulin signaling in peripheral tissues and the liver [130].

The fenugreek seed extracts demonstrated potent antioxidative potential in diabetic animals. The seed powder stabilized glucose homeostasis, oxidative stress, and free radical metabolism in alloxan induced diabetic rats, preventing liver degenerative and early nephritic changes in diabetic rats [131]. The combination of vanadate and fenugreek in many studies was found to be more potent in reversing disturbed antioxidant levels and peroxidative damage in diabetic animals. Fenugreek plays a significant role in reversing the diabetic state at the cellular level by ameliorating the key glycolytic, gluconeogenic, and lipogenic enzymes, thereby partially preventing structural abnormalities in the liver and kidney of diabetic animals [132,133]. Fenugreek seed at a daily dose of 2 g/kg body weight significantly inhibited the cataract development in alloxan diabetic rats [134]. The fenugreek seed powder or its soluble dietary fiber fraction has been shown to significantly reduce atherogenic lipids in animals and humans [128,135]. The defatted seed fraction, for example, significantly lowered basal cholesterol levels in normal dogs and declined cholesterolemia in diabetic hypercholesterolemic dogs [128]. Steroid saponins and diosgenin are found to be responsible for these actions.

Many clinical studies have investigated its hypoglycemic and hypocholesterolemic actions in normal and diabetic human subjects. In a randomized control trial [135], 100 g/day of defatted seed powder administered for 10 days in two equal doses to IDDM patients through diet supplementation during lunch and dinner significantly reduced fasting blood glucose and improved glucose tolerance. Serum total cholesterol, LDL, and Very low density lipoprotein (VLDL) cholesterol, triglycerides, and urinary glucose were also reduced significantly, whereas the HDL cholesterol remained unchanged. In a recent double blind placebo controlled study with newly diagnosed mild to moderate type 2 diabetes, adjunct use of fenugreek seeds improved glycemic control and decreased insulin resistance and showed a favorable effect on triglyceridemia [136]. Administration of twice daily doses of 2.5 g of fenugreek for a 3 months period to NIDDM patients with coronary artery disease (CAD) significantly decreased blood lipids without affecting HDL cholesterol, whereas a similar treatment to NIDDM patients without CAD significantly reduced both fasting and postprandial blood glucose [137]. The addition of powdered fenugreek seed (15 g) soaked in water to NIDDM patients significantly reduced the postprandial glucose levels following the meal tolerance test. The plasma insulin did not show a statistical difference and fenugreek had no effect on lipid levels 3 h following the meal test [138]. Many other human trials on fenugreek seeds also reported improved blood sugar control and lipid profile. A standard battery of tests recommended by USFDA using fenugreek seed extract containing 40% of 4-hydroxyisoleucine revealed that the use of fenugreek is safe [139]. Its long history of usage as a medicine and food flavoring agent further establishes its safety.

21.8   COCCINIA INDICA

Coccinia indica is a climbing perennial herb distributed throughout India. It is popularly known as Ivy gourd and has a long history in the ancient Indian medical system for its use in the treatment of diabetes [140]. In a comparative evaluation of hypoglycemic medicinal plants selected from indigenous folk medicines, the Ayurveda, Unani, and Siddha systems of medicines that include many of the other plants described in this review, the ethanolic extract of *C. indica* was found to be the most active one [141]. A number of animal studies confirm its hypoglycemic activity.

Oral administration of *C. indica* leaf ethanolic extract (200 mg/kg body weight) for 45 days improved significantly the antioxidant status in serum, liver, and kidney of STZ-induced diabetic rats and the effect was significantly better than glibenclamide [142]. *Coccinia indica* has been reported to act like insulin in diabetic patients and corrected the elevated levels of enzymes, glucose-6-phosphatase, and lactate dehydrogenase (LDH) in the glycolytic pathway, and restored the lipoprotein lipase (LPL) activity in the lipolytic pathway—along with the control of hyperglycemia in diabetics. The efficacy of *C. indica* was proven by a randomized controlled study on uncontrolled and untreated type 2 diabetic patients. Administration of freeze dried powder preparation from the leaves of the plant at a dose of 900 mg twice a day for six weeks showed marked improvement in the glucose tolerance to values close to normal, while none of the subjects from the placebo group exhibited significant improvement [143].

21.9   SALACIA OBLONGA

Salacia oblonga, which is native to regions of India and Sri Lanka, has traditionally been known as a cure for diabetes. The petroleum ether extract of *S. oblonga* root bark exhibited antidiabetic and antioxidative activity in STZ-induced diabetic rats [144]. In a comparative study, *Salacia* health tea exhibited inhibitory effect for the longest duration compared to mulberry, guava, gymnema, banaba, etc [145]. Other animal studies confirm its hypoglycemic activity. *S. oblonga* extract tended to lower postprandial glycemia and significantly reduced the postprandial insulin response, insulinemia, in healthy adults [146]. It binds to intestinal enzymes called alpha-glucosidase that break down carbohydrates in the body and potentially inhibits their activity [146]. In a double blind randomized crossover study conducted on 39 healthy adults, *S. oblonga* reduced insulin and blood glucose levels by 29 and 23 percent, respectively [146]. In a similar study, *Salacia reticulata* was also found to be an effective and safe treatment for type 2 diabetes. *S. oblonga* hot water extract at a 10 fold higher dose than that proposed for human intake did not result in any clinical chemistry or histopathologic indications of toxic effects in animal studies [147].

21.10   VIJAYSAR

Pterocarpus marsupium is more commonly known as Vijaysar or Indian Kino tree and grows more abundantly in hilly regions of Central and Peninsular India. The wood and bark have been used as an antidiabetic and antidiarrheic in traditional Indian medicine [148]. Practitioners of the Indian system of medicine are of the view that the heartwood rather than the bark of *P. marsupium* is useful for treatment of diabetic patients. *Pterocarpus marsupium* was found to show hypoglycemic and antihyperglycemia actions in rats [141] and rabbits [149]. Its antidiabetic activity has been further supported by other animal experiments. Marsupsin and pterostilbene, two pure components of *P. marsupium*, exhibited blood glucose lowering effects comparable to metformin in hyperglycemic rats [150]. Epicatechin, another active principle in water extract has been reported to have insulinogenic as well as insulin-like properties. In a second flexible multi centric open study with newly diagnosed or untreated NIDDM patients, *P. marsupium* significantly reduced both fasting and postprandial blood glucose levels by 32 and 45 mg/dl at the 12th week from the initial means of 151 and 216 mg/dl respectively [151]. An aqueous extract of *P. marsupium* exerted better anti-cataract effect than fenugreek by decreasing the opacity index in alloxan-induced diabetic rats [134].

21.11   CINNAMON

Cinnamon is the inner bark from the shoots of *Cinnamomum zeylanicum* that grows abundantly in India, China, and Ceylon. Cinnamon is aromatic and one of the predominant spices in most of the cultures around the world. In recent years, scientists attributed strong antioxidant, antidiabetic, and cholesterol maintaining properties to cinnamon. Cinnamon bark reduced the enhanced biochemical and hematological parameters of STZ-induced diabetes in rats and may provide benefit against diabetic-related conditions [152].

Cinnamon was the most active product among a large number of herb, spice, and medicinal extracts that were tested in a rat epididymal adipocyte assay for the insulin-dependent utilization of glucose. One study indicated that cinnamon extract was shown to prevent the development of insulin resistance induced by a high fructose diet in male Wistar rats at least in part by enhancing insulin signaling [153]. According to a University of California press release dated April 12, 2004, and a news article authored by Judy Foreman in The Boston Globe dated August 24, 2004, bioactive compounds such as proanthocyanidin, extracted from cinnamon, act like insulin and potentiate insulin activity by stimulating insulin receptor kinase and inhibiting PTP-1B, a phosphatase that inactivates insulin receptors [154]. A randomized and placebo-controlled study demonstrated that intake of 1, 3, or 6 g of cinnamon per day reduces serum glucose, triglyceride, LDL cholesterol, and total cholesterol in people with type 2 diabetes, and suggests that the inclusion of cinnamon in the diet of people with type 2 diabetes will reduce risk factors associated with diabetes and cardiovascular diseases [155]. In a controlled human clinical trial, the cinnamon extract was well tolerated, and side effects were found be minimal [156].

21.12 TURMERIC

Turmeric is *Curcuma longa*, a famous herb known in Ayurveda and Chinese medicine for thousands of years for its amazing healing properties. The fleshy underground rhizomes are a regular condiment in food items like Indian curries and other Asian cuisine and they contain medicinally important yellow pigments called curcumins. Many potential health benefits of turmeric were attributed to these compounds. Turmeric or curcumin significantly reduced blood sugar, hemoglobin, and glycosylated hemoglobin in alloxan-induced diabetic rats [157]. Curcumin was also shown to have hypoglycemic effects in humans [158]. Turmeric and curcumin reduced oxidative stress encountered by diabetic rats, as indicated by lower levels of thiobarbeturic acid reactive substances (TBARS) [159]. Curcumin as a potential antioxidant may be effective in attenuating many diabetes related complications. Curcumin exhibited direct anti-angiogenic activity in both in vitro and in vivo models [160]. Studies indicate that it may inhibit angiogenesis by preventing proliferation and migration of endothelial cells. The in vivo angioinhibitory effect of curcumin was corroborated by the downregulation of proangiogenic genes such as angiopoietin 1 and 2, and VEGF, etc., in EAT, NIH3T3, and endothelial cells treated with curcumin [160].

In fact, curcumin, and its derivatives demonstrated significant inhibition of basic fibroblast growth factor (bFGF) mediated corneal neovascularization in mice [161]. *C. longa* was reported to enhance cutaneous wound healing in both normal and diabetic rats [162], normal guinea pigs, and genetically diabetic mice [163]. Curcumin exhibited a differential regulatory effect on transforming growth factor β1 (TGF-β1), its receptors, and inducible nitric oxide synthase (iNOS)—the important factors in the healing process and the accelerated healing of wounds in both normal and dexamethasone impaired healing wounds. The curcumin treatment resulted in the enhanced expression of transforming growth factor TGF-beta1 and its receptor TGF-beta tIIrc, and the macrophages in the curcumin treated wound bed showed the enhanced expression of TGF-beta 1 mRNA [162]. Curcumin supplementation in STZ-induced diabetic rats brought about significant beneficial modulation of renal lesions associated with diabetes [164]. The antioxidant and anti-angiogenic actions of curcumin may thus be exploited to develop a suitable drug candidate for the treatment of diabetic retinopathy and other diabetic complications, especially the chronic diseases that are associated with extensive neovascularization.

21.13 ALOE

Aloe (*Aloe vera*) is a succulent cactus-like plant, originally indigenous to Africa and is prominent among more than 300 species that belong to this family. It was widely regarded as a master healing plant in folklore medicine. Studies suggest that *A. vera* extracts maintain glucose homeostasis in experimental diabetes rats by controlling the carbohydrate metabolizing enzymes [165]. *Aloe vera* leaf pulp extract showed hypoglycemic activity on IDDM and NIDDM rats; the effectiveness for type II diabetes was comparable with glibenclamide [166].

The ethanolic extract from *A. vera* leaf gel showed potential antioxidant properties and was reported to play a role in ameliorating the oxidative stress observed in diabetes [167]. Aloe gel extract has a protective effect comparable to glibenclamide against the hepatotoxicity [168] produced by diabetes, and also showed protective effect against mild damage in kidney tissue caused by type-II diabetes. Aloe increases blood flow to wounded areas and stimulates fibroblasts, the skin cells responsible for wound healing and which enhance the process of wound healing in diabetic rats [169]. Consistant with its efficient wound healing beneficial effect, *A. vera* gel demonstrated angiogenic activity both in vitro and in vivo. *A. vera* gel exhibited angiogenic activity on calf pulmonary artery endothelial (CPAE) cells [170]. The most active fraction from the gel increased the CPAE cell proliferation and enhanced the mRNA expression of urokinase-type plasminogen activator, matrix metalloproteinase-2 (MMP-2), and membrane type MMP in CPAE cells [170]. β-Sitosterol a constituent of Aloe, stimulated new vessel formation (neovascularization) in gerbil brains damaged by ischaemia/reperfusion in a dose-dependent fashion and also enhanced the expressions of proteins related to angiogenesis, namely von Willebrand factors, VEGF, VEGF receptor Flk-1, and blood vessel matrix laminin [171].

Recent scientific evidence suggests that *Aloe* might be useful as a treatment for type 2 diabetes. Evidence from human clinical trials complement the mounting evidence from animal studies [166,172] that Aloe can improve blood sugar control. A single-blind, placebo-controlled trial on people with diabetes exhibited significantly greater improvements in blood sugar levels among those given *A. vera* juice over the 2-week treatment period [173]. A combination of glibenclamide and *A. vera* juice was evaluated in diabetes mellitus patients who failed to respond to the oral diabetes drug glibenclamide in a single-blind, placebo-controlled study. The group taking glibenclamide and aloe showed definite improvements in blood sugar levels over 42 days as compared to those taking glibenclamide and placebo [174].

21.14 GUDUCHI

Guduchi (*Tinospora cordifolia*) is a large deciduous climbing shrub distributed throughout the tropical Indian subcontinent. It is widely used in the Ayurvedic system of medicine as an anti-inflammatory, antiarthritic, antiallergic, and antidiabetic agent [175]. It is also a prominent ingredient in Ayurvedic Rasayanas to improve the immune system and body resistance. A large number of in vivo experiments support its use for the treatment of diabetes mellitus. The daily administration of either alcoholic or aqueous extract of *T. cordifolia* attenuates blood glucose levels [141,176,177] and enhances glucose tolerance [178] in normal and alloxan- or STZ-induced hyperglycemic rats and rabbits. The alcoholic extract enhanced total hemoglobin and hexokinase and lowered hepatic glucose-6-phosphatase and serum acid phosphatase, alkaline phosphatase, and lactate dehydrogenase in diabetic rats [177]. It has shown hepatoprotective and immunomodulatory properties in a randomized controlled clinical study [179] on the surgical outcome in patients with malignant obstructive jaundice.

The ethanolic extract of *T. cordifolia* root restored antioxidant defense status in diabetic rats. Elevated thiobarbituturic acid reactive substances in the liver, kidney, and brain—with a decreased concentration in the heart—and also a reduced concentration of glutathione and the diminished activities of SOD and catalase in liver, kidney, brain, and heart were all normalized in diabetic rats after 6 weeks of daily oral administration of 100 mg of alcoholic guduchi extract per kg body weight [180]. Intraperitoneal administration of Guduchi extract inhibited melanoma tumor-induced angiogenesis in vivo models and differentially regulated the elevation of proinflammatory cytokines such as interleukin-1beta (IL-1beta), IL-6, and TNF-α and growth factors VEGF in the serum of the angiogenesis-induced animals. It also increased the production of anti-angiogenic factors interleukin-2 (IL-2) and the tissue inhibitor of metalloproteinase-1 (TIMP-1) in the treated animals [181]. Consistent with its antiangiogenesis activity, *T. cordifolia* alcoholic extract moderately prevented the development of murine alloxan diabetic cataract [44]. The *T. cordifolia* root extract also exerted hypolipidemic actions in addition to hypoglycemic activity in alloxan diabetic rats [176,177]. Guduchi may thus have a role in minimizing the complications of diabetes.

21.15 CONCLUSIONS

Herbal treatments continue to be a source of major therapeutic interventions among many cultures. Some of these commonly used natural diabetic medicines often have a long history of usage in folk medicine and traditional systems. The popularity of these commonly used herbal medicines varies among people of different ethnicities, although some of them have gained universal appeal. Analogous to a fair number of currently marketed drugs, whose development was based on their use in traditional medicine and folk remedies, plant derived products may also hold promise for new drugs for diabetes [182]. The ethanobotanical search for new antidiabetic drugs is gaining prominence [183]. Metformin, a modern pharmaceutical drug used in conventional medicine was derived from a flowering plant *Galega officinalis*. *G. officinalis*, commonly known as Gouts Rue or French Lilac, was known as a traditional remedy for diabetes [184].

Diet is the most important element in the control of diabetes. Lately, herbal products as natural dietary supplements in functional foods have attained a much broader appeal for the prevention and management of several diseases. Several traditional herbal products, such as bitter melon and fenugreek, with purported antidiabetic activity have been taken as components of a normal diet in many cultures, conferring value addition to these products. Both type 1 and type 2 diabetes are associated with the higher risk of developing chronic microvascular, retinal, renal, and neuro-pathic complications, erectile dysfunction, Carpal Tunnel Syndrome, and accelerated aging—depending on the extent and duration of hyperglycemia. The conception and progression of these complications is a slow process that spreads over many years and phases. Improved glycemic control not only deters the onset of complications but also slows down their progression. A large number of herbal supplements have been found to be quite useful in reducing these symptoms of diabetes.

Although the foregoing covers only botanicals with reasonable scientific evidence, more than a thousand plant species have been alleged to offer some form of benefit for the treatment of diabetes or its complications. Supplementation with primrose oil, a source of γ-linolenic acid, has been found in double-blind study to improve nerve function and to relieve symptoms of pain in diabetic neuropathy [185]. α-Lipoic acid was found to exert a similar effect against diabetic neuropathy in addition to improving insulin sensitivity [186]. Acetyl-L-carnitine improves nerve regeneration and vibratory perception, and relieves pain in patients with chronic diabetic neuropathy [187]. Topical application of creams containing 0.025%–0.075% capsaicin was shown to relieve symptoms of diabetic neuropathy in double-blind trials [188]. Several minerals are essential co-factors for signaling intermediates of insulin signaling and key enzymes of glucose metabolism. Diabetes tends to alter the mineral levels in diabetic patients [189]. Chromium supplements, for example, significantly improve glucose tolerance in both type 1 and type 2 diabetic patients [190]. Magnesium, zinc and, vanadium have shown similar effects in people with diabetes during human clinical trials.

Although clinical studies have attributed significant antidiabetic effects to many herbs, potential flaws in the study design, such as lack of large scale trials, lack of randomization, inadequate placebo or control groups, and inadequate trial duration unfortunately limit the general acceptability of the study results. However, their usage in folk and traditional medicine for ages, may well be construed as an indication of their general safety. Many of the herbal supplements covered thus far warrant further study. Plant products may be helpful when used under medical supervision as an adjunct to standard treatment to stabilize, reduce, or eliminate medication requirements; or correct nutritional deficiencies associated with diabetes. Future research on plant medicine may provide new options to improve diabetes and to treat diabetic complications.

REFERENCES

1. National diabetes statistics, *National Institute of Diabetes and Digestive and Kidney Diseases*, 2005, http://diabetes.niddk.nih.gov/dm/pubs/statistics/index.htm#7
2. Deborah, J. C., Rustam, R., and Richard, D., Vascular risk: diabetes, *Vasc. Med.*, 9, 307–310, 2004.
3. Swanston-Flatt, S. K., Flatt, P. R., Day, C., and Bailey, C. J., Traditional dietary adjuncts for the treatment of diabetes mellitus, *Proc. Nutr. Soc.*, 50 (3), 641–651, 1991.
4. Shashtri, K. M. and Chaturvedi, G. N., *Charak Samhita Nidan: 11 Hindi Commentary*, 5th ed., Chaukhamba Sanskriti Sansthan, Varanasi, 1977. (Part I and II)

5. Swami, S. S. T., *The Ayurveda Encyclopedia, Natural Secrets to Healing, Prevention and Longevity*, Sri Satguru Publications, New Delhi, India, pp. 423–427, 1998.

6. *American Academy of Family Physicians*, familydoctor.org, website, http://familydoctor.org/x2038.xml?#5, 1999.

7. Yeh, G. Y., Eisenberg, D. M., Kaptchuk, T. J., and Phillips, R. S., Systematic review of herbs and dietary supplements for glycemic control in diabetes, *Diabetes Care*, 26 (4), 1277–1294, 2003.

8. Martin, A., Komada, M. R., and Sane, D. C., Abnormal angiogenesis in diabetes mellitus, *Med. Res. Rev.*, 23 (2), 117–145, 2003.

9. Stitt, A. W., McGoldrick, C., Rice-McCaldin, A., McCance, D. R., Glenn, V. J., Hsu, D. K., Liu, F.-T., Thorpe, S. R., and Gardiner, A. T., Impaired retinal angiogenesis in diabetes role of advanced glycation end products and Galectin-3, *Diabetes*, 54, 785–794, 2005.

10. Eugene, L. B., Marvin, A. M., Paul, S. L. D., George, A., and Gerald, W. S., The effect of diabetes on endothelin, interleukin-8 and vascular endothelial growth factor-mediated angiogenesis in rats, *Clin. Sci.*, 103, 424S–429S, 2002.

11. Wadhwa, S., Therapeutic myocardial angiogenesis: a ray of hope for patients unsuitable for CABG/PTCA, *J. Indian Acad. Clin. Med.*, 1 (3), 252–256, 2000.

12. Zenere, B. M., Arcaro, G., Saggiani, F., Rossi, L., Muggeo, M., and Lechi, A., Noninvasive detection of functional alterations of the arterial wall in IDDM patients with and without microalbuminuria, *Diabetes Care*, 18, 975–982, 1995.

13. Michael, S. G, Chen, J., and Brodsky, S., Workshop: endothelial cell dysfunction leading to diabetic nephropathy, focus on nitric oxide, *Hypertension*, 37, 744–748, 2001.

14. Kalfa, T. A., Gerritsen, M. E., Carlson, E. C., Binstock, A. J., and Tsilibary, E. C., Altered proliferation of retinal microvascular cells on glycated matrix, *Invest. Ophthalmol. Vis. Sci.*, 36, 2358–2367, 1995.

15. Yamagashi, S., Fujimori, H., Yonekura, H., Yamamoto, Y., and Yamamoto, H., Advanced glycation end products inhibit prostacyclin production and induce plasminogen activator inhibitor-1 in human microvascular endothelial cells, *Diabetologia*, 41, 1435–1441, 1998.

16. Craven, P., DeRubertis, F., and Melhem, M., Nitric oxide in diabetic nephropathy, *Kidney Int.* 52, S46–S53, 1997; Diedrich, D., NO in diabetic nephropathy, In *Nitric Oxide and the Kidney*, Goligorsky, M. and Gross, S. Eds., Chapman and Hall, New York, pp. 349–367, 1997.

17. Baynes, J. W. and Thorpe, S. R., Role of oxidative stress in diabetic complications: a new perspective on an old paradigm, *Diabetes*, 48 (1), 1–9, 1999.

18. Schmidt, A. M., Yan, S. D., Wautier, J. L., and Stern, D., Activation of receptor for advanced glycation end products: a mechanism for chronic vascular dysfunction in diabetic vasculopathy and atherosclerosis, *Circ. Res.*, 84 (5), 489–497, 1999.

19. Idris, I., Gray, S., and Donnelly, R., Comment, *Diabetologia*, 44 (6), 657–658, 2001. Idris, I., Gray, S., and Donnelly, R., Protein kinase C activation: isozyme-specific effects on metabolism and cardiovascular complications in diabetes, *Diabetologia*, 44 (6), 659–673, 2001.

20. Beks, P. J., Mackaay, A. J., de Neeling, J. N., de Vries, H., Bouter, L. M., and Heine, R. J., Peripheral arterial disease in relation to glycemic level in an elderly Caucasian population: the Hoorn study, *Diabetologia*, 38 (1), 86–96, 1995.

21. Adler, A. I., Stevens, R. J., Neil, A., Stratton, I. M., Boulton, A. J., and Holman, R. R., UKPDS 59: hyperglycemia and other potentially modifiable risk factors for peripheral vascular disease in type 2 diabetes, *Diabetes Care*, 25 (5), 894–899, 2002.

22. *American Heart Association, 2005, website*, http://www.s2mw.com/heartofdia-betes/resistance.html

23. Hayden, M. R. and Tyagi, S. C., Vasa vasorum in plaque angiogenesis, metabolic syndrome, type 2 diabetes mellitus, and atheroscleropathy: a malignant transformation, *Cardiovasc. Diabetol.*, 3 (1), 1–16, 2004.

24. Witte, M. B., Kiyama, T., and Barbul, A., Nitric oxide enhances experimental wound healing in diabetes, *Br. J. Surg.*, 89, 1594–1605, 2002.

25. Ruggenenti, P. and Remuzzi, G., Nephropathy of type-2 diabetes mellitus, *J. Am. Soc. Nephrol.*, 9, 2157–2169, 1998.

26. Mogensen, C. E., Microalbuminuria, blood pressure and diabetic renal disease: origin and development of ideas, *Diabetologia*, 42, 263–285, 1999.

27. *MedlinePlus medical encyclopedia*, 2004, http://www.nlm.nih.gov/medlineplus/ency/article/000494.htm

28. Vinik, A. I., Park, T. S., Stansberry, K. B., and Pittenger, G. L., Diabetic neuropathies, *Diabetologia*, 43, 957–973, 2000.

29. The Diabetes Control and Complications Trial Research Group, The effect of intensive diabetes therapy on the development and progression of neuropathy, *Ann. Int. Med.*, 122, 561–568, 1995.

30. Taniyama, Y., Morishita, R., Hiraoka, K., Aoki, M., Nakagami, H., Yamasaki, K., Matsumoto, K., Nakamura, T., Kaneda, Y., and Ogihara, T., Therapeutic angiogenesis induced by human hepatocyte growth factor gene in rat diabetic hind limb ischemia model: molecular mechanisms of delayed angiogenesis in diabetes, *Circulation*, 104 (19), 2344–2350, 2001.

31. Rathi, S. S., Grover, J. K., and Vats, V., The effect of *Momordica charantia* and *Mucuna pruriens* in experimental diabetes and their effect on key metabolic enzymes involved in carbohydrate metabolism, *Phytother. Res.*, 16 (3), 236–243, 2002.

32. Dandona, P., Aljada, A., and Bandyopadhyay, A., Inflammation: the link between insulin resistance, obesity and diabetes, *Trends Immunol.*, 25 (1), 4–7, 2004.

33. Leslie Taylor, N. D., *The Healing Power of Rainforest Herbs*, Square One Publishers, Inc., New York, 2004.

34. Duke, J. A., *CRC Handbook of Medicinal Herbs*, CRC Press, Boca Raton, FL, pp. 315–316, 1985.

35. Grover, J. K. and Yadav, S. P., Pharmacological actions and potential uses of *Momordica charantia*: a review, *J. Ethnopharmacol.*, 93 (1), 123–132, 2004.

36. Raman, A. and Lau, C., Anti-diabetic properties and phytochemistry of *Momordica charantia* L. (*Curcurbitaceae*), *Phytomedicine*, 2, 349–362, 1996.

37. Bailey, C. J., Day, C., Turner, S. L., and Leatherdale, B. A., Cerasee, a traditional treatment for diabetes. Studies in normal and streptozotocin diabetic mice, *Diab. Res.*, 2 (2), 81–84, 1985.

38. McCarty, M. F., Does bitter melon contain an activator of AMP-activated kinase? *Med. Hypotheses*, 63 (2), 340–343, 2004.

39. Rotshteyn, Y. and Zito, S. W., Application of modified in vitro screening procedure for identifying herbals possessing sulfonylurea-like activity, *J. Ethnopharmacol.*, 93 (2–3), 337–344, 2004.

40. Virdi, J., Sivakami, S., Shahani, S., Suthar, A. C., Banavalikar, M. M., and Biyani, M. K., Antihyperglycemic effects of three extracts from *Momordica charantia*, *J. Ethnopharmacol.*, 88 (1), 107–111, 2003.

41. Ahmed, I., Adeghate, E., Cummings, E., Sharma, A. K., and Singh, J., Beneficial effects and mechanism of action of *Momordica charantia* juice in the treatment of streptozotocin-induced diabetes mellitus in rat, *Mol. Cell Biochem.*, 261 (1–2), 63–70, 2004.

42. Sitasawad, S. L., Shewade, Y., and Bhonde, R., Role of bitter gourd fruit juice in stz-induced diabetic state in vivo and in vitro, *J. Ethnopharmacol.*, 73 (1–2), 71–79, 2000.

43. Raza, H., Ahmed, I., John, A., and Sharma, A. K., Modulation of xenobiotic metabolism and oxidative stress in chronic streptozotocin-induced diabetic rats fed with *Momordica charantia* fruit extract, *J. Biochem. Mol. Toxicol.*, 14 (3), 131–139, 2000.

44. Rathi, S. S., Grover, J. K., Vikrant, V., and Biswas, N. R., Prevention of experimental diabetic cataract by Indian Ayurvedic plant extracts, *Phytother. Res.*, 16 (8), 774–777, 2002.

45. Srivastava, Y., Venkatakrishna-Bhatt, H., Verma, Y., and Prem, A. S., Retardation of retinopathy by *Momordica charantia L.*, (bitter gourd) fruit extract in alloxan diabetic rats, *Indian J. Exp. Biol.*, 25, 571–572, 1987.

46. Srivastava, Y., Venkatakrishna-Bhatt, H., and Verma, Y., Effects of *Momordica charantia* L. pomous aqueous extract on cataractogenesis in murrin alloxan diabetic, *Pharmacol. Res. Commun.*, 20 (3), 201–209, 1988.

47. Grover, J. K., Vats, V., Rathi, S. S., and Dawar, R., Traditional Indian anti-diabetic plants attenuate progression of renal damage in streptozotocin-induced diabetic mice, *J. Ethnopharmacol.*, 76 (3), 233–238, 2001.

48. Ahmed, I., Lakhani, M. S., Gillett, M., John, A., and Raza, H., Hypotriglyceridemic and hypocholesterolemic effects of anti-diabetic *Momordica charantia* (karela) fruit extract in streptozotocin-induced diabetic rats, *Diab. Res. Clin. Pract.*, 51 (3), 155–161, 2001.

49. Welihinda, J. and Karunanayake, E. H., Extra-pancreatic effects of *Momordica charantia* in rats, *J. Ethnopharmacol.*, 17 (3), 247–255, 1986.

50. Welihinda, J., Arvidson, G., Gylfe, E., Hellman, B., and Karlsson, E., The insulin-releasing activity of the tropical plant *Momordica charantia*, *Acta Biol. Med. Ger.*, 41 (12), 1229–1240, 1982.

51. Miura, T., Itoh, C., Iwamoto, N., Kato, M., Kawai, M., Park, S. R., and Suzuki, I., Hypoglycemic activity of the fruit of the *Momordica charantia* in type 2 diabetic mice, *J. Nutr. Sci. Vitaminol. (Tokyo)*, 47 (5), 340–344, 2001.

52. Ng, T. B., Wong, C. M., Li, W. W., and Yeung, H. W., A steryl glycoside fraction from *Momordica charantia* seeds with an inhibitory action on lipid metabolism in vitro, *Biochem. Cell Biol.*, 64 (8), 766–771, 1986.

53. Ng, T. B., Wong, C. M., Li, W. W., and Yeung, H. W., Isolation and characterization of a galactose binding lectin with insulinomimetic activities. From the seeds of the bitter gourd *Momordica charantia* (Family *Cucurbitaceae*), *Int. J. Pept. Protein Res.*, 28 (2), 163–172, 1986.

54. Kirti, S., Kumar, V., Nigam, P., and Srivastava, P., Effect of *Momordica charantia* (karela) extract on blood and urine sugar in diabetes mellitus—study from a diabetic clinic, *Clinician*, 46 (1), 26–29, 1982.

55. Leatherdale, B. A., Panesar, R. K., Singh, G., Atkins, T. W., Bailey, C. J., and Bignell, A. H., Improvement in glucose tolerance due to *Momordica charantia* (karela), *Br. Med. J., (Clin. Res. Ed.)*, 282 (6279), 1823–1824, 1981.

56. Akhtar, M. S., Trial of *Momordica charantia* L. (karela) powder in patients with maturity onset diabetes, *J. Pak. Med. Assoc.*, 32, 106–107, 1982.

57. Srivastava, Y., Venkatakrishna-Bhatt, H., Verma, Y., Venkaiah, K., and Raval, B., Antidiabetic and adaptogenic properties of *Momordica charantia* extract: an experimental and clinical evaluation, *Phytother. Res.*, 7, 285–289, 1993.

58. Welihinda, J., Karunanayake, E. H., Sheriff, M. H., and Jayasinghe, K. S., Effect of *Momordica charantia* on the glucose tolerance in maturity onset diabetes, *J. Ethnopharmacol.*, 17 (3), 277–282, 1986.

59. Ahmad, N., Hassan, M. R., Halder, H., and Bennoor, K. S., Effect of *Momordica charantia* (Karolla) extracts on fasting and postprandial serum glucose levels in NIDDM patients, *Bangladesh Med. Res. Counc. Bull.*, 25 (1), 11–13, 1999.

60. Baldwa, V. S., Bhandari, C. M., Pangaria, A., and Goyal, R. K., Clinical trial in patients with diabetes mellitus of an insulin-like compound obtained from a plant source, *Upsala J. Med. Sci.*, 82, 39–41, 1977.

61. Grover, J. K. and Gupta, S. R., Hypoglycemic activity of seeds of *Momordica charantia*, *Eur. J. Pharmacol.*, 183, 1026–1027, 1990.

62. Tongia, A., Tongia, S. K., and Dave, M., Phytochemical determination and extraction of *Momordica charantia* fruit and its hypoglycemic potentiation of oral hypoglycemic drugs in diabetes mellitus (NIDDM), *Indian J. Physiol. Pharmacol.*, 48 (2), 241–244, 2004.

63. Chopra, R. N., *Indigenous Drugs of India*, 2nd ed., Art Press, Calcutta, India, p. 336, 1958.

64. Liu, H. M., Kiuchi, F., and Tsuda, Y., Isolation and structure elucidation of gymnemic acids, antisweet principles of *Gymneme sylvestre Chem. Pharm. Bull.*, 40 (6), 1366–1375, 1992.

65. Agarwal, S. K., Singh, S. S., Verma, S., Lakshmi, V., Sharma, A., and Kumar, S., Chemistry and medicinal uses of *Gymnema sylvestre* (Gur-Mar) leaves-A review, *Indian Drugs*, 37, 354–360, 2000.

66. Ye, W., Liu, X., Zhang, Q., Che, C. T., and Zhao, S., Antisweet saponins from *Gymnema Sylvestre*, *J. Nat. Prod.*, 64 (2), 232–235, 2001.

67. Hooper, D., An examination of the leaves of *Gymnema sylvestre Nature*, 35, 565–567, 1887.

68. Hellekant, G., Ninomiya, Y., and Danilova, V., Taste in chimpanzees. III: labeled-line coding in sweet taste, *Physiol. Behav.*, 65 (2), 191–200, 1998.

69. Mhasker, K. S. and Caius, J. F., A study of Indian medicinal plants. II, *Gymnema sylvestre* R.Br., *Indian J. Med. Res. Memoirs*, 16, 2–75, 1930.

70. Persaud, S. J., Al-Majed, H., Raman, A., and Jones, P. M., *Gymnema Sylvestre* stimulates insulin release in vitro by increased membrane permeability, *J. Endocrinol.*, 163 (2), 207–212, 1999, and references there in.

71. Shanmugasundaram, E. R., Gopinath, K. L., Shanmugasundaram, K. R., and Rajendran, V. M., Possible regeneration of the islets of Langerhans in streptozotocin-diabetic rats given *Gymnema Sylvestre* leaf extracts, *J. Ethnopharmacol.*, 30 (3), 265–279, 1990.

72. Shanmugasundaram, K. R., Panneerselvam, C., Samudram, P., and Shanmugasundaram, E. R., Enzyme changes and glucose utilization in diabetic rabbits: the effect of *Gymnema Sylvestre* R.Br., *J. Ethnopharmacol.*, 7 (2), 205–234, 1983.

73. Yeh, G. Y., Kaptchuk, T. J., Eisenberg, D. M., and Phillips, R. S., Systematic review of herbs and dietary supplements for glycemic control in diabetes, *Diabet. Care*, 26 (4), 1277–1294, 2003.

74. Sugihara, Y., Nojima, H., Matsuda, H., Murakami, T., Yoshikawa, M., and Kimura, I., Antihyperglycemic effects of gymnemic acid IV, a compound derived from *Gymnema sylvestre* leaves in streptozotocin-diabetic mice, *J. Asian Nat. Prod. Res.*, 2, 321–327, 2000.

75. Hirata, S., Terasawa, H., Katou, T., and Imoto, T., The inhibitory effects of novel substances, which were extracted from *Gymnema sylvestre* leaves and partially purified by affinity chromatography, on glucose absorption in the small intestines of rats, *J. Yonago Med. Ass.*, 43, 397–404, 1992.

76. Hirata, S., Abe, T., and Imoto, T., Effects of crude gymnemic acid on the oral glucose tolerance test in human being, *J. Yonago Med. Ass.*, 43, 392–396, 1992.

77. Shanmugasundaram, K. R., Panneerselvam, C., Samudram, P., and Shanmugasundaram, E. R., The insulinotropic activity of *Gymnema Sylvestre* R.Br. An Indian medical herb used in controlling diabetes mellitus, *Pharmacol. Res. Commun.*, 13 (5), 475–486, 1981.

78. Baskaran, K., Ahamath, B. K., Shanmugasundaram, K. R., and Shanmugasundaram, E. R., Antidiabetic effect of a leaf extract from *Gymnema Sylvestre* in non-insulin-dependent diabetes mellitus patients, *J. Ethnopharmacol.*, 30 (3), 295–300, 1990.

79. Shanmugasundaram, E. R., Rajeswari, G., Baskaran, K., Rajesh Kumar, B. R., Shanmugasundaram, K. R., and Ahmath, B. K., Use of *Gymnema Sylvestre* leaf extract in the control of blood glucose in insulin-dependent diabetes mellitus, *J. Ethnopharmacol.*, 30 (3), 281–294, 1990.

80. Joffe, D. J. and Freed, S. H., Effect of extended release *Gymnema Sylvestre* leaf extract alone or in combination with oral hypoglycemics or insulin regimens for type 1 and type 2 diabetes, *Diab. Control Newslett.*, 76 (1), 1–4, 2001.

81. Bishayee, A. and Chatterjee, M., Hypolipidemic and antiatherosclerotic effects of oral *Gymnema sylvestre* R.Br. leaf extract in albino rats fed on a high fat diet, *Phytother. Res.*, 8, 118–120, 1994.

82. Nakamura, Y., Tsumura, Y., Tonogai, Y., and Shibata, T., Fecal steroid excretion is increased in rats by oral administration of Gymnemic acids contained in *Gymnema sylvestre* leaves, *J. Nutr.*, 129, 1214–1222, 1999.

83. Ogawa, Y., Sekita, K., Umemura, T., Saito, M., Ono, A., Kawasaki, Y., Uchida, O., Matsushima, Y., Inoue, T., and Kanno, J., *Gymnema Sylvestre* leaf extract: a 52-week dietary toxicity study in Wistar rats, *Shokuhin Eiseigaku Zasshi*, 45 (1), 8–18, 2004.

84. Bertoni, M. S., Le Kaa He-e Sa nature et ses properietes, *Ancient. Paraguayos*, 1 (5), 1–14, 1905.

85. Bridel, M. and Lavielle, R., Sur le principe sucre des feuilles de kaa-he-e (*Stevia rebaundiana* B), *Compt. Rend. Acad. Sci.*, Parts 192, 1123–1125, 1931.

86. Leung, A. Y. and Foster, S., *Encyclopedia of Common Natural Ingredients Used in Foods, Drugs, and Cosmetics* 2nd ed., Wiley, New York, pp. 478–480, 1996.

87. Raskovic, A., Gavrilovic, M., Jakovljevic, V., and Sabo, J., Glucose concentration in the blood of intact and alloxan-treated mice after pretreatment with commercial preparations of *Stevia rebaudiana* (Bertoni), *Eur. J. Drug Metab. Pharmacokinet.*, 29 (2), 87–90, 2004.

88. Raskovic, A., Jakovljevic, V., Mikov, M., and Gavrilovic, M., Joint effect of commercial preparations of *Stevia rebaudiana* Bertoni and sodium monoketocholate on glycemia in mice, *Eur. J. Drug Metab. Pharmacokinet.*, 29 (2), 83–86, 2004.

89. Yamamoto, N. S., Kelmer Bracht, A. M., Ishii, E. L., Kemmelmeier, F. S., Alvarez, M., and Bracht, A., Effect of steviol and its structural analogues on glucose production and oxygen uptake in rat renal tubules, *Experientia*, 41 (1), 55–57, 1985.

90. Jeppesen, P. B., Gregersen, S., Poulsen, C. R., and Hermansen, K., Stevioside acts directly on pancreatic beta cells to secrete insulin: actions independent of cyclic adenosine monophosphate and adenosine triphosphate-sensitive K+-channel activity, *Metabolism*, 49 (2), 208–214, 2000.

91. Abudula, R., Jeppesen, P. B., Rolfsen, S. E., Xiao, J., and Hermansen, K., Rebaudioside a potently stimulates insulin secretion from isolated mouse islets: studies on the dose-, glucose-, and calcium-dependency, *Metabolism*, 53 (10), 1378–1381, 2004.

92. Hubler, M. O., Bracht, A., and Kelmer-Bracht, A. M., Influence of stevioside on hepatic glycogen levels in fasted rats, *Res. Commun. Chem. Pathol. Pharmacol.*, 84 (1), 111–118, 1994.

93. Jeppesen, P. B., Gregersen, S., Alstrup, K. K., and Hermansen, K., Stevioside induces antihyperglycemic, insulinotropic and glucagonostatic effects in vivo: studies in the diabetic Goto-Kakizaki (GK) rats, *Phytomedicine*, 9 (1), 9–14, 2002.

94. Oviedo, C. A., Fronciani, C., Moreno, R., and Maas, L. C., Accion hipoglicemiante de la *Stevia rebaudiana* Bertoni, *Excerpta Medica*, 209, 92–93, 1971, Seventh Congress of the International diabetes.

95. Gregersen, S., Jeppesen, P. B., Holst, J. J., and Hermansen, K., Antihyperglycemic effects of stevioside in type 2 diabetic subjects, *Metabolism*, 53 (1), 73–76, 2004.

96. Curi, R., Alvarez, M., Bazotte, R. B., Botion, L. M., Godoy, J. L., and Bracht, A., Effect of *Stevia rebaudiana* on glucose tolerance in normal adult humans, *Braz. J. Med. Biol. Res.*, 19 (6), 771–774, 1986.

97. Oliveira-Filho, R. M., Uehara, O. A., Minetti, C. A., and Valle, L. B., Chronic administration of aqueous extract of *Stevia rebaudiana* (Bert.) Bertoni in rats: endocrine effects, *Gen. Pharmacol.*, 20 (2), 187–191, 1989.

98. Sekihashi, K., Saitoh, H., and Sasaki, Y., Genotoxicity studies of Stevia extract and steviol by the comet assay, *J. Toxicol. Sci.*, 27 (suppl. 1), 1–8, 2002.

99. Kinghorn, A. D. and Soejarto, D. D., Current status of stevioside as a sweetening agent for human use, In *Economic and Medicinal Plant Research, Volume 1*, Wagner, H., Hikino, H., and Farnsworth, N. R. Eds., Academic Press, New York, pp. 1–51, 1985.

100. Soejarto, D. D., Compadre, C. M., Medon, P. J., Kameth, S. K., and Kinghorn, A. D., Potential sweetening agents of plant origin II. Field search for sweet tasting *Stevia* species, *Econ. Bot.*, 37, 71–79, 1983.

101. Quisumbing, E., *Medicinal Plants of the Philippines*, Katha Publishing, Quezon city, pp. 640–642, 1978.

102. Murakami, C., Myoga, K., Kasai, R., Ohtani, K., Kurokawa, T., Ishibashi, S., Dayrit, F., Padolina, W. G., and Yamasaki, K., Screening of plant constituents for effect on glucose transport activity in Ehrlich ascites tumour cells, *Chem. Pharm. Bull. (Tokyo)*, 41 (12), 2129–2131, 1993.

103. Hayashi, T., Maruyama, H., Kasai, R., Hattori, K., Takasuga, S., Hazeki, O., Yamasaki, K., and Tanaka, T., Ellagitannins from *Lagerstroemia speciosa* as activators of glucose transport in fat cells, *Planta. Med.*, 68 (2), 173–175, 2002.

104. Hattori, K., Sukenobu, N., Sasaki, T., Takasuga, S., Hayashi, T., Kasai, R., Yamasaki, K., and Hazeki, O., Activation of insulin receptors by lagerstroemin, *J. Pharmacol. Sci.*, 93 (1), 69–73, 2003.

105. Unno, T., Sugimoto, A., and Kakuda, T., Xanthine oxidase inhibitors from the leaves of *Lagerstroemia speciosa* (L.) Pers, *J. Ethnopharmacol.*, 93 (2–3), 391–395, 2004.

106. Hosoyama, H., Sugimoto, A., Suzuki, Y., Sakane, I., and Kakuda, T., Isolation and quantitative analysis of the alpha-amylase inhibitor in *Lagerstroemia speciosa* (L.) Pers (Banaba), *Yakugaku Zasshi*, 123 (7), 599–605, 2003.

107. Kakuda, T., Sakane, I., Takihara, T., Ozaki, Y., Takeuchi, H., and Kuroyanagi, M., Hypoglycemic effect of extracts from *Lagerstroemia speciosa* L. leaves in genetically diabetic KK-AY mice, *Biosci. Biotechnol. Biochem.*, 60 (2), 204–208, 1996.

108. Miura, T., Itoh, Y., Kaneko, T., Ueda, N., Ishida, T., Fukushima, M., Matsuyama, F., and Seino, Y., Corosolic acid induces GLUT4 translocation in genetically type 2 diabetic mice, *Biol. Pharm. Bull.*, 27 (7), 1103–1105, 2004.

109. Liu, F., Kim, J., Li, Y., Liu, X., Li, J., and Chen, X., An extract of *Lagerstroemia speciosa* L. has insulin-like glucose uptake-stimulatory and adipocyte differentiation-inhibitory activities in 3T3-L1 cells, *J. Nutr.*, 131 (9), 2242–2247, 2001.

110. Suzuki, Y., Unno, T., Ushitani, M., Hayashi, K., and Kakuda, T., Antiobesity activity of extracts from *Lagerstroemia speciosa* L. leaves on female KK-Ay mice, *J. Nutr. Sci. Vitaminol. (Tokyo)*, 45 (6), 791–795, 1999.

111. Matsuyama, F., Method for inhibiting increase of blood sugar level or lowering blood sugar level with a Lagerstroemia extract, United States Patent, 6,485,760 dtd., November 26, 2002.

112. Ikeda, Y., Chen, J.-T., and Matsuda, T., Effectiveness and safety of banabamin tablet containing extract from banaba in patients with mild type 2 diabetes, *Jpn. Pharmacol. Ther.*, 27, 829–835, 1999.

113. Judy, W. V., Hari, S. P., Stogsdill, W. W., Judy, J. S., Naguib, Y. M., and Passwater, R., Antidiabetic activity of a standardized extract (Glucosol) from *Lagerstroemia speciosa* leaves in type II diabetics. A dose-dependence study, *J. Ethnopharmacol.*, 87 (1), 115–117, 2003.

114. Nadkarni, K. M., Ed., 3rd ed. *Indian Materia Medica*, Vol. 1, Popular Prakashan Private Limited, Mumbai, pp. 516–518, 1982.

115. Prince, P. S., Menon, V. P., and Pari, L., Hypoglycaemic activity of *Syzigium cumini* seeds: effect on lipid peroxidation in alloxan diabetic rats, *J. Ethnopharmacol.*, 61 (1), 1–7, 1998.

116. Prince, P. S., Kamalakkannan, N., and Menon, V. P., Antidiabetic and antihyperlipidemic effect of alcoholic *Syzigium cumini* seeds in alloxan-induced diabetic albino rats, *J. Ethnopharmacol.*, 91 (2–3), 209–213, 2004.

117. Kedar, P. and Chakrabarti, C. H., Effects of jambolan seed treatment on blood sugar, lipids and urea in streptozotocin-induced diabetes in rabbits, *Indian J. Physiol. Pharmacol.*, 27, 135–140, 1983.

118. Sharma, S. B., Nasir, A., Prabhu, K. M., Murthy, P. S., and Dev, G., Hypoglycemic and hypolipidemic effect of ethanolic extract of seeds of *Eugenia jambolana* in alloxan-induced diabetic rabbits, *J. Ethnopharmacol.*, 85 (2–3), 201–206, 2003.

119. Ravi, K., Sivagnanam, K., and Subramanian, S., Anti-diabetic activity of *Eugenia jambolana* seed kernels on streptozotocin-induced diabetic rats, *J. Med. Food*, 7 (2), 187–191, 2004.

120. Achrekar, S., Kaklij, G. S., Pote, M. S., and Kelkar, S. M., Hypoglycemic activity of *Eugenia jambolana* and Ficus bengalensis: mechanism of action, *In Vivo*, 5 (2), 143–147, 1991.

121. Ravi, K., Ramachandran, B., and Subramanian, S., Effect of *Eugenia jambolana* seed kernel on antioxidant defense system in streptozotocin-induced diabetes in rats, *Life Sci.*, 75 (22), 2717–2731, 2004.

122. Agency for Healthcare Research and Quality (AHRQ), AHRQ Evidence Reports, number 41, Ayurvedic Interventions for Diabetes Mellitus: A Systematic Review, 2001, http://www.ncbi.nlm.nih.gov/books/bv.fcgi?rid=hstat1.section.95526.

123. Nande, C. V., Kale, P. M., Wagh, S. Y., Antarkar, D. S., and Vaidya, A. B., Effect of jambu fruit pulp (*Eugenia jambolana* Lam) on blood sugar levels in healthy volunteers and diabetics, *J. Res. Ayur. Siddha*, IV (1–4), 1–5, 1983.

124. Kohli, K. R. and Singh, R. H., A clinical trial of NIDDM, *J. Res. Ayur. Sidha*, 14 (3–4), 89–97, 1993.

125. Ratsimamanga, S. U., *Eugenia jambolana*: Madagascar, Monograph, Malagasy Institute of Applied Research (IMRA), Madagascar (http://tcdc.undp.org/sie/experiences/vol7/Eugenia%20Jambolana_Madagascar.pdf).

126. British Herbal Medicine Association., *British Herbal Pharmacopoeia* (BHP), Keighley, U.K., 1983.

127. Ribes, G., Sauvaire, Y., Baccou, J. C., Valette, G., Chenon, D., Trimble, E. R., and Loubatieres-Mariani, M. M., Effects of fenugreek seeds on endocrine pancreatic secretions in dogs, *Ann. Nutr. Metab.*, 28 (1), 37–43, 1984.

128. Valette, G., Sauvaire, Y., Baccou, J. C., and Ribes, G., Hypocholesterolemic effect of fenugreek seeds in dogs, *Atherosclerosis*, 50 (1), 105–111, 1984.

129. Devi, B. A., Kamalakkannan, N., and Prince, P. S., Supplementation of fenugreek leaves to diabetic rats. Effect on carbohydrate metabolic enzymes in diabetic liver and kidney, *Phytother. Res.*, 17 (10), 1231–1233, 2003.

130. Broca, C., Breil, V., Cruciani-Guglielmacci, C., Manteghetti, M., Rouault, C., Derouet, M., Rizkalla, S., et al., Insulinotropic agent ID-1101 (4-hydroxyisoleucine) activates insulin signaling in rat, *Am. J. Physiol. Endocrinol. Metab.*, 287 (3), E463–E471, 2004.

131. Thakran, S., Siddiqui, M. R., and Baquer, N. Z., *Trigonella foenum graecum* seed powder protects against histopathological abnormalities in tissues of diabetic rats, *Mol. Cell Biochem.*, 266 (1–2), 151–159, 2004.

132. Raju, J., Gupta, D., Rao, A. R., Yadava, P. K., and Baquer, N. Z., *Trigonella foenum graecum* (fenugreek) seed powder improves glucose homeostasis in alloxan diabetic rat tissues by reversing the altered glycolytic, gluconeogenic and lipogenic enzymes, *Mol. Cell Biochem.*, 224 (1–2), 45–51, 2001.

133. Gupta, D., Raju, J., and Baquer, N. Z., Modulation of some gluconeogenic enzyme activities in diabetic rat liver and kidney: effect of antidiabetic compounds, *Indian J. Exp. Biol.*, 37 (2), 196–199, 1999.

134. Vats, V., Yadav, S. P., Biswas, N. R., and Grover, J. K., Anti-cataract activity of *Pterocarpus marsupium* bark and *Trigonella foenum-graecum* seeds extract in alloxan diabetic rats, *J. Ethnopharmacol.*, 93 (2–3), 289–294, 2004.

135. Sharma, R. D., Raghuram, T. C., and Rao, N. S., Effect of fenugreek seeds on blood glucose and serum lipids in type I diabetes, *Eur. J. Clin. Nutr.*, 44 (4), 301–306, 1990.

136. Gupta, A., Gupta, R., and Lal, B., Effect of *Trigonella foenum-graecum* (fenugreek) seeds on glycemic control and insulin resistance in type 2 diabetes mellitus: a double blind placebo controlled study, *J. Assoc. Phys. India*, 49, 1057–1061, 2001; *J. Assoc. Phys. India*, 49, 1055–1056, 2001.

137. Bordia, A., Verma, S. K., and Srivastava, K. C., Effect of ginger (*Zingiber officinale* Rosc.) and fenugreek (*Trigonella foenumgraecum* L.) on blood lipids, blood sugar and platelet aggregation in patients with coronary artery disease, *Prostaglandins Leukot Essent Fatty Acids*, 56 (5), 379–384, 1997.

138. Madar, Z., Abel, R., Samish, S., and Arad, J., Glucose-lowering effect of fenugreek in non-insulin dependent diabetics, *Eur. J. Clin. Nutr.*, 42 (1), 51–54, 1988.
139. Flammang, A. M., Cifone, M. A., Erexson, G. L., and Stankowski, L. F. Jr., Genotoxicity testing of a fenugreek extract, *Food Chem. Toxicol.*, 42 (11), 1769–1775, 2004.
140. Chopra, R. N., Chopra, I. L., Handa, K. L., and Kapur, L. D., *Indigenous Drugs of India*, 2nd ed., UN Dhar and Sons Pvt Ltd, Calcutta, 1958.
141. Kar, A., Choudhary, B. K., and Bandyopadhyay, N. G., Comparative evaluation of hypoglycemic activity of some Indian medicinal plants in alloxan diabetic rats, *J. Ethnopharmacol.*, 84 (1), 105–108, 2003.
142. Venkateswaran, S. and Pari, L., Effect of *Coccinia indica* leaves on antioxidant status in streptozotocin-induced diabetic rats, *J. Ethnopharmacol.*, 84 (2–3), 163–168, 2003.
143. Khan, A. K., Akhtar, S., and Mahtab, H., Treatment of diabetes mellitus with *Coccinia indica*, *Br. Med. J.*, 280 (6220), 1044, 1980.
144. Krishnakumar, K., Augusti, K. T., and Vijayammal, P. L., Hypoglycaemic and anti-oxidant activity of *Salacia oblonga* Wall. extract in streptozotocin-induced diabetic rats, *Indian J. Physiol. Pharmacol.*, 43 (4), 510–514, 1999.
145. Toshiki, M., Yukako, Y., Hironori, M., and Mitsuaki, S., Suppression of glucose absorption by various health teas in rats, *Yakugaku Zasshi*, 124 (4), 217–223, 2004.
146. Heacock, P. M., Hertzler, S. R., Williams, J. A., and Wolf, B. W., Effects of a medical food containing an herbal alpha-glucosidase inhibitor on postprandial glycemia and insulinemia in healthy adults, *J. Am. Diet Assoc.*, 105 (1), 65–71, 2005.
147. Heacock, P. M., Hertzler, S. R., Williams, J. A., Wolf, B. W., and Weisbrode, S.E, Safety evaluation of an extract from *Salacia oblonga, Food Chem. Toxicol.*, 41 (6), 867–874, 2003.
148. Kirtikar, K. R. and Basu, B. D., In *Indian Medicinal Plants*, Blatter, E., Cailes, J. F., and Mhaskar, K. S. Eds., Singh and Singh, Delhi, India, p. 2135, 1975.
149. Shah, D. S., A preliminary study of the hypoglycemic action of heartwood of *Pterocarpus marsupium* roxb, *Indian J. Med. Res.*, 55, 166–168, 1967.
150. Manickam, M., Ramanathan, M., Jahromi, M. A., Chansouria, J. P., and Ray, A. B., Antihyperglycemic activity of phenolics from *Pterocarpus marsupium*, *J. Nat. Prod.*, 60 (6), 609–610, 1997.
151. Indian Council of Medical Research (ICMR), Collaborating Centres, New Delhi, Flexible dose open trial of Vijayasar in cases of newly-diagnosed non-insulin-dependent diabetes mellitus, *Indian J. Med. Res.*, 108, 24–29, 1998. Indian Council of Medical Research (ICMR), Erratum in *Indian J. Med. Res.*, 108, 253, 1998.
152. Onderoglu, S., Sozer, S., Erbil, K. M., Ortac, R., and Lermioglu, F., The evaluation of long-term effects of cinnamon bark and olive leaf on toxicity induced by streptozotocin administration to rats, *J. Pharm. Pharmacol.*, 51 (11), 1305–1312, 1999.
153. Qin, B., Nagasaki, M., Ren, M., Bajotto, G., Oshida, Y., and Sato, Y., Cinnamon extract prevents the insulin resistance induced by a high-fructose diet, *Horm. Metab. Res.*, 36 (2), 119–125, 2004.
154. Imparl-Radosevich, J., Deas, S., Polansky, M. M., Baedke, D. A., Ingebritsen, T. S., Anderson, R. A., and Graves, D. J., Regulation of PTP-1 and insulin receptor kinase by fractions from cinnamon implications for cinnamon regulation of insulin signaling, *Horm. Res.*, 50 (3), 177–182, 1998.
155. Khan, A., Safdar, M., Ali Khan, M. M., Khattak, K. N., and Anderson, R. A., Cinnamon improves glucose and lipids of people with type 2 diabetes, *Diabetes Care*, 26 (12), 3215–3218, 2003.

156. Nir, Y., Potasman, I., Stermer, E., Tabak, M., and Neeman, I., Controlled trial of the effect of cinnamon extract on Helicobacter pylori, *Helicobacter*, 5 (2), 94–97, 2000.

157. Arun, N. and Nalini, N., Efficacy of turmeric on blood sugar and polyol pathway in diabetic albino rats, *Plant Foods Hum. Nutr.*, 57 (1), 41–52, 2002.

158. Srinivasan, M., Effect of curcumin on blood sugar as seen in a diabetic subject, *Indian J. Med. Sci.*, 26, 269–270, 1972.

159. Bagchi, D. and Preuss, H. G., Eds., *Phytopharmaceuticals in Cancer Chemoprevention*, CRC Press, Boca Raton, pp. 349–388, 2005 chap. 23

160. Gururaj, A. E., Belakavadi, M., Venkatesh, D. A., Marme, D., and Salimath, B. P., Molecular mechanisms of antiangiogenic effect of curcumin, *Biochem. Biophys. Res. Commun.*, 297 (4), 934–942, 2002.

161. Arbiser, J. L., Klauber, N., Rohan, R., van Leeuwen, R., Huang, M. T., Fisher, C., Flynn, E., and Byers, H. R., Curcumin is an in vivo inhibitor of angiogenesis, *Mol. Med.*, 4 (6), 376–383, 1998.

162. Mani, H., Sidhu, G. S., Kumari, R., Gaddipati, J. P., Seth, P., and Maheshwari, R. K., Curcumin differentially regulates TGF-beta1, its receptors and nitric oxide synthase during impaired wound healing, *Biofactors*, 16 (1–2), 29–43, 2002.

163. Sidhu, G. S., Mani, H., Gaddipati, J. P., Singh, A. K., Seth, P., Banaudha, K. K., Patnaik, G. K., and Maheshwari, R. K., Curcumin enhances wound healing in streptozotocin-induced diabetic rats and genetically diabetic mice, *Wound Repair Regen.*, 7 (5), 362–374, 1999.

164. Suresh Babu, P. and Srinivasan, K., Amelioration of renal lesions associated with diabetes by dietary curcumin in streptozotocin diabetic rats, *Mol. Cell Biochem.*, 181 (1–2), 87–96, 1998.

165. Rajasekaran, S., Sivagnanam, K., Ravi, K., and Subramanian, S., Hypoglycemic effect of *Aloe vera* gel on streptozotocin-induced diabetes in experimental rats, *J. Med. Food*, 7 (1), 61–66, 2004.

166. Okyar, A., Can, A., Akev, N., Baktir, G., and Satlupinar, N., Effect of *Aloe vera* leaves on blood glucose level in type I and type II diabetic rat models, *Phytother. Res.*, 15, 157–161, 2001.

167. Rajasekaran, S. and Subramanian, S., Modulatory effects of *Aloe vera* leaf gel extract on oxidative stress in rats treated with streptozotocin, *J. Pharm. Pharmacol.*, 57 (2), 241–246, 2005.

168. Can, A., Akev, N., Ozsoy, N., Bolkent, S., Arda, B. P., Yanardag, R., and Okyar, A., Effect of *Aloe vera* leaf gel and pulp extracts on the liver in type-II diabetic rat models, *Biol. Pharm. Bull.*, 27 (5), 694–698, 2004.

169. Davis, R. H., Leitner, M. G., and Russo, J. M., *Aloe vera*. A natural approach for treating wounds, edema, and pain in diabetes, *J. Am. Podiatr. Med. Assoc.*, 78 (2), 60–68, 1988.

170. Lee, M. J., Lee, O. H., Yoon, S. H., Lee, S. K., Chung, M. H., Park, Y. I., Sung, C. K., Choi, J. S., and Kim, K. W., In vitro angiogenic activity of *Aloe vera* gel on calf pulmonary artery endothelial (CPAE) cells, *Arch. Pharm. Res.*, 21 (3), 260–265, 1998.

171. Choi, S., Kim, K. W., Choi, J. S., Han, S. T., Park, Y. I., Lee, S. K., Kim, J. S., and Chung, M. H., Angiogenic activity of beta-sitosterol in the ischaemia/reperfusion-damaged brain of Mongolian gerbil, *Planta. Med.*, 68 (4), 330–335, 2002.

172. Ajabnoor, M. A., Effect of aloes on blood glucose levels in normal and alloxan diabetic mice, *J. Ethnopharmacol.*, 28, 215–220, 1990.

173. Yongchaiyudha, S., Rungpitarangsi, V., Bunyapraphatsara, N., and Chokechaijaroenpom, O., Antidiabetic activity of *Aloe vera L.* juice I. Clinical trial in new cases of diabetes mellitus, *Phytomedicine*, 3, 241–243, 1996.

174. Bunyapraphatsara, N., Yongchaiyudha, S., Rungpitarangsi, V., and Chokechaijaroenpom, O., Antidiabetic activity of *Aloe vera L.* juice II. Clinical trial in diabetes mellitus patients in combination with glibenclamide, *Phytomedicine*, 3, 245–248, 1996.

175. Singh, S. S., Pandey, S. C., Srivastava, S., Gupta, V. S., Patro, B., and Ghosh, A. C., Chemistry and medicinal properties of *Tinospora cordifolia* (Guduchi), *Indian J. Pharmacol.*, 35, 83–91, 2003.

176. Wadood, N., Wadood, A., and Shah, S. A., Effect of *Tinospora cordifolia* on blood glucose and total lipid levels of normal and alloxan-diabetic rabbits, *Planta. Med.*, 58 (2), 131–136, 1992.

177. Stanely, P., Prince, M., and Menon, V. P., Hypoglycemic and other related actions of *Tinospora cordifolia* roots in alloxan-induced diabetic rats, *J. Ethnopharmacol.*, 70 (1), 9–15, 2000.

178. Gupta, S. S., Varma, S. C. L., Garg, V. P., and Rai, M., Antidiabetic effect of *Tinospora cordifolia* effect on fasting blood sugar levels and glucose tolerance and adrenaline induced hyperglycemia, *Indian J. Exp. Biol.*, 55, 733–745, 1967.

179. Rege, N., Bapat, R. D., Koti, R., Desai, N. K., and Dahanukar, S., Immunotherapy with *Tinospora cordifolia*: a new lead in the management of obstructive jaundice, *Indian J. Gastroenterol.*, 12 (1), 5–8, 1993.

180. Prince, P. S., Padmanabhan, M., and Menon, V. P., Restoration of antioxidant defense by ethanolic *Tinospora cordifolia* root extract in alloxan-induced diabetic liver and kidney, *Phytother. Res.*, 18 (9), 785–787, 2004.

181. Leyon, P. V. and Kuttan, G., Effect of *Tinospora cordifolia* on the cytokine profile of angiogenesis-induced animals, *Int. Immunopharmacol.*, 4 (13), 1569–1575, 2004.

182. Gray, A. M. and Flatt, P. R., Nature's own pharmacy: the diabetes perspective, *Proc. Nutr. Soc.*, 56, 507–517, 1997.

183. Oubre, A. Y., Carlson, T. J., King, S. R., and Reaven, G. M., From plant to patient: an ethanobotanical approach to the identification of new drugs for the treatment of NIDDM, *Diabetologia*, 40, 614–617, 1997.

184. Pandey, V. N., Rajagopalan, S. S., and Chowdhary, D. P., An effective Ayurvedic hypoglycemic formulation, *J. Res. Ayur. Siddha*, XVI, 1–14, 1995.

185. Jamal, G. A. and Carmichael, H., The effect of gamma-linolenic acid on human diabetic peripheral neuropathy: a double-blind placebo-controlled trial, *Diab. Med.*, 7, 319–323, 1990.

186. Ziegler, D., Schatz, H., Conrad, F., Gries, F. A., Ulrich, H., and Reichel, G., Effects of treatment with the antioxidant alpha-lipoic acid on cardiac autonomic neuropathy in NIDDM patients. A 4-month randomized controlled multicenter trial (DEKAN Study), *Diabetes Care*, 20, 369–373, 1997.

187. Sima, A. A., Calvani, M., Mehra, M., and Amato, A., Acetyl-L-carnitine improves pain, nerve regeneration, and vibratory perception in patients with chronic diabetic neuropathy: an analysis of two randomized placebo-controlled trials, *Diabetes Care*, 28 (1), 89–94, 2005.

188. The Capsaicin Study Group, Treatment of painful diabetic neuropathy with topical capsaicin. A multicenter, double-blind, vehicle-controlled study, *Arch. Int. Med.*, 151, 2225–2229, 1991.

189. Caroline, D., Traditional plant treatments for diabetes mellitus: pharmaceutical foods, *Br. J. Nutr.*, 80, 5–6, 1998.

190. Anderson, R. A., Chromium, glucose intolerance and diabetes, *J. Am. Coll. Nutr.*, 17, 548–555, 1998.

22 Obesity and Angiogenesis

Debasis Bagchi, Shirley Zafra-Stone,
Chandan K. Sen, and Manashi Bagchi

CONTENTS

22.1 SUMMARY

Obesity, a worldwide epidemic, is associated with an increased risk for diverse cardiovascular diseases including atherosclerosis, thrombosis, hypertension, hyperinsulinemia, insulin resistance, type 2 diabetes, and cancer. Obesity is characterized by an excess of fat mass caused by adipocyte hypertrophy and hyperplasia. Angiogenesis, the formation of new blood vessels from pre-existing vessels, is one rate-limiting factor that supports the development of adipose tissue during obesity.

Angiogenesis plays a major role in the regaining of body weight during nutritional health recovery. During weight regain after a food-restricted diet, angiogenic activity in the adipose tissue sharply increases because of induction of vascular endothelial growth factorvascular endothelial growth factor vascular endothelial growth factor (VEGF). Leptin, a circulating hormone encoded by the Ob gene, maintains body weight homeostasis by regulating food intake and energy expenditure. Recently, leptin has been found to generate a growth signal involving a tyrosine kinase-dependent intracellular pathway and promote angiogenic processes via activation of the leptin receptor (Ob-R) in endothelial cells (EC). On the other hand, inhibitors of angiogenesis significantly reduce adipose tissue mass by cutting off the blood supply to the expanding mass of tissue. Current developments demonstrate potent anti-angiogenic properties of novel natural nutrients such as edible berries. Such anti-angiogenic nutrients offer a significant opportunity to stifle adipose tissue growth and manage obesity.

22.2   INTRODUCTION

Obesity is now a worldwide epidemic [1–3]. The prevalence of overweight and obesity in the U.S.A. and other industrialized and developing countries is increasing exponentially [1–6]. Approximately half a billion of the world's population is now considered overweight [Body mass index (BMI) 25–29 kg/m^2] or obese (BMI > 30 kg/m^2) [1–3,9]. Genetic predisposition, inadequate energy expenditure, increased caloric intake, environmental and social factors, and a sedentary lifestyle represent major contributors to obesity [1–9].

Angiogenesis is defined as unwanted growth or the formation of new blood vessels by capillary sprouting from preexisting vessels [10–13]. It has been well demonstrated that development of metabolic syndrome including obesity and cancer are linked with angiogenesis, and the cause of these pathological conditions can be related [10–13].

Adipogenesis, the development of adipose tissue, is intrinsically associated with angiogenesis. In other words, the expanding adipose tissue in adults represents one of the few sites of active angiogenesis [10]. The development and maturation of adipose tissue vascularity are critical to the function of adipose tissue [10,11]. Adipocytes induce EC-specific mitogens and angiogenic factors such as VEGF, monobutyrin, and leptin [10]. These angiogenic factors promote EC proliferation and differentiate within the fat tissue [11]. The interaction between adipocytes and endothelium is presumed to be involved in the development and maintenance of adipose tissue [10–12]. EC secrete basic fibroblast growth factors, platelet-derived EC growth factor, and other soluble matrix-bound preadipocyte mitogens, which stimulate preadipocyte replication and promote differentiation [10–12]. Newly formed adipose tissue depends on continued angiogenesis for further growth [10]. The interaction between microvascular EC and preadipocytes promotes the expansion of adipose tissue [11,13,14].

22.3   ADIPOKINES, ADIPOKINOMES, INFLAMMATION, AND WHITE ADIPOSE TISSUE

Conventionally, white adipose tissue or fat was known to provide thermal and mechanical insulation to the body [12,15]. In addition, it serves as an energy storage depot for mammals [11]. The white adipose tissue has now emerged as a dynamic, multifunctional endocrine organ involved in a wide range of physiological and metabolic processes [10–12]. The tissue acts as a central reservoir for lipid storage, including deposition and release of fatty acids [10–13,15].

White adipocytes secrete several major hormones, most importantly leptin and adiponectin, and a diverse range of signaling mediators, collectively known as "adipokines" or "adipocytokines" [12–15]. The total number of adipokines is now well over fifty, and their main functional categories include appetite and energy balance, immunity, insulin sensitivity, angiogenesis, homeostasis, lipid metabolism, and blood pressure, [11,15]. Furthermore, there is a growing list of adipokines involved in inflammation, Tumor necrosis factor alpha and Interleukins (TNFα, IL-1β, IL-6, IL-8, IL-10, transforming growth factor-β, nerve growth factor) and the acute-phase response (plasminogen activator inhibitor-1, haptoglobin, serum amyloid A) [10–14]. Production of these inflammatory adipokines by adipose tissue is increased in obesity.

Elevated circulating levels of several acute-phase proteins and inflammatory adipokines in the obese has led to the hypothesis that obesity is characterized by a state of chronic low-grade inflammation, and that this links directly to the symptoms of insulin resistance and metabolic syndrome [10,12]. The term "adipokinome" refers to proteins involved in lipid metabolism, insulin sensitivity, the alternative complement system, vascular homeostasis, regulation of blood pressure and energy balance, and angiogenesis [11,12,15]. It is, however, unclear as to the extent to which adipose tissue contributes quantitatively to the elevated circulating levels of these factors in obesity, and whether there is a generalized or local state of inflammation [12,15]. Overall, increased production of inflammatory cytokines and acute-phase proteins by the adipose tissue in obesity relates primarily to localized events within the expanding fat depots [11]. Hypoxia, consequent to the rapid expansion of white fat mass ahead of angiogenesis, could be a key trigger for the inflammation-related events in white adipose tissue in obesity [12–15]. Adipose-derived inflammatory cytokines may lead to a direct stimulation of angiogenic factors, such as VEGF and leptin, as well as through the activation of the transcription factor hypoxia inducible factor-1 (HIF-1), the key and central controller of the cellular response to hypoxia [13,15].

22.4   ROLE OF LEPTIN, THE PRODUCT OF OB GENE, IN ANGIOGENESIS

Leptin, an adipocyte-derived 16,000 molecular weight cytokine-like hormone, plays a pivotal role in the regulation of body weight and obesity, as well as on the control of food consumption, sympathetic nervous system activation, thermogenesis, and proinflammatory immune responses [16]. Because angiogenesis and adipogenesis are integrally linked during embryonic development of adipose tissue mass, Bouloumie et al. [16] evaluated the regulatory role of leptin in the growth of the vasculature using human umbilical venous endothelial cells (HUVEC) and porcine aortic EC. Leptin, via activation of the endothelial Ob-R, generates a growth signal involving a tyrosine kinase-dependent intracellular pathway and promotes angiogenic processes [11,16]. This leptin-mediated stimulation of angiogenesis is viewed as a prime event in the development of obesity [10,11,16]. Leptin also contributes to the modulation of growth under physiological and pathophysiological conditions in other tissues [10–12,16].

22.5   ANTI-ANGIOGENIC THERAPEUTIC STRATEGIES TO REGULATE OBESITY

Rupnick et al. [17] explored the co-relationship between adipose tissue growth, angiogenesis, and leptin levels in Ob/Ob mice, which rapidly accumulate adipose tissue and develop spontaneous obesity because of a lack of functional leptin. The authors also examined EC proliferation, apoptotic cell death in adipose tissue, and metabolic consequences in these mice following treatment with anti-angiogenic inhibitor TNP-470, a synthetic analog of fumagillin, which is also an inhibitor of EC proliferation in vitro [17]. TNP-470-treated Ob/Ob mice dose-dependently lost body weight and adipose tissue weight as compared to the control animals. Obesity-associated hyperglycemia, serum glucose levels, and appetite level were also reduced. TNP-470-treated mice also had the lowest average lean mass [17]. A number of angiogenesis inhibitors including endostatin, Bay12-9566 (a matrix metalloproteinase inhibitor), thalidomide, angiostatin, and TNP-470, were tested on Ob/Ob mice, and all treatment groups gained less or lost weight compared to the control animals [17]. Mice from other obesity models also yielded similar results. The authors also treated 3T3-L1 preadipocytes with different concentrations of TNP-470 to evaluate whether suppression of preadipocyte proliferation contributes to this effect, and the results demonstrate that both endothelial-mediated mechanism and preadipocyte suppression are involved in this pathophysiology [17]. Cell proliferation and apoptosis were assessed in epididymal fat sections from TNP-470-treated and control Ob/Ob mice; TNP-470 decreased EC proliferation and increased apoptosis. Furthermore, the basal metabolic rate (as measured by oxygen consumption, VO2) was also increased in TNP-470-treated Ob/Ob mice [17].

Brakenhielm et al. [18] evaluated the effect of TNP-470 on high-fat diet-fed C57Bl/6 wt and Ob/Ob mice. Systemic administration of TNP-470 prevented obesity in high caloric diet-fed C57Bl/6 wt mice as well as in genetically leptin-deficient Ob/Ob mice. This containment of obesity in mice was accompanied by a reduction of vascularity in the adipose tissue [18]. TNP-470 selectively affected the growth of adipose tissue as measured by the ratio between total fat and lean body mass, as well as decreased serum levels of low-density lipoprotein cholesterol [18]. Furthermore, TNP-470 increased insulin sensitivity, as demonstrated by reduced insulin levels, suggesting the therapeutic role of a potent angiogenesis inhibitor in the prevention of type 2 diabetes [10,18]. The investigators demonstrated that adipose tissue growth is dependent on angiogenesis, and that potent anti-angiogenic compounds may serve as novel therapeutic agents for the prevention of obesity and symptoms of metabolic syndrome [18].

Voros et al. [19] studied the intricate aspects of the development of vasculature and mRNA expression of 17 pro and anti-angiogenic factors during adipose tissue development in nutritionally-induced or genetically-predisposed murine obesity models. Male C57Bl/6 mice were maintained either on a standard food diet (SFD) or a high-fat diet (HFD), and male Ob/Ob mice were maintained on a SFD over a period of 15 weeks. Ob/Ob mice and male C57BL/6 mice on a HFD had significantly larger subcutaneous and gonadal fat pads, accompanied by significantly higher blood content, increased total blood vessel volume, and a higher number of proliferating cells [19]. Fat pad growth was accompanied by increased vascularization. mRNA and protein levels of angiopoietin-1 were downregulated, whereas those of thrombospondin-1 were upregulated in the developing adipose tissue in both obesity models. Angiopoietin-1 mRNA levels correlated negatively with adipose tissue weight in the early phase of nutritionally-induced obesity, as well as in genetically-predisposed obesity [19]. Placental growth factor and angiopoietin-2 expression were increased in subcutaneous adipose tissue of Ob/Ob mice, and thrombospondin-2 expression was increased in both subcutaneous and gonadal fat pads. No changes were observed in the mRNA levels of VEGF-A isoforms VEGF-B, VEGF-C, VEGF receptor-1, -2, and -3, and neuropilin-1 [19].

Neels et al. [13] examined the angiogenic process using the 3T3-F442A model of adipose tissue development. These investigators subcutaneously implanted 3T3-F442A preadipocytes into athymic Balb/c nude mice to study the neovascularization of developing adipose tissue [13]. These cells developed into highly vascularized fat pads over the next 14–21 days, and these fat pads were morphologically similar to normal subcutaneous adipose tissue. Histological studies demonstrated that a new microvasculature comes up as early as five days after cell implantation, and real-time quantitative real time-polymerase chain reaction (RT–PCR) analyses exhibited that the expression of EC markers and adipogenesis markers had simultaneous and parallel increases during fat pad development [13]. Thus, neovasculature originates by sprouting from larger host-derived blood vessels that run parallel to peripheral nerves, and the endothelial progenitor cells play a minor role in this process [13].

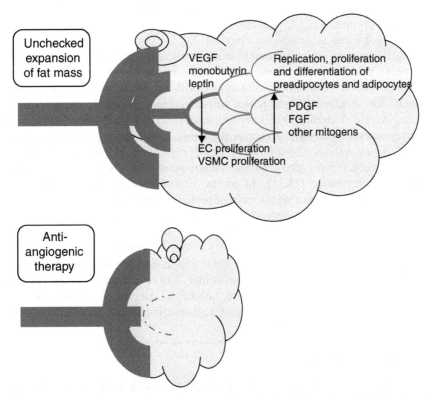

FIGURE 22.1 This demonstrates the key factors involved in adipose tissue development and the influence of anti-angiogenic therapy. Abbreviations: VSMC, Vascular smooth muscle cells; PDGF, Platelet-derived growth factor; FGF, Fibroblast growth factor.

Figure 22.1 demonstrates the key factors involved in adipose tissue development and the influence of anti-angiogenic therapy.

22.6   ANTI-ANGIOGENIC NUTRIENTS

Tumor vascularization is a key step in the development of solid tumors and the vast majority of pharmaceutical activity surrounding angiogenesis relates to the development of therapeutic strategies to destroy existing tumor vasculature and to prevent neovascularization [11–14]. Thus, anti-angiogenic approaches to treating cancer represent a prime area in vascular tumor biology [11–14]. Catechin derivatives exhibited novel anti-angiogenic properties in three different in vitro bioassays with concentrations ranging from 1.56 to 100 μM [20,21]. Epigalloca-techin-3-gallate (EGCG) exhibited the most potent anti-angiogenic activity in all the three assays among all the catechins tested. EGCG inhibited the binding of VEGF, a key regulator of tumor angiogenesis, to HUVEC in a concentration-dependent manner [20,21]. Resveratrol, a novel phytoalexin found in grapes, red wine, and diverse functional foods, and quercetin exhibited anti-angiogenic properties in a concentration-dependent manner (6–100 μM) [22]. Pure flavonoids including rutin, kaempferol, ferrulic, acid, genistein, fisetin, and luteolin also exhibited novel anti-angiogenic properties [22,23]. These studies demonstrate that functional foods are effective in limiting angiogenis in vivo [20–23].

OptiBerry, a blend of six edible berry extracts including blueberry, bilberry, cranberry, elderberry, raspberry seeds, and strawberry extracts was recently developed in our laboratories [23,24]. OptiBerry possessed high antioxidant capacity, low cytotoxicity and potent anti-angiogenic properties. With respect to anti-angiogenic properties, OptiBerry was superior to individual berry anthocyanins, underscoring the synergistic power of the blend. OptiBerry significantly inhibited both H_2O_2- as well as TNFα-induced VEGF expression by human keratinocytes [23,24]. Matrigel assay using human microvascular EC showed that OptiBerry impaired angiogenesis [23,24]. In an in vivo model, OptiBerry significantly inhibited basal monocyte chemoattractant protein-1 (MCP-1) and inducible nuclear transcription factor-kappa beta (NF-κβ) transcription [24]. Endothelioma cells pretreated with OptiBerry showed a diminished ability to form hemangioma and markedly decreased growth of neoplastic vasculature by more than 50% in 129 P3 mice [24]. These studies highlight the novel anti-angiogenic and anti-carcinogenic potential of select natural food constituents including catechin derivatives, curcumin (comprising about 40% of the spice turmeric), selected flavonoids including quercetin, rutin, kaempferol, ferrulic acid, genistein, fisetin, and luteolin, resveratrol, and berry anthocyanins [20–24].

22.7   CONCLUSIONS

Overweight and obesity pose epidemic threats to human health, especially in the Western world. Healthy diet, proper nutrition and exercise represent the fundamental pre-requisites to fight such threat. Emergent studies reveal that the growth of adipose tissue may be regulated by inhibiting the development of vasculature that feeds the fat mass. Anti-Angiogenic therapeutics show significant promise to manage undesired expansion of fat mass. Anti-Angiogenic nutrients provide a safe and potentially effective strategy to fight obesity. Several polyphenolic antioxidants, especially berry anthocyanins, possess potent anti-angiogenic properties and should be considered as counter measures for obesity in humans.

REFERENCES

1. Wyatt, S. B., Winters, K. P., and Dubbert, P. M., Overweight and obesity: prevalence, consequences, causes of a growing public health problem, *Am. J. Med. Sci.*, 331, 166, 2006.
2. Brownell, K. D., Obesity: understanding and treating a serious, prevalent, and refractory disorder, *J. Consult. Clin. Psychol.*, 50, 820, 1992.
3. Bray, G. A., Obesity, In *Present Knowledge in Nutrition*, Ziegler, E. E. and Filer, L. J. Jr. Eds., ILSI Press, Washington, DC, pp. 19–32, 1996.
4. Guterman, L., Obesity problem swells worldwide, *The Chronicle of Higher Education*, A18, March 8, 2002.
5. U.S. Department of Health and Human Services, *The Surgeon General's Call to Action to Prevent and Decrease Over Weight and Obesity 2001*, U.S. General Printing Office, Washington, DC, 2001.
6. World Health Organization, Report: Controlling the global obesity epidemic. Available at www.who.int/nut/obs.htm. Updated August 15, 2003.
7. Campbell, I., The obesity epidemic: can we turn the tide?, *Heart*, 89, 35, 2003.
8. Critser, G., *Fat Land: How Americans Became the Fattest People in the World*, Houghton Mifflin, Boston, MA, p. 232, 2003. Critser, G., *N. Engl. J. Med.* 348, 2161, 2003.
9. Jequier, E., Pathways to obesity, *Int. J. Obes. Relat. Metab. Disord.*, 260, S12, 2002.
10. Wasserman, F., The development of adipose tissue, In *Handbook of Physiology*, Renold, A. E. and Cahill, G. F. Eds., Vol. 5, American Society of Physiology, Washington, DC, pp. 87–100, 1965.
11. Hausman, G. J. and Richardson, R. L., Adipose tissue angiogenesis, *J. Anim. Sci.*, 82, 925, 2004.
12. Dallabrida, S. M., Zurakowski, D., Shih, S. C., Smith, L. E., Folkman, J., Moulton, K. S., and Rupnick, M. A., Adipose tissue growth and regression are regulated by angiopoietin-1, *Biochem. Biophys. Res. Commun.*, 311, 563, 2003.
13. Neels, J. G., Thinnes, T., and Loskutoff, D. J., Angiogenesis in an in vivo model of adipose tissue development, *FASEB J.*, 18, 983, 2004.
14. Mandrup, S. and Lane, M. D., Regulating adipogenesis, *J. Biol. Chem.*, 272, 5367, 1997.
15. Trayhurn, P. and Wood, I. S., Adipokines: Inflammation and pleiotropic role of white adipose tissue, *Br. J. Nutr.*, 92, 347, 2004.
16. Bouloumie, A., Drexler, H. C. A., Lafontan, M., and Busse, R., Leptin, the product of Ob gene, promotes angiogenesis, *Circ. Res.*, 83, 1059, 1998.
17. Rupnick, M. A., Panigrahy, D., Zhang, C.-Y., Dallabrida, S. M., Lowell, B. B., Langer, R., and Folkman, M. J., Adipose tissue mass can be regulated through the vasculature, *Proc. Natl Acad. Sci. (U.S.A.)*, 99, 10730, 2002.
18. Brakenhielm, E., Cao, R., Gao, B., Angelin, B., Cannon, B., Parini, P., and Cao, Y., Angiogenesis inhibitor, TNP-470, prevents diet-induced and genetic obesity in mice, *Circ. Res.*, 94, 1579, 2004.

19. Voros, G., Maquoi, E., Demeulemeester, D., Clerx, N., Collen, D., and Lijnen, H. R., Modulation of angiogenesis during adipose tissue development in murine models of obesity, *Endocrinology*, 146, 4545, 2005.

20. Annabi, B., Lee, Y. T., Martel, C., Pilorget, A., Bahary, J. P., and Beliveau, R., Radiation induced-tubulogenesis in endothelial cells is antagonized by the anti-angiogenic properties of green tea polyphenol ($-$) epigallocatechin-3-gallate, *Cancer Biol. Ther.*, 2, 642, 2003.

21. Sartippour, M. R., Shao, Z. M., Heber, D., Beatty, P., Zhang, L., Liu, C., Ellis, L., Liu, W., Go, V. L., and Brooks, M. N., Green tea inhibits vascular endothelial growth factor (VEGF) induction in human breast cancer cells, *J. Nutr.*, 132, 2307, 2002.

22. Oak, M. H., El Bedoui, J., and Schini-Kerth, V.B, Antiangiogenic properties of natural polyphenols from red wine and green tea, *J. Nutr. Biochem.*, 16, 1, 2005.

23. Roy, S., Khanna, S., Alessio, H. M., Bagchi, D., Bagchi, M., and Sen, C. K., Antiangiogenic properties of berry nutrients, *Free Radic. Res.*, 36, 1023, 2002.

24. Bagchi, D., Sen, C. K., Bagchi, M., and Atalay, M., Antiangiogenic, antioxidant and anti-carcinogenic properties of a novel anthocyanin-rich berry extract formula, *Biochemistry*, 69, 75, 2004.

Part IV

Angiogenesis, Functional, and Medicinal Foods

23 Screening Functional Foods as Inhibitors of Angiogenesis Biomarkers

Jack N. Losso

CONTENTS

23.1   INTRODUCTION

The formation of new blood vessels in humans occurs via two distinct pathways, vasculogenesis and angiogenesis. Vasculogenesis is restricted to embryonic development and involves the differentiation of progenitor cells into endothelial and their subsequent organization into vascular structures. Angiogenesis, the formation of new blood vessels from preexisting capillaries, occurs both during embryogenesis and vasculogenesis. Physiological angiogenesis is a precisely regulated process that is associated with the growth, development, reproduction, and repair of wounded tissues in the body.

Physiological angiogenesis occurs as brief burst of capillary blood vessel growth that lasts days or weeks [1]. Once formed, the vascular network is a stable system that regenerates slowly. In a healthy body, endothelial cells undergo about 0.01% turnover every three years. Pregnancy, the development of a child in the mother's womb, the healing of a wound, exercise, and the menstrual cycle of a woman are typical examples of physiological angiogenesis where new blood vessels are formed in response to signals generated during tissue growth and repair. The angiogenic process "switches off" after achieving the appropriate biological end point (e.g., healed wound or menstrual cycle).

Pathological angiogenesis, however, continues unabated and paves the way to aberrant tissue growth. Tumor cells have an absolute requirement for a persistent supply of new blood vessels to nourish their growth and facilitate metastasis. Angiogenesis is the hallmark of several degenerative diseases [2] (Figure 23.1). Biochemical and genetic data have supported the hypothesis of angiogenesis-dependent tumor and angiogenic-dependent non-neoplastic diseases [3–8].

The prevascular phase is quiescent [9]. Once the switch to an angiogenic phenotype has occurred, avascular tumor cells can then acquire their own blood

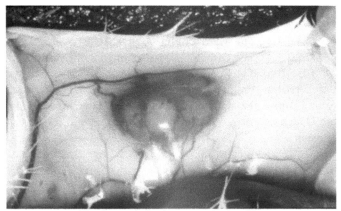

Blood vessels feed a tumour in the skin

FIGURE 23.1 In vivo angiogenesis showing blood vessels feeding a tumor in a mouse skin. (From Drs. M. Achen and S. Stacker, Ludwig Institute, Angiogenesis Laboratory, Melbourne, Australia. With permission.)

supply to support a rapid rate of growth. The angiogenic phenotype of tumors is regulated by local balance between pro-angiogenic and anti-angiogenic factors.

The "angiogenic switch" has been recognized for its role in the progression and complications of several angiogenic diseases such as cancer, diabetic retinopathy, nephropathy and angiopathy, atherosclerosis, HIV, AIDS, bowl disease, multiple sclerosis, chronic inflammation, and arthritis [2,9,10]. Activated endothelial cells that form the inner lining of the vascular tube are the primary targets of angiogenesis inhibitors, mostly because they are, by far, the best understood components of the neovasculature [2,11]. However, pericytes that form the outer sheath around the endothelium participate early and play an important role in the process of angiogenesis sprouting by providing survival signals for endothelial cells and promoting capillary maturation [11,12].

There are several proteins known to activate new blood vessel growth. These include vascular endothelial growth factor (VEGF), the acidic and basic fibroblast growth factors (aFGF and bFGF), matrix metalloproteinases (MMPs), angiogenin, epidermal growth factor (EGF), placental growth factor (PlGF), platelet derived growth factor (PDGF), urokinase-type plasminogen activator, angiopoietin, integrins, and tumor necrosis factor alpha (TNF-α) to name a few. The expression of proangiogenic proteins such as VEGF and others can be induced by several factors including defects in the mitochondria, hsp90, hypoxia, activation of oncogenes, suppression of tumor suppressor genes, enzyme receptors, adhesion molecules, metals, oncogenes, and proteasome [13–15]. However, most of these biomarkers are molecules secreted by the tumor cells to stimulate endothelial cells; they have been identified in the invasive and proliferative stages of angiogenic diseases and few of these factors are proving clinically useful for healthy life survival [16]. Tumors, themselves, secrete anti-angiogenic factors such as angiostatin and endostatin that are proteolytic fragments of the large precursor proteins plasminogen and collagen XVIII, respectively [17]. Angiostatin is a naturally occurring polypeptide derived from the cleavage of a larger protein, plasminogen. Endostatin is a polypeptide derived from the enzymatic hydrolysis of type XVIII collagen. Other endogenous anti-angiogenic factors include uroglobin and pigment epithelium derived factor (PEDF) which are found either in the tumor environment or in the body.

23.2 DIETARY STRATEGIES AGAINST PATHOLOGICAL ANGIOGENESIS

There are several advantages to anti-angiogenic therapy over conventional therapies. Because angiogenesis in healthy individuals is normally restricted and only about 0.01% of adult endothelial cells undergo the process of angiogenesis at any given time, the side effects of inhibitors of angiogenesis on normal tissues should be negligible [9]. These inhibitors affect tumor growth indirectly by targeting the steps involved in the process leading to the formation of new blood vessels and are mainly cytostatic [18].

A variety of inhibitors of angiogenesis and growth factor-targeted agents have been evaluated, but have so far shown little promise [19]. However, targeting endothelial cells with anti-angiogenic molecules in combination with standard chemotherapy has recently shown promise in the treatment of different types of advanced cancer, more so than the use of anti-angiogenic molecules alone [20,21]. Potential factors involved in the slow progress in anti-angiogenic therapy include among others (1) genetic alterations such as p53, (2) redundancy of tumor cell secreted growth factors, (3) the impact of the tumor micro-environment such as hsp90, hypoxia, and alteration of the HIF-1α pathway, and (4) the advanced stage of the disease, while preclinical data are often collected using animals with relatively small tumors [13,22].

Most chronic angiogenic diseases are complex and involve multistep enzymatic activities that may span over years or decades. And when the symptoms of the angiogenic disease erupt, it is often too late to reverse the cascade of enzymatic reactions to the initial step of homeostasis. Because drugs are designed to cure and not prevent diseases, sustained surveillance of the body against chronic degenerative diseases requires sustained use of non-drug bioactive compounds with the ability to delay or prevent the onset or progression of chronic diseases.

The symptoms or manifestation of angiogenic diseases are often associated with multiple biomarkers for which a single inhibitor may not exist. Food is a daily necessity and consumed over the lifespan of an individual. Functional foods that can prevent the early steps of angiogenic diseases offer an opportunity to prevent disease onset and progression. Anti-angiogenic functional foods will likely be most effective when used as an integral part of a disease prevention program that would also involve identifying proangiogenic factors in the food we eat. In this chapter, we discuss recent advances and limitations in screening methods for functional food inhibitors of angiogenic disease biomarkers.

23.3 SCREENING ANTI-ANGIOGENIC BIOACTIVE COMPOUNDS

Activated endothelial cells are the primary targets of angiogenic inhibitors [9]. Angiogenesis inhibition is not designed to directly attack the cancer cells, rather the targets are normal processes controlled by *normal* cells (such as the cells that form blood vessels). Cancer cells have a high rate of mutation that often renders chemotherapy ineffective. Anti-angiogenic compounds are immune to the high mutation rate characteristics of cancer cells. It is also recognized that pericytes participate early and play an important role in the process of angiogenic sprouting by providing survival signals for endothelial cells and promoting capillary maturation [12]. As a result, pericytes may also be good targets of angiogenesis inhibitors. Whereas VEGF is critical for endothelial cells, PDGF is required for pericyte viability [11]. It takes two—endothelial cells and pericytes—to make blood vessels [23]. Like endothelial cells, pericytes may be involved in diseases such as diabetic microangiopathy, tissue fibrosis, cancer, atherosclerosis, and Alzheimer's disease [23,24].

Screening functional foods for anti-angiogenic and vascular targeting activity will help identify and validate their appropriate molecular targets and optimize their dose scheduling and combination strategies, if necessary. Angiogenesis screening assays include in vitro and in vivo methods. Each model has limitations, but it is essential to select a model that is easy, reproducible, quantitative, cost-effective, permits rapid analysis, and is the most appropriate for the objectives to be investigated. Among the in vitro assays, the endothelial cell assays that look at proliferation, migration, and tube formation as indicators of the ability of a compound to inhibit angiogenesis are the simplest.

23.3.1 IN VITRO ANGIOGENESIS ASSAYS

23.3.1.1 Cell Proliferation Assay

In vitro models of angiogenesis generally use endothelial cells from large vessels such as the human umbilical vein endothelial cells (HUVEC) and the bovine aortic endothelial (BAE) cells, or from small vessels such as the human dermal microvascular endothelial cells (HuDMECs). The assays use markers of cell division to assess cell proliferation and are divided into classes: those that analyze cell cycle kinetics and those that measure net cell number. Most of these assays are rapid and quantifiable.

Assays that analyze cell cycle kinetics include (1) DNA synthesis (by thymidine incorporation using cell labeling reagent 5-bromo-2′-deoxyuridine (BrdU)), (2) staining cells with a DNA-binding dye (followed by colorimetric evaluation of bound dye, (3) ATP levels measurement (using luciferase), tetrazolium salt 3-(4,5-dimethylthiazol-2-yl)-2,5-diphenyltetrazolium bromide) (MTT), and DNA-binding followed by flow cytometric determination of DNA content of individual cells (e.g., using propidium iodide) which may provide the endothelial cells' response to an inhibitor or stimulator of proliferation [25]. Assays that measure net cell number use the haemocytometer or the electronic Coulter counter. However, all endothelial cells are not alike and demonstrate structural, organ-associated, and phenotype differences [25,26].

In the existing established vasculature, endothelial cells are quiescent and have a turnover of about 0.01% every three years. But endothelial cells used in the laboratory are proliferative, cannot be used without passing them, and may lose attributes found in vivo. The assay may be rapid and quantifiable, but may not represent the complexity of physiological interactions that occur in vivo. It is suggested that interactions of bioactive compounds with endothelial be studied using endothelial cells from more than one source, followed by in vivo evaluation of the bioactive compound.

23.3.1.2 Cell Migration

The principle of the assay is to create a gradient of a chemotactic agent such as VEGF or bFGF and allow cells to migrate through a membrane towards the chemotactic agent. The assay uses the Boyden Chamber which consists of an upper and lower wells separated by a membrane filter to evaluate cell migration. The membrane, which is typically 3–12 μm (pores of lower size, e.g., 0.3 μm can be used for lympocytes) serves as a barrier to discriminate migrating cells from non-migrating cells and is coated with a matrix protein such as fibronectin or collagen at the bottom outside. An angiogenesis stimulator such as VEGF or FGF is added to the medium in the lower chamber. The cells will cross the membrane filter and move down toward the chemotactic stimulus in the lower chamber. Cell movement in the lower chamber can be recorded after 6–24 h. An inhibitor of angiogenesis when incubated with cancer cells in the upper chamber will block endothelial cell chemotaxis. Improved variations of the Boyden Chambers are commercially available as kits (Chemicon, Becton Dickinson) some of which study cell migration by chemotaxis (movement of cells in a direction controlled by exposure to a gradient of a diffusible chemical such as growth factors), haptotaxis (control of cell migration from a less adherent to a more adherent surface), or chemokinesis (response of a cell to a chemical that causes the cell to make some kind of change in its movement by speeding it up, slowing it down, or changing its direction). These new improvements offer some advantages such as the use of colorimetric or fluorescence reading over traditional chambers.

23.3.1.3 Tube or Cord Formation

Tube or cord formation measures the ability of endothelial cells to rapidly form two- or three-dimensional structures. Tube formation is a multistep process involving cell adhesion, migration, differentiation, and growth. The assay involves monitoring the induction or inhibition of tube formation by exogenous signals. Endothelial cells seeded onto plastic culture dishes precoated with adhesive proteins such as type I collagen or fibrin, or loaded on top of a Matrigel, rapidly form capillary-like structures. Tube formation on collagen or fibrin is reasonably faithful to the in vivo situation, and the formation of tight junctions can be confirmed by electron microscopy [26]. Matrigel, a solid gel of basement proteins prepared from the Engelbreth Holm-Swarm (EHS) mouse tumor, whose primary component is laminin, can evoke endothelial cell tube formation within 24 h. The extent of tube assembly in various endothelial cells, e.g., HUVEC, bovine capillary endothelial cells (BCEC), bovine aortic endothelial cells (BAEC), and many others can be monitored within days or weeks [27]. Three-dimensional configuration closely mimics events such as cell proliferation, migration, and tubulogenesis that occur during angiogenesis in vivo and also allow the measuring of the bioavailability of angiogenic or anti-angiogenic factors (Figure 23.2).

23.3.1.4 The Aortic Ring Assay

The aortic ring assay is a widely used, sensitive, and reproducible ex vivo model that closely bridges the gap between the in vitro and in vivo models. Isolated mouse or rat aorta (explant) are cultured in a collagen or fibrin matrix such as Matrigel in the presence of growth factors. The explant is monitored, over 7–14 days, for outgrowth of endothelial (and other) cells in the absence or presence of a test compound. Quantification by pixel counts can be achieved by using endothelium-selective reagents such as fluorescein-labeled BSL-I or by measuring the length and abundance of vessel-like extensions from the explant [26]. The aortic ring assay simulates in vivo angiogenesis because the endothelial cells are not in a proliferative state at the time of explantation and have not been preselected by passaging. However, because angiogenesis is mostly a vascular event, the aortic ring assay is less than an ideal choice.

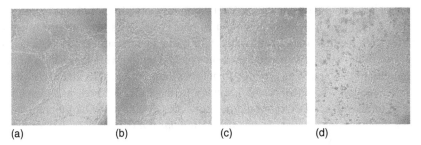

(a)   (b)   (c)   (d)

FIGURE 23.2 Inhibition of VEGF-induced endothelial cell tube formation by auraptene. Endothelial cells with VEGF (10 ng/ml) (a), in the presence of 100 nM of auraptene (b), 300 nM of auraptene (c) and 500 nM of auraptene (d).

23.3.2 HIGH THROUGHPUT TECHNOLOGICAL APPROACH TO IDENTIFY BIOMARKERS OF ANGIOGENESIS

23.3.2.1 cDNA Microarray and RT–PCR

cDNA microarray technology allows comparison of the mRNA expression of large numbers of genes between normal and diseased tissues and can quantitatively detect altered gene expression associated with the pathology or the altered biology of angiogenic disease cells. To circumvent the technical problems inherent in the cDNA microarray technique, and because of its performance, accuracy, sensitivity, wide dynamic range, and high throughput capacity for nucleic acid quantification, quantitative real-time RT–PCR is being applied to genes involved in pathways (cell cycling, signal transduction, apoptosis, angiogenesis, and metastasis). Gene expression profiling, coupled with quantitative RT–PCR, Northern blot, and Western blot analyses of genes in samples from diseased tissues, in comparison to genes in samples from normal tissues, has been used to characterize the gene expression patterns that are induced by anti-angiogenic bioactive compounds [28–30]. Gene array and proteomic techniques were used to investigate the response of colorectal cancer cells to butyrate [31]. Smirnov et al. [32] used RT–PCR to identify a set of biomarker genes that can be used to monitor circulating endothelial cells in the peripheral blood of cancer patients. Microarray technologies such as the Affymetrix Genechip® platform facilitate rapid measurement of the expression levels of thousands of transcripts in a single experiment and allow comparison of expression patterns across many samples.

23.3.2.2 Digital Differential Display

Digital Differential Display (DDD) is a data mining tool from the Cancer Genome Anatomy Project (CGAP) used to predict solid tumor- and organ-specific genes from the expressed sequence tag (EST) database. Differential display was developed by integrating PCR and DNA sequencing by gel electrophoresis. In essence, DDD works by systematically amplifying the $3'$ termini of eukaryotic mRNA by reverse transcription-PCR, using one of the three anchored oligo-dT primers (that is, the run of Ts ending with a C, G, or A) in combination with a set of short primers of arbitrary sequences. The DDD method uses pools of cDNAs from groups of cDNA libraries from specific tissues and is useful for initial identification of cDNAs that are expressed preferentially in particular tissues or frequently overexpressed in tumors. The quality of the input data can significantly influence the results and usefulness of the DDD method. To enhance the accuracy of the DDD method, a large number (several thousand) of cDNAs should be sequenced from each cDNA library that is used in the DDD comparison. More information on DDD can be obtained by visiting http://www.ncbi.nlm.nih.gov/ncicgap/ (accessed August 10, 2006).

23.3.2.3 In Silico Northern Blotting

In silico data filtering allows the use of a nucleotide array for the analysis of well-controlled in vitro biological models and the correlation in silico with less precise human disease tissue data sets [33]. Starting with thousands of genes from in vitro models of endothelial cells, Gerritsen et al. [33] identified more than 50% of genes that had never been previously associated with angiogenesis, and with further filtering and using an in silico comparison to tissue data sets the authors identified smaller subsets of genes that would be potential targets for angiogenesis.

Human kallikreins are a cluster of 15 serine protease genes located in the chromosomal band 19q13.4, a non-randomly rearranged region in many solid tumors, including pancreatic cancer. Yousef et al. [34] used the SAGE and EST databases to perform an in silico analysis of kallikrein gene expression in normal and cancerous pancreatic and colon tissues and cell lines using virtual Northern blotting (VNB), DDD, and X-profiler. One kallikrein, the KLK6 was found to be expressed in five out of six cancer libraries and its level of expression was higher in cancer cells than normal cells. Three kallikrein genes, KLK6, 8, and 10 are overexpressed in colon cancer compared to the normal colon, while one kallikrein, KLK1, is downregulated.

23.3.2.4 Pitfalls of In Vitro Angiogenesis Assays

Understanding the process of angiogenesis and its regulation requires identification and use of models for the study of vascular development and regulation that are truly representative of the in vivo complexity of the vasculature. Whereas in vitro models facilitate the ascertainment of specific information regarding endothelial cell biology, they do not recreate the microenvironment of an intact organism [26]. These methods cannot be considered conclusive, and active compounds should be tested in other complex assays such as the in vivo assays, angiogenic dependent tumor growth, and metastasis [35].

23.3.3 In Vivo Angiogenesis Assays

The chick chorioallantoic membrane (CAM), corneal assay, sponge implant models such as the Matrigel plug assay, and conventional tumor models are classical in vivo angiogenesis assays.

23.3.3.1   The Chick Chorioallantoic Membrane Assay

The CAM assay represents an in vivo vascular network where a chorioallantoic membrane surrounds the developing chicken embryo. As the embryo develops, it becomes vascularized. New vessels form from the existing ones. An inhibitor of angiogenesis will cause a regression of the vessels. The CAM assay is quite simple and inexpensive to perform. The major drawback of the CAM assay is that the presence of an already well-developed vascular network at the time of the assay makes it difficult to distinguish between newly formed capillaries and previously existing ones. However, the disappearance of capillaries can easily be seen (Figure 23.3). Another major drawback is the need to analyze several fertilized eggs before drawing an acceptable conclusion.

23.3.3.2   The Corneal Pocket Assay

The cornea is an avascular tissue and for that reason the assay is considered as one of the best angiogenesis assays. Originally developed in rabbit cornea, the assay can nowadays be performed in mice. Angiogenic or anti-angiogenic activity of a bioactive compound is determined as follows. The bioactive compound is encapsulated with a proangiogenic substance (i.e., bFGF or VEGF) in polymer pellets such as Hydron [36]. A single pellet is implanted and advanced toward the temporal corneal limbus within 0.5–1.0 mm. A veterinary ophthalmic antibiotic ointment is applied to the eye to prevent infection. Angiogenesis or antiangiogenesis is measured after 4–6 days by examining new vessels penetrating from the limbus into the corneal stroma. The cornea is perfused with India ink and analyzed by computer image analysis, or angiogenesis can be monitored using a stereomicroscope in the case of mice. The method is very reliable as it allows monitoring angiogenesis in an avascular system. Although reliable, this method is not practical for screening because the surgical procedure is technically demanding. Few animals can undergo operation on a daily basis, the pocket for introducing the test compound

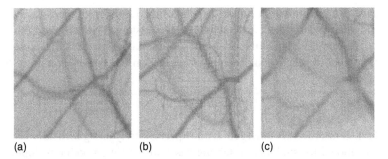

(a)   (b)   (c)

FIGURE 23.3 Thymoquinone inhibits angiogenesis in vivo. Photographs of developing CAMs incubated with 50 µM of thymoquinone showing control (a), day 1 (b) and day 2 (c).

is restricted, inflammation of the tissues is difficult to avoid, and the technique is more expensive than the CAM assay [25,26,37].

23.3.3.3 The Matrigel Plug Assay

Matrigel is a complex mixture of proteins and growth factors such as laminin, collagen type IV, heparan sulfate, fibrin, EGF, TGF-b, PDGF, and IGF-1, and proteolytic enzymes (plasminogen, tPA, MMPs) that occur normally in Engelbreth Holm-Swarm (EHS) mouse tumors [37,38]. The mixture is a liquid at 4°C and forms a gel at 37°C. In this assay, the Matrigel (500 µl) on ice is mixed with either a growth factor alone or cancer cells at 10^6 cells, or with a growth factor (or cancer cells) and the potential anti- or pro-angiogenic compound to be tested. The mixture is injected s.c. into a 4–8 week-old mouse at sites near the abdominal midline or into the flank. The number of injections allows each animal to simultaneously receive a positive control plug (growth factor and heparin), a negative control plug (heparin plus buffer), and a plug containing the treatment to be tested (growth factor such as FGF-2, heparin, and anti-or proangiogenic compound to be tested). All treatments are tested in triplicates. Animals are sacrificed 5–7 days after injection. The Matrigel plugs are recovered and analyzed for hemoglobin levels using Drabkin's solution (Sigma) according to the manufacturer's instructions for the quantitation of blood vessel formation. The concentration of hemoglobin is calculated from a known amount of hemoglobin assayed in parallel [39]. The advantage of Matrigel is that it provides a natural environment that mimics angiogenesis in vivo. However, it is expensive.

23.3.3.4 The Orthotopic Models in Nude Mice

Surgical orthotopic implantation (SOI) was conceived as a clinically-accurate rodent model that allows human tumors of all the major types of human cancer to reproduce clinically-like tumor growth and metastasis in the transplanted nu/nu mice of both sexes [40]. The model allows accurate expression of the clinical features of human cancer in nude mice. SOI technology involves microsurgical transplantation of tumor fragments orthotopically [40]. SOI was developed in part because subcutaneous implantation of tumors precludes the tumor from metastasizing. The anti-angiogenic and antimetastatic activities of genistein, TNP-470, and Batimastat have been demonstrated in SOI models [41–44]. A major drawback to the use of SOI is the high level of surgical skill required to perform the transplantation of tumors into the mouse.

23.3.3.5 The Zebrafish

The zebrafish has recently emerged as an experimentally and genetically accessible alternative model for studying vertebrate vascular formation both developmentally and in adult tissues. There are several advantages to using zebrafish in angiogenesis research. Zebrafish have a short breeding cycle (every 1–2 weeks), and a short generation time (3 months) that gives a high number of progenies (100's/clutch). Zebrafish embryos are transparent vertebrates that develop outside the mother's body, and the fish change from an egg to a well-developed transparent embryo within 24 h. Thus, the entire process, as well as the blood vessels, can be watched under a stereomicroscope, with the fish embryos developing organs that are similar to those in humans, such as central nervous system, pancreas, and thymus—and they quickly form blood vessels and beating hearts. The 2–3 day assay time with zebrafish is similar to the time required for a cell-based assay, i.e., low cost, whereas a mouse tumor assay requires at least 4 weeks. Additionally, a single dosing regimen is used compared to a daily dosing for a mouse. And finally, zebrafish serve as an in vivo toxicological model, allowing the use of enough vertebrates to obtain good replicates and easily test statistically significant numbers of animals. The inhibition of angiogenesis in zebrafish mimics the effects observed in mice and humans because genes such as VEGF, Flk-1, Tie-1, and Tie-2 have been identified in zebrafish and have been shown to have the same function as in mammals [35,37,45]. Methods for visualizing blood vessels in the zebrafish include (1) whole mount in situ hybridization, (2) detection of endogenous alkaline phosphatase activity, (3) microangiography, and (4) transgenic zebrafish with fluorescent blood vessels [46–49].

23.3.3.6 Pitfalls of In Vivo Assays

These assays are complex, expensive, may require surgical skills, in the case of the orthotopic mouse model, and often are unsuitable for the quantitative screening of a large number of compounds. However, the activity of a potential anti-angiogenic or proangiogenic compound can be confirmed.

23.4 SCREENING ANTI-ANGIOGENIC OR PRO-ANGIOGENIC FUNCTIONAL FOODS

Several naturally occurring bioactive compounds such as carbohydrates, proteins, peptides, amino acids, lipids, vitamins, and phenolics have been identified as inhibitors of angiogenesis (Table 23.1). Since angiogenesis occurs at high levels during fetal development, the menstrual cycle, and in wound healing, anti-angiogenic compounds might be expected to interfere with these processes, but they should not harm most normal dividing cells.

Diet contributes to more than 40% of chronic angiogenic diseases. Most major angiogenic chronic diseases are genetic, diet, or environmental-related diseases. Whereas efforts are being made to identify foods with inhibitory activity against excessive angiogenesis, similar efforts must be made to identify proangiogenic compounds in foods. Screening tests for anti-angiogenic functional foods should reveal the stimulatory effect of pro-angiogenic foods. The identification of stimulators of angiogenesis should also be part of the effort of industry to remove these undesirable compounds found in foods.

TABLE 23.1

Mechanism of Inhibition of Angiogenesis by Functional Foods

Food Source	Bioactive Compound	Mechanisms of Inhibition	References
Green tea, Black tea	Catechins (EGCG, ECG)	↓EC proliferation, Pro-MMP-2, MMP-1, MMP-2, MMP-9, elastase, urokinase	[50,51]
Oolong tea	Catechins (theasinensin A, theaflavin derivatives)	Apoptosis, ↑caspase, Nitric oxide synthase	[52]
Turmeric	Curcumin	Inhibits MMPs, ↓EC proliferation. Apoptosis, LOX, ODC, COX-2	[53]
Soybean	Isoflavones (genistein, daidzein)	↓EC proliferation, block uPA, EC migration and proliferation+↓EGF,	[54,55]
Grapes, nuts	Resveratrol	Inhibits EC activation, FGF, VEGF, wound healing, MMP-9, COX-1&-2 TNF-alpha, MAP kinase	[56,57]
Citrus	apigenin	↓TNF, IL-1, E-selectin expression, inflammation, HIF-1 allpha, MMPs	[58,59]
Grape seeds	Proanthocyanidins + resveratrol	↑VEGF expression (↑wound healing), Neutrophil proteases	[60]
Tangerine peel oils	Polymethoxylated flavones (Tangeretin + Nobiletin)	Apoptosis, Inhibition of cell proliferation, ↓ TNF, restoration of E-cadherin, MMP-2, MMP-9, COX-2	[61–63]
Citrus, Onion	Quercetin	Inhibits inflammation, Lipoxygenase, aromatase, PKC	[54,64,65]
Citrus	Naringenin	COX-2 , iNOS, Inhibits IkB kinase	[66]
Citrus	Nobiletin	↓invasion, migration, and adhesion ↓gene expression of MMP-9, MMP-2	[63,67]
Beer hop	Humulone	↓COX-2,↓EC proliferation ↓VEGF production	[68]
Walnut, raspberries, strawberries	Ellagic acid	Activates p53, Triggers p21, ↓regulate IGI-II, ODC, protein kinase	[69,70]
Strawberries, red raspberries	Gallic acid	Trypsin, chymotrypsin,	[71,72]
Fenugreek	Saponins	MMPs, TNF-alpha, proliferation, apoptosis	[73]
Almond hull	Betulinic acid	COX-2 , Inhibits bFGF	[74]
Licorice	Aqueous exytract, isoliquiritin	↓tube formation, VEGF	[75,76]
Ginseng	Ginsenosides	↓COX-2, enhance wound healing, ↓VEGF and bFGF-induced angiogenesis, MMP-2 and MMP-9	[77]
Fruits & vegetables	Perillyl alcohol, geranial, farnesol	↓geranyltransferase, farnesyltransferase, ↓cell proliferation, PKC,	[78]
Citrus	Pectin	Binds tumor surface lectins, ↓NFkB	[79,80]

(continued)

TABLE 23.1 *(Continued)*

Food Source	Bioactive Compound	Mechanisms of Inhibition	References
Cereals	Vitamin B3	↓HIF-1 alpha	[81]
Mammalian milk	Vitamin D3	↓EC proliferation, VEGF, EGF, MMPs, ↑TGF	[82,83]
Oilseeds, milk	Vitamin E	↓VEGF,TGF-aplha, IL-8,VE-cadherin	[84,85]
Buttermilk, resistant starch	Butyrate	↓EC proliferation, HIF-1 alpha, histone deacetylase, MMPs, ↑apoptosis,	[86,87]
Pepper	Capsaicin	↓COX-2, wound proteases, VEGF	[88,89]
Milk fat, beef, lamb	Conjugated linoleic acid	↓VEGF, bFGF-induced proliferation	[90,91]

Table 23.1 lists anti-angiogenic functional foods, the angiogenesis-associated enzymes they inhibit, the mechanism(s) of inhibition, good sources of the anti-angiogenic compound, and references for more information.

Abbreviations: ↑, enhances; ↓, downregulates; ??, questionable; β-Lg f, beta-lactoglobulin fragment; COX, cyclooxygenase; COX-1, cyclooxygenase-1; COX-2, cyclooxygenase-2; LOX, lipoxygenase; MMP, metalloproteinase; MMP-1, metalloproteinase-1; MMP-2, metalloproteinase-2; MMP-9, metalloproteinase-9; MAP, mitogen activated protein; PKC, protein kinase; ODC, ornithine decarboxylase; EC, endothelial cell; IL-1, interleukin-1; IL-8, interleukin-8; TGF, transforming growth factor; ACE, angiotensin converting enzyme; VE-cadherin, vascular endothelial-cadherin; EGF, epidermal growth factor; HIF-1, hypoxia-inducible factor-1 alpha; TNF, tumor necrosis factor; uPA, urokinase plasminogen; IGF-II, insulin growth factor-II.

23.5 FUTURE DIRECTIONS

The last few years have been a period of great progress in identifying bioactive compounds in foods that may have inhibitory activities against excessive angiogenesis. Although more scientific and epidemiological research are needed before making specific recommendations for anti-angiogenic functional foods, there is ample evidence to suggest that dietary bioactive compounds with inhibitory activity against angiogenesis should be recommended, and this often translates into consuming a diet rich in a variety of fruits, legumes, nuts, vegetables, and whole grains.

23.6 CONCLUSION

Biomarkers can be major targets for the early detection of chronic degenerative diseases and identification of disease risk. The relevance of angiogenesis inhibition to functional foods development is without doubt unquestioned. Our understanding of the basic biology of many angiogenic diseases is far from complete. We cannot predict with full certainty which functional food will work efficiently and which individuals will benefit fully from these bioactive compounds. But, as the field of molecular functional foods develops, and as our understanding of the molecular biology of disease, biochemistry, and pharmacology progresses, there will be a corresponding increase in the number of novel disease-associated biomarkers that will require the attention of food scientists and nutritionists.

REFERENCES

1. Folkman, J. and Kalluri, R., Tumor angiogenesis, In *Cancer Medicine*, Kufe, D., Pollock, R., Welchselbaum, R., Bast, R., Gansler, T., Holland, J., and Frei, E. Eds., BC Decker, Hamilton, CA, pp. 161–194, 2003.
2. Folkman, J., Angiogenesis in cancer, vascular, rheumatoid and other disease, *Nat. Med.*, 1 (1), 27–31, 1995.
3. Carmeliet, P., Angiogenesis in life, disease and medicine, *Nature*, 438 (7070), 932–936, 2005.
4. Yamamoto, Y., Maeshima, Y., Kitayama, H., Kitamura, S., Takazawa, Y., Sugiyama, H., Yamasaki, Y., and Makino, H., Tumstatin peptide, an inhibitor of angiogenesis, prevents glomerular hypertrophy in the early stage of diabetic nephropathy, *Diabetes*, 53 (7), 1831–1840, 2004.
5. Ferrara, N., Gerber, H. P., and LeCouter, J., The biology of VEGF and its receptors, *Nat. Med.*, 9 (6), 669–676, 2003.
6. Viloria-Petit, A., Miquerol, L., Yu, J. L., Gertsenstein, M., Sheehan, C., May, L., Henkin, J. et al., Contrasting effects of VEGF gene disruption in embryonic stem cell-derived versus oncogene-induced tumors, *EMBO J.*, 22 (16), 4091–4102, 2003.
7. Inoue, K., Chikazawa, M., Fukata, S., Yoshikawa, C., and Shuin, T., Frequent administration of angiogenesis inhibitor TNP-470 (AGM-1470) at an optimal biological dose inhibits tumor growth and metastasis of metastatic human transitional cell carcinoma in the urinary bladder, *Clin. Cancer Res.*, 8 (7), 2389–2398, 2002.
8. Carmeliet, P. and Jain, R. K., Angiogenesis in cancer and other diseases, *Nature*, 407 (6801), 249–257, 2000.
9. Folkman, J., Role of angiogenesis in tumor growth and metastasis, *Semin. Oncol.*, 29 (6), 15–18, 2002 (suppl. 16).
10. Hanahan, D. and Folkman, J., Patterns and emerging mechanisms of the angiogenic switch during tumorigenesis, *Cell*, 86 (3), 353–364, 1996.
11. Ozerdem, U. and Stallcup, W. B., Pathological angiogenesis is reduced by targeting pericytes via the NG2 proteoglycan, *Angiogenesis*, 7 (3), 269–276, 2004.
12. Wilkinson-Berka, J. L., Babic, S., De Gooyer, T., Stitt, A. W., Jaworski, K., Ong, L. G., Kelly, D. J., and Gilbert, R. E., Inhibition of platelet-derived growth factor promotes pericyte loss and angiogenesis in ischemic retinopathy, *Am. J. Pathol.*, 164 (4), 1263–1273, 2004.
13. Neckers, L., Chaperoning oncogenes: Hsp90 as a target of geldanamycin, *Handb. Exp. Pharmacol.*, 172, 259–277, 2006.
14. Papetti, M. and Herman, I. M., Mechanisms of normal and tumor-derived angiogenesis, *Am. J. Physiol. Cell Physiol.*, 282 (5), C947–C970, 2002.
15. Semenza, G. L., HIF-1 and tumor progression: pathophysiology and therapeutics, *Trends Mol. Med.*, 8 (4), S62–S67, 2002 (suppl.).
16. Coradini, D. and Daidone, M. G., Biomolecular prognostic factors in breast cancer, *Curr. Opin. Obstet. Gynecol.*, 16 (1), 49–55, 2004.
17. Folkman, J., Angiogenesis, *Annu. Rev. Med.*, 57, 1–18, 2006.
18. Gasparini, G., The rationale and future potential of angiogenesis inhibitors in neoplasia, *Drugs*, 58 (1), 17–38, 1999.

19. Modlin, I. M., Latich, I., Kidd, M., Zikusoka, M., and Eick, G., Therapeutic options for gastrointestinal carcinoids, *Clin. Gastroenterol. Hepatol.*, 4 (5), 526–547, 2006.

20. Gille, J., Anti-Angiogenic cancer therapies get their act together: current developments and future prospects of growth factor- and growth factor receptor-targeted approaches, *Exp. Dermatol.*, 15 (3), 175–186, 2006.

21. Wakelee, H. A. and Schiller, J. H., Targeting angiogenesis with vascular endothelial growth factor receptor small-molecule inhibitors: novel agents with potential in lung cancer, *Clin. Lung Cancer*, 7 (suppl. 1), S31–S38, 2005.

22. Shannon, A. M., Bouchier-Hayes, D. J., Condron, C. M., and Toomey, D., Tumor hypoxia, chemotherapeutic resistance and hypoxia-related therapies, *Rev. Cancer Treat.*, 29 (4), 297–307, 2003.

23. Gerhardt, H. and Betsholtz, C., Endothelial–pericyte interactions in angiogenesis, *Cell Tissue Res.*, 314 (1), 15–23, 2003.

24. Wilkinson-Berka, J. L., Diabetes and retinal vascular disorders: role of the reninangiotensin system, *Expert Rev. Mol. Med.*, 23 (6(15)), 1–18, 2004.

25. Staton, C. A., Stribbling, S. M., Tazzyman, S., Hughes, R., Brown, N. J., and Lewis, C. E., Current methods for assaying angiogenesis in vitro and in vivo, *Int. J. Exp. Pathol.*, 85 (5), 233–248, 2004.

26. Auerbach, R., Lewis, R., Shinners, B., Kubai, L., and Akhtar, N., Angiogenesis assays: a critical overview, *Clin. Chem.*, 49 (1), 32–40, 2003.

27. Vailhe, B., Vittet, D., and Feige, J. J., In vitro models of vasculogenesis and angiogenesis, *Lab. Invest.*, 81 (4), 439–452, 2001.

28. Zhang, W., Chuang, Y. J., Jin, T., Swanson, R., Xiong, Y., Leung, L., and Olson, S. T., Anti-Angiogenic antithrombin induces global changes in the gene expression profile of endothelial cells, *Cancer Res.*, 66 (10), 5047–5055, 2006.

29. Loboda, A., Jazwa, A., Jozkowicz, A., Molema, G., and Dulak, J., Angiogenic transcriptome of human microvascular endothelial cells: effect of hypoxia, modulation by atorvastatin, *Vascul. Pharmacol.*, 44 (4), 206–214, 2006.

30. Ambra, R., Rimbach, G., de Pascual, T. S., Fuchs, D., Wenzel, U., Daniel, H., and Virgili, F., Genistein affects the expression of genes involved in blood pressure regulation and angiogenesis in primary human endothelial cells, *Nutr. Metab. Cardiovasc. Dis.*, 16 (1), 35–43, 2006.

31. Williams, E. A., Coxhead, J. M., and Mathers, J. C., Anti-cancer effects of butyrate. Use of microarray technology to investigate mechanisms, *Proc. Nutr. Soc.*, 62 (1), 107–115, 2003.

32. Smirnov, D. A., Foulk, B. W., Doyle, G. V., Connelly, M. C., Terstappen, L. W., and O'Hara, S. M., Global gene expression profiling of circulating endothelial cells in patients with metastatic carcinomas, *Cancer Res.*, 66 (6), 2918–2922, 2006.

33. Gerritsen, M. E., Soriano, R., Yang, S., Ingle, G., Zlot, C., Toy, K., Winer, J., Draksharapu, A., Peale, F., Wu, T. D. et al., In silico data filtering to identify new angiogenesis targets from a large in vitro gene profiling data set, *Physiol. Genomics*, 10 (1), 13–20, 2002.

34. Yousef, G. M., Borgono, C. A., Popalis, C., Yacoub, G. M., Polymeris, M. E., Soosaipillai, A., and Diamandis, E. P., In silico analysis of kallikrein gene expression in pancreatic and colon cancers, *Anticancer Res.*, 24 (1), 43–51, 2004.

35. Taraboletti, G. and Giavazzi, R., Modelling approaches for angiogenesis, *Eur. J. Cancer*, 40 (6), 881–889, 2004.

36. Presta, M., Rusnati, M., Belleri, M., Morbidelli, L., Ziche, M., and Ribatti, D., Purine analogue 6-methylmercaptopurine riboside inhibits early and late phases of the angiogenesis process, *Cancer Res.*, 59 (10), 2417–2424, 1999.

37. Losso, J. N. and Bawadi, H. A., Anti-Angiogenic proteins, peptides, and amino acids, In *Nutraceutical Proteins and Peptides in Health and Disease*, Mine, Y. and Shahidi, F. Eds., CRC Press, Boca Raton, FL, pp. 191–216, 2005.

38. Lawley, T. J. and Kubota, Y., Induction of morphologic differentiation of endothelial cells in culture, *J. Invest. Dermatol.*, 93 (2), 59S–61S, 1989.

39. Passaniti, A., Taylor, R. M., Pili, R., Guo, Y., Long, P. V., Haney, J. A., Pauly, R. R., Grant, D. S., and Martin, G. R., A simple, quantitative method for assessing angiogenesis and anti-angiogenic agents using reconstituted basement membrane, heparin, and fibroblast growth factor, *Lab. Investig.*, 67 (4), 519–528, 1992.

40. Hoffman, R. M., Orthotopic metastatic (MetaMouse) models for discovery and development of novel chemotherapy, *Methods Mol. Med.*, 111, 297–322, 2005.

41. Wang, Y., Raffoul, J. J., Che, M., Doerge, D. R., Joiner, M. C., Kucuk, O., Sarkar, F. H., and Hillman, G. G., Prostate cancer treatment is enhanced by genistein in vitro and in vivo in a syngeneic orthotopic tumor model, *Radiat. Res.*, 166 (1 Pt 1), 73–80, 2006.

42. Singh, A. V., Franke, A. A., Blackburn, G. L., and Zhou, J. R., Soy phytochemicals prevent orthotopic growth and metastasis of bladder cancer in mice by alterations of cancer cell proliferation and apoptosis and tumor angiogenesis, *Cancer Res.*, 66 (3), 1851–1858, 2006.

43. Zervox, E. E., Franz, M. G., Salhab, K. F., Shafii, A. E., Menendez, J., Gower, W. R., and Rosemurgy, A. S., Matrix metalloproteinase inhibition improves survival in an orthotopic model of human pancreatic cancer, *J. Gastrointest. Surg.*, 4 (6), 614–619, 2000.

44. Kanai, T., Konno, H., Tanaka, T., Matsumoto, K., Baba, M., Nakamura, S., and Baba, S., Effect of angiogenesis inhibitor TNP-470 on the progression of human gastric cancer xenotransplanted into nude mice, *Int. J. Cancer*, 71 (5), 838–841, 1997.

45. Cross, L. M., Cook, M. A., Lin, S., Chen, J. N., and Rubinstein, A. L., Rapid analysis of angiogenesis drugs in a live fluorescent zebrafish assay, *Arterioscler. Thromb. Vasc. Biol.*, 23 (5), 911–912, 2003.

46. Fouquet, B., Weinstein, B. M., Serluca, F. C., and Fishman, M. C., Vessel patterning in the embryo of the zebrafish: guidance by notochord, *Dev. Biol.*, 183 (1), 37–48, 1997.

47. Serbedzija, G. N., Flynn, E., and Willett, C. E., Zebrafish angiogenesis: a new model for drug screening, *Angiogenesis*, 3 (4), 353–359, 1999.

48. Liao, W., Bisgrove, B. W., Sawyer, H., Hug, B., Bell, B., Peters, K., Grunwald, D. J., and Stainier, D. Y., The zebrafish gene cloche acts upstream of a flk-1 homologue to regulate endothelial cell differentiation, *Development*, 124, 381–389, 1997.

49. Lawson, N. D. and Weinstein, B. M., In vivo imaging of embryonic vascular development using transgenic zebrafish, *Dev. Biol.*, 248 (2), 307–318, 2002.

50. Garbisa, S., Sartor, L., Biggin, S., Salvato, B., Benelli, R., and Albini, A., Tumor gelatinases and invasion inhibited by the green tea flavonol epigallocatechin-3-gallate, *Cancer*, 91 (4), 822–832, 2001.

51. Lamy, S., Gingras, D., and Beliveau, R., Green tea catechins inhibit vascular endothelial growth factor receptor phosphorylation, *Cancer Res.*, 62 (2), 381–385, 2002.

52. Pan, M. H., Liang, Y. C., Lin-Shiau, S. Y., Zhu, N. Q., Ho, C. T., and Lin, J. K., Induction of apoptosis by the oolong tea polyphenol theasinensin A through cytochrome c release and activation of caspase-9 and caspase-3 in human U937 cells, *J. Agric. Food Chem.*, 48 (12), 6337–6346, 2000.

53. Shishodia, S., Sethi, G., and Aggarwal, B. B., Curcumin: getting back to the roots, *Ann. N.Y. Acad. Sci.*, 1056, 206–217, 2005.

54. Lambert, J. D., Hong, J., Yang, G. Y., Liao, J., and Yang, C. S., Inhibition of carcinogenesis by polyphenols: evidence from laboratory investigations, *Am. J. Clin. Nutr.*, 81 (1), 284S–291S, 2005 (suppl.).

55. Su, S. J., Yeh, T. M., Chuang, W. J., Ho, C. L., Chang, K. L., Cheng, H. L., Liu, H. S., Cheng, H. L., Hsu, P. Y., and Chow, N. H., The novel targets for antiangiogenesis of genistein on human cancer cells, *Biochem. Pharmacol.*, 69 (2), 307–318, 2005.

56. Brakenhielm, E., Cao, R., and Cao, Y., Supression of angiogenesis, tumor growth, and wound healing by resveratrol, a natural compound in red wine and grapes, *FASEB J.*, 15 (10), 1798–1800, 2001.

57. Delmas, D., Lancon, A., Colin, D., Jannin, B., and Latruffe, N., Resveratrol as a chemopreventive agent: a promising molecule for fighting cancer, *Curr. Drug Targets*, 7 (4), 423–442, 2006.

58. Osada, M., Imaoka, S., and Funae, Y., Apigenin suppresses the expression of VEGF, an important factor for angiogenesis, in endothelial cells via degradation of HIF-1alpha protein, *FEBS Lett.*, 575 (1–3), 59–63, 2004.

59. Kim, M. H., Flavonoids inhibit VEGF/bFGF-induced angiogenesis in vitro by inhibiting the matrix-degrading proteases, *J. Cell Biochem.*, 89 (3), 529–538, 2003.

60. Khanna, S., Roy, S., Bagchi, D., Bagchi, M., and Sen, C. K., Upregulation of oxidant-induced VEGF expression in cultured keratinocytes by a grape seed proanthocyanidin extract, *Free Radical Biol. Med.*, 31 (1), 38–42, 2001.

61. Datla, K. P., Christidou, M., Widmer, W. W., Rooprai, H. K., and Dexter, D. T., Tissue distribution and neuroprotective effects of citrus flavonoid tangeretin in a rat model of Parkinson's disease, *Neuroreport*, 12 (17), 3871–3875, 2001.

62. Hirano, T., Abe, K., Gotoh, M., and Oka, K., Citrus flavone tangeretin inhibits leukaemic HL-60 cell growth partially through induction of apoptosis with less cytotoxicity on normal lymphocytes, *Br. J. Cancer*, 72 (6), 1380–1388, 1995.

63. Manthey, J. A. and Guthrie, N., Antiproliferative activities of citrus flavonoids against six human cancer cell lines, *J. Agric. Food Chem.*, 50 (21), 5837–5843, 2002.

64. Kim, J. D., Liu, L., Guo, W., and Meydani, M., Chemical structure of flavonols in relation to modulation of angiogenesis and immune-endothelial cell adhesion, *J. Nutr. Biochem.*, 17 (3), 165–176, 2006.

65. Jackson, J. K., Higo, T., Hunter, W. L., and Burt, H. M., The antioxidants curcumin and quercetin inhibit inflammatory processes associated with arthritis, *Inflamm. Res.*, 55 (4), 168–175, 2006.

66. Raso, G. M., Meli, R., Di Carlo, G., Pacilio, M., and Di Carlo, R., Inhibition of inducible nitric oxide synthase and cyclooxygenase-2 expression by flavonoids in macrophage J774A, *Life Sci.*, 68 (8), 921–931, 2001.

67. Ishiwa, J., Sato, T., Mimaki, Y., Sashida, Y., Yano, M., and Ito, A., A citrus flavonoid, nobiletin, suppresses production and gene expression of matrix metalloproteinase 9/gelatinase B in rabbit synovial fibroblasts, *J. Rheumatol.*, 27 (1), 20–25, 2000.

68. Shimamura, M., Hazato, T., Ashino, H., Yamamoto, Y., Iwasaki, E., Tobe, H., Yamamoto, K., and Yamamoto, S., Inhibition of angiogenesis by humulone, a bitter acid from beer hop, *Biochem. Biophys. Res. Commun.*, 289 (1), 220–224, 2001.

69. Labrecque, L., Lamy, S., Chapus, A., Mihoubi, S., Durocher, Y., Cass, B., Bojanowski, M. W., Gingras, D., and Beliveau, R., Combined inhibition of PDGF and VEGF receptors by ellagic acid, a dietary-derived phenolic compound, *Carcinogenesis*, 26 (4), 821–826, 2005.

70. Losso, J. N., Bansode, R. R., Trappey, A. 2nd, Bawadi, H. A., and Truax, R., In vitro anti-proliferative activities of ellagic acid, *J. Nutr. Biochem.*, 15 (11), 672–678, 2004.

71. Bagchi, D., Sen, C. K., Bagchi, M., and Atalay, M., Anti-Angiogenic, antioxidant, and anti-carcinogenic properties of a novel anthocyanin-rich berry extract formula, *Biochemistry, (Mosc)*, 69 (1), 75–80, 2004.

72. Liu, Z., Schwimer, J., Liu, D., Greenway, F. L., Anthony, C. T., and Woltering, E. A., Black raspberry extract and fractions contain angiogenesis inhibitors, *J. Agric. Food Chem.*, 53 (10), 3909–3915, 2005.

73. Aggarwal, B. B. and Shishodia, S., Molecular targets of dietary agents for prevention and therapy of cancer, *Biochem. Pharmacol.*, 71 (10), 1397–1421, 2006.

74. Kwon, H. J., Shim, J. S., Kim, J. H., Cho, H. Y., Yum, Y. N., Kim, S. H., and Yu, J., Betulinic acid inhibits growth factor-induced in vitro angiogenesis via the modulation of mitochondrial function in endothelial cells, *Jpn. J. Cancer Res.*, 93 (4), 417–425, 2002.

75. Kobayashi, S., Miyamoto, T., Kimura, I., and Kimura, M., Inhibitory effect of isoliquiritin, a compound in licorice root, on angiogenesis in vivo and tube formation in vitro, *Biol. Pharm. Bull.*, 18 (10), 1382–1386, 1995.

76. Sheela, M. L., Ramakrishna, M. K., and Salimath, B. P., Angiogenic and proliferative effects of the cytokine VEGF in Ehrlich ascites tumor cells is inhibited by Glycyrrhiza glabra, *Int. Immunopharmacol.*, 6 (3), 494–498, 2006.

77. Yue, P. Y., Wong, D. Y., Wu, P. K., Leung, P. Y., Mak, N. K., Yeung, H. W., Liu, L. et al., The angiosuppressive effects of 20(R)-ginsenoside Rg(3), *Biochem. Pharmacol.*, 72 (4), 437–445, 2006.

78. Loutrari, H., Hatziapostolou, M., Skouridou, V., Papadimitriou, E., Roussos, C., Kolisis, F. N., and Papapetropoulos, A., Perillyl alcohol is an angiogenesis inhibitor, *J. Pharmacol. Exp. Ther.*, 311 (2), 568–575, 2004.

79. Hayashi, A., Gillen, A. C., and Lott, J. R., Effects of daily oral administration of quercetion chalcone and modified citrus pectin on implanted colon-25 tumor growth in balb-c mice, *Altern. Med. Rev.*, 5 (6), 546–552, 2000.

80. McCarty, M. F. and Block, K. I., Toward a core nutraceutical program for cancer management, *Integr. Cancer Ther.*, 5 (2), 150–171, 2006.

81. Wouters, B. G., Weppler, S. A., Koritzinsky, M., Landuyt, W., Nuyts, S., Theys, J., Chiu, R. K., and Lambin, P., Hypoxia as a target for combined modality treatments, *Eur. J. Cancer*, 38 (2), 240–257, 2002.

82. Iseki, K., Tatsuta, M., Uehara, H., Iishi, H., Yano, H., Sakai, N., and Ishiguro, S., Inhibition of angiogenesis as a mechanism for inhibition by 1alpha-hydroxyvitamin D3 and 1,25-dihydroxyvitamin D3 of colon carcinogenesis induced by azoxymethane in Wistar rats, *Int. J. Cancer*, 81 (5), 730–733, 1999.

83. Shokravi, M. T., Marcus, D. M., Alroy, J., Egan, K., Saornil, M. A., and Albert, D. M., Vitamin D inhibits angiogenesis in transgenic murine retinoblastoma, *Invest. Ophthalmol. Vis. Sci.*, 36 (1), 83–87, 1995.

84. Miyazawa, T., Tsuzuki, T., Nakagawa, K., and Igarashi, M., Anti-Angiogenic potency of vitamin E, *Ann. N.Y. Acad. Sci.*, 1031 , 401–404, 2004.

85. Tang, F. Y. and Meydani, M., Green tea catechins and vitamin E inhibit angiogenesis of human microvascular endothelial cells through suppression of IL-8 production, *Nutr. Cancer*, 41 (1–2), 119–125, 2001.

86. Zgouras, D., Wachtershauser, A., Frings, D., and Stein, J., Butyrate impairs intestinal tumor cell-induced angiogenesis by inhibiting HIF-1alpha nuclear translocation, *Biochem. Biophys. Res. Commun.*, 300 (4), 832–838, 2003.

87. Myzak, M. C. and Dashwood, R. H., Histone deacetylases as targets for dietary cancer preventive agents: lessons learned with butyrate, diallyl disulfide, and sulforaphane, *Curr. Drug Targets*, 7 (4), 443–452, 2006.
88. Surh, Y. J., Antitumor promoting potential of selected spice ingredients with antioxidative and anti-inflammatory activities: a short review, *Food Chem. Toxicol.*, 40 (8), 1091–1097, 2002.
89. Min, J. K., Han, K. Y., Kim, E. C., Kim, Y. M., Lee, S. W., Kim, O. H., Kim, K. W., Gho, Y. S., and Kwon, Y. G., Capsaicin inhibits in vitro and in vivo angiogenesis, *Cancer Res.*, 64 (2), 644–651, 2004.
90. Moon, E. J., Lee, Y. M., and Kim, K. W., Anti-Angiogenic activity of conjugated linoleic acid on basic fibroblast growth factor-induced angiogenesis, *Oncol. Rep.*, 10 (3), 617–621, 2003.
91. Wang, L. S., Huang, Y. W., Sugimoto, Y., Liu, S., Chang, H. L., Ye, W., Shu, S., and Lin, Y. C., Effects of human breast stromal cells on conjugated linoleic acid (CLA) modulated vascular endothelial growth factor-A (VEGF-A) expression in MCF-7 cells, *Anticancer Res.*, 25 (6B), 4061–4068, 2005.

24 Role of Edible Berry Anthocyanins in Angiogenesis

*Manashi Bagchi, Shirley Zafra-Stone,
Jack N. Losso, Chandan K. Sen, Sashwati Roy,
Soumyadipta Hazra, and Debasis Bagchi*

CONTENTS

Abstract

Nutrition is a major tool in human health and disease prevention. Edible berries are rich in natural anthocyanin antioxidants. The pharmacologic and therapeutic properties of berry anthocyanins have been demonstrated to exert significant chemopreventive, anti-angiogenic and anti-cancer properties. A significant number of studies have also shown that berry anthocyanins are novel cardioprotectants, beneficial in reducing age-associated oxidative stress, improving neuronal and cognitive brain function and ocular health, and protecting genomic deoxyribonucleic acid (DNA) integrity. Angiogenesis, a natural process of blood vessel growth, is a double-edged sword that can either be protective or destructive. Although therapeutic angiogenesis takes place for wound healing and for restoring blood flow in ischemic tissues/organs, unwanted growth of blood vessels may lead to the formation of varicose vein, tumor or cancer metastases—and therapeutic intervention requires an anti-angiogenic process.

A healthy body controls angiogenesis through angiogenesis-stimulating growth factors or angiogenesis inhibitors. The diverse health benefits of edible berry anthocyanins have been reviewed. A synergistic formula, OptiBerry, was developed by combining wild blueberry, bilberry, cranberry, elderberry, raspberry seeds, and strawberry. This formula exhibited high antioxidant efficacy [oxygen radical absorbance capacity (ORAC) value] and cellular uptake, minimal cytotoxicity and novel anti-angiogenic properties. OptiBerry significantly inhibited both hydrogen peroxide (H_2O_2)- as well as tumor necrosis factor alpha (TNFα)-induced vascular endothelial growth factor (VEGF) expression, a key regulator of tumor angiogenesis, by human keratinocytes. A Matrigel assay using human microvascular endothelial cells (EC) showed that OptiBerry impaired angiogenesis. In an in vitro model of angiogenesis, OptiBerry significantly inhibited basal monocyte chemoattractant protein-1 (MCP-1) and inducible nuclear transcription factor-kappa beta (NF-$\kappa\beta$) transcriptions. Endothelioma (EOMA) cells pretreated with OptiBerry showed a diminished ability to form hemangioma and markedly decreased tumor growth by more than 50% in 129 P3 mice. Results demonstrated the novel antioxidant, anti-angiogenic and anti-carcinogenic potential of OptiBerry. A review of a large number of studies exhibited the health benefits of berry anthocyanins in cardiovascular dysfunction, cerebral ischemic stroke, diabetes, and healthy vision—and therapeutic angiogenesis causes intervention. Overall, berry anthocyanins favorably regulate angiogenesis in promoting human health and disease prevention.

24.1 INTRODUCTION

Epidemiological data support the association between the high intake of fresh fruits and vegetables and reduced risks of chronic degenerative diseases including cancer [1]. There are numerous mechanistic interpretations that demonstrate how consumption of fresh fruits and vegetables might delay or prevent the onset of chronic degenerative diseases and induce chemoprevention. It has been well documented that fresh fruits and vegetables are rich sources of a large number of diverse nutrients, including vitamins/antioxidants, trace minerals/micronutrients, phytosterols, novel enzymes, dietary fiber, and miscellaneous biologically active chemoprotectants [1,2]. Phytochemicals derived from fresh fruits, nuts, and seeds exert a broad spectrum of medicinal, pharmacological, and therapeutic properties. Fruits and vegetable consumption reduce the risk of certain forms of cancer [2–4]. These natural products can possess complementary and overlapping or identical mechanisms of potential disease-preventive action, including novel antioxidant, anti-bacterial, anti-viral, and anti-angiogenic properties, modulation of detoxification enzymes, stimulation of the immune system, reduction of platelet aggregation, and favorable reduction of cholesterol biosynthesis, hypertension, and hormone metabolism [2–4]. Research shows that almost a third of all cancer events may be prevented by changes in diet [2–5].

24.2   ANTIOXIDANTS IN HUMAN HEALTH AND DISEASE PREVENTION

Oxidative stress and tissue damage have been implicated in more than one hundred disease conditions in animals and humans, including atherosclerosis, arthritis, ischemia, and reperfusion injury of many tissues, central nervous system (CNS) injury, gastrointestinal injury, initiation and promotion of multistage carcinogenesis, and Acquired Immuno Deficiency Syndrome (AIDS) [4–6]. Antioxidants are potent scavengers of free radicals and function as inhibitors of both the initiation and promotion/propagation/transformation stages of tumor promotion/carcinogenesis as well as neoplastic processes, and thus protect cells against oxidative damage [1,4]. The consumption of edible fruits and vegetables has been demonstrated to prevent the occurrence of a number of diseases in animals and humans. Vegetables, fruits, and their seeds are rich sources of vitamins C and E, β-carotene, or protease inhibitors, natural chemoprotectants, that might protect the organism against free radical-induced injury and diseases [1–6].

24.3   ANTHOCYANINS AS NOVEL ANTIOXIDANTS

A growing body of evidence demonstrates the diverse health benefits and protective effects of anthocyanins present in various fruits and vegetables [1,7,8]. Anthocyanins are integral constituents of fruits and vegetables, in particular berries; they provide pigmentation and serve as natural antioxidants [1,7,8]. Numerous therapeutic benefits attributable to anthocyanins include antioxidant protection and maintaining Deoxyribonucleic acid (DNA) integrity. Anthocyanins act as potent anti-inflammatory and anti-mutagenic agents, and provide cardioprotection by maintaining vascular permeability [7,8]. Dietary consumption of anthocyanins has been shown to improve the overall antioxidant defense status [1,7]. Anthocyanins are effective in reducing advancing age-induced oxidative stress and reversing neuronal and behavioral changes [7,8].

Anthocyanins, found in high concentrations in the berries, are capable of inhibiting the initiation of carcinogenesis, as well as inhibiting tumor formation [1,7,9]. In fact, edible berries are rich in both anthocyanins and flavonoid glycosides, which is responsible for the red, violet, purple, and blue color of the fruits [7,8]. Anthocyanins can be identified in human blood plasma after consumption of berries [9].

24.3.1 CHEMISTRY OF ANTHOCYANINS

Anthocyanins are a group of natural antioxidants widely distributed in fruits, seeds, and vegetables, and can be identified in human plasma. Anthocyanins have two absorbance peaks at 270–280 nm and 510–540 nm [10].

The natural color of anthocyanins differs from other naturally occurring flavonoid antioxidants because of their ability for electron delocalization and to form resonance structures by changes in pH [11–13]. Anthocyanins exist in an aqueous phase in a mixture of four molecular species (Figure 24.1). Their relative color depends upon pH. At pH 1–3, the flavylium cation is red colored, at pH 5 the colorless carbinol pseudo base (pb) is generated, and at pH 7–8 the blue purple quinoidal base (qb) is formed. The high levels of anthocyanins in berries contribute to their powerful antioxidant activity [14,15]. The positively charged oxygen atom in the anthocyanin molecule makes it a more potent hydrogen-donating antioxidant compared to oligomeric proanthocyanidins (OPCs) and other flavonoids [7].

Anthocyanins are the integral components and natural colorants in red wines; they provide the color and contribute to the powerful antioxidant properties. The 3-glucoside anthocyanins: delphinidin, cyanidin, petunidin, and malvidin are

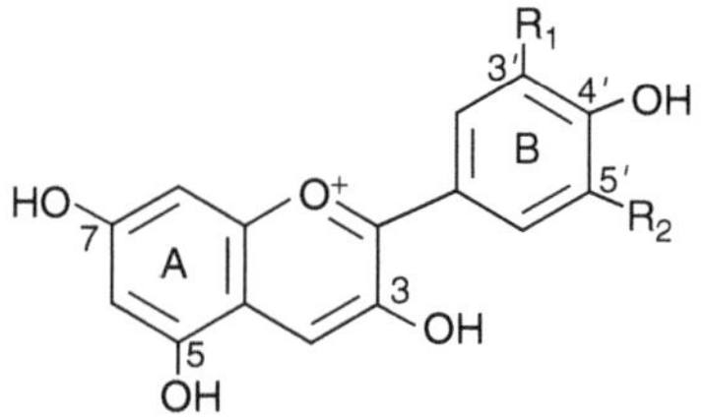

R_1 = H	R_2 = H	: Pelargonidin (Pg)
R_1 = OH	R_2 = H	: Cyanidin (Cy)
R_1 = OH	R_2 = OH	: Delphinidin (Dp)
R_1 = OCH_3	R_2 = H	: Peonidin (Pn)
R_1 = OCH_3	R_2 = OH	: Petunidin (Pt)
R_1 = OCH_3	R_2 = OCH_3	: Malvidin (Mv)

FIGURE 24.1 Structures of anthocyanins.

present in red wines. However, malvidin 3-glucoside, malvidin 3-glucoside acetate, and malvidin 3-glucoside coumarate are the most abundant constituents [16,17].

24.3.2 ABSORPTION AND BIOAVAILABILITY OF ANTHOCYANINS

Various biological and pharmacological activities of anthocyanins have been reported using crude fruit extracts that are rich in anthocyanins. The intake of anthocyanins in humans has been estimated to be 180–215 mg/day in the U.S.A., which is much higher than the intake (23 mg/day) of other flavonoids, including quercetin, kaempferol, myricetin, apigenin, and luteolin. The direct evidence of the absorption of anthocyanins in humans was obtained by combining an octadecylsilane (ODS) solid-phase extraction procedure for plasma sample preparation and an high performance liquid chromatography (HPLC) system with diode array for anthocyanin separation and detection [18]. Anthocyanin-like compounds have been found in human urine. Consumption of 500 ml of blueberry juice or cranberry juice by healthy female subjects increased plasma phenolic content and antioxidant capacity [19].

Anthocyanins can be identified in human blood plasma using a protocol of a restricted access phase that involves removal of the proteins and enrichment of the anthocyanins. Spray-dried elderberry juice demonstrated antioxidant activities in vitro. The anthocyanins in elderberry (*Sambucus nigra*) could be analyzed quantitatively in the blood sample. The limit that this quantification reached was 0.5 ng/ml. In this experiment, maximum concentration in the blood (35 mg/ml) was observed after 1 h with a quick decay [9].

Another study assessed whether consumption of 500 ml of blueberry juice or cranberry juice by healthy female subjects increased plasma phenolic content and antioxidant capacity. Nine volunteers consumed 500 ml of blueberry juice, cranberry juice, or a sucrose solution (control) after fasting overnight. Blood and urine samples were obtained up to 4 h after consumption of the juices. Consumption of cranberry juice resulted in a significant reduction of plasma potassium nitrosodisulphonate and Fe(III)-2,4,6-tri(2-pyridyl)-s-triazine, with measurements of antioxidant capacity attaining a maximum after 60–120 min. This corresponded to a 30% increase in vitamin C and a small but significant increase in total phenols in the plasma. Consumption of blueberry juice had no such effects [19].

24.3.3 ANTIOXIDANT ACTIVITY OF ANTHOCYANINS

Anthocyanin has been found to be a potent antioxidant compared to classic antioxidants such as butylated hydroxyanisole (BHA), butylated hydroxytoulene (BHT) and α-tocopherol (vitamin E). This natural agent, in addition to imparting color to the food, might prevent autooxidation of lipids as well as lipid peroxidation in biological systems [20].

Endothelial dysfunction has been proposed to play an important role in the initiation and development of vascular disease and anthocyanins improve endothelial function. The enrichment of endothelial cells (EC) with elderberry anthocyanins conferred significant protective effects of EC against different oxidative stressors including hydrogen peroxide H_2O_2, 2,2′-azobis(2-amidinopropane) dihydrochloride (AAPH), and $FeSO_4$/ascorbic acid. These results demonstrated that vascular EC can incorporate anthocyanins into the membrane and cytosol, conferring significant protective effects against oxidative insult, which may have important implications on preserving EC functions and preventing vascular diseases [21].

In another study, the potential antioxidant properties of blueberry anthocyanins were investigated in vitro and in vivo using red blood cell (RBC) resistance to reactive oxygen species (ROS) as the model. In vitro incubation with anthocyanins or hydroxycinnamic acids (HCA) (0.5 and 0.05 mg/ml) was found to significantly enhance RBC resistance to H_2O_2 (100 μM)-induced ROS production. A similar protective effect was also observed in vivo following oral supplementation in rats at 100 mg/ml. Only the anthocyanin-fed group had significant protection at 6- and 24-h post-supplementation. But it was not consistent with the measured plasma levels of anthocyanins. Indeed, plasma concentrations were highest after 1 h, declining considerably after 6 h and not detected after 24 h. The difference in absorption between anthocyanins and HCA is likely to have contributed to the observed difference in their abilities to afford protection to RBC. This protection represents a positive role following dietary consumption of blueberries against ROS formation within RBC in vivo [22].

In a related study, researchers investigated the responses in serum total antioxidant capacity following consumption of strawberries (240 g), spinach (294 g), red wine (300 ml), or vitamin C (1250 mg) in eight elderly women. Total antioxidant capacity was determined using different methods: an oxygen radical absorbance capacity (ORAC) assay, a trolox equivalent antioxidant capacity (TEAC) assay, and a ferric reducing ability (FRAP) assay. The results showed that the total antioxidant capacity of serum determined as ORAC, TEAC, and FRAP, using the area under the curve, increased significantly by 7%–25% during the 4-h period following consumption of red wine, strawberries, vitamin C, or spinach [23]. The total antioxidant capacity of urine determined as ORAC increased ($p < 0.05$) by 9.6%, 27.5%, and 44.9% for strawberries, spinach, and vitamin C, respectively, during the 24-h period following these treatments. The plasma vitamin C level after the strawberry drink, and the serum urate level after the strawberry and spinach treatments also increased significantly. However, the increased vitamin C and urate levels could not fully account for the increased total antioxidant capacity in serum following the consumption of strawberries, spinach, or red wine. It was concluded that the consumption of strawberries, spinach, or red wine, which are rich in antioxidant phenolic compounds, can increase the serum antioxidant capacity in humans [23].

In vitro enzymatic and non-enzymatic polyunsaturated fatty acid peroxidation was significantly inhibited in a dose-dependent manner by purified anthocyanin, a deep-red color pigment from a carrot cell culture. The kinetics showed that anthocyanin is a noncompetitive inhibitor of lipid peroxidation [20].

An extensive study was conducted on six edible berry extracts including wild blueberry, wild bilberry, cranberry, elderberry, raspberry seed, and strawberry. A novel synergistic combination of these six berry anthocyanins was developed, OptiBerry, which demonstrated maximal ORAC value in conjunction with high cellular uptake and low cytotoxicity, as demonstrated by a lactate dehydrogenase leakage assay [24]. In a more recent study, investigators demonstrated OptiBerry's unique antioxidant potential in a whole body scenario using a state-of-the-art electron paramagenetic resonance (EPR) technique (Sen and Bagchi, manuscript in preparation [25]).

24.4 ANGIOGENESIS

The term "angiogenesis" describes the growth of new blood vessels, an important natural process that takes place in the animal or human body, contributing to both health and diseased conditions [26]. Angiogenesis is a double-edge sword; it can be protective or destructive. Although angiogenesis occurs in the healthy body for healing wounds and for restoring blood flow to tissues after ischemic injury or insult, unwanted growth of blood vessels may lead to varicose vein, tumor formation, and cancer metastases—and an anti-angiogenic approach can serve as a therapeutic intervention [26]. Angiogenesis also takes place in many diseased conditions including rheumatoid arthritis, psoriasis, scleroderma, placental growth, and embryo implantation, and in three common causes of blindness: diabetic retinopathy, retrolental fibroplasia, and neovascular glaucoma (in fact, diseases of the eye are almost always accompanied by vascularization) [26,27].

Additionally, the process of wound angiogenesis has many features in common with tumor angiogenesis [28]. Dermal wound healing, other than the most superficial injury, cannot occur without angiogenesis [29,30]. In all wound healing processes, damaged vasculature need to be repaired, and increased local cell activity for the healing process requires an increased supply of nutrients from the bloodstream [29,30]. Furthermore, the EC which form the lining of the blood vessels are important in themselves as organizers and regulators of healing [29–31].

24.4.1 ANGIOGENESIS CASCADE

The healthy body controls angiogenesis through angiogenesis-stimulating growth factors or angiogenesis inhibitors. Angiogenesis proceeds concurrently with the formation of new tissue (granulation tissue) that typically begins about 4 days post-wounding [29]. It is stimulated by the chemicals (soluble factors) released by wounded tissue [30]. The resulting processes are tightly regulated by cell–cell interactions, cell–extra cellular matrix (ECM) interactions, and cell–soluble factor interactions [31].

A blood capillary consists of a hollow tube lined with EC. The outside of the tube is covered with a layer known as the basement membrane, a major component of which is collagen IV, and which also contains fibronectin and proteoglycans (compounds consisting mainly of polysaccharides but also containing protein) [31].

Angiogenesis begins with degradation of the basement membrane, followed by migration of EC out of the vessel. These cells then form a tube which "sprouts" from the old capillary and is extended further into the wound space as the cells behind the leading tip begin to proliferate. The tips of such tubes can branch and eventually join up with other sprouts to form a closed loop through which blood can flow [32,33].

The sprouting process begins again from these new vessels, until the wound space is permeated by a network of new capillaries. The cells of the capillaries first synthesize themselves into a provisional covering containing fibronectin and proteoglycans, and then finally form a true basement membrane [32,33].

As the granulation tissue matures, most of its vessels begin to disappear [40]. The EC begin to undergo programmed cell death (apoptosis) and are removed from the tissue by scavenging macrophages [27–34].

24.4.2 ANGIOGENESIS, VASCULAR ENDOTHELIAL GROWTH FACTOR AND BERRY ANTHOCYANINS

Anti-angiogenic approaches to prevent and treat cancer represent a priority area in investigative tumor biology [35,36]. Vascular endothelial growth factor (VEGF) plays a crucial role for the vascularization of tumors [24]. The vasculature in adult skin remains normally quiescent. However, skin retains the capacity for brisk initiation of angiogenesis during inflammatory skin diseases such as psoriasis and skin cancers [35–37].

The effects of multiple berry extracts on inducible VEGF expression by human HaCaT keratinocytes were evaluated [24]. Six individual berry extracts (wild blueberry, bilberry, cranberry, elderberry, raspberry seed, and strawberry) and OptiBerry, a novel combination of six edible berries, were investigated. All berry samples and OptiBerry demonstrated significant inhibition of both H_2O_2- as well as

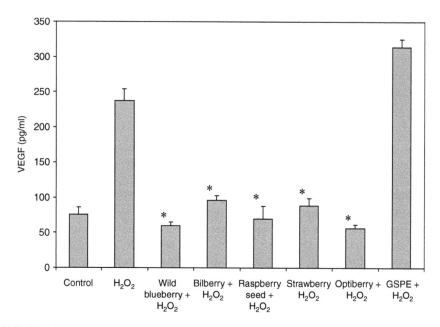

FIGURE 24.2 Individual berry samples and a combination of berry extracts inhibited H_2O_2-induced expression of vascular endothelial growth factor (VEGF). HaCaT cells were seeded at density 0.45×10^6/well/3 ml. After 24 h, growth media was changed to serum-free Ruswell Park Memorial Institute medium (RPMI) and 50 µg/ml berry samples were added. After 12 h, cells were challenged with 150 µM H_2O_2. After 12 h of activation with H_2O_2, media was collected for enzyme-linked immunosorbent assay (ELISA). *Represents $p < 0.05$, higher in response to H_2O_2 treatment; *Represents lower compared to H_2O_2 treated cells. Mean $\pm$ SD of three experiments.

TNFα-induced VEGF expression by human keratinocytes, while OptiBerry exhibited the greatest effect [24]. Figure 24.2 demonstrates the inhibitory effects of wild blueberry, bilberry, raspberry seed, strawberry, OptiBerry, and grape seed proanthocyanidin extract (GSPE) against H_2O_2-induced expression of VEGF. The berries demonstrated significant inhibitory effects, while OptiBerry exhibited the maximum inhibition. GSPE exhibited no inhibitory effect [24].

Figure 24.3 provides an overview of the inhibitory effects of wild blueberry, bilberry, raspberry seed, strawberry, OptiBerry, and GSPE on TNFα-induced expression of VEGF. Wild blueberry exhibited the maximum inhibitory effect, while no significant difference was observed among raspberry seed, strawberry, and OptiBerry. GSPE exhibited no inhibitory effect.

Antioxidants such as GSPE, with comparable ORAC, or α-tocopherol did not influence inducible VEGF expression, suggesting that the observed effect of berry extracts was not dependent on their antioxidant property alone [24]. On the contrary, pure flavonoids such as ferrulic acid, catechin, and rutin shared the ability to suppress oxidant-inducible VEGF expression. Thus, results demonstrate that berry anthocyanin structural moiety is responsible for the superior effect on inducible VEGF expression and release [24].

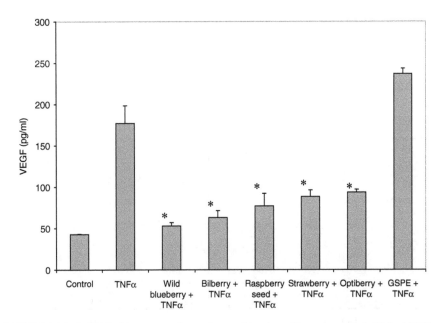

FIGURE 24.3 Individual berry samples and a combination of berry extracts inhibited TNFα-induced expression of vascular endothelial growth factor (VEGF). HaCaT cells were seeded at density 0.45×10^6/well/3 ml. After 24 h, growth media was changed to serum-free RPMI and 50 μg/ml berry samples were added. After 12 h, cells were challenged with 25 ng/ml TNFα. After 12 h of activation with TNFα, media was collected for ELISA. *Represents $p < 0.05$, higher in response to TNFα treatment; *Represents lower compared to TNFα treated cells. Mean ± SD of three experiments.

24.4.3 MATRIGEL ASSAY DEMONSTRATING ANTI-ANGIOGENIC EFFECT OF OPTIBERRY

Matrigel assay, a highly reliable approach to study in vitro angiogenesis, is based on the differentiation of EC to form capillary structures on a basement membrane matrix. Matrigel is a matrix of a mouse basement membrane neoplasm that represents a complex mixture of basement membrane proteins including type IV collagen, entactin/nidogen, proteoheparan sulfate, and other growth factors. Matrigel assays induce EC to differentiate, as shown by morphological changes and by the reduction in proliferation, which offers a convenient model to study biochemical changes associated with angiogenesis [38]. Human dermal micro-vascular EC and an in vitro angiogenesis kit (Chemicon International, Inc., Temecula, CA, U.S.A.) are ideal for this assay.

Briefly, ECMatrix is a solid gel of basement proteins prepared from the Engelbreth-Holm-Swarm (EHS) mouse tumor. Approximately 100 μl diluted ECMatrix (10×) solution is taken in a 96-well tissue culture plate and incubated at 37°C for at least an hour to allow the matrix solution to solidify. Human dermal microvascular EC (5000 cells/well), harvested in the presence and absence of chemicals to determine their anti-angiogenic potential, were added on top of the solidified matrix solution and maintained in an incubator at 37°C overnight. Endothelial tube formation is monitored and digitally photographed under an inverted light microscope at 20× magnification. In a recent study, OptiBerry, a novel combination of six edible berry anthocyanins, impaired angiogenesis in this model [24] (Figure 24.4).

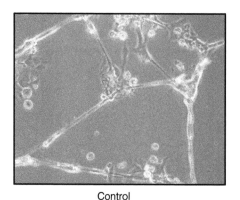
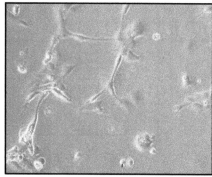

Control	Optiberry

FIGURE 24.4 Anti-angiogenic property of OptiBerry in vitro. An in vitro model of angiogenesis was used employing Matrigel and human dermal microvascular endothelial cells (EC). Human dermal microvascular EC (0.5×10^5 cells/per well) were seeded onto 4-well plates precoated with Matrigel. After 48 h of seeding, OptiBerry sample was added (50 mg/ml). Endothelial tube formation is observed and digitally photographed under an inverted light microscope at 20–100× (20× shown) magnification.

24.4.4 ANTI-ANGIOGENIC PROPERTY OF OPTIBERRY IN A MODEL OF HEMANGIOMA

Hemangiomas are the most common infancy tumors, which occur in approximately 1:100 normal newborns; however, in premature infants weighing less than 1000 g, the incidence rises to 1:5 live births [39]. The hemangioma is characterized by rapid growth during the first year of life (proliferative phase) followed by a decline in growth (involutional phase) over the next 5–6 years, with complete regression of the lesion in 90% of affected individuals by age 9 (the involuted phase). Hemangiomas arise from clonal expansion of a single EC precursor [40]. Approximately 5% of hemangiomas cause serious tissue damage, while 1%–2% of all hemangiomas are life threatening. Proliferating hemangiomas are highly angiogenic and, thus, experimentally, hemangiomas represent a powerful model to study in vivo angiogenesis.

The presence of macrophages is associated with proliferating hemangiomas. The CC (characterized by a cys-cys sequence motif) chemokine MCP-1 (monocyte chemoattractant protein-1) has been shown to be responsible for recruiting macrophages to infection or inflammation sites. MCP-1 is acknowledged as a major accessory facilitating angiogenesis and a direct role of MCP-1 on angiogenesis has been recently demonstrated, and antagonists to MCP-1 are considered to be anti-angiogenic [41,42].

The endothelioma (EOMA) cell line has also been well characterized in the literature. In a recent study, the EOMA cell line derived from the spontaneously arising hemangioma was used. EOMA cells were treated with OptiBerry for 12 or 24 h [43]. High baseline luciferase activity suggests elevated levels of basal MCP-1 transcription in these EOMA cells. Cells were activated with TNFα (400 IU/ml) for 12 h. Pretreatment of the EOMA cells with OptiBerry significantly inhibited basal MCP-1 transcription. MCP-1 transcription is mediated by several transcription factors among which NF-κβ is a key player. TNFα significantly induced NF-κβ transcription, while NF-κβ activity was significantly reduced in OptiBerry-treated EOMA cells [43].

Eight week old 129P3/J mice (Jackson Laboratories, Bar Harbor, ME, U.S.A.) were anesthetized and injected subcutaneously with 100 µl of EOMA cell suspension (5×10^6 cells) with or withour OptiBerry pretreatment. The tumor was harvested one week after the injection. Injection of the EOMA cells pretreated with OptiBerry did not result in hemangioma formation in all mice. Although the OptiBerry treated group did test positive for the presence of hemangioma, the average mass of such tumor growth was below 50% as compared to the untreated control group [43] (Figure 24.5). Histological analysis demonstrated markedly decreased infiltration of macrophages in hemangioma of treated mice compared to controls [42,43].

The underlying mechanistic aspects of the antitumor activity by berry anthocyanins are not entirely clear. Several different mechanisms have been suggested. Edible berry anthocyanins have been shown to inhibit cellular transformation and our study exhibited the potent inhibitory effect on inducible VEGF expression. Furthermore, antioxidant property alone may not account for the total observed anti-angiogenic effect. OptiBerry, the novel combination of six berry

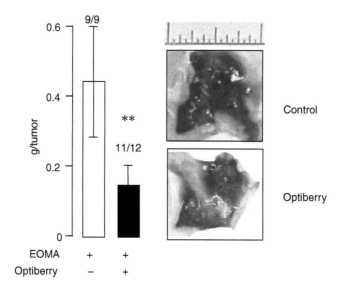

FIGURE 24.5 Hemangioma mass, incidence, and appearance. Data collected one week after subcutaneous injection of OptiBerry treated or untreated (control) endothelioma (EOMA) cells to 129P3/J mice. Mean + SD. **Lower mass compared to control group. $p < 0.01$. Photo: top, control; bottom, OptiBerry-treated. Bar graph represents the average mass of hemangioma. Scale = 1 in.

anthocyanins, significantly inhibits basal MCP-1 and inducible NF-κβ transcriptions, and EOMA cells pretreated with OptiBerry showed a diminished ability to form hemangioma [42,43]. Histological analysis exhibited markedly decreased infiltration of macrophages in the hemangioma of treated mice compared to control animals [42,43]. Thus, berry anthocyanins may exert novel chemoprevention and anti-angiogenic properties by several mechanistic pathways.

24.5 ANTI-CANCER POTENTIAL OF EDIBLE BERRY ANTHOCYANINS

Edible berry anthocyanins have demonstrated potent chemopreventive, anti-inflammatory, anti-carcinogenic, and anti-mutagenic properties. Fruit extracts from blueberry, bilberry, and cranberry exhibited potent anticancer activities in in vitro screening tests [17]. Strawberries and raspberries have exhibited potential cancer chemopreventive properties in vivo [44]. Fruit extracts of four *Vaccinium* species (lowbush blueberry, bilberry, cranberry, and lingonberry) exhibited potential anticarcinogenic activity, as evaluated by in vitro screening tests, by inhibiting the induction of orinithine decarboxylase—the rate limiting enzyme in polyamine synthesis [17].

24.5.1 Freeze Dried Strawberries as Inhibitors of Esophageal Cancer

Esophageal squamous cell carcinoma in humans occurs worldwide with a variable geographic distribution. Environmental factors thought to play a role in the development of esophageal cancer include the use of tobacco, alcoholic beverages, and highly salty pickled foods. Freeze dried strawberries have potential inhibitory effects against esophageal cancer [45]. Freeze dried strawberries have the potential to prevent N-nitrosomethylbenzylamine (NMBA)-induced esophageal cancer in the F344 rat esophagus. The tumor multiplicity was reduced in a dose-dependent manner in rats fed either 5% strawberries or 10% strawberries, and DNA adduct formation was significantly inhibited [45].

The potent cancer chemopreventive activity of freeze dried berries, including strawberries and raspberries, is well documented. Anthocyanins and ellagic acid are integral constituents in these berries, which exhibit novel anti-carcinogenic effects. In another study, freeze dried extracts of strawberries (*Fragara ananassa*) or black raspberry (*Rubus ursinus*) extracts, along with ellagic acid, were analyzed for anti-progression activity in the Syrian hamster embryo cell transformation model. These cells were treated with these extracts and ellagic acid and/or benzo[a]pyrene (B[a]P) for 7 days [44]. Ellagic acid and these extracts produced a dose-dependent decrease in progression/transformation compared with B[a]P treatment only. Ellagic acid and selected fractions were further examined using a 24 h co-treatment with B[a]P or a 6 day treatment following 24 h with B[a]P. Ellagic acid showed inhibitory ability in both protocols. Selected fractions significantly reduced B[a]P-induced progression/transformation only when co-treated with B[a]P for 24 h [44]. These results suggest that a methanol extract from strawberries and black raspberries displays chemopreventive activity. The possible mechanism by which these methanol fractions inhibited cell transformation appears to involve interference of uptake, activation, detoxification of B[a]P, or intervention of DNA binding and DNA repair [44].

24.5.2 Black Raspberries Inhibit Azoxymethane-Induced Colon Cancer

The effect of lyophilized black raspberries (LBR) was investigated on azoxymethane (AOM)-induced colon tumors in male Fischer 344 rats. AOM was injected (15 mg/kg body wt i.p.) once per week for 2 weeks. At 24 h after the final injection, AOM-treated rats began consuming diets containing 0, 2.5, 5, or 10% (wt/wt) LBR. Rats were sacrificed after 9 and 33 weeks of LBR feeding for aberrant crypt foci (ACF) enumeration and tumor analysis. ACF multiplicity decreased 36%, 24%, and 21% ($p < 0.01$ for all groups) in the 2.5%, 5%, and 10% LBR groups, respectively, relative to the AOM-only group. Total tumor multiplicity declined 42%, 45%, and 71% ($p < 0.05$ for all groups). A decrease in tumor burden (28%, 42%, and 75%) was observed in all LBR groups. Adenocarcinoma multiplicity decreased 28%, 35%, and 80% ($p < 0.01$) in the same treatment groups. These results indicate that LBR inhibits several measures of AOM-induced colon carcinogenesis in the Fischer 344 rat [46].

Another study was conducted to assess anti-initiation activity of LBR, which also included a 30-week tumorigenicity bioassay, quantification of DNA adducts, and NMBA metabolism. Feeding 5% and 10% LBR for 2 weeks prior to NMBA treatment (0.25 mg/kg, weekly for 15 weeks) and throughout the 30-week bioassay, significantly reduced tumor multiplicity (39% and 49%, respectively). At 25 weeks, both 5% and 10% LBRs significantly reduced tumor incidence (54% and 46%, respectively), tumor multiplicity (62% and 43%, respectively), proliferation rates, and preneoplastic lesion development. At 35 weeks, only 5% LBRs significantly reduced tumor incidence and multiplicity, proliferation indices and preneoplastic lesion formation. These findings demonstrate that dietary administration of LBR inhibited events of NMBA induced esophageal tumorigenesis in the F344 rats during initiation and post-initiation [47].

24.5.3 RASPBERRIES AND THEIR ANTI-PROLIFERATIVE EFFECT IN LIVER CANCER CELLS

The potency of four fresh raspberry varieties (Heritage, Kiwigold, Goldie, and Anne) was investigated. The color of the raspberry juice correlated well with the total phenolic, flavonoid, and anthocyanin contents of the raspberry [48]. The anti-proliferative effect was also increased with the color intensity of the juice. The proliferation of HepG human liver cancer cells was significantly inhibited in a dose-dependent manner after exposure to the raspberry extracts. The extract equivalent to 50 mg of Goldie, Heritage, and Kiwigold fruit inhibited the proliferation of those cells by 89.4 ± 0.1, 88 ± 0.2, and 87.6 ± 1.0, respectively. Anne had the lowest antiproliferative activity but still exhibited a significant inhibition of 70.3 ± 1.2 with an extract equivalent to 50 mg of fruit [48].

24.5.4 NATURAL ANTHOCYANINS INHIBIT COLORECTAL CARCINOGENESIS

Studies also demonstrated the inhibition of colorectal carcinogenesis by anthocyanin in rats. Hagiwara et al. (2001) demonstrated that purple corn color (PCC), a natural anthocyanin supplement, suppressed the lesion development and proliferation of 1,2-dimethylhydrazine-induced colorectal carcinogenesis significantly in male F344 rats. PCC supplementation suppressed the multiplicities of both colorectal adenomas and carcinomas [49].

24.5.5 EPIDERMAL GROWTH FACTOR (EGF) RECEPTOR INHIBITORY EFFECT OF ANTHOCYANINS

The aglycons of the most abundant anthocyanins in food, cyanidin, and delphinidin were found to inhibit the growth of human tumor cells in vitro in the micromolar range. The aglycons preferentially inhibited the growth of the human vulva carcinoma cell line A-431, overexpressing the epidermal growth-factor receptor (EGFR). In intact cells, the influence of anthocyanin treatment on downstream signaling cascades was investigated by measuring the phosphorylation of the transcription factor Elk-1. A-431 cells were transiently transfected with a luciferase reporter gene. Thus, the anthocyanidins are potent inhibitors of the EGFR, shutting off downstream signaling cascades. These effects contribute substantially to the growth-inhibitory properties of these natural anthocyanins from edible berries [50].

24.5.6 ESTROGENIC ACTIVITY OF ANTHOCYANINS IN MCF-7 AND BG-1 BREAST CANCER CELLS

Anthocyanins are known to have estrogenic activity. Anthocyanidins showed estrogenic-inducible cell proliferation in two cell lines (MCF-1 and BG-1). Anthocyanidin treatment resulted in a reduction of estrodiol-induced cell proliferation in the above cell lines. Furthermore, anthocyanidin may exert estrogenic activity that may play a role in altering the development of hormone dependent breast cancer [51].

24.6 CARDIOVASCULAR DYSFUNCTIONS, ANGIOGENESIS, AND BERRY ANTHOCYANINS

Cardiovascular disease is a major killer of older men and women in the U.S.A. Therapeutic angiogenesis describes an emerging field of cardiovascular medicine whereby new blood vessels are induced to grow to supply oxygen and nutrients to cardiac and skeletal muscle rendered ischemic as a result of progressive atherosclerosis or arteriosclerosis. This promotes collateral growth of blood vessels that are free from plaques or blockages, for the purpose of treating/preventing cardiac ischemia as well as peripheral arterial and coronary artery disease. This therapy may utilize angiogenic growth factor protein or genes to promote endogenous collateral vessels in the ischemic myocardium [52–54].

Anthocyanins act as a super cardioprotectant by maintaining vascular permeability and demonstrating potential anti-inflammatory properties. Anthocyanins reduce platelet aggregation in vitro and offer vascular protection comparing favorably with other cardioprotective drugs. Hypertension, atherosclerosis, and arteriosclerosis can reduce the flexibility of arterial walls, which contributes to poor blood flow and plaque formation. Rat aortas exposed to anthocyanin-enriched blueberry extract in vitro exhibited relaxation caused by endothelium-generated nitric oxide [55]. In another study, treatment of rats with anthocyanosides of *Vaccinium myrtillus* for 12 days before the induction of hypertension kept the blood-brain barrier permeability normal and limited the increase in vascular permeability in the skin and the aortic wall [56].

A commercial *V. myrtillus* (bilberry) extract enhanced relaxation of calf aortas in vitro that had been exposed to adrenalin. The dose response was nearly linear as the extract concentration was increased from 25 to 100 µg/ml. Anthocyanins and ascorbic acid inhibited contractile responses in calf aortas in the presence of histamine and angiotensin II. This effect was not observed when indomethacin or lysine acetylsalicylate was added. In a separate study, hamsters that were given oral doses (10 mg/10 g body weight) of a commercial product containing 36% bilberry anthocyanosides for two or four weeks exhibited better capillary perfusion and fewer sticking leukocytes in the capillaries [56,57].

24.7 CEREBRAL ISCHEMIC STROKE, ANGIOGENESIS, AND BERRY ANTHOCYANINS

Cerebral ischemic stroke is a leading cause of human death and disability causing paralysis, sensory, and motor damage, as well as behavioral and neurological dysfunctions. Inadequate clinical intervention is available to prevent ischemic damage and restore lost function in stroke victims. Extensive research has focused on protective maneuvers against ischemia-induced cell death, especially on potential strategies for promoting tissue repair and functional recovery in the damaged post-ischemic brain [58]. Current studies on angiogenesis may contribute to cell survival, re-oxygenation, and functional recovery of the area of neurological insult caused by ischemia and reperfusion-induced oxidative stress. Additionally, stem cell transplantation has emerged in the last few years as a potential therapy for ischemic stroke because of their capability of differentiating into multiple cell types and the possibility that they may provide trophic support for cell survival, tissue repair, and functional recovery [58]. Furthermore, stem cell transplantation is potential therapy for ischemic stroke because of their ability to differentiate multiple cell types and provide support for cell survival tissue repair and functional recovery [58].

Strawberry supplementation was shown to enhance striatal muscarinic receptor sensitivity, and this appeared to be reflected in the reversal of cognitive behavioral deficits [59]. Berries also enhance dopamine release in the brain, which improves the brain cells' ability to communicate. Brain function declines with advancing age, but research has shown that eating blueberries can reverse age-related decline in brain function.

Studies show that diets supplemented with strawberries or blueberries reverse age-induced declines in β-adrenergic receptor function in cerebellar Purkinje neurons [2]. Blueberries were shown to prevent or reverse age-related declines in cerebellar noradrenergic receptor function [2]. Brain functions such as balance, coordination, short-term memory and information retrieval can be impaired with advancing age. An animal study recently provided some evidence that dietary anthocyanins can protect the brain from oxidative injury. Six to eight month old F344 rats were fed diets containing vitamin E, aqueous blueberry extract, and dried strawberry extract or dried spinach extract. After eight weeks on the diet, the rats were subjected to 48 h of 100% oxygen-induced damage similar to that found in aged rats. All antioxidant diets prevented decreases in nerve growth factor in the basal forebrain and other adverse effects were reduced. In another study, rats were given intraperitoneal injections of bilberry anthocyanins (200 mg/kg/day) for five days had significantly more T3 in their brains than rats given only the solvent (26% alcohol). T3 enters the brain by a specific transport in the capillaries; therefore, anthocyanins may mediate T3 transport at the capillary level. Bilberry-treated animals exhibited superior memory, better vision and better control of sensory input [60].

24.8 DIABETES, OCULAR HEALTH, ANGIOGENESIS, AND BERRY ANTHOCYANINS

The leaves and fruit of *V. myrtillus* (containing 25% anthocyanidins) have been used for centuries in Europe to ameliorate the symptoms of diabetes. Alterations in the capillary filtration of biological macromolecules are well documented in diabetics, and anthocyanins have been shown to be effective against experimentally-induced capillary hyperfiltration [61]. The high levels of glucose in the blood of diabetics trigger many deteriorative events in the body. Medical complications of diabetes include microangiopathy, cataracts, blindness due to retinopathy, neuropathy, decreased resistance to infections and hyperlipidemia. The capillary walls in diabetic patients thicken due to collagen and glycoprotein deposits. The thickened capillaries are less flexible and more susceptible to blockage, leading to atherosclerosis.

Rat aorta smooth muscle cells incorporated less radio-labeled amino acids when cultured with *V. myrtillus* anthocyanidins (VMA), suggesting one mechanism by which bilberry maintains normal capillary structure. Flavonoids inhibited aldose reductase, an enzyme that converts sugars to sugar alcohols and is implicated with diabetic complications such as neuropathy, retinopathy, and heart disease. Boniface and colleagues (1986) also reported positive results for diabetic patients treated with 500–600 mg of VMA for eight to 33 months. Based upon this series of experiments, it was recommended that diabetic patients be given 500 mg VMA daily, split in two doses, for a period of several months. An aqueous alcohol extract of *V. myrtillus* leaves produced a 26% reduction in plasma glucose levels in rats made diabetic by the drug streptozotocin. Plasma triglycerides decreased in proportion with the amount of bilberry leaf extract given to rats (1.2 or 3.0 g/kg body weight) fed a hyperlipidemic diet. Consumption of blueberries with breakfast cereals and other whole grain foods could add another level of protection against the onset of diabetes [61].

24.9 HEALTHY VISION

Bilberry is believed to improve eyesight, particularly night vision, and was extensively used by British Air Force Pilots during World War I and II. The berry anthocyanins have been shown to improve circulation to the eye and retina, and demonstrate healing effects on diabetic retinopathy, macular degeneration, cataracts, and retinitis pigmentosa [62]. This health benefit is the primary reason for the product's popularity in Japan and Korea, where it is used to relieve eye strain caused by excessive computer use. Since carotenoids with vitamin A activity are found in *Vaccinium* species, some of the benefits pertaining to vision are attributable to these compounds. A double-blind, placebo-controlled study showed that oral doses of anthocyanins are important for generation of visual purple, which helps to convert light into electrical signals for the brain. Adapto-electroretinograms (AERG) of two sets of six subjects were made before treatment at one and three hours post-administration. Subjects given the bilberry adapted to the light within

6.5 min, compared with 9 min for the control group. Bilberry extracts appear to benefit vision in several ways, including improving night vision by enhanced generation of retinal pigments, increasing circulation within the capillaries of the retina, decreasing macular degeneration and diabetic retinopathy, and improving or preventing glaucoma and cataracts [62]. The anthocyanosides are typically concentrated in a 25% standardized extract. In a report of 50 patients with senile cataracts, a combination of bilberry extract standardized to contain 25% anthocyanosides (180 mg twice daily) and vitamin E in the form of dl-tocopheryl acetate (100 mg twice daily) administered for four months stopped the progression of cataracts in 96% of the subjects treated ($n=25$) compared to 76% in the control group ($n=25$) [63].

In a human study, fourteen diabetic or hypertensive outpatients with vascular retinopathy underwent therapy with *V. myrtillus* (bilberry) anthocyanosides (160 mg b.i.d.) or a placebo ($n=20$) for one month. The study was double-blind. At the end of the month, patients placebo-treated received the active drug for one additional month. Ophthalmoscopic and fluoroangiographic findings recovered before and after treatments showed an improvement ranging from 77% to 90% of anthocyanosides-treated patients. No other side effects or adverse drug reactions were recorded. Anthocyanosides appear to be safe and effective therapy for diabetic or hypertensive vascular retinopathy [64].

In another study, fifty patients, 21 men and 29 women (mean age 67 years, range 48–81) suffering from mild senile cortical cataract (62 eyes) underwent therapy with Vitamin E plus *V. myrtillus* (bilberry) anthocyanosides (FAR-1, 2 tabs b.i.d.) for 4 months. The study was a randomized, double-blind, placebo-controlled trial. FAR-1 was able to stop lens opacity progress in 97% of the eyes. No adverse drug reactions were recorded [14].

Long-term administration of antioxidants can inhibit the development of the early stages of diabetic retinopathy, and the mechanism by which this action occurs warrants further investigation. Supplementation with anthocyanins can offer an achievable and inexpensive adjunct therapy to inhibit the development of retinopathy and can normalize hyperglycemia [65].

24.10 CONCLUSION

Anthocyanins are common components of fruits and vegetables, in particular berries, and they provide pigmentation (color) and serve as natural antioxidants. In vitro and in vivo studies have shown that berry anthocyanins possess potent antioxidant and many potential health benefits, including cardiovascular, anti-carcinogenic potential, antidiabetic properties, improved brain function and mental clarity, healthy vision, urinary tract health, and dermal health. Recent studies have examined and demonstrated the potential cancer chemopreventive activity of freeze dried berries or strawberries and raspberries. Certainly, more studies with human subjects are needed to verify the broad spectrum of health benefits of anthocyanins. OptiBerry, a novel synergistic blend of wild blueberry, bilberry, cranberry, elderberry, raspberry seed, and strawberry anthocyanin extracts, was shown to be safer and more potent than all other combinations of berry extracts tested [24,25]. It demonstrated superior bioavailability, anti-oxidative, and anti-angiogenic properties. It is worthwhile to emphasize that angiogenesis is related to disorders such as cancer and inflammation. Taken together, a broad spectrum of mechanistic interpretations support the concept that edible berries high in anthocyanin content may serve as a potent anticarcinogenic and antimutagenic supplement to promote a wide range of health benefits.

REFERENCES

1. Heber, D., Vegetables, fruits and phytoestrogens in the prevention of diseases, *J. Postgrad. Med.*, 50, 145–149, 2004.
2. Bickford, P. C., Gould, T., Briederick, L., Chadman, K., Pollock, A., Young, D., Shukitt-Hale, B., and Joseph, J., Antioxidant-rich diets improve cerebellar physiology and motor learning in aged rats, *Brain Res.*, 866, 211–217, 2000.
3. Halliwell, B., Role of free radicals in the neurodegenerative diseases: therapeutic implications for antioxidant treatment, *Drugs Aging*, 18, 685–716, 2001.
4. Evans, P. and Halliwell, B., Micronutrients: oxidant/antioxidant status, *Br. J. Nutr.*, 85 (2), S67–S74, 2001.
5. Aruoma, O. I., Nutrition and health aspects of free radicals and antioxidants, *Food Chem. Toxicol.*, 32, 671–683, 1994.
6. Bagchi, D., Bagchi, M., Stohs, S. J., Das, D. K., Ray, S. D., Kuszynski, C. A., Joshi, S. S., and Preuss, H. G., Free radicals and grape seed proanthocyanidin extract: importance in human health and disease prevention, *Toxicology*, 148, 187–197, 2000.
7. Kong, J. M., Chia, L. S., Goh, N. K., Chia, T. F., and Brouillard, R., Analysis and biological activities of anthocyanins, *Phytochemistry*, 64, 923–933, 2003.
8. Prior, R. L., Fruits and vegetables in the prevention of cellular oxidative damage, *Am. J. Clin. Nutr.*, 78 (3), 570S–578S, 2003.
9. Murkovic, M., Adam, U., and Pfannhauser, W., Analysis of anthocyane glycosides in human serum, *Fresenius J. Anal. Chem.*, 366, 379–381, 2000.
10. Karakaya, S., Bioavailability of phenolic compounds, *Crit. Rev. Food Sci. Nutr.*, 44, 453–464, 2004.
11. Amouretti, M., Therapeutic value of *Vaccinium myrtillus* anthocyanosides in an internal medicine department, *Therapeutique*, 48, 579–581, 1972.
12. Andriambeloson, E., Magnier, C., Haan-Archipoff, G., Lobstein, A., Anton, R., Beretz, A., Stoclet, J. C., and Andriantsitohaina, R., Natural dietary polyphenolic compounds cause endothelium-dependent vasorelaxation in rat thoracic aorta, *J. Nutr.*, 128, 2324–2333, 1998.
13. Bettini, V., Fiori, A., Martino, R., Mayellaro, F., and Ton, P., Study of the mechanism whereby anthocyanosides potentiate the effect of catecholamines on coronary vessels, *Fitoterapia*, 56, 67–72, 1985.
14. Bravetti, G. O., Fraboni, E., and Maccolini, E., Preventive medical treatment of senile cataract with vitamin E and *Vaccinium myrtillus* anthocynosides: clinical evaluation, *Annali di Ottalmologia e Clinica Oculistica*, 115, 109–116, 1989.
15. Camire, M. E., Bilberries and blueberries as functional foods and nutraceuticals, In *Herbs, Botanicals and Teas*, Technomic Publishing Company, Lancaster, PA, pp. 289–319, 2000.

16. Bickford, P. C., Shukitt-Hale, B., and Joseph, J., Effects of aging on cerebellar noradrenergic function and motor learning: nutritional interventions, *Mech. Ageing Dev.*, 111, 141–154, 1999.
17. Bomser, J., Madhavi, D. L., Singletary, K., and Smith, M. A., In vitro anticancer activity of fruit extracts from *Vaccinium* species, *Planta Med.*, 62, 212–216, 1996.
18. Cao, G. and Prior, R. L., Anthocyanins are detected in human plasma after oral administration of an elderberry extract, *Clin. Chem.*, 45, 574–576, 1999.
19. Pedersen, C. B., Kyle, J., Jenkinson, A. M., Gardner, P. T., McPhail, D. B., and Duthie, G. G., Effects of blueberry and cranberry juice consumption on the plasma antioxidant capacity of healthy female volunteers, *Eur. J. Clin. Nutr.*, 54, 405–408, 2000.
20. Narayan, M. S., Naidu, K. A., Ravishankar, G. A., Srinivas, L., and Venkataraman, L. V., Antioxidant effect of anthocyanin on enzymatic and non-enzymatic lipid peroxidation, *Prostaglandins Leukot. Essent. Fatty Acids*, 60, 1–4, 1999.
21. Youdim, K. A., Martin, A., and Joseph, J. A., Incorporation of the elderberry anthocyanins by endothelial cells increases protection against oxidative stress, *Free Radic. Biol. Med.*, 29, 51–60, 2000.
22. Youdim, K. A., Shukitt-Hale, B., MacKinnon, S., Kalt, W., and Joseph, J. A., Polyphenolics enhance red blood cell resistance to oxidative stress: in vitro and in vivo, *Biochim. Biophys. Acta*, 1523, 117–122, 2000.
23. Cao, G., Russel, R. M., Lischner, N., and Prior, R. L., Serum antioxidant capacity is increased by consumption of strawberries, spinach, red wine or vitamin C in elderly women, *J. Nutr.*, 128, 2383–2390, 1998.
24. Roy, S., Khanna, S., Alessio, H. M., Vider, J., Bagchi, D., Bagchi, M., and Sen, C. K., Antiangiogenic property of edible berries, *Free Radic. Res.*, 36, 1023–1031, 2002.
25. Bagchi, D. and Sen, C. K., Manuscript in preparation.
26. Auerbach, R., Auerbach, W., and Polakowski, L., Assays for angiogenesis: a review, *Pharmacol. Ther.*, 51, 1–11, 1991.
27. BenEzra, D., Ocular circulation and neovascularization, *Documenta Opthalmologica Proceedings Series No. 50*, Nijhoff/Junk, Hingham, MA, 1987.
28. Whalen, G. F. and Zetter, B. R., Angiogenesis, In *Wound Healing: Biochemical and Physical Aspects*, Cohen, I. K. Ed., Saunders, Philadelphia, PA, 1992.
29. Clark, R. A. F., Overview of wound repair, In *The Molecular and Cellular Biology of Wound Repair*, Clark, R. A. F. Ed. 2nd ed., Plenum Press, New York, 1996.
30. Madri, J. A., Sankar, S., and Romanic, A. M., Angiogenesis, In *The Molecular and Cellular Biology of Wound Repair*, Clark, R. A. F. Ed. 2nd ed., Plenum Press, New York, 1996.
31. Alberts, B., Bray, D., Lewis, J., Raff, M., Roberts, K., and Watson, J. D., *Molecular Biology of the Cell*, 1st ed., Garland, New York, p. 709, 1983.
32. Madri, J. A., Sankar, S., and Romanic, A. M., Angiogenesis, In *The Molecular and Cellular Biology of Wound Repair*, Clark, R. A. F. Ed. 2nd ed., Plenum Press, New York, pp. 18–19, 1996.
33. Arnold, F. and West, D. C., Angiogenesis in wound healing, *Pharmacol. Ther.*, 52, 407–422, 1991.
34. Ausprunk, D. H., Falterman, K., and Folkman, J., The sequence of events in the regression of corneal capillaries, *Lab. Investig.*, 38, 284–294, 1978.
35. Giavazzi, R. and Taraboletti, G., Angiogenesis and angiogenesis inhibitors in cancer, *Forum*, 9, 261–272, 1999.
36. Griffioen, A. W. and Molema, G., Angiogenesis: potentials for pharmacologic intervention in the treatment of cancer, cardiovascular diseases, and chronic inflammation, *Pharmacol. Rev.*, 52, 237–268, 2000.

37. Detmar, M., The role of VEGF and thrombospondins in skin angiogenesis, *J. Dermatol. Sci.*, 24, S78–S84, 2000.

38. Ponce, M. L., Nomizu, M., and Kleinman, H. K., *FASEB J.*, 15, 1389–1397, 2001.

39. Folkman, J., Mulliken, J., and Ezekowitz, R., In *Surgery of Infants and Children: Scientific Principles and Practices*, Oldham, K. Ed., Lippincott-Raven, Philadelphia, PA, pp. 569–584, 1997.

40. Boye, E., Yu, Y., Paranya, G., Mulliken, J., Olsen, B., and Bischoff, J., Clonality and altered behavior of endothelial cells from hemangiomas, *J. Clin. Investig.*, 107, 745–752, 2001.

41. Salcedo, R., Ponce, M. L., Young, H. A., Wasserman, K., Ward, J. M., Kleinman, H. K., Oppenheim, J. J., and Murphy, W. J., *Blood*, 96, 34–40, 2000.

42. Atalay, M., Gordillo, G., Roy, S., Rovin, B., Bagchi, D., Bagchi, M., and Sen, C. K., Antiangiogenic property of edible berry in a model of hemangioma, *FEBS Lett.*, 544, 252–257, 2003.

43. Bagchi, D., Sen, C. K., Bagchi, M., and Atalay, M., Antiangiogenic, antioxidant, and anti-carcinogenic properties of a novel anthocyanin-rich berry extract formula, *Biochemistry (Mosc)*, 69, 75–80, 2004.

44. Xue, H., Aziz, R. M., Sun, N., Cassady, J. M., Kamendulis, L. M., Xu, Y., Stoner, G. D., and Klaunig, J. E., Inhibition of cellular transformation by berry extracts, *Carcinogenesis*, 22, 351–356, 2001.

45. Stoner, G. D., Kresty, L. A., Carlton, P. S., Siglin, J. C., and Morse, M. A., Isothiocyanates and freeze dried strawberries as inhibitors of esophageal cancer, *Toxicol. Sci.*, 52, 95–100, 1999.

46. Harris, G. K., Gupta, A., Nines, R. G., Kresty, L. A., Habib, S. G., Frankel, W. L., LaPerle, K., Gallaher, D. D., Schwartz, S. J., and Stoner, G. D., Effects of lyophilized black raspberries on azoxymethane-induced colon cancer and 8-hydroxy-2'-deoxyguanosine levels in the Fischer 344 rat, *Nutr. Cancer*, 40, 125–133, 2001.

47. Kresty, L. A., Morse, M. A., Morgan, C., Carlton, P. S., Lu, J., Gupta, A., Blackwood, M., and Stoner, G. D., Chemoprevention of esophageal tumorigenesis by dietary administration of lyophilized black raspberries, *Cancer Res.*, 61, 6112–6119, 2001.

48. Liu, M., Li, X. Q., Weber, C., Lee, C. Y., Brown, J., and Liu, R. H., Antioxidant and antiproliferative activities of raspberries, *J. Agric. Food Chem.*, 50, 2926–2930, 2002.

49. Hagiwara, A., Miyashita, K., Nakanishi, T., Sano, M., Tamano, S., Kadota, T., Koda, T. et al., Pronounced inhibition by a natural anthocyanin, purple corn color, of 2-amino-1-methyl-6-phenylimidazo[4,5-b]pyridine (PhIP)-associated colorectal carcinogenesis in male F344 rats pretreated with 1,2-dimethylhydrazine, *Cancer Lett.*, 171, 17–25, 2001.

50. Meiers, S., Kemeny, M., Weyand, U., Gastpar, R., von Angerer, E., and Marko, D., The anthocyanidins cyanidin and delphinidin are potent inhibitors of the epidermal growth-factor receptor, *J. Agric. Food Chem.*, 49, 958–962, 2001.

51. Schmitt, E. and Stopper, H., Estrogenic activity of naturally occurring anthocyanidins, *Nutr. Cancer*, 41, 145–149, 2001.

52. Hughes, G. C. and Annex, B. H., Angiogenic therapy for coronary artery and peripheral arterial disease, *Expert Rev. Cardiovasc. Ther.*, 3, 521–535, 2005.

53. Angus, J. A., Ward, J. E., Smolich, J. J., and McPherson, G. A., Reactivity of canine isolated epicardial collateral coronary arteries. Relation to vessel structure, *Circ. Res.*, 69, 1340–1352, 1991.

54. Banai, S., Shweiki, D., Pinson, A., Chandra, M., Lazarovici, G., and Keshet, E., Upregulation of vascular endothelial growth factor expression induced by myocardial ischaemia: implications for coronary angiogenesis, *Cardiovasc. Res.*, 28, 1176–1179, 1994.

55. Zaragoza, F., Iglesias, I., and Benedi, J., Comparative study of the anti-aggregation effects of anthocyanosides and other agents, *Arch. Farmacol. Toxicol.*, 11, 183–188, 1985.

56. Detre, Z., Jellinek, H., Miskulin, M., and Robert, A. M., Studies on vascular permeability in hypertension: action of anthocyanosides, *Clin. Physiol.-Biochem.*, 4, 143–149, 1986.

57. Kadar, A., Robert, L., Miskulin, M., Tixier, J. M., Brechemier, D., and Robert, A. M., Influence of anthocyanoside treatment on the cholesterol-induced atherosclerosis in the rabbit, *Paroi Arterielle*, 5, 187–205, 1979.

58. Wei, L., Keogh, C. L., Whitaker, V. R., Theus, M. H., and Yu, S. P., Angiogenesis and stem cell transplantation as potential treatments of cerebral ischemic stroke, *Pathophysiology*, 12, 47–62, 2005.

59. Joseph, J. A., Shukitt-Hale, B., Denisova, N. A., Bielinski, D., Martin, A., McEwen, J. J., and Bickford, P. C., Reversals of age-related declines in neuronal signal transduction, cognitive, and motor behavior deficits with blueberry, spinach, or strawberry dietary supplementation, *J. Neurosci.*, 19, 8114–8121, 1999.

60. Saija, A., Princi, P., D'Amico, N., De Pasquale, R., and Costa, G., Effect of *Vaccinium myrtillus* anthocyanins on triiodothyronine transport into brain in the rat, *Pharmacol. Res.*, 22, 59–60, 1990.

61. Cignarella, A., Nastasi, M., Cavalli, E., and Puglisi, L., Novel lipid-lowering properties of *Vaccinium myrtillus* L. leaves, a traditional antidiabetic treatment, in several models of rat dyslipidaemia: a comparison with ciprofibrate, *Thromb Res.*, 84, 311–322, 1996.

62. Camire, M. E., Bilberries and blueberries as functional foods and nutraceuticals, In *Herbs, Botanicals, and Teas*, Technomic Publishing Company, Lancaster, PA, pp. 289–319, 2000.

63. Head, K. A., Natural therapies for ocular disorders, part two: cataracts and glaucoma, *Altern. Med. Rev.*, 6, 141–166, 2001.

64. Perossini, M., Guidi, G., Chiellini, S., and Siravo, D., Diabetic and hypertensive retinopathy therapy with *Vaccinium myrtillus* anthocianosides (Tegens) double-blind placebo-controlled clinical trial, *Annali di Ottalmologia e Clinica Oculistica*, 12, 1173–1190, 1987.

65. Kowluru, R. A., Tang, J., and Kern, T. S., Abnormalities of retinal metabolism in diabetes and experimental galactosemia. VII. Effect of long-term administration of antioxidants on the development of retinopathy, *Diabetes*, 50, 1938–1942, 2001.

25 Redox Regulation of Angiogenesis: Anti-Angiogenic Properties of Edible Berries and Its Significance in a Clinical Setting

Gayle M. Gordillo, Debasis Bagchi, and Chandan K. Sen

CONTENTS

25.1 INTRODUCTION

Current work from our laboratory supports the finding that angiogenesis in vivo is subject to redox control [1]. Numerous mechanisms implicated in angiogenesis are redox-sensitive [1,2]. This new insight into angiogenic mechanisms points towards

549

the therapeutic potential of redox-based strategies to enhance or arrest angiogenesis. While enhancing angiogenesis is desired in wound healing [1,3–12], anti-angiogenic strategies are useful to arrest tumor growth. In this chapter, we will discuss the anti-angiogenic potential of nutritional antioxidants such as edible berries.

Nutrition is a major tool in health preservation and disease prevention. Dietary approaches are considered to be significantly safer than medical drugs because the anti-angiogenic ingredient is present in foods with proven safety records. In the United States, consumer demand for dietary supplements has risen steadily over the past few years. Over $12 billion dollars worth of "nutraceuticals" were purchased in the U.S. in 1998 alone [13]. The U.S. government has responded to the increasing consumer demand for dietary supplements by specifying that a portion of the NIH budget be allocated toward research on alternative medicine [14]. The therapeutic property of edible berries has been long known [15]. More recently, it has been observed that edible berries may have potent chemopreventive properties [16–20]. Berries are rich in anthocyanins, flavonoid glycosides, responsible for the red, violet, purple, and blue color of the fruit. Dietary consumption of anthocyanin improves the overall antioxidant defense status of human plasma [21–23]. On one hand, the search is on for specific medical drugs that would efficiently limit angiogenesis [24,25]. On the other hand, diet-based approaches to limit angiogenesis are being actively explored as well [16,17,26–32].

Anthocyanins are flavonoid glycosides widely present in edible berries and which possess strong antioxidant capacities [33,34]. The ability of endothelial cells to incorporate anthocyanins into the membrane and cytosol, where they maintain their antioxidant functions, has been documented in vitro [35]. Oral intake of berry anthocyanins enhances serum antioxidant status in humans in vivo [22,23].

Additionally, anthocyanins have been shown to have anti-angiogenic properties [32,36–38]. Tumor growth is an angiogenesis-dependent process with increasing evidence that many elements of angiogenesis regulation are redox sensitive. The combination of antioxidant and redox-independent anti-angiogenic qualities of anthocyanins found in edible berry extracts has led to increased interest in these types of food for potential use as anticancer agents. Little progress has been made to evaluate anthocyanins in clinical trials. This may be attributable to the inability to determine the active components of berry extracts and the bioavailability of the active ingredients in an in vivo setting. Indeed, the need for more in vivo studies to refine the anticancer activity of dietary agents has been highlighted [16]. This article will discuss both the potential mechanisms of anthocyanins in cancer chemoprevention and a model for evaluating their anti-angiogenic properties in vivo.

25.2   CELLULAR MECHANISMS

Activator protein-1 (AP-1) plays a critical role in the process of tumor transformation [39,40], and anthocyanins inhibit AP-1 activity and the transformation process [41,42]. Specifically, increased AP-1 activity is involved in the carcinogenesis phases of tumor promotion and progression both in vitro and in vivo [43–48]. Hou et al. showed that inducible AP-1 activity and neoplastic transformation can be blocked by anthocyanidins, which are anthocyanins lacking a sugar moiety, and that the orthohydroxyphenyl group on the B-ring is critical for the prevention of tumor promotion. They showed that the anthocyanidin delphinidin was the most potent inhibitor of tumor transformation in the TPA-induced transformation of mouse JB6 cells. Delphinidin blocked multiple steps in the mitogen-activated protein (MAP) kinase signaling cascade thought to play a role in the neoplastic process. Delphinidin inhibited the phosphorylation of TPA-inducible MAPK/ERK kinase (MEK), its downstream target ERK, and SAPK/ERK kinase 1 (SEK1) together with its downstream target JNK, as well as phosphorylation of c-Jun [42]. JNK-1 activation causes c-Jun phosphorylation and leads to activation of the transcription factor AP-1 [49,50].

Activator protein-1 has known binding sites on the monocyte chemoattractant protein-1 (MCP-1) promoter [51,52], and constitutive activation of c-Jun has been shown to result in MCP-1 expression in endothelial cells [53]. Monocyte chemoattractant protein-1 is expressed by endothelial cell neoplasms in humans [54], as well as in breast [55,56], bladder [57], and ovarian cancers [58], and melanoma [59]. Expression of MCP-1 correlates with tumor proliferation for endothelial cell neoplasms [54] and breast cancer [60,61]. Breast cancer mortality correlates with MCP-1 levels [56,60]. Thus, inhibition of JNK-1 resulting in decreased AP-1 formation may have significant clinical impact for neoplasms known to express MCP-1. We have demonstrated that MCP-1 is actually required for endothelial neoplasm growth in vivo [62].

25.3   ENDOTHELIAL CELL NEOPLASMS AS AN EXPERIMENTAL MODEL

25.3.1   ENDOTHELIAL CELL NEOPLASMS AS A MODEL OF ANGIOGENESIS

Vascular tumors, specifically endothelial cell neoplasms, are an excellent model that can be used to study the effects of anti-angiogenic agents in vivo. Endothelial cell neoplasms are highly angiogenic, as the growth of these lesions entails endothelial cell proliferation, formation of luminal channels communicating with host vessels, and creation of perfused vascular spaces. Judah Folkman has referred to the growth of endothelial cell neoplasms as "a relatively pure form of angiogenesis." [63] The fact that humans with proliferating hemangiomas, a benign endothelial cell neoplasm, can have urinary basic fibroblast growth factor (bFGF) levels elevated 25–50 fold higher than normal individuals attests to the degree of angiogenic activity associated with endothelial cell neoplasms [64]. Halting the growth of these lesions requires the arrest of angiogenesis that is inherent in this process and is a direct indication of the anti-angiogenic capabilities of the agent being tested. Several investigators have recognized this concept and used endothelial cell neoplasm models to evaluate the anti-angiogenic effects of the thiol antioxidant N-acetylcysteine, batimistat, IL-12, angiostatin, and AGM-1470 [65–69].

25.3.2 MURINE MODELS OF ENDOTHELIAL CELL NEOPLASMS

There are two murine models of endothelial cell neoplasms generated by subcutaneous injection of endothelial cells. One uses murine endothelial cells transformed with the middle T antigen of the murine polyoma virus. The other uses endothelial cells derived from a spontaneously arising hemangioendothelioma (HE) [70–74]. The endothelial cells that are virally transformed are on a mixed major histocompatability complex (MHC) background (H-2^d/H-2^b) [74]. The endothelial cells (EOMA) derived from the spontaneously arising HE are from the 129/J strain, which is commercially available (now called the 129P3/J) with a defined H-2^b MHC background. EOMA cells have also been well characterized with regard to endothelial cell phenotype, protein expression, response to inhibitors of angiogenesis, and even with regard to the development of the Kasabach–Merritt syndrome [67,68,70,75–79]. Mice injected with EOMA cells develop HE within four days with 100% efficiency. The fact that the EOMA cell model is relatively well characterized with a defined MHC background in a commercially available murine strain makes it the preferred model [80].

25.3.3 CLINICAL RELEVANCE OF THE ENDOTHELIAL CELL NEOPLASM MODEL

Any results obtained with this model have direct clinical applicability. Hemangioendothelioma are vascular neoplasms of borderline/intermediate malignancy [81]. They do not metastasize; however, in humans, the mortality rate ranges from 12 to 24% [82]. This lesion is also associated with the development of Kasabach–Merritt syndrome, a consumptive coagulopathy with a 20%–30% mortality rate [83,84]. The fact that mice injected with EOMA cells develop Kasabach–Merritt syndrome is a good indicator of how closely this model mimics the human condition [70]. The overall incidence of endothelial cell neoplasms in children is approximately 10%, and although many of these lesions eventually involute, they have serious or life-threatening complications in 5%–10% of affected children (Figure 25.1), while 50% of all lesions leave permanent residual deformity [85]. Many children and their parents are forced to accept a period of deformity, as shown in Figure 25.2, until involution occurs because of the high risk side effect profiles of current treatment regimens (high-dose steroids or interferon-alpha). Potential complications from steroid or interferon alpha therapy include: fevers, irritability, growth disturbance, immunosuppression, and spastic diplegia (a form of cerebral palsy). Therefore, these therapies are usually reserved for patients with life, limb, or vision-threatening complications and the need for less toxic diet-based treatments is acute.

25.3.4 ADVANTAGES OF USING DIETARY CHEMOPREVENTIVE STRATEGIES

Dietary chemoprevention represents an attractive approach to this problem, especially given the young age of the patient population. Dietary supplements could be taken orally, which is certainly advantageous compared to the daily injections required with interferon therapy. Presumably, the side effect profile would

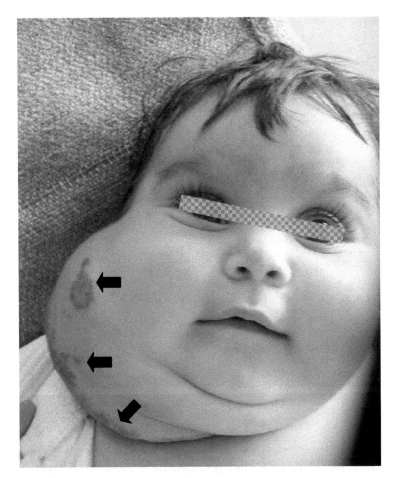

FIGURE 25.1 Six month old female with a large life-threatening vascular neoplasm involving the face, neck, and chest and which is compressing internal airway structures.

be minimal and certainly much more favorable than that associated with the current treatment options. Indeed, the EOMA model has proven efficacy in demonstrating the anti-angiogenic properties of edible berry extracts and for exploring the feasibility of dietary chemoprevention for endothelial cell neoplasms [86].

25.4 ANTI-ANGIOGENIC EFFECTS OF EDIBLE BERRY EXTRACTS

25.4.1 In Vitro Findings

Anthocyanins are flavanoid glycosides with antioxidant properties capable of inhibiting a number of processes involved in tumor angiogenesis. Vascular endothelial growth factor (VEGF) is a critical growth factor required for tumor

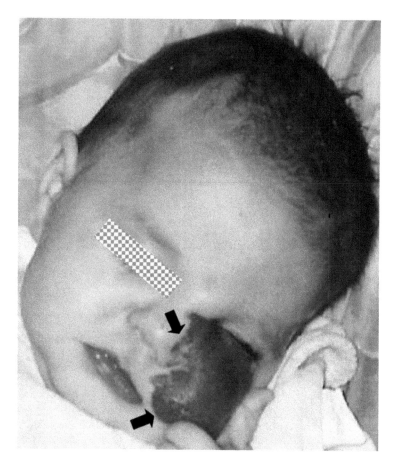

FIGURE 25.2 Three month old female with vascular neoplasm covering the entire left cheek.

angiogenesis and its expression is known to be redox sensitive [3]. Pure flavanoids, such as ferrulic acid, catechin, and rutin, inhibit H_2O_2-inducible VEGF expression, and the combination of multiple flavanoids present in berry extracts seems to have a more potent effect on inhibiting both H_2O_2 and tumor necrosis factor-α (TNF-α) inducible VEGF [32]. We have demonstrated that the expression of MCP-1 in EOMA cells is redox sensitive [80] and that MCP-1 is required for the HE tumor development that occurs with EOMA cell injections in mice [62]. In vitro experiments have demonstrated that berry extracts inhibit inducible MCP-1 expression in EOMA cells and that this effect is mediated by the inhibition of nuclear factor-κB (NF-κB), a redox sensitive transcription factor [86].

25.4.2 IN VIVO EVIDENCE OF FEASIBILITY

The HE model has been productively used to document the anti-angiogenic effects of berry extracts. The incidence of HE formation was significantly reduced [86] in mice injected with EOMA cells treated with blueberry extract or an OptiBerry mix [32]. Mice that received EOMA cells treated with berry extracts—and went on to develop HE—had lesions that were significantly smaller than observed in control mice [86]. This study supports the feasibility to experimentally manipulate EOMA cells in vitro and then observe the effects of that manipulation in vivo. It is also the only reported in vivo model to examine the effects of berry extracts on angiogenesis and tumor growth. There are several studies that demonstrate the ability of other dietary antioxidants to limit tumor growth or vessel development by inhibiting angiogenesis, but most use immunodeficient mice engrafted with human tumor cells [69,87–89]. The host immune response is a significant issue that cannot be overlooked as investigators accumulate the kind of animal data necessary to lead to successful outcomes in clinical trials.

25.5   SUMMARY

Redox-based strategies have the potential to modulate angiogenesis by both inducing or inhibiting it. Dietary chemopreventive strategies using antioxidant edible berries aimed at inhibiting angiogenesis represent an alluring option to treat a variety of neoplasms clinically. While it is unlikely that one approach will fit all cancers, it is imperative to develop models that permit experimental manipulation to identify optimal formulations of berry extracts. As in vivo models become more standardized, the rate of data accumulation and the ability to define how berries affect signal transduction, gene expression, and cell behavior will be enhanced. This will increase the likelihood of success in clinical trials.

ACKNOWLEDGMENTS

Supported by K08GM06696 and RO1GM69589.

REFERENCES

1. Roy, S., Khanna, S., Nallu, K., Hunt, T. K., and Sen, C. K., Dermal wound healing is subject to redox control, *Mol. Ther.*, 13, 211–220, 2006.
2. Maulik, N., Redox signaling of angiogenesis, *Antioxid. Redox Signal.*, 4, 805–815, 2002.
3. Sen, C. K., Khanna, S., Babior, B. M., Hunt, T. K., Ellison, E. C., and Roy, S., Oxidant-induced vascular endothelial growth factor expression in human keratinocytes and cutaneous wound healing, *J. Biol. Chem.*, 277, 33284–33290, 2002.
4. Roy, S., Clark, C. J., Mohebali, K., Bhatt, U., Wallace, W. A., Nahman, N. S., Ellison, E. C., Melvin, W. S., and Sen, C. K., Reactive oxygen species and EGR-1 gene expression in surgical postoperative peritoneal adhesions, *World J. Surg.*, 28, 316–320, 2004.
5. Patel, V., Chivukula, I. V., Roy, S., Khanna, S., He, G., Ojha, N., Mehrotra, A., Dias, L. M., Hunt, T. K., and Sen, C. K., Oxygen: from the benefits of inducing VEGF expression to managing the risk of hyperbaric stress, *Antioxid. Redox Signal.*, 7, 1377–1387, 2005.

6. Fries, R. B., Wallace, W. A., Roy, S., Kuppusamy, P., Bergdall, V., Gordillo, G. M., Melvin, W. S., and Sen, C. K., Dermal excisional wound healing in pigs following treatment with topically applied pure oxygen, *Mutat. Res.*, 579, 172–181, 2005.

7. Sen, C. K., The general case for redox control of wound repair, *Wound Repair Regen.*, 11, 431–438, 2003.

8. Gordillo, G. M. and Sen, C. K., Revisiting the essential role of oxygen in wound healing, *Am. J. Surg.*, 186, 259–263, 2003.

9. Sen, C. K., Khanna, S., Gordillo, G., Bagchi, D., Bagchi, M., and Roy, S., Oxygen, oxidants, and antioxidants in wound healing: an emerging paradigm, *Ann. N.Y. Acad. Sci.*, 957, 239–249, 2002.

10. Sen, C. K., Khanna, S., Babior, B. M., Hunt, T. K., Ellison, E. C., and Roy, S., Oxidant-induced vascular endothelial growth factor expression in human keratinocytes and cutaneous wound healing, *J. Biol. Chem.*, 277, 33284–33290, 2002.

11. Sen, C. K., Khanna, S., Venojarvi, M., Trikha, P., Ellison, E. C., Hunt, T. K., and Roy, S., Copper-induced vascular endothelial growth factor expression and wound healing, *Am. J. Physiol. Heart Circ. Physiol.*, 282, H1821–H1827, 2002.

12. Khanna, S., Roy, S., Bagchi, D., Bagchi, M., and Sen, C. K., Upregulation of oxidant-induced VEGF expression in cultured keratinocytes by a grape seed proanthocyanidin extract, *Free Radic. Biol. Med.*, 31, 38–42, 2001.

13. Zeisel, S., Health-regulation of "nutraceuticals", *Science*, 285, 1853–1855, 1999.

14. Couzin, J., Alternative medicine beefed-up NIH center probes unconventional therapies, *Science*, 282, 2175–2176, 1998.

15. Ofek, I., Goldhar, J., Zafriri, D., Lis, H., Adar, R., and Sharon, N., Anti-*Escherichia coli* adhesin activity of cranberry and blueberry juices, *N. Engl. J. Med.*, 324, 1599, 1991.

16. Colic, M. and Pavelic, K., Molecular mechanisms of anticancer activity of natural dietetic products, *J. Mol. Med.*, 78, 333–336, 2000.

17. Kresty, L. A., Morse, M. A., Morgan, C., Carlton, P. S., Lu, J., Gupta, A., Blackwood, M., and Stoner, G. D., Chemoprevention of esophageal tumorigenesis by dietary administration of lyophilized black raspberries, *Cancer Res.*, 61, 6112–6119, 2001.

18. McCarty, M. F., Current prospects for controlling cancer growth with non-cytotoxic agents—nutrients, phytochemicals, herbal extracts, and available drugs, *Med. Hypotheses*, 56, 137–154, 2001.

19. She, Q. B., Bode, A. M., Ma, W. Y., Chen, N. Y., and Dong, Z., Resveratrol-induced activation of p53 and apoptosis is mediated by extracellular-signal-regulated protein kinases and p38 kinase, *Cancer Res.*, 61, 1604–1610, 2001.

20. Xue, H., Aziz, R. M., Sun, N., Cassady, J. M., Kamendulis, L. M., Xu, Y., Stoner, G. D., and Klaunig, J. E., Inhibition of cellular transformation by berry extracts, *Carcinogenesis*, 22, 351–356, 2001. [erratum appears in *Carcinogenesis*; 22 (5), 831–833]

21. Cao, G., Muccitelli, H. U., Sanchez-Moreno, C., and Prior, R. L., Anthocyanins are absorbed in glycated forms in elderly women: a pharmacokinetic study, *Am. J. Clin. Nutr.*, 73, 920–926, 2001.

22. Cao, G. and Prior, L., Anthocyanins are detected in human plasma after oral administration of an elderberry extract, *Clin. Chem.*, 45, 574–576, 1999.

23. Cao, G., Russell, R. M., Lischner, N., and Prior, R. L., Serum antioxidant capacity is increased by consumption of strawberries, spinach, red wine or vitamin C in elderly women, *J. Nutr.*, 128, 2383–2390, 1998.

24. Giavazzi, R. and Taraboletti, G., Angiogenesis and angiogenesis inhibitors in cancer, *Forum*, 9, 261–272, 1999.

25. Griffioen, A. W. and Molema, G., Angiogenesis: potentials for pharmacologic intervention in the treatment of cancer, cardiovascular diseases, and chronic inflammation, *Pharmacol. Rev.*, 52, 237–268, 2000.

26. Fotsis, T., Pepper, M. S., Aktas, E., Breit, S., Rasku, S., Adlercreutz, H., Wahala, K., Montesano, R., and Schweigerer, L., Flavonoids, dietary-derived inhibitors of cell proliferation and in vitro angiogenesis, *Cancer Res.*, 57, 2916–2921, 1997.

27. Fotsis, T., Pepper, M. S., Montesano, R., Aktas, E., Breit, S., Schweigerer, L., Rasku, S., Wahala, K., and Adlercreutz, H., Phytoestrogens and inhibition of angiogenesis, *Baillieres Clin. Endocrinol. Metab.*, 12, 649–666, 1998.

28. Hayashi, A., Gillen, A. C., and Lott, R., Effects of daily oral administration of quercetin chalcone and modified citrus pectin on implanted colon-25 tumor growth in Balb-c mice, *Altern. Med. Rev.*, 5, 546–552, 2000.

29. Hisa, T., Kimura, Y., Takada, K., Suzuki, F., and Takigawa, M., Shikonin, an ingredient of *Lithospermum erythrorhizon*, inhibits angiogenesis in vivo and in vitro, *Anticancer Res.*, 18, 783–790, 1998.

30. Jiang, C., Agarwal, R., and Lu, J., Anti-Angiogenic potential of a cancer chemopreventive flavonoid antioxidant, silymarin: inhibition of key attributes of vascular endothelial cells and angiogenic cytokine secretion by cancer epithelial cells, *Biochem. Biophys. Res. Commun.*, 276, 371–378, 2000.

31. Paper, D., Natural products as angiogenesis inhibitors, *Planta Med.*, 64, 686–695, 1998.

32. Roy, S., Khanna, S., Alessio, H. M., Vider, J., Bagchi, D., Bagchi, M., and Sen, C. K., Anti-Angiogenic property of edible berries, *Free Radic. Res.*, 36, 1023–1031, 2002.

33. Pool-Zobel, B. L., Bub, A., Schroder, N., and Rechkemmer, G., Anthocyanins are potent antioxidants in model systems but do not reduce endogenous oxidative DNA damage in human colon cells, *Eur. J. Nutr.*, 38, 227–234, 1999.

34. Kahkonen, M. P., Hopia, A. I., and Heinonen, M., Berry phenolics and their antioxidant activity, *J. Agric. Food Chem.*, 49, 4076–4082, 2001.

35. Youdim, K. A., Martin, A., and Joseph, J. A., Incorporation of the elderberry anthocyanins by endothelial cells increases protection against oxidative stress, *Free Radic. Biol. Med.*, 29, 51–60, 2000.

36. Bagchi, D., Sen, C., Bagchi, M., and Atalay, M., Anti-Angiogenic, antioxidant, and anti-carcinogenic properties of a novel anthocyanin-rich berry extract formula, *Biochemistry (Moscow)*, 69, 75–80, 2004.

37. Chow, L. M., Chui, C. H., Tang, J. C., Lau, F. Y., Yau, M. Y., Cheng, G. Y., Wong, R. S. et al., Anti-Angiogenic potential of Gleditsia sinensis fruit extract, *Int. J. Mol. Med.*, 12, 269–273, 2003.

38. Liu, Z., Schwimer, J., Liu, D., Greenway, F., Anthony, C., and Woltering, E., Black raspberry extract and fractions contain angiogenesis inhibitors, *J. Agric. Food Chem.*, 53, 3909–3915, 2005.

39. Li, J. J., Dong, Z., Dawson, M. I., and Colburn, N. H., Inhibition of tumor promoter-induced transformation by retinoids that transrepress AP-1 without transactivating retinoic acid response element, *Cancer Res.*, 56, 483–489, 1996.

40. Dong, Z., Birrer, M. J., Watts, R. G., Matrisian, L. M., and Colburn, N. H., Blocking of tumor promoter-induced AP-1 activity inhibits induced transformation in JB6 mouse epidermal cells, *Proc. Natl Acad. Sci. U.S.A.*, 91, 609–613, 1994.

41. Ding, M., Lu, Y., Bowman, L., Huang, C., Leonard, S., Wang, L., Vallyathan, V., Castranova, V., and Shi, X., Inhibition of AP-1 and neoplastic transformation by fresh apple peel extract, *J. Biol. Chem.*, 279, 10670–10676, 2004.

42. Hou, D. X., Kai, K., Li, J. J., Lin, S., Terahara, N., Wakamatsu, M., Fujii, M., Young, M. R., and Colburn, N., Anthocyanidins inhibit activator protein 1 activity and cell transformation: structure–activity relationship and molecular mechanisms, *Carcinogenesis*, 25, 29–36, 2004.

43. Chen, N., Nomura, M., She, Q. B., Ma, W. Y., Bode, A. M., Wang, L., Flavell, R. A., and Dong, Z., Suppression of skin tumorigenesis in c-Jun NH(2)-terminal kinase-2-deficient mice, *Cancer Res.*, 61, 3908–3912, 2001.

44. Young, M. R., Li, J. J., Rincon, M., Flavell, R. A., Sathyanarayana, B. K., Hunziker, R., and Colburn, N., Transgenic mice demonstrate AP-1 (activator protein-1) transactivation is required for tumor promotion, *Proc. Natl Acad. Sci. U.S.A.*, 96, 9827–9832, 1999.

45. Dumont, J. A., Bitonti, A. J., Wallace, C. D., Baumann, R. J., Cashman, E. A., and Cross-Doersen, D. E., Progression of MCF-7 breast cancer cells to antiestrogen-resistant phenotype is accompanied by elevated levels of AP-1 DNA-binding activity, *Cell Growth Differ.*, 7, 351–359, 1996.

46. Risse-Hackl, G., Adamkiewicz, J., Wimmel, A., and Schuermann, M., Transition from SCLC to NSCLC phenotype is accompanied by an increased TRE-binding activity and recruitment of specific AP-1 proteins, *Oncogene*, 16, 3057–3068, 1998.

47. Dong, Z., Crawford, H. C., Lavrovsky, V., Taub, D., Watts, R., Matrisian, L. M., and Colburn, N. H., A dominant negative mutant of jun blocking 12-*O*-tetradecanoyl-phorbol-13-acetate-induced invasion in mouse keratinocytes, *Mol. Carcinog.*, 19, 204–212, 1997.

48. Saez, E., Rutberg, S. E., Mueller, E., Oppenheim, H., Smoluk, J., Yuspa, S. H., and Spiegelman, B. M., c-Fos is required for malignant progression of skin tumors, *Cell*, 82, 721–732, 1995.

49. Minden, A., Lin, A., Smeal, T., Derijard, B., Cobb, M., Davis, R., and Karin, M., c-Jun N-terminal phosphorylation correlates with activation of the JNK subgroup but not the ERK subgroup of mitogen-activated protein kinases, *Mol. Cell. Biol.*, 14, 6683–6688, 1994.

50. Ventura, J. J., Kennedy, N. J., Lamb, J. A., Flavell, R. A., and Davis, R. J., c-Jun NH(2)-terminal kinase is essential for the regulation of AP-1 by tumor necrosis factor, *Mol. Cell. Biol.*, 23, 2871–2882, 2003.

51. Martin, T., Cardelli, P., Parry, G., Felts, K., and Cobb, R., Cytokine induction of monocyte chemoattractant protein-1 gene expression in human endothelial cellsdepends upon the cooperative action of NF-kappaB and AP-1, *Eur. J. Immunol.*, 27, 1091–1097, 1997.

52. Sica, A., Wang, J., Collota, F., DeJana, E., Montovani, A., Larsen, C., Zachariae, C., and Matsushima, K., Monocyte chemotactic and activating factor gene expression induced in endothelial cells by IL-1 and tumor necrosis factor, *J. Immunol.*, 144, 3034–3038, 1990.

53. Wang, N., Verna, L., Hardy, S., Forsayeth, J., Zhu, Y., and Stemerman, M. B., Adenovirus-mediated overexpression of c-Jun and c-Fos induces intercellular adhesion molecule-1 and monocyte chemoattractant protein-1 in human endothelial cells, *Arterioscler. Thromb. Vasc. Biol.*, 19, 2078–2084, 1999.

54. Isik, F., Rand, R., Gruss, J., Benjamin, D., and Alpers, C., Monocyte chemoattractant protein-1 mRNA expression in hemangiomas and vascular malformations, *J. Surg. Res.*, 61, 71–76, 1996.

55. Saji, H., Koike, M., Yamori, T., Saji, S., Seiki, M., Matsushima, K., and Toi, M., Significant correlation of monocyte chemoattractant protein-1 expression with neovascularization and progression of breast carcinoma, *Cancer*, 92, 1085–1091, 2001.

56. Ueno, T., Toi, M., Saji, H., Muta, M., Bando, H., Kuroi, K., Koike, M., Inadera, H., and Matsushima, K., Significance of macrophage chemoattractant protein-1 in macrophage recruitment, angiogenesis, and survival in human breast cancer, *Clin. Cancer Res.*, 6, 3282–3289, 2000.

57. Amann, B., Pearbo, F., Wirger, A., Hugenschmidt, H., and Schultze-Seeman, W., Urinaryl levels of monocyte chemoattractant protein-1 correlate with tumor stage and grade in patients with bladder cancer, *Br. J. Urol.*, 82, 118–121, 1998.

58. Hefler, L., Tempfer, C., Heinze, G., Mayerhofer, K., Breitenecker, G., Leodolter, S., Reinthaller, A., and Kainz, C., Monocyte chemoattractant protein-1 serum levels in ovarian cancer patients, *Br. J. Cancer*, 81, 855–859, 1999.

59. Torisu, H., Ono, M., Kiryu, H., Furue, M., Ohmoto, Y., Nakayama, J., Nishioka, Y., Sone, S., and Kuwano, M., Macrophage infiltration correlates with tumor stage and angiogenesis in human malignant melanoma: possible involvement of TNFalpha and IL-1alpha, *Int. J. Cancer*, 85, 182–188, 2000.

60. Leek, R. D., Lewis, C. E., Whitehouse, R., Greenall, M., Clarke, J., and Harris, A. L., Association of macrophage infiltration with angiogenesis and prognosis in invasive breast carcinoma, *Cancer Res.*, 56, 4625–4629, 1996.

61. Bingle, L., Brown, N., and Lewis, C., The role of tumor-associated macrophages in tumor progression: implications for new anticancer therapies, *J. Pathol.*, 196, 254–265, 2002.

62. Gordillo, G., Onat, D., Stockinger, M., Roy, S., Atalay, M., Beck, M., and Sen, C., A key angiogenic role of monocyte chemoattractant protein-1 in hemangioendothelioma proliferation, *Am. J. Physiol. (Cell)*, 287, C866–C873, 2004.

63. Folkman, J., Mulliken, J., and Ezekowitz, R., Angiogenesis and hemangiomas, In *Surgery of Infants and Children: Scientific Principles and Practice*, Oldham, K. Ed., Lippincott-Raven, Philadelphia, PA, pp. 569–584, 1997.

64. Takahashi, K., Mulliken, J., Kozakewich, H., Rogers, R., Folkman, J., and Ezekowitz, R., Cellular markers that distinguish the phases of hemangioma during infancy and childhood, *J. Clin. Invest.*, 93, 2357–2364, 1994.

65. Wang, C., Quevedo, M. E., Lannutti, B. J., Gordon, K. B., Guo, D., Sun, W., and Paller, A. S., In vivo gene therapy with interleukin-12 inhibits primary vascular tumor growth and induces apoptosis in a mouse model, *J. Invest. Dermatol.*, 112, 775–781, 1999.

66. Taraboletti, G., Garofalo, A., Belotti, D., Drudis, T., Borsotti, P., Scanziani, E., Brown, P. D., and Giavazzi, R., Inhibition of angiogenesis and murine hemangioma growth by batimastat, a synthetic inhibitor of matrix metalloproteinases, *J. Natl Cancer Inst.*, 87, 293–298, 1995.

67. Lannutti, B., Gately, S., Quevedo, M., Soff, G., and Paller, A., Human angiostatin inhibits murine hemangioendothelioma tumor growth in vivo, *Cancer Res.*, 57, 5277–5280, 1997.

68. O'Reilly, M., Brem, H., and Folkman, J., Treatment of murine hemangioendotheliomas with the angiogenesis inhibitor AGM-1470, *J. Pediatr. Surg.*, 30, 325–330, 1995.

69. Albini, A., Morini, M., D'Agostini, F., Ferrari, N., Campelli, F., Arena, G., Noonan, D. M., Pesce, C., and De Flora, S., Inhibition of angiogenesis-driven Kaposi's sarcoma tumor growth in nude mice by oral *N*-acetylcysteine, *Cancer Res.*, 61, 8171–8178, 2001.

70. Warner, E. D., Hoak, J. C., and Fry, L., Hemangioma, thrombocytopenia, and anemia. The Kasabach–Merritt syndrome in an animal model, *Arch. Pathol. Lab. Med.*, 91, 523–528, 1971.

71. Hoak, J. C., Warner, E. D., Cheng, H. F., Fry, G. L., and Hankenson, R. R., Hemangioma with thrombocytopenia and microangiopathic anemia (Kasabach–Merritt syndrome): an animal model, *J. Lab. Clin. Med.*, 77, 941–950, 1971.

72. Bautch, V. L., Toda, S., Hassell, J. A., and Hanahan, D., Endothelial cell tumors develop in transgenic mice carrying polyoma virus middle T oncogene, *Cell*, 51, 529–537, 1987.

73. Montesano, R., Pepper, M., Mohle-Steinlein, U., Risau, W., Wagner, E., and Orci, L., Increased proteolytic activity is responsible for the aberrant morphogenetic behavior of endotehlial cells expressing middle T oncogene, *Cell*, 62, 435–445, 1990.

74. Taraboletti, G., Belotti, D., Dejana, E., Montovani, A., and Giavazzi, R., Endothelial cell migration and invasiveness are induced by a soluble factor produced by murine endothelioma cells transformed by polyoma virus middle T oncogene, *Cancer Res.*, 53, 3812–3816, 1993.

75. Obeso, J., Weber, J., and Auerbach, R., A hemangioendothelioma-derived cell line: its use as a model for the study of endothelial cell biology, *Lab. Invest.*, 63, 259–269, 1990.

76. O'Reilly, M., Boehm, T., Shing, Y., Fukai, N., Vasios, G., Lane, W., Flynn, E., Birkhead, J., Olsen, B., and Folkman, J., Endostatin: an endogenous inhibitor of angiogenesis and tumor growth, *Cell*, 88, 277–285, 1997.

77. Felbor, U., Dreier, L., Bryant, R., Ploegh, H., Olsen, B., and Mothes, W., Secreted cathepsin L generates endostatin from collagen XVIII, *EMBO J.*, 19, 1187–1194, 2000.

78. Sage, H. and Bornstein, P., Endothelial cells from umbilical vein and hemangioendothelioma secrete basement mebrane largely to the exclusion of interstitial procollagens, *Arteriosclerosis*, 2, 27–36, 1982.

79. Wei, W., Moses, M., Wiederschain, D., Arbiser, J., and Folkman, J., The generation of endostatin is mediated by elastase, *Cancer Res.*, 59, 6052–6056, 1999.

80. Gordillo, G. M., Atalay, M., Roy, S., and Sen, C. K., Hemangioma model for in vivo angiogenesis: inducible oxidative stress and MCP-1 expression in EOMA cells, *Methods Enzymol.*, 352, 422–432, 2002.

81. Enzinger, F. and Weiss, S., *Soft Tissue Tumors*, C.V. Mosby, St. Louis, MO, 1983.

82. Esterly, N., Cutaneous hemangiomas, vascular stains and malformations, and associated syndromes, *Curr. Probl. Dermatol.*, 7, 65–108, 1995.

83. Sarkar, M., Mulliken, J., Kozakewich, H., Robertson, R., and Burrows, P., Thrombocytopenic coagulopathy (Kasabach–Merritt Phenomenon) is associated with Kaposiform Hemangioendothelioma and not common infantile hemangioma, *Plast. Reconstr. Surg.*, 100, 1377–1386, 1997.

84. Zuckerberg, L., Nickoloff, B., and Weiss, S., Kaposiform hemangioendothelioma of infancy and childhood: an aggressive neoplasm associated with Kasabach–Merritt syndrome and lymphangiomatosis, *Am. J. Surg. Pathol.*, 17, 321–328, 1993.

85. Paller, A. S., Responses to anti-angiogenic therapies, *J. Investig. Dermatol. Symp. Proc.*, 5, 83–86, 2000.

86. Atalay, M., Gordillo, G., Roy, S., Rovin, B., Bagchi, D., Bagchi, M., and Sen, C. K., Anti-Angiogenic property of edible berry in a model of hemangioma, *FEBS Lett.*, 544, 252–257, 2003.

87. Shklar, G. and Schwartz, J. L., Vitamin E inhibits experimental carcinogenesis and tumor angiogenesis, *Eur. J. Cancer, Part B, Oral Oncol.*, 32B, 114–119, 1996.

88. Marikovsky, M., Thiram inhibits angiogenesis and slows the development of experimental tumors in mice, *Br. J. Cancer*, 86, 779–787, 2002.

89. Malafa, M. P., Fokum, F. D., Smith, L., and Louis, A., Inhibition of angiogenesis and promotion of melanoma dormancy by vitamin E succinate, *Ann. Surg. Oncol.*, 9, 1023–1032, 2002.

26 A Novel Nutrient Mixture Containing Ascorbic Acid, Lysine, Proline, and Green Tea Extract Inhibits Critical Parameters in Angiogenesis

M. Waheed Roomi, V. Ivanov, T. Kalinovsky,
A. Niedzwiecki, and M. Rath

CONTENTS

26.1   INTRODUCTION

Angiogenesis, the formation of new capillaries from existing blood vessels, is essential for progressive growth, invasion, and metastasis of solid tumors. Over 2500 scientific reports demonstrate the dependency of tumor growth on angiogenesis [1]. Angiogenesis not only allows the tumor to increase in size, but it also provides a route for metastasis to distal sites in the body. The degree of vascularization in a tumor has been correlated with the metastatic potential and prognosis of the disease [2].

The regulation of angiogenesis is achieved through a balance of pro- and anti-angiogenic stimuli. Two major factors driving angiogenesis are matrix metallopro-teinases (MMPs) that degrade ECM, and vascular endothelial growth factor (VEGF), a stimulatory factor for cell migration. The prevention of ECM degradation through the inhibition of MMP activity, in particular MMP-2 (gelatinase A) and MMP-9 (gelatinase B), has been shown to be a promising therapeutic approach to blocking the invasion process that occurs during angiogenesis and tumor progression. Vascular endothelial growth factor is specific and critical for blood vessel formation, and is one of the most powerful stimulators of angiogenesis.

Blood vessels local to the tumor respond to the malignant cells' elaboration of VEGF and fibroblast growth factor (FGF), inducing local blood vessels to sprout branches to feed the metastases. This causes small micrometastases to grow beyond the 2-mm size, which is functionally dormant, and become a threat to the patient when rapid growth causes local damage [3]. Vascular endothelial growth factor is secreted by tumor cells and promotes the proliferation of endothelial cells by binding to cell surface receptors, as well as the migration toward the tumor. Since endothelial cells can communicate directly with tumor cells by producing growth-promoting factors, the interrelationship between endothelial and tumor cells and the imbalance between angiogenic factors and angiogenic inhibitors can promote tumor vascularization. Other stimulating factors include angiopoeitn-1, epidermal growth factor (EGF), tumor necrosis factor alpha (TNF-α), interleukin (IL)-1, IL -6, IL-8, and platelet derived growth factor (PDGF).

Physiologically, angiogenesis is suppressed by one or more of the known endogenous inhibitors, such as angiostatin, endostatin, thrombobspondin, and tissue inhibitors of metalloproteinases. Angiostatin, a fragment of plasminogen endogenously produced by tumors, is found naturally in significant amounts in the circulation of patients with primary tumors; angiostatin levels can control metastatic cell proliferation until a primary tumor is removed [4,5]. However, when primary tumors are surgically removed, the endogenous levels of angiostatin and other inhibitors decreases and micrometastases that have previously seeded elsewhere in the body are allowed to grow. Angiostatin, however, has a short half-life of the peptide, requiring continuous administration. Efforts have been made to identify other anti-angiogenic agents as potential cancer treatments.

Earlier work by Rath and Pauling [6] defined common pathomechanisms for all cancers, the destruction of ECM as a precondition for cancer cell invasion, metastasis, and angiogenesis, and suggested intervention through natural inhibitors of plasmin-induced proteolysis, such as lysine and its analogues. The prevention of

ECM degradation through the inhibition of MMP activity, in particular MMP-2 (gelatinase A) and MMP-9 (gelatinase B), has been shown to be a promising therapeutic approach to blocking the invasion process that occurs during angiogenesis and tumor progression.

The identification of novel angiogenic inhibitors that target both proliferating endothelial and tumor cells and MMP inhibitors may, therefore, lead to the therapeutic regulation of tumor growth. Most angiogenic inhibitors also act as anti-invasive or antimetastatic agents. Recently, several MMP inhibitors and anti-angiogenic agents have been developed. An increasing number of clinical trials are testing the therapeutic efficacy and tolerance of angiogenic agents, targeting MMPs, angiogenic growth factors, and their receptors [7].

Our previous work confirmed the direction described by Rath and Pauling [6] and resulted in identifying a novel formulation of lysine, proline, ascorbic acid, and EGCG-enriched green tea extract (NM) that has shown significant anti-cancer activity against a large number of cancer cell lines—blocking cancer cells' growth, tissue invasion and MMPs' activity both in vitro and in vivo [8–14]. The aim of this study was to determine the effectiveness of this novel nutrient formulation as an inhibitor of angiogenesis using both in vitro and in vivo models.

26.2 COMPOSITION OF THE NUTRIENT MIXTURE

The nutrient mixture (NM) is composed of the following relative amounts of components: Vitamin C (as ascorbic acid and as Mg, Ca, and palmitate ascorbate) 700 mg, L-lysine 1000 mg, L-proline 750 mg, L-arginine 500 mg, N-acetyl cysteine 200 mg, standardized green tea extract 1000 mg (green tea extract derived from green tea leaves was obtained from U.S. Pharma Lab). The certificate of analysis indicates the following characteristics: total polyphenol 80%, catechins 60%, EGCG 35%, and caffeine 1.0% (80% polyphenol), selenium 30 μg, copper 2 mg, and manganese 1 mg.

We formulated and tested NM because we were looking at the multiple effects of cancer inhibition at different stages of cancer progression and metastasis. For example, the ECM integrity is dependent upon adequate collagen formation; the amino acids lysine and proline are necessary for formation of collagen chains, and ascorbic acid is essential for the hydroxylation reaction. Manganese and copper are also essential for collagen formation. Ascorbic acid has also been shown to inhibit cell division and growth through production of hydrogen peroxide [15]. Green tea extract has been shown to be a promising agent in controlling angiogenesis, metastasis, and other aspects of cancer [16]. N-acetyl cysteine has been observed to inhibit MMP-9 activity [17] and the invasive activities of tumor cells [18], as well as endothelial tissue invasion [7]. Selenium has been shown to interfere with MMP expression and tumor invasion [19], as well as the migration of endothelial cells through ECM [7,18]. Since arginine is a precursor of nitric oxide (NO), any deficiency of arginine can limit the production of NO, which has been shown to predominantly act as an inducer of apoptosis, as in breast cancer cells [20].

Based on the evidence available in literature and our own research, we hypothesized that a combination of ascorbic acid, lysine, proline, green tea extract, arginine, *N*-acetyl cysteine, selenium, copper, and manganese would work synergistically. For example, we found that a combination of ascorbic acid, lysine, and proline used with EGCG enhanced the anti-invasive activity of 20 μg/ml EGCG to that of 50 μg/ml [21]. Thus by including nutrients like *N*-acetyl cysteine, arginine, selenium, manganese, and copper, in addition to ascorbic acid, proline, lysine, and EGCG, we could obtain a significant reduction in cell invasion at a much lower concentration of EGCG.

The presence of an adequate blood supply is required for the growth and metastasis of malignant tumors; thus, inhibition of tumor-induced angiogenesis represents a promising approach for cancer therapy. A number of in vivo and in vitro models have been developed facilitating the study of angiogenesis. Using various in vivo and in vitro models, we demonstrated that the nutrient mixture of lysine, proline, ascorbic acid, and green tea extract had anti-angiogenic properties.

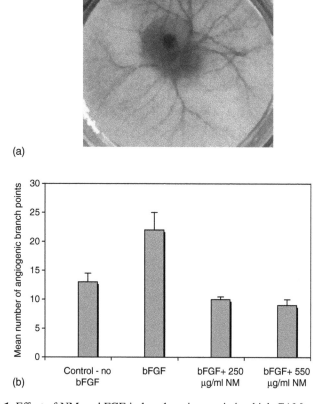

FIGURE 26.1 Effect of NM on bFGF-induced angiogenesis in chick CAM assay.

26.3 THE EFFECT OF NM ON SURROGATE MODELS FOR ANGIOGENESIS

Angiogenesis induces a release of various angiogenic factors, among them bFGF. We applied the chick embryo chorioallantoic membrane (CAM) assay to test the effects of NM, as this is a comprehensive in vitro system in tissue that incorporates all angiogenic processes in one mode. This assay utilizes a microenvironment in which angiogenesis naturally occurs and provides a good model for evaluation of systemically administered antagonists. In addition, it allows selection of inhibitors of angiogenesis that interfere with new blood vessel development without affecting pre-existing vessels.

26.3.1 CAM STUDY

The chick CAM angiogenesis assay was performed essentially as described by Brooks, et al. [22]. Briefly, the CAMs of 10-day old chick embryos were separated from the shell membrane. Filter discs previously coated with cortisone acetate were saturated with 15 µl of recombinant bFGF at a concentration of 1.0 µg/ml. The embryos were allowed to incubate for a total of 24 h. The embryos were next treated with a single I.V. injection of NM (250 or 500 µg/embryo) in a total volume of 100 µl. At the end of a 3-day incubation period, the embryos were sacrificed and the CAMs were ressected and washed. The number of branching angiogenic blood vessels were counted within the confines of the filter discs for each CAM for each experimental condition. The nutrient mixture caused a significant ($P < 0.50$) reduction (from 22 to 10 blood vessel branch points within the confined region of the filter disc) in bFGF-induced angiogenesis as compared to no treatment (bFGF only) ((Figure 26.1) [23]. The number of blood vessel branch points is relative to the number of newly sprouting angiogenic vessels.

26.3.2 IN VIVO MOUSE MATRIGEL PLUG ASSAY

The anti-angiogenic effects of NM observed in the CAM study were congruent with our in vivo mouse Matrigel study, which showed that NM, included as a component of a diet, strongly suppressed bFGF-induced angiogenesis in C57BL/6J female mice. To investigate the anti-angiogenic potential of NM, an extract of basement membrane proteins (Matrigel) impregnated with bFGF, an inducer of neovascularization, was injected subcutaneously into C57BL/6J female mice; Passaniti [24] found that a subcutaneous injection of Matrigel, supplemented with angiogenic factors, into C57BL/6J mice reconstituted into a gel and supported an intense vascular response.

The mouse Matrigel plug assay was performed as described by Passaniti, et al. [24]. Nutrient mixture 5 mg/ml and bFGF 400 ng/ml in PBS were mixed with Matrigel in proportions not exceeding 1% of the total volume of Matrigel. A mixture of 0.5 ml Matrigel with bFGF with NM was injected s.c into four C57BL/6J female mice and the mixture of 0.5 ml Matrigel with bFGF in vehicle were injected s.c. into another group of four C57BL/6J female mice. After seven days, mice were

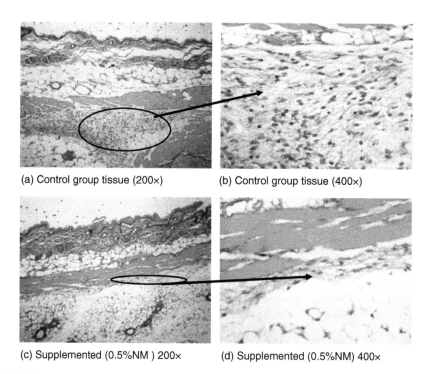

(a) Control group tissue (200×)   (b) Control group tissue (400×)

(c) Supplemented (0.5%NM) 200×   (d) Supplemented (0.5%NM) 400×

FIGURE 26.2 Effect of NM on bFGF-induced vessel growth in C57BL/6J female mice.

sacrificed, skin was excised, fixed, and stained with H&E and by the Masson-Trichrome method—and representative photographs were taken. The test group of mice received NM in the injection mixture and the control mice received just the vehicle. After seven days, red blood cells were abundant within the lumen of numerous vessels in the control mice (Figure 26.2a and Figure 26.2b). In contrast, NM strongly suppressed the bFGF-stimulated angiogenesis in supplemented mice (Figure 26.2c and Figure 26.2d) [23].

26.4 THE EFFECT OF NM ON HUMAN OSTEOSARCOMA MNNG–HOS CELLS IN VIVO

We also tested the nutrient mixture on tumor growth in vivo by implanting human osteosarcoma MNNG–HOS cells in athymic male nude mice and treating one group of mice with nutrient-supplemented (NM 0.5%) Purina mouse chow and the other group with unsupplemented Purina mouse chow.

Human osteosarcoma cells MNNG–HOS (ATCC, Rockville, MD) were maintained in MEM culture, supplemented with 10% fetal bovine serum, 100 U/ml penicillin, and 100 μg/ml streptomycin. The media and sera used were obtained from ATCC, and antibiotics (penicillin and streptomycin) were from Gibco BRL, Long Island, NY. At near confluence, the cultured cells were detached by trypsinizing, washed with PBS, and diluted and emulsified to a concentration of

3×10^6 cells in 0.2 ml PBS and 0.1 ml Matrigel (BD Bioscience, Bedford, MA) for inoculation.

Male athymic nude mice (NCr–nu/nu), approximately six weeks of age on arrival, (Simonsen Laboratories, Gilroy, CA) were maintained in microinsulator cages under pathogen-free conditions on a 12-hour light/12-hour dark schedule for a week. All animals were cared for in accordance with institutional guidelines for the care and use of experimental animals. After housing for a week, the mice were inoculated with 3×10^6 human osteosarcoma MNNG–HOS cells in 0.2 ml of PBS and 0.1 ml of Matrigel. After implantation, the 12 mice were randomly divided into two groups of six mice. From day one, mice from Group A were fed a Purina mouse chow diet and those in Group B were fed a Purina mouse chow diet supplemented with 0.5% NM. After four weeks, the mice were sacrificed, tumors were excised, measured, weighed, fixed in 10% (v/v) buffered formalin, and processed for histology.

Tissue samples were fixed in 10% buffered formalin. All tissues were embedded in paraffin and cut at 4–5 μm. Sections were deparaffinized through xylene and a graduated alcohol series to water, and incubated for 5 min in aqueous 3% hydrogen peroxide to block endogenous peroxidase. Histological sections were stained with hematoxylin and eosin (H&E) stain for evaluation using a standard light microscope. Immunochemical studies were performed on formalin-fixed, paraffin-embedded sections. Standard immunohistochemical staining procedures were used for staining antibodies. After deparaffinization and appropriate epitope retrieval, the sections were incubated with primary antibody. Detection was by biotinylated goat anti-mouse antibodies followed by streptavidin conjugated to horseradish peroxidase with the use of diaminobenzidine as the chromogen. Polyclonal rabbit anti-human antibodies used for MMP-9 and VEGF were obtained from Santa Cruz Biotechnology, Inc., CA, and from Sigma.

The results provided further evidence of the significant cancer protecting effect of this specific nutrient mixture in nude mice. Nude mice on nutrient-supplemented diet with implants of the highly metastatic human osteosarcoma MNNG–HOS cells developed significantly smaller tumors (by 53%, $p = 0.0001$) and less vascular ones than did the tumors grown in the control group of nude mice [23] (see Figure 26.3a through Figure 26.3d). Furthermore, immunohistochemical analysis of these tumors demonstrated lower VEGF staining, an indicator of angiogenesis, in nutrient-supplemented animals compared to the control group of mice. This was accompanied by lower cytoplasmic staining for MMP-9 in tumor tissues taken from nutrient-supplemented mice than those fed the standard diet, indicating decreased matrix degradation (Figure 26.4) [23].

26.5 THE EFFECT OF NM ON IN VITRO STUDIES IN HUMAN OSTEOSARCOMA U2OS CELLS

We also tested the effect of the nutrient mixture on tumor cells in culture, with an eye towards the invasive and angiogenic potential. Human osteosarcoma cells U2OS (ATCC) were grown in McCoy medium supplemented with 10% fetal bovine serum,

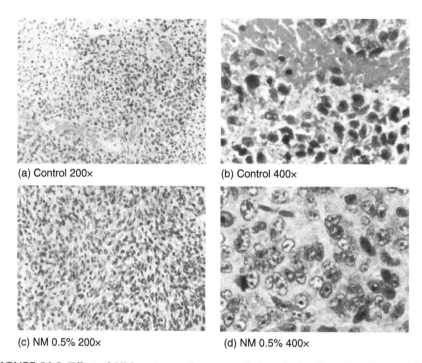

(a) Control 200×   (b) Control 400×

(c) NM 0.5% 200×   (d) NM 0.5% 400×

FIGURE 26.3 Effect of NM on tumor tissue vascularity of athymic nude mice receiving osteosarcoma MNNG xenografts.

penicillin (100 U/ml), and streptomycin (100 mg/ml). Cells were incubated with 1 ml of media at 37°C in a tissue culture incubator equilibrated with 95% air and 5% CO_2. At near confluence, the cells were treated with the NM dissolved in media and tested at 0, 10, 100, 500, and 1000 μg/ml in triplicate at each dose. A group of cells were also treated with PMA 200 ng/ml. The plates were then returned to the incubator.

26.6 EFFECT OF NM ON HUMAN OSTEOSARCOMA

26.6.1 U2OS MMP ACTIVITY: GELATINASE ZYMOGRAPHY

Matrix metalloproteinase expression in conditioned media was determined by gelatinase zymography. Gelatinase zymography was performed in 10% polyacrylamide precast Novex gel (Invitrogen Corporation) in the presence of 0.1% gelatin. Culture media (20 μl) was loaded and SDS-PAGE was performed with a tris–glycine SDS buffer. After electrophoresis, the gels were washed with 5% Triton X-100 for 30 min. The gels were then incubated for 24 h at 37°C in the presence of 50 mM Tris–HCl, 5 mM $CaCl_2$, 5 μM $ZnCl_2$, and pH 7.5, and then stained with Coomassie Blue R 0.5% for 30 min and destained. Protein standards were run concurrently and approximate molecular weights were determined. Zymography demonstrated secretion of MMP-2 and MMP-9 by human osteosarcoma U2OS cells

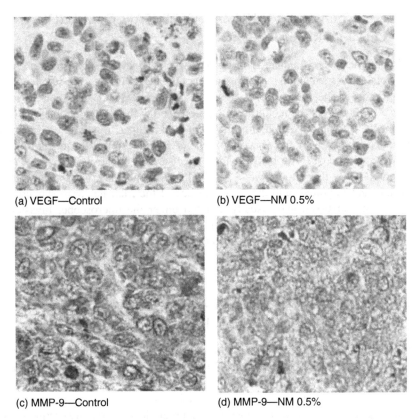

(a) VEGF—Control

(b) VEGF—NM 0.5%

(c) MMP-9—Control

(d) MMP-9—NM 0.5%

FIGURE 26.4 Effect of NM on tumor tissue of athymic nude mice receiving osteosarcoma MNNG xenografts: immunohistochemistry.

with significantly increased MMP-9 activity in PMA (200 ng/ml) treated osteosarcoma cells. The nutrient mixture inhibited the expression of both MMPs in a dose-dependent fashion with virtual total inhibition at 500 μg/ml concentration (Figure 26.5) [23].

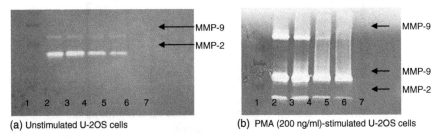

(a) Unstimulated U-2OS cells

(b) PMA (200 ng/ml)-stimulated U-2OS cells

1-Markers, 2-Control, 3–7 Nutrient Mixture 10, 50, 100, 500, 1000 μg/ml

FIGURE 26.5 Effect of NM on MMP secretion by human osteosarcoma U2OS cells.

26.7   EFFECT OF NM ON HUMAN OSTEOSARCOMA U2OS CELL MATRIGEL INVASION

Invasion studies were conducted using Matrigel™ (Becton Dickinson) inserts in 24-well plates. Suspended in medium, osteosarcoma U2OS cells were supplemented with nutrients, as specified in the design of the experiment and seeded on the insert in the well. Thus, both the medium on the insert and in the well contained the same supplements. The plates with the inserts were then incubated in a culture incubator equilibrated with 95% air and 5% CO_2 for 24 h. After incubation, the media from the wells were withdrawn. The cells on the upper surface of the inserts were gently scrubbed away with cotton swabs. The cells that had penetrated the Matrigel membrane and migrated onto the lower surface of the Matrigel were stained with Hematoxylin and Eosin and visually counted under the microscope. Invasion of osteosarcoma cells U2OS through Matrigel was reduced by 74% at 50 µg/ml and totally inhibited at 100 µg/ml ($p = 0.003$) (see Figure 26.6) [23].

26.8   EFFECT OF NM ON HUMAN OSTEOSARCOMA U2OS SECRETION OF VEGF, IL-6, IL-8, FGF, AND TGFB

Nutrient mixture was also found to inhibit the secretion of proangiogenic factors by osteosarcoma U2OS: VEGF, IL-6, IL-8, bFGF, and TGF-β (Table 26.1) [23]. These factors, which have been identified in various settings of physiologic and pathologic angiogenesis, affect the endothelium directly or indirectly by activation of surrounding cells to produce other factors with proangiogenic activity or modulation of receptors/receptor activities [25]. Conditioned media were collected after confluent cell culture incubation for 24 h in serum-free medium with the indicated supplements. Triplicate samples were pooled and the level of respective protein was measured in duplicate using an immunoassay kit (BioSource International and Quantikine, R&D) according to the manufacturer's protocol. Values are expressed as mean in percentage units of unstimulated control.

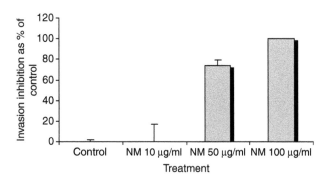

FIGURE 26.6 Effect of NM on human osteosarcoma U2OS invasion and migration through Matrigel.

TABLE 26.1
Effect of the Nutrient Mixture (NM) on Angiogenesis Mediator Secretion
by Osteosarcoma U2OS Cells Treated with PMA (200 ng/ml)

Pro-Angio-genic Factor	Factor Level in PMA-Treated U2OS Cells as % of Untreated Control	Factor Level in PMA and + NM (100 µg/ml)-Treated U2OS Cells as % of Untreated Control	NM Reduction of Factor Secretion (%)
VEGF	172	100	72
Interleukin-6	560	100	460
Interleukin-8	4800	300	4500
TGF-β	140	29	111
FGF	56	11	45

26.9   EFFECT OF THE NUTRIENT MIXTURE ON ENDOTHELIAL CELLS

Tumor growth is dependent upon angiogenesis and vascular remodeling. Endothelial cell activation is the first process to occur under pathological and physiological conditions. To initiate formation of new capillaries, endothelial cells of existing blood vessels must degrade the underlying basement membrane and invade the stroma of neighboring tissue. Therefore, we also investigated the effect of the NM on endothelial cell angiogenic parameters.

Human umbilical vein endothelial cells (HUVECs) obtained from ATCC were grown in M199 media supplemented with 10% FBS, 50 µg/ml endothelial cell growth supplement, 100 µg/ml heparin, and 20 mM Hepes in 24-well tissue culture plates (Costar, Cambridge, MA). Cells were incubated with 1 ml of media at 37°C in a tissue culture incubator equilibrated with 95% air and 5% CO_2. At near confluence, the cells were treated with the NM dissolved in media and tested at 0, 10, 100, 500, and 1000 µg/ml in triplicate at each dose. The plates were then returned to the incubator.

26.10   EFFECT OF NM ON HUVEC CAPILLARY TUBE FORMATION

The inhibitory effect of NM on capillary tube formation (in vitro differentiation) was assessed by exposure to NM of HUVECs on Matrigel. Unpolymerized Matrigel was placed in wells (300 µl/well) of a 24-well tissue culture plate and allowed to polymerize for an hour at 37°C. Human umbilical vein endothelial cells were plated in triplicate at a density of 100,000 cells per well with NM at 0, 10, 50, 100, 500, and 1000 µg/ml concentration. After 24 h of incubation in 5% CO_2 humidified atmosphere at 37°C, endothelial cells were washed with PBS, fixed, and stained with H&E. Tube morphology was observed and photomicrographs were taken. The

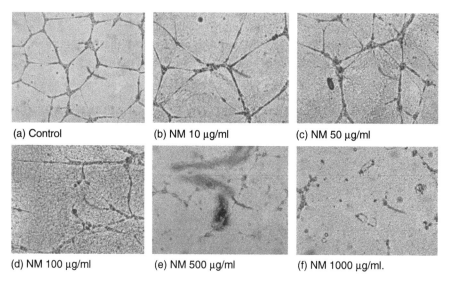

(a) Control   (b) NM 10 µg/ml   (c) NM 50 µg/ml

(d) NM 100 µg/ml   (e) NM 500 µg/ml   (f) NM 1000 µg/ml.

FIGURE 26.7   Effect of the nutrient mixture (NM) on HUVEC capillary tube formation assay (H&E, 100x).

inhibitory effect of NM on capillary tube formation (in vitro differentiation) was assessed by exposure to NM of HUVECs on Matrigel.

After 22 h, the control showed HUVEC-formed hollow tubes on Matrigel in contrast to the NM treated HUVECs, which demonstrated dose-dependent inhibition of capillary tube formation with complete disruption of capillary tubes at 1000 µg/ml NM (Figure 26.7a through Figure 26.7f H&E 100x) [26]. This ability to disrupt the integrity of preformed tubes indicates that NM may not only prevent but also regress new blood vessels. Capillary tube formation in the basement membrane-like matrix of Matrigel requires that endothelial cells adhere to and move on the extracellular matrix, a process dependent mainly on various integrins [27]. Capillary tube formation is a complex process requiring cell-matrix interactions and inter-cellular communications, as well as cell motility. The inhibitory effect of NM on HUVEC vascular network formation in light of the other in vitro studies indicates NM inhibits attachment, migration, and invasion of endothelial cells.

26.11   EFFECT OF NM ON HUVEC PROLIFERATION: MTT ASSAY 24 H

Nutrient effects on HUVEC proliferation were evaluated by an MTT assay, a colorimetric assay based on the ability of viable cells to reduce a soluble yellow tetrazolium salt [3-(4,5-dimethylthiazol-2-yl) 2,5-diphenyl tetrazolium bromide] (MTT) to a blue formazan crystal by mitochondrial succinate dehydrogenase activity. After MTT addition (0.5 mg/ml), the plates were covered and returned to the 37°C incubator for 2 h, the optimal time for formazan product formation.

Following incubation, the supernatant was carefully removed from the wells, the formazan product was dissolved in 1 ml DMSO, and absorbance was measured at 570 nm in a Bio Spec 1601, Shimadzu spectrometer. The OD_{570} of the DMSO solution in each well was considered to be proportional to the number of cells. The OD_{570} of the control (treatment without supplement) was considered 100%. The nutrient mixture had no significant effect on HUVEC proliferation [26].

26.12   EFFECT OF NM ON HUVEC MORPHOLOGY (H&E)

Human umbilical vein endothelial cell morphology was unchanged even at the highest concentration (Figure 26.8) [26].

26.13   EFFECT OF NM ON HUVEC MMP EXPRESSION: GELATINASE ZYMOGRAPHY

Matrix metalloproteinase expression in conditioned media was determined by gelatinase zymography. Gelatinase zymography was performed in 10% polyacrylamide precast Novex gel (Invitrogen Corporation) in the presence of 0.1% gelatin. Culture media (20 μl) was loaded and SDS-PAGE was performed with a tris–glycine SDS buffer. After electrophoresis, the gels were washed with 5% Triton X-100 for 30 min. The gels were then incubated for 24 h at 37°C in the presence of 50 mM Tris–HCl, 5 mM $CaCl_2$, 5 μM $ZnCl_2$, and pH 7.5, and then stained with Coomassie Blue R 0.5% for 30 min and destained. Protein standards were run concurrently and approximate molecular weights were determined.

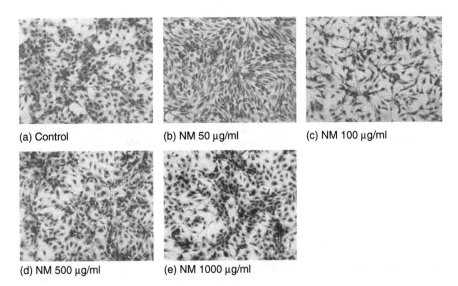

(a) Control   (b) NM 50 μg/ml   (c) NM 100 μg/ml

(d) NM 500 μg/ml   (e) NM 1000 μg/ml

FIGURE 26.8 Effect of the nutrient mixture (NM) on HUVEC morphology.

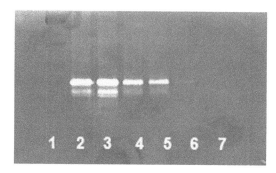

FIGURE 26.9 Effect of NM on HUVEC MMP-2 secretion; Legend: 1-Markers, 2-Control, 3–7 NM 10, 50, 100, 500, 1000µg/ml.

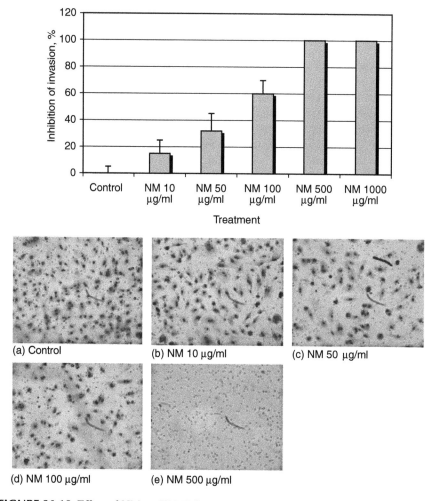

(a) Control

(b) NM 10 µg/ml

(c) NM 50 µg/ml

(d) NM 100 µg/ml

(e) NM 500 µg/ml

FIGURE 26.10 Effect of NM on HUVEC Matrigel invasion.

Zymography demonstrated expression of MMP-2 by HUVECs. Nutrient mixture inhibited the expression of MMP-2 in a dose-dependent fashion with virtual total inhibition at 500 µg/ml concentration (Figure 26.9) [26].

26.14   EFFECT OF NM ON HUVEC MATRIGEL INVASION

Invasion studies were conducted using Matrigel™ (Becton Dickinson) inserts in 24-well plates. Suspended in medium, HUVECs were supplemented with nutrients, as specified in the design of the experiment, and seeded on the insert in the well. Thus, both the medium on the insert and in the well contained the same supplements. The plates with the inserts were then incubated for 24 h in a culture incubator equilibrated with 95% air and 5% CO_2. After incubation, the media from the wells were withdrawn. The cells on the upper surface of the inserts were gently scrubbed away with cotton swabs. The cells that had penetrated the Matrigel membrane and migrated onto the lower surface of the Matrigel were stained with Hematoxylin

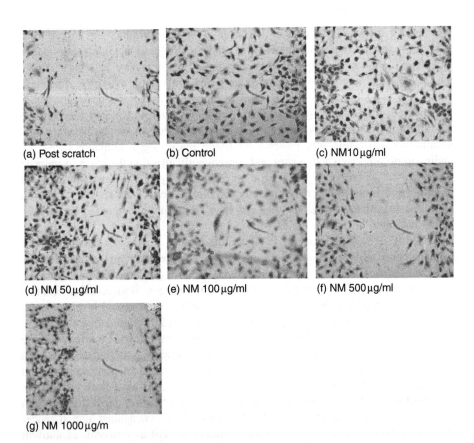

(a) Post scratch   (b) Control   (c) NM 10 µg/ml

(d) NM 50 µg/ml   (e) NM 100 µg/ml   (f) NM 500 µg/ml

(g) NM 1000 µg/m

FIGURE 26.11 Effect of NM on HUVEC migration: scratch test.

and Eosin and visually counted under the microscope. Invasion of HUVECs through Matrigel was reduced by 60% at 100 μg/ml and totally inhibited at 500 μg/ml ($p < 0.0001$) (Figure 26.10) [26].

26.15   EFFECT OF NM ON HUVEC MIGRATION

Human umbilical vein endothelial cell migration was investigated by making a 2 mm wide single uninterrupted scratch from top to bottom of the culture plates at near confluence. Culture plates were washed with PBS and incubated with NM in medium and tested at 0, 10, 50, 100, 500, and 1000 μg/ml in triplicate at each dose for 24 h. The cells were washed with PBS, then fixed and stained with H&E. Photomicrographs 100x were taken. Nutrient mixture reduced HUVEC migration by a scratch test in a dose-dependent fashion with total inhibition at 500 μg/ml concentration, as shown in Figure 26.11 [26].

26.16   DISCUSSION

A promising therapeutic approach to cancer is through simultaneous targeting of universal pathomechanisms involved in cancer progression. Inhibiting matrix degradation and optimizing ECM structure and its integrity can curb cancer invasiveness and angiogenesis. Control of ECM proteolytic activity provides an opportunity to address common mechanisms of metastasis, angiogenesis, and tumor growth. Rath and Pauling [6] suggested targeting plasmin-mediated mechanisms with nutritional components, such as lysine and its analogues, which interfere with plasminogen activation into plasmin by tissue plasminogen activator (tPA), by binding to plasminogen active sites and consequently affecting the plasmin-induced MMP activation cascade. Expression of urokinase-type plasminogen activator inhibitor type one by human prostate carcinoma cells was shown to inhibit angiogenesis and metastasis to lung and liver in an athymic mouse model [28]. Lysine-mediated effects on the ECM include increased connective tissue strength and stability. It is well known that the synthesis and structure of collagen fibrils depends upon hydroxylation of proline and lysine residues in collagen fibers, catalyzed by ascorbic acid. Lysine importance in collagen structure and synthesis relates to it being the most abundant amino acid in collagen. Both ascorbic acid and lysine are not produced in the human body; therefore, sub-optimal levels of these nutrients are possible in various pathological stages and through deficient diets.

The inhibitory effects of the individual nutrients composing the nutrient mixture have been reported in both clinical and experimental studies. Ascorbic acid has been reported to have cytotoxic, antimetastatic, and anti-angiogenic actions on malignant cell lines [29–32]; in addition, low levels of ascorbic acid have been reported in cancer patients [33–35]. Green tea extract is a potent anticancer agent that has been reported to have antitumorigenic and anti-angiogenic effects against human cancer cell lines [36–41]. However, individual nutrients are not as powerful as nutrient synergy. Our previous studies demonstrated that the synergistic anticancer effect of ascorbic acid, proline, lysine, and EGCG on several cancer cell lines in tissue culture studies was greater than that of the individual nutrients [21].

Because of the growing number of agents tested for anti-angiogenic properties, the U.S. National Cancer Institute (NCI) uses a classification system that categorizes anti-angiogenic drugs into five groups: (1) agents that inhibit endothelial cells directly; (2) agents that block activators of angiogenesis; (3) agents that block ECM breakdown; (4) agents that inhibit endothelial-specific integrin signaling; and (5) agents with non-specific mechanisms of action. Many MMP inhibitors are also anti-angiogenic agents. Unfortunately, many compounds have limited efficacy, due to problems of delivery and penetration and moderate effects on the tumor cells, accompanied by severe toxicity and damage to healthy tissues. In addition, the activity of these compounds is mainly limited by the development of drug resistance [42].

Tumor cells are rapidly changing because of their genetic instability, heterogeneity, and high rate of mutation. We postulate that cancer development and progression can be controlled only through a multi-targeted approach, in contrast to the application of agents selected for a highly specific metabolic target. The results of this and our previous studies indicate that such multifaceted approaches targeting the mechanisms that are common to all types of cancer cells can be achieved through nutrient synergy. We demonstrated the inhibitory effects of this nutrient synergy on cancer cell invasion, proliferation, and angiogenesis executed through various mechanisms. We have shown that this nutrient mixture affects MMPs, can induce apoptosis, and modulates the effects of various growth factors.

While clinical studies are necessary to better determine the efficacy of nutrient therapy in both cancer prevention and treatment, the results of these studies suggest that the formulation of green tea extract, lysine, proline, and ascorbic acid is an excellent candidate for adjunctive therapeutic use in the treatment of metastatic cancer, as it inhibits MMP expression and invasion and angiogenesis without cytotoxic effects.

REFERENCES

1. Folkman, J., What is the evidence that tumors are angiogenesis dependent?, *J. Natl. Cancer Inst.*, 82, 4–6, 1990.
2. Fang, W., Li, H., Kong, L., Niu, G., Gao, Q., Zhou, K., Zheng, J., and Wu, B., Role of matrix metalloproteinases (MMPs) in tumor invasion and metastasis: serial studies on MMPs and TIMPs, *Bejing D Xue Bao*, 4, 441–443, 2003.
3. Wahl, M. R., Moser, T. L., and Pizzo, S. V., Angiostatin and antiangiogenic therapy in human disease, *The Endocrine Society*, 59, 73–104, 2004.
4. Canfield, A. E., Schor, A. M., Schor, S. L., and Grant, M. E., The biosynthesis of extracellular matrix components by bovein retinal endothelial cells displaying distinctive morphological phenotypes, *Biochem. J.*, 235, 375–383, 1986.
5. Cao, Y., Veitonmaki, N., Keough, K., Cheng, H., Lee, L. S., and Zurakowski, D., Elevated levels of angiostatin and plasminogen/plasmin in cancer patients, *Intl. J. Mol. Med.*, 5, 547–551, 2000.

6. Rath, M. and Pauling, L., Plasmin-induced proteolysis and the role of apoprotein(a), lysine and synthetic analogs, *Ortho. Med.*, 7, 17–23, 1992.

7. Tossetti, F., Ferrari, N., De Flora, S., and Albini, A., Angioprevention: angiogenesis is a common and key target for cancer chemopreventative agent, *FASEB J.*, 16, 2–14, 2002.

8. Roomi, M. W., Ivanov, V., Kalinovsky, T., Niedzwiecki, A., and Rath, M., Antitumor effect of nutrient synergy on human osteosarcoma cells U2OS, MNNG–HOS and Ewing's sarcoma SK-ES.1, *Oncol. Rep.*, 13 (2), 253–257, 2005.

9. Roomi, M. W., Ivanov, V., Kalinovsky, T., Niedzwiecki, A., and Rath, M., In vitro and in vivo antitumorigenic activity of a mixture of lysine, proline, ascorbic acid and green tea extract on human breast cancer lines MDA MB-231 and MCF-7, *Med. Oncol.*, 22 (2), 129–138, 2005.

10. Roomi, M. W., Ivanov, V., Kalinovsky, T., Niedzwiecki, A., and Rath, M., Synergistic effect of combination of lysine, proline, arginine, ascorbic acid, and epigallocatechin gallate on colon cancer cell line HCT 116, *JANA*, 7 (2), 40–43, 2004.

11. Roomi, M. W., Ivanov, V., Kalinovsky, T., Niedzwiecki, A., and Rath, M., Antitumor effect of a combination of lysine, proline, arginine, ascorbic acid, and green tea extract on pancreatic cancer cell line MIA PaCa-2, *Int. J. Gastrointest. Cancer*, 35 (2), 97–102, 2005.

12. Roomi, M. W., Ivanov, V., Kalinovsky, T., Niedzwiecki, A., and Rath, M., In vivo antitumor effect of ascorbic acid, lysine, proline and green tea extract on human colon cancer cell HCT 116 xenografts in nude mice: evaluation of tumor growth and immunohistochemistry, *Oncol. Rep.*, 13 (3), 421–445, 2005.

13. Roomi, M. W., Ivanov, V., Kalinovsky, T., Niedzwiecki, A., and Rath, M., In vivo antitumor effect of ascorbic acid, lysine, proline and green tea extract on human prostate cancer PC-3 xenografts in nude mice: evaluation of tumor growth and immunohistochemistry, *In Vivo*, 19 (1), 179–183, 2005.

14. Roomi, M. W., Ivanov, V., Kalinovsky, T., Niedzwiecki, A., and Rath, M., Modulation of *N*-methyl–*N*-nitrosourea induced mammary tumors in Sprague–Dawley rats by combination of lysine, proline, arginine, ascorbic acid and green tea extract, *Breast Cancer Res.*, 7, R291–R295, 2005.

15. Maramag, C., Menon, M., Balaji, K. C., Reddy, P. G., and Laxmanan, S., Effect of vitamin C on prostate cancer cells in vitro: effect of cell number, viability and DNA synthesis, *Prostate*, 32, 188–195, 1997.

16. Hare, Y., *Green Tea: Health Benefits and Applications*, Marcel Dekker, New York, Basel, 2001.

17. Kawakami, S., Kageyama, Y., Fujii, Y., Kihara, K., and Oshima, H., Inhibitory effects of *N*-acetyl cysteine on invasion and MMP-9 production of T24 human bladder cancer cells, *Anticancer Res.*, 21, 213–219, 2001.

18. Morini, M., Cai, T., Aluigi, M. G., Noonan, D. M., Masiello, L., De Flora, S., D'Agostini, F., Albini, A., and Fassina, G., The role of *N*-acetyl cysteine in the prevention of tumor invasion and angiogenesis, *Int. J. Biol. Markers*, 14, 268–271, 1999.

19. Yoon, S. O., Kim, M. M., and Chung, A. S., Inhibitory effects of selenite on invasion of HT 1080 tumor cells, *J. Biol. Chem.*, 276, 20085–20092, 2001.

20. Cooke, J. P. and Dzau, V. J., Nitric oxide synthase: role in the genesis of vascular disease, *Annu. Rev. Med.*, 48, 489–509, 1997.

21. Roomi, M. W., Ivanov, V., Niedzwiecki, A., and Rath, M., Synergistic antitumor effect of ascorbic acid, lysine, proline, and epigallocatechin gallate on human fibrosarcoma cells HT-1080, *Ann. Cancer Res. Ther.*, 12 (1–2), 6–11, 2004.

22. Brooks, P. C., Montgomery, A. M., and Cheresh, D. A., Use of the 10-day-old chick embryo model for studying angiogenesis, *Meth. Mol. Biol.*, 129, 257–269, 1997.
23. Roomi, M. W., Roomi, N., Ivanov, V., Kalinovsky, T., Niedzwiecki, A., and Rath, M., Inhibitory effect of a mixture containing ascorbic acid, lysine, proline, and green tea extract on critical parameters in angiogenesis, *Oncol. Rep*, 14(4), 807–815, 2005.
24. Passaniti, A., Taylor, R. M., Phili, R., Guo, Y., Long, P. V., Haney, J. A., Pauly, R. R., Grant, D. S., and Martin, G. R., A simple, quantitative method for assessing angiogenesis and antiangiogenesis agents using reconstituted basement membrane, heparin, and fibroblast growth factor, *Lab. Invest.*, 67, 519–528, 1992.
25. Yoshida, S., Ono, M., Shono, T., Izumi, H., Ishibashi, T., Suzuki, H., and Kuwano, M., Involvement of interleukin-8, vascular endothelial growth factor, and basic fibroblast growth factor in tumor necrosis factor α-dependent angiogenesis, *Mol. Cell. Biol.*, 17, 4015–4023, 1997.
26. Roomi, M. W., Ivanov, V., Kalinovsky, T., Niedzwiecki, A., and Rath, M., Antiangiogenic Effects of a Nutrient Mixture on Human Umbilical Vein Endothelial Cells, *Oncol. Rep*, 14(6), 1399–1404, 2005.
27. Friedlander, M., Brooks, P. C., Schaffer, R. W., Kincaid, C. M., Varner, J. A., and Cheresh, D. A., Definition of two angiogenic pathways by distinct α integrins, *Science*, 270, 1500–1502, 1995.
28. Soff, G. A., Sanderowiz, J., Gately, S., Verrusio, E., Weiss, I., Brem, S., and Kwaan, H. C., Expression of plasminogen activator inhibitor type 1 by human prostate carcinoma cells inhibits primary tumor growth, tumor-associated angiogenesis, and metastasis to lung, liver in an athymic mouse model, *J. Clin. Investig.*, 96, 2593–2600, 1995.
29. Koh, W. S., Lee, S. J., Lee, H., Park, C., Park, M. H., Kim, W. S., Yoon, S. S. et al., Differential effects and transport kinetics of ascorbate derivatives in leukemic cell lines, *Anticancer Res.*, 8, 2487–2493, 1998.
30. Roomi, M. W., House, D., Eckert-Maksic, M., Maksic, Z. B., and Tsao, C. S., Growth suppression of malignant leukemia cell line in vitro by ascorbic acid (vitamin C) and its derivatives, *Cancer Lett.*, 122, 93–99, 1998.
31. Naidu, K. A., Karl, R. C., Naidu, K. A., and Coppola, D., Antiproliferative and proapoptotic effect of ascorbyl stearate in human pancreatic cancer cells: association with decreased expression of insulin-like growth factor 1 receptor, *Dig. Dis. Sci.*, 48 (1), 230–237, 2003.
32. Ashino, H., Shimamura, M., Nakajima, H., Dombou, M., Kawanaka, S., Oikav, T., Iwaguchi, T., and Kawashima, S., Novel function of ascorbic acid as an angiostatic factor, *Angiogenesis*, 6 (4), 259–269, 2003.
33. Anthony, H. M. and Schorah, C. J., Severe hypovitaminosis C in lung-cancer patients: The utilization of vitamin C in surgical repair and lymphocyte related host resistance, *Br. J. Cancer*, 46, 354–367, 1982.
34. Nunez, C., Ortiz, D. E., Apodaca, Y., and Ruiz, A., Ascorbic acid in the plasma and blood cells of women with breast cancer. The effect of consumption of food with an elevated content of this vitamin, *Nutr. Hosp.*, 10, 368–372, 1995.
35. Kurbacher, C. M., Wagner, U., Kolster, B., Andreotti, P. E., Krebs, D., and Bruckner, H. W., Ascorbic acid (vitamin C) improves the antineoplastic activity doxorubicin, cisplatin, and paclitaxel in human breast carcinoma cells in vitro, *Cancer Lett.*, 103 (2), 183–189, 1996.
36. Jung, Y. D., Kim, M. S., Shin, B. A., Chay, K. O., Ahn, B. W., Liu, W., Bucana, C. D., Gallick, G. E., and Ellis, L. M., EGCG, a major component of green tea, inhibits tumor growth by inhibiting VEGF induction in human colon carcinoma cells, *Br. J. Cancer*, 84 (6), 844–850, 2001.

37. Koima-Yuasa, A., Hua, J. J., Kennedy, D. O., and Matsui-Yuasa, I., Green tea extract inhibits angiogenesis of human umbilical vein endothelial cells through reduction of expression of VEGF receptors, *Life Sci.*, 73 (10), 1299–1313, 2003.
38. Valcic, S., Timmermann, B. N., Alberts, D. S., Wachter, G. A., Krutzsch, M., Wymer, J., and Guillen, M., Inhibitory effects of six green tea catechins and caffeine on the growth of four selected human tumor cell lines, *Anticancer Drugs*, 7, 461–468, 1996.
39. Mukhtar, H. and Ahmed, N., Tea polypheonols: prevention of cancer and optimizing health, *Am. J. Clin. Nutr.*, 71, 1698S–1720S, 2000.
40. Yang, G. Y., Liao, J., Kim, K., Yurkow, E. J., and Yang, C. S., Inhibition of growth and induction of apoptosis in human cancer cell lines by tea polyphenols, *Carcinogenesis*, 19, 611–616, 1998.
41. Metz, N., Lobstein, A., Schneider, Y., Gosse, F., Schleiffer, R., Anton, R., and Raul, F., Suppression of azoxymethane-induced preneoplastic lesions and inhibition of cyclooxygenase-2 activity in the colonic mucosa of rats drinking a crude green tea extract, *Nutr. Cancer*, 38 (1), 60–64, 2000.
42. Griffioen, A. W. and Molema, G., Angiogenesis: potential for pharmacologic intervention in the treatment of cancer, cardiovascular diseases, and chronic inflammation, *Pharm. Rev.*, 52 (2), 237–268, 2000.

27 Angiogenesis and Chinese Medicinal Foods

Anthony Y. H. Woo, Y. Zhao, R. Zhang, C. Zhou, and Christopher H. K. Cheng

CONTENTS

27.1 ANGIOGENESIS AND THE DISEASE PROCESS

Angiogenesis is the growth of new blood vessels sprouting from existing vessel beds responsible for formation of the vasculature. Angiogenesis plays a major role in the initiation, progression, and prognosis of a number of diseases. Recently, angiogenesis has become a hot topic of research because of its potential applications in the treatment of these pathological conditions. These studies can be categorized into two aspects: (1) studies on the inhibition of angiogenesis and (2) studies on the promotion of angiogenesis.

The excessive growth of blood vessels is related to the initiation and progression of many diseases such as cancer, diabetes, rheumatoid arthritis, ocular fundus diseases, psoriasis, and neurodegenerative diseases. In cancer, for example, the formation of new blood vessels is pivotal to the growth and spreading of the tumor. The vessels supply nutrients to the tumor to support its growth. In addition, as the vessel walls of newly formed blood vessels are thinner, tumor cells can pass through the walls more easily, thus allowing metastasis to occur via the blood circulation. Therefore, by inhibiting angiogenesis, blood supply to the tumor can be restricted and the objective of "starving" the tumor cells can be achieved.

There are several advantages to treating these pathological conditions by targeting the newly formed blood vessels. First, compared with conventional chemotherapeutic agents which are cytotoxic to both the tumor and normal tissues, anti-angiogenic treatment enjoys a higher specificity because only the already turned-on angiogenic blood vessels are targeted and existing blood vessels are spared. Second, in contrast to conventional chemotherapeutic agents, where delivery of high enough concentrations of the drug to the inner part of the tumor always poses a problem, higher efficacy can be achieved in anti-angiogenic treatment by using a lower drug dose because the drug can act on the endothelial cells directly exposed to the blood circulation. Side effects associated with a high drug dose would be avoided.

On the other hand, promotion of angiogenesis could be beneficial in some clinical conditions where blood supply is inadequate. In bone fracture and trauma for example, local promotion of angiogenesis can improve blood supply to the traumatic tissues and accelerate the healing process. In ischemic diseases like myocardial infarction, development of new blood vessels in the infarct region can replenish blood supply to the infarct tissues and thus can reduce tissue injury.

27.2 RECENT PROGRESS IN THE STUDIES OF CHINESE MEDICINAL FOODS ON ANGIOGENESIS

27.2.1 INHIBITION OF ANGIOGENESIS

Several Chinese medicinal foods have been identified to possess therapeutic effects on angiogenesis-related diseases. The effects of some of them on the anti-angiogenic treatment of cancer are particularly promising. They are described as follows.

27.2.1.1 Laminarin Sulfate

Laminarin, also known as brown algal starch, is a neutral polysaccharide widely distributed in brown algae. It is particularly abundant in the geni *Ecklonia* and *Laminaria*. Laminarin sulfate (LAMS), the sulfonated product of laminarin, possesses several reported bioactivities. LAMS has been shown to inhibit basic fibroblast growth factor (bFGF) binding and bFGF-stimulated proliferation of fetal bovine heart endothelial cells, inhibit tube formation by endothelial cells and vascularisation of the chick embryo chorioallantic membrane (CAM), and inhibit the growth of the murine RIF-1 tumor [1]. Xu et al. [2] have also used the chick embryo CAM method to observe the anti-angiogenic effect of LAMS. Compared with the positive control suramin, LAMS possesses a higher potency in inhibiting angiogenesis but is more toxic.

27.2.1.2 Ginsenoside Rg3

The root of *Panax ginseng* C.A. Mey (ginseng root) is commonly used in Chinese medicine to treat cancer. It contains a number of saponins in which ginsenoside Rg3 (Figure 27.1) possesses antitumor effects. Pun et al. [3] have used a severe combined immunodeficient (SCID) mouse model of peritoneal ovarian cancer to study the mechanism of the antitumor effect of ginsenoside Rg3. After treatment with ginsenoside Rg3, the mRNA, and protein levels of vascular endothelial growth factor (VEGF) and the microvessel density (MVD) in the tumor tissues were assayed by reverse transcription-polymerase chain reaction (RT-PCR), enzyme-linked immunosorbent assay (ELISA) and immunohistochemistry, respectively. The results showed that no peritoneal fluid was produced in the tumor-inoculated SCID mice treated with ginsenoside Rg3. The VEGF mRNA level in the tumor tissues of the drug-treated group was significantly lower than that of the blank group or the control group. The MVD in the tumor tissue of the drug-treated group was also significantly decreased. These results suggest that ginsenoside Rg3 inhibits angiogenesis by downregulating VEGF mRNA and protein expression in the tumor tissues, thereby inhibiting the growth and metastasis of the tumor.

27.2.1.3 Polysaccharide K

Polysaccharide K (PSK) or Krestin is the intrasporal polysaccharide of the mushroom *Coriolus vesicolor* (L. ex Fr.) Quél (family: Polyporaceae). For years, PSK has been used as a complementary therapy to treat cancer, as it could stimulate the immune system which is compromised in conventional chemotherapy [4]. Recent studies showed that PSK could also inhibit tumor growth by suppressing tumor-induced angiogenesis. Kanoh et al. [5] have demonstrated the inhibitory

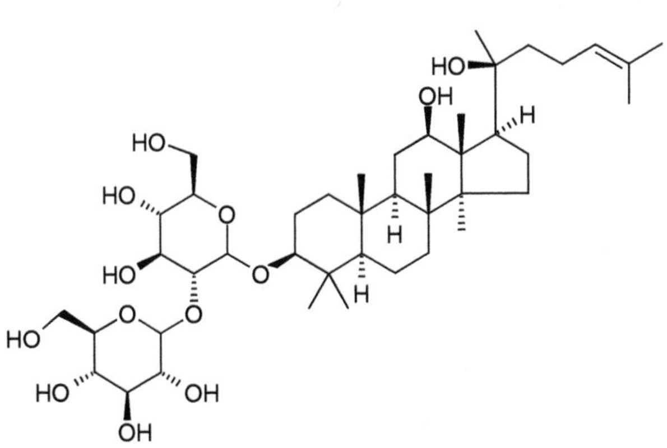

FIGURE 27.1 Ginsenoside Rg3.

effects of PSK on MH-134 murine hepatocarcinoma-induced angiogenesis in mice. In another study, PSK was found to inhibit tumor-induced angiogenesis in skin tissues of Balb/c mice inoculated with mammary adenocarcinoma cells [6]. Tumor growth and transforming growth factor-β1 (TGF-β1) expression in the tumor of the treatment group was also suppressed. It was suggested that the anti-angiogenic effect of PSK might be due to the suppression of TGF-β1 level. Polysaccharide K was also shown to inhibit carcinogen-induced mammary carcinogenesis in Sprague Dawley rats [7]. Histological and immunohistological studies showed that PSK decreased the MVD and the expression of VEGF and Ras protein suggesting that its anti-cancer effect may be related to the inhibition of angiogenesis.

27.2.1.4 Kanglaite Injection

Kanglaite (KLT) injection is a fat emulsion of anti-cancer substances extracted from the seed of *Coix lacryma-jobi* L. var. *ma-yuen* (Roman.) Stapf. (synonyms: coix seed or job's tear) for intra-arterial or intravenous injection. It is used in China for the treatment of colon cancer and kidney cancer. Jiang et al. [8] have studied the effects of KLT on angiogenesis using a serum-free aorta culture model. Kanglaite was demonstrated to exhibit a significant inhibitory effects on angiogenesis. Treatment of the aorta with 10 μL/mL KLT resulted in sparse blood vessel formation. Vessel growth also quickly entered into the regression phase. Inhibition of angiogenesis was suggested to be responsible for, in part at least, the anti-cancer mechanisms of KLT.

27.2.1.5 Ursolic Acid

Ursolic acid (Figure 27.2) is a triterpene widely found in plants. For example, it is present in the fruits and the leaves of *Viccinium vitis-idaea* L. (cowberry), and the leaves of *Paulownia tomentosa* (Thunb.) Steud. (foxglove tree) and *Ligustrum lucidum* Ait. (ligustrum or glossy privet). Ursolic acid exhibits a wide spectrum of biological activities. It has been discovered to antagonize the actions of carcinogens and oncogens, and inhibit the growth of many malignant tumor cells including P-388 and L1210 human leukemia cells, and A-549 human lung adenocarcinoma cells.

In 1995, Sohn et al. [9] studied the anti-angiogenic effects of ursolic acid by the chick embryo CAM method. They discovered that ursolic acid was effective in

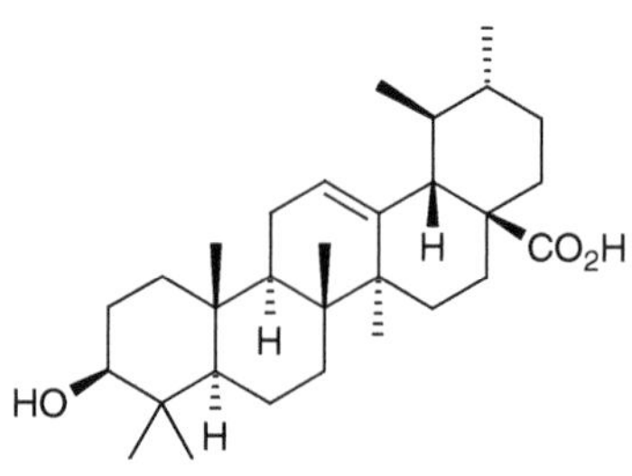

FIGURE 27.2 Ursolic acid.

inhibiting angiogenesis at 4 nmol/egg, and the ID_{50} was determined to be 10 nmol/ egg at which no cytotoxic effect was observed. The ID_{50} of ursolic acid on endothelial cell proliferation was determined to be 5 μmol/L. This anti-angiogenic effect might explain the antitumor actions of ursolic acid.

27.2.1.6 Curcumin

Curcumin (Figure 27.3) is the major antitumor ingredient isolated from the *Curcuma* species. Herbs from this genus are well known for their effects "to quicken blood and to transform stasis" (to activate blood circulation) in traditional Chinese medicine (TCM), and they are commonly applied in the treatment of cancer. Studies have shown that curcumin could significantly reduce angiogenesis as determined by the chick embryo CAM method, and 20 μmol/L of curcumin was sufficient to induce apoptosis in SMMC-7221 hepatocarcinoma cells [10].

27.2.1.7 Tea Catechins

Recent studies have shown that green tea extract could lower the expression of VEGF receptors on endothelial cells and subsequently inhibit angiogenesis [11]. The results of three different in vitro experimental models of human endothelial cell growth, migration, and tube formation have been utilized to demonstrate the anti-angiogenic effects of the catechins, viz. (−)-epicatechin, (−)-epicatechin 3-gallate, (−)-epigallocatechin, and (−)-epigallocatechin 3-gallate (EGCg) (Figure 27.4) [12]. The effective concentrations of these compounds were between 1.56 and 100 μM in which EGCg was the most potent in the three experimental models. Inhibition of angiogenesis by tea catechins may be one of the factors that contribute to the lower incidence of cancer in individuals with habitual consumption of tea.

27.2.2 PROMOTION OF ANGIOGENESIS

Promotion of angiogenesis, also known as therapeutic angiogenesis, is a recent approach to treat certain conditions where local blood supply to tissues is impaired. This can be achieved by agents that increase the number of microangium, thereby promoting blood flow [13]. Western medicine has made great progress in the area of antiangiogenesis. However, there is no major breakthrough in the development of treatment regimens through the promotion of angiogenesis. On the other hand, a

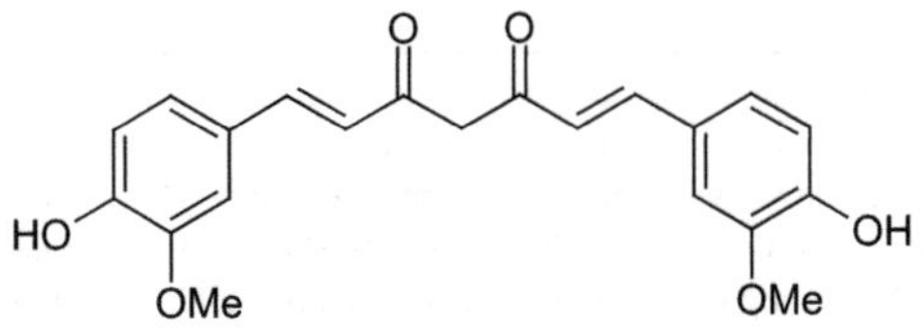

FIGURE 27.3 Curcumin.

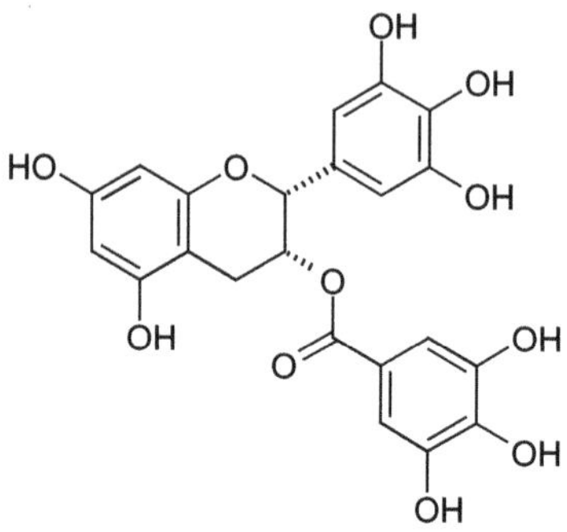

FIGURE 27.4 (−)-Epigallocatechin-3-gallate.

number of TCM theories describe the actions of certain Chinese medicinal foods as "quickening blood and transforming stasis," "benefiting qi and supplementing blood," "tonifying kidney and regulating menstruation," "filling channels and engendering blood," and "moving qi and unblocking vessels."* These provide some clues on the search for angiogenic principles in Chinese medicinal foods.

In addition, the recent use of these materials in the treatment of angiogenesis impairment-related conditions has achieved some promising results, and some Chinese medicinals and formulae have been demonstrated to exhibit definitive effects in promoting angiogenesis [14]. Dai et al. [15] have suggested the possibility of using Chinese medicinal materials with angiogenic properties to treat coronary heart diseases. This suggestion is mainly based on experimental evidence of drug actions on the number and density of microvessels identified histologically in bone fracture models or in bone marrow lesion models in animals. Some of these effective formulae are described below.

27.2.2.1 Ligustrum Impregnating Decoction

Except in the case of tumor and recovery from trauma, the female endometrium is the only human tissue that still maintains an active process of new blood vessel formation during adulthood. The regulation of new vessel formation and breakdown in the endometrium during the menstrual cycle is thus crucial for human fertility. A study has been conducted to investigate the effect of a traditional Chinese formula, Nue-jing-yun-yu-tang (Ligustrum impregnating decoction), used to treat sterility in

---

* Some of the terms used in TCM do not have the same meaning as in modern medicine. For example, "kidney" has extensive functions not limited to excretion. In addition, some TCM terms such as qi (vital energy of the body) do not have synonyms in modern medicine. The phases "quickening blood and transforming stasis" and "moving qi and unblocking vessels" have the meaning of activating blood circulation. The phases "benefiting qi and supplementing blood" and "filling channels and engendering blood" have the meaning of increasing the blood supply. And the phrase "tonifying kidney and regulating menstruation" has the meaning of regulating the menstrual cycle and promoting fertility.

women, on angiogenesis [16]. This formula is composed of 14 ingredients including Semen Cuscutae (cuscuta seed), Fructus Ligustri Lucidi (ligustrum fruit), Fructus Lycii (Chinese wolfberry), Radix Angelicae Sinensis (danggui root), and Radix Salviae Miltiorrhizae (salvia root).

A sero-pharmacological approach was adopted to study the effect of the formula on promoting angiogenesis in women with ovulation-deficient sterility. The abilities of the serum samples to promote angiogenesis were determined by the chick embryo CAM method. Results showed that ingestion of the decoction for three consecutive menstrual cycles resulted in a significant increase in the angiogenic activities of the serum samples as compared to the serum samples before treatment. This result suggests that the promotion of angiogenesis may contribute towards the efficacy of this formula in treating sterility.

27.2.2.2   Kidney-Tonifying and Menstruation Regulating Formula

Bao-shen-tiao-jing-fang (Kidney-tonifying and menstruation regulating formula) is composed of ingredients including Placenta Hominis (human placenta), danggui root, cuscuta seed, Radix Morindae Officinalis (morinda root), and Faeces Trogopterori (flying squirrel's droppings). This formula has a definitive effect in the chick embryo CAM model of blood vessel formation, as observed in a 120% increase in the number of vessels formed. The effects of the formula on the expression of VEGF, VEGF receptor, bFGF/FGF, platelet-derived growth factor receptor-α, and epidermal growth factor receptor in the human endometrial tissues during implantation were studied by immunohistochemistry before and after drug administration [17]. The results showed that the expression levels of these 5 parameters, as well as the total number of cells positively stained for these factors, had increased significantly, suggesting that the formula has an effect in promoting blood vessel formation in vivo.

27.2.2.3   Kidney-Tonifying and Blood-Engendering Formula

In a series of studies using the chick embryo CAM model, Zhang, Wang, and others [18,19] have demonstrated an increase in the angiogenic effects of the serum of mice administered with a kidney-tonifying and blood-engendering formula containing Fructus Corni (cornus fruit), Radix Rehmanniae Conquita (cooked rehmannia root), Chinese wolfberry, Plastrum Testudinis (tortoise plastron), and danggui root as the major ingredients. Oral treatment of old age golden hamsters with physiological kidney deficiency increased the expression of bFGF and VEGF in the uterine tissues and the number of blood vessels in the endometrium. The angiogenic activity of the formula may explain its beneficial effect on patients suffering from kidney deficiency.

27.2.2.4   Blood-Quickening Formula

Using a rabbit bilateral radial fracture model, Wang et al. [20] have found that a kidney-tonifying and blood-quickening formula shortened the time of the

TABLE 27.1
Summary of the Inhibitory and Stimulatory Effects of Chinese Medicinal Foods on Angiogenesis

Inhibitors of Angiogenesis	Effects	Promoters of Angiogenesis	Effects
Laminarin sulfate	Antitumor: murine RIF-1 tumor Anti-angiogenic: I: CAM, EC growth and tube formation, ↓ bFGF binding	Ligustrum impregnating decoction Kidney-tonifying and menstruation regulating formula	Treatment of sterility, Angiogenic: P: sero-CAM Angiogenic: ↑ VEGF, ↑ VEGFR, ↑ bFGF/FGF, ↑ PDGFR-α, ↑ EGFR in endometrium
Ginsenoside Rg3	Antitumor: ovarian cancer Anti-angiogenic: ↓ VEGF, ↓ MVD	Kidney-tonifying and blood-engendering formula	Kidney-tonifying, Angiogenic: P: sero-CAM ↑ VEGF, ↑ bFGF/FGF in endometrium
Polysaccharide K	Antitumor: breast cancer, ↓ Ras Anti-angiogenic: ↓ TGF-β1, ↓ VEGF, ↓ MVD	Blood-quickening formula	Bone fracture healing, Angiogenic: ↑ vessel regeneration
Kanglaite injection	Antitumor: colon caner, kidney cancer Anti-angiogenic: I: aorta culture	Pulse-engendering and bone-forming formula	Ischemic bone-lesion healing, Angiogenic: ↑ blood flow, ↑ VEGF in bone tissue
Ursolic acid	Antitumor: leukemia, lung cancer Anti-angiogenic: I: CAM	Qi-benefiting and stasis-transforming formula	Ischemic bone-lesion healing, Angiogenic: ↑ blood vessel, ↑ VEGF
Curcumin	Antitumor: liver cancer Anti-angiogenic: I: CAM		
Tea catechins	Anti-angiogenic: I: EC growth, migration, and differentiation, ↓ VEGF		

Abbreviations: bFGF, basic fibroblast growth factor; CAM, chick embryo chorioallantic membrane assay; EC, endothelial cell; EGFR, epidermal growth factor receptor; EMC, extracellular matrix components; I, inhibition; MVD, microvessel density; P, promotion; PDGFR-α, platelet-derived growth factor receptor-α; TGF-β1, transforming growth factor-β1; VEGFR, VEGF receptor.

organization of hematoma, promoted local growth of vascular endothelial cells and fibroblast cells, enhanced regeneration of the microangium, and subsequently shortened the time of fracture healing. Zhang et al. [21] have studied the effects of the external application of a blood-quickening and stasis-transforming formula in a bilateral tibial fracture model in rabbits. The results showed that the formula promoted vasodilation and vessel regeneration, and increased blood supply to the fracture regions. It was postulated that the formula could speed up the blood vessel formation process in the fracture regions to ensure sufficient blood supply during the healing process, thus facilitating fracture healing.

27.2.2.5   Pulse-Engendering and Bone-Forming Formula

The clinical effect of Sheng-mai-cheng-gu-fang (pulse-engendering and bone-forming formula) on ischemic necrosis of the femoral head was reported [22]. Emission-computed tomography and magnetic resonance imaging revealed a significant increase in blood flow to the femoral head in the treatment group. The authors have also studied the effects of the formula by in situ hybridization in a rabbit steroid-induced femoral head necrosis model. They have found that the formula significantly increased the VEGF mRNA expression level in the endothelial cells and osteoblasts, and subsequently enhanced the activities of these cells in promoting blood vessel formation as well as new bone formation and remodeling.

27.2.2.6   Qi-Benefiting and Stasis-Transforming Formula

In a study on ischemic degeneration of the intervertebral disc, Wang et al. [23] have observed the effects of Yi-qi-hua-yu-fang (qi-benefiting and stasis-transforming formula), composed of ingredients including Radix Astragali (astragalus root), Moschus (musk), and Rhizoma Chuanxiong (chuanxiong rhizome) on the cartilage end-plate of ischemic cervical intervertebral discs in rats. The formula was found to significantly increase the number of blood vessel sprouts and the expression level of VEGF in the cartilage end-plate, promote the repair and regeneration of blood vessels, and increase nutrient supply to the cervical intervertebral discs.

A summary of these inhibitory and stimulatory effects of Chinese medicinal foods on angiogenesis is tabulated in Table 27.1.

27.3   CONCLUSION

It appears that most of the current studies on the inhibition of angiogenesis by Chinese medicinal foods are restricted to cancer treatment. As some other diseases apart from cancer are also related to excessive angiogenesis, studies of the therapeutic values of the anti-angiogenic Chinese medicinal foods on these other diseases are necessary. Future research in this direction is highly warranted.

As far as the development of therapeutic angiogenesis is concerned, Chinese medicinal foods appear to have great potential. Chinese medicinal foods have been demonstrated to exhibit promising results in the promotion of angiogenesis in vivo and results on the efficacy of some formulae in the treatment of ischemic diseases and bone fracture have been obtained. However, research regarding the active principles and the mechanisms of the angiogenic activities in Chinese medicinal foods is still in its infancy.

On the other hand, investigators in other parts of the world have conducted extensive studies on the molecular mechanisms of angiogenesis, identifying the biological factors, signal transduction processes, and control mechanisms involved. Future research should focus on understanding the mechanisms of the angiogenic activities of these effective Chinese medicinal foods and their isolated active principles. At present, clinical research on the promotion of angiogenesis is generally lacking. Investigation in this field will have important implications in the development of prophylactic and therapeutic regimes for disorders caused by insufficient blood supply.

REFERENCES

1. Hoffman, R., Paper, D. H., Donaldson, J., and Vogl, H., Inhibition of angiogenesis and murine tumor growth by laminarin sulphate, *Br. J. Cancer*, 73, 1183–1186, 1996.
2. Xu, Z., Li, F., and Wang, H., Anti-angiogenic activity and antitumor effect of laminarin sulfate [Chinese], *Chin. Tradit. Herb. Drugs*, 30, 551–553, 1999.
3. Pun, Z., Ye, D., Xie, X., Chen, H., and Lu, W., Antiangiogenesis of ginsenoside Rg3 in severe combined immunodeficient mice with human ovarian carcinoma [Chinese], *Chin. J. Obstet. Gynecol.*, 34, 227–230, 2002.
4. Toi, M., Hattori, T., Akagi, M., Inokuchi, K., Orita, K., Sugimachi, K., Dohi, K. et al., Randomized adjuvant trial to evaluate the addition of tamoxifen and PSK to chemotherapy in patients with primary breast cancer, *Cancer*, 70, 2475–2483, 1992.
5. Kanoh, T., Matsunaga, K., Saito, K., and Fuji, T., Suppression of in vivo tumor-induced angiogenesis by the protein-bound polysaccharide PSK, *In Vivo*, 8, 247–250, 1994.
6. Wu, A., Lin, F., Kung, L., Liu, S., and Zhang, Z., Experimental study of IPPV in the inhibition of tumor-induced angiogenesis and the growth of transplanted mammary adenocarcinoma [Chinese], *Chin. J. Gen. Surg.*, 16, 124, 2001.
7. Wu, A., Lin, F., Kung, L., Liu, S., and Li, H., Effects of IPPV on atypical proliferation of mammary glands and angiogenesis [Chinese], *Chin. J. Exp. Surg.*, 18, 372, 2001.
8. Jiang, X., Zhang, L., Xu, Z., and Guo, C., Effect of coix seed injection on angiogenesis [Chinese], *Tumor (Shanghai)*, 20, 313–314, 2000.
9. Sohn, K. H., Lee, H. Y., Chung, H. Y., Young, H. S., Yi, S. Y., and Kim, K. W., Antiangiogenic activity of triterpene acids, *Cancer Lett.*, 94, 213–218, 1995.
10. Ding, Z., Gao, C., Chen, N., Yuan, W., and Wo, X., Effects of curcumin on antiangiogenesis and induction of SMMC-7221 cell apoptosis [Chinese], *Chin. Pharmacol. Bull.*, 19, 171–173, 2003.
11. Kojima-Yuasa, A., Hau, J. J., Kennedy, D. O., and Matsui-Yuasa, I., Green tea extract inhibits angiogenesis of human umbilical vein endothelial cells through reduction of expression of VEGF receptors, *Life Sci.*, 73, 1299–1313, 2003.
12. Kondo, T., Ohta, T., Igura, K., Hara, Y., and Kaji, K., Tea catechins inhibit angiogenesis in vitro, measured by human endothelial cell growth, migration and tube formation, through inhibition of VEGF receptor binding, *Cancer Lett.*, 180, 139–144, 2002.

13. Durairaj, A., Mehra, A., Singh, R. P., and Faxon, D. P., Therapeutic angiogenensis, *Cardiol. Rev.*, 8, 279–287, 2000.

14. Wang, G., Yi, C., and Zhou, H., Research progress in Chinese medicine and angiogenesis [Chinese], *Chin. J. Microcirc.*, 13, 33–35, 2003.

15. Dai, R. and Li, Y., Ischemic blood vessels producing in coronary heart disease and traditional Chinese medicine [Chinese], *Chin. J. Integr. Tradit. West Med.*, 20, 163–164, 2000.

16. Xia, Y., Cai, L., and Zhang, S., Sero-pharmacological study of Nujing Yuyu Tang on the chicken chorioallantic membrane angiogenesis model [Chinese], *Study J. Trad. Chin. Med.*, 21, 531–534, 2003.

17. Zheng, S., Liu, X., Zhang, Z., Shen, M., Kan, G., and Wang, J., Effect of prescription of tonifying the kidney to regulate menstruation on angiogenic factors and their receptors of human endometrium during implantation [Chinese], *Chin. J. Bas. Med. Trad. Chin. Med.*, 8, 384–386, 2002.

18. Zhang, S., Wu, Z., Wang, L., Wang, H., and Wang, J., Application of chicken chorioallantoic membrane as a model for study of effects of Chinese medicine on angiogenesis [Chinese], *Chin. J. Bas. Med. Trad. Chin. Med.*, 5 (5), 16–19, 1999.

19. Wang, L., Lu, X., and Wu, Z., Effect of Bushenshengxueyao on bFGF, VEGF expression in golden hamster's uterus tissue [Chinese], *Chin. J. Bas. Med. Trad. Chin. Med.*, 6, 795–798, 2000.

20. Wang, W., Xu, L., and Yu, H., Experimental study of fracture healing promoted by Chinese medicinal drugs of invigorating the kidney and activating blood circulation [Chinese], *Chin. J. Bas. Med. Trad. Chin. Med.*, 4 (9), 35–38, 1998.

21. Zhang, J., Chen, L., and Jiang, W., Effect of external application of Chinese medicinal herbs on reconstruction of microangium in fracture healing [Chinese], *Chin. J. Orthop. Trauma.*, 13, 86–87, 2000.

22. Fan, Y., Xu, C., He, W., Fang, B., Li, X., Liu, S., Wang, H., and Yuan, H., Regulatory effect of Shengmai Chenggu capsule on vascular endothelial cell function in steriod-induced avascular necrosis of femoral head [Chinese], *Chin. J. Bas. Med. Trad. Chin. Med.*, 8, 675–677, 2002.

23. Wang, Y., Shi, Q., Zhou, C., Lan, F., Shen, P., Liu, M., and Xu, Y., The effect of Yiqi Huayu Fang on the blood of the cartilage end-plate [Chinese], *Chin. J. Trad. Med. Traum. Orthop.*, 10 (4), 1–4, 2002.

28 Disposition and Metabolism of Dietary Flavonoids

Min Jung Kang and Dong-Hyun Kim

CONTENTS

28.1 INTRODUCTION

Flavonoids occur widely in the natural diet in such items as fruits, vegetables, teas, wine, nuts, seeds, and flowers and are important constituents of the human diet. Flavonoids are a diverse group of natural products, and variation in the heterocyclic C-ring gives rise to over 4000 chemically unique flavonoids that have been identified in plants sources. On average, the daily Western diet contains approximately 1 g of mixed flavonoids, a quantity that could provide pharmacologically significant concentrations in body fluids and tissues. The biological effects of flavonoids include the reduction of cardiovascular disease risk, the inhibition of hepatocytic autophagy, antiviral activity, anticlastogenic effects, an anti-inflammatory analgesic effect, and an antiischemic effect.

There has been considerable interest in studying the metabolism and disposition of these compounds to better understand the biological fate of flavonoids. Absorption is an important unsolved problem because of the limited data available about it. Glycosidic flavonoids are considered non-absorbable and only free flavonoids without sugar molecules are thought to pass through the gut well. Absorption depends on the flavonoids. Quercetin is not absorbed in humans and rutin is poorly absorbed, whereas procyanidins are well absorbed in mice. After absorption, the subsequent metabolism is quite well known in animal studies before excretion into urine and bile. Another important aspect of metabolism is that their structural similarities can result in metabolic interconversion. Since metabolism might transform one class of flavonoid into another, such as the conversion of galangin to kaemferol, new pharmacological activity might result due to metabolism. This chapter describes the disposition and metabolism of flavonoids, which exist abundantly in functional foods.

28.2 CHEMISTRY AND CLASSES OF FLAVONOIDS

The basis of flavonoids' structure is the nucleus, which consists of three phenolic rings referred to as A-, B- and C-rings (Figure 28.1). The C-ring C is varied with a heterocyclic pyran, pyrone, or 4-oxo-pyrone, which yields flavonols, flavones, or flavanols and flavanones.

The chemical structure of the flavonoids depends on structural class, degree of hydroxylation, other substitutions, conjugations, and the degree of poymerization. In plants, they are relatively resistant to heat, oxygen, dryness, and a moderated degree of acidity—but can be modified by light. Photostability of the flavonoids depends on the modification of the hydroxyl group attached to C-3 of ring C. The absence or glycosylation of this hydroxyl group results in high photostability of the molecule [1,2].

Flavonoids are divided into several classes, according to their oxidation level on the C-ring, which include anthocyanidins, flavanols (catechins), flavones, flavonols, flavanones, and isoflavonoids. The main classes of flavonoids are shown in Figure 28.2.

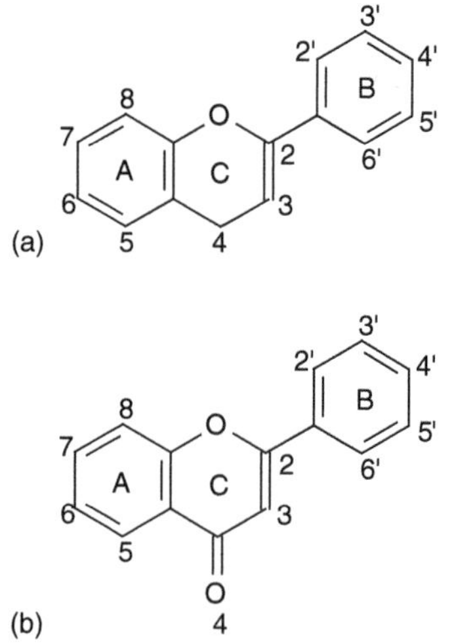

FIGURE 28.1 Flavan nuclus (a) and 4-oxo-flavonoid nucleus (b).

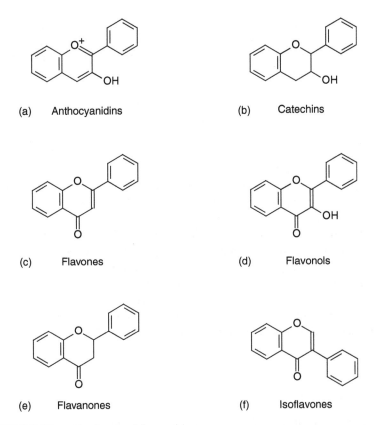

(a)   Anthocyanidins   (b)   Catechins

(c)   Flavones   (d)   Flavonols

(e)   Flavanones   (f)   Isoflavones

FIGURE 28.2 The main classes of flavonoids.

The structures of flavonoids differ greatly within the major classifications and substitutions, including glycosylation, hydrogenation, hydroxylation, malonylation, methylation, and sulfation. The pattern of conjugation, glycosylation, or methylation, which is very complex, can modify the hydrophilicity of the molecule and its biological properties, and markedly increase the molecular weight of the flavonoids. Flavonoid molecules lacking sugar moieties are referred to as aglycone, whereas flavonoid molecules with sugar moieties are called flavonoids as glycosides.

28.3   DIETARY SOURCES OF FLAVONOIDS

Fruits, vegetables, and beverages such as tea and red wine are especially rich sources of flavonols [3–6]. Tea, onions, and apples are the most predominant food sources of flavonols in The Netherlands [7], Denmark [8], and the United States [9]. Dietary sources can differ significantly depending on the country of origin [4]. Green tea was the predominant source of these five flavonoids in Japan, whereas fruit and vegetables are reported to be the most important dietary sources in Finland [4].

In parts of Greece [10] and the middle east [11], where the traditional diet is consumed, edible wild greens and fermented foods are the main staple diets. These foods have been shown to contain significant levels of flavonoids with values greater than those of red wine and black tea. Variations of dietary sources also can occur among regions within the same country. The flavonol quercetin was the main compound found in some fruits and vegetables [12]. However, the flavonols myricetin and kaempferol, and the flavones apigenin and luteolin were detected in many foods. High concentrations of quercetin can be found in onions in such forms as quercetin-3-glucoside, quercetin-4'-glucoside, quercetin-3,4'-diglucoside, and isorhamnetin-4'-glucoside.

Anthcyanins are reddish pigments that are widely distributed in many plant materials such as strawberries, blueberries, black currants, and grapes. Tea beverages are rich sources of flavonoids, predominantly catechins. Red wines contain flavonoids that can vary in content between 1 and 3 g/l, the majority of which is usually flavanols as monomers and oligomers. However, anthocyanins can comprise 20%–80% of the flavonoids content [13]. Berries represent another rich plant sources of dietary anthocyanins. Intake of anthocyanins from berry sources has been estimated to be 180–215 mg per day among the U.S. population [1]. This figure far exceeds the estimated intake of 23 mg per day of other flavonoids such as luteolin, kaempferol, myricetin, apigenin, and quercetin [5].

28.4   ABSORPTION AND METABOLISM

Deglycosylation of flavonoids glycosides has been proposed as the first stage of metabolism. The aglycones can be absorbed from the small intestine. However, most polyphenols are also present in food in the form of esters or glycosides that cannot be absorbed in their native form. Therefore, the substances have to be hydrolyzed or deglycosylated by enzymes or bacteria. During the course of absorption, polyphenols are conjugated in the small intestine and later in the liver. This process mainly includes methylation, sulfation, and glucuronidation. This is a metabolic detoxication process common to various classes of xenobiotics that restricts their potential toxic effects and facilitates their biliary and urinary elimination by increasing their hydrophilicity. The conjugation mechanisms are highly efficient, and aglycones are generally either absent in the blood or present in low concentrations after consumption of nutritional doses.

Circulating polyphenols are conjugated derivatives that are extensively bound to albumin. Polyphenols are able to penetrate tissues, particularly those in which they are metabolized, but their ability to accumulate within specific target tissues needs to be further investigated. Polyphenols and their derivatives are eliminated chiefly in urine and bile. Polyphenols are secreted via the biliary route into the duodenum, where they are subjected to the action of bacterial enzymes, especially β-glucuronidase, in the distal segments of the intestine, after which they may be reabsorbed. This enterohepatic recycling may lead to a longer presence of polyphenols within the body.

In foods, all flavonoids except flavanols are found in glycosylated forms, and glycosylation influences absorption. The fate of glycosides in the stomach is not clear. Experiments using surgically treated rats in which absorption was restricted to the stomach showed that absorption at the gastric level is possible for some flavonoids, such as quercetin and daidzein, but not for their glycosides [14,15]. Most of the glycosides probably resist acid hydrolysis in the stomach and thus arrive intact in the duodenum [16]. This has been shown in humans for quercetin glycosides: maximum absorption occurs 0.5–0.7 h after ingestion of quercetin 4′-glucoside and 6–9 h after ingestion of the same quantity of rutin (quercetin-3β-rutinoside). The bioavailability of rutin is only 15%–20% of that of quercetin 4′-glucoside [17,18]. Similarly, absorption of quercetin is more rapid and efficient after the ingestion of onions, which are rich in glucosides, than after the ingestion of apples containing both glucosides and various other glycosides [19].

In the case of quercetin glucosides, absorption occurs in the small intestine, and the efficiency of absorption is higher than that for the aglycone itself [20,21]. The underlying mechanism by which glucosylation facilitates quercetin absorption has been partly elucidated. Hollman et al. suggested that glucosides could be transported into enterocytes by the sodium-dependent glucose transporter SGLT1 [20]. They could then be hydrolyzed inside the cells by a cytosolic β-glucosidase [22]. Another pathway involves the lactase phloridzine hydrolase, a glucosidase of the brush border membrane of the small intestine, that catalyzes extracellular hydrolysis of some glucosides, followed by diffusion of the aglycone across the brush border [23].

Isoflavone glycosides present in soya products can also be deglycosylated by β-glucosidases from the human small intestine [22,23]. However, the effect of glucosylation on absorption is less clear for isoflavones than for quercetin. Aglycones present in fermented soya products were shown to be better absorbed than were the glucosides ingested from soybeans [24]. Dose or matrix effect may explain the difference in absorption observed in this first study. Setchell et al. [25] showed that when pure daidzein, genistein, or their corresponding 7-glucosides were administered orally to healthy volunteers, a tendency toward greater bioavailability was observed with the glucosides, as measured from the area under the curve of the plasma concentrations: 2.94, 4.54, 4.52, and 4.95 µg h/mL for daidzein, genistein, daidzin, and genistin, respectively. However, in another human study, peak plasma concentrations were markedly higher after aglycone ingestion than after glucoside ingestion.

Glycosylation does not influence the nature of the circulating metabolites. Intact glycosides of quercetin, daidzein, and genistein were not recovered in plasma or urine after ingestion as pure compounds or from complex food [26–29]. For flavanones, only trace amounts of glycosides have been detected in human urine, corresponding to 0.02% of the administered dose of naringin [30]. The glycoside of anthocyanins can be directly converted into glucuronides by a uridine diphosphate (UDP) glucose dehydrogenase, as suggested by Wu and his colleagues [31].

Caco-2 cells as a model of absorption in the small intestine showed that only the dimer and trimers of flavanols are able to cross the intestinal epithelium [32]. Some procyanidin dimers are poorly absorbed in humans. The procyanidin dimer B2 was detected in the plasma of volunteers after ingestion of a cocoa beverage; however, the maximal plasma concentration that was reached 2 h after ingestion was much lower than that reached after a roughly equivalent intake of epicatechin (0.04 compared with 6.0 µmol/L) [33].

Despite the scarcity of studies performed on the bioavailability of hydroxycinnamic acids, when ingested in free form, these compounds are rapidly absorbed from the small intestine and are conjugated and, in particular, glucuronidated in the same way that flavonoids are [34,35]. However, these compounds are naturally esterified in plant products, and this impairs their absorption. Human tissues (intestinal mucosa, liver) and biological fluids (plasma, gastric juice, duodenal fluid) do not possess esterase capable of hydrolyzing chlorogenic acid to release caffeic acid [36–38].

Existing data do not suggest a marked effect of the various diet components on polyphenol bioavailability. The absorption of quercetin, catechin, and resveratrol in humans was recently shown to be broadly equivalent when these polyphenols were administered in three different matrices: white wine, grape juice, and vegetable juice [39].

Day et al. [40] used human small intestine and liver cell-free extracts to see whether there is glucosidase activity toward flavonoid glycosides. Some but not all flavonoid glycosides were hydrolyzed by the small intestine and liver extracts. Lactase phlorizin hydrolase, a 1-β-glycosidase found on the brush border of the mammalian small intestine, can deglycosylate dietary flavonoid glycosides, which suggests a possible role for the enzyme in flavonoid metabolism [41].

After absorption, flavonoids are bound to albumin and transported to the liver via the portal vein [42]. The liver seems to be the chief organ involved in flavonoid metabolism; however, the intestinal mucosa and/or kidneys must not be ruled out [43]. Flavonoids and their derivatives may undergo reactions such as hydroxylations, methylations, and reductions. Manach et al. [44] identified 3' methylquercetin (isorhamnetin) in human plasma after consumption of various fruits and vegetables. After ingestion of a single dose of quercetin glucosides, concentrations of isorhamnetin in plasma peak shortly after the quercetin concentration peak, suggesting that quercetin glucosides can be methylated into isorhamnetin immediately after absorption [45].

Conjugation reactions with glucuronic acid or sulfate seem to be the most common type of metabolic pathways for the flavonoids [46]. For instance, glucuronidation of flavonols occurs in human microsomes (UGT1A9) [51] and liver [47], and UDP-glucuronosyltransferase plays an important role in this process [48]. Glucuronidation, O-methylation, and O-methyl-glucuronidation occur in flavonoid metabolism in the small intestine and O-methyl-transferases were involved in this process [49]. After ingestion of an extract of Gingko biloba in humans, no flavonoid metabolites could be detected in blood [50]. Urine samples contained detectable amounts of substituted benzoic acids; however, no phenylacetic acids or phenylpropionic acid derivatives were found, thus indicating that extensive metabolism of the flavonoids occurred [50].

28.5   BACTERIAL DEGRADATION

Flavonoid metabolism in humans has been reported to also depend on the participation of intestinal microflora [46,51–53]. Flavonoids that are not absorbed in the small intestine can be metabolized by colonic microflora into aglycones and phenolic acids. These in turn may be absorbed from the colon [54–56]. This has been suggested to take place in the lower part of the ileum and the cecum [51]. Flavonoid metabolism produces a series of phenolic compounds that have been identified as aromatic acids. They form by detachment of the A-ring from the residual flavonoid molecule and the opening of the heterocyclic C-ring.

McDonald et al. [56] reported that the flavonoids quercitrin and rutin are bacterially hydrolyzed in the human gut to quercetin, whereas robinin is converted into kaempferol. Bokkenheuser et al. [51] identified flavonoid glycoside-hydrolyzing enzymes in fecal flora cultures grown under anaerobic conditions. They recovered three enzyme-producing strains that, using β-glucosidases, α-rhamnosidases, or β-galactosidases were capable of converting rutin to quercetin. They showed that at least some of the bacterial glycosidases are capable of cleaving glycosidic bonds and flavonoid–saccharide bonds in the gut. In addition, four Clostridia strains, recovered from human feces, were identified to have the enzymes that cleave the C-ring of flavonoids [52]. Increasing evidence shows that human intestinal bacteria are equipped with a vast selection of hydrolytic enzymes capable of degrading numerous dietary flavonoid glycoside compounds.

28.6   PLASMA CONCENTRATIONS

Plasma concentrations reached after flavonoid consumption vary highly according to their nature and the food source. They are on the order of 0.3–0.75 µmol/L after consumption of 80–100 mg quercetin equivalent administered in the form of apples, onions, or meals rich in plant products [17,19,26]. When ingested in the form of green tea (0.1–0.7 µmol/L for an intake of 90–150 mg), cocoa (0.25–0.7 µmol/L for an intake of 70–165 mg) [57–60], or red wine (0.09 µmol/L for an intake of 35 mg) [61], catechin and epicatechin are as effectively absorbed as is quercetin. The maximum concentrations of hesperetin metabolites determined in plasma 5–7 h after consumption of orange juice were 1.3–2.2 µmol/L for an intake of 130–220 mg [62,63].

Naringenin in grapefruit juice appears to be absorbed even better: a peak plasma concentration of 6 µmol/L is obtained after ingestion of 200 mg. In contrast, plasma concentrations of anthocyanins are very low: peak concentrations, which occur between 30 min and 2 h after consumption, are on the order of a few tens of nanomoles per liter for an intake of ≈10–200 mg anthocyanins [64–66]. Similarly, the intake of ≈25 mg secoisolariciresinol diglucoside in the form of linseed produces only a slight increase (30 nmol/L) in plasma lignan concentrations, and this increase occurs gradually between 9 and 24 h [67]. Isoflavones are certainly the best absorbed flavonoids: plasma concentrations of 1.4–4 µmol/L are obtained between 6 and 8 h in adults who consume relatively low quantities of soya derivatives supplying 50 mg isoflavones [68–70].

28.7   TISSUE UPTAKE

When single doses of radiolabeled polyphenols (quercetin, epigallocatechin gallate, quercetin 4′-glucoside, resveratrol) are given to rats or mice killed 1–6 h later, radioactivity is mainly recovered in the blood and tissues of the digestive system, such as the stomach, intestine, and liver [71–74]. However, polyphenols have been detected by high pressure liquid chromatography (HPLC) analysis in a wide range of tissues in mice and rats, including the brain [75,76], endothelial cells [77], heart, kidney, spleen, pancreas, prostate, uterus, ovary, mammary gland, testes, bladder, bone, and skin [72,78–80]. The concentrations obtained in these tissues ranged from 30 to 3000 ng aglycone equivalents/g tissue depending on the dose administered and the tissue considered. The time of tissue sampling may be of great importance because we have no idea of the kinetics of penetration and elimination of polyphenols in the tissues.

It is still difficult to say whether some polyphenols accumulate in specific target organs. A few studies seem to indicate that some cells may readily incorporate polyphenols by specific mechanisms. The endothelium is likely to be one of the primary sites of flavonoid action. Microautoradiography of mice tissues after the administration of radiolabeled epigallocatechin gallate or resveratrol indicated that radioactivity is unequally incorporated into the cells of organs [72,74].

The nature of the tissue metabolites may be different from that of blood metabolites because of the specific uptake or elimination of some of the tissue metabolites, or because the metabolism occurred in the cells. Youdim et al. [81] showed that the uptake of flavanone glucuronides by rat and mouse brain endothelial cultured cells is much lower than that of their corresponding aglycones. In rats fed a genistein-supplemented diet, the fraction of genistein present in the aglycone form was much more important in several tissues than in blood. It accounted for >50% of the total genistein metabolites in the mammary gland [82], uterus, and ovary, and 100% in the brain [78] and prostate [83], whereas it represented only 8% of the total plasma metabolites.

Only two studies reported data on polyphenol concentrations in human tissues. The first study measured phytoestrogens in human prostate tissue. Surprisingly, the study showed significantly lower prostatic concentrations of genistein in men with benign prostatic hyperplasia than in those with a normal prostate, whereas plasma genistein concentrations were higher in men with benign prostatic hyperplasia [84]. In addition, concentrations of enterodiol and enterolactone were higher in prostatic tissue than in plasma, whereas the opposite was true for daidzein, genistein, and equol.

In the other study, equol concentrations in women who ingested isoflavones were found to be higher in breast tissue than in serum, whereas genistein and daidzein were more concentrated in serum than in breast tissue. Note that very high equol concentrations have been obtained in breast tissue, and these concentrations are equivalent to 6 μmol/L for an intake of 110 mg of its precursor, daidzein [85]. These initial studies show that plasma concentrations are not directly correlated with concentrations in target tissues and that the distribution between blood and tissues differs between the various polyphenols. This raises the question of whether plasma concentrations are accurate biomarkers of exposure.

28.8 ELIMINATION

Flavonoid glucuronides and sulfates are polar, water-soluble compounds that are apparently readily excreted by mammals in the urine and bile. When excreted in bile, flavonoids are passed into the duodenum and metabolized by intestinal bacteria, which results in the production of fragmentation products or the hydrolysis of glucurono- or sulfoconjugates [43]. The resulting metabolites that are released may be reabsorbed and enter an enterohepatic cycle. Substitution on the flavonoid molecule, degree of polarity, and molecular weight determine the extent of biliary excretion [53]. Flavonoids also are eliminated by renal excretion after conjugation in the liver [43].

The half-life of elimination also can be affected and plasma levels of quercetin have been detected up to 48 h after consumption of flavonols. This implies that repeated intakes of onion quercetin glycosides lead to a build-up of the concentration in plasma.

Biliary excretion seems to be a major pathway for the elimination of genistein, epigallocatechin gallate, and eriodictyol [86,87]. But the biliary excretion studies in human have never been examined yet. Urinary excretion has often been determined in human studies. The total amount of metabolites excreted in urine is roughly correlated with maximum plasma concentrations. It is quite high for flavanones from citrus fruit (4%–30% of intake), especially for naringenin from grapefruit juice [30,62,63,88,89], and is even higher for isoflavones: the percentages excreted are 16%–66% for daidzein and 10%–24% for genistein [68,70,71,90,91]. It may appear surprising that plasma concentrations of genistein are generally higher than those of daidzein despite the higher urinary excretion of daidzein, but this can be explained by the efficient biliary excretion of genistein. Certain metabolites of anthocyanins may still be unidentified as a result of analytic difficulties with these unstable compounds. Urinary excretion of flavonols accounts for 0.3%–1.4% of the ingested dose for quercetin and its glycosides [17,19,20], but reaches 3.6% when purified glucosides are given in a hydroalcoholic solution to fasted volunteers [92]. Urinary recovery is 0.5%–6% for some tea catechins [57,93], 2%–10% for red wine catechin [94], and up to 30% for cocoa epicatechin [95]. For caffeic and ferulic acids, relative urinary excretion ranges from 5.9 to 27% [38,96].

The exact half-lives of polyphenols in plasma have rarely been calculated with great precision but are on the order of 2 h for compounds such as anthocyanins [66] and 2–3 h for flavanols [97–100], except for epigallocatechin gallate, which is eliminated more slowly, probably because of higher biliary excretion or greater complexing with plasma proteins, as described for galloylated compounds. The half-lives of isoflavones and quercetin are on the order of 4–8 [103–105] and 11–28 [17,19] hours, respectively. This suggests that maintenance of high plasma concentrations of flavonoid metabolites could be achieved with regular and frequent consumption of plant products. For instance, heavy consumption of onions favors accumulation of quercetin in plasma [106]. For compounds such as tea catechins with rapid absorption and a short half-life, repeated intakes must be very close together in time to obtain an accumulation of metabolites in plasma [101,102,107].

Otherwise, plasma concentrations regularly fluctuate after repeated ingestions, and no final accumulation occurs [108].

28.9 MASS SPECTROMETRY IN THE STRUCTURAL ANALYSIS OF FLAVONOIDS

For the investigation of metabolism and the disposition of flavonoids, it is important to have access to rapid and reliable methods for the analysis and identification of these natural polyphenolic compounds in all their many forms. Modern mass spectrometric techniques are very well suited to the analysis of flavonoids in plants and foodstuffs, and play a key role in the analysis, as they can provide significant structural information on small quantities of pure samples as well as on mixtures. Liquid chromatography coupled to mass spectrometry (LC/MS) represents a very powerful tool for the analysis of natural products. The mass spectrometer is a universal detector that can achieve very high sensitivity and provide information on the molecular mass and on structural features. More detailed structural information can subsequently be obtained by resorting to tandem mass spectrometry (MS/MS) in combination with collision-induced dissociation (CID).

With regard to the structural characterization of flavonoids, information can be obtained on (1) the aglycone moiety, (2) the types of carbohydrates (mono-, di-, tri-, or tetrasaccharides, and hexoses, deoxyhexoses or pentoses) or other substituents present, (3) the stereochemical assignment of terminal monosaccharide units, (4) the sequence of the glycan part, (5) interglycosidic linkages, and (6) attachment points of the substituents to the aglycone.

Structural information can be obtained from the chromatographic retention times. For the C_{18}- or C_8-RP columns generally used, the more polar compounds are eluted first. Thus, retention times are inversely correlated with increasing glycosylation, whereas acylation, methylation, or prenylation have the opposite effect, although the position of glycosylation [109] or methylation [110] can have a significant effect on the retention time. Flavanones precede flavonols, which in turn precede flavones for compounds with an equivalent substitution pattern. For isomeric compounds that differ in the structure of the saccharide residues, rutinosides elute ahead of neohesperidosides, galactosides ahead of glucosides [111], glucosides ahead of arabinosides, and arabinosides ahead of rhamnosides [112,113]. One should therefore be aware that the linkage position also can have an influence on retention.

Mass spectrometry has proven to be a very powerful technique in the analysis of flavonoids owing to its high sensitivity and the possibility of coupling with different chromatographic techniques, e.g., GC/MS, CE/MS, and especially LC/MS because it allows both qualitative and quantitative determinations. Structural information can be obtained on the flavonoid aglycone part, the types of carbohydrates, or other substituents present, the stereochemical structure of terminal monosaccharide units, the sequence of the glycan part, interglycosidic linkages, and attachment points of the substituents to the aglycone. Many of the structural characterization methods discussed are not restricted to flavonoid glycosides, but can also be applied to other glycosylated secondary plant metabolites (e.g., saponins, iridoid glycosides,

chromon glycosides). Although some of the procedures described were established using older soft ionization techniques (e.g., Fast-atom bombardment [FAB] or Liquid-phase secondary ion mass spectrometry [LSIMS]), they also hold for the newer techniques (e.g., Electrospray ionization [ESI] or Atmosphere–Pressure ionization [APCI]) because the fragmentation behavior of the molecular ion species generated with these techniques (e.g., $[M+H]^+$, $[M+Na]^+$ and $[M-H]^-$) are essentially the same.

With the recent developments of user-friendly LC/MS instrumentation that incorporates an ion trap or a quadrupole time-of-flight analyzer and allows low-energy CID, it can be envisaged that methods based on the latter technique will be more widely applied in the future.

The MS methods available at present rarely provide complete structural information, so that the search for novel MS approaches to the characterization of flavonoids and the implementation of new technologies to make the present methods faster, easier, and more selective should continue. It is fair to state, however, that MS in combination with other spectroscopic techniques [116], e.g., Ultraviolet (UV) and Nuclear Magnetic Resonance (NMR), or directly coupled in an LC UV MS NMR system [114,115] is a most powerful technique in the identification of complex unknown polyphenolic compounds of plant origin.

REFERENCES

1. Kehnau, J., The flavonoids, a class of semi-essential food components; their role in human nutrition, *World Rev. Nutr. Diet*, 24, 117–191, 1976.
2. Dunford, C. L., Smith, G. J., Swinny, E. E., and Markham, K. R., The fluorescence and photostabilities of naturally occurring isoflavones, *Photochem. Photobiol. Sci.*, 2 (5), 611–615, 2003.
3. Crozier, A., Lean, M. E. J., McDonald, M. S., and Black, C., Quantitative analysis of the flavonoid content of commercial tomatoes, onions, lettuce, and celery, *J. Agric. Food Chem.*, 45, 590–595, 1997.
4. Hertog, M. G. L., Kromhout, D., Aravanis, C., Blackburn, H., Buzina, R., Fidanza, F., Giampaoli, S. et al., Flavonoid intake and long term risk of coronary heart disease and cancer in the Seven Countries Study, *Arch. Int. Med.*, 155, 381–386, 1995.
5. Hertog, M. G. L., Hollman, P. C. H., and van de Putte, B., Content of potentially anticarcinogenic flavonoids of tea infusions, wines and fruit juices, *J. Agric. Food Chem.*, 41, 1242–1246, 1993.
6. McDonald, M., Hughes, M. J., Michael, B., Lean, E. J., Matthews, D., and Crozier, A., Survey of the free and conjugated myricetin and quercetin content of red wines of different geographical origins, *J. Agric. Food Chem.*, 46, 368–375, 1998.
7. Hertog, M., Hollman, P. C. H., Katan, M. B., and Kromhout, D., Intake of potentially anticarcinogenic flavonoids and their determinants in adults in The Netherlands, *Nutr. Cancer*, 20, 21–29, 1993.
8. Justesen, U., Knuthsen, P., and Leth, T., Determination of plant polyphenols in Danish foodstuffs by HPLC–UV and LC–MS detection, *Cancer Lett.*, 114, 165–167, 1997.
9. Rimm, E. B., Katan, M. B., Ascherio, A., Stampfer, M. J., and Willett, W. C., Relation between intake of flavonoids and risk for coronary heart disease in male health professionals, *Ann. Int. Med.*, 125, 384–389, 1996.

10. Trichopoulou, A., Vasilpoulou, E., Hollman, P., Chamalides, Ch., Foufa, E., Kaloudis, Tr., Kromhout, D. et al., Nutritional composition and flavonoid content of edible wild greens and green pies: a potential rich source of antioxidant nutrients in the Mediterranean diet, *Food Chem.*, 70, 319–323, 2000.

11. Miean, K. H. and Mohamed, S., Flavonoid (myricetin, quercetin, kaempferol, luteolin, and apigenin) content of edible tropical plants, *J. Agric. Food Chem.*, 49 (6), 3106–3112, 2001.

12. Aherne, S. A. and O'Brien, N. M., Dietary flavonols: chemistry, food content, and metabolism, *Nutrition*, 18, 75–81, 2002.

13. Glories, Y., Anthocyanins and tannins from wine: organoleptic properties, *Prog. Clin. Biol. Res.*, 280, 123–134, 1988.

14. Crespy, V., Morand, C., Besson, C., Manach, C., Demigne, C., and Remesy, C., Quercetin, but not its glycosides, is absorbed from the rat stomach, *J. Agric. Food Chem.*, 50, 618–621, 2002.

15. Piskula, M. K., Yamakoshi, J., and Iwai, Y., Daidzein and genistein but not their glucosides are absorbed from the rat stomach, *FEBS Lett.*, 447, 287–291, 1999.

16. Gee, J. M., Du Pont, M. S., Rhodes, M. J. C., and Johnson, I. T., Quercetin glucosides interact with the intestinal glucose transport pathway, *Free Radic. Biol. Med.*, 25, 19–25, 1998.

17. Graefe, E. U., Wittig, J., Mueller, S., Riethling, A. K., Uehleke, B., Drewelow, B., Pforte, H., Jacobasch, G., Derendorf, H., and Veit, M., Pharmacokinetics and bioavailability of quercetin glycosides in humans, *J. Clin. Pharmacol.*, 41, 492–499, 2001.

18. Hollman, P. C., Bijsman, M. N., van Gameren, Y., Cnossen, E. P., de Vries, J. H., and Katan, M. B., The sugar moiety is a major determinant of the absorption of dietary flavonoid glycosides in man, *Free Radic. Res.*, 31, 569–573, 1999.

19. Hollman, P. C., van Trijp, J. M., Buysman, M. N., van der Gaag, M. S., Mengelers, M. J., de Vries, J. H., and Katan, M. B., Relative bioavailability of the antioxidant flavonoid quercetin from various foods in man, *FEBS Lett.*, 418, 152–156, 1997.

20. Hollman, P. C. H., Devries, J. H. M., Vanleeuwen, S. D., Mengelers, M. J. B., and Katan, M. B., Absorption of dietary quercetin glycosides and quercetin in healthy ileostomy volunteers, *Am. J. Clin. Nutr.*, 62, 1276–1282, 1995.

21. Morand, C., Manach, C., Crespy, V., and Remesy, C., Quercetin 3-*O*-beta-glucoside is better absorbed than other quercetin forms and is not present in rat plasma, *Free Radic. Res.*, 33, 667–676, 2000.

22. Day, A. J., DuPont, M. S., Ridley, S., Rhodes, M., Rhodes, M. J., Morgan, M. R., and Williamson, G., Deglycosylation of flavonoid and isoflavonoid glycosides by human small intestine and liver β-glucosidase activity, *FEBS Lett.*, 436, 71–75, 1998.

23. Day, A. J., Canada, F. J., Diaz, J. C., Kroon, P. A., Mclauchlan, R., Faulds, C. B., Plumb, G. W., Morgan, M. R., and Williamson, G., Dietary flavonoid and isoflavone glycosides are hydrolysed by the lactase site of lactase phlorizin hydrolase, *FEBS Lett.*, 468, 166–170, 2000.

24. Hutchins, A. M., Slavin, J. L., and Lampe, J. W., Urinary isoflavonoid phytoestrogen and lignan excretion after consumption of fermented and unfermented soy products, *J. Am. Diet. Assoc.*, 95, 545–551, 1995.

25. Setchell, K. D., Brown, N. M., Desai, P., Zimmer-Nechemias, L., Wolfe, B. E., Brashear, W. T., Kirschner, A. S., Cassidy, A., and Heubi, J. E., Bioavailability of pure isoflavones in healthy humans and analysis of commercial soy isoflavone supplements, *J. Nutr.*, 131, 1362S–1375S, 2001.

26. Manach, C., Morand, C., Crespy, V., Demigne, C., Texier, O., Regerat, F., and Remesy, C., Quercetin is recovered in human plasma as conjugated derivatives which retain antioxidant properties, *FEBS Lett.*, 426, 331–336, 1998.

27. Sesink, A. L., O'Leary, K. A., and Hollman, P. C., Quercetin glucuronides but not glucosides are present in human plasma after consumption of quercetin-3-glucoside or quercetin-4′-glucoside, *J. Nutr.*, 131, 1938–1941, 2001.

28. Wittig, J., Herderich, M., Graefe, E. U., and Veit, M., Identification of quercetin glucuronides in human plasma by high-performance liquid chromatography–tandem mass spectrometry, *J. Chromatogr. B*, 753, 237–243, 2001.

29. Setchell, K. D., Brown, N. M., Zimmer-Nechemias, L., Brashear, W. T., Wolfe, B. E., Kirschner, A. S., and Heubi, J. E., Evidence for lack of absorption of soy isoflavone glycosides in humans, supporting the crucial role of intestinal metabolism for bioavailability, *Am. J. Clin. Nutr.*, 76, 447–453, 2002.

30. Ishii, K., Furuta, T., and Kasuya, Y., Mass spectrometric identification and high-performance liquid chromatographic determination of a flavonoid glycoside naringin in human urine, *J. Agric. Food Chem.*, 48, 56–59, 2000.

31. Wu, X., Cao, G., and Prior, R. L., Absorption and metabolism of anthocyanins in elderly women after consumption of elderberry or blueberry, *J. Nutr.*, 132, 1865–1871, 2002.

32. Déprez, S., Mila, I., Huneau, J.-F., Tomé, D., and Scalbert, A., Transport of proanthocyanidin dimer, trimer and polymer across monolayers of human intestinal epithelial Caco-2 cells, *Antioxid. Redox Signal.*, 3, 957–967, 2001.

33. Holt, R. R., Lazarus, S. A., Sullards, M. C., Zhu, Q. Y., Schramm, D. D., Hammerstone, J. F., Fraga, C. G., Schmitz, H. H., and Keen, C. L., Procyanidin dimer B2 [epicatechin-(4beta-8)-epicatechin] in human plasma after the consumption of a flavanol-rich cocoa, *Am. J. Clin. Nutr.*, 76, 798–804, 2002.

34. Choudhury, R., Srai, S. K., Debnam, E., and Rice-Evans, C. A., Urinary excretion of hydroxycinnamates and flavonoids after oral and intravenous administration, *Free Radic. Biol. Med.*, 27, 278–286, 1999.

35. Cremin, P., Kasim-Karakas, S., and Waterhouse, A. L., LC/ES–MS detection of hydroxycinnamates in human plasma and urine, *J. Agric. Food Chem.*, 49, 1747–1750, 2001.

36. Gonthier, M. P., Verny, M. A., Besson, C., Remesy, C., and Scalbert, A., Chlorogenic acid bioavailability largely depends on its metabolism by the gut microflora in rats, *Nutrition*, 133 (6), 1853–1859, 2003.

37. Olthof, M. R., Hollman, P. C. H., and Katan, M. B., Chlorogenic acid and caffeic acid are absorbed in humans, *J. Nutr.*, 131, 66–71, 2001.

38. Rechner, A. R., Spencer, J. P., Kuhnle, G., Hahn, U., and Rice-Evans, C. A., Novel biomarkers of the metabolism of caffeic acid derivatives in vivo, *Free Radic. Biol. Med.*, 30, 1213–1222, 2001.

39. Goldberg, D. M., Yan, J., and Soleas, G. J., Absorption of three wine-related polyphenols in three different matrices by healthy subjects, *Clin. Biochem.*, 36, 79–87, 2003.

40. Day, A. J., DuPont, M. S., Ridley, S., Rhodes, M., Rhodes, M. J., Morgan, M. R., and Williamson, G., Deglycosylation of flavonoid and isoflavonoid glycosides by human small intestine and liver β-glucuronidase activity, *FEBS Lett.*, 436, 71–75, 1998.

41. Day, A. J., Canada, F. J., Diaz, J. C., Kroon, P. A., Mclauchlan, R., Faulds, C. B., Plumb, G. W., Morgan, M. R., and Williamson, G., Dietary flavonoid and isoflavone glycosides are hydrolysed by the lactase site of lactase phlorizin hydrolase, *FEBS Lett.*, 468, 166–170, 2000.

42. Manach, C., Texier, O., Morand, C., Crespy, V., Regerat, F., Demigne, C., and Remesy, C., Comparison of the bioavailability of quercetin and catechin in rats, *Free Radic. Biol. Med.*, 27, 1259–1266, 1999.

43. Hackett, A., The metabolism of flavonoid compounds in mammals, In *Plant Flavonoids in Biology and Medicine: Biochemical, Pharmacological, and Structure–Activity Relationships*, Cody, V., Middleton, E., and Harborne, J. B. Eds., Alan R Liss, New York, p.177, 1986.

44. Manach, C., Morand, C., Crespy, V., Demigne, C., Texier, O., Regerat, F., and Remesy, C., Quercetin is recovered in human plasma as conjugated derivatives which retain antioxidant properties, *FEBS Lett.*, 426 (3), 331–336, 1998.

45. Olthof, M. R., Hollman, P. C., Vree, T. B., and Katan, M. B., Bioavailabilities of quercetin-3-glucoside and quercetin-4′-glucoside do not differ in humans, *J. Nutr.*, 130, 1200–1203, 2000.

46. Kühnau, J., The flavonoids, a class of semi-essential food components: their role in human nutrition, *World Rev. Nutr. Diet*, 24, 117–191, 1976.

47. Yilmazer, M., Stevens, J. F., and Buhler, D. R., In vitro glucuronidation of xanthohumol, a flavonoid in hop and beer, by rat and human liver microsomes, *FEBS Lett.*, 91 (3), 252–256, 2001.

48. Oliveira, E. J. and Watson, D. G., In vitro glucuronidation of kaempferol and quercetin by human UGT-1A9 microsomes, *FEBS Lett.*, 471, 1–6, 2000.

49. Kuhnle, G., Spencer, J. P., Schroeter, H., Shenoy, B., Debnam, E. S., Srai, S. K., Rice-Evans, C., and Hahn, U., Epicatechin and catechin are O-methylated and glucuronidated in the small intestine, *Biochem. Biophys. Res. Commun.*, 277, 507–512, 2000.

50. Pietta, P. G., Gardana, C., and Mauri, P. L., Identification of Gingko biloba flavonol metabolites after oral administration to humans, *J. Chromatogr. A*, 693, 249–255, 1997.

51. Bokkenheuser, V. D., Shackleton, C. H., and Winter, J., Hydrolysis of dietary flavonoid glycosides by strains of intestinal bacteroides from humans, *Biochem. J.*, 248, 953–956, 1987.

52. Winter, J., Moore, L. H., Dowell, V. R. Jr., and Bokkenheuser, V. D., C-ring cleavage of flavonoids by human intestinal bacteria, *Appl. Environ. Microbiol.*, 55, 1203–1208, 1989.

53. Griffiths, L. A., Mammalian metabolism of flavonoids, In *The Flavonoids: Advances in Research*, Harborne, J. H. and Mabry, T. J. Eds., Chapman and Hall, London, p. 681, 1982.

54. Manach, C., Morand, C., Crespy, V., Demigne, C., Texier, O., Regerat, F., and Remesy, C., Quercetin is recovered in human plasma as conjugated derivatives which retain antioxidant properties, *FEBS Lett.*, 426, 331–336, 1998.

55. Hollman, P. C., de Vries, J. H., van Leeuwen, S. D., Mengelers, M. J., and Katan, M. B., Absorption of dietary quercetin glycosides and quercetin in healthy ileostomy volunteers, *Am. J. Clin. Nutr.*, 62 (6), 1276–1282, 1995.

56. McDonald, I. A., Mader, J. A., and Bussard, R. G., The role of rutin and quercitrin in stimulating flavonol glycosidase activity by cultured cell-free microbial preparations of human feces and saliva, *Mutat. Res.*, 122 (2), 95–102, 1983.

57. Lee, M. J., Wang, Z. Y., Li, H., Chen, L., Sun, Y., Gobbo, S., Balentine, D. A., and Yang, C. S., Analysis of plasma and urinary tea polyphenols in human subjects, *Cancer Epidemiol. Biomarkers Prev.*, 4, 393–399, 1995.

58. Unno, T., Kondo, K., Itakura, H., and Takeo, T., Analysis of (−)-epigallocatechin gallate in human serum obtained after ingesting green tea, *Biosci. Biotechnol. Biochem.*, 60, 2066–2068, 1996.

59. Rein, D., Lotito, S., Holt, R. R., Keen, C. L., Schmitz, H. H., and Fraga, C. G., Epicatechin in human plasma: in vivo determination and effect of chocolate consumption on plasma oxidation status,, *J. Nutr.*, 130, 2109S–2114S, 2000.

60. Wang, J. F., Schramm, D. D., Holt, R. R., Ensunsa, J. L., Fraga, C. G., Schmitz, H. H., and Keen, C. L., A dose–response effect from chocolate consumption on plasma epicatechin and oxidative damage, *J. Nutr.*, 130 (8S suppl.), 2115S–2119S, 2000.

61. Donovan, J. L., Bell, J. R., Kasim-Karakas, S., German, J. B., Walzem, R. L., Hansen, R. J., and Waterhouse, A. L., Catechin is present as metabolites in human plasma after consumption of red wine, *J. Nutr.*, 129, 1662–1668, 1999.

62. Manach, C., Morand, C., Gil-Izquierdo, A., Bouteloup-Demange, C., and Remesy, C., Bioavailability in humans of the flavanones hesperidin and narirutin after the ingestion of two doses of orange juice, *Eur. J. Clin. Nutr.*, 57, 235–242, 2003.

63. Erlund, I., Meririnne, E., Alfthan, G., and Aro, A., Plasma kinetics and urinary excretion of the flavanones naringenin and hesperetin in humans after ingestion of orange juice and grapefruit juice, *J. Nutr.*, 131, 235–241, 2001.

64. Miyazawa, T., Nakagawa, K., Kudo, M., Muraishi, K., and Someya, K., Direct intestinal absorption of red fruit anthocyanins, cyanidin-3-glucoside and cyanidin-3,5-diglucoside, into rats and humans, *J. Agric. Food Chem.*, 47, 1083–1091, 1999.

65. Matsumoto, H., Inaba, H., Kishi, M., Tominaga, S., Hirayama, M., and Tsuda, T., Orally administered delphinidin 3-rutinoside and cyanidin 3-rutinoside are directly absorbed in rats and humans and appear in the blood as the intact forms, *J. Agric. Food Chem.*, 49, 1546–1551, 2001.

66. Cao, G., Muccitelli, H. U., Sanchez-Moreno, C., and Prior, R. L., Anthocyanins are absorbed in glycated forms in elderly women: a pharmacokinetic study, *Am. J. Clin. Nutr.*, 73, 920–926, 2001.

67. Nesbitt, P. D., Lam, Y., and Thompson, L. U., Human metabolism of mammalian lignan precursors in raw and processed flaxseed, *Am. J. Clin. Nutr.*, 69, 549–555, 1999.

68. Xu, X., Wang, H.-J., Murphy, P. A., Cook, L., and Hendrich, S., Daidzein is a more bioavailable soymilk isoflavone than is genistein in adult women, *J. Nutr.*, 124, 825–832, 1994.

69. King, R. A. and Bursill, D. B., Plasma and urinary kinetics of the isoflavones daidzein and genistein after a single soy meal in humans, *Am. J. Clin. Nutr.*, 67, 867–872, 1998.

70. Watanabe, S., Yamaguchi, M., Sobue, T., Takahashi, T., Miura, T., Arai, Y., Mazur, W., Wahala, K., and Adlercreutz, H., Pharmacokinetics of soybean isoflavones in plasma, urine and feces of men after ingestion of 60 g baked soybean powder (kinako), *J. Nutr.*, 128, 1710–1715, 1998.

71. Ueno, I., Nakano, N., and Hirono, I., Metabolic fate of [^{14}C]quercetin in the ACI rat, *Jpn J. Exp. Med.*, 53, 41–50, 1983.

72. Suganuma, M., Okabe, S., Oniyama, M., Tada, Y., Ito, H., and Fujiki, H., Wide distribution of [^{3}H](−)-epigallocatechin gallate, a cancer preventive tea polyphenol, in mouse tissue, *Carcinogenesis*, 19, 1771–1776, 1998.

73. Mullen, W., Graf, B. A., Caldwell, S. T., Hartley, R. C., Duthie, G. G., Edwards, C. A., Lean, M. E., and Crozier, A., Determination of flavonol metabolites in plasma and tissues of rats by HPLC-radiocounting and tandem mass spectrometry following oral ingestion of [2-(14)C]quercetin-4′-glucoside, *J. Agric. Food Chem.*, 50, 6902–6909, 2002.

74. Vitrac, X., Desmouliere, A., Brouillaud, B., Krisa, S., Deffieux, G., Barthe, N., Rosenbaum, J., and Merillon, J. M., Distribution of [^{14}C]-*trans*-resveratrol, a cancer chemopreventive polyphenol, in mouse tissues after oral administration, *Life Sci.*, 72, 2219–2233, 2003.

75. Datla, K. P., Christidou, M., Widmer, W. W., Rooprai, H. K., and Dexter, D. T., Tissue distribution and neuroprotective effects of citrus flavonoid tangeretin in a rat model of Parkinson's disease, *Neuroreport*, 12, 3871–3875, 2001.

76. Abd El Mohsen, M. M., Kuhnle, G., Rechner, A. R., Schroeter, H., Rose, S., Jenner, P., and Rice-Evans, C. A., Uptake and metabolism of epicatechin and its access to the brain after oral ingestion, *Free Radic. Biol. Med.*, 33, 1693–1702, 2002.

77. Youdim, K. A., Martin, A., and Joseph, J. A., Incorporation of the elderberry anthocyanins by endothelial cells increases protection against oxidative stress, *Free Radic. Biol. Med.*, 29, 51–60, 2000.

78. Chang, H. C., Churchwell, M. I., Delclos, K. B., Newbold, R. R., and Doerge, D. R., Mass spectrometric determination of genistein tissue distribution in diet-exposed Sprague–Dawley rats, *J. Nutr.*, 130, 1963–1970, 2000.

79. Kim, S., Lee, M. J., Hong, J., Li, C., Smith, T. J., Yang, G. Y., Seril, D. N., and Yang, C. S., Plasma and tissue levels of tea catechins in rats and mice during chronic consumption of green tea polyphenols, *Nutr. Cancer*, 37, 41–48, 2000.

80. Coldham, N. G. and Sauer, M. J., Pharmacokinetics of [(14)C]Genistein in the rat: gender-related differences, potential mechanisms of biological action, and implications for human health, *Toxicol. Appl. Pharmacol.*, 164, 206–215, 2000.

81. Youdim, K. A., Dobbie, M. S., Kuhnle, G., Proteggente, A. R., Abbott, N. J., and Rice-Evans, C., Interaction between flavonoids and the blood–brain barrier: in vitro studies, *J. Neurochem.*, 85, 180–192, 2003.

82. Fritz, W. A., Coward, L., Wang, J., and Lamartiniere, C. A., Dietary genistein: perinatal mammary cancer prevention, bioavailability and toxicity testing in the rat, *Carcinogenesis*, 19, 2151–2158, 1998.

83. Wang, J., Eltoum, I. E., and Lamartiniere, C. A., Dietary genistein suppresses chemically induced prostate cancer in Lobund–Wistar rats, *Cancer Lett.*, 186, 11–18, 2002.

84. Hong, S. J., Kim, S. I., Kwon, S. M., Lee, J. R., and Chung, B. C., Comparative study of concentration of isoflavones and lignans in plasma and prostatic tissues of normal control and benign prostatic hyperplasia, *Yonsei Med. J.*, 43, 236–241, 2002.

85. Maubach, J., Bracke, M. E., Heyerick, A., Depypere, H. T., Serreyn, R. F., Mareel, M. M., and De Keukeleire, D., Quantitation of soy-derived phytoestrogens in human breast tissue and biological fluids by high-performance liquid chromatography, *J. Chromatogr. B*, 784, 137–144, 2003.

86. Sfakianos, J., Coward, L., Kirk, M., and Barnes, S., Intestinal uptake and biliary excretion of the isoflavone genistein in rats, *J. Nutr.*, 127, 1260–1268, 1997.

87. Kohri, T., Nanjo, F., Suzuki, M., Seto, R., Matsumoto, N., Yamakawa, M., Hojo, H. et al., Synthesis of (−)-[4-^{3}H]epigallocatechin gallate and its metabolic fate in rats after intravenous administration, *J. Agric. Food. Chem.*, 49, 1042–1048, 2001.

88. Fuhr, U. and Kummert, A. L., The fate of naringin in humans: a key to grapefruit juice–drug interactions? *Clin. Pharmacol. Ther.*, 58, 365–373, 1995.

89. Lee, Y. S. and Reidenberg, M. M., A method for measuring naringenin in biological fluids and its disposition from grapefruit juice by man, *Pharmacology*, 56, 314–317, 1998.

90. Richelle, M., Pridmore-Merten, S., Bodenstab, S., Enslen, M., and Offord, E. A., Hydrolysis of isoflavone glycosides to aglycones by beta-glycosidase does not alter

plasma and urine isoflavone pharmacokinetics in postmenopausal women, *J. Nutr.*, 132, 2587–2592, 2002.

91. Zhang, Y., Hendrich, S., and Murphy, P. A., Glucuronides are the main isoflavone metabolites in women, *J. Nutr.*, 133, 399–404, 2003.

92. Olthof, M. R., Hollman, P. C. H., Vree, T. B., and Katan, M. B., Bioavailabilities of quercetin-3-glucoside and quercetin-4′-glucoside do not differ in humans, *J. Nutr.*, 130, 1200–1203, 2000.

93. Yang, B., Arai, K., and Kusu, F., Determination of catechins in human urine subsequent to tea ingestion by high-performance liquid chromatography with electrochemical detection, *Anal. Biochem.*, 283, 77–82, 2000.

94. Donovan, J. L., Kasim-Karakas, S., German, J. B., and Waterhouse, A. L., Urinary excretion of catechin metabolites by human subjects after red wine consumption, *Br. J. Nutr.*, 87, 31–37, 2002.

95. Baba, S., Osakabe, N., Yasuda, A., Natsume, M., Takizawa, T., Nakamura, T., and Terao, J., Bioavailability of (−)-epicatechin upon intake of chocolate and cocoa in human volunteers, *Free Radic. Res.*, 33, 635–641, 2000.

96. Jacobson, E. A., Newmark, H., Baptista, J., and Bruce, W. R., A preliminary investigation of the metabolism of dietary phenolics in humans, *Nutr. Rep. Int.*, 28, 1409–1417, 1983.

97. Hollman, P. C. H. and Arts, I. C. W., Flavonols, flavones and flavanols—nature, occurrence and dietary burden, *J. Food Sci. Agric.*, 80, 1081–1093, 2000.

98. Lee, M. J., Maliakal, P., Chen, L., Meng, X., Bondoc, F. Y., Prabhu, S., Lambert, G., Mohr, S., and Yang, C. S., Pharmacokinetics of tea catechins after ingestion of green tea and (−)-epigallocatechin-3-gallate by humans: formation of different metabolites and individual variability, *Cancer Epidemiol. Biomarkers Prev.*, 11, 1025–1032, 2002.

99. Bell, J. R. C., Donovan, J. L., Wong, R., Waterhouse, A. L., German, J. B., Walzem, R. L., and Kasim-Karakas, S. E., (+)-Catechin in human plasma after ingestion of a single serving of reconstituted red wine, *Am. J. Clin. Nutr.*, 71, 103–108, 2000.

100. Richelle, M., Tavazzi, I., Enslen, M., and Offord, E. A., Plasma kinetics in man of epicatechin from black chocolate, *Eur. J. Clin. Nutr.*, 53, 22–26, 1999.

101. Yang, C. S., Chen, L., Lee, M. J., Balentine, D., Kuo, M. C., and Schantz, S. P., Blood and urine levels of tea catechins after ingestion of different amounts of green tea by human volunteers, *Cancer Epidemiol. Biomarkers Prev.*, 7, 351–354, 1998.

102. Okuda, T., Mori, K., and Hatano, T., Relationship of the structures of tannins to the binding activities with hemoglobin and methylene blue, *Chem. Pharm. Bull. (Tokyo)*, 33, 1424–1433, 1985.

103. Cassidy, A., Hansley, B., and Lamuela-Raventos, R. M., Isoflavones, lignans and stilbenes—origins, metabolism and potential importance to human health, *J. Sci. Food Agric.*, 80, 1044–1062, 2000.

104. Setchell, K. D., Faughnan, M. S., Avades, T., Zimmer-Nechemias, L., Brown, N. M., Wolfe, B. E., Brashear, W. T., Desai, P., Oldfield, M. F., Botting, N. P., and Cassidy, A., Comparing the pharmacokinetics of daidzein and genistein with the use of ^{13}C-labeled tracers in premenopausal women, *Am. J. Clin. Nutr.*, 77, 411–419, 2003.

105. Shelnutt, S. R., Cimino, C. O., Wiggins, P. A., and Badger, T. M., Urinary pharmacokinetics of the glucuronide and sulfate conjugates of genistein and daidzein, *Cancer Epidemiol. Biomarkers Prev.*, 9, 413–419, 2000.

106. Moon, J. H., Nakata, R., Oshima, S., Inakuma, T., and Terao, J., Accumulation of quercetin conjugates in blood plasma after the short-term ingestion of onion by women, *Am. J. Physiol. Regul. Integr. Comp. Physiol.*, 279, R461–R467, 2000.

107. Warden, B. A., Smith, L. S., Beecher, G. R., Balentine, D. A., and Clevidence, B. A., Catechins are bioavailable in men and women drinking black tea throughout the day, *J. Nutr.*, 131, 1731–1737, 2001.

108. van het Hof, K. H., Wiseman, S. A., Yang, C. S., and Tijburg, L. B., Plasma and lipoprotein levels of tea catechins following repeated tea consumption, *Proc. Soc. Exp. Biol. Med.*, 220, 203–209, 1999.

109. Harborne, J. B. and Boardley, M., Use of high-performance liquid chromatography in the separation of flavonol glycosides and flavonol sulphates, *J. Chromatogr.*, 299, 377, 1984.

110. Greenham, J., Harborne, J. B., and Williams, C. A., Identification of lipophilic flavones and flavonols by comparative HPLC, TLC and UV spectral analysis, *Phytochem. Anal.*, 14 (2), 100–118, 2003.

111. Robards, K. and Antolovich, M., Analytical chemistry of fruit bioflavonoids: a review, *Analyst*, 122, 11R, 1997.

112. Harborne, J. B. and Boardley, M., Use of high-performance liquid chromatography in the separation of flavonol glycosides and flavonol sulphates, *J. Chromatogr.*, 299, 377, 1984.

113. Schieber, A., Keller, P., Streker, P., Klaiber, I., and Carle, R., Detection of isorhamnetin glycosides in extracts of apples (Malus domestica cv. 'Brettacher') by HPLC–PDA and HPLC–APCI–MS/MS, *Phytochem. Anal.*, 13 (2), 87–94, 2002.

114. Hansen, S. H., Jensen, A. G., Cornett, C., Bjørnsdottir, I., Taylor, S., Wright, B., and Wilson, I. D., High-performance liquid chromatography on-line coupled to high-field NMR and mass spectrometry for structure elucidation of constituents of *Hypericum perforatum* L., *Anal. Chem.*, 71, 5235, 1999.

115. Lommen, A., Godejohann, M., Venema, D. P., Hollman, P. C. H., and Spraul, M., Application of directly coupled HPLC–NMR–MS to the identification and confirmation of quercetin glycosides and phloretin glycosides in apple peel, *Anal. Chem.*, 72 (8), 1793–1797, 2000.

116. Olthof, M. R., Hollman, P. C. H., Vree, T. B., and Katan, M. B., Bioavailabilities of quercetin-3-glucoside and quercetin-4′-glucoside do not differ in humans, *J. Nutr.*, 130, 1200–1203, 2000.

29 Functional Foods: Probiotics

Maria Saarela

CONTENTS

29.1 INTRODUCTION

In EU countries healthiness is one of the most frequently mentioned reasons behind food choices [1]. A trend where people want to manage their health through the foods they eat is continuously strengthening, and, as a result, food products that communicate healthy messages are increasing on the market. The European market of functional food is dominated by gut-health products, especially probiotics—live microorganisms which when administered in adequate amounts confer a health benefit on the host [2,3].

Only a few years ago, consumer awareness about the health benefits of probiotics was fairly low: in a study in 2001, only 44%–48% of European consumers (including Finnish and Danish people) were aware of the health effects of probiotics [4]. In a more recent study, probiotic products were already familiar to consumers: 88% of Finnish consumers included into the study recognized probiotic/gut-friendly juice, but only 33% of them used it [5]. However, when the consumption of both probiotic yoghurt and juice was surveyed, consumption was clearly more common: 64% of respondents [6]. In the United States, consumer awareness about probiotics is still low, with only about 20% of grocery shoppers showing an awareness of the term, while roughly 70% do not know any health benefits associated with probiotics [7].

A rising number of probiotic-containing products are nowadays available on the market. During the past few years, probiotic bacteria have been increasingly

TABLE 29.1
Examples of Probiotic Foods on European Market

Product Types and Trade Names	Probiotic Microorganisms in the Products as Stated by the Manufacturer (Starter Microbes are not Listed)
Non-drinkable fermented milks (such as yoghurt, cheese, quark, viili) Bifisoft, Bifidus, Bighurt, Biofit, BiofardePlus, Biola, Biologic Bifidus, Cultura Dofilus, Dujat Bio Aktiv, Ekologisk jordgubbs youghurt, Fit & Aktiv, Fjällyoughurt, Fysiq, Gaio Dofilus, Gefilac, Gefilus, Lc1, Probiotisches Joghurt, ProViva, RELA, Verum, Vifit Vitamel, Vitality, Weight Watchers, Yogosan Milbona	*Lactobacillus acidophilus*/"acidophilus" bacteria, *L. acidophilus* LA5, *Lactobacillus* LGG, *Lactobacillus casei* (incl. F19), *Lactobacillus johnsonii*, *Lactobacillus plantarum* 299v, *Lactobacillus reuteri*, *Lactobacillus rhamnosus* 271 and LB21 *Lactococcus lactis* L1A *Bifidobacteria/Bifidobacterium*, "bifidus Bb12," "bifidus," "*L. bifidus*," "*Lb. bifidus*," *Bifidobacterium* BB12, *Bifidobacterium lactis*, *Bifidobacterium bifidum*
Drinkable fermented milks (including cultured buttermilk, yoghurt drink, dairy drink) A-fil, Actimel, Aktifit, AB-piimä, Bella Vita, Bifidus, Biofit, Biola, Casilus, Cultura, Emmifit, Everybody, Fit & Aktiv, Fundo, Gaio, Gefilac, Gefilus, Kaiku Actif, Lc1go, LGG+, Onaka, Öresundsfil, Philura, Probiotic drink, ProViva, Pro.x, Verum, Vikt-Väktarna, Vitality, Vive+, Yakult, Yoco acti-vit	*Lactobacillus acidophilus*/"acidophilus bacteria," *L. acidophilus* LA5, *L. casei* (incl. F19, 431, Immunitas, Shirota), "casei," *Lactobacillus* LGG, *L. johnsonii*, *L. rhamnosus* (incl. 271 and LB21), *L. reuteri*, *L. plantarum* 299v, "*Lactobacillus Fortis*" *Lactococcus lactis* L1A *Bifidobacterium*/"bifidus" bacteria, "*Lb. bifidus*," *B. lactis*, *Bifidobacterium* BB12, *B. longum* BB536
Non-fermented dairy products (milk, ice cream) Gefilus, God Hälsa, RELA, Vivi Vivo	*Lactobacillus* LGG, *L. plantarum* 299v, *L. reuteri*
Non-dairy foods (fruit and berry juices and drinks, recovery drinks, cereal-based drinks and snacks) BioGrain, Cultura, Dexal PRO Gefilac, Gefilus, ProViva, ProViva Active, RELA, Vifit Vitamel, Yosa	*Lactobacillus* LGG, *L. plantarum* 299v, *L. reuteri*, *L. acidophilus* (incl. LA5), *L. casei* F19 *Bifidobacterium*, *B. lactis* (BB12)

Information obtained mainly from www-pages. Invalid names are indicated by quotes. Starter microbes are not listed.

included in foods, mainly in yoghurts and fermented milks, but more lately in non-dairy foods such as fruit and berry juices, as well. (Table 29.1). Europe has traditionally had a strong position on the probiotic food market (along with Japan), a position that is expressed today as a wide range of probiotic food products available for consumers (examples are listed in Table 29.1) [8,9]. Daily-dose dairy drinks have been the largest growing probiotic product type on the European market [10].

29.2 HEALTH BENEFITS OF PROBIOTICS

Inadequate consumer understanding of probiotic foods—in spite of the improving situation—still poses a challenge to the potential growth of this market sector. To eliminate these hurdles and to enhance the attractiveness of probiotic foods, it is important to inform consumers about the scientific evidence of the health and nutritional benefits of probiotics. This has to be done clearly, understandably, and truthfully [6]. The scientific back-up for probiotic health claims has significantly advanced during the past few years, mainly due to the improved quality of human intervention trials, but also due to the increasing number of studies carried out. Most commonly, probiotic microbes have been given to humans in the form of macroencapsulated lyophilized powders. More recently, however, the number of trials performed with real probiotic food products (e.g., fermented or non-fermented dairy products or fruit/berry juices) has been increasing. This trend towards studying the effects of probiotic foods and not just probiotic bacteria is necessary to enhance consumer trust in these products—and also to give more credibility to the health-benefiting effects of probiotic food products.

The types of bacteria studied for their probiotic potential include *Lactobacillus* sp. (*Lactobacillus acidophilus, Lactobacillus reuteri, Lactobacillus casei, Lactobacillus johnsonii, Lactobacillus plantarum, Lactobacillus rhamnosus*), *Bifidobacterium* sp. (*Bifidobacterium bifidum, Bifidobacterium infantis, Bifidobacterium animalis/lactis, Bifidobacterium longum, Bifidobacterium breve*), *Enterococcus faecalis, Escherichia coli,* and *Bacillus cereus* [11]. Of the large number of probiotic bacterial strains studied, only relatively few have so far been involved in high-quality human intervention trials. *Lactobacillus rhamnosus* GG (LGG) is the most extensively clinically studied strain in humans [12–14] , followed by others such as *L. johnsonii* (acidophilus) LJ-1 (La1, LC-1), *B. animalis* subsp. *lactis* Bb-12, *L. casei* Immunitas (DN-114001), *L. casei* Shirota, and *L. plantarum* DSM 9843 (299v) (Table 29.2). The most frequently used probiotic daily dose has been 10^{10}–10^{11} CFU (colony forming units), while the duration of the trials has largely varied from a few days up to one year depending on the nature of the condition targeted (acute vs. chronic, and prevention vs. treatment) (Table 29.2).

The health claims attributed to probiotics are diverse, including alleviation of lactose intolerance symptoms, treatment of viral and antibiotic associated diarrhea, treatment of infant gastroenteritis, reduction of the symptoms of antibiotic treatment of *Helicobacter pylori*, alleviation of atopic dermatitis symptoms in children, prevention of the risk of allergy in infancy, alleviation of symptoms of IBD (inflammable bowel disease) and IBS (irritable bowel syndrome), enhancement of the immune response, and positive effects on superficial bladder and cervical cancer. There are also several potential effects that have not been adequately established, including lowering the activity of rheumatoid arthritis, prevention of urinary tract infection, reduction in the risk of respiratory infections, etc., [11,13,15,16] (Table 29.2).

The target population groups for probiotic intervention studies have included children (also infants), adults and the elderly. By far, the most frequently studied group thus far has been adults, and especially "healthy" adults. In healthy adults, the

TABLE 29.2
Clinical Effects of Some Probiotic Strains

Probiotic Strain	Daily Dose, CFU (Number of Studies)/Duration	Potential Clinical Effects in Humans
Lactobacillus rhamnosus GG (ATCC 53103)	10^{10}–10^{11} (21[a]) 10^9 (3) 10^8 (3) Also heat-killed 2 days—15 months	Lowering fecal enzyme activities Reduction of antibiotic-associated diarrhea in children Treatment and prevention of rotavirus and acute diarrhea in children Treatment of relapsing *Clostridium difficile* diarrhea Reduction of symptoms of antibiotic treatment of *H. pylori* Modulation of immune response Alleviation of atopic dermatitis symptoms in children Prevention of the risk of allergy in infancy Reduction of the severity of respiratory infection in children
Lactobacillus johnsonii (acidophilus) LJ-1 (La1, LC-1)	10^{10} (2) 10^9 (2) 10^8 vs. 10^9 (1) Also supernatant & live vs. heat-killed 3–4 weeks	Modulation of intestinal flora Immune enhancement Down-regulation of *H. pylori* infection and gastritis Decrease of *H. pylori* colonization in children
Bifidobacterium animalis subsp. lactis Bb-12	10^{11} (1) 10^{10} (4) 10^9 (4) 10^8 (2) 3 weeks—6 months	Prevention of traveller's diarrhea Treatment of viral diarrhea including rotavirus diarrhea Modulation of intestinal flora Improvement of constipation Modulation of immune response Alleviation of atopic dermatitis symptoms in children
Lactobacillus casei Immunitas (DN-114001)	10^{10}(6) Days—6 months	Reduction of the duration, severity, and incidence of diarrhea in children Reduction of the duration of winter infections in elderly Lowering fecal enzyme activities Treatment of HAM/TPS Maintaining immune competence in aging

(continued)

TABLE 29.2 *(Continued)*

Probiotic Strain	Daily Dose, CFU (Number of Studies)/Duration	Potential Clinical Effects in Humans
		Enhancing the therapeutic effect of antibiotics on *H. pylori* eradication
Lactobacillus casei Shirota	10^{11} (2) 10^{10} (4) Also heat-killed 3 week—1 year	Modulation of intestinal flora Lowering fecal enzyme activities Positive effects on superficial bladder cancer and cervical cancer Immune response modulation Adjunctive therapy of chronic constipation
Lactobacillus plantarum DSM 9843 (299v)	10^{10} (5) 10^8 (1) 3 days—6 weeks	Modulation of intestinal flora Increase in fecal short-chain fatty acid content Lowering of fibrinogen and cholesterol levels Reducing the symptoms of IBS Reduction in cardiovascular disease risk factors in smokers Attenuation of systemic inflammatory response in critically ill patients

[a] Indicates the minimum number of studies performed with the dose (the table is not all inclusive and in some cases exact information was not readily available). Only studies showing positive effects are included.

Source: From Hawrelak, J. A., Whitten, D. L., and Myers, S. P., *Digestion*, 72, 51–56, 2005; Saarela, M., Mogensen, G., Fonden, R., Mättö, J., and Mattila-Sandholm, T., *J. Biotech.*, 84, 197–215, 2000; Bergonzelli, G. E., Blum, S., Brüssow, H., and Corthésy-Theulaz, I., *Digestion*, 72, 57–68, 2005; Guandalini, S., *J. Clin. Gastroenterol.*, 40, 244–248, 2006; McFarland L. V., *Am. J. Gastroenterol.*, 101, 812–822, 2006; Sazawal, S., Hiremath, G., Dhingra, U., Malik, P., Deb, S., and Black, R. E., *Lancet Infect. Dis.*, 6, 374–382, 2006.

aim of the intervention study has usually been to look at the fecal recovery of the probiotic strain, its persistence in the gastrointestinal (GI) tract, the effect on GI microbiota and immune responses, and the possible side-effects of consumption. These types of studies are more focused on the safety aspects of consumption than on the actual health effects. In addition to healthy adults, adults with various acute or chronic infections, or other usually chronic conditions, have been targeted.

In children the most commonly studied patient group has been subjects with acute diarrhea followed by those with atopic disease or those colonised with *H. pylori*. In the elderly group, healthy, constipated, and those with antibiotic

TABLE 29.3

Target Groups for Human Intervention Studies with Probiotic *Lactobacilli* and *Bifidobacteria*

Children/Infants	Adults	Elderly
Acute diarrhea	Healthy[a]	Healthy[a]
Healthy[a]	*H. pylori* treatment side effects	Incidence and severity of
Atopic disease and immune	Antibiotic treatment side	winter infections
functions	effects	Antibiotic associated diarrhea
H. pylori colonization	Traveller's diarrhea	Constipation
Severity of respiratory infection	Irritable bowel syndrome	
Infections in child care centers	Inflammatory bowel disease	
	Immune functions in diseased	
	subjects	
	Constipation	
	Arthritis	
	Prevention of UTI	
	Prophylaxis for vaginitis	
	Cancer prevention	
	Irradiation side effects	
	Cardiovascular disease risk	
	factors	

[a] These studies mainly look at fecal recovery of the strain; also persistence, tolerance, effects on microbiota and immune responses (in elderly especially optimisation of immunity and gut function).

Source: From Hawrelak, J. A., Whitten, D. L., and Myers, S. P., *Digestion*, 72, 51–56, 2005; Saarela, M., Mogensen, G., Fonden, R., Mättö, J., and Mattila-Sandholm, T., *J. Biotech.*, 84, 197–215, 2000; Bergonzelli, G. E., Blum, S., Brüssow, H., and Corthésy-Theulaz, I., *Digestion*, 72, 57–68, 2005; Guandalini, S., *J. Clin. Gastroenterol.*, 40, 244–248, 2006; McFarland L. V., *Am. J. Gastroenterol.*, 101, 812–822, 2006; Sazawal, S., Hiremath, G., Dhingra, U., Malik, P., Deb, S., and Black, R. E., *Lancet Infect. Dis.*, 6, 374–382, 2006.

associated diarrhea have been most commonly included (Table 29.3). Due to the rapidly aging population, this group will become a more important target group for both pro- and prebiotic interventions in the future [17]. In the elderly, the possibility of averting or delaying age-associated degenerative diseases with functional foods is important. Similarly, in the adult population, combating lifestyle-related diseases is of concern, whereas in children/infants both prevention/treatment of diarrhea and allergy symptoms are important targets for functional foods.

29.3 PROBIOTICS AND ANGIOGENESIS

Angiogenesis is an essential process during tissue healing after injury or inflammation. On the other hand, it is also involved in cancer progression.

Angiogenesis may also function in host defence, since angiogenins show microbicidal activity and can thus play a role in epithelial host defence by inhibiting the access of microbes to the gut epithelium [18].

Lactobacillus acidophilus (ATCC 4356 and 43121) supernatants have been shown to be able to stimulate the inflammatory stage of tissue repair, TNF-α production, and angiogenesis in mice [19]. Furthermore the same *L. acidophilus* supernatants were shown to stimulate the proliferation of embryonic cells [20]. Other lactic acid bacteria, *Bifidobacterium adolescentis* and *Lactococcus lactis*, have been applied in another approach—as a delivery system of endostatin (an angiogenesis inhibitor) to tumors in mice and rats, respectively [21,22].

29.4   SAFETY OF PROBIOTICS

Lactobacillus and *Bifidobacterium* species are common but non-dominant members of the indigenous microbiota of the human GI-tract [23,24]. Like other members of the indigenous microbiota of the orogastrointestinal tract, they occasionally cause opportunistic infections in humans. Indigenous bifidobacteria and lactobacilli have been connected to certain dental infections, bacteremia, endocarditis, and rare cases of other infections [25,26].

Because probiotic consumption may involve ingestion of large numbers of viable bacterial cells (daily dose 10^9–10^{11} CFU), safety aspects of probiotic consumption are of the utmost importance. In the assessment of probiotic safety, several approaches including studies on the intrinsic properties and pharmaco-kinetics of the probiotic strain, and on interactions between the probiotic strain and the host are possible. Data on survival of probiotics within the GI-tract, their translocation and colonization properties, and the fate of probiotic-derived components is important for the safety evaluation of probiotic ingestion. These characteristics are strain specific; for example, it is known that survival of probiotic strains through the GI-tract varies largely [27]. O'Brien et al. [28] listed several metabolic activities that can be considered in probiotic safety evaluations. These included biogenic amine production, bile salt hydroxylase, D- vs. L-lactate production, mucin degradation, nitroreductase, and β-glucuronidase. However, the relevance of probiotic metabolic activities detected in vitro often remains speculative, as these activities are also present among the members of the normal GI microbiota [29]. An additional safety consideration is the possible transferability of antibiotic resistance genes between probiotics and the members of the GI microbiota. Lactobacilli and bifidobacteria display a wide range of antibiotic resistances naturally, but in most cases antibiotic resistance is not of the transmissible type, and therefore does not usually create a safety concern [25].

To our knowledge, there are currently only few relevant reports of probiotic bacterium found in an infection. A *L. rhamnosus* strain indistinguishable from *L. rhamnosus* GG has been isolated from a liver abscess from an elderly lady with a history of hypertension and diabetes mellitus [30]. In another case a probiotic *L. rhamnosus* strain (strain or product specifications were not given) was suggested to have caused endocarditis in an elderly male [31]. In a more recent publication [32]

L. rhamnosus GG was detected in two septic patients (6-week-old infant and 6-year-old child) after enteral administration of the probiotic following antibiotic therapies (including vancomycin). Predisposing factors for *Lactobacillus* infection include diabetes and prior treatment with antibiotics inactive against the genus (e.g., vancomycin) [25].

Probiotic products have been safely consumed in large quantities for a long time in Europe and in Japan. However, the above findings indicate that sporadic localized infections in mainly immunocompromised patients may occur and show that no zero risk can be attributed to the consumption of living microbes. It has to be noted that isolating a probiotic strain in an infection site does not necessarily mean that there is a direct cause–effect relationship between the isolate and the disease, because the growth of the probiotic strain in the site may be a secondary effect.

In Finland *Lactobacillus* strains isolated from bacteremia have been systematically characterized since 1989. Although the yearly consumption of probiotic products containing lactobacilli (especially *L. rhamnosus* GG) has increased during the last 10 years (at the time of the investigation 3×10^{11} CFU per capita), the incidence of lactobacillemia has not increased [33]. In more recent papers by Salminen et al. [34,35], both the demographic and clinical charcteristics of patients with *Lactobacillus* bacteremia and the *Lactobacillus* isolates were thoroughly analyzed and characterized. The extensive follow-up period of *Lactobacillus* bacteremia cases indicates that the risk of serious infection by one single probiotic strain is very low. Simultaneously, it is difficult to estimate all the health benefits the same probiotic strain has implemented, also in immunocompromised patients.

Assessing the risks of probiotic consumption can be a very expensive and time-consuming task. While considering the risk of probiotic consumption, we have to keep in mind that lactic acid bacteria have been globally consumed in a myriad of fermented food varieties (milk, meat, vegetable, and cereal products) for a very long time without an indication that they could be generally harmful to the consumers' health.

29.5 PROBIOTIC FOODS ON EUROPEAN MARKET—THE IMPORTANCE OF QUALITY

The most commonly used probiotics in foods in Europe are lactobacilli (e.g., *L. rhamnosus*, *L. paracasei*, *L. acidophilus*) and bifidobacteria (mainly *B. animalis* subsp. *lactis*) (Table 29.1). Typically, probiotics have been incorporated into dairy products, mainly fermented ones such as yogurt (also drink), cheese, quark, cultured buttermilk and dairy drink. Other probiotic food products include milk, ice cream, fruit, and berry juices and drinks, recovery drinks, cereal-based drinks, and snacks (Table 29.1). The far from complete, yet lengthy list of trade names of probiotic food products in Table 29.1 illustrates the importance of this market in Europe.

Probiotic functionality

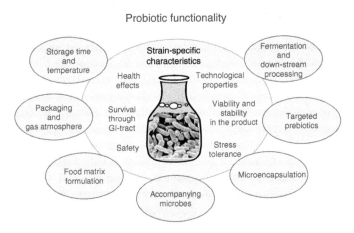

FIGURE 29.1 Factors important for probiotic functionality.

Good viability is generally considered a prerequisite for optimal probiotic functionality [36]; therefore, probiotic products should contain high enough levels of the specific probiotic strain(s) throughout storage and during consumption. Several factors, such as strain characteristics, food matrix, temperature (e.g., during storage), pH, and accompanying microbes affect the viability of a probiotic strain (Figure 29.1).

Concerns about the quality of probiotic products have been broadly expressed. For quality assessment, especially regarding the numbers and identification of living bacteria in the products, several studies investigating the microbiology of commercial probiotic foods have been published [37–41]. Mislabeling is commonly detected in commercial probiotic food products: the products often contain bacterial species not indicated on the label, unacceptably low numbers of the added probiotic, or an erroneous name is given for the probiotic bacterium [37–39,41]. According to Temmerman et al. [38] the numbers of viable probiotic cells were 10^5–10^9 CFU/ml in dairy products and <1–10^6 CFU/g in supplements. In the study of Coeuret et al. [39] dairy products contained 10^4–10^7 CFU/g of probiotic cells, whereas, according to Gueimonde et al. [40], the corresponding range was 10^4–10^8 CFU/ml. These studies indicate that—unfortunately for the consumer—achieving the effective daily dose of living cells (often considered to be 10^9 CFU although most clinical studied have been performed with a daily dose of 10^{10}–10^{11} CFU) in a reasonable portion size of the product is often not possible.

A restricted variability seems to exist among commercial *Lactobacillus* and especially *Bifidobacterium* strains. In the studies of Fasoli et al. [37], Temmerman et al. [38], and Gueimonde et al. [40] *B. animalis* subsp. *lactis* was the only *Bifidobacterium* species detected regardless of what was claimed on the food product label. In the study of Masco et al. [41], few products with *B. breve* and *B. longum* were additionally identified. The popularity of *B. animalis* subsp. *lactis* is due to its superior technological properties (e.g., tolerance to oxygen and low pH) compared to other *Bifidobacterium* species [42–44].

29.6 PRODUCTION OF PROBIOTIC FOODS

The most common food carriers for probiotics are fermented milks (Table 29.1). There are different ways in which probiotics can be incorporated into these types of products. The most typical way is to add the probiotic culture to starters as DVI (direct vat inoculation) culture. Since milk fermentation rarely occurs in conditions optimal for probiotics, the probiotic does not usually grow notably during a mixed fermentation with starters. To promote probiotic growth, fermentation can be performed in two separate batches: one batch of milk is fermented with the probiotic (in optimal conditions for it) and another batch with starters. After fermentation, the two products are combined to generate the final probiotic fermented milk product. A third way of producing probiotic fermented milk products is to use the probiotic culture alone as the fermenting starter (e.g., milk drink Yakult) [45].

In non-fermented foods, the growth of probiotic cultures is usually considered undesirable. This is due to the fact that the end products of probiotic metabolism (e.g., organic acids) can create off-flavors into the product. Therefore, the probiotic is added to these types of products as a concentrated culture without any propagating steps. Non-dairy foods especially are considered to be a challenging environment for probiotics: often the ingredients, pH, and oxygen content are more unfavorable to probiotic viability in non-dairy matrices [36,46,47].

29.7 CONCLUSIONS

During the past few years, research activity in the field of health benefiting microbes—probiotics—has increased rapidly. The quality of research has improved; this is especially notable when we compare the human intervention trials performed 10 years ago and now. However, although probiotic products have gained increasing popularity and their health effects have been assiduously studied, research, and development of processes aimed at improving the technological properties of probiotics are still inadequate. Unfortunately, this manifests as common quality deficiencies (both qualitative and quantitative) in the probiotic products on the market. Solving these quality problems and guaranteeing the consumer an effective daily dose of probiotics is a challenge to industry, but is essential for the future success of probiotic products. Technological improvements are especially important when aimed at the diversified application of probiotics in novel and non-traditional products.

In the future, the special needs of different consumer groups will become more and more important in functional food product development. The numbers of elderly people are increasing and this group is a strong potential customer for probiotic products. Therefore, averting or delaying age-associated degenerative diseases and combating lifestyle-related diseases should be major research areas in probiotic efficacy studies in the future.

REFERENCES

1. Lappalainen, R., Kearney, J., and Gibney, M., A pan EU survey of consumer attitudes to food, nutrition and health: an overview, *Food Qual. Prefer.*, 9, 467–478, 1998.
2. Joint FAO/WHO working group report on drafting guidelines for the evaluation of probiotics in food, *Guidelines for the evaluation of probiotics in food*, 2002.
3. Menrad, K., Market and marketing of functional food in Europe, *J. Food Eng.*, 56, 181–188, 2003.
4. Bech-Larsen, T., Grunert, K. G., and Poulsen, J. B., *The Acceptance of Functional Foods in Denmark, Finland and the United States*, MAPP working paper 73, The Aarhus School on Business, 2001.
5. Urala, N. and Lähteenmäki, L., Attitudes behind consumers' willingness to use functional foods, *Food Qual. Prefer.*, 15, 793–803, 2004.
6. Urala, N., Arvola, A., and Lähteenmäki, L., Strength of health-related claims and their perceived advantage, *Int. J. Food Sci. Technol.*, 38, 815–826, 2003.
7. Sloan, E., Top 10 functional food trends, *Food Technol.*, 4, 22–40, 2004.
8. Halliwell, D. E., Hands up for probiotics! The mechanism for health benefits is unclear, but interest remains strong, *World Food Ingred.*, March, 46–50, 2002.
9. Hilliam, M., Healthier dairy, *World Food Ingred.*, September, 52–54, 2004.
10. Mellentin, J., The key trends in functional foods 2006, In *New Nutrition Business*, The Centre for Food and Health Studies, London, 2006.
11. Alvarez-Olmos, M. I. and Oberhelman, R. A., Probiotic agents and infectious diseases: a modern perspective on a traditional therapy, *Clin. Infect. Dis.*, 32, 1567–1576, 2001.
12. Marteau, P. R., de Vrese, M., Cellier, C. J., and Schrezenmeir, J., Protection form gastrointestinal diseases with the use of probiotics, *Am. J. Clin. Nutr.*, 73 (suppl.), 430S–436S, 2001.
13. Reid, G., Jass, J., Sebulsky, M. T., and McCormick, J. K., Potential uses of probiotics in clinical practice, *Clin. Microbiol. Rev.*, 16, 658–672, 2003.
14. Hawrelak, J. A., Whitten, D. L., and Myers, S. P., Is *Lactobacillus rhamnosus* GG effective in preventing the onset of antibiotic-associated diarrhea: a systematic review, *Digestion*, 72, 51–56, 2005.
15. Isolauri, E., Salminen, S., and Ouwehand, A. C., Probiotics, *Best Pract. Res. Clin. Gastroenterol.*, 18, 299–313, 2004.
16. Saxelin, M., Tynkkynen, S., Mattila-Sandholm, T., and de Vos, W., Probiotic and other functional microbes: from markets to mechanisms, *Curr. Opin. Biotechnol.*, 16, 1–8, 2005.
17. Hamilton-Miller, J. M., Probiotics and prebiotics in the elderly, *Postgrad. Med. J.*, 80, 447–451, 2004.
18. Hooper, L. V., Stappenbeck, T. S., Hong, C. V., and Gordon, J. I., Angiogenins: a new class of microbicidal proteins involved in innate immunity, *Nat. Immunol.*, 4, 269–273, 2003.
19. Halper, J., Leshin, L. S., Lewis, S. J., and Li, W. I., Wound healing and angiogenic propreties of supernatants from *Lactobacillus* cultures, *Exp. Biol. Med.*, 228, 1329–1337, 2003.
20. Li, W. I., Brackett, B. G., and Halper, J., Culture supernatant of *Lactobacillus acidophilus* stimulates proliferation of embryonic cells, *Exp. Biol. Med.*, 230, 494–501, 2005.
21. Li, X., Fu, G. F., Fan, Y. R., Liu, W. H., Liu, X. J., Wang, J. J., and Xu, G. X., *Bifidobacterium adolescentis* as a delivery system of endostatin for cancer gene therapy: selective inhibitor of angiogenesis and hypoxic tumor growth, *Cancer Gene Ther.*, 10, 105–111, 2003.

22. Li, W. and Li, C. B., Effect of oral *Lactococcus lactic* containing endostatin in 1,2-dimethylhydrazine-induced colon tumor in rats, *World J. Gastroenterol.*, 11, 7242–7247, 2005.

23. Sghir, A., Gramet, G., Suau, A., Rochet, V., Pochart, P., and Dore, J., Quantification of bacterial groups within human fecal flora by oligonucleotide probe hybridization, *Appl. Environ. Microbiol.*, 66, 2263–2266, 2000.

24. Walter, J., Hertel, C., Tannock, G. W., Lis, C. M., Munro, K., and Hammes, W. P., Detection of *Lactobacillus Pediococcus, Leuconostoc*, and *Weissella* species in human feces using group-specific PCR primers and denaturing gradient gel electrophoresis, *Appl. Environ. Microbiol.*, 67, 2578–2585, 2001.

25. Saarela, M., Mättö, J., and Mattila-Sandholm, T., Safety aspects of *Lactobacillus* and *Bifidobacterium* species originating from human orogastrointestinal tract or from probiotic products, *Microb. Ecol. Health Dis.*, 14, 233–240, 2002.

26. Cannon, J. P., Lee, T. A., Bolanos, J. T., and Danziger, L. H., Pathogenic relevance of *Lactobacillus*: a retrospective review of over 200 cases, *Eur. J. Clin. Microbiol. Infect. Dis.*, 24, 31–40, 2005.

27. Marteau, P. and Shanahan, F., Basic aspects and pharmacology of probiotics: an overview of pharmacokinetics, mechanisms of action and side-effects, *Best Pract. Res. Clin. Gastroenterol.*, 17, 725–740, 2003.

28. O'Brien, J., Crittenden, R., Ouwehand, A. C., and Salminen, S., Safety evaluation of probiotics, *Trends Food Sci. Technol.*, 10, 418–424, 1999.

29. Borriello, S. P., Hammes, W. P., Holzapfel, W., Marteau, P. M., Schrezenmeier, J., Vaara, M., and Valtonen, V., Safety of probiotics that contain lactobacilli or bifidobacteria, *Clin. Infect. Dis.*, 36, 775–780, 2003.

30. Rautio, M., Jousimies-Somer, H., Kauma, H., Pietarinen, I., Saxelin, M., Tynkkynen, S., and Koskela, M., Liver abscess due to a *Lactobacillus rhamnosus* strain indistinguishable from *L. rhamnosus* strain GG, *Clin. Infect. Dis.*, 28, 1160–1161, 1999.

31. Mackay, A. D., Taylor, M. B., Kibbler, C. C., and Hamilton-Miller, J. M. T., *Lactobacillus* endocarditis caused by a probiotic organism, *Clin. Microbiol. Infect.*, 6, 290–292, 1999.

32. Land, M. H., Rouster-Stevens, K., Woods, C. R., and Cannon, M. L., *Lactobacillus* sepsis associated with probiotic therapy, *Pediatrics*, 115, 178–181, 2005.

33. Salminen, M. K., Tynkkynen, S., Rautelin, H., Saxelin, M., Vaara, M., Ruutu, P., Sarna, S., Valtonen, V., and Järvinen, A., *Lactobacillus* bacteremia during a rapid increase in probiotic use of *Lactobacillus rhamnosus* GG in Finland, *Clin. Infect. Dis.*, 35, 1155–1160, 2002.

34. Salminen, M. K., Rautelin, H., Tynkkynen, S., Poussa, T., Saxelin, M., Valtonen, V., and Järvinen, A., *Lactobacillus* bacteremia, clinical significance, and patient outcome, with special focus on probiotic *L. rhamnosus* GG, *Clin. Infect. Dis.*, 38, 62–69, 2004.

35. Salminen, M. K., Rautelin, H., Tynkkynen, S., Poussa, T., Saxelin, M., Valtonen, V., and Järvinen, A., *Lactobacillus* bacteremia, species identification, and antimicrobial susceptibility of 85 blood isolates, *Clin. Infect. Dis.*, 42, 35–44, 2006.

36. Saarela, M., Mogensen, G., Fonden, R., Mättö, J., and Mattila-Sandholm, T., Probiotic bacteria: safety, functional and technological properties, *J. Biotechol.*, 84, 197–215, 2000.

37. Fasoli, S., Marzotto, M., Rizzotti, L., Rossi, F., Dellaglio, F., and Torriani, S., Bacterial composition of commercial probiotic products as evaluated by PCR–DGGE analysis, *Int. J. Food Microbiol.*, 82, 59–70, 2003.

38. Temmerman, R., Scheirlinck, I., Huys, G., and Swings, J., Culture-independent analysis of probiotic products by denaturating gradient gel electrophoresis, *Appl.*

Environ. Microbiol., 69, 220–226, 2003; Temmerman, R., Pot, B., Huys, G., and Swings, J., Identification and antibiotic susceptibility of bacterial isolates from probiotic products, *Int. J. Food Microbiol.*, 81, 1–9, 2003.

39. Coeuret, V., Gueguen, M., and Vernoux, J. P., Numbers and strains of lactobacilli in some probiotic products, *Int. J. Food Microbiol.*, 97, 147–156, 2004.

40. Gueimonde, M., Delgado, S., Mayo, B., Ruas-Madiedo, P., Margolles, A., and de los Reyes-Gavilán, C. G., Viability and diversity of probiotic *Lactobacillus* and *Bifidobacterium* populations included in commercial fermented milks, *Food Res. Int.*, 37, 839–850, 2004.

41. Masco, L., Huys, G., De Brandt, E., Temmerman, R., and Swings, J., Culture-dependent and culture-independent qualitative analysis of probiotic products claimed to contain bifidobacteria, *Int. J. Food Microbiol.*, 102, 221–230, 2005.

42. Truelstrup Hansen, L., Allan-Wojtas, P. M., Jin, Y. L., and Paulson, A. T., Survival of Ca-alginate microencapsulated *Bifidobacterium* spp. in milk and simulated gastrointestinal conditions, *Food Microbiol.*, 29, 35–45, 2002.

43. Matsumoto, M., Ohishi, H., and Benno, Y., H^+-ATPase activity in *Bifidobacterium* with special reference to acid tolerance, *Int. J. Food Microbiol.*, 93, 109–113, 2004.

44. Mättö, J., Malinen, E., Suihko, M. L., Alander, M., Palva, A., and Saarela, M., Genetic heterogeneity and technological properties of intestinal *Bifidobacteria*, *J. Appl. Microbiol.*, 97, 459–470, 2004.

45. Leporanta, K., Valio Ltd, personal communication.

46. Svensson, U., Industrial perspectives, In *Probiotics—A Critical Review*, Tannock, G. W. Ed., Horizon Scientific Press, Wymondham, UK, pp. 57–64, 1999.

47. Fonden, R., Saarela, M., Mättö, J., and Mattila-Sandholm, T., Lactic acid bacteria in functional dairy products, In *Functional Dairy Products*, Mattila-Sandholm, T. and Saarela, M. Eds., CRC Press Woodhead Publishing Limited, England, pp. 244–262, 2003.

48. Bergonzelli, G. E., Blum, S., Brüssow, H., and Corthésy-Theulaz, I., Probiotics as a treatment strategy for gastrointestinal diseases, *Digestion*, 72, 57–68, 2005.

49. Guandalini, S., Probiotics for children: use in diarrhea, *J. Clin. Gastroenterol.*, 40, 244–248, 2006.

50. McFarland, L. V., Meta-analysis of probiotics for the prevention of antibiotic associated diarrhea and the treatment of *Clostridium difficile* disease, *Am. J. Gastroenterol.*, 101, 812–822, 2006.

51. Sazawal, S., Hiremath, G., Dhingra, U., Malik, P., Deb, S., and Black, R. E., Efficacy of probiotics in prevention of acute diarrhea: a meta-analysis of masked, randomised, placebo-controlled trials, *Lancet Infect. Dis.*, 6, 374–382, 2006.

30 Potential Anti-Cancer Effects of Shark Cartilage

Kenji Sato, Masumi Suganuma, and Kazuhiro Shichinohe

CONTENTS

30.1 BACKGROUND

A tumor frequently induces new capillary vessels to obtain nutrients from the host. It is referred to as "tumor-induced angiogenesis" and is a critical process for tumor growth. Tumor-induced angiogenesis is believed to be a good target for cancer therapy. In addition, some diseases such as rheumatoid arthritis also require angiogenesis for the progression of disease [1,2]. Folkman and his coworkers first demonstrated that animal cartilage, avascular tissue, has an ability to suppress tumor-induced angiogenesis [3]. Thereafter, Langer and associates have demonstrated that shark cartilage or its extract implanted near a tumor can also suppress the tumor-induced angiogenesis and tumor growth [4]. Because the anti-angiogenesis activity in the cartilage was co-purified with collagenase (matrix metalloprotease: MMP) inhibitory activity, it has been suggested that the MMP inhibitory activity may play a significant role in the suppression of angiogenesis and cancer invasion and metastasis [5–7]. However, the mechanisms involved in the inhibition of tumor-induced angiogenesis by cartilage have not been elucidated in detail until now.

On the basis of these findings, Lane and his coworkers have proposed using shark cartilage powder for cancer therapy by intrarectal and, thereafter, oral administrations [8,9]. Their proposal, however, was not supported by well-designed clinical trials and animal experiments. Nevertheless, the myth of shark cartilage therapy has been prevalent worldwide, even though there are some discouraging results on the anti-cancer effects of shark cartilage powder [10]. Recently, another group has prepared a formulized water extract of shark cartilage and suggested an anti-cancer effect by animal experiments [11,12] and human trials [13,14]. To confirm the anti-cancer effect of shark cartilage, some possible candidates responsible for the anti-cancer effect have been proposed through laboratory studies. In this chapter, recent knowledge on the anti-cancer effect of shark cartilage is reviewed, and we discuss the possible mechanism of the potential anti-cancer effect of shark cartilage through oral administration.

30.2   HUMAN STUDY

Lane and his coworkers prepared dry shark cartilage powder for use in cancer therapy. Twenty-one patients suffering from inoperative cancers were administered this preparation (30 g/day) intrarectally by using a retention enema. They reported that more than half of the patients responded positively to this treatment, and stated that the administration of shark cartilage powder can inhibit tumor-induced angiogenesis and consequently suppress tumor growth [9]. However, the number of subjects and follow-up period were limited in their study. Then, a randomized placebo-control study using a corresponding shark cartilage powder was conducted. In that study, patients were orally given dry shark cartilage powder (1 g/kg/day) for 12 weeks. Oral administration of this dose of the dry shark cartilage powder frequently caused serious adverse events such as nausea, vomiting, and constipation. On the basis of tumor size, no significant beneficial effect was observed [10]. However, there are some anecdotal reports suggesting anti-cancer effects of shark cartilage powder.

Shark cartilage powder contains large amount of calcium salts that may have no beneficial activity for cancer patients and may cause serious adverse events. Since that time, efforts have been focused on extracting the "active component" from shark cartilage. More recently, some kinds of water extract of shark cartilage have been prepared in laboratory and industrial scales. Oral administration of the extract may not cause the severe side effects observed in the whole cartilage preparation. A phase II clinical trial suggested that a high dose (240 ml/day) of oral administration of a formularized water extract of shark cartilage shows a survival benefit in renal cell carcinoma [14]. Currently, phase III clinical trials against some types of cancer are in progress.

30.3   ANIMAL STUDIES

Before the trial by Lane and co-workers [9], a preliminary animal experiment was carried out, which was published as a personal communication in a review article by Lane [8]. As popular interest in the anti-cancer effect of shark cartilage has increased, some animal studies have been undertaken to evaluate the potential anti-cancer effect of shark cartilage and its extract.

By using streptozotocin-induced murine renal cell carcinoma, the effect of orally administered powdered shark cartilage (0.6% in diet) on carcinogenesis and cancer progression was examined [15]. The authors concluded that oral administration of the crude shark cartilage cannot abolish the chemo-induced carcinogenesis, but can delay the progression of cancer. Similarly, the present authors also found that prefeeding of a commercially available shark cartilage powder (1% in diet) for three weeks before implantation of Ehrich asetic sarcoma can suppress growth of the tumor in mice and extend the survival period, as shown in Figure 30.1, whereas no beneficial effect was observed when shark cartilage was simultaneously given with the implantation of Ehrich asetic sarcoma.

In this experiment, the present authors observed increased inhibitory activity against MMP-9 in the serum of cancer-bearing mice receiving the shark cartilage compared to those that received the control diet (Figure 30.2). Because MMP-9 can degrade type IV and V collagens, which are main constituents of basal lamina and pericellular connective tissue, it has been believed to play a significant role in cancer invasion, metastasis, and tumor-induced angiogenesis. Thus, the increased inhibitory activity against MMP-9 might, at least partially, account for the suppression of cancer growth by the oral administration of shark cartilage powder. For the anti-angiogenesis activity, Gonzares et al. [16] demonstrated that oral administration of shark cartilage powder can suppress basic fibroblast growth factor

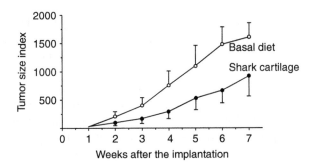

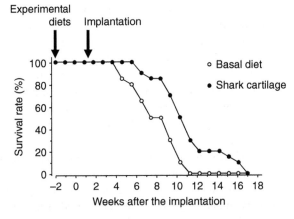

FIGURE 30.1 Effect of orally administered shark cartilage powder (1% in diet) on growth of Ehrich asetic sarcoma and survival period of mice. Shark cartilage diet was given from three weeks before the implantation.

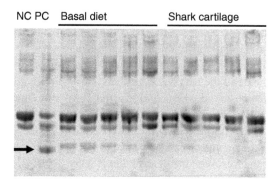

FIGURE 30.2 Increased inhibitory activity against MMP-9 in serum of mice implanted with Ehrich asetic sarcoma by oral administration of shark cartilage powder. Negative control (NC); substrate type V collagen only, positive control (PC); type V collagen was reacted with human fibrosarcoma MMP-9. Serums (1 μL) from cancer-bearing mice given basal and shark cartilage (1%) diets were added to the reaction mixture. Degradation product of type V collagen by MMP-9 (arrow) decreased more intensively by addition of serum from mice given shark cartilage diet.

(bFGF)-induced angiogenesis. They mentioned a possibility that the bFGF-induced signal transmission might be inhibited by the oral administration of shark cartilage. On the other hand, the migration and organization of endothelial cells into capillary structures depends on the collagenase activity of MMPs. Then, inhibition of MMPs, if it occurred in their model, also could suppress bFGF-induced angiogenesis.

These animal studies also support the finding that oral administration of shark cartilage powder can moderately suppress the progression of cancer, possibly by inhibition of angiogenesis or MMPs; however, a relatively high dose (1–1.2 g/kg body weight) is necessary. Although animals can tolerate relatively high dose loadings of dry shark cartilage powder, these doses, as described above, may be too high for human therapy. To reduce the dose, some types of water extract of shark cartilage have been prepared and their anti-cancer effect has been evaluated on animal models. A Canadian company has prepared a water extract from raw shark cartilage. According to their patent [17], an ultrafiltration technique was used to concentrate molecules between 50 and 500,000 Da. Thus, their preparation might contain a low level of high molecular weight proteoglycan. Oral administration of this type of extract also can suppress cancer progression in animal models [11,12] as well as in humans [13,14]. The company also reported that this preparation has in vitro anti-angiogenesis and collagenase-inhibitory activities [12,18–20]; however, the effect of the oral administration of their preparation on either anti-angiogenesis or collagenase-inhibitory activities in tissues has not been demonstrated.

The present author and coworkers have prepared another type of extract from a byproduct of the fishery industry. Shark fin has been used as an ingredient in high-grade Chinese cuisine. The fin was solar-dried and rehydrated in water at 50°C–60°C and then skinned. Transparent fibers on both sides of fin cartilage are used for Chinese cuisine. The remaining fin cartilage is a low-valued byproduct (Figure 30.3).

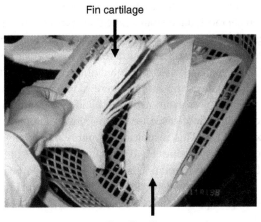

FIGURE 30.3 Photograph of shark fin cartilage.

We found that a significant inhibitory activity against MMPs-2 and 9 could be extracted with water from the dried shark fin cartilage. This inhibitory activity in the water extract can be precipitated by the addition of three volumes of ethanol, which consist of high molecular weight ($>500,000$) proteoglycan with inhibitory activity

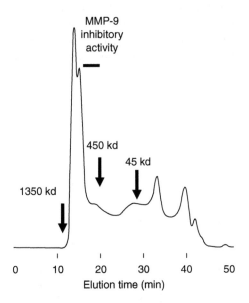

FIGURE 30.4 Fractionation of water extract of shark fin cartilage by size exclusion chromatography (Superdex 200). MMP-9 inhibitory activity was eluted in high molecular weight fraction indicated with bar.

(Figure 30.4). The anti-cancer effect of the ethanol precipitate was evaluated by using hamsters with chemo-induced pancreatic duct cancer in the method of Iki et al. [21], which is a good experimental model for human pancreatic cancer.

After the carcinogenic treatment, the animals were given experimental diets containing 0%, 0.2%, and 0.4% of the ethanol precipitate. After receiving the experimental diets for 50 days, the animals were sacrificed and the incidence of the carcinoma in the pancreas was examined. The incidence of advanced carcinoma in the pancreas decreased in a dose-response manner through the oral administration of shark cartilage extract [22]. This response is roughly comparable with that of the oral administration of a synthetic MMP inhibitor (OPB-3206; 3S-[4-(N-hydro-xyamino)-2R-isobutylsuccinyl]amino-1-methoxy-3, 4-dihydrocarbostyril) in the dose of 0.1% in the same animal model [21]. In the serum of the hamster that received the ethanol precipitate, an increased level of MMP-9-inhibitory activity was also observed.

On the other hand, no significant inhibition of BrU incorporation to cancer cells was observed in the animal receiving the ethanol precipitate, indicating no significant cytotoxic effect on cancer cells [22]. The ethanol precipitate was also prepared from a water extract of the commercially available powdered shark cartilage (a kind gift from Lane-lab Far East), which was same preparation used in the experiment shown in Figure 30.1 and Figure 30.2.

As shown in Figure 30.5, this ethanol precipitate also suppressed growth of Ehrich asetic sarcoma in mice, using a smaller dose (0.4% in diet) than the crude shark cartilage powder (1%), and without the three weeks of prefeeding that was necessary in the case of crude shark cartilage powder (Figure 30.1). The increased MMP-9 inhibitory activity in the serum of cancer-bearing mice was also observed. On the basis of these studies, at least part of "the anti-cancer component" in shark cartilage, including commercial products, can be extracted with water, which would enable a reduction in the dose for oral administration. In addition, some of "the

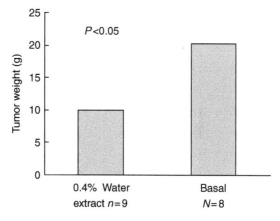

FIGURE 30.5 Tumor size of mice receiving basal diet and water extract of the shark cartilage powder (0.4%) after seven weeks.

anti-cancer component(s)" in shark cartilage are stable against heat (50°C–60°C) and 75% ethanol treatments.

30.4 LABORATORY WORK: POSSIBLE MECHANISMS OF THE ANTI-CANCER EFFECTS OF SHARK CARTILAGE

Implanted shark cartilage or its extract near the tumor on a rabbit cornea has been demonstrated to inhibit tumor-induced angiogenesis [4]. In addition, physiological angiogenesis on a chick embryo is similarly suppressed [6]. These early studies clearly demonstrate that shark cartilage has moderate anti-angiogenesis inhibitory activity. To identify component(s) responding to the anti-angiogenetic activity, shark cartilage extract was fractionated. As described above, the anti-angiogenesis activity was co-fractionated with the inhibitory activity against collagenases [6]. It has been suggested that collagenase inhibitors, such as the tissue inhibitor of matrix melalloprotese (TIMP), might be responsible for anti-angiogenesis activity. In fact, the presence of TIMP-1, -2, and -3-like proteins in shark cartilage has been reported.

Recently, we isolated a high molecular type of MMP-inhibitor from a dried shark cartilage (Figure 30.4). Cellulose acetate electrophoresis revealed that the inhibitor fraction consists of chondroichin sulfate C and protein conjugate—a high molecular weight proteoglycan. This type of MMP-inhibitor is heat-stable. As described above, the proteoglycan fraction without TIMP activity can suppress the growth of Ehrich asetic sarcoma in mice (Figure 30.1) and the progression of chemo-induced pancreatic duct cancer [22]. The presence of MMP-inhibitors in shark cartilage powder and its extract might lead to a hypothesis that the MMP inhibitors in shark cartilage might be absorbed and subsequently inhibit the MMPs on and near the tumor. As shown in Figure 30.6, animal and human serums show approximately 10 times higher MMP-2 and -9 inhibitory activities than a formulized water extract of shark cartilage. Thus, it is unlikely that the inhibitor(s) in shark cartilage directly

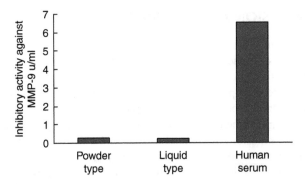

FIGURE 30.6 MMP-9 inhibitory activity of water extract of a commercially available shark cartilage (powder and liquid types) in comparison with human serum. Powder type was suspended in water to 1% (w/v) and supernatant was used. MMP-9 activity was evaluated by SDS–PAGE. One milliunit of the activity was defined as yielding degradation product to give 10% on the basis of staining intensity on SDS–PAGE.

affect the inhibitor level in tissue by oral administration of the extract, even when assuming that the shark inhibitor survives proteolytic digestion in the gastrointestinal duct and is absorbed with activity. However, as mentioned above, the inhibitory activity against MMP-9 increased by oral administration of shark cartilage and its water-extract. This phenomenon could be interrupted, as the oral ingestion of shark cartilage might enhance induction of endogenous MMP-inhibitors, or the metabolic alteration of the shark cartilage compounds might lead to derivatives with enhanced inhibitory activity. Although there is no direct evidence, these hypotheses seem to be attractive.

Because angiogenesis is specifically and strongly induced by vascular endothelial growth factor (VEGF) and bFGF, inhibition of angiogenesis induced by VEGF and bFGF has been examined [12,16]. Recently, Beliveau et al. demonstrated that the binding of VEGF to an endothelial cell surface and the phosphorisation of a VEGF-dependent receptor *v.e.g.f.*recepter-2; (VEGFR-2) or fetal liver kinase-1; FLK-1) are inhibited by the addition of a formulated shark cartilage extract to the cell culture medium [20]. However, a relatively high dosage of shark cartilage-derived protein (10–100 µg/mL) was required to show the inhibition. Generally, it might be difficult to suppose that most of the protein in shark cartilage might survive digestion and be adsorbed with activity. Furthermore, their in vitro experiment revealed that the shark cartilage protein (200 µg protein/mL) cannot inhibit the binding of bFGF, another stimulator for angiogenesis, to endothelial cells, whereas oral administration of shark cartilage can inhibit bFGF-induced angiogenesis. These facts suggest that shark cartilage contains some compounds that modestly inhibit the VEGF-dependent signal transmission, which, however, could not be directly linked to the potential anti-angiogenesis and anti-cancer effects by oral administration of shark cartilage. To solve these problems, the effect of orally administered shark cartilage extract on the inhibitory activity against the signal transmission in serum or tissue extracts should be examined.

30.5   CONCLUSION AND FUTURE PROSPECTS

Conventionally, anti-cancer effects have been evaluated on the basis of the shrinkage of tumors. On the basis of a limited numbers of animal and human studies, the disappearance (complete response) or extensive shrinkage (partial response) of tumors could not be anticipated solely by the oral administration of shark cartilage-based products. On the other hand, the delay of cancer progression, prolonged survival period, improvement of quality of life, and so on might be anticipated. On the basis of these criterion, the potential beneficial effect of shark cartilage-based products should be confirmed by further human trial.

For application to humans, extraction and concentration of the active component, with its beneficial activity, should be done, as the oral administration of crude cartilage preparations in high doses (approximately 1 g/kg body weight) frequently caused serious adverse events due to presence of large amounts of calcium salts. Until now, on the basis of in vitro and exo vivo assay systems, relatively low molecular proteinaous components and high molecular proteoglycan have been proposed to be possible candidates for the anti-cancer effect. In some cases, these extracts have been directly applied to the assay systems to evaluate the anti-cancer effect; however, they could not be directly linked to the biological response by oral administration. To break it, biomarker responding to the oral administration of shark cartilage-based product should be detected. The increased inhibitory activity against MMP-9 in the serum may be one good biomarker for the evaluation of antitumor activity, as MMP-9 plays a significant role in cancer progression.

REFERENCES

1. Folkman, J., What is the evidence that tumors are angiogenesis dependent? *J. Natl Cancer Inst.*, 82, 4–6, 1989.
2. Folkman, J., Angiogenesis-dependent diseases, *Semin. Oncol.*, 28, 536–542, 2001.
3. Berm, H. and Folkman, J., Inhibition of tumor angiogenesis mediated by cartilage, *J. Exp. Med.*, 141, 427–439, 1975.
4. Lee, A. and Langer, R., Shark cartilage contains inhibitors of tumor angiogenesis, *Science*, 221, 1185–1187, 1983.
5. Lee, A. K., van Beuzekom, M., Glowacki, J., and Langer, R., Inhibitors, enzymes and growth factors from shark cartilage, *Comp. Biochem. Physiol.*, 78B, 609–616, 1984.
6. Oikawa, T., Ashino-Fuse, H., Shimamura, M., Koide, U., and Iwaguchi, T., A novel angiogenetic inhibitor derived from Japanese shark cartilage (I). Extraction and estimation of inhibitory activities toward tumor and embryonic angiogenesis, *Cancer Lett.*, 51, 181–186, 1990.
7. Mose, M. A. and Langer, R., A metalloproteinase inhibitor as an inhibitor of neovascularization, *J. Cell. Biochem.*, 47, 230–235, 1991.
8. Lane, I. W., Shark cartilage: its potential medical application, *J. Adv. Med.*, 4, 263–271, 1991.
9. Lane, I. W. and Contreras, E. Jr., Jr., High rate of bioactivity (reduction in gross tumor size) observed in advanced cancer patients treated with shark cartilage material, *J. Naturopathic Med.*, 3, 8–88, 1992.
10. Miller, D. R., Anderson, G. T., Stark, J. J., Granick, J. L., and Richardson, D., Phase I/II trial of the safety and efficiency of shark cartilage in the treatment of advanced cancer, *J. Clin. Oncol.*, 16, 3649–3655, 1998.
11. Weber, M. H., Lee, J., and Orr, F. W., The effect of Neovastat (AE-941) on an experimental metastatic bone tumor model, *Int. J. Oncol.*, 20, 299–303, 2002.
12. Dupont, E., Falardeau, P., Mousa, S. A., Dimitriadou, V., Pepin, M. C., Wang, T., and Alaoui-Jamali, M. A., Antiangiogenic and antimetastatic properties of Neovastat (AE-941), an orally active extract derived from cartilage tissue, *Clin. Exp. Metastasis*, 19, 145–153, 2002.
13. Falardeau, P., Champagne, P., Poyet, P., Hariton, C., and Dupont, E., Neovastat, a naturally occurring multifunctional anti-angiogenic drug, in phase III clinical trial, *Semin. Oncol.*, 28, 620–625, 2001.
14. Batist, G., Patenaude, F., Champagne, P., Croteasu, D., Levinton, C., Harton, C., Escudier, B., and Dupont, E., Neovastat (AE-941) in refractory renal cell carcinoma patients: report of a phase II trial with two dose levels, *Ann. Oncol.*, 13, 1259–1263, 2002.

15. Barber, R., Delahunt, B., Grebe, S. K. G., Davis, P. F., Thornton, A., and Slim, G., Oral shark cartilage does not abolish carcinogenosis but delays tumor progression in a murine model, *Anticancer Res.*, 21, 1065–1070, 2001.

16. Gonzalez, R. P., Soares, F. dos S. D., Farias, R. F., Pessoa, C., Leyva, A., Viana, G.S.de B., and Moraes, M. O., Demonstration of inhibitory effect of oral shark cartilage on bFGF-induced angiogenesis in the rabbit cornea, *Biol. Pharm. Bull.*, 24, 151–154, 2001.

17. Dupont, E., Brazeau, P., and Juneau, C., Extracts of shark cartilage having an angiogenenic activity and an effect on tumor progression: process of making thereof, *United States Patent*, #5, 618–925, 1995.

18. Gringras, D., Renaud, A., Mousseau, N., and Beliveau, R., Shark cartilage extracts as anti-angiogenic agents: smart drink or bitter pills, *Cancer Metastasis Rev.*, 19, 83–86, 2000.

19. Gingras, D., Renaud, A., Mousseau, N., Beaulieu, E., Kachra, Z., and Beliveau, R., Matrix protease inhibition by AE-941, a multifunctional angiogenetic compound, *Anticancer Res.*, 21, 145–156, 2001.

20. Beliveau, R., Gingras, D., Kruger, E. A., Lamy, S., Sirois, P., Simard, B., Sirois, M. G. et al., The antiangiogenic agent Neovastat (AE-941) inhibits vascular endothelial growth factor-mediated biological effects, *Clin. Cancer Res.*, 8, 1242–1250, 2002.

21. Iki, K., Tstsumi, M., Kido, A., Sakitani, H., Takahama, M., Yoshimoto, M., Motoyama, M., Tatsumi, K., Tsunoda, T., and Konishi, Y., Expression of matrix metalloproteinase 2 (MMP-2), membrane-type I MMP and tissue inhibitor of metalloproteinase 2 and activation of proMMP-2 in pancreatic duct adenocarcinomas in hamsters treated with *N*-nitrosobis(2-oxopropyl)amine, *Carcinogenesis*, 20, 1323–1329, 1999.

22. Murata, N., Effects of bovine and shark cartilage water extracts on pancreatic ductal carcinogenesis in hamster, *J. Nara Med. Assoc.*, 53, 241–252, 2002.

31 Physiological Effects of Eicosapentaenoic Acid (EPA) and Docosahexaenoic Acid (DHA)—A Review

Narayan Bhaskar and K. Miyashita

CONTENTS

31.1 INTRODUCTION

Lipids play a major role in human nutrition and health while accounting for roughly 40% of the total calories in most of the industrialized countries. They reduce the bulk of the diet as they provide a concentrated source of energy. Added to this, lipids are indispensable from the human diet as they form the major sources of essential fatty acids and contain fat-soluble substances like vitamins and carotenoids. Plants and animals, including those from the aquatic environ, account for most of the dietary lipids.

Vegetable oils contribute 70% of the global oil production while marine oils contribute about 2% of the total with the remainder being contributed by fats from land-based animals. The per capita consumption of marine oils is approximately 1 g per person per day. However, in countries like Japan and Ghana it varies between 2.1 and 8.4 [1]. Furthermore, marine lipids serve a more important role in human nutrition than previously realized.

Fatty acid composition of lipids varies depending on the source. For instance, marine lipids generally contain a wider range of fatty acids than that of their terrestrial counterparts including both plants and animals (Table 31.1). As can be seen in Table 31.1, polyunsaturated fatty acids (PUFAs) of the linolenic acid family (n-3 fatty acids) are typical of marine lipids, whereas PUFAs belonging to the linoleic acid family (n-6 fatty acids) are predominant in vegetable oils. Like vegetable oils, the fats derived from terrestrial animal sources largely consist of PUFAs belonging to the linoleic acid family. Marine lipids contain substantial amounts of long chain PUFAs and monounsaturated fatty acids. Added to this, the metabolic effects of these long chain PUFAs and monounsaturated fatty acids (MUFAs), especially in human nutrition, has attracted major interest among biochemists and nutritionists alike.

As mentioned earlier, n-3 fatty acids are significant components of marine lipids and are not present in other foods. The main effects of n-3 fatty acids in human health can be divided into three main categories: (1) Their essentiality in specific organs; (2) their significant role in lowering blood lipids; and (3) their role as precursors for mediating biochemical and physiological responses. Eicosapentae-noic acid (EPA, C20:5) and docosahexaenoic acid (DHA, C22:6) are the two typical fatty acids from marine lipids. These two long chain PUFAs have been shown to cause significant biochemical and physiological changes in the body. Most of these changes exert positive influence on human nutrition and health. Apart from this, there are also some controversies pertaining to the functioning of these n-3 PUFAs and the extent of their requirement by the body.

This brief review is mainly concerned with the physiological effects and potential therapeutic uses of EPA and DHA vis-à-vis their less desirable features. An effort is also made to outline the factors responsible for these effects along with those factors that affect the quality of these long chain PUFAs. Because work on marine lipids has been reviewed time and again, this review mainly focuses on more recent studies on the physiological and related effects of EPA and DHA.

31.2   EPA AND DHA—BASIC FACTS AND SOURCES

Eicosapentaenoic acid and DHA are two important n-3 fatty acids. The other important n-3 fatty acids are α-linolenic acid (ALA, C18:3) and docosapentaenoic acid (DPA, C22:5). Structures of these n-3 fatty acids along with some typical n-6 fatty acids are presented in Figure 31.1. As can be seen, they are basically differentiated by the position of the first double bond counting from the methyl end of the carbon chain. Another significant factor is the spatial configuration of these long chain PUFAs, especially EPA and DHA. This can have a direct bearing—on the membrane fluidity, membrane receptor sites, substrate binding and a host of

TABLE 31.1
Fatty Acid Composition of Lipids from Selected Animals and Plants

	Salmon	Seal	Whale	Beef	Pork	Chicken	Olive	Soybean	L. japonica	A. spicifera	C. vulgaris
Saturated											
12:0	—	—	0.1	—	—	—	0	0.1	0.1	0.6	—
14:0	6.7	3.3	5.1	3.2	1.6	1.3	tr	0.2	20.1	6.7	0.4
16:0	13.7	6.5	12.0	26.9	27.1	26.7	12.0	10.0	20.7	40.3	15.4
18:0	1.2	0.4	1.7	13.0	13.8	7.1	2.3	4.0	0.4	1.4	6.2
Monounsaturated											
16:1	10.0	17.6	11.8	6.3	3.4	7.2	1.0	0.2	4.8	1.4	1.2
18:1	15.2	30.2	30.6	42.0	43.8	39.8	72.0	25.0	10.8	12.6	34.3
20:1	13.2	13.6	12.1	tr	0.7	0.6	0	0.2	0.1	0.3	—
22:1	17.4	6.0	9.7	—	—	—	0	tr	0.1	—	—
Polyunsaturated (linoleic family)											
18:2	1.4	0.8	1.3	2.0	7.4	13.5	11.0	52.0	6.4	1.4	9.7
20:4	0.6	0.3	0.3	1.0	tr	0.7	—	—	9.1	10.2	Tr
Polyunsaturated (linolenic family)											
18:3	1.2	0.2	0.5	1.3	0.9	0.7	0.7	7.4	5.6	0.8	1.9
20:5	7.0	5.5	3.7	tr	tr	—	—	—	—	6.2	3.2
22:5	1.1	4.2	0.2	tr	tr	tr	—	—	—	—	3.1
22:6	6.5	7.5	4.1	—	—	—	—	—	—	—	20.9

Source: From Paul, A. A. and Southgate, D. A. T., *McCane and Widdowson's the Composition of Foods*, 4th ed., Elsevier, Amsterdam, 1978; Jangaard, P. M., Ackman, R. G., and Burgher, R. D., *Can J. Biochem. Physiol.*, 41, 2543–2548, 1963; Ackman, R. G., Eaton, C. A., and Jangaard, P. M., *Can J. Biochem. Physiol.*, 45, 1513–1517, 1965; Nisizawa, K., *Jpn Seaweed Assoc.*, 44–50, 2002; Tokusoglu, O. and Unal, M. K., *J. Food Sci.*, 68, 1144–1148, 2003.

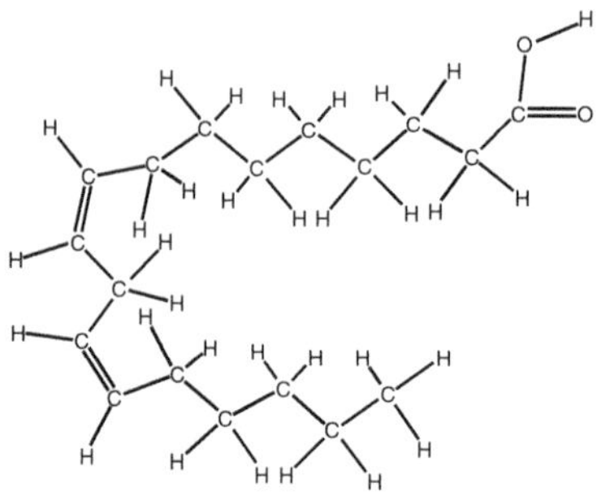

LA, linoleic acid, 18:2ω6 (18:2n-6)
(contains 18 carbon atoms and 2 double bonds
or unsaturation sites)

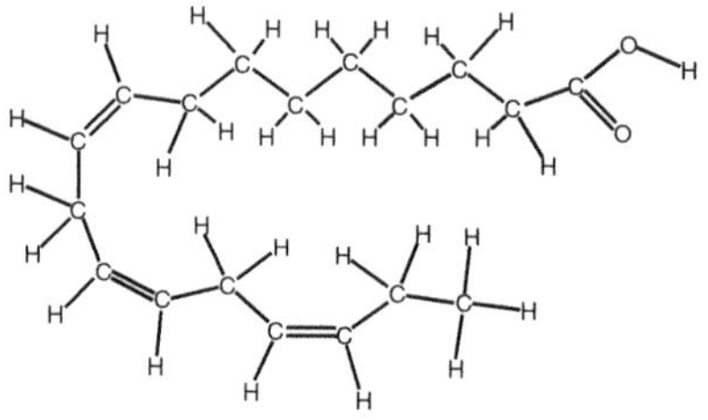

ALA, alpha-linolenic acid, 18:3ω3 (18:3n-3)
(contains 18 carbon atoms and 3 double bonds
or unsaturation sites)

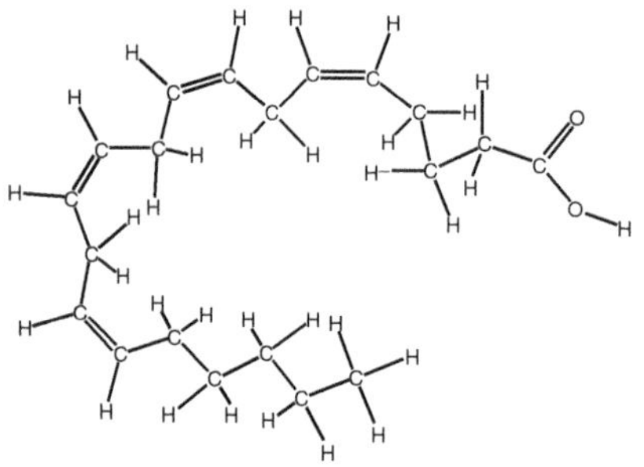

AA, arachidonic acid, 20:4ω6 (20:4n-6)
(contains 20 carbon atoms and 4 double bonds
or unsaturation sites)

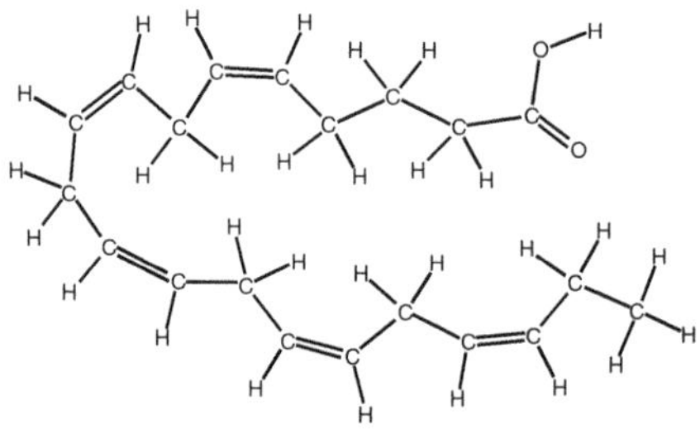

EPA, eicosapentaenoic acid, 20:5ω3 (20:5n-3)
(contains 20 carbon atoms and 5 double bonds
or unsaturation sites)

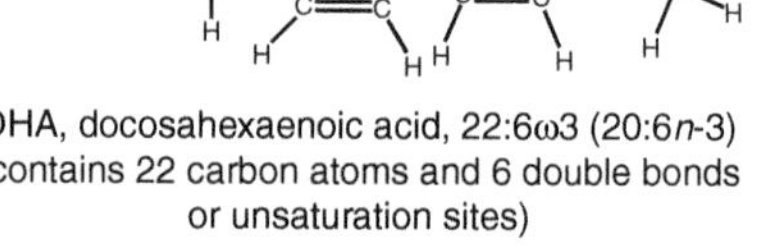

DHA, docosahexaenoic acid, 22:6ω3 (20:6n-3)
(contains 22 carbon atoms and 6 double bonds
or unsaturation sites)

FIGURE 31.1 Chemical structures of some fatty acids of linoleic (n-6) and linolenic (n-3) family.

other functions—once incorporated into the membranes of land-based animals through diet [7].

Both EPA and DHA are basically derived from ALA through elongation and desaturation. Unicellular phytoplankton and seaweed are major sources of EPA and

DHA. They are accumulated in fish and other marine animals that consume algae. They get passed on to other species through the food chain. Thus, the major sources of EPA and DHA in the human diets are marine products. Although terrestrial plants can produce small to moderate amounts of ALA, the mammalian systems cannot elongate and desaturate this fatty acid into EPA or DHA upon consumption [7]. Persons consuming a normal diet were found to have low tissue levels of EPA and DHA, while those who consume fish have higher levels in their tissues [8].

31.3 PHYSIOLOGICAL EFFECTS OF EPA AND DHA

In this section, the physiological effects of dietary EPA and DHA, both in terms of beneficial and less desirable effects have been reviewed.

31.3.1 CARDIOPROTECTIVE EFFECTS

The role of *n*-3 fatty acids including EPA and DHA in alleviating cardiovascular disease has been shown in several in vivo and in vitro experiments apart from several clinical trials [9]. Although saturated fats and cholesterol are pathogenic for coronary heart disease (CHD), *n*-3 fatty acids, especially EPA and DHA from marine foods, are protective and prevent death from CHD by a variety of mechanisms [10].

This unique property of *n*-3 fatty acids first came to light in the investigations of the eating habits and CHD incidence in the Eskimos of Greenland. Greenland's Eskimos, who consume a diet very high in fat from seal, whales, and fish had a very low incidence of CHD [11,12]. This was further supported by the fact that the diet of this community was rich in EPA and DHA. These *n*-3 fatty acids prevent CHDs through a variety of actions. They prevent arrhytmias, act as precursors to prostaglandins and leukotrienes, prevent inflammation (anti-inflammatory), prevent thrombosis (anti-thrombotic), act as hypolipidemics and inhibit atherosclerosis [13]. The various mechanisms by which EPA and DHA prevent/modify cardiovascular diseases have been thoroughly reviewed by several authors [7,10,14–16] and they are summarized in Table 31.2.

31.3.2 ESSENTIALITY IN MEMBRANE STRUCTURE AND RELATED FUNCTIONS

Fatty acids including DHA and EPA form major constituents of phospholipids, which in turn are major molecules in most biological membranes. Furthermore, *n*-3 PUFA are important for structure and function [28,29]. Docosahexaenoic acid is an important constituent of the membrane phospholipids of important neural structures, including the brain and retina, usually occupying the sn-2 position of the phospholipid moiety. It is specifically concentrated in the brain and retina of humans and other mammals.

Several researchers [30–33] have reported the importance of DHA as a prominent fatty acid in several phospholipid molecules of the brain and retina. It is essential for brain function [32–35] and vision [34,36]. It has also been reported that

TABLE 31.2
Cardioprotective Effects Associated with EPA and DHA along with Possible Mechanisms of Action

Effect	Mechanism
Strong anti-arrhythmic effect	Reduction in malignant ventricular arrhythmias via enrichment of cardiac lipids thereby preventing the development of ventricular tachycardia and fibrillation
Increase in heart rate variability	By increasing parasympathetic tone, inhibition or alteration of cytokine levels, altering the levels of mitogens and other factors
Antithrombotic effect	Via inhibition of thromboxane A2 in the arachidonic acid cascade, reducing platelet activity, enhancing the production of prostacyclins (pro-vasodialatory) and by lowering postprandial lipemia thereby lowering the pro-coagulant activated factor VII
Inhibitory effect on atherosclerosis	By regulating plasma cholesterol concentrations, by inhibiting monocyte migration into the plaque and through stimulation of endothelial production of nitric oxide
Hypolipidemic effect	Lowers plasma cholesterol and triacylglycerol concentrations by inhibiting triacylglycerol and very low density lipoproteins (VLDL) synthesis in the liver, and by stimulating the synthesis of membrane phospholipids. This also aids in preventing obesity related problems
Anti-inflammatory effect	Through inhibition of smooth cell proliferation, altered eicosanoid synthesis (especially PGE2 and LTB4) and by reducing the expression of cell adhesion molecules. By salvaging cardiomyocytes from hypoxia/re-oxygenation induced damage

Source: From Gordon, D. T. and Ratliff, V., *Advances in Seafood Biochemistry*, Flick, G. J. and Martin, R. E., Eds., Technomic Publ., Lancaster 1992; Connor, W. E., *Am. J. Clin. Nutr.*, 71 (suppl.), 171S–175S, 2000; Holub, B. J., *Can Med. Assoc. J.*, 166, 608–615, 2002; Goodnight, S. H. Jr., Harris, W. S., Connor, W. E., and Illingworth, D. R., *Arteriosclerosis*, 2, 87–113, 1982; Phillipson, B. E., Rothrock, D. W., Connor, W. E., Harris, W. S., and Illingworth, D. R., *N. Engl. J. Med.*, 312, 1210–1216, 1985; Weiner, B. H., Ockene, I. S., Levine, P. H. et al., *Circulation*, 315, 841–846, 1993; Davis, H. R., Bridenstine, R. T., Vesselinovitch, D., and Wissler, R. W., *Arteriosclerosis*, 7, 441–449, 1987; Haris, W. S., Connor, W. E., Alam, N., and Illingworth, D. R., *J. Lipid Res.*, 29, 1451–1460, 1988; Shimokawa, H. and Vanhoutte, P. M., *Am. J. Physiol.*, 256, H968–H973, 1989; Benner, K. G., Sasaki, A., Gowen, D. R., Weaver, A., and Connor, W. E., *Lipids*, 25, 534–540, 1990; Harris, W. S., Connor, W. E., Illingworth, D. R., Rothrock, D. W., and Foster, D. M., *J. Lipid Res.*, 31, 1549–1558, 1990; Larsen, L. F., Bladjberg, E. M., Jespersen, J., and Marckmann, P., *Arterioscler. Thromb. Vasc. Biol.*, 17, 2904–2909, 1997; Nasa, Y., Hayashi, M., Sasaki, H., Hayashi, J., and Takeo, S., *Jpn J. Pharmacol.*, 77, 137–146, 1998; James, M. J., Gibson, R. A., and Cleland, L. G., *Am. J. Clin. Nutr.*, 71 (suppl.), 343S–348S, 2000.

the time during fetal development and until the completion of biochemical points in the brain and retina after birth are the two critical periods during which this essential fatty acid is acquired [9]. There are reports of lower intelligence quotients in formula fed infants than those fed human milk [37]. This was attributed to the fact that formula-fed infants had lower brain DHA concentrations as compared to the human milk-fed infants [38,39].

TABLE 31.3
Essentiality of DHA in Retinal and Cell Membrane Functions

Retinal functions
    Photosensitivity of retina (dynamics of rhodopsin movement)
    Kinetics of career mediated transport
    Activity of membrane bound enzymes
    Properties of membrane receptors and membrane fluidity
Cell membranes/Central nervous system (CNS)
    Influences the structure of cell membranes and functioning of CNS as nearly one third of fatty acids in
    ethanolamine and serum phosphoglycerides are DHA
    Its association with $5'$-nucleotidase activity [46] has a structural role in brain functioning
    DHA acts as an anti-apoptotic factor in neuronal survival [32]

Source: From Gordon, D. T. and Ratliff, V., *Advances in Seafood Biochemistry*, Flick, G. J. and Martin, R. E., Eds., Technomic Publ., Lancaster, 1992.

It has also been reported that all the fatty acids including DHA get transferred across the placenta into fetal blood [40]. Eicosapentaenoic acid and DHA in maternal adipose tissue can be mobilized as free fatty acids bound to albumin and be made available for the developing fetus via transport. In several studies involving monkeys, low DHA concentrations in plasma and red blood cells in the infant at birth has been reported to be directly related to the DHA deficient maternal diet [40,41]. The deficiency of these important *n*-3 fatty acid gets manifested in blood and in tissue biochemistry [32,41,42].

In experiments involving human subjects, it has been reported that pregnant woman administered with fish oil or sardines showed higher DHA concentrations in both maternal plasma and red blood cells, and in cord blood plasma and red blood cells at the time of birth [43]. In another study involving healthy male volunteers, it was found that the administration of dietary EPA and DHA resulted in a 3 fold increase of EPA/DHA level in plasma, platelet, and mononuclear cell phospholipids [44]. It was noticed, in an in vitro trial with mice, that ethyl esters of EPA and DHA can modify spleen phospholipid fatty acid composition [45]. The retinal and cell membrane functions in which DHA can be considered essential are summarized as shown in Table 31.3.

It can be concluded that once membrane phospholipids have adequate concentrations of DHA or EPA, there would be an avid retention of these fatty acids in the brain and the retina even if the subsequent diet is deficient in these fatty acids [9].

31.3.3 Effect on Hypertension, Cancer, and Diabetes

Several studies involving both human subjects and animals with hypertension have revealed the importance of the long chain *n*-3 fatty acids, mainly EPA and DHA, in alleviating blood pressure [47–49]. In an experiment involving human subjects with hypertension and diabetes, it was found that higher DHA levels in blood platelet membranes resulted in significantly lower diastolic blood pressure and higher heart rate variability as compared to the control group that had low DHA levels in the blood platelet membranes [50]. Eicosapentaenoic acid and DHA differ in their effect on blood pressure as observed by Mori et al. [49] in a study involving double-blind, randomized placebo controlled human subjects. According to this study, DHA had a very significant effect on heart rate and blood pressure, while EPA had no significant effect on blood pressure or heart rate. It has been proven in a study with rats that n-3 PUFA deficiency in the perinatal period results in increased blood pressure later in life [51]. This study implies that an adequate intake of n-3 fatty acids in humans, including DHA at an early stage in life, may prevent increased blood pressure in later life [15].

Compared to blood pressure, the etiology of cancer is complicated and multi-factorial, apart from being uncertain. However, a strong positive relationship between dietary fat intake or body fat and the manifestation of cancer has been clearly established [52,53]. The fact that higher levels of n-6 fatty acids favor the development of tumors while their n-3 counterparts reduce or protect against tumor development has been thoroughly reviewed by Cave [54]. Docosahexaenoic acid and EPA inhibit cyclocoxygenase thereby reducing the amount of prostaglandins and increasing the lipoxygenase activity. This in turn results in higher production of hydoxyeicosatrienoic acids (HETE) and leukotriene B4 (LTB4) which are suggested to retard the process of cancerous cells overtaking a tissue [7]. Both DHA and EPA aid in the prevention of cancerous cell growth. However, in an experimental animal model, DHA was found to be more effective than EPA in inhibiting transcription factor activator protein 1 (AP-1), which has been implicated in the development of cancer [55].

Impaired glucose production and functions later in life leading to impaired glucose metabolism characterize adult on-set type II diabetes. Eicosapentaenoic acid was found to effectively increase the ability of the erythrocytes to clear glucose in EPA administered type II diabetes subjects [56]. It is speculated that incorporation of this n-3 fatty acid into erythrocyte cell membranes could change the structural or morphological characteristics of the cells to allow for greater glucose entry and subsequent metabolism [7].

Furthermore, another experiment involving chemically-induced diabetes in rats by Strolein et al. [57] supported the findings of Popp-Snijders et al. [56] that EPA and DHA allow tissues to more efficiently absorb and metabolize glucose in the absence of insulin. In all, these experiments suggest that the physical attributes of EPA and DHA, in particular in cell membranes, play an important role in regulating cellular metabolism [58].

31.3.4 EFFECT ON NEUROPSYCHIATRIC DISORDERS AND AUTO IMMUNE DISEASES

In recent years, several case-control studies and clinical trials have shown that long chain n-3 PUFA, mainly EPA and DHA, play a very important role in neuropsychiatry performance [15]. By effectively regulating the plasma/serum cholesterol that is associated with an increased risk of depression and suicide, these long chain PUFA aid in the prevention of neuropsychiatric disorders [59]. Several authors [60,61] have reviewed the association of abnormalities of C20 and C22 PUFA such as AA, EPA, and DHA with that of attention deficit/hyperactivity disorder (ADHD). Depressive patients were found to have decreased levels of long chain n-3 PUFA in the serum phospholipids and cholesterol esters [62,63], and the same has been noticed in schizophrenic patients as well [64,65]. Serum cholesterol levels were strongly associated with violent behavior in 20 psychiatric patients [66]. All these suggest that the use of cholesterol lowering drugs could lead to a change in mood [15].

The role of EPA and DHA in reducing the risk of neuropsychiatric disorders could be attributed to their effect on neurotransmitter receptor and G-proteins via effects on the bio-physical properties of cell membranes and secondary messengers—and protein kinases [59,67]. It is worth mentioning that EPA is able to reverse the phospholipid abnormalities in patients with schizophrenia via inhibition of phospholipase A2 (an enzyme that is specific to phospholipids (PL) and removes the PUFA from the sn-2 position of membrane PLs) or by activating a fatty acid Coenzyme A ligase [64].

Autoimmune diseases are a direct result of specific immune system aberrations [7]. These diseases affect different organs and are characterized by different clinical manifestations. However, the common factor is that the body harbors an overzealous immune system that attacks normal body cells in addition to foreign bodies (antigens) or diseased cells. The effectiveness of these long chain n-3 PUFA on some immunologic and inflammatory disorders [68,69], perhaps by influencing the cytokine and leukotriene generation [70–72], has been well documented. The involvement of EPA and DHA with the immune system and ultimately with autoimmune diseases [73,74], along with the comparative effect in relation to n-6 fatty acids [75], has been reviewed. These long chain n-3 fatty acids allow the suppressor T-cells (cells that are responsible for regulating the immune reaction) to function properly. If these are not regulated properly, then the helper T-cells may function out of control, thereby manifesting in autoimmune diseases [76]. However, in the case of autoimmune diseases, EPA and DHA may be effective only in larger doses and hence can be considered more as drugs rather than prophylactics or therapeutics [7,76].

31.3.5 Less Desirable Effects of EPA and DHA

The possible side effects vis-à-vis the benefits of an increased uptake of n-3 fatty acids including EPA and DHA has been the subject of review by several authors [1,7,17,77]. These long chain n-3 PUFA can alter gene expression by down-regulating proteoglycan degrading enzymes, inflammation inducible cytokines [78], and fatty acid synthase [79]. As mentioned earlier, EPA and DHA can have a beneficial effect in the case of autoimmune diseases. However, hypothetically, large intakes of these fatty acids could weaken the defense mechanisms against infections or malignancies, although there are no clinical data available on this as yet [77]. Although EPA and DHA exert a positive influence in preventing cancer related

TABLE 31.4
Less Desirable Effects Attributed to EPA and DHA

Reduced platelet count and long bleeding times
Increased production and accumulation of erucic acid (C22:1) that may lead to heart lesions
Higher vitamin E requirements to counter possible oxidation problems
Higher intake of vitamin A and D there by manifesting in problems of hypervitaminosis
Higher chances for consuming oxidation products (peroxides and aldehyde)

Source: From Carroll, K. K. and Woodward, C. J. H., *Marine Biogenic Lipids, Fats and Oils Vol. II*, Ackman, R. G., Ed., 435–456, CRC Press, Florida, 1989; Gordon, D. T. and Ratliff, V., *Advances in Seafood Biochemistry*, Flick, G. J. and Martin, R. E., Eds., Technomic Publ., Lancaster, 1992; Eritsland, J., *Am. J. Clin. Nutr.*, 71, 197–201, 2000.

manifestations, they might even lead to the development of undesirable effects as they are highly unsaturated (and in the absence of an antioxidant) thereby leading to the increased intake of harmful peroxides and aldehydes [17,77]. Since EPA and DHA have an antithrombotic effect, they might cause enhanced clotting time which may have an adverse effect at the time of severe injuries. The less desirable effects of EPA and DHA are summarized in Table 31.4 below.

31.4 SAFETY RELATED ISSUES OF EPA AND DHA

Several researchers have emphasized the safety related issues surrounding long chain *n*-3 fatty acids including EPA and DHA. This topic has been the subject of thorough reviews by various authors [1,77] and related strategies for public policies regarding the same [80]. The major safety-related aspect associated with EPA and DHA concerns that of their auto-oxidation potential. On exposure to oxidant stress, they can be easily attacked by free radicals and oxidized into peroxides [81], which in turn can result in a variety of products such as aldehydes, ketones, and cyclic peroxides [82] that could be cytotoxic or genotoxic [83]. Thus, it becomes important to encourage the reasonable intake of antioxidants when consuming a PUFA rich diet. Because EPA and DHA could influence bleeding time and associated clotting time, it may have serious consequences in individuals who have inherited or acquired hemorrhagic problems. Therefore, individual medical history has to be considered while consuming these long chain *n*-3 fatty acids either as functional foods or in the form of PUFA rich diets.

The increased intake of these long chain *n*-3 fatty acids should be at the expense of saturated fats because a substantial addition of energy dense PUFAs to the habitual diet could result in obesity related problems and associated metabolic disorders [77,84]. Also, commercially available EPA and DHA rich products should mention the quality and quantity of fatty acids apart from the above, along with any other toxic substance present in the product [17]. Apart from these considerations, the major issues related with EPA and DHA intake could be: what should be the optimum intake in each serving? What should be the balance between LA, EPA, and

DHA? What should be the ratio of total n-6 fatty acids to n-3 fatty acids? What should be the correct balance between EPA and DHA? And, should these be incorporated into the diet or can they be taken as supplements? These issues need to be resolved before drawing any conclusions regarding the safety of these n-3 fatty acids.

31.5 FUTURE AREAS OF RESEARCH

There are several areas in which research related to EPA and DHA needs to be done. The important one among them is to develop a public policy on the usage of these substances either as supplements, recommended daily intakes, or as therapeutics. A policy regarding labeling requirements for any of the functional foods or nutraceuticals that have these as ingredients also needs to be spelled out.

Furthermore, additional scientific evidence needs to be gathered on the pro- or anti-cancerous effects of EPA and DHA. Additionally, the association of EPA and DHA with autommune and inflammatory diseases, along with some of the gene related disorders, needs to be studied further before drawing any conclusions. And there should be a collective effort from nutritionists, food technologists, and medical professionals to develop dietary guidelines for the consumption of these beneficial substances. All these, apart from revealing both the beneficial and less desirable effects, would provide a clearer picture of the value of these long chain n-3 fatty acids in the prevention and treatment of various chronic diseases.

31.6 SUMMARY AND CONCLUSIONS

There is an increased interest in the nutrition and health-related aspects of EPA and DHA as a result of the evidence that has come from scientific findings that these fatty acids could prevent or alleviate several chronic diseases or health-related problems. Some of the apparent benefits and some less desirable effects have been outlined in this review. In summary, it can be said that EPA and DHA carry more beneficial effects than less desirable effects. In other words, the risk of adverse effects seems to be small when compared to the numerous health benefits afforded by them.

REFERENCES

1. Carroll, K. K. and Woodward, C. J. H., Nutrition and human health aspects of marine oils and lipids, In *Marine Biogenic Lipids, Fats and Oils Vol. II*, Ackman, R. G. Ed., CRC Press, Florida, pp. 435–456, 1989.
2. Paul, A. A. and Southgate, D. A. T., *McCane and Widdowson's the Composition of Foods*, 4th ed., Elsevier, Amsterdam, 1978.
3. Jangaard, P. M., Ackman, R. G., and Burgher, R. D., Component of fatty acids of the blubber fat from the common or harbor seal Phoca vitulina concolor de Kay, *Can J. Biochem. Physiol.*, 41, 2543–2548, 1963.

4. Ackman, R. G., Eaton, C. A., and Jangaard, P. M., Lipids of the fin whale (*Balaenoptera physalus*) from North Atlantic waters I. Fatty acid composition of whole blubber and blubber sections, *Can J. Biochem. Physiol.*, 45, 1513–1517, 1965.

5. Nisizawa, K., Seaweeds Kaiso—bountiful harvest from the seas, *Jpn. Seaweed Assoc.*, 44–50, 2002.

6. Tokusoglu, O. and Unal, M. K., Biomass nutrient profile of three micro-algae: *Spirulina platensis*, *Chlorella vulgaris* and *Isochrisis galbana*, *J. Food Sci.*, 68, 1144–1148, 2003.

7. Gordon, D. T. and Ratliff, V., The implications of omega 3 fatty acids in human health, In *Advances in Seafood Biochemistry*, Flick, G. J. and Martin, R. E. Eds., Technomic Publ., Lancaster, pp. 69–98, 1992.

8. Kinsella, J. E., *Seafoods and Fish Oils in Human Health and Disease*, Marcel Dekker Inc., New York, 1987.

9. Connor, W. E., Importance of *n*-3 fatty acids in health and diseases, *Am. J. Clin. Nutr.*, 71 (suppl.), 171S–175S, 2000.

10. Kang, J. X. and Leaf, A., Antiarrhythmic effects of polyunsaturated fatty acids, *Circulation*, 94, 1774–1780, 1996.

11. Bang, H. O. and Dyerberg, J., The composition of food consumed by Greenland Eskimos, *Acta Med. Scand.*, 200, 69–73, 1973.

12. Dyerberg, J. and Bang, H. O., Haemostatic function and platelet polyunsaturated fatty acids in Eskimos, *Lancet*, 2, 433–435, 1979.

13. Connor, W. E., Fatty acids and heart diseases, In *Nutrition and Disease Update: Heart Disease*, Kritchevsky, D. and Carroll, K. K. Eds., AOCS, Champaign, IL, pp. 7–42, 1979.

14. Holub, B. J., Clinical nutrition: 4 Omega-3 fatty acids in cardiovascular care, *Can Med. Assoc. J.*, 166, 608–615, 2002.

15. Li, D., Bode, O., Drummond, H., and Sinclair, A. J., Omega-3 (*n*-3) fatty acids, In *Lipids for Functional Foods and Nutraceuticals*, Gunstone, F. D. Ed., The Oily Press, New York, pp. 225–262, 2003.

16. Calder, P. C., Long chain *n*-3 fatty acids and inflammation: potential application in surgical and trauma patients, *Braz. J. Med. Biol. Res.*, 36, 433–466, 2003.

17. Goodnight, S. H. Jr., Harris, W. S., Connor, W. E., and Illingworth, D. R., Polyunsaturated fatty acids, hyperlipedimia and thrombosis, *Arteriosclerosis*, 2, 87–113, 1982.

18. Phillipson, B. E., Rothrock, D. W., Connor, W. E., Harris, W. S., and Illingworth, D. R., Reduction of plasma lipids, lipoproteins, and apoproteins by dietary fish oils in patients with hypertriglyceridemia, *N. Engl. J. Med.*, 312, 1210–1216, 1985.

19. Weiner, B. H., Ockene, I. S., Levine, P. H., Cuenoud, H. F., Fisher, M., Johnson, B. F., Daoud, A. S. et al., Inhibition of atherosclerosis by cod liver oil in a hyperlipidemic swine model, *N. Engl. J. Med.*, 315, 841–846, 1993.

20. Davis, H. R., Bridenstine, R. T., Vesselinovitch, D., and Wissler, R. W., Fish oil inhibits development of atherosclerosis in rhesus monkeys, *Arteriosclerosis*, 7, 441–449, 1987.

21. Haris, W. S., Connor, W. E., Alam, N., and Illingworth, D. R., The reduction of postprandial triglyceridemia in humans by dietary *n*-3 fatty acids, *J. Lipid Res.*, 29, 1451–1460, 1988.

22. Shimokawa, H. and Vanhoutte, P. M., Dietary omega-3 fatty acids and endothelium-dependent relaxations in porcine coronary arteries, *Am. J. Physiol.*, 256, H968–H973, 1989.

23. Benner, K. G., Sasaki, A., Gowen, D. R., Weaver, A., and Connor, W. E., The differential effect of eicosapentaenoic acid and oleic acid in lipid synthesis and VLDL secretion in rabbit hepatocytes, *Lipids*, 25, 534–540, 1990.

24. Harris, W. S., Connor, W. E., Illingworth, D. R., Rothrock, D. W., and Foster, D. M., Effect of fish oil on VLDL triglyceride kinetics in man, *J. Lipid Res.*, 31, 1549–1558, 1990.

25. Larsen, L. F., Bladjberg, E. M., Jespersen, J., and Marckmann, P., Effects of dietary fat quality and quantity on postprandial activation of blood coagulation factor VII, *Arterioscler. Thromb. Vasc. Biol.*, 17, 2904–2909, 1997.

26. Nasa, Y., Hayashi, M., Sasaki, H., Hayashi, J., and Takeo, S., Long term supplementation with eicosapentaenoic acid salvages cardiomyocytes from hypoxia/reoxygenation-induced injury in rats fed with fish-oil deprived diet, *Jpn. J. Pharmocol.*, 77, 137–146, 1998.

27. James, M. J., Gibson, R. A., and Cleland, L. G., Dietary polyunsaturated fatty acids and inflammatory mediator production, *Am. J. Clin. Nutr.*, 71 (suppl.), 343S–348S, 2000.

28. Holman, R. T., Biological activities of and requirements for polyunsaturated fatty acids, *Prog. Chem. Fats Other Lipids*, 9, 611–680, 1968.

29. Mitchell, D. C., Gawrisch, K., Litman, B. J., and Salem, N. Jr., Why is docosahexaenoic acid essential for nervous system function? *Biochem. Soc. Trans.*, 26, 365–370, 1998.

30. Lin, D. S., Connor, W. E., Anderson, G. J., and Neuringer, M., The effects of dietary *n*-3 fatty acids upon the phospholipid molecular species of monkey brain, *J. Neurochem.*, 55, 1200–1207, 1990.

31. Lin, D. S., Anderson, G. J., Connor, W. E., and Neuringer, M., The effects of dietary *n*-3 fatty acids upon the phospholipid molecular species of monkey retina, *Invest. Opthalmol. Visc. Sci.*, 35, 794–803, 1994.

32. Salem, N. Jr., Litman, B., Kim, H. Y., and Gawrisch, K., Mechanisms of action of docosahexaenoic acid in the nervous system, *Lipids*, 36, 945–959, 2001.

33. Crawford, M. A., Bloom, M., Cunnane, S., Holmsen, H., Ghebremeskel, K., Parkington, J., Schimdt, W., Sinclair, A. J., and Leigh-Broadhurst, C., *World Rev. Nutr. Diet,*, 88, 6–17, 2001.

34. Salem, N. and Ward, G. R., Are omega-3 fatty acids essential nutrients for mammals?, *World Rev. Nutr. Diet*, 72, 128–147, 1993.

35. Greiner, R. S., Moriguchi, T., Hutton, A., Slotnick, B. M., and Salem, N., Rats with low levels of brain docosahexaenoic acid show mpaired performance in olfactory based and spatial learning tasks, *Lipids*, 34 (suppl.), S239–S243, 1999.

36. Weisinger, H. S., Vingrys, A. J., Bui, B. V., and Sinclair, A. J., Effects of dietary *n*-3 fatty acid deficiency and repletion in the guinea pig retina, *Invest. Opthalmol. Vis. Sci.*, 40, 327–338, 1999.

37. Lucas, A., Morley, R., Cole, T. J., Lister, G., and Leeson-Payne, C., Breast milk and subsequent intelligence quotient in children born pre-term, *Lancet*, 339, 261–264, 1992.

38. Farquharson, J., Cockburn, F., Partick, W. A., Jamieson, E. C., and Logan, R. W., Infant cerebral cortex phospholipid fatty acid composition and diet, *Lancet*, 340, 810–813, 1992.

39. Makrides, M., Neumann, M. A., Byard, R. W., Simmer, K., and Gibson, R. A., Fatty acid composition of brain, retina and erythrocytes in breast- and formula fed infants, *Am. J. Clin. Nutr.*, 60, 189–194, 1994.

40. Ruyle, M., Connor, W. E., Anderson, G. J., and Lowenson, R. I., Placental transfer of essential fatty acids in humans: Venous arterial difference for docosahexaenoic acid in fetal umbilical erythrocytes, *Proc. Natl Acad. Sci. U.S.A*, 87, 7902–7906, 1990.

41. Neuringer, M., Connor, W. E., Van Petten, C., and Barstad, L., Dietary omega-3 fatty acid deficiency and visual loss in infant rhesus monkeys, *J. Clin. Invest.*, 73, 272–276, 1984.

42. Crawford, M. A., Bloom, M., Leigh-Broadhurst, C., Schmidt, W. F., Cunnane, S. C., Galli, C., Ghebremeskel, K., Linseisen, F., Lloyd-Smith, J., and Parkington, J., Evidence for the unique function of docosahexaenoic acid (DHA) during the evolution of the modern hominid brain, *Lipids*, 34, S39–S47, 2000.

43. Connor, W. E., Lowensohn, R., and Hatcher, L., Increased docosahexaenoic acid levels in human new born infants by the administration of sardines and fish oils during pregnancy, *Lipids*, 31 (suppl.), S183–S187, 1996.

44. Mantizioris, E., Cleland, L. G., Gibson, R. A., Neumann, M. A., Demasi, M., and James, M. J., Biochemical effects of a diet containing foods enriched with *n*-3 fatty acids, *Am. J. Clin. Nutr.*, 72, 42–48, 2000.

45. Robinson, D. R., Zu, L. L., Knoell, C. T., Tateno, S., and Olesiak, W., Modification of spleen phospholipid fatty acid composition by dietary fish oil and *n*-3 fatty acid ethyl esters, *J. Lipid Res.*, 34, 1423–1434, 1993.

46. Holman, R. T., Johnson, S. B., and Hatch, T. F., A case of human linolenic acid deficiency involving neurological abnormalities, *Am. J. Clin. Nutr.*, 35, 617–623, 1982.

47. Morris, M. C., Sack, F., and Rosner, B., Does fish oil lower blood pressure? A meta analysis of controlled trials, *Circulation*, 88, 523–533, 1993.

48. Bao, D. Q., Mori, T. A., Burke, V., Puddey, I. B., and Beilin, L. J., Effects of dietary fish and weight reduction on ambulatory blood pressure in overweight hypertensives, *Hypertension*, 32, 710–717, 1998.

49. Mori, T. A., Bao, D. Q., Burke, V., Puddey, I. B., and Beilin, L. J., Docosahexaenoic acid but not eicosapentaenoic acid lowers ambulatory blood pressure and heart rate in humans, *Hypertension*, 34, 253–260, 1999.

50. Christensen, J. H., Skou, H. A., Madsen, T., Torring, I., and Schmidt, E. B., Heart rate variability and *n*-3 polyunsaturated fatty acids in patients with diabetes mellitus, *J. Intern. Med.*, 249, 545–552, 2001.

51. Weisinger, H. S., Armitage, J. A., Sinclair, A. J., Vingrys, A. J., Burns, P. L., and Weisinger, R. S., Perinatal omega-3 fatty acid deficiency affects blood pressure later in life, *Nat. Med.*, 7, 258–259, 2001.

52. NRC (National Research Council), Assembly of Life Sciences, *Diet, Nutrition and Cancer*, National Academy Press, Washington, DC, 1982.

53. Carroll, K. K., Experimental studies on dietary fat and cancer in relation to epidemiological data, In *Dietary Fat and Cancer—Progress in Chemical and Biological Research Vol. 222*, Ip, C., Birt, D. F., Rogers, A. E., and Mettlin, C. Eds., Alan RL Inc., New York, 1986.

54. Cave, W. T. Jr., Omega-3 polyunsaturated fatty acids in rodent models of breast cancers, *Breast Cancer Res. Treat.*, 46, 239–246, 1997.

55. Liu, G., Bibus, D. M., Bode, A. M., Ma, W. Y., Holman, R. T., and Dong, Z., Omega 3 but not polyunsaturated fat content of canned meats commonly available in Australia, *Food Austr.*, 54, 311–315, 2001.

56. Popp-Snijders, C., Schouten, J. A., Heine, R. J., and Van der Veen, E. A., Dietary supplementation of omega-3 polyunsaturated fatty acids improves insulin sensitivity in non-insulin dependent diabetes, *Net. J. Med.*, 29, 74–79, 1986.

57. Strolien, L. H., Kraegen, E. W., Chrisholm, D. J., Ford, G. L., Bruce, D. G., and Pacoe, W. S., Fish oil prevents insulin resistance induced by high fat feeding in rats, *Science*, 237, 885–888, 1987.

58. Kamada, T., Yamashita, T., Baba, Y., Kai, M., Setoyania, S., Chuman, Y., and Otsuji, S., Dietary sardine oil increases erythrocyte membrane fluidity in diabetic patients, *Diabetes*, 35, 604–610, 1986.

59. Hibbeln, J. R. and Salem, N. Jr., Dietary polyunsaturated fatty acids and depression: when cholesterol does not satisfy, *Am. J. Clin. Nutr.*, 62, 1–9, 1995.

60. Richardson, A. J. and Puri, B. K., The potential role of fatty acids in attention-deficit/hyperactivity disorder. Prostaglandins leukotrienes essent, *Fatty Acids*, 63, 79–87, 2000.

61. Kidd, P. M., Attention deficit/hyperactivity disorder (ADHD) in children: rationale for its integrative management, *Altern. Med. Rev.*, 5, 402–428, 2000.

62. Adams, P. B., Lawson, S., Sanigorski, A., and Sinclair, A. J., Arachidonic acid to eicosapentaenoic acid ratio in blood correlates positively with clinical symptoms of depression, *Lipids*, 31 (suppl.), S157–S161, 1996.

63. Maes, M., Christophe, A., Delanghe, J., Altamura, C., Neels, H., and Meltzer, H. Y., Lowered omega-3 polyunsaturated fatty acids in serum phospholipids and cholesteryl esters of depressed patients, *Psychiatr. Res.*, 85, 275–291, 1999.

64. Richardson, A. J., Easton, T., and Puri, B. K., Red cell and plasma fatty acid changes accompanying symptom remission in a patient with schizophrenia treated with eicosapentaenoic acid, *Eur. Neropsychopharmocol.*, 10, 189–193, 2000.

65. Assies, J., Liverse, R., Vreken, P., Wanders, R. J., Dingemans, P. M., and Linszen, D. H., Significantly reduced docosahexaenoic and docosapentaenoic acid concentrations in erythrocyte membrane from schizophrenic patients compared with a carefully matched control group, *Biol. Psych.*, 49, 510–522, 2001.

66. Mufti, R. M., Balon, R., and Arfken, C. L., Low cholesterol and violence, *Psychiatr. Serv.*, 49, 214–221, 1998.

67. Edwards, R. H. and Peet, M., Essential fatty acid intake in relation to depression, In *Phospholipid Spectrum Disorder in Psychiatry*, Peet, M., Glenn, I., and Horrobin, D. F. Eds., Marius Press, Lancashire, 1999.

68. Heide, V. J. J., Bilo, H. J. G., Donker, J. M., Wilmink, J. M., and Tegzess, A. M., Effect of dietary fish oil on renal function and rejection in cyclosporine-treated recipients of renal transplants, *N. Engl. J. Med.*, 329, 769–773, 1993.

69. Belluzzi, A., Brignola, C., Campieri, M., Pera, A., Boschi, S., and Miglioli, M., Effect of an enteric coated fish oil preparation on relapses in Crohn's disease, *N. Engl. J. Med.*, 334, 1557–1560, 1996.

70. Lee, T. H., Hoover, R. L., Williams, J. D., Sperling, R. I., Ravalese, J. III., Spur, B. W., Robinson, D. R., Corey, E. J., Lewis, R. A., and Austen, K. F., Effect of dietary enrichment with eicosapentaenoic and docosahexaenoic acids on in vitro neutrophil and monocyte leukotriene generation and neutrophil function, *N. Engl. J. Med.*, 312, 1217–1224, 1985.

71. Endres, S., Gorbani, R., Kelly, V. E., Georgilis, K., Lonnemann, G., van der Meer, J. W., Cannon, J. G. et al., The effect of dietary supplementation with *n*-3 polyunsaturated fatty acids on the synthesis of interleukin-1 and tumor necrosis factor by mononuclear cells, *N. Engl. J. Med.*, 320, 265–271, 1989.

72. De Caterina, R., Cybulsky, M. I., Clinton, S. K., Gimbrone, M. A., and Libby, P., The omega-3 fatty acid docosahexaenoate reduces cytokine induced expression of proatherogenic and proinflammatory proteins in human endothelial cells, *Arteriosclr. Thromb.*, 14, 1829–1836, 1994.

73. Gurr, M. I., The role of lipids in the regulation of the immune system, *Prog. Lipid Res.*, 22, 257–287, 1983.

74. Johnston, P. V., Dietary fat, eicosanoids and immunity, *Adv. Lipid Res.*, 21, 103–141, 1985.

75. Harbige, L. S., Fatty acids, the immune response, and auto-immunity: a question of *n*-6 essentiality and the balance between *n*-6 and *n*-3, *Lipids*, 38, 323–342, 2003.

76. Lands, W. E. M., *Fish and Human Health*, Academic Press, New York, pp. 63–82, 1985.

77. Eritsland, J., Safety considerations of polyunsaturated fatty acids, *Am. J. Clin. Nutr.*, 71, 197–201, 2000.

78. Curtis, C. L., Hughes, C. E., Flannery, C. R., Little, C. B., Harwood, J. L., and Caterson, B., *n*-3 fatty acids specifically modulate catabolic factors involved in articular cartilage degradation, *J. Biol. Chem.*, 275, 721–724, 2000.

79. Pegorier, J. P., Regulation of gene expression by fatty acids, *Curr. Opin. Clin. Nutr. Metab. Care*, 1, 329–334, 1998.

80. Simopoulos, A. P., *n*-3 fatty acids and human health: defining strategies for public policy, *Lipids*, 36 (suppl.), S83–S89, 2001.

81. Halliwell, B. and Chirico, S., Lipid peroxidation: its mechanism, measurement and significance, *Am. J. Clin. Nutr.*, 57 (suppl.), 715S–725S, 1993.

82. Dargel, S., Lipid peroxidation-a common pathogenic mechanism? *Exp. Toxicol. Pathol.*, 44, 169–181, 1992.

83. Esterbauer, H., Cytotoxicity and genotoxicity of lipid oxidation products, *Am. J. Clin. Nutr.*, 57 (suppl.), 779S–786S, 1993.

84. NCPEP, The expert panel. Report of the national cholesterol education program expert panel on detection, evaluation and treatment of high blood cholesterol in adults, *Arch. Intern. Med.*, 148, 36–39, 1988.

32   Marine Polysaccharides and Angiogenesis: Modulation of Angiogenesis by Fucoidans

Shinji Soeda, Satoru Koyanagi, and Hiroshi Shimeno

CONTENTS

32.1   INTRODUCTION

In Japan, brown seaweeds such as kombu, wakame, and mozuku have been a mainstay of the traditional diet for several thousand years. The prefecture of Okinawa, whose inhabitants enjoy some of the highest life expectancies in Japan, also happens to have one of the highest per capita consumption rates of kombu (1 g per person per day). In contrast, the average per capita consumption rate of kombu in Japan is approximately 0.5 g per day. On Okinawa, the death rate from cancer is the lowest of all the prefectures in Japan. The longevity and low cancer rates may be caused in part by a substance called fucoidan.

Fucoidans are a unique class of sulfated fucans isolated from brown seaweeds (Phaeophyceae). The composition of each varies with the species such as Chorda filum [1], Ascophyllum nodosum [2], and Fucus evanescens [3], but they always contain L-fucose and sulfate ester groups, with small proportions of D-xylose, D-galactose, and D-glucuronic acid [4]. However, they may also contain additional sugars such as D-mannose and D-glucose [5]. Because it is commercially available, a fucoidan prepared from Fucus vesiculosus has commonly been used in biological studies of fucoidans.

Earlier data suggested that this fucoidan's structure consists of $(1 \rightarrow 2)$-linked, 4-O-sulfated fucopyranose residues. It is now thought that the main chain of the fucoidan contains $(1 \rightarrow 3)$-linked fucopyranose residues with some branches at the O-2 or O-4 positions, and that some of the residues are 4-O-sulfated [6]. Despite extensive studies of algal fucoidans, their precise structures remain uncertain. The uncertainty is caused largely by lack of strict molecular symmetry and the presence of minor sugar components and, occasionally, protein contaminants [4].

However, algal fucoidans have potent anticoagulant [4,7], antitumor [8–12], fibrinolytic [13,14], antiviral [15–19], anti-inflammatory [20,21], and contraceptive [22,23] activities. Fucoidans also act as ligands for P- and L-selectins and block cell–cell binding [20,21,24–28]. These biological activities may vary depending on the fucoidans' structural peculiarities, molecular weights, and degrees of sulfation [4].

Angiogenesis (new blood vessel formation), which is essential for several physiological or pathological processes, can be affected by the biological actions of fucoidans. This chapter focuses the fucoidans' effects on angiogenesis, and their potential use in treating ischemic limb disease and malignant tumors. We also review the potential uses of chitin and chitosan for promoting wound healing and cancer treatment (for the molecular characteristics and other medical applications that have been uncovered, see a review [29]).

32.2 THE EFFECT OF FUCOIDAN ON BASIC FIBROBLAST GROWTH FACTOR-INDUCED ANGIOGENESIS

Basic fibroblast growth factor (FGF-2), a member of the heparin-binding growth factor family, plays a major angiogenic role in vascular wound repair and collateral vessel formation [30,31]. FGF-2 induces many endothelial cell modifications involved in angiogenesis: increases in production of urokinase-type plasminogen activator (u-PA), tissue-type plasminogen activator (t-PA), and their receptors that participate in the degradation of basement membrane; stimulation of cell proliferation and migration during vessel elongation; and increases in the expression of cell surface proteins such as integrins and adhesion molecules [32]. Angiogenesis depends on the interaction between cell surface proteins and extracellular matrix proteins; α_6 and β_1 integrin subunits [33] and PECAM-1 [34] were identified as endothelial cell–cell adhesion molecules.

The biological functions of FGF-2 are mediated through its binding to FGF-2 receptors that exhibit intracellular tyrosine kinase activity [35]. To bind to the receptors, FGF-2 needs the help of heparan sulfate proteoglycans (HSPG) that are low-affinity FGF-2 receptors present on the cell surfaces and surrounding extracellular matrix [36]. The highly charged macromolecules consist of different core proteins with covalently linked heparan sulfate chains (HS) of varying monosaccharide sequences. Sydecans-1, -2, and -4, as well as glypican-1, possess HS which carry out this mediation. The amino acid residues in the FGF-2: Asp^{28}, Arg^{121}, Lys^{126}, and Gln^{135} interact with the 2-O-sulfated iduronic acid and glucosamine N-sulfate residues as well as other carboxyl groups within heparin oligosaccharides [37,38].

As illustrated in Figure 32.1, the binding of HS or heparin to FGF-2 leads to the dimerization of the growth factor without significant conformational alterations and results in a direct interaction with FGF-2 receptors [37–40]. The receptors are dimerized transiently, and the resulting dimer formation facilitates phosphorylation of cytoplasmic tails, assembly of intracellular components, and mitogenesis [41,42]. In addition to the promotion of FGF-2 binding to the receptor, endothelial HSPG protect FGF-2 from proteolytic degradation and serve as reservoirs that release the growth factor after HSPG degradation [43]. HSPG also induce internalization of FGF-2 in Chinese hamster ovary cells [44]. This suggests that the functions of FGF-2 in endothelium may be partly regulated by the availability of extracellular glycosaminoglycans and that exogenous polysaccharides may modulate the endothelial angiogenic response.

In this section, we describe the effect of two fucoidans on FGF-2-induced angiogenesis. The present data are from the studies done by the INSERM groups in Paris, France [45–47].

32.2.1 FUCOIDANS USED IN THESE STUDIES

Unless otherwise indicated, two fucoidans were used in all the studies mentioned below. One [45] is a natural fucoidan extracted and purified from the brown seaweed Ascophyllum nodosum, as previously described by Colliec et al. [48]. This fucoidan is homogenous and has an average molecular weight of $16,000 \pm 3000$. Its chemical composition is 55.7% fucose, 6.3% uronic acid, and 29% SO_3Na. Its anticoagulant activity is 9.2 IU/mg, as determined using the activated partial thromboplastin time assay with heparin as standard. We designate this fucoidan as natural fucoidan. The other [46,47] is a low-molecular weight (LMW) fucoidan obtained by radical processing of Ascophyllum nodosum fucoidan [49]. This LMW fucoidan has an average molecular weight of 7000 ± 2000. Its chemical composition is 35% fucose, 3% uronic acid and 34% SO_3Na. The amount of this fucoidan required for an activated partial thromboplastin time of 80 s (control, 40 s) is 25 µg/ml, suggesting that it has a low affinity for thrombin. We designate this fucoidan as LMW fucoidan.

32.2.2 EFFECTS OF THESE FUCOIDANS ON FGF-2-INDUCED ANGIOGENESIS

Matou et al. [45] have tested the effect of natural fucoidan on FGF-2-induced proliferation, surface protein expression, and differentiation of human umbilical

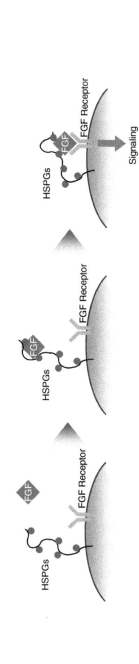

FIGURE 32.1 A model of the FGF-2 binding to its receptors. The binding of HS in HSPG to FGF-2 leads to the dimerization of growth factor and results in a direct interaction with FGF-2 receptors. The receptors are transiently dimerized, and the resulting dimer formation facilitates intracellular signal transduction.

vein endothelial cells (HUVEC). This fucoidan did not modulate mitogenic activity of FGF-2, but it did cause a significant increase in FGF-2-induced tube formation in HUVEC. The cell surface proteins, such as u-PA receptors, $\alpha_v\beta_3$ integrin, α_6 and β_1 integrin subunits, and PECAM-1 are known to be modulated by FGF-2 and are involved in tube formation [33,34,50,51]. The natural fucoidan (10 µg/ml) alone increased the expression of α_6 integrin subunits, whereas heparin did not. The increase in α_6 integrin subunits was smaller than that observed with FGF-2 alone. In the presence of FGF-2, the natural fucoidan enhanced the expressions of α_6, β_1, and PECAM-1, while it reduced that of $\alpha_v\beta_3$ integrin. The most striking effect of this fucoidan on the expression of α_6 integrin subunits and tube formation was stopped by adding monoclonal antibodies against α_6. Furthermore, monoclonal antibodies against FGF-2 strongly inhibited the effect of natural fucoidan plus FGF-2 on the expression of α_6. These data suggest that the natural fucoidan enhances FGF-2-induced angiogenesis by promoting the FGF-2-induced expression of endothelial surface proteins (mainly α_6).

Like heparin, natural fucoidan shows a potent inhibitory effect on the arterial smooth muscle cells in rats [52]. This fucoidan's effect is applicable to the prevention of restenosis. Compounds which have antithrombotic and anticoagulant activities, and which inhibit proliferation of smooth muscle cells, are candidates as drugs to treat restenosis. Such drugs will be more effective if they promote angiogenesis so as to restore the endothelial monolayers. Although the results need to be confirmed through in vivo angiogenesis assays, they [45] suggest natural fucoidan is a possible drug candidate for the prevention of restenosis.

Because most natural fucoidans have high-molecular weight, which makes it difficult to use them in vivo, a LMW fucoidan was prepared for possible medical applications [49]. Chabut et al. [46] compared the effect of a LMW fucoidan on FGF-2-induced tube formation in HUVEC with that of a LMW heparin (5 kDa). In a standard Matrigel assay system, both the LMW fucoidan and LMW heparin promoted FGF-2-induced tube formation in HUVEC. The LMW fucoidan also caused the HUVEC to branch and form numerous tubes with closed areas. The authors also demonstrated that both LMW fucoidan and LMW heparin potentiate FGF-2-induced tubular morphogenesis through the enhancement of expression of α_6 integrin, but not of β_1. Vascular tube formation may result from a finely tuned balance between proliferation, migration, and differentiation of endothelial cells. However, the results suggest that tube formation in HUVEC is independent of the cell's proliferation, because the enhanced expression of α_6 integrin triggered differentiation but not proliferation [46].

This process is analogous to that in a myoblast cell system [53], in which the α_6 integrin regulates differentiation, while the β_1 subunit triggers proliferation. A direct link between α_6 integrin expression and vascular tube formation was also confirmed by the use of antibodies against α_6: antibodies against FGF-2 receptor had no effect on the enhancement of α_6 integrin expression. Interestingly, removing HS from HUVEC with heparitinases stopped the FGF-2-induced expression of α_6 integrin. Adding unfractionated heparin partly restored stimulation of the α_6 expression, but neither LMW fucoidan nor LMW heparin could. A simple explanation is that both LMW polysaccharides cannot present FGF-2 to the FGF-2 receptor. However, the FGF-2 receptor seemed dispensable for the FGF-2-induced α_6 integrin expression. Expression of α_6 had an absolute requirement for HS, but the mechanism of action seemed independent of the FGF-2 receptor. Thus, HS mediate directly the FGF-2-induced differentiation in HUVEC [46]. Overall, LMW fucoidan and LMW heparin may act as chaperones to shuttle FGF-2 to HS, as has been suggested for other heparin-like molecules [54].

LMW fucoidan highly consists of fucose sulfate [55]. A fucosylated chondroitin sulfate extracted from sea cucumber has been demonstrated to potentiate FGF-2-induced tubular morphogenesis [56]. Branched sulfated fucoses constitute the key motif in this activity, and thus fucose seems critical for FGF-2-induced HUVEC differentiation. In conclusion, the LMW fucoidan can enhance FGF-2-induced tubular morphogenesis through the overexpression of α_6 integrin, which is HS-dependent [46].

In animal experiments, the LMW fucoidan prevents arterial thrombus growth with less of the hemorrhaging caused by heparin [57], and it inhibits smooth muscle cell proliferation [52]. After myocardial infarction, endothelial cells increase the expression of HSPG specifically at iscemic sites [58]. Chabut et al. [46] conclude that, because of the antithrombotic and potential angiogenic properties, LMW fucoidan might be developed into a drug that can initiate endothelial cell differentiation in the revascularization of ischemic areas after myocardial infarction.

32.2.3 THE THERAPEUTIC POTENTIAL OF THE LMW FUCOIDAN FOR REVASCULARIZATION IN ANIMAL MODELS

Critical limb ischemia is estimated to develop in approximately 500–1000 individuals per million per year (2nd European consensus document on chronic critical leg ischemia). In a large proportion of these patients, the anatomic extent and the distribution of arterial occlusive disease make the patients unsuitable for operative or percutaneous revascularization. No pharmacological treatment has been shown to favorably affect the natural history of critical limb ischemia. Recently, successful vascular endothelial growth factor (VEGF) gene therapy to treat the patients with critical limb ischemia has been reported [59]. However, in some cases, such direct administration of the VEGF gene alone may not be sufficient to treat impaired neo-vascularization.

Studies by Luyt et al. [47] suggest that the LMW fucoidan may have potential to treat limb ischemia, because of its ability to promote revascularization. Luyt et al. investigated the fucoidan's effect on stimulated critical hindlimb ischemia in rats. The LMW fucoidan enhanced FGF-2-induced (3H) thymidine incorporation in cultured rat smooth muscle cells. Its intravenous injection in rats significantly increased the stromal-derived factor (SDF)-1 levels in plasma. SDF-1 is a heparin-binding cytokine [60] involved in the process of angiogenesis [61], where it regulates endothelial cell branching morphogenesis [62]. On the other hand, FGF-2, and VEGF upregulate the expression of SDF-1 receptors on endothelial cells [63]. Thus, SDF-1 mobilization may be one of the molecular effectors of therapeutic revascularization.

Sweeney et al. [64], using the natural fucoidan, found that it helped increase plasma SDF-1 levels. Their study also indicated that intravenous injection of the fucoidan increases the plasma levels of matrix metalloprotease-9 [64]. However, LMW fucoidan did not increase the protease levels in plasma [47]. That substances belonging to the same polysaccharide family vary greatly in their effects may be caused by the degree of sulfation or variation in molecular weight.

In rats [47], the LMW fucoidan or FGF-2 used separately caused similar improvements in residual muscle blood flow, compared to the control. Used together, they created even more significant improvement in tissue blood flow, and apparently caused some muscle regeneration. The capillary density count increased from 9.6 ± 0.7 capillaries/muscle section in untreated ischemic controls to 14.3 ± 0.9 in rats with the LMW fucoidan alone, to 14.5 ± 0.9 with FGF-2 alone, and to 19.1 ± 0.9 in combination ($p < .001$). The study by Luyt et al. [47] indicates that the LMW fucoidan potentiates FGF-2 activity, mobilizes SDF-1, and facilitates angiogenesis in the rat model. Thus, the LMW fucoidan alone or with FGF-2 may aid the VEGF promotion of neovascularization without antithrombin side effects.

32.3 OVERSULFATION OF NATURAL FUCOIDAN AND ITS ANTI-ANGIOGENIC AND ANTITUMOR ACTIVITIES

Highly regulated and transient angiogenesis is necessary for embryonic development, wound healing, and corpus luteum formation. As described in the above section, therapeutic angiogenesis is an effective treatment for patients with critical limb ischemia or with myocardial ischemia. However, uncontrolled and persistent angiogenesis occurs in several pathological states such as diabetic retinopathy, rheumatoid arthritis, and tumor progression.

As illustrated in Figure 32.2, the growth of tumors requires angiogenesis to supply nutrients and oxygen. Tumors use the newly formed blood vessels as conduits to disseminate tumor cells. Because the growth and metastasis of malignant tumors are dependent upon their angiogenetic potencies, a novel anticancer treatment has been developed in which tumors are regressed by prolonged inhibition of angiogenesis. A variety of anti-angiogenic agents are currently undergoing clinical trials for dormancy therapy of cancers [65].

Tumor-induced angiogenesis is modulated by cell-produced factors that have mitogenic and chemotactic effects on vascular endothelial cells. Several endothelial growth factors such as FGF-2, transforming growth factor (TGF)-β, platelet-derived growth factor (PDGF), and VEGF have been identified as angiogenic factors [66–68]. In particular, the expression of VEGF in tumor cells is thought to play a key role in tumor-induced angiogenesis [69,70]. Because tumor vascularity and metastasis vary almost directly with the expression of VEGF and its receptors, the targeting of VEGF or VEGF receptors may be an effective strategy for anti-angiogenic therapy [71,72].

We have found that an oversulfated fucoidan (OSF), a natural fucoidan modified by oversulfation, reduces the invasiveness of Lewis lung carcinoma (3LL) cells in vitro [73], inhibits tube formation by HUVEC [74,75], and suppresses the growth

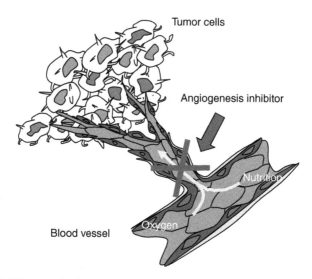

FIGURE 32.2 The growth of tumors requires new blood vessel formation (angiogenesis) to supply nutrients and oxygen. Inhibition of angiogenesis regresses tumor growth. It is one of several promising anticancer treatments.

of tumor cells implanted in mice by preventing tumor-induced angiogenesis [76]. One of the mechanisms by which OSF prevents the tumor's angiogenesis is the inhibition of VEGF-induced phosphorylation (activation) of VEGF receptor-2 by binding to VEGF itself [76].

32.3.1   OVERSULFATION OF NATURAL FUCOIDAN

For our study, we modified a natural fucoidan obtained from Fucus vesiculosus. The fucoidan was purchased from Sigma Chemical Co., St. Louis, MO, U.S.A. and modified by mixing it with sulfur trioxide–trimethylamine complex in the presence of dimethylformamide [77]. Sulfation followed at 50°C for 24 h. The yield of OSF was 54%; the sulfate content was estimated to be 56.7% (that of the natural fucoidan was estimated to be 32.6%). Electrophoretic analysis revealed that the 100–130 kDa form of OSF was nearly homogeneous [78].

32.3.2   EFFECTS OF OSF ON TUBE FORMATION BY HUVEC

OSF at concentrations of 10–25 µg/ml significantly inhibited tube formation by HUVEC on Matrigel, while natural fucoidan did not [74]. OSF also inhibited the migration of HUVEC, which is critical for angiogenesis. Increasing t-PA release from HUVEC enhances the formation of plasmin. Plasmin degrades extracellular components such as fibronectin and laminin and thereby promotes cell migration. Furthermore, OSF, but not natural fucoidan, increased the release of plasminogen activator inhibitor-1 (PAI-1), which is the primary inhibitor of u-PA and t-PA.

This suggests that the balance between plasminogen activation and its inhibition shifts to inhibition in the presence of OSF. The predominance of the PAI-1 levels in medium may reduce plasmin formation, thereby reducing the activation of pro-collagenases. Incubation of HUVEC with OSF significantly decreased the collagenase activity in medium.

Matrigel is prepared from murine Engelbreth–Holm–Swarm sarcoma and contains t-PA, laminin, type IV collagen, and various angiogenic factors including FGF-2. These factors might affect the tube formation by HUVEC. Therefore, additional experiments were done to test the potential effect of OSF on angiogenesis in highly purified collagen media [75]. Unlike natural fucoidan, OSF inhibited FGF-2-induced cell migration and tube formation of HUVEC, and promoted an FGF-2-induced release of PAI-1, but not of t-PA.

The increase in PAI-1 release may be part of the mechanism(s) by which OSF inhibits migration and tube formation of HUVEC. OSF promoted the binding of FGF-2 to HUVEC surfaces and followed enhanced activation (phosphorylation) of the HUVEC FGF-2 receptors. In heparitinase-treated HUVEC, both FGF-2 binding and PAI-1 release were decreased by OSF. These results suggest that the fucoidan derivative can promote the binding of FGF-2 to HS molecules on HUVEC and facilitate FGF-2 binding to the receptors displaying tyrosine kinase activity (see, Section 32.2). Heparitinase treatment largely decreases the cell surface HS. The HS molecules may be too few to act as the acceptors of FGF-2, causing the FGF-2 to bind to the more numerous OSF molecules, and preventing their transfer to the acceptor molecule HS.

In contrast to our results, Matou et al. [45] and Chabut et al. [46] showed that LMW fucoidan enhances FGF-2-induced tube formation by HUVEC on Matrigel. The discrepancy may depend on the difference between the molecular natures of the fucoidans used. In other studies, hyaluronans hampered tubular morphogenesis, whereas the oligosaccharides stimulated proliferation and tube formation of endothelial cells [79]. Collen et al. [80] observed that, LMW heparin inhibits the stimulation of angiogenesis induced by FGF-2, VEGF, or TNF-α, whereas unfractionated heparin enhances it. Chemically derived dextrans enhance the angiogenic effect of FGF-2 [81], while chemically modified heparin-like compounds inhibit FGF-2-induced endothelial cell proliferation and angiogenesis [82]. Thus, the effects of polysaccharides on angiogenesis are largely affected by their chemical structure, molecular weight, and the model used for the assays.

32.3.3 THE EFFECTS OF OSF ON TUMOR CELL INVASION

Tumor cell invasion is a complex phenomenon that involves the disruption of basement membrane barriers and the penetration of cells into normal adjacent tissues. The invading cells first attach themselves to basement membranes. The anchored tumor cells next secret u-PA, which converts plasminogen to plasmin. Plasmin degrades noncollagenous matrix proteins, such as laminin and fibronection, and activates procollagenase secreted from the tumor cells.

OSF inhibited the invasion of 3LL cells through a Matrigel-coated filter and their adhesion to laminin; however, it did not prevent their adhesion to fibronectin or to type IV collagen [73]. These two effects were more potent than those of natural fucoidan. Both OSF and natural fucoidan bound to 56- and 62-kDa fragments of laminin, suggesting that laminin has binding site(s) for these fucoidans. 3LL cells have 67 kDa laminin receptors on their cell surfaces. YIGSR (Tyr-Ile-Gly-Ser-Arg), a synthetic laminin pentapeptide, specifically binds to the 67 kDa receptors and inhibits experimental metastasis [83]. Our results [73] demonstrated that the laminin attachment of both YIGSR-pretreated and non-treated cells was equally inhibited by OSF or natural fucoidan. Therefore, the 67 kDa receptor-binding site (YIGSR) on the laminin molecule is not found on the fucoidan binding site. The release of u-PA from 3LL cells was increased by adding laminin. OSF reduced the increase in u-PA activity. Therefore, the inhibitory effect of OSF on 3LL cell invasion may result from its inhibition of physical interaction between tumor cells and laminin, followed by suppression of the laminin-induced increase in u-PA release.

The fucoidans' ability to inhibit the tumor cell invasion depended on the degree of sulfation. Natural fucoidan itself is a negatively charged sulfated polysaccharide. For a better understanding of the relationship between fucoidan structure and its function on invasion and metastasis of tumor cells, we prepared an aminated fucoidan and examined its effect on the invasion of 3LL cells through a Matrigel-coated filter [84]. The purified fucoidan was aminated in epichlorohydrin and 30% ammonia water at 40°C for 90 m [85]. The positively charged fucoidan, unlike OSF, promoted tumor cell invasion, demonstrating that the biological activities of polysaccharides are largely affected by their molecular weights and the degrees of electrical charge.

32.3.4   ANTI-ANGIOGENIC AND ANTITUMOR ACTIVITIES OF OSF

VEGF is a potent selective cytokine that acts on endothelial cells to promote the formation and spreading of blood vessels [86,87]. The angiogenic action of VEGF is mediated through VEGF receptor-2 (also known as KDR) and the downstream phosphatidylinositol $3'$-kinase (PI-3K)/Akt signaling pathway [88]. The binding of VEGF to VEGF receptor-2 induces conformational changes in the receptor, followed by dimerization and autophosphorylation of the tyrosine residues [89]. Thereafter, consecutive intracellular signal transduction events are initiated and culminate in VEGF-mediated angiogenesis. Five isoforms of human VEGF mRNA have been produced from a single gene by means of alternative splicing [90]. The best characterized VEGF is $VEGF_{165}$, a 165-amino acid form that induces angiogenesis and blood vessel permeability in vivo and displays a mitogenic activity restricted to vascular endothelial cells [91].

Both OSF and natural fucoidan inhibit VEGF's binding to its receptor, thus slowing or inhibiting angiogenesis in tumor masses [76]. OSF inhibited $VEGF_{165}$-induced proliferation and migration of HUVEC more effectively than natural fucoidan. As shown in the upper panel in Figure 32.3, treatment of HUVEC with $VEGF_{165}$ resulted in extensive phosphorylation of VEGF receptor-2, when compared with the control (lane 1 versus lane 2). However, the co-presence of natural fucoidan (lane 3) or OSF (lane 4) decreased the amount of the phosphorylated VEGF receptor-2. The effect of OSF was more potent than that of

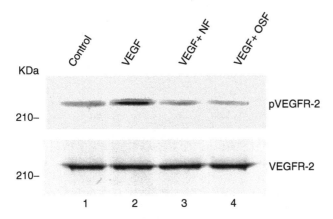

FIGURE 32.3 Effects of natural fucoidan and OSF on the VEGF$_{165}$-induced autophosphorylation of VEGF receptor-2. (Cited from Koyanagi, S., Tanigawa, N., Nakagawa, H., Soeda S., and Shimeno, H., *Biochem. Pharmacol.*, 65, 173–179, 2003. With permission.)

natural fucoidan. These results suggest that both fucoidans inhibit the VEGF$_{165}$-induced proliferation and migration of HUVEC by preventing VEGF$_{165}$-mediated signal transduction.

VEGF$_{165}$ binds to heparin through a site separated from VEGF's receptor-binding domain [92]; the formation of the complex is thought to facilitate the binding of VEGF$_{165}$ to VEGF receptor-2. Thus, both fucoidans may have the ability to modulate VEGF$_{165}$ binding to vascular endothelial cells before or after the complex has been formed. Both fucoidans prevented the binding of (^{125}I) VEGF to HUVEC; the cause appeared to be the interaction of fucoidans with VEGF$_{165}$. Surface plasmon resonance analyses of the VEGF$_{165}$ binding to fucoidans showed that the affinities (K_d) of natural fucoidan and OSF for immobilized VEGF$_{165}$ were 368 nM and 52 nM, respectively. The result suggests that OSF and natural fucoidan bind directly to VEGF$_{165}$, thereby preventing VEGF$_{165}$ from binding to the receptor. The difference in the fucoidans' molecular structure may affect the efficiency of fucoidan–VEGF complex formation.

Natural fucoidan or OSF (5 mg/kg each, i.v.) was administrated to mice, in which a membrane including Sarcoma 180 cells had been implanted. As shown in Figure 32.4a, 7 days after implanting the tumor cells, neovascularization from surrounding blood vessels was observed in the control mice in the region adjacent to the implanted chamber. On the other hand, neovascularization in mice given natural fucoidan or OSF was clearly suppressed. Figure 32.4b shows the effects of natural fucoidan and OSF on the growth of Lewis lung carcinoma or B16 melanoma cells inoculated into the right hind footpads of mice. The growth of these tumor cells was clearly suppressed by the repetitive administration of each fucoidan (5 mg/kg, i.v.). These results suggest that the antitumor action of fucoidan is caused, at least in part, by its anti-angiogenic potency and that effect varies directly with the fucoidan molecule's degree of sulfation.

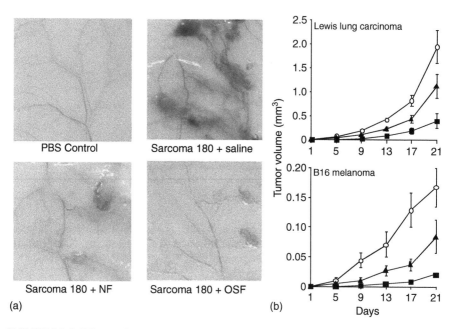

(a)   (b)

FIGURE 32.4 Effects of natural fucoidan and OSF on angiogenesis and tumor growth in mice. (a) Photographs showing the neovascularization induced by Sarcoma 180 cells packed in membrane chambers in mice. The control chamber contained phosphate-buffered saline (PBS) instead of Sarcoma 180 cells. (b) The tumor cells (1.5×10^6) were inoculated into the right hind footpads of mice. The mice were injected i.v. with a single daily dose of natural fucoidan or OSF (5 mg/kg each) starting 3–5 days after inoculation with the tumor cells. The tumor volume was estimated according to the following formula: tumor volume $(cm^3) = 4\pi(xyz)/3$, where x, y, and z are the three perpendicular diameters of the tumor. $\blacktriangle$, natural fucoidan; $\blacksquare$, OSF; $\bigcirc$, saline (control). Each point is the mean $\pm$ SEM of 7–11 mice. (Cited from Koyanagi, S., Tanigawa, N., Nakagawa, H., Soeda S., and Shimeno, H., *Biochem. Pharmacol.*, 65, 173–179, 2003. With permission.)

Recently, Hamma-Kourbali et al. [93] have shown that carboxymethyl dextran benzylamide (CMDB7) inhibits the mitogenic effect of $VEGF_{165}$ on HUVEC by preventing the $VEGF_{165}$-induced phosphorylation of VEGF receptor-2 and consequently the cell proliferation arrest. CMDB7 mimics some properties of heparin, such as the interactions with angiogenic growth factors including FGF-2 [94,95]. In vivo study demonstrated that MCF-7 ras xenograft growth in nude mice is blocked by CMDB7 treatment [96]. Furthermore, CMDB7 is capable of inhibiting by 80% the incidence of lung micrometastasis from breast carcinoma MDA-MB435 cell implants in the foot pads of nude mice [97]. The molecule's capabilities may be largely caused by its inhibiton of $VEGF_{165}$-induced angiogenesis. The dextran derivative acts by displacing heparin from $VEGF_{165}$ to $VEGF_{165}$ receptor complexes, but not by direct interactions with VEGF receptor-2 [93]. CMDB7 is non-cytotoxic dextran and, unlike fucoidans, has no anticoagulant activity. The selective biological activity of this compound make it a good choice for increasing the efficiency of conventional anticancer treatments. OSF acts in a way similar to CMDB7 on the binding of $VEGF_{165}$ to VEGF receptor-2 [76]. However, because it shows anticoagulant activity, OSF at present may not be an ideal choice for anticancer treatment [77,78]. Further fractionation and purification of the OSF used in our studies may partly solve this problem. In conclusion, varying the structure of natural fucoidans, including minimizing their molecular weight, open avenues for the creation of new drugs to fight cancer.

32.4 CHITIN AND CHITOSAN: THEIR EFFECTS ON ANGIOGENESIS

Chitin, poly-β-$(1 \rightarrow 4)$-N-acetyl D-glucosamine is the most abundant renewable organic skeletal component of invertebrates. It serves as a "glue" for the chemical components making up the delicate wings of insects and the hard exoskeletons of crustaceans such as crabs and shrimps. Chitin is manufactured commercially in Japan and the U.S.A. from crustacean exoskeletons; crab, lobster, shrimp, prawn, and krill are the best sources. Chitosan is manufactured by hydrolyzing chitin's N-acetyl linkages in an aqueous 40%–50% sodium hydroxide solution at 110°C–120°C for several hours.

Chitin and chitosan are being evaluated in a number of biomedical applications. These include wound healing and dressing, antitumor uses, dialysis membranes, fibers for digestible sutures, liposome stabilization agents, serum cholesterol level reducers, hypocholesteremic, and hemostatic agents, and others [29]. In this section, we describe the effects of chitin, chitosan, and their derivatives on wound healing and tumor angiogenesis.

Sulfated chitin derivatives (SCM-chitin III) are reported to inhibit B16-BL6 cell-induced angiogenesis in mice [98]. SCM-chitin III, when injected into the tumor site on day 1 or 3 after tumor inoculation, caused a marked decrease in the angiogenic response without affecting the tumor growth. In contrast, carboxymethyl chitin and heparin had no effect. SCM-chitin III also inhibited the migration of endothelial cells through Matrigel and the haptotactic migration to fibronectin-substrate. The sulfated chitin derivative did not directly affect the viabilities or the growths of tumor cells and endothelial cells. These results suggest that the SCM-chitin III's antitumor effect is its ability to inhibit tumor-induced angiogenesis. Qin et al. [99] prepared a water-soluble LMW chitosan by hydrolyzing chitosan with hemicellulase. Intraperitoneal injections of LMW chitosan or its N-acetylated product inhibited the growth of Sarcoma 180 cells inoculated in mice. Oral administration was also effective. However, sulfate groups on SCM-chitin III [98] may play an important role in anti-angiogenic activity.

Angiogenesis is an essential component of wound healing [100]. Chitin and chitosan as wound healing accelerators have recently been studied [101,102]. Ueno et al. [101] made experimental open skin wounds on the dorsal side in dogs. They applied cottonfiber-type chitosan for 15 days and evaluated the process of wound healing by using histological and immunohistochemical techniques. The cottonfiber-type chitosan obtained is composed of fibers 2–20 mm in length,

20–50 µm in width, and 3–15 µm in thickness, and has an apparent specific gravity of 0.1–0.2 g/cm^3. By day 3, chitosan treatment had induced extensive infiltration of polymorphonuclear (PMN) cells. The treatment also increased the production of type III collagen. In the control group, large numbers of mitotic cells were seen on day 6, and in the chitosan-treated group on day 3. These results suggest that chitosan accelerates infiltration of PMN cells early in wound healing and promotes collagen production by fibroblast.

Cho et al. [102] compared the effect of a water-soluble chitin (WSC) on wound healing in rats with that of chitin and chitosan. Powdered chitin, chitosan, WSC, or a WSC solution was embedded in incisions made on the backs of rats. WSC was more efficient than chitin or chitosan as a wound-healing accelerator. The wound treated with WSC solution was completely re-epithelialized; the granulation tissues were nearly replaced by fibrosis, and the hair follicles were almost healed after 7 days. Also, the skin treated with WSC solution had the highest tensile strength, and the arrangement of collagen fibers in the skin was similar to normal skins. The efficacy of chitosan for wound healing may be caused by the following: accelerating PMN cell infiltration into the wound area; increasing effusion which forms thick fibrin and promotes the migration of fibroblasts into the wound area; and stimulating macrophage migration, fibroblast proliferation, and collagen III production [101]. WSC may be more potent in these wound healing processes than chitin and chitosan [102]. However, no direct evidence was found that chitosan and WSC accelerate angiogenesis during wound healing [101,102]. Because the process of healing does not occur in the absence of angiogenesis [100], the chitosan and WSC may play direct or indirect roles in the acceleration of angiogenesis in these experimental models. Current biomedical applications of chitins and chitosans in wound healing are reviewed by Muzzarelli et al. [103].

32.5 CONCLUSIONS

Fucoidans have been shown, in vitro and in vivo, to have properties useful for the treatment of diseases. If fucoidans are to be used clinically, it may be necessary to modify them so that they only exert their angiogenic or anti-angiogenic (antitumor) effects. Future clinical studies should involve the development of extremely LMW and synthetic fucoidan-like materials with a suitable number of sulfate groups and sugar units.

Chitin's and chitosan's biological activities and mechanical and physical properties make them an attractive bio-polymer. Some chitins and chitosans can accelerate wound healing. The possibilities for their clinical uses in treating various diseases appear great. However, as with fucoidans, further pharmaceutical innovations will be needed to realize their potential. Finally, a better understanding of the chemistry and biochemistry of these marine polysaccharides might provide a basis for the development of new therapeutic agents to overcome a number of diseases such as critical limb ischemia, cancers, diabetic retinopathy, and rheumatoid arthritis.

REFERENCES

1. Chizhov, A. O., Dell, A., Morris, H. R., Haslam, S. M., McDowell, R. A., Shashkov, A. S., Nifantev, N. E., Khatuntseva, E. A., and Usov, A. I., A study of fucoidan from the brown seaweed *Chorda filum*, *Carbohydr. Res.*, 320, 108–119, 1999.

2. Marais, M. F. and Joseleau, J. P., A fucoidan fraction from *Ascophyllum nodosum*, *Carbohydr. Res.*, 336, 155–159, 2001.

3. Bilan, M. I., Grachev, A. A., Ustuzhanina, N. E., Shashkov, A. S., Nifantiev, N. E., and Usov, A. I., Structure of a fucoidan from the brown seaweed *Fucus evanescens*, *Carbohydr. Res.*, 337, 719–730, 2002.

4. Nagumo, T. and Nishino, T., Fucan sulfates and their anticoagulant activities, In *Polysaccharides in Medical Applications*, Dumitriu, S. Ed., Marcel Dekker, New York, pp. 545–574, 1996.

5. Duarte, M. E. R., Cardoso, M. A., Noseda, M. D., and Cerezo, A. S., Structural studies on fucoidans from the brown seaweed Sargassum stenophyllum, *Carbohydr. Res.*, 333, 281–293, 2001.

6. Patankar, M. S., Oehninger, S., Barnett, T., Williams, R. L., and Clark, G. F., A revised structure for fucoidan may explain some of its biological activities, *J. Biol. Chem.*, 268, 21770–21776, 1993.

7. Chaubet, F., Chevolot, L., Jozefonvicz, J., Durand, P., and Boisson-Vidal, C., Relationships between chemical characteristics and anticoagulant activity of low molecular weight fucans from marine algae, In *Bioactive Carbohydrate Polymers*, Paulsen, B. S. Ed., Kluwer Academic, Netherlands, pp. 59–84, 2000.

8. Ito, H., Noda, H., Amano, H., Zhuaug, C., Mizuno, T., and Ito, H., Antitumor activity and immunological properties of marine algal polysaccharides, especially fucoidan, prepared from *Sargassum thunbergii* of Phaeophyceae, *Anticancer Res.*, 13, 2045–2052, 1993.

9. Zhuang, C., Mizuno, T., and Ito, H., Antitumor active fucoidan from the brown seaweed, umitoranoo (*Sargassum thunbergii*), *Biosci. Biotechnol. Biochem.*, 59, 563–567, 1995.

10. Ito, H., Noda, H., Amano, H., and Ito, H., Immunological analysis of inhibition of lung metastases by fucoidan (GIV-A) prepared from brown seaweed *Sargassum thuunbergii*, *Anticancer Res.*, 15, 1937–1947, 1995.

11. Riou, D., Colliec-Jouault, C., Pinczon du Sel, D., Bosch, S., Siavoshian, S., LeBert, V., Tomasoni, C., Singuin, C., Durand, P., and Roussakis, C., Antitumor and antiproliferative effects of a fucan extracted from *Ascophyllum nodosum* against a non-small cell bronchopulmonary carcinoma line, *Anticancer Res.*, 16, 1213–1218, 1996.

12. Murayama, H., Tamauchi, H., Hashimoto, M., and Nakano, T., Antitumor activity and immune response of Mekabu fucoidan extracted from Sporophyll of Undaria pinnatifida, *In Vivo*, 17, 245–249, 2003.

13. Doctor, V. M., Hill, C., and Jackson, G. J., Effect of fucoidan during activation of human plasminogen, *Thromb. Res.*, 79, 237–247, 1995.

14. Nishino, T., Yamauchi, T., Horie, M., Nagumo, T., and Suzuki, H., Effects of a fucoidan on the activation of plasminogen by u-PA and t-PA, *Thromb. Res.*, 99, 623–634, 2000.

15. McClure, M. O., Moore, J. P., Blanc, D. F., Scotting, P., Cook, G. M., Keynes, R. J., Weber, J. N., Davies, D., and Weiss, R. A., Investigations into the mechanism by which sulfated polysaccharides inhibit HIV infection *in vitro*, *AIDS Res. Hum. Retroviruses*, 8, 19–26, 1992.

16. Hoshino, T., Hayashi, T., Hayashi, K., Hamada, J., Lee, J. B., and Sankawa, U., An antivirally active sulfated polysaccharide from *Sargassum horneri* (TURNER) C. AGARDH, *Biol. Pharm. Bull.*, 21, 730–734, 1998.

17. Feldman, S. C., Reynaldi, S., Stortz, C. A., Cerezo, A. S., and Damont, E. B., Antiviral properties of fucoidan fractions from *Leathesia difformis*, *Phytomedicine*, 6, 335–340, 1999.

18. Schaeffer, D. J. and Krylov, V. S., Anti-HIV activity of extracts and compounds from algae and cyanobacteria, *Ecotoxicol. Environ. Saf.*, 45, 208–227, 2000.

19. Preeprame, S., Hayashi, K., Lee, J. B., Sankawa, U., and Hayashi, T., A novel antivirally active fucan derived from an edible brown alga, *Sargassum horneri*, *Chem. Pharm. Bull. (Tokyo)*, 49, 484–485, 2001.

20. Preobrazhenskaya, M. E., Berman, A. E., Mikhailov, V. I., Ushakova, N. A., Semenov, A. V., Usov, A. I., Nifant'ev, N. E., and Bovin, N. V., Fucoidan inhibits leukocyte recruitment in a model peritoneal inflammation in rat and blocks interaction of P-selectin with its carbohydrate ligand, *Biochem. Mol. Biol. Int.*, 43, 443–451, 1997.

21. Teixeira, M. M. and Hellewell, P. G., The effect of the selectin binding polysaccharide fucoidan on eosinophil recruitment in vivo, *Br. J. Pharmacol.*, 120, 1059–1066, 1997.

22. Mahony, M. C., Oehninger, S., Clark, G. F., Acosta, A. A., and Hodgen, G. D., Fucoidan inhibits the zona pellucida-induced acrosome reaction in human spermatozoa, *Contraception*, 44, 657–665, 1991.

23. Mahony, M. C., Clark, G. F., Oehninger, S., Acosta, A. A., and Hodgen, G. D., Fucoidan binding activity and its localization on human spermatozoa, *Contraception*, 48, 277–289, 1993.

24. Nasu, T., Fukuda, Y., Nagahira, K., Kawashima, H., Noguchi, C., and Nakanishi, T., Fucoidan, a potent inhibitor of L-selectin function, reduces contact hypersensitivity reaction in mice, *Immunol. Lett.*, 59, 47–51, 1997.

25. Chauvet, P., Bienvenu, J. G., Theoret, J. F., Latour, J. G., and Merhi, Y., Inhibition of platelet–neutrophil interactions by fucoidan reduces adhesion and vasoconstriction after acute arterial injury by angioplasty in pigs, *J. Cardiovasc. Pharmacol.*, 34, 597–603, 1999.

26. Ostergaard, C., Yieng-Kow, R. V., Benfield, T., Frimodt-Moller, N., Espersen, F., and Lundgren, J. D., Inhibition of leukocyte entry into the brain by the selectin blocker fucoidan decreases interleukin-1 (IL-1) levels but increases IL-8 levels in cerebrospinal fluid during experimental pneumococcal meningitis in rabbits, *Infect. Immun.*, 68, 3153–3157, 2000.

27. Frenette, P. S. and Weiss, L., Sulfated glycans induce rapid hematopoietic progenitor cell mobilization: evidence for selectin-dependent and independent mechanisms, *Blood*, 96, 2460–2468, 2000.

28. Bojakowski, K., Abramczyk, P., Bojakowska, M., Zwolinska, A., Przybylski, J., and Gaciong, Z., Fucoidan improves the renal blood flow in the early stage of renal ischemia/reperfusion injury in the rat, *J. Physiol. Pharmacol.*, 52, 137–143, 2001.

29. Hon, D. N. S., Chitin and chitosan: medical applications, In *Polysaccharides in Medical Applications*, Dumitriu, S. Ed., Marcel Dekker, New York, pp. 631–649, 1996.

30. Watanabe, E., Smith, D. M., Sun, J., Smart, F. W., Delcarpio, J. B., Roberts, T. B., Van Meter, C. H., and Claycomb, W. C., Effect of basic fibroblast growth factor on angiogenesis in the infarcted porcine heart, *Basic Res. Cardiol.*, 93, 30–37, 1998.

31. Detillieux, K. A., Sheikh, F., Kardami, E., and Cattini, P. A., Biological activities of fibroblast growth factor-2 in the adult myocardium, *Cardiovasc. Res.*, 57, 8–19, 2003.
32. Bikfalvi, A., Klein, S., Pintucci, G., and Rifkin, D. B., Biological roles of fibroblast growth factor-2, *Endocr. Rev.*, 18, 26–45, 1997.
33. Davis, G. E. and Camarillo, C. W., Regulation of endothelial cell morphogenesis by integrins, mechanical forces, matrix guidance pathways, *Exp. Cell Res.*, 216, 113–123, 1995.
34. DeLisser, H. M., Christofidou-Solomidou, M., Strieter, R. M., Burdick, M. D., Robinson, C. S., Wexler, R. S., Kerr, J. S. et al., Involvement of endothelial PECAM-1/CD31 in angiogenesis, *Am. J. Pathol.*, 151, 671–677, 1997.
35. Fantl, W. J., Johnson, D. E., and Williams, M. T., Signalling by receptor tyrosine kinases, *Annu. Rev. Biochem.*, 62, 453–481, 1993.
36. Rosenberg, R. D., Shworak, N. W., Liu, J., Schwartz, J. J., and Zhang, L., Heparan sulfate proteoglycans of the cardiovascular system, *J. Clin. Investig.*, 99, 2062–2070, 1997.
37. Faham, S., Hileman, R. E., Fromm, J. R., Linhardt, R. J., and Rees, D. C., Heparin structure and interactions with basic fibroblast growth factor, *Science*, 271, 1116–1120, 1996.
38. Ornitz, D. M., Herr, A. B., Nilsson, M., Westman, J., Svahn, C. M., and Waksman, G., FGF binding and FGF receptor activation by synthetic heparan-derived di- and trisaccharides, *Science*, 268, 432–436, 1995.
39. Yayon, A., Klagsbrun, M., Esko, J. D., Leder, P., and Ornitz, D. M., Cell surface, heparin-like molecules are required for binding to basic fibroblast growth factor to its high affinity receptor, *Cell*, 64, 841–848, 1991.
40. Kan, M., Wang, F., Xu, J., Crabb, J. W., Hou, J., and McKeehan, W. L., An essential heparin-binding domain in the fibroblast growth factor receptor kinase, *Science*, 259, 1918–1921, 1993.
41. Krufka, A., Guimond, S., and Rapraeger, A. C., Two hierarchies of FGF-2 signaling in heparin: mitogenic stimulation and high-affinity binding/receptor transphosphorylation, *Biochemistry*, 35, 11131–11141, 1996.
42. Ueno, H., Gunn, M., Dell, K., Tseng, A. J., and Williams, L., A truncated form of fibroblast growth factor receptor 1 inhibits signal transduction by multiple types of fibroblast growth factor receptor, *J. Biol. Chem.*, 267, 1470–1476, 1992.
43. Saksela, O., Moscatelli, D., Sommer, A., and Rifkin, D. B., Endothelial cell-derived heparan sulfate binds basic fibroblast growth factor and protects it from proteolytic degradation, *J. Cell Biol.*, 107, 743–751, 1988.
44. Rhogani, M. and Moscatelli, D., Basic fibloblast growth factor is internalized through both receptor-mediated and heparan sulfate-mediated mechanisms, *J. Biol. Chem.*, 267, 22156–22162, 1992.
45. Matou, S., Helley, D., Chabut, D., Bros, A., and Fischer, A.-M., Effect of fucoidan on fibroblast growth factor-2-induced angiogenesis in vitro, *Thromb. Res.*, 106, 213–221, 2002.
46. Chabut, D., Fischer, A. M., Colliec-Jouault, S., Laurendeau, I., Matou, S., Le Bonniec, B., and Helley, D., Low molecular weight fucoidan and heparin enhance the basic fibroblast growth factor-induced tube formation of endothelial cells through heparan sulfate-dependent α6 overexpression, *Mol. Pharmacol.*, 64, 696–702, 2003.
47. Luyt, C. E., Meddahi-Pellé, A., Ho-Tin-Noe, B., Colliec-Jouault, S., Guezennec, J., Louedec, L., Prats, H. et al., Low molecular-weight fucoidan promotes therapeutic revascularization in a rat model of critical hindlimb ischemia, *J. Pharmacol. Exp. Ther.*, 305, 24–30, 2002.

48. Colliec, S., Fischer, A. M., Tapon-Bretaudiere, J., Boisson, C., Durand, P., and Jozefonvicz, J., Anticoagulant properties of a fucoidan fraction, *Thromb. Res.*, 64, 143–154, 1991.

49. Nardella, A., Chaubet, F., Boisson-Vidal, C., Blondin, C., Durand, P., and Jozefonvicz, J., Anticoagulant low molecular weight fucans produced by radical process and ion exchange chromatography of high molecular weight fucans extracted from the brown seaweed *Ascophyllum nodosum*, *Carbohydr. Res.*, 289, 201–208, 1996.

50. Mignatti, P., Mazzieri, R., and Rifkin, D. B., Expression of urokinase receptor in vascular endothelial cells is stimulated by basic fibroblast growth factor, *J. Cell Biol.*, 113, 1193–1201, 1991.

51. Klein, S., Giancotti, F., Presta, M., Albelda, S. M., Buck, C. A., and Riflin, D. B., Basic fibroblast growth factor mudulates integrin expression in microvascular endothelial cells, *Mol. Biol. Cell*, 4, 973–982, 1993.

52. Logeart, D., Prigeant-Richard, S., Jozefonvicz, J., and Letourmeur, D., Fucan, a sulfated polysaccharide extracted from brown seaweeds, inhibits vascular SMC proliferation: part I. Comparison with heparin for the antiproliferativeactivity, binding and internalization, *Eur. J. Cell Biol.*, 74, 385–390, 1997.

53. Sastry, S. K., Lakonishok, M., Wu, S., Truong, T. Q., Huttenlocher, A., Turner, C. E., and Horwitz, A. F., Quantitative changes in integrin and focal adhesion signaling regulate myoblast cell cycle withdrawal, *J. Cell Biol.*, 144, 1295–1309, 1999.

54. Liekens, S., Leali, D., Neyts, J., Esnouf, R., Rusnati, M., Dell'Era, P., Maudgal, P. C., De Clercq, E., and Presta, M., Modulation of fibroblast growth factor-2 receptor binding, signaling and mitogenic activity by heparin–mimicking polysulfonated compounds, *Mol. Pharmacol.*, 56, 204–213, 1999.

55. Chevolot, L., Mulloy, B., Ratiskol, J., Foucault, A., and Colliec-Jouault, S., A disaccharide repeat unit is the major structure in fucoidans from tow species of brown algae, *Carbohydr. Res.*, 330, 529–535, 2001.

56. Tapon-Bretaudiére, J., Chabut, D., Zierer, M., Matou, S., Helley, D., Bros, A., Mourao, P. A., and Fischer, A. M., A fucosylated chondroitin sulfate from echinoderm modulates in vitro fibroblast growth factor-2-dependent angiogenesis, *Mol. Cancer Res.*, 1, 96–102, 2002.

57. Millet, J., Colliec-Jouault, S., Mauray, S., Theveniaux, J., Sternberg, C., Boisson Vidal, C., and Fischer, A. M., Antithrombotic and anticoagulant activities of a low molecular weight fucoidan by subcutaneous route, *Thromb. Haemost.*, 81, 391–395, 1999.

58. Kojima, T., Takagi, A., Maeda, M., Segawa, T., Shimizu, A., Yamamoto, K., Matsushita, T., and Saito, H., Plasma levels of syndecan-4 (ryudocan) are elevated in patients with acute myocardial infarction, *Thromb. Haemost.*, 85, 793–799, 2001.

59. Baumgartner, I., Pieczek, A., Manor, O., Blair, R., Kearney, M., Walsh, K., and Isner, J. M., Constitutive expression of phVEGF$_{165}$ after intramuscular gene transfer promotes collateral vessel development in patients with critical limb ischemia, *Circulation*, 97, 1114–1123, 1998.

60. Lortat-Jacob, H., Grosdidier, A., and Imberty, A., Structural diversity of heparan sulfate binding domains in chemokines, *Proc. Natl Acad. Sci. U.S.A.*, 99, 1229–1234, 2002.

61. Mirshahi, F., Pourtau, J., Li, H., Muraine, M., Trochon, V., Legrand, E., Vannier, J., Soria, J., Vasse, M., and Soria, C., SDF-1 activity on microvascular endothelial cells: consequences on angiogenesis in in vitro and in vivo models, *Thromb. Res.*, 99, 587–594, 2000.

62. Salvucci, O., Yao, L., Villalba, S., Sajewicz, A., Pittaluga, S., and Tosato, G., Regulation of endothelial cell branching morphogenesis by endogenous chemokine stromal-derived factor-1, *Blood*, 99, 2703–2711, 2002.

63. Salcedo, R., Wasserman, K., Young, H. A., Grimm, M. C., Howard, O. M., Anver, M. R., Kleinman, H. K., Murphy, W. J., and Oppenheim, J. J., Vascular endothelial growth factor and basic fibroblast growth factor induce expression of CXCR4 on human endothelial cells: in vivo neovascularization induced by stromal-derived factor-1α, *Am. J. Pathol.*, 154, 1125–1135, 1999.

64. Sweeney, E., Lortat-Jacob, H., Priestley, G. V., Nakamoto, B., and Papayannopoulou, T., Sulfated polysaccharides increase plasma level of SDF-1 in monkeys and mice: involvement in mobilization of stem/progenitor cells, *Blood*, 99, 44–51, 2002.

65. Nelson, N. J., Inhibitors of angiogenesis enter phase III testing, *J. Natl Cancer Inst.*, 90, 960–963, 1998.

66. Montesano, R., Vassalli, J. D., Baird, A., Guillemin, R., and Orci, L., Basic fibroblast growth factor induces angiogenesis in vitro, *Proc. Natl Acad. Sci. U.S.A.*, 83, 7297–72301, 1986.

67. Roberts, A. B., Sporn, M. B., Assoian, R. K., Smith, J. M., Roche, N. S., Wakefield, L. M., Heine, U. I. et al., Transforming growth factor type β: rapid induction of fibrosis and angiogenesis in vivo and stimulation of collagen formation in vitro, *Proc. Natl Acad. Sci. U.S.A.*, 83, 4167–4177, 1986.

68. Ishikawa, F., Miyazono, K., Hellman, U., Hannes, D., Wernstadt, C., Hagiwara, K., Usuki, K., Takaku, F., Risau, W., and Heldin, C. H., Identification of angiogenic activity and the cloning and expression of platelet-derived endothelial cell growth factor, *Nature*, 338, 557–562, 1989.

69. Dvorak, H. F., Sioussat, T. M., Brown, L. F., Berse, B., Nagy, J. A., Sotrel, A., Manseau, E. J., Van de Water, L., and Senger, D. R., Distribution of vascular permeability factor (vascular endothelial growth factor) in tumors: concentration in tumor blood vessels, *J. Exp. Med.*, 174, 1275–1278, 1991.

70. Asano, M., Yukita, A., Matsumoto, T., Kondo, S., and Suzuki, H., Inhibition of tumor growth and metastasis by an immnoneutralizing monoclonal antibody to human vascular endothelial growth factor/vascular permeability factor 121, *Cancer Res.*, 55, 5296–5301, 1995.

71. Weidner, N., Semple, J. P., Welch, W. R., and Folkman, J., Tumor angiogenesis and metastasis: correlation in invasive breast carcinoma, *N. Engl. J. Med.*, 324, 1–8, 1991.

72. Takahashi, Y., Kitadai, Y., Bucana, C. D., Cleary, K. R., and Ellis, L. M., Expression of vascular endothelial growth factor and its receptor, KDR, correlates with vascularity, metastasis, and proliferation of human colon cancer, *Cancer Res.*, 55, 3964–3968, 1995.

73. Soeda, S., Ishida, S., Shimeno, H., and Nagamatsu, A., Inhibitory effect of oversulfated fucoidan on invasion through reconstituted basement membrane by murine lewis lung carcinoma, *Jpn J. Cancer Res.*, 85, 1144–1150, 1994.

74. Soeda, S., Shibata, Y., and Shimeno, H., Inhibitory effect of oversulfated fucoidan on tube formation by human vascular endothelial cells, *Biol. Pharm. Bull.*, 20, 1131–1135, 1997.

75. Soeda, S., Kozako, T., Iwata, K., and Shimeno, H., Oversulfated fucoidan inhibits the basic fibroblast growth factor-induced tube formation by human umbilical vein endothelial cells: its possible mechanism of action, *Biochim. Biophys. Acta*, 1497, 127–134, 2000.

76. Koyanagi, S., Tanigawa, N., Nakagawa, H., Soeda, S., and Shimeno, H., Oversulfation of fucoidan enhances its antiangiogenic and antitumor activities, *Biochem. Pharmacol.*, 65, 173–179, 2003.
77. Soeda, S., Sakaguchi, S., Shimeno, H., and Nagamatsu, A., Fibrinolytic and anticoagulant activities of highly sulfated fucoidan, *Biochem. Pharmacol.*, 43, 1853–1858, 1992.
78. Soeda, S., Ohmagari, H., Shimeno, H., and Nagamatsu, A., Preparation of oversulfated fucoidan fragments and evaluation of their antithrombotic activities, *Thromb. Res.*, 72, 247–256, 1993.
79. Rahmanian, M., Pertoft, H., Kanda, S., Christofferson, R., Claesson-Welsh, L., and Heldin, P., Hyaluronan oligosaccharides induce tube formation of a brain endothelial cell line in vitro, *Exp. Cell Res.*, 237, 223–230, 1997.
80. Collen, A., Smorenburg, S. M., Peters, E., Lupu, F., Koolwijk, P., Van Noorden, C., and van Hinsbergh, V. W., Unfractionated and low molecular weight heparin affect fibrin structure and angiogenesis *in vitro*, *Cancer Res.*, 60, 6196–6200, 2000.
81. Desgranges, P., Barritault, D., Caruelle, J. P., and Tardieu, M., Transmural endothelialization of vascular prostheses is regulated *in vitro* by fibloblast growth factor 2 and heparin-like molecule, *Int. J. Artif. Organs*, 20, 589–598, 1997.
82. Miao, H., Ornitz, D. M., Aingorn, E., Ben-Sasson, S. A., and Vlodavsky, I., Modulation of fibloblast growth factor-2 receptor binding, dimerization, signaling, and angiogenic activity by a synthetic heparin–mimicking polyanionic compound, *J. Clin. Investig.*, 99, 1565–1575, 1997.
83. Iwamoto, Y., Robey, F. A., Graf, J., Sasaki, M., Kleinman, H. K., Yamada, Y., and Martin, G. R., YIGSR, a synthetic laminin pentapeptide, inhibits experimental metastasis formation, *Science*, 238, 1132–1134, 1987.
84. Soeda, S., Ishida, S., Honda, O., Shimeno, H., and Nagamatsu, A., Aminated fucoidan promotes the invasion of 3LL cells through reconstituted basement membrane: its possible mechanism of action, *Cancer Lett.*, 85, 133–138, 1994.
85. Soeda, S., Ohmagari, Y., Shimeno, H., and Nagamatsu, A., Preparation of aminated fucoidan and its evaluation as an antithrombotic and antilipemic agent, *Biol. Pharm. Bull.*, 17, 784–788, 1994.
86. Neufeld, G., Cohen, T., Gengrinovitch, S., and Poltorak, Z., Vascular endothelial growth factor (VEGF) and its receptors, *FASEB J.*, 13, 9–22, 1999.
87. Petrova, T. V., Makinen, T., and Alitalo, K., Signaling *via* vascular endothelial growth factor receptors, *Exp. Cell Res.*, 253, 117–130, 1999.
88. Gerber, H. P., McMurtrey, A., Kowalski, J., Yan, M., Keyt, B. A., Dixit, V., and Ferrara, N., Vascular endothelial growth factor regulates endothelial cell survival through the phosphatidylinositol 3'-kinase/Akt signal transduction pathway: requirement for Flk-1/KDR activation, *J. Biol. Chem.*, 273, 30336–30343, 1998.
89. Waltenberger, J., Claesson-Welsh, L., Siegbahn, A., Shibuya, M., and Heldin, C. H., Different signal transduction properties of KDR and Flt 1, two receptors for vascular endothelial growth factor, *J. Biol. Chem.*, 269, 26988–26995, 1994.
90. Ferrara, N. and Davis-Smith, T., The biology of vascular endothelial growth factor, *Endocr. Rev.*, 18, 4–25, 1997.
91. Ferrara, N., Houck, K., Jakeman, L., and Leung, D. W., Molecular and biological properties of the vascular endothelial growth factor family of proteins, *Endocr. Rev.*, 13, 18–32, 1992.

92. Dougher, A. M., Wasserstrom, H., Torley, L., Shridaran, L., Westdock, P., Hileman, R. E., and Fromm, J. R., Identification of a heparin binding peptide on the extracellular domain of the KDR VEGF receptor, *Growth Factors*, 14, 257–268, 1997.

93. Hamma-Kourbali, Y., Vassy, R., Starzec, A., Meuth-Metzinger, V. L., Oudar, O., Bagheri-Yarmand, R., Perret, G., and Crépin, M., Vascular endothelial growth factor 165 (VEGF$_{165}$) activities are inhibited by carboxymethyl benzylamide dextran that competes for heparin binding to VEGF$_{165}$ and VEGF$_{165}$ KDR complexes, *J. Biol. Chem.*, 276, 39748–39754, 2001.

94. Bagheri-Yarmand, R., Liu, J. F., Ledoux, D., Morere, J. F., and Crépin, M., Inhibition of human breast epithelial HBL 100 cell proliferation by a dextran derivative (CMDB7): interference with the FGF-2 autocrine loop, *Biochem. Biophys. Res. Commun.*, 239, 424–428, 2001.

95. Bagheri-Yarmand, R., Kourbali, Y., Mabilat, C., Morere, J. F., Martin, A., Lu, H., Soria, C., Jozefonvicz, J., and Crépin, M., The suppression of fibroblast growth factor 2/fibroblast growth factor 4-dependent tumour angiogenesis and growth by the anti-growth factor activity of dextran derivative (CMDB7), *Br. J. Cancer*, 78, 111–118, 1998.

96. Bagheri-Yarmand, R., Kourbali, Y., Morere, J. F., Jozefonvicz, J., and Crépin, M., Inhibition of MCF-7 ras tumor growth by carboxymethyl benzylamine dextran: blockage of the paracrine effect and receptor binding of transforming growth factor beta and platelet-derived growth factor, *Cell Growth Differ.*, 9, 497–504, 1998.

97. Bagheri-Yarmand, R., Kourbali, Y., Rath, A. M., Vassy, R., Martin, A., Jozefonvicz, J., Soria, C., Lu, H., and Crépin, M., Carboxymethyl benzylamide dextran blocks angiogenesis of MDA-MB435 breast carcinoma xenografted in fat pad and its lung metastasis in nude mice, *Cancer Res.*, 59, 507–510, 1999.

98. Murata, J., Saiki, I., Makabe, T., Tsuta, Y., Tokura, S., and Azuma, I., Inhibition of tumor-induced angiogenesis by sulfated chitin derivatives, *Cancer Res.*, 51, 22–26, 1991.

99. Qin, C., Du, Y., Xiao, L., Li, Z., and Gao, X., Enzymic preparation of water-soluble chitosan and their antitumor activity, *Int. J. Biol. Macromol.*, 31, 111–117, 2002.

100. Arnold, F. and West, D. C., Angiogenesis in wound healing, *Pharmacol. Ther.*, 52, 407–422, 1991.

101. Ueno, H., Yamada, H., Tanaka, I., Kaba, N., Matsuura, M., Okumura, M., Kadosawa, T., and Fujinaga, T., Accelerating effects of chitosan for healing at early phase of experimental open wound in dogs, *Biomaterials*, 20, 1407–1414, 1999.

102. Cho, Y.-W., Cho, Y.-N., Chung, S.-H., Yoo, G., and Ko, S.-W., Water-soluble chitin as a wound healing accelerator, *Biomaterials*, 20, 2139–2145, 1999.

103. Muzzarelli, R. A., Mattioli-Belmonte, M., Pugnaloni, A., and Biagini, G., Biochemistry, histology and clinical uses of chitins and chitosans in wound healing, *EXS*, 87, 251–264, 1999.

33 Development and Delivery of Anti-Angiogenic Functional Food Products: Opportunities and Challenges

Jack N. Losso

CONTENTS

33.1 INTRODUCTION

Angiogenesis inhibition has been approved for the treatment of cancer and macular degeneration by the Food and Drug Administration in the United States and several other countries including China [1–3]. Pharmaceutical and biomedical companies, as well food scientists and nutritionists, are investigating naturally occurring bioactive compounds in foods also known as "functional foods" as inhibitors of pathological angiogenesis. These compounds work slowly over time and have a history of usage in many rural communities around the world where the incidence of some angiogenic diseases is low [4,5]. But most dietary anti-angiogenic compounds such as genistein, hydroxytyrosol, resveratrol, lutein, isothiocyanates, monoterpenes, Bowman–Birk inhibitor, quinones, terpenoids, lactoferrin, phycocyanins, curcumin, and many others are either secondary metabolites or found in trace amounts in their naturally occurring sources. At the present time, bioseparation and solvent extraction techniques involving the isolation and concentration of the anti-angiogenic compounds are the standard isolation protocols.

33.2 BIOTECHNOLOGY AS A STRATEGY TO DEVELOP ANTI-ANGIOGENIC FUNCTIONAL FOODS

Plant-and animal-derived functional foods are poised to become the next major commercial development in biotechnology. Because of their long history of use by humans, functional foods offer enormous advantages in terms of product safety and acceptability that may not be matched by some synthetic products on the market. Plants represent a safer production system for anti-angiogenic functional foods than animal bioreactors because plants do not harbor human viral pathogens.

However, despite the promised benefits, the development and commercialization of anti-angiogenic functional food products is overshadowed by the uncertain regulatory terrain with regard to regulatory approval, adaptation of good manufacturing practice regulations, intellectual property, and others. Numerous professional organizations, academic research entities, and regulatory bodies such as the National Academy of Sciences, American Medical Association, Food and Agriculture Organization of the United Nations, Council for Agricultural Science and Technology, National Center for Food and Agricultural Policy, Pew Charitable Trusts, Society of Toxicology, International Life Science Institute, and World

Health Organization support the concept of using agricultural and food biotechnology to improve human health. Large-scale production using enzymes, fermentation, in vitro fertilization, tissue culture, or genes isolated from engineered microorganisms or whole cells could promote a new era of anti-angiogenic functional foods. Corn, soybean, cotton, tomato, rice, potato, and rapeseed (grown for canola oil) that have been modified to resist insects or increase herbicide tolerance are being produced from modern biotechnology. It is possible at the present level of knowledge to insert agronomic traits that increase the levels of anti-angiogenic factors in agricultural crops used for human food.

33.2.1 FERMENTATION AND ENZYMOLOGY

Fermented foods are consumed in every country. Fermented Japanese soy sauce contains an inhibitor of angiotensin-converting enzyme I which was identified as N-[N-(3-amino-3-carboxypropyl)-3-amino-3-carboxyl]-azetidine-2-carboxylic acid [6]. Tempeh is a traditional Indonesian fermented soybean food. A method for producing γ-aminobutyric-enriched tempeh using anaerobic incubation of soybean with *Rhizopus* has been reported [7]. This type of tempeh lowers blood pressure in hypertensive rats. Similarly, tempeh-enriched in isoflavone (daidzein, glycetin, and genistein) can be prepared by fermentation by adding soybean germ to soybean cotyledon [8]. Isoflavones isolated from tempeh inhibited cell proliferation and Ets 1 expression in vitro and angiogenesis in vivo using the chorioallantoic membrane (CAM) assay [9].

Miso, a traditional Japanese fermented soybean product contains several anti-angiogenic bioactive compounds such as the isoflavones, terpenoids, and protease inhibitors [10]. Angiotensin converting enzyme I (ACEI) is an angiogenic factor and its inhibition has been associated with the inhibition of angiogenesis progression [11]. Ca-enriched *Enterococcus faecalis* CECT 5728-fermented milk contains high levels of ACEI [12]. Two peptides with sequences, PYVRYL and LVYPFTGPIPN, with potent inhibitory activity against ACEI were isolated in commercial kefir made from caprine milk [13]. Long-term administration of the fermented milk to spontaneous hypertensive rats caused a significant decrease in both systolic and diastolic blood pressure (SBP and DBP), comparable to the drug captopril, but the SBP and DBP increased when the food was removed. Ca-enriched *E. faecalis* CECT 5728-fermeneted milk can be studied as a functional food for individuals with high blood pressure conditions.

Isomers of conjugated linoleic acid (CLA) such as t10,c12-CLA and c9,t11-CLA, which occur as natural products of fermentation in ruminant dairy and beef products, inhibit angiogenesis in vitro and in vivo [14,15]. Diet formulation that leads to increased rumen outflow of vaccenic acid and delta9-desaturase activity, both the substrate and the enzyme that catalyze the biosynthesis of CLA, can increase the content of CLA in milk fat [16].

Consumption of 450 g/d of yogurt-containing live cultures may stimulate IFN-γ synthesis, and the latter inhibits tumor growth by inducing responsive tumor cells to generate anti-angiogenic IFNγ-inducible protein-10 (chemokine IP-10) [17]. Certain strains of *Lactobacilli* such as *Lactobacillus casei* strain Shirota,

Lactobacillus rhamnosus, and *Lactobacillus lactis* stimulate IL-12 production; the latter is a strong inducer of NK cells and anti-angiogenic compounds [18,19]. Lactic acid bacteria such as *L. lactis* subsp *cremoris* or *L. lactis* subsp *lactis* can also generate anti-angiogenic bioactive compounds. Yogurt as food can induce the production of IL-10, CD4 + T cells and downregulate the inflammatory response induced by carcinogens. Bioactive peptides from enzymatic hydrolysis in fermented milk proteins include, among others, lactoferricin, casein phosphopeptides, β-lactorphin, and lactokinins [20]. The anti-angiogenic activities of lactoferricin have been established [11]. It is reasonable to speculate that casokinins and lactokinins that are ACEI are potent anti-angiogenic peptides.

Osteoporosis, the thinning of bone tissue and loss of bone density over time, is an angiogenic disease that occurs when the body fails to form enough new bone, or when too much old bone is reabsorbed by the body-or both. The disease affects more than 10 million Americans and the incidence of the disease is on the rise. Atherosclerosis is associated with the thickening or hardening of the coronary arteries and affects more than 10 million Americans. In addition to this group, over half the U.S. population over the age of 50 has an elevated risk of developing osteoporosis [21]. Vitamin K2 is approved as a therapeutic agent for the treatment of osteoporosis [22,23]. Kawashima et al. [23] reported that vitamin K2, a naturally occurring quinone, up to a daily dose of 100 mg/kg of body weight of rabbit prevented both the progression of atherosclerosis and cholesterol deposition in the aorta of hypercholesterolemic rabbits.

The Rotterdam Study examined more than 7500 men and women age 55 years and over who lived in a defined district of Rotterdam. The study assessed the occurrence of diseases of the elderly and sought to clarify their determinants [24]; it followed the subjects in Rotterdam to establish a relationship between dietary intake of phylloquinone (vitamin K1) and menaquinone (vitamin K2) to aortic calcification and coronary heart disease [24]. The study concluded that adequate intake of vitamin K2 was inversely related to severe aortic calcification, whereas no protective effect was observed with vitamin K1 intake. Fermented soybeans (Natto), a traditional Japanese food, contain more than 100 times as much vitamin K2 as various cheeses. Thus it is conceivable that natto soybean could be fermented with *Bacillus subtilis* (Natto) strain MH-1 for the isolation and purification of vitamin K2 to fortify food products such as cheese, salad dressings, chocolate, or other types of food commonly preferred by postmenopausal women and older men. Natto soybean is a specialty soybean for the production of natto and is grown in Arkansas (U.S.) and in Ontario (Canada). Menaquinone is also produced by the intestinal flora, but the level may vary from one individual to another and absorption seems to be limited [25].

The low incidence of breast cancer in Asian women who maintain their native diet has been associated with the high consumption of soy products of which isoflavones such as genistein, daidzein, and glycitein appear to play a major role. However, these isoflavones are poorly soluble in water and less bioavailable. Li et al. [26] used maltosyltransferase from *Thermotoga maritima* to produce transglycosylated daidzin products with improved water solubility and bioavailability.

Kimchi, the Korean fermented cabbage, contains bioactive compounds including beta-sisterol which inhibit the Ras-dependent oncogenic signaling pathway [27].

Doenjang is a soy-based soup consumed daily in Korean households. It contains higher levels of isoflavones than regular soy, tofu or soymilk. Kimchi and doenjang have antihypertensive and anti-angiogenic activities [28].

Lutein and xanthophylls, 3,3'-dihydroxy-α-carotene and 3,3'-dihydroxy-β-carotene, have been identified and recognized by various interdisciplinary studies as the major dietary carotenoids present in the human retina and capable of delaying the onset of macular degeneration [29]. Age-related macular degeneration (AMD) is a major cause of blindness in the Western world. Risk factors include age, family history, exposure to sunlight, complications of diabetes, and high cholesterol levels. There are therapies, but no curative treatments; most AMD patients progress to legal blindness. As standard therapies are limited, there is recognition in the scientific community that nutrition with antioxidants and carotenoids is a meaningful preventative approach against the onset or progression of AMD [30].

Lutein is also a potential anti-angiogenic, anti-cancer and immune enhancer bioactive compound [31,32]. Sweet potato leaves are excellent sources [33]. Corn is an economically viable source of lutein [34]. Microalgae such as *Dunaliella barwadil*, *Chlorella zofingiensis*, *Chlorella protothecoide*, *Muriellopsis* can produce large amounts of lutein and zeaxanthin under specified conditions [30]. Similarly, the green alga *Dunaliella salina*, the blue green alga *Phormidium laminosum*, *Flavobacterium multivorum*, and yeast *Phaffia rhodozyma* accumulate zeaxanthin under specified growth conditions [30].

Solid state (or substrate) fermentation (SSF) offers several biotechnological advantages such as higher fermentation productivity, higher end-concentration of products, higher product stability, lower catabolic repression, cultivation of microorganisms specialized for water-insoluble substrates or mixed cultivation of various fungi, and lower demand on sterility due to the low water activity used in SSF [35]. Solid state (or substrate) fermentation has been used at laboratory and pilot scales for the production of anti-angiogenic functional food such as gallic acid (from tannin-rich mixed substrates using filamentous fungi) and chlorogenic acid (from cranberry pomace) [36,37].

Epidemiological evidence suggests that a high intake of resistant starch and non-starch polysaccharides (NSP) protects against colorectal cancer [38]. The mechanisms underlying the protection of resistant starch against colon cancer are thought to be mediated by the short-chain butyrate, which is present in the colonic lumen in millimolar concentrations as a result of bacterial fermentation of carbohydrates that have resisted digestion in the small intestine. The molecular mechanism underlying the anti-angiogenic activity of butyrate included the inhibition of HIF-1α protein levels, increased apoptosis, reduced proliferation, anti-inflammatory activity, enhanced immunosurveillance, and decreased levels of angiogenic factors such as vascular endothelial growth factor (VEGF) mRNA and matrix metalloproteinases [39–41]. However, the intestinal flora may vary from one individual to another. Individuals who are physiologically impaired may not be able to produce these bioactive compounds de novo. The bioavailability of butyrate was reported to be very low, suggesting that these compounds may be good for local colonic health. Large scale controlled bacterial fermentation may help produce these short chain fatty acids for incorporation into food formulations.

33.2.2 Recombinant and Genetic Engineering Technologies

Chlorogenic acid is a major hydroxycinnamic acid in potato, sunflower seeds, tomato, eggplant, apples, and coffee. It scavenges free radicals, protects LDL against peroxidation, inhibits carcinogenesis in vitro, stimulates the nuclear translocation of NF-E2-related factor (Nrf2) and the induction of GSTA1 antioxidant response element (ARE)-mediated GST activity, and inhibits DNA damage in vitro [42]. Because of its high bioavailability and potency, overexpression of the enzyme hydroxycinnamoyl-CoA quinate:hydroxycinnamoyl transferase that synthesizes chlorogenic acid in tomato caused transgenic tomato to accumulate a two-fold increase in soluble chlorogenic acid with no side effects on the levels of other phenolics in the tomato [43].

Because of the potential health enhancing activities of flavonoids, there is increasing interest in developing alternative food sources rich in flavonoids. Tomatoes (*Lycopersicon esculentum* Mill.) including a wide variety of processed tomato food products (e.g., ketchup, pasta sauce, tomato puree, etc.) are one of the major vegetables in human diets. Tomato contains trace amounts of naringenin, rutin, and kaempferol in other parts of the plant such as the peel and leaf tissues. High levels of quercetin accumulated in tomato fruit when accession LA1926 for *Lycospericon pennellii v. puberulum* was crossed with cultivated *L. esculentum* [44]. Germplasm of crops that express genes of the flavonols biosynthesis in the fruit flesh can be surveyed for the high traits to create a platform for the production of a high flavonoid tomato.

Plants are a major source of vitamins needed for human health. Recent findings show that vitamin K2 deficiency is associated with atherosclerosis and many other angiogenic diseases [45]. As the technologies to design transgenic plants with high levels of vitamin A, C, and E already exist, metabolic engineering for plants with high levels of vitamin K may be desired in the near future. Coenzyme Q10 is used for the treatment of certain cardiovascular conditions and has been shown to delay Parkinson's disease [46]. While low levels of coenzyme Q10 can be obtained from animal products, including cheese and chicken legs, high levels of coenzyme Q10 can be obtained from soybean by fermentation [47] and metabolic engineering [48].

Omega-3 fatty acids, naturally occurring nutrients of proven health benefit to infants and adults, inhibit angiogenesis in vivo [49,50]. cDNA clones encoding the enzyme that synthesize stearidonic acid, $\Delta 6$, $\Delta 12$, and $\Delta 15$ fatty acid desaturases were characterized from commercially grown fungus (*Mortierella alpina*) and canola (*Brassica napus*). Overexpression of the fatty acid desaturases in canola caused seed oils to accumulate stearidonic acid up to 23% of the oil by weight, bringing the total omega-3 content in the seed lipids to over 60% of the fatty acids [51]. Omega-3 fatty acids could become an alternative or an adjuvant treatment for hemangiomas by slowing down their rapid proliferation phase through anti-angiogenic and anti-tumoral effects [49].

Conjugated linoleic acid has a number of health benefits such as anti-adipogenic, anti-carcinogenic, anti-atherogenic, and immune modulation effects in animal models, which are all related to angiogenesis. *Propioniumbacterium acnes* has been cloned in Eischeria coli to produce pure 10,12-CLA isomers in large quantities [52].

Lactoferrin and its fragments lactoferricin inhibit angiogenesis [53]. Transgenic mammary glands for the delivery of high content of lactoferrin in whey are needed.

33.2.3   ANTI-ANGIOGENIC FOOD PRODUCT DEVELOPMENT: OPPORTUNITIES AND CHALLENGES

33.2.3.1   Opportunities

Carmeliet [54] suggested that angiogenesis is implicated in more than 70 diseases with a worldwide potential market of 500 million patients. The opportunity to develop anti-angiogenic functional foods already exists and some products have been suggested [5,55,56]. It is well known that most consumers prefer taste over nutrition. To bring anti-angiogenic functional foods on the market, we need to establish safety and dose-response of anti-angiogenic functional foods, provide adequate processing and storage conditions, address children with bitter foods as early as possible in their lives, reformulate food to contain anti-angiogenic bioactive compounds, remove proangiogenic compounds from our foods, change the way we do sensory evaluation, and involve medical doctors so that they can prescribe functional foods.

The efficacy of anti-angiogenic functional foods may be realized by individuals who understand that food should be consumed in its matrix rather than picking and choosing individual bioactive compounds. Most "good for you anti-angiogenic foods" are bitter. For anti-angiogenic functional foods to make a successful inroad into the food chains in the western nations, it will require an early adaptation to the taste of anti-angiogenic functional foods. People will have to appreciate and cope with the bitterness of functional foods rather than circumventing it, because very often it is the bitter compound that is the bioactive principle in the concoction. However, the food industry in general has an aversion to high-risk research and development, and often prefers to rely on familiar technologies when introducing new products knowing that only a few will likely succeed. Functional food development is a long-term effort that requires a high cost and early usage. A typical example of industry impatience with a long-term shot is the removal of Intelligence Cuisine-a range of clinically-proven functional foods for diabetes and other specific health problems introduced by Campbell Soup in 1997–1998 and pulled out within a year because the business expectations were not met. Now, it appears that Campbell Soup was simply ahead of many other manufacturers-and functional foods for diabetes and other specific diseases are already on the U.S. market.

33.2.3.2   Challenges

Although an increased intake of a specific food may improve health status in some cases, chronic consumption of large amounts of one specific food may be detrimental. Ethyl carbamate, which is mainly present in fermented foods and beverages, can be found at levels between 0 and 16.2 ppb in kimchi [28]. Considering the large daily intake of kimchi in Korean families, this may be a problem in long-term uses at high levels. Downregulation of angiogenic stimulators may have several downstream negative implications including reduced VEGF activity and angiogenic activity associated with vessel growth; so care should be exercised for ingestion of anti-angiogenic functional foods in cases like pregnancy or even wound healing. Feeding 200 mg of chocolate rich in catechins and theobromine to 2-month old Balb/c mice (an equivalent of 200 g of chocolate per person) decreased the relative length of limbs and thigh bones in four-week old progeny and decreased VEGF content of offspring femoral bones [57]. This is a challenging area for food scientists and an opportunity for medical doctors and pharmacologists to work with food formulators so that 2020 vision that includes healthy people can be met by all.

33.3   DELIVERY OF ANTI-ANGIOGENIC FUNCTIONAL FOODS

33.3.1   INTRODUCTION

Oral ingestion remains the major route for consuming functional foods and nutraceuticals. Functional foods like any other food material ingested must be dissolved, transferred to the stomach for digestion, pass the gut wall for absorption, enter the portal circulation, and pass through the liver before gaining access to the systemic circulation. Despite the advantages associated with oral delivery of bioactive compounds, which include easy compliance, ingested functional foods have to overcome several physical, biochemical, and efflux barriers before reaching their target organs or tissue.

The first such biochemical hurdle involves the saliva in the mouth. The biochemical barriers can be primarily attributed to digestive enzymes and other reagents in the saliva that breakdown whatever is introduced in the mouth. The second and greatest biochemical barrier encompasses the acidic conditions and the proteolytic enzyme pepsin in the stomach. The pH of an empty stomach is between 1.4 and 2.1, and, in the presence of food, the pH value can increase to nearly 4.0. In the stomach, unprotected proteins are denatured or degraded; some become useless while others, through their peptides, may become bioactive compounds for disease prevention. The pancreas releases serine endopeptidases (trypsin, α-chymotrypsin, elastase) and exopeptidases (carboxypeptidase A and B) which further the proteolysis of polypeptides.

After the stomach, the food reaches the small intestine. The entire small intestine is 5 m (15 ft.) and ingested material typically resides in the small intestine between 2 and 4 h. Enzymes in the brush border membrane, cytoplasm, and lysosomes also contribute to the pre-systemic degradation of peptides. Ingested peptides and proteins are metabolized in the intestine to be absorbed as amino acids. However, it is desirable for most protein or peptide functional foods to be transported intact to the systemic circulation and target tissue.

The third biochemical hurdle is the hepatic first-pass metabolism degradation. Physical hurdles include: the mucus layer of the epithelial layer, the phospholipid bilayer of the plasma membrane, and the tight junctions. The mucus layer may restrict access of large molecules such as peptides and proteins to the epithelial surface. The phospholipid bilayer of the plasma membrane allows the transport of lipid-soluble molecules by passive diffusion, but prevents the movement of large hydrophilic molecules across the epithelial cells. The tight junctions are selectively permeable to certain small hydrophilic molecules (ions, nutrients) and control diffusion of solutes through the paracellular route while forming a barrier to the transport of large molecules through the intercellular spaces. The efflux systems which include P-glycoprotein (P-gp) in combination with intracellular metabolism may also reduce the bioavailability of certain bioactive dietary compounds such as peptides.

Other barriers include the blood–brain and retina barriers which restrict certain molecules from entering the brain and eye. The limitations associated with the hurdles mentioned above result in low bioavailability and a short half-life that require frequent ingestion of the bioactive compound. Also, because of the many biotransformations such as deglycosylation, glucuronization, and sulfation that the bioactive compound may undergo in the GI tract or the liver (hepatic degradation) and which may result in bioactivation or loss of bioactivity, in vitro data may not correlate with in vivo data. Therefore, one possible approach to improve the bioavailability of functional foods while maintaining a high level of efficacy against the target (as seen in vitro) is to facilitate and enhance the delivery of bioactive compounds to the circulation by bypassing the barriers mentioned above.

Similarly, efforts must be made to enhance the absorption of bioactive compounds that require biotransformation before being active. All this should take effect when the safety of bioactive compounds at levels higher than physiological levels has been established with in vivo data in humans. The transmucosal and intestinal (oral and gut mucosa) routes are considered for functional food delivery and are discussed below.

Among the transmucosal routes, the oral mucosa offers an attractive option for systemic delivery of functional foods. The oral mucosa consists of: (i) the sublingual area which includes the floor of the mouth below the tongue; (ii) the buccal area which involves the lining of the cheeks, the gingival, and palates; and (iii) the oral area which includes the palate, gingival, and surface of the tongue-and the membrane lining the roof of the mouth. Within the adsorptive surfaces of oral mucosa, the buccal and gingival mucosa and cheek linings are considered for enhanced systemic delivery of functional foods. Various factors including the pathophysiological state and surface area of the mucosa as well as the properties of both the bioactive compound and the delivery system can increase or decrease the efficacy of the buccal delivery.

The advantages of the oral mucosa for systemic delivery of functional foods include: (i) large surface area of the buccal mucosa; (ii) some highly permeable and extensively vascularized networks; (iii) the ability of the retromandibular veins to drain oral mucosa directly into the internal jugular vein while avoiding first-pass clearance; (iv) good blood flow; (v) the ability to rapidly remove the bioactive compound should the food be health threatening; (vi) the mucosa is both hydrophilic and lipophilic in nature; (vii) an attractive route for functional foods that undergo first-pass effect, such as many phenolics; and (viii) an attractive alternative for individuals who experience nausea, vomiting, dysphagia or gastrointestinal discomfort [58].

Salivary glands secrete mucus that is negatively charged at physiological pH of 5.8–7.4.

33.3.2 ABSORPTION ENHANCEMENT OF FUNCTIONAL FOODS THROUGH BUCCAL MUCOSA

The presence of hydrophilic and lipophilic regions in the oral mucosa suggests the existence of two routes for the absorption of functional foods through the buccal mucosa: the paracellular (between the cells) for hydrophilic molecules, and the transcellular (across the cells) for lipophilic molecules (Figure 33.1). To deliver a broad range of desirable functional foods such as polypeptides and polysaccharides across the buccal mucosa, transmucosal and transdermal penetration enhancers that reversibly and safely alter the permeability restriction of the buccal mucosa may be employed to overcome its barrier properties. These large molecules need to bypass the hepatic first-pass metabolism and degradation in the intestine. The absorption enhancers must be non-toxic, reactively inert, non-irritant, non-allergenic, and capable of reversing the tissue to its normal integrity upon removal of the absorption enhancer. Absorption enhancers include bile salts, surfactants, fatty acids and derivatives, chelators, inclusion complexes, nanoparticles, and other naturally occurring compounds. Technologies employed to improve oral absorption of functional foods may include complexation, solid dispersion techniques, micro-emulsions, and nanoparticles. The mechanism of action including the advantages and limitations of each absorption enhancer (technology) with respect to functional foods are presented below.

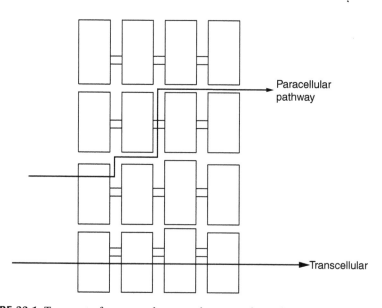

FIGURE 33.1 Transport of compounds across the mucosal membrane.

33.3.2.1 Complexation

There are 3 natural cyclodextrins (CDs), α-, β-, and γ-CDs containing 6, 7, or 8 D-(+) glucopyranose units attached by α-[1,4] glucosidic bonds. Cyclodextrin with 9 glucose units (δ-CD or cyclomaltose) has been reported [59]. The MW of α-, β-, γ-, and δ-CDs are 972, 1135, 1297, and 1459 Da, respectively. Cyclodextrins use their inclusion cavities to retain small hydrophobic molecules, mask the bitterness of bioactive compounds, and enhance their bioavailabilty. The solubilities (g/100 ml) of CDs are 14.5, 1.85, 23.2, and 8.19 for α-, β-, γ-, and δ-CDs, respectively. The inner cavities of CD are lipophilic while the outer surfaces are hydrophilic. Cyclodextrins form noncovalent inclusion complexes with guest molecules. The cavity of α-CD (4.7–5.3 Å) is insufficient for many bioactive compounds and this CD is undergoing regulatory approval in the U.S.A. β-CD (inner cavity, 6.0–6.5 Å) is poorly water-soluble and nephrotoxic. γ-CD (inner cavity, 7.5–8.3 Å) is expensive. Both β-and γ-CDs have GRAS status in the U.S. All CDs are approved in Japan, and β-CD is a food additive in Europe. In general, nonionic molecules are better complex forming agents than ionic forms. Increased temperature decreases guest/CD complexation due to reduction in van der waals and hydrophobic forces. Spray-drying (vitamin D_2 complexed with β-CD) and freeze drying are common methods for complexation [59] but are less efficient. Chemically modified CD derivatives (such as hydroxypropyl-CD, carboxymethyl-CD, hydroxyethyl-CD, and sulfobutyl-CD) are commercially available and most of these CD derivatives are approved for use in most countries.

    Cyclodextrins, because of their ability either to complex guest molecules or to act as carriers, can serve as efficient and precise delivery agents for functional foods to targeted sites for a necessary period of time. The application of CDs in oral functional food delivery involves increased absorption via buccal and sublingual mucosa, taste masking, enhanced bioavailability and permeability by increasing the free functional food availability at the absorptive surface, resulting in high plasma concentration of the functional food. At the mucosa surface, CD makes the guest functional food available for partitioning into the membrane without disrupting the lipid layers of the barrier. Inclusion complexation with a lipid of human erythrocytes, perturbation of membrane integrity, and ability to remove cholesterol were suggested to contribute to CD-induced increased bioavailability [59]. For oral delivery, all CDs are considered as nontoxic due to lack of CD absorption through the gastrointestinal tract [59]. Hydrophilic derivatives of β-CD improve the aqueous solubility and dissolution rate of poorly soluble drugs and are used in immediate release-type formulation. Hydrophobic CD derivatives retard the dissolution rate of water-soluble guests from vehicles and are used in prolonged release-type formulation.

33.3.2.2 Chelators

Chelators interfere with divalent cations. EDTA, sodium citrate, and chitosan are among potential absorption enhancers for functional foods. At acidic pH, chitosan is a positively charged polymer which enhances bioavailability by ionic interactions with negative charge on the mucosal surface. Because of its bioadhesive nature, chitosan can increase the retention and bioavailability of bioactive compounds at the buccal mucosa surface. However, chitosan is not FDA approved for use in foods. Other positively charged (cationic) compounds include poly-arginine and lysine which act by the same mechanism as positively charged polymers.

33.3.2.3 Surfactants and Other Permeation Enhancers

Several dietary bioactive compounds such as curcumin are poorly water-soluble and bioavailable. Incorporation of the lipophilic compounds into surfactants such as polysorbate 80, lecithin, and lysolecithin may improve the oral absorption of poorly water-soluble functional foods. Surfactants may enhance the absorption of functional foods through buccal mucosa by perturbation of intercellular lipids and protein domain integrity. Several anionic (sodium laurate), cationic (cetylpyridium chloride), and nonionic (span, fatty acid esters) surfactants are available as absorption enhancers. Lipid extraction and interactions with proteinaceous domains are the main mechanisms by which surfactants and bile salts improve buccal permeability. Enhancement in permeability may be achieved at a concentration above the surfactant critical micelle concentration (CMC) where the surfactants extract the mucosal lipids to affect its permeability characteristics [60]. However, it appears that the lipophilicity and permeation pathway of the molecule play a significant role in enhancing its passage through the buccal mucosa. Surfactants can also enhance the permeability of polar molecules such as caffeine that would traverse the buccal mucosa through the paracellular route. Some surfactants would cause irritation of the buccal mucosa and may not be recommended for functional foods. Bile salts aid in the absorption of lipids and lipid soluble vitamins. The most abundant of the bile salts in humans are cholate and deoxycholate that are conjugated with either glycine or taurine to give glycocholate or taurocholate, respectively. Bile salts, for instance, may cause irritation and damage of mucosal tissue, but the effects may be reversible within 24 h [61]. Whereas their use in drug absorption has been extensively investigated and approved, their use for functional food delivery needs to be studied.

Fatty acids and their esters (oleic, caprylic, lauric, 5% cod liver oil extract, caprate, and fatty esters of sucrose) disrupt lipid bilayers in the buccal membranes, denature proteins, inactivate enzymes, swell tissues, increase the fluidity of phospholipids domains, and, as a result, enhance the permeation of lipophilic compounds. Nonionic surfactants such as fatty acid esters are less irritating to buccal mucosa than are bile salts or ionic surfactants [58]. A synergistic buccal permeation enhancement is observed when surfactants are combined with fatty acids and their esters are combined with polyethylene glycol 200. Caffeine permeates the buccal mucosa predominantly through the paracellular route. Karaya gum proved to be better than guar gum, but concentrations greater than 50% w/w may be required to provide zero-order suitable sustained release [62].

Other buccal permeation enhancers include menthol, cetyl pyridinium chloride, aprotinin, limonene, sulfoxides, lecithin, and lysolecithin, but their long term safety needs to be established. Chewing gums, gummy bears, bioadhesive tablets, and candies form an intimate contact with the buccal mucosa and offer tremendous opportunities for the delivery of functional foods through the buccal mucosa. Stay Alert®, a caffeine chewing gum, is available on the market to help increase alertness in some people.

33.3.3 Colonic Delivery of Anti-Angiogenic Functional Foods

33.3.3.1 Polysaccharide-Based Encapsulation

Encapsulation techniques using polysaccharides for delivery of bioactive compounds for colon-based diseases have been reported [63]. Pectin, chitosan, guar gum, dextrans, alginates, CDs, inulin, starch/amylase, and chondroitin sulfate have all been suggested for the delivery of functional foods to the colon to circumvent the acidity of the stomach. Modified pectins such as methoxylated pectin, Zn-pectate, and pectin–chitosan complex were the most promising from this polysaccharide [63]. Chitosan is soluble at acidic pH. Chitosan–tripolyphosphate hydrogel beads, chitosan salts such as glutamic and aspartic salts, chitosan–pectin, and chitosan–locust bean gum complex have been designed and proven to enhance colonic delivery and, in the case of chitosan locust bean gum complex, enhanced bioavailability was observed [64].

Guar gum, from the seeds of *Cyamopsis tetragonolobus*, has shown excellent biocompatibility for colonic functional food delivery. Bioactive compounds in guar gum layered tablets, guar gum crosslinked with trisodium trimetaphosphate or guar gum complexed with pectin, sustain the acidity of the stomach and can be released in the colon in the presence of galactomannase [63]. Dextrans from *Leuconostoc mesenteroids*, dextran-podophyllotoxin, and dextran sulfate are water soluble and can slowly release bioactive compounds against colon cancer [63]. Calcium-alginate-lactose or calcium-alginate-lactose-chitosan complex can have a controlled induction and release of bioactive compounds in the colon [65]. Cyclodextrins and CD-conjugates can be designed for the release of anti-colon cancer bioactive compounds. Resistant starch is supposed to provide additional butyrate for colon health, provided that the individual is physiologically capable of hydrolyzing the starch. Enzymatically or chemically modified resistant starch such as esterified, acetylated, succinylated, or carboxymethylated can deliver bioactive compounds to the colon. Amylose–ethylcellulose complex can provide controlled released of bioactive compounds in the gut. Controlled crosslinked chondroitin sulfate is another food grade carrier for colonic delivery of functional foods that warrants investigation [63]. Whereas these polysaccharides can mostly deliver to the colon, most of them do not deliver into the circulation and would not be a good choice for systemic delivery of functional foods. Liposomes containing lactoferrin were a more effective anti-inflammatory than conventional lactoferrin in the intestinal site, which is regarded as an active site of orally administered lactoferrin [66].

33.3.3.2 Cyclodextrins

Cyclodextrins are slightly absorbed in the upper gastrointestinal tract, but are fermented into small saccharides by colonic microbial flora. Cyclodextrins are good for delayed release colonic functional food delivery that may alleviate any potential side effect that may result from the release of a large amount of bioactive compound at the site of action. Cyclodextrins can also be used as excipients for functional foods. Most functional foods are bitter and bitterness increases tremendously when these compounds are concentrated for enhanced bioactivity. Green tea tablets and fenugreek capsules are bitter, as are many other compounds good for health. Cyclodextrins can be used to mask their taste in solutions. Among the CDs, β-CD is better than γ-CD or α-CD in suppressing bitter taste [67].

33.3.4 INTESTINAL ABSORPTION ENHANCEMENT AND BIOAVAILABILITY OF FUNCTIONAL FOODS

33.3.4.1 Introduction

While the pharmaceutical industry invests great sums of money in the development of drugs that reach their target, but which are often beyond the gastrointestinal tract, the food industry has for many years settled on the idea of "from farm to table" instead of "from farm to the gut and beyond." As a result, funding for food bioavailability is abysmal. The topic of bioavailability has only received attention in recent years and funding is still not adequate.

Systemic delivery of functional food penetration through the oral mucosa can be categorized into: (i) paracacellular, for hydrophilic compounds; (ii) transcellular, for lipophilic bioactive compounds, and this transport can be subcategorized into passive and carrier-mediated components; and (iii) carrier-mediated transporters, a transport mode used by amino acids, nicotinamide, thiamine, glutathione, and glucose co-transporters such as SGLT1, GLUT1, GLUT2, and GLUT3 [68]. However, the oral mucosa forms a barrier to entry of foreign substances and can hinder the absorption of desirable functional foods.

Direct targeting to the systemic circulation requires that the bioactive compound is absorbed through the gastrointestinal tract. Hydrophilic bioactive compounds of medium to high molecular weight have poor bioavailability. Their transport across the intestinal barrier occurs through the paracellular pathway that occupies less than 0.1% of the total surface area of the intestinal epithelium and the presence of tight junctions between epithelial cells [69,70]. Reduced membrane permeability is the main cause of poor bioactive compound bioavailability. Technologies to increase intestinal absorption would enhance functional food bioavailability and probably reduce inter-and intrasubject variability in plasma concentrations, and thus enhance their preventative effects. The following technologies can enhance bioactive compounds absorption. Surfactants have been described in the previous section on buccal delivery.

33.3.4.2 Medium Chain Fatty Acids

Medium chain fatty acids such as sodium caprylate and caprate enhance the permeability of hydrophilic compounds by decreasing lipid membrane fluidity or interacting with the hydrophilic domains of membrane and by inducing dilations of tight junctions [69]. Lipoidal adjuvants that include the sodium salts of medium-chain fatty acid, such as sodium caprylate (C8), caprate (C10), and laurate (C12) were shown to be good absorption enhancers for bioactive compounds having the molecular weight of about 10,000 Da, with C12 being the most effective enhancer [71]. When evaluating the bioavailability of functional foods, the effect on Cmax and the area under the curve (AUC) are the indicators most commonly measured. Glycyrrhizin bioavailability was enhanced 60-fold [72].

33.3.4.3 Liposomes

Liposome-based systems are used to enhance efficacy or ameliorate the toxicity of certain drugs. Liposomes designed for long circulation times combine stable encapsulation of the active compound within the liposome and surface modified formulations containing polyethylene glycol that allow liposomes to target diseased tissues while resisting recognition and uptake by reticuloendothelial system cells [73]. Phospholipid-based liposomes may assist in the formulation of poorly-soluble bioactive compounds, provide a slow-release vehicle to achieve pharmacokinetic profiles that maximize the preventative effect of the bioactive compound, or behave as long-circulating nanoparticulates that allow the bioactive compound to accumulate in the disease tissue. The colloidal nature and foreignness of the lipid microspheres enable these vesicles to be dispatched to the liver as soon as they get into systemic circulation. Silymarin, the major source of the bioactive flavonoid silibin, was complexed with phosphatidylcholine and used against human ovarian cancer [74]. Free silibin levels were found to be 7.0 ± 5.3 μg/ml and 183.5 ± 85.9 ng/g tissue in the plasma and tumor samples, respectively. Liposomal curcumin suppressed growth, and induced apoptosis of human pancreatic cells in vitro and in vivo by down-regulating the NF-κB machinery, providing a biochemical rationale for the treatment of patients suffering from pancreatic carcinoma with encapsulated curcumin for systemic delivery [75]. But most conventional liposomes (a combination of phospholipids with or without cholesterol, and possibly other lipids) are rapidly cleared from the circulation by macrophage cells in the liver, spleen, and lung

Mixed micelles containing bile salts, and fatty acids can enhance the absorption of poorly water-soluble bioactive compounds to a greater extent relative to nonmicellar and simple micellar systems [76].

33.3.4.4 Self-Emulsifying Delivery Systems

Several dietary bioactive compounds such as curcumin are poorly water-soluble and bioavailable. Incorporation of the lipophilic compounds into lipid vehicles may improve the oral absorption of poorly water-soluble functional foods. A self-microemulsifying delivery system is an isotropic mixture of an oil, surfactant, co-surfactant (or solubilizer), and the bioactive compound of interest. Olive oil or medium chain triglyceride (as oils), polyoxyethylene [20] sorbitan coconut oil ester (as surfactant), and propylene glycol (as co-surfactant) can be used to formulate microemulsions containing functional foods. A microemulsion is a thermo-dynamically stable solution of 1–100 nm droplets composed of water, oil, and surfactant. The microemulsion can be ingested as a soft capsule containing an oil solution of a poorly-soluble bioactive compound and a surfactant. The principle of a self-emulsifying delivery system is its ability to form nanosize fine oil-in-water or water-in-oil microemulsions under gentle agitation in an aqueous environment such as the digestive motility of the stomach and upper small intestine; it mimics the bile salt micelles, thereby enhancing the absorption of the bioactive compound. The micellar properties of the microemulsions, the nano range size of the droplets, and the very high surface area to volume ratio enhance the bioavailability and plasma concentration of the poorly water-soluble bioactive compound [77–79]. Knowledge of the bioactive compound solubility in various components of the mixture, the area of self-emulsifying region in a phase diagram, and droplet size distribution following self-emulsification are important factors when formulating a self-emulsifying delivery system.

33.3.4.5   Lectin- and M Cells-Mediated Absorption

Lectins, such as lectins from tomato fruit (*L. esculentum*) have been touted as potential absorption enhancers by endocytosis through cytoadhesison [80]. Similar lectins which compete with bacteria for receptors on the GI tract are found in garlic [81]. Lectins, which bind to certain sugars on the cell membrane, can increase bioadhesion and functional food absorption.

M cells have a high transcytotic capacity for mucosal delivery of health enhancing bioactive compounds [82] to underlying lymphoid cells. Liposomes can be delivered to M cells. Lectins have been touted as potential M cell delivery vehicles but additional knowledge is needed to investigate the benefits of mucosal delivery of functional foods via the M cell route.

33.3.4.6   Protein Transduction Domain

Protein transduction domains are small proteins characterized by the presence of numerous positively charged lysine or arginine residues. Protamine, a 35 amino acid protein, is the best food protein that can be used as the source of a protein transduction domain. Protamine has a repeat of eight arginine in a sequence that makes this protein a valuable source for the delivery of macromolecules inside the cytoplasm [83].

33.3.4.7   Nanoparticles

Nanoparticles have been recognized as stable systems suitable for providing targeted drug delivery and enhancing the efficacy and bioavailability of poorly soluble drugs. However, nanoparticles have low loading and limited entrapment efficiency which limit their efficacy. Inclusion complexation of bioactive compounds with CDs, such as β-CD or HP-β-CD, prior to its entrapment in nanospheres increased the drug loading into the nanospheres and its subsequent delivery to target tissues [59]. The safety of nanotechnology in human needs to be established. Non-targeted nanoparticles can stick to the microvasculature of the liver and spleen [84]. Nanoparticles are double-edged swords that need to be managed closely because they are more toxic than larger particles of the same materials, and animal studies have shown an increase in lung inflammation, oxidative stress, and inflammatory cytokine production [84]. Hazards and exposure data are lacking. There are no real regulations covering the development of nanoparticles. The size and surface properties of nanoparticles are very important in understanding their toxicity. It is not known if these particles can cross the blood–brain barrier and cause damage to the brain.

33.3.4.8 Protease Inhibitors

P-glycoprotein is a 170 kDa inducible glycoprotein transporter that belongs to the family of ATP-binding cassette (ABC) transporter proteins. P-glycoprotein is found in normal human tissues, such as the gastrointestinal tract, liver, kidneys, placenta, testis, and blood–brain barrier where it serves as an efflux pump for the absorption, distribution, and excretion of many bioactive compounds including drugs. The protein is also responsible for the excretion of toxins from tissues [85]. P-glycoprotein is involved in herb–drug interactions and its modulation can enhance bioavailability. For instance, interaction with garlic-warfarin, ginseng-warfarin, or ginko-aspirin may lead to spontaneous bleeding [86]. In cancer cells, the overexpression of P-gp, after disease progression following chemotherapy, is considered one of the major obstacles to successful cancer chemotherapy. Because of its ability to exclude cytotoxic drugs, efforts are being made to find safe and reversible inhibitors of P-gp in cancer treatment. Several functional foods such as curcumin, ginsenosides, bergamottin, epigallocatechin gallate, silymarin, quercetin, piperine, vincristine, vinblastine, and other alkaloids have been identified as modulators of P-gp and their use can enhance absorption [87]. Functional foods that inhibit P-gp and tumor cells should be studied because of their ability to prevent resistance and inhibit tumor growth. Cytochrome P450s, particularly CYP3A4, are also a key to altered systemic drug delivery. Grapefruit contains epoxybergamottin which is an inhibitor of CYP3A4, and, as a result, it enhances the absorption of drugs that are subject to first-pass intestinal/hepatic metabolism-and this may lead to serious side effects. Functional food inhibitors of CYP3A4 may enhance plasma and tissue concentration of the same and other functional foods. Similarly, bioactive compound inhibitors of multiple resistance proteins (MRPs) would enhance bioavailability and tissue concentration of functional foods that may lead to toxicity. Whereas inhibition of P-gp, MRPs, and CYP3A4 may lead to toxicity, stimulation of these efflux proteins may lead to decreased bioactive compound efficacy.

Piperine, an alkaloid (1-peperyl piperidine), is a vanilloid naturally occurring at a concentration between 1 and 9% in *Piperaceae* [88]. Co-ingestion of bioactive compounds such as coenzyme Q10, theophylline, beta-carotene, vitamin B6, vitamin C, and selenium by mouth with piperine at a concentration as low as 5 mg significantly increased plasma concentrations of each one of these compounds without inhibiting xenobiotic drug metabolizing enzymes [88]. A similar increase in bioavailability was observed when epigallocatechin-3-gallate was orally adminis-tered to mice in the presence of piperine [89]. At a high dose, such as 50 μmol/L, the mechanism of enhanced absorption includes inhibition of xenobiotic metabolizing enzymes such as P-gp or cytochrome P450 [90]. At a low dose, such as 5 mg of piperine, the mechanism by which piperine enhances the serum levels of nutrients and non-nutrients appears to be non-specific and localized directly in the gastrointestinal tract, and may include increased micelle formation, blood supply, and modification of the epithelial cell wall.

33.3.5 DIRECT TARGETING OF ENDOTHELIAL CELLS

Endothelial cells that line tumor vessels selectively express a high level of a number of markers that are characteristics of angiogenesis and are good targets for angiogenesis inhibition. The markers include VEGF, certain integrins ($\alpha v \beta 3$, $\alpha v \beta 5$, and $\alpha 5 \beta 1$), aminopeptidase N, NG2 proteoglycan (specific to pericytes), PDGF-β receptor, and anionic phospholipids [91,92]. Targeting VEGFR is insufficient to permanently halt angiogenesis in vivo [2], but sufficient to decrease interstitial fluid pressure and produce a functionally normalized vascular network [93]. Angiogenesis inhibition can also be achieved by targeting pericytes or mural cells that form an outer sheath around the endothelium, promote vessel stabilization, and are intimately involved in angiogenesis [94]. PlGF appears to be a key player in pathological angiogenesis but not physiological angiogenesis [95]. Combined inhibition of PDGFRβ, PlGF, RTKIs, and VEGFRs can lead to the successful inhibition of angiogenesis. Functional food inhibitors of the markers mentioned above are needed.

33.3.6 DIRECT TARGETING OF INTRACELLULAR TISSUES

Successful inhibition of pathological angiogenesis will depend on the ability of inhibitors to effectively block multiple angiogenic pathways. Direct targeting of intracellular tissues refers to the direct delivery of functional foods to specific organelles (organ tropism) and compartments (cytoplasm, endo-lysosomes, mitochondria, and nucleus) within the cell. By delivering a bioactive compound at or near its receptor site, one can achieve a higher bioavailability of the functional food at the site of action, better health enhancing effect, and reduced unwanted side effects. Intracellular delivery of low molecular weight bioactive compounds is beset with low intracellular retention. Low molecular weight lipophilic molecules can diffuse across the cell, but they tend to diffuse back or efflux out of the cell rapidly either when the concentration gradient outside the cell is removed or due to the presence of transporters such as P-gp or MRP. Macromolecules have poor permeation and can enter the cell through endocytosis; once inside the cell, these macromolecules are delivered to the lysosomes where they are degraded.

  Liposomes and nanoparticles can penetrate into phagosomes or lysosomes. When the site of action is located in the cytoplasm, such as the receptors for terpenoids, bioactive compounds such as diosgenin, ginsenoides, and yamogenin should be delivered to and target the cytoplasm. Most macromolecules that enter the cell are delivered to the lysosomes where they are mostly degraded. It is important for the carrier of the bioactive compound to deliver it to the cytoplasm and not the lysosomes.

Mitochondrial targeting is important for the delivery of antioxidant molecules to mitochondria to prevent oxidative damage and the delivery of compounds that target mtDNA mutations which cause neural or muscular degenerative diseases, obesity, diabetes, and normal aging [96]. The membrane potential of mitochondria is very negative inside the organelle (-130 to -150 mV). Many cancer cells have even higher mitochondrial membrane potential than non-transformed cells. These properties of cancer cell mitochondria are being used to selectively target the mitochondria of cancer cells with compounds such as lipophilic cations or protein transduction domains [96]. Lipophilic cations, such as cationic lipids, and protein transduction domains, such as low molecular weight protamine peptide, can be used to kill the mitochondria of cancer cells and carry antioxidants and other health enhancing molecules to the mitochondria. Nuclear targeting occurs through the pore complex (NPC) which is present in the nuclear membrane [96]. The NPC is permeable to molecules smaller than 40–45 kDa. Nuclear targeting of functional foods can block the action of a nuclear protein from reaching its target cellular compartment. For instance, functional foods such as ellagic acid, genistein, diosgenin, capsaicin, and quercetin that inhibit the nuclear export of the transcription factor NF-kappaB are promising anti-cancer bioactive compounds [4]. Well designed nanoparticles can penetrate the lysosomes, nucleus, and mitochondria and have the potentials to deliver anti-angiogenic functional foods to these organelles.

33.4　CONCLUSION

Anti-angiogenic functional foods have been identified. Most of the bioactive compounds are found in very small concentration in their natural sources. Fermentation, enzymology, and recombinant and genetic engineering technologies can assist in large scale production of these compounds for human use. Anti-angiogenic functional foods can be delivered through buccal mucosa or gastrointestinal mucosa using naturally occurring absorption enhancers that reversibly and safely open or inactivate the barriers (paracellular or transcellular pathways, P-gp, CYPs, or MDRs) to absorption. The critical questions to be asked are:

1. Is it possible to consume anti-angiogenic functional foods and achieve levels in plasma and target tissues that are likely to inhibit excessive or insufficient angiogenesis with no side effects?
2. Is it, at the present time, possible to demonstrate the in vivo efficacy of anti-angiogenic functional foods?
3. Is it possible for the medical and pharmaceutical communities to embrace the idea of disease prevention centered around a well-designed and balanced diet?

REFERENCES

1. Folkman, J., Angiogenesis, *Annu. Rev. Med.*, 57, 1–18, 2006.
2. Carmeliet, P., Angiogenesis in life, disease and medicine, *Nature*, 438, 932–936, 2005.
3. Ferrara, N. and Kerbel, R. S., Angiogenesis as a therapeutic target, *Nature*, 438, 967–974, 2005.
4. Aggarwal, B. B. and Shishodia, S., Molecular targets of dietary agents for prevention and therapy of cancer, *Biochem. Pharmacol.*, 7 (10), 1397–1421, 2006.
5. Losso, J. N., Preventing chronic diseases with anti-angiogenic functional foods, *Food Technol.*, XX (6), 78–86, 2002.
6. Kataoka, S., Functional effects of Japanese style fermented soy sauce (shoyu) and its components, *J. Biosci. Bioeng.*, 100 (3), 227–234, 2005.
7. Aoki, H., Furuya, Y., Endo, Y., and Fujimoto, K., Effect of gamma-aminobutyric acid-enriched tempeh-like fermented soybean (GABA-Tempeh) on the blood pressure of spontaneously hypertensive rats, *Biosci. Biotechnol. Biochem.*, 67 (8), 1806–1808, 2003.
8. Nakajima, N., Nozaki, N., Ishihara, K., Ishikawa, A., and Tsuji, H., Analysis of isoflavone content in tempeh, a fermented soybean, and preparation of a new isoflavone-enriched tempeh, *J. Biosci. Bioeng.*, 100 (6), 685–687, 2005.
9. Kiriakidis, S., Hogemeier, O., Starcke, S., Dombrowski, F., Hahne, J. C., Pepper, M., Jha, H. C., and Wernert, N., Novel tempeh (fermented soyabean) isoflavones inhibit in vivo angiogenesis in the chicken chorioallantoic membrane assay, *Br. J. Nutr.*, 93 (3), 317–323, 2005.
10. Minamiyama, Y., Takemura, S., Yoshikawa, T., and Okada, S., Fermented grain products, production, properties and benefits to health, *Pathophysiology*, 9 (4), 221–227, 2003.
11. Losso, J. N., Targeting angiogenesis with functional foods and nutraceuticals, *Trends Food Sci. Technol.*, 14, 455–468, 2003.
12. Miguel, M., Muguerza, B., Sanchez, E., Delgado, M. A., Recio, I., Ramos, M., and Aleixandre, M. A., Changes in arterial blood pressure in hypertensive rats caused by long-term intake of milk fermented by *Enterococcus faecalis* CECT 5728, *Br. J. Nutr.*, 94 (1), 36–43, 2005.
13. Quiros, A., Hernandez-Ledesma, B., Ramos, M., Amigo, L., and Recio, I., Angiotensin-converting enzyme inhibitory activity of peptides derived from caprine kefir, *J. Dairy Sci.*, 88 (10), 3480–3487, 2005.
14. Moon, E. J., Lee, Y. M., and Kim, K. W., Anti-angiogenic activity of conjugated linoleic acid on basic fibroblast growth factor-induced angiogenesis, *Oncol Rep.*, 10(3), 617–621, 2003.
15. Masso-Welch, P. A., Zangani, D., Ip, C., Vaughan, M. M., Shoemaker, S., Ramirez, R. A., and Ip, M. M., Inhibition of angiogenesis by the cancer chemopreventive agent conjugated linoleic Acid, *Cancer Res.*, 62 (15), 4383–4389, 2002.
16. Lock, A. L. and Bauman, D. E., Modifying milk fat composition of dairy cows to enhance fatty acids beneficial to human health, *Lipids*, 39 (12), 1197–1206, 2004.
17. Coughlin, C. M., Salhany, K. E., Gee, M. S., LaTemple, D. C., Kotenko, S., Ma, X., Gri, G. et al., Tumor cell responses to IFNgamma affect tumorigenicity and response to IL-12 therapy and Antiangiogenesis, *Immunity*, 9 (1), 25–34, 1998.
18. Van de Water, J., Yogurt and immunity: the health benefits of fermented milk products that contain lactic acid bacteria, In *Handbook of Fermented Functional Foods*, Farnworth, E. R. Ed., CRC Press, Boca Raton, FL, 2003.

19. Kerbel, R. S. and Hawley, R. G., Interleukin 12: newest member of the antiangiogenesis club, *J. Natl Cancer Inst.*, 87 (8), 557–559, 1995.

20. Matar, C., LeBlanc, J. G., Martin, L., and Perdigon, G., Biologically active peptides released in fermented milk: role and functions, In *Handbook of Fermented Functional Foods*, Farnworth, E. R. Ed., CRC Press, Boca Raton, FL, 2003.

21. Blanchette, J., Kavimandan, N., and Peppas, N. A., Principles of transmucosal delivery of therapeutic agents, *Biomed. Pharmacother.*, 58 (3), 142–151, 2004.

22. Katsuyama, H., Ideguchi, S., Fukunaga, M., Fukunaga, T., Saijoh, K., and Sunami, S., Promotion of bone formation by fermented soybean (Natto) intake in premenopausal women, *J. Nutr. Sci. Vitaminol.*, 50 (2), 114–120, 2004.

23. Kawashima, H., Nakajima, Y., Matubara, Y., Nakanowatari, J., Fukuta, T., Mizuno, S., Takahashi, S., Tajima, T., and Nakamura, T., Effects of vitamin K2 (menatetrenone) on atherosclerosis and blood coagulation in hypercholesterolemic rabbits, *Jpn J. Pharmacol.*, 75 (2), 135–143, 1997.

24. Geleijnse, J. M., Vermeer, C., Grobbee, D. E., Schurgers, L. J., Knapen, M. H., Van der Meer, I. M., Hofman, A., and Witteman, J. C., Dietary intake of menaquinone is associated with a reduced risk of coronary heart disease: the Rotterdam study, *J. Nutr.*, 134 (11), 3100–3105, 2004.

25. Groenen-van Dooren, M. M., Ronden, J. E., Soute, B. A., and Vermeer, C., Bioavailability of phylloquinone and menaquinones after oral and colorectal administration in vitamin K-deficient rats, *Biochem. Pharmacol.*, 50, 797–801, 1995.

26. Li, D., Park, J. H., Park, J. T., Park, C. S., and Park, K. H., Biotechnological production of highly soluble daidzein glycosides using *Thermotoga maritima* maltosyltransferase, *J. Agric. Food Chem.*, 52 (9), 2561–2567, 2004.

27. Park, K. Y., Cho, E. J., Rhee, S. H., Jung, K. O., Yi, S. J., and Jhun, B. H., Kimchi and an active component, beta-sitosterol, reduce oncogenic H-Ras(v12)-induced DNA synthesis, *J. Med. Food*, 6 (3), 151–156, 2003.

28. Kwon, H. and Kim, Y. K. L., Korean fermented foods: Kimchi and Doenjang, In *Handbook of Fermented Functional Foods*, Farnworth, E. R. Ed., CRC Press, Boca Raton, FL, 2003.

29. Bernstein, P. S., Zhao, D. Y., Wintch, S. W., Ermakov, I. V., McClane, R. W., and Gellermann, W., Resonance Raman measurement of macular carotenoids in normal subjects and in age-related macular degeneration patients, *Ophthalmology*, 109 (10), 1780–1787, 2002.

30. Bhosale, P. and Bernstein, P. S., Microbial xanthophylls, *Appl. Microbiol. Biotechnol.*, 68 (4), 445–455, 2005.

31. Nishino, H., Murakosh, M., Ii, T., Takemura, M., Kuchide, M., Kanazawa, M., Mou, X. Y. et al., Carotenoids in cancer chemoprevention, *Cancer Metastasis Rev.*, 21 (3–4), 257–264, 2002.

32. Chew, B. P., Brown, C. M., Park, J. S., and Mixter, P. F., Dietary lutein inhibits mouse mammary tumor growth by regulating angiogenesis and apoptosis, *Anticancer Res.*, 23 (4), 3333–3339, 2003.

33. Menelaou, E., Khachatryan, A., Losso, J. N., Cavalier, M., and Labonte, D., Lutein content in sweet potato leaves, *HortSci.*, in press.

34. Jones, S. T., Aryana, K., and Losso, J. N., Stability of lutein in cheddar cheese, *J. Dairy Sci.*, 88 (5), 1661–1700, 2005.

35. Holker, U., Hofer, M., and Lenz, J., Biotechnological advantages of laboratory-scale solid-state fermentation with fungi, *Appl. Microbiol. Biotechnol.*, 64 (2), 175–186, 2004.

36. Mukherjee, G. and Banerjee, R., Evolutionary operation-factorial design technique for optimization of conversion of mixed agroproducts into gallic acid, *Appl. Biochem. Biotechnol.*, 118 (1–3), 33–46, 2004.

37. Zheng, Z. and Shetty, K., Solid-state bioconversion of phenolics from cranberry pomace and role of Lentinus edodes beta-glucosidase, *J. Agric. Food Chem.*, 48 (3), 895–900, 2000.

38. Williams, E. A., Coxhead, J. M., and Mathers, J. C., Anti-cancer effects of butyrate: use of micro-array technology to investigate mechanisms, *Proc. Nutr. Soc.*, 62 (1), 107–115, 2003.

39. Miki, K., Unno, N., Nagata, T., Uchijima, M., Konno, H., Koide, Y., and Nakamura, S., Butyrate suppresses hypoxia-inducible factor-1 activity in intestinal epithelial cells under hypoxic conditions, *Shock*, 22 (5), 446–452, 2004.

40. Zgouras, D., Becker, U., Loitsch, S., and Stein, J., Modulation of angiogenesis-related protein synthesis by valproic acid, *Biochem. Biophys. Res. Commun.*, 316 (3), 693–697, 2004.

41. Gururaj, A. E., Belakavadi, M., and Salimath, B. P., Antiangiogenic effects of butyric acid involve inhibition of VEGF/KDR gene expression and endothelial cell proliferation, *Mol. Cell. Biochem.*, 243 (1–2), 107–112, 2003.

42. Yu, X. and Kensler, T., Nrf2 as a target for cancer chemoprevention, *Mutat. Res.*, 591 (1–2), 93–102, 2005.

43. Niggeweg, R., Michael, A. J., and Martin, C., Engineering plants with increased levels of the antioxidant chlorogenic acid, *Nat. Biotechnol.*, 22 (6), 746–754, 2004.

44. Willits, M. G., Kramer, C. M., Prata, R. T., De Luca, V., Potter, B. G., Steffens, J. C., and Graser, G., Utilization of the genetic resources of wild species to create a nontransgenic high flavonoid tomato, *J. Agric. Food Chem.*, 53 (4), 1231–1236, 2005.

45. Yoshiji, H., Kuriyama, S., Noguchi, R., Yoshii, J., Ikenaka, Y., Yanase, K., Namisaki, T. et al., Amelioration of carcinogenesis and tumor growth in the rat liver by combination of vitamin K2 and angiotensin-converting enzyme inhibitor via anti-angiogenic activities, *Oncol. Rep.*, 15 (1), 155–159, 2006.

46. Shults, C. W., Oakes, D., Kieburtz, K., Beal, F., Haas, R., Plumb, S., Juncos, J. L. et al., Effects of coenzyme Q_{10} in early Parkinson disease: evidence of slowing of the functional decline, *Arch. Neurol.*, 59 (10), 1541–1550, 2002.

47. Kurowska, E. M., Dresser, G., Deutsch, L., Bassoo, E., and Freeman, D. J., Relative bioavailability and antioxidant potential of two coenzyme Q10 preparations, *Ann. Nutr. Metab.*, 47 (1), 16–21, 2003.

48. Choi, J. H., Ryu, Y. W., and Seo, J. H., Biotechnological production and applications of coenzyme Q10, *Appl. Microbiol. Biotechnol.*, 68 (1), 9–15, 2005.

49. Sterescu, A. E., Rousseau-Harsany, E., Farrell, C., Powell, J., David, M., and Dubois, J., The potential efficacy of omega-3 fatty acids as anti-angiogenic agents in benign vascular tumors of infancy, *Med. Hypotheses*, 66 (6), 1121–1124, 2006.

50. Hardman, W. E., $(n-3)$ fatty acids and cancer therapy, *J. Nutr.*, 134 (suppl. 12), 3427S–3430S, 2004.

51. Ursin, V. M., Modification of plant lipids for human health: development of functional land-based omega-3 fatty acids, *J. Nutr.*, 133 (12), 4271–4274, 2003.

52. Hornung, E., Krueger, C., Pernstich, C., Gipmans, M., Porzel, A., and Feussner, I., Production of (10E,12Z)-conjugated linoleic acid in yeast and tobacco seeds, *Biochim. Biophys. Acta*, 1738 (1–3), 105–114, 2005.

53. Shimamura, M., Yamamoto, Y., Oikawa, T., Hazato, T., Tsuda, H., and Iigo, M., Bovine lactoferrin inhibits tumor-induced angiogenesis, *Int. J. Cancer*, 111 (1), 111–116, 2004.

54. Carmeliet, P., Written in the blood, *Nature*, 440, 710–711, 2006.
55. Bagchi, D., Sen, C. K., Bagchi, M., and Atalay, M., Anti-angiogenic, antioxidant, and anti-carcinogenic properties of a novel anthocyanin-rich berry extract formula, *Biochemistry (Mosc)*, 69 (1), 75–80, 2004.
56. Atalay, M., Gordillo, G., Roy, S., Rovin, B., Bagchi, D., Bagchi, M., and Sen, C. K., Anti-angiogenic property of edible berry in a model of hemangioma, *FEBS Lett.*, 544 (1–3), 252–257, 2003.
57. Skopinski, P., Skopinska-Rozewska, E., Sommer, E., Chorostowska-Wynimko, J., Rogala, E., Cendrowska, I., Chrystowska, D., Filewska, M., Bialas-Chromiec, B., and Bany, J., Chocolate feeding of pregnant mice influences length of limbs of their progeny, *Pol. J. Vet. Sci.*, 6 (3), 57–59, 2003.
58. Birudaraj, R., Mahalingam, R., Li, X., and Jasti, B. R., Advances in buccal drug delivery, *Crit. Rev. Ther. Drug Carrier Syst.*, 22 (3), 295–330, 2005.
59. Challa, R., Ahuja, A., Ali, J., and Khar, R. K., Cyclodextrins in drug delivery: an updated review, *AAPS Pharm. Sci. Tech.*, 6 (2), E329–E357, 2005.
60. Nicolazzo, J. A., Reed, B. L., and Finnin, B. C., Assessment of the effects of sodium dodecyl sulfate on the buccal permeability of caffeine and estradiol, *J. Pharm. Sci.*, 93 (2), 431–440, 2004.
61. Senel, S. and Hincal, A. A., Drug permeation enhancement via buccal route: possibilities and limitations, *J. Control Release*, 72 (1–3), 133–144, 2001.
62. Park, C. R. and Munday, D. L., Evaluation of selected polysaccharide excipients in buccoadhesive tablets for sustained release of nicotine, *Drug Dev. Ind. Pharm.*, 30 (6), 609–617, 2004.
63. Kosaraju, S. L., Colon targeted delivery systems: review of polysaccharides for encapsulation and delivery, *Crit. Rev. Food Sci. Nutr.*, 45 (4), 251–258, 2005.
64. Raghavan, C. V., Muthulingam, C., Jenita, J. A., and Ravi, T. K., An in vitro and in vivo investigation into the suitability of bacterially triggered delivery system for colon targeting, *Chem. Pharm. Bull.*, 50 (7), 892–895, 2002.
65. Takeuchi, H., Yasuji, T., Yamamoto, H., and Kawashima, Y., Spray-dried lactose composite particles containing an ion complex of alginate-chitosan for designing a dry-coated tablet having a time-controlled releasing function, *Pharm. Res.*, 17 (1), 94–99, 2000.
66. Ishikado, A., Imanaka, H., Takeuchi, T., Harada, E., and Makino, T., Liposomalization of lactoferrin enhanced it's anti-inflammatory effects via oral administration, *Biol. Pharm. Bull.*, 28 (9), 1717–1721, 2005.
67. Funasaki, N., Kawaguchi, R., Hada, S., and Neya, S., Ultraviolet spectroscopic estimation of microenvironments and bitter tastes of oxyphenonium bromide in cyclodextrin solutions, *J. Pharm. Sci.*, 88 (8), 759–762, 1999.
68. Song, Y., Wang, Y., Thakur, R., Meidan, V. M., and Michniak, B., Mucosal drug delivery: membranes, methodologies, and applications, *Crit. Rev. Ther. Drug Carrier Syst.*, 21 (3), 195–256, 2004.
69. Cano-Cebrian, M. J., Zornoza, T., Granero, L., and Polache, A., Intestinal absorption enhancement via the paracellular route by fatty acids, chitosans and others: a target for drug delivery, *Curr. Drug Deliv.*, 2 (1), 9–22, 2005.
70. Salama, N. N., Eddington, N. D., and Fasano, A., Tight junction modulation and its relationship to drug delivery, *Adv. Drug Deliv. Rev.*, 58 (1), 15–28, 2006.
71. Higaki, K., Yata, T., Sone, M., Ogawara, K., and Kimura, T., Estimation of absorption enhancement by medium-chain fatty acids in rat large intestine, *Res. Commun. Mol. Pathol. Pharmacol.*, 109 (3–4), 231–240, 2001.

72. Sasaki, K., Yonebayashi, S., Yoshida, M., Shimizu, K., Aotsuka, T., and Takayama, K., Improvement in the bioavailability of poorly absorbed glycyrrhizin via various non-vascular administration routes in rats, *Int. J. Pharm.*, 265 (1–2), 95–102, 2003.

73. Park, J. W., Benz, C. C., and Martin, F. J., Future directions of liposome- and immunoliposome-based cancer therapeutics, *Semin. Oncol.*, 31 (6 suppl.13), 196–205, 2004.

74. Gallo, D., Giacomelli, S., Ferlini, C., Raspaglio, G., Apollonio, P., Prislei, S., Riva, A., Morazzoni, P., Bombardelli, E., and Scambia, G., Antitumour activity of the silybin-phosphatidylcholine complex IdB 1016 against human ovarian cancer, *Eur. J. Cancer*, 39 (16), 2403–2410, 2003.

75. Li, L., Braiteh, F. S., and Kurzrock, R., Liposome-encapsulated curcumin: in vitro and in vivo effects on proliferation, apoptosis, signaling, and angiogenesis, *Cancer*, 104 (6), 1322–1331, 2005.

76. Dangi, J. S., Vyas, S. P., and Dixit, V. K., Effect of various lipid-bile salt mixed micelles on the intestinal absorption of amphotericin-B in rat, *Drug Dev. Ind. Pharm.*, 24 (7), 631–635, 1998.

77. Iwanaga, K., Kushibiki, T., Miyazaki, M., and Kakemi, M., Disposition of lipid-based formulation in the intestinal tract affects the absorption of poorly water-soluble drugs, *Biol. Pharm. Bull.*, 29 (3), 508–512, 2006.

78. Araya, H., Tomita, M., and Hayashi, M., The novel formulation design of self-emulsifying drug delivery systems (SEDDS) type O/W microemulsion III: the permeation mechanism of a poorly water soluble drug entrapped O/W microemulsion in rat isolated intestinal membrane by the using chamber method, *Drug Metab. Pharmacokinet.*, 21 (1), 45–53, 2006.

79. Subramanian, N., Ray, S., Ghosal, S. K., Bhadra, R., and Moulik, S. P., Formulation design of self-microemulsifying drug delivery systems for improved oral bioavailability of celecoxib, *Biol. Pharm. Bull.*, 27 (12), 1993–1999, 2004.

80. Woodley, J. F., Lectins for gastrointestinal targeting—15 years on, *J. Drug Target*, 7 (5), 325–333, 2000.

81. Kaku, H., Goldstein, I. J., Van Damme, E. J., and Peumans, W. J., New mannose-specific lectins From garlic (*Allium sativum*) and ramsons (*Allium ursinum*) bulbs, *Carbohydr. Res.*, 229 (2), 347–353, 1992.

82. Clark, M. A., Jepson, M. A., and Hirst, B. H., Exploiting M cells for drug and vaccine delivery, *Adv. Drug Deliv. Rev.*, 50 (1–2), 81–106, 2001.

83. Park, Y. J., Chang, L. C., Liang, J. F., Moon, C., Chung, C. P., and Yang, V. C., Nontoxic membrane translocation peptide from protamine, low molecular weight protamine (LMWP), for enhanced intracellular protein delivery: in vitro and in vivo study, *FASEB J.*, 19 (11), 1555–1557, 2005.

84. Brower, V., Is nanotechnology ready for primetime?, *J. Natl Cancer Inst.*, 98 (1), 9–11, 2006.

85. Leonard, G. D., Fojo, T., and Bates, S. E., The role of ABC transporters in clinical practice, *Oncologist*, 8 (5), 411–424, 2003.

86. Pal, D. and Mitra, A. K., MDR- and CYP3A4-mediated drug-herbal interactions, *Life Sci.*, 78 (18), 2131–2145, 2006.

87. Zhou, S., Lim, L. Y., and Chowbay, B., Herbal modulation of P-glycoprotein, *Drug Metab. Rev.*, 36 (1), 57–104, 2004.

88. Badmaev, V., Majeed, M., and Prakash, L., Piperine derived from black pepper increases the plasma levels of coenzyme Q10 following oral supplementation, *J. Nutr. Biochem.*, 11 (2), 109–113, 2000.

89. Lambert, J. D., Hong, J., Kim, D. H., Mishin, V. M., and Yang, C. S., Piperine enhances the bioavailability of the tea polyphenol (−)-epigallocatechin-3-gallate in mice, *J. Nutr.*, 134 (8), 1948–1952, 2004.

90. Bhardwaj, R. K., Glaeser, H., Becquemont, L., Klotz, U., Gupta, S. K., and Fromm, M. F., Piperine, a major constituent of black pepper, inhibits human P-glycoprotein and CYP3A4, *J. Pharmacol. Exp. Ther.*, 302 (2), 645–650, 2002.

91. Ruoslahti, E., Vascular zip codes in angiogenesis and metastasis, *Biochem. Soc. Trans.*, 32 (Pt3), 397–402, 2004.

92. Duda, D. G., Antiangiogenesis and drug delivery to tumors: bench to bedside and back, *Cancer Res.*, 66 (8), 3967–3970, 2006.

93. Tong, R. T., Boucher, Y., Kozin, S. V., Winkler, F., Hicklin, D. J., and Jain, R. K., Vascular normalization by vascular endothelial growth factor receptor 2 blockade induces a pressure gradient across the vasculature and improves drug penetration in tumors, *Cancer Res.*, 64 (11), 3731–3736, 2004.

94. Ozerdem, U. and Stallcup, W. B., Pathological angiogenesis is reduced by targeting pericytes via the NG2 proteoglycan, *Angiogenesis*, 7 (3), 269–276, 2004.

95. Luttun, A., Autiero, M., Tjwa, M., and Carmeliet, P., Genetic dissection of tumor angiogenesis: are PlGF and VEGFR-1 novel anti-cancer targets?, *Biochim. Biophys. Acta*, 1654 (1), 79–94, 2004.

96. Panyam, J. and Labhasetwar, V., Targeting intracellular targets, *Curr. Drug Deliv.*, 1 (3), 235–247, 2004.

Index

Milton Keynes UK
Ingram Content Group UK Ltd.
UKHW021936071024
449327UK00022B/1825